STOCHASTIC LEARNING AND OPTIMIZATION

A SENSITIVITY-BASED APPROACH

STOCHASTIC LEARNING AND OPTIMIZATION

A Sensitivity-Based Approach

With 119 Figures, 27 Tables, and 212 Problems

Xi-Ren Cao
Hong Kong University of Science and Technology

Xi-Ren Cao, PhD, Professor
Hong Kong University of Science and Technology
Clear Water Bay, Kowloon
Hong Kong
email: eecao@ust.edu.hk

ISBN-13: 978-1-4419-4222-7 e-ISBN-13: 978-0-387-69082-7

Printed on acid-free paper.

9 8 7 6 5 4 3 2 1

springer.com

To the memory of my parents
Cao Yun Jiu and Guo Wen Ying

Preface

Performance optimization is very important in the design and operation of modern engineering systems in many areas, including communications (Internet and wireless), manufacturing, robotics, and logistics. Most engineering systems are too complicated to be modelled, or the system parameters cannot be easily identified. Therefore, learning techniques have to be utilized.

A Brief Description of Learning and Optimization

Learning and optimization of stochastic systems is a multi-disciplinary area that has attracted wide attention from researchers in many disciplines including control systems, operations research, and computer science. Areas such as perturbation analysis (PA) in discrete event dynamic systems (DEDSs), Markov decision processes (MDPs) in operations research, reinforcement learning (RL) in computer science, neuro-dynamic programming (NDP), identification, and adaptive control (I&AC) in control systems, share a common goal: to make the "best decision" to optimize a system's performance.

Different areas take different perspectives and have different formulations for the problems with the same goal. *This book provides an overview of these different areas, PA, MDPs, RL, and I&AC, with a unified framework based on a sensitivity point of view. It also introduces new approaches and proposes new research topics and directions with this sensitivity-based framework.*

Roughly speaking, with RL, we learn how to make decisions to improve a system's performance by observing and analyzing the system's current behavior; the structure and the parameters of the system may not be known and even may not need to be estimated. PA estimates the derivatives of a system's performance with respect to the system's parameters by observing and analyzing the system's behavior. Optimization is achieved by combining performance derivative estimation and other optimization techniques such as stochastic approximation. MDPs provide a theoretical foundation for performance optimization of systems with a Markov model [21, 216]. In adaptive control, the system behavior is described by differential or difference equations; when the system parameters are unknown, they have to be identified using the observed data. Adaptive control together with identification achieves the same goal as learning and optimization.

The goal of these research areas is the same: to find a policy that optimizes a system's performance, by using the information "learned" by observing or analyzing the system's behavior. Given a system's status or history, a policy determines an action to be applied to the system, which controls the system evolution. In some cases, policies depend on continuous parameters and the policy space is continuous; in other cases, the policy space is discrete and usually contains a huge number of policies.

A Sensitivity-Based View

Recent research indicates that the various disciplines in learning and optimization can be explained from a unified point of view based on the performance sensitivities in the policy space [56]. The fundamental elements of learning and optimization are two types of performance sensitivity formulas, one for performance derivatives at any policy in the policy space, the other, for performance differences between any two policies in the policy space. With these two types of sensitivity formulas, existing results in the various areas and their relations can be derived or explained in a simple and intuitive way, new approaches can be introduced, and the average, discounted, and other performance criteria can be treated in the same way.

The unified framework is based on a few simple and fundamental facts: Naturally, by observing and analyzing a system's behavior under one policy, we cannot know the system's performance under other policies, if no structural information about the system is known; and we can only compare the performance of two policies at a time. The question is, with these fundamental limitations, how can we achieve our goal of performance optimization by using as little information about the system structure as possible and with as little computation effort as possible?

Thinking along this direction, we find that two things can be done: First, if the policy space is continuous, *with some knowledge about the system structure (e.g., queueing or Markov) and by the PA principles, we may estimate the performance derivatives at a policy along any direction in the policy space by observing/analyzing the behavior of the system under this policy [70, 62, 69].* This leads to the performance derivative formula; all the quantities required to calculate the derivative along a given direction can be obtained by analyzing the sample paths of the current policy. The performance derivative formula forms the basis for PA and the "policy gradient" approach that was proposed recently in the RL research community.

Second, if the policy space is discrete, the performance difference formula forms the basis for optimization. The difference formula compares the performance of the system under two policies. However, unlike the derivative formula, the difference formula involves quantities for both policies and it is not possible to know the performance of another policy, or the difference in the performance of a system under two policies, by observing or analyzing only the behavior of the system under one policy. Fortunately, *by the particular factorized form of the performance difference formula, under some structural conditions, we can always use the information learned from observing/analyzing the system behavior under a policy to find another policy under which the performance of the system is better, if such better policies exist.* This leads to policy iteration: learn from a policy to find another better policy, and learn from this better policy to find an even better policy, and so on. Thus, the performance difference formula forms the basis for policy-iteration type approaches to performance optimization. We will show that the results in I&AC can also be derived using this principle, which also provides a learning-based perspective to the area.

The fundamental quantity in the two sensitivity formulas (and thus in the two types of optimization approaches, the gradient-based and the policy-iteration-based) is the performance potential, which has a clear physical meaning: It measures the "potential" contribution of a state to the system performance. The difference of the potentials of two states measures the effect of changing from one state to the other on the system performance. Such a change from one state to the other is called a perturbation in PA (or simply called a "jump" on a sample path). In RL, many efficient algorithms (e.g., TD(λ) and Q-learning) have been developed for estimating the potential and its variant Q-factor and their values for the optimal policies.

The physical interpretation of the potentials leads to the fundamental principle in PA: The effect of any change in a system's structure or parameters can be decomposed into the effects of many jumps among states (or many perturbations). With this principle, we can use the potentials as building blocks to construct new sensitivity formulas by first principles for many problems that do not fit into the standard formulation in the existing literature [59]. Since, as explained, such sensitivity formulas serve as the basis for learning and optimization, the sensitivity construction approach opens up a new direction: New learning and optimization schemes can be developed based on these new sensitivity formulas, and special system features can be utilized.

One of the approaches developed based on sensitivity construction is called the *event-based optimization* where actions can be taken only when some events happen. This approach utilizes the special feature of a system captured by events. Policy depends on events, rather than on states. An event is defined as a set of state transitions and, therefore, an event occurring in the present contains some information about the next state, i.e., the future. In many modern engineering systems in information technology, such information is accessible before actions are taken, and the standard Markov model does not capture this special property. Thus, in some cases, event-based policies may perform better than state-based ones. Furthermore, the number of events usually scales to the system size, which is much smaller than that of the states, which grows exponentially with the size of the system. Thus, under some conditions, this approach may provide a possibility to overcome or to alleviate a computational difficulty: the curse of dimensionality. In addition, many existing approaches, such as partially observed MDPs (POMDPs), state and time aggregation, hierarchical control (hybrid systems), options, and singular perturbation, can be treated as special cases of the event-based optimization by defining different events to capture the special features of these different problems.

The Unique Features of This Book

Compared with other books in the area of learning and optimization, this book is unique in the following aspects.

1. The book covers various disciplines in learning and optimization, including PA, MDPs, RL, and I&AC, with a unified framework based on a sensitivity perspective in the policy space. Many results can be explained with the two types of fundamental sensitivity formulas in a simple way.
2. We emphasize physical interpretations rather than mathematics. With the intuitive physical explanations, we propose to construct new sensitivity formulas with performance potentials as building blocks. The physical intuition may provide insights that complement to other existing approaches.
3. With the unified framework and the construction approach, we introduce the recently-developed event-based optimization approach; this approach opens up a research direction in overcoming/alleviating the curse of dimensionality issue by utilizing the system's special features.
4. The performance difference-based approach is applied to all the MDP problems, including ergodic and multi-chain systems, average and discounted performance criteria, and even bias optimality and nth-bias optimality. It is shown that the nth-bias optimal policies eventually lead to the Blackwell optimal policies. This approach provides a simple, intuitively clear, and comprehensive presentation of all these problems in MDPs in a unified way. This presentation of MDPs is unique in existing books.

The Contents of This Book

Chapter 1 presents an introduction, which consists of an overview of the different disciplines in learning and optimization and a discussion of the event-based approach. This chapter serves as a road map for the book. The rest of the book consists of three parts. Part I, consisting of Chapters 2 to 7, describes how the sensitivity point of view in policy spaces leads to the main concepts and results in PA, MDPs, RL, and I&AC. Part II, consisting of Chapters 8 and 9, presents the recent developments in event-based learning and optimization with this sensitivity point of view. Part III consists of three appendices that provide the mathematical background required for this book.

Part I starts with PA in Chapter 2. We derive the performance derivative formulas, by using performance potentials or realization factors as building blocks, for Markov systems and queueing systems. The sample-path-based sensitivity point of view in PA is the core of the unified approach of this book. In Chapter 3, we discuss performance potentials and develop sample-path-based algorithms for estimating potentials and performance derivatives, as well as for performance optimization with the potentials. In Chapter 4, we show how policy iteration for both uni-chain and multi-chain MDPs can be easily derived from the performance difference formulas; this approach applies in the same way to both average and discounted criteria, as well as bias optimality, etc. We also define and solve the nth-bias optimality problem with the same approach. On-line policy iteration algorithms are developed in Chapter 5 with the potentials estimated from the sample paths. Chapter 6 presents basic results of RL, which is essentially a combination of stochastic approximation and the sample-path-based estimation of the potentials and their variants Q-factors. In Chapter 7, we show that the on-line policy iteration approach can be applied to I&AC problems, including linear systems and some non-linear systems.

In Part II, Chapter 8 presents the event-based optimization approach. This approach provides a possible way to address the difficult issue of the curse of dimensionality by utilizing particular system structures; in some cases event-based policies may perform better than state-based ones. The construction of sensitivity formulas with performance potentials as building blocks for general problems is presented in Chapter 9.

How to Use This Book

This book provides, in a unified way, good introductory materials for graduate students and engineers who wish to have an overview of learning and optimization theory, the related methodologies in different disciplines, including PA, MDPs, RL, I&AC, and stochastic approximation, and their relations. The new perspective presented in this book is helpful in finding new research topics. Thus, the book is useful to researchers in these areas who wish to find some motivation and to promote inter-disciplinary collaborations. In addition, engineers, in particular those in information technology, may find the ideas and methodologies introduced in this book useful in their practical applications.

The chapters and sections marked with asterisks "*" are supplementary reading material and can be omitted by first-time readers. Each chapter contains a considerable number of problems that may help students to enhance their understanding of the main contents. Some of the problems are summaries of past research topics and might be difficult. These are also marked with asterisks. Solutions to the problems are available upon request and can be found on my website http://www.ee.ust.hk/~eecao.

An earlier version of this book was used as the textbook for graduate courses at the Hong Kong University of Science and Technology and Tsinghua University in Beijing, China. A suggested time table for a course in a fourteen-week term (three hours per week) is as follows.

Chapters	Sections		Hours	Weeks
A-C	A.1 - C.2		3	1
1	1.1 - 1.4		2	2/3
2	2.1		4	4/3
	2.2		1	1/3
	2.4		3	1
3	3.1 - 3.3		3	1
Review			1	1/3
4	4.1		3	1
	4.2		3	1
5	5.1 - 5.2		3	1
6	6.1 - 6.4		4	4/3
7	7.1 - 7.3		2	2/3
8	8.1 - 8.5		5	5/3
9	9.1 - 9.2		1	1/3
Review			1	1/3
Examination			3	1
		Total	42	14

The following are some suggestions and comments about the contents covered in each chapter:

0. The contents covered in the appendices are the prerequisite of the course. Three hours are not enough to cover the details in the three appendices. In a brief review, we may focus more on probability and the theory on Markov chains, which are closely related to the main concepts presented in the book. Appendix B is mainly related to Chapter 4, and Appendix C is mainly related to Section 2.4. Some results can be reviewed when the main texts are taught.
1. For students with a background in control, Section 1.1 on policies can be taught fast. Sections 1.2-1.3 are intended to give an overview of the different disciplines, and they should be revisited after studying Part I to get a better picture.
2. The main part of Chapter 2 is Sections 2.1 and 2.4.
3. Section 3.2 is relatively new in the literature.
4. Sections 4.1 and 4.2 cover the main ideas and methodologies. If time permits, we may cover the main results in Section 4.3 without going through the proofs.
5. The proof in Section 5.2.3 is interesting, but it is a bit technical and requires some careful thinking.
6. In Chapter 6, we emphasize on the intuitions behind the development of recursive algorithms, with the principles in stochastic approximation, and we do not intend to provide proofs for these algorithms. The algorithms for performance derivative estimates are new research topics in recent years.

7. In Chapter 7, it is easy to convince students that a control system can be modelled as an MDP. The extension of MDP from a discrete state space to a continuous state space is of no conceptual difficulty. We may cover only the LQ problem as an example.
8. Section 8.1 provides a nice overview of the event-based optimization approach. If we wish to avoid studying the tedious mathematical formulation, we may study the two examples to obtain a clear picture of the approach.
9. Section 9.2 provides the basic ideas for the construction of the performance difference formula. Other sections illustrate the flexibility of this approach and are for additional reading.

The following is a suggested time table for a course in a nine-week term (three hours per week).

Chapters	Sections		Hours	Weeks
A-C	A.1 - C.2		1.5	1/2
1	1.1 - 1.4		1.5	1/2
2	2.1		4	4/3
3	3.1 - 3.3		2	2/3
4	4.1		3	1
	4.2.1		2	2/3
5	5.1 - 5.2		2	2/3
6	6.1 - 6.4		3	1
7	7.1 - 7.3		2	2/3
8	8.1 - 8.5		3	1
9	9.1 - 9.2		1	1/3
Examinations			2	2/3
		Total	27	9

The additional suggestions are as follows.

1. We do not have time to cover PA of queueing systems in Section 2.4, PA-based optimization of queueing systems in Section 3.3, etc.
2. In Chapter 4, we may only briefly introduce the concept of the nth bias and the problem of the nth-bias optimality.
3. Event-based optimization can be introduced via examples.

Acknowledgements

A large portion of the book is based on my research on learning and optimization conducted over the past two to three decades, starting early 1980s when I was at Harvard University and working on PA. Many people have helped me in my research in various ways during this period.

I would like to express my sincere thanks to Prof. Yu-Chi Ho for his continuing support and encouragement; his insights and inspiration have made a significant impact on my research. I would like to thank the following people for joint works and/or insightful discussions in various periods on topics that are related to the materials covered in this book: K. J. Åström, T. Başar, A. G. Barto, D. P. Bertsekas, R. W. Brockett, C. G. Cassandras, H. F. Chen, A. Ephremides, H. T. Fang, E. A. Feinberg, M. C. Fu, P. Glasserman, W. B. Gong, X. P. Guo, B. Heidergott, P. V. Kokotovic, F. L. Lewis, L. Ljung, D. J. Ma, S. I. Marcus, S. P. Meyn, G. Ch. Pflug, L. Qiu, Z. Y. Ren, J. Si, R. Suri, J. N. Tsitsiklis, B. Van Roy, P. Varaiya, A. F. Veinott, Y. Wardi, Y. W. Wan, and J. Y. Zhang. I also wish to thank those people who have carefully read parts of the early draft of this book and provided useful comments regarding the presentations of the book and corrected typos: F. Cao, H. F. Chen, T. W. Chen, X. P. Guo, Q. L. Li, Y. J. Li, D. Y. Shi, L. Xia, Y. K. Xu, and J. Y. Zhang. In addition, I particularly appreciate the tedious work of Y. K. Xu and J. Y. Zhang in making Latex files for many pictures in the book. I also thank V. Unkefer for her technical editing of most part of this book and thank J. Q. Shen for drawing the figure for the book cover. Of course, all errors remain my responsibility. My sincere thanks also go to Harvard University, Digital Equipment Corporation, U.S.A., and the Hong Kong University of Science and Technology for providing me with financial support as well as excellent research environment during the past years.

Finally, I wish to express my sincere appreciation to my wife, Mindy Wang Cao, for her continuing support and understanding under all circumstance in the past years.

Hong Kong
April 2007

Xi-Ren Cao
The Hong Kong University
of Science and Technology
eecao@ust.hk

Contents

A journey of a thousand miles begins with a single step.

Lao-tzu, Chinese philosopher
(604 BC - 531 BC)

1 Introduction

1.1 An Overview of Learning and Optimization

Performance optimization plays an important role in the design and operation of modern engineering systems in many areas, including communications (Internet and wireless networks), manufacturing, logistics, robotics, and bio-informatics. Most engineering systems are too complicated to be analyzed, or the parameters of the system models cannot be easily obtained. Therefore, learning techniques have to be applied.

The goal of learning and optimization is to make the "best" decisions to optimize, or to improve, the performance of a system based on the information obtained by observing and analyzing the system's behavior. A system's behavior is usually represented by a model, or by the sample paths (also called *trajectories*) of the system. A *sample path* is a record of the operation history of a system.

1.1.1 Problem Description

In this book, we mainly study stochastic dynamic systems. A dynamic system evolves as time passes. It is generally easier to explain the ideas with a discrete time model, in which time takes discrete values denoted as $l = 0, 1, 2, \ldots$. In addition to its dynamic nature, a stochastic system is always subject to random influences caused by noise or other uncertainties.

States, Actions, and Observations

To study the system behavior, we need to describe precisely the system's status. A system's status at any time $l = 0, 1, \ldots$ can be represented by a quantity called the system's *state* at time l, denoted as X_l, $l = 0, 1, \ldots$. The *state space* (i.e., the set of all states) is denoted as $\mathcal{S}$, which may be either discrete or continuous. A *sample path* of a system is a record of state history denoted as $\boldsymbol{X} = \{X_0, X_1, \ldots\}$. In stochastic dynamic systems, X_l, $l = 0, 1, \ldots$, are random variables (may be multi-dimensional random vectors). A system's dynamic behavior is then represented by its sample paths. We denote a "finite-length" sample path as $\boldsymbol{X}_l := \{X_0, X_1, \ldots, X_l\}$.

In this book, the word "*state*" is used in a strict sense that a sample path $\boldsymbol{X}$ is a Markov chain. This means that *given the current state X_l, the system's future behavior $\{X_{l+1}, X_{l+2}, \ldots\}$ is independent of its past history $\{X_0, X_1, \ldots, X_{l-1}\}$*, $l = 1, 2, \ldots$ (see Appendix A.2 for more details). This is called the *Markov property*. Intuitively, a state completely captures the system's current status in regard to its future evolution.

In optimization problems, at any time l, we can apply an *action*, denoted as $A_l \in \mathcal{A}$, $l = 0, 1, \ldots$, where $\mathcal{A}$ is an action space. In most cases, $\mathcal{A}$ contains a finite number of actions, but in general it may contain infinitely many actions, or even be a continuous space.

The actions $A_0, A_1, \ldots$ may affect the evolution of the system. With the Markov model, the actions control the transition probabilities of the state process. If action $\alpha \in \mathcal{A}$ is taken at time l (i.e., $A_l = \alpha$), then the transition probabilities at time l are denoted as $p^{\alpha}(X_{l+1}|X_l)$, $X_{l+1}, X_l \in \mathcal{S}$, $l = 0, 1, \ldots$.

Because the actions affect the system behavior, the operation history of a system should include the actions. Let $\boldsymbol{A}_{l-1} := \{A_0, A_1, \ldots, A_{l-1}\}$ denote an action history with a finite length and $\boldsymbol{A} := \{A_0, A_1, \ldots\}$ denote an infinitely long action history. Taking the actions into consideration, we denote a sample path as $\boldsymbol{H} := (\boldsymbol{X}, \boldsymbol{A})$, or $\boldsymbol{H}_l := (\boldsymbol{X}_l, \boldsymbol{A}_{l-1})$.

In many cases, the system's state cannot be exactly observed, and we can only observe a random variable Y_l at time l that is related to X_l, $l = 0, 1, \ldots$. The observation history is denoted as $\boldsymbol{Y} := \{Y_0, Y_1, \ldots\}$, or $\boldsymbol{Y}_l := \{Y_0, Y_1, \ldots, Y_l\}$. In such cases, we say that the system is partially observable. The information history up to time l is $\boldsymbol{H}_l := (\boldsymbol{Y}_l, \boldsymbol{A}_{l-1})$. When $Y_l = X_l$, for all $l = 0, 1, \ldots$, we say that the system is completely observable. In such cases, we have $\boldsymbol{H}_l = (\boldsymbol{X}_l, \boldsymbol{A}_{l-1})$. Note that even for partially observable systems, we reserve the word "sample path" for $\boldsymbol{H}_l = (\boldsymbol{X}_l, \boldsymbol{A}_{l-1})$, or $\boldsymbol{H} = (\boldsymbol{X}, \boldsymbol{A})$.

Here are some examples of the observations Y_l: When the observations of states contain additive noise, we have $Y_l = X_l + \epsilon_l$, with $E(Y_l) = X_l$ if $E(\epsilon_l) = 0$. When the states of the Markov chain $\boldsymbol{X}$ are aggregated, Y_l may be an aggregation state. In the event-based optimization approach (see Chapters 9 and 8), Y_l may represent an event at time l, and the event may contains information about the state transition at the time instant.

Rewards and Performance Measures

Associated with each sample path $\boldsymbol{H}_L = (\boldsymbol{X}_L, \boldsymbol{A}_{L-1})$, there is a *reward* (or cost; we use the word "reward" in most cases in this book) denoted as $\eta_L(\boldsymbol{H}_L)$. Because the states $\boldsymbol{X}_L$ and the actions $\boldsymbol{A}_{L-1}$ are generally random, $\eta_L(\boldsymbol{H}_L)$ is usually a random variable. For finite-length problems, $\eta_L(\boldsymbol{H}_L)$ represents the total reward received when the system is going through the sample path $\boldsymbol{H}_L$. The *performance measure* η (or simply called the *performance*) is defined as the mean of the rewards

$$\eta = E[\eta_L(\boldsymbol{H}_L)]. \tag{1.1}$$

For sample paths with infinitely long lengths, the *performance measure* η is defined as the limit of the mean rewards

$$\eta = \lim_{L\to\infty} E[\eta_L(\boldsymbol{H}_L)], \tag{1.2}$$

in which we assume that both the expectation and limit exist. In this case, $\eta_L(\boldsymbol{H}_L)$ usually represents the average reward per step received by the system during the operation.

The reward η_L and the performance η may take different forms in different problems. In an optimization problem with the Markov model, there is a reward function denoted as $f(i,\alpha)$, $i \in \mathcal{S}$, $\alpha \in \mathcal{A}$. At time l, if the system is in state i and action $\alpha \in \mathcal{A}$ is taken, then the system receives a reward of $f(i,\alpha)$. For a sample path with a finite length L, the *total reward* received on this sample path is

$$\eta_L = \sum_{l=0}^{L-1} f(X_l, A_l).$$

The performance measure is the expected total reward $\eta = E(\eta_L)$. For problems with infinitely long sample paths, we consider the average

$$\eta_L = \frac{1}{L}\sum_{l=0}^{L-1} f(X_l, A_l),$$

and the performance measure is (when it exists)

$$\eta = \lim_{L\to\infty} E(\eta_L) = \lim_{L\to\infty} \frac{1}{L} E\left\{\sum_{l=0}^{L-1} f(X_l, A_l)\right\}, \tag{1.3}$$

which is called the *long-run average reward* in the literature. Note that in general the expectations in (1.1), (1.2), and (1.3) depend on the initial states or the initial state distributions.

For ergodic Markov chains, the long-run average reward is

$$\eta := \lim_{L\to\infty} \frac{1}{L}\sum_{l=0}^{L-1} f(X_l, A_l), \qquad \text{w.p.1},$$

which does not depend on the initial state.

Other performance measures, such as the discounted reward, exist and will be discussed in other chapters. For simplicity, a performance measure is also referred to as a performance.

The Problem Description

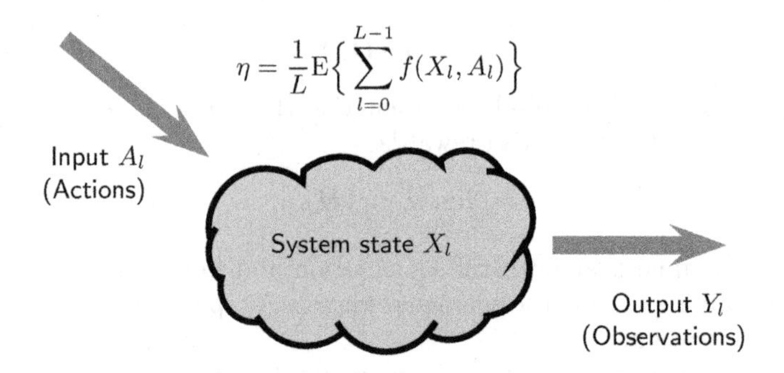

Fig. 1.1. A Model of Learning and Optimization

A general description of the learning and optimization problem is illustrated by Figure 1.1. In the figure, the shaded area represents a stochastic dynamic system. The system is essentially a black box and it can only interact with the outside through its inputs and outputs. The inputs provide a vehicle to intervene or to control the operation of the system, and/or to affect the reward of the operation. For example, in a Markov system (whose behavior can be modelled as a Markov chain), the inputs may be the actions A_l that determine the system's state transition probabilities at time l, $l = 0, 1, \ldots$. In other cases, an input can also control the system operation modes, or tune the values of system parameters, etc. In this terminology, setting different values for system parameters is viewed as taking different actions. It is usually assumed that the available actions are known to us (e.g., we know that we can accept or reject a packet in a communication system, or we can tune the rate of a transmission line to θ megabit/second).

The outputs provide a window for observing the system. That is, the outputs are the observations. In partially observable Markov systems, the output at time l is the observation Y_l, $l = 0, 1, \ldots$; and in completely observable systems, the output at time l is the state X_l, $l = 0, 1, \ldots$.

Associated with every system, there is a performance measure η. In the figure, as an example, η is taken to be the mean of the average reward for L steps; L is usually a very large integer or infinity.

The goal of an optimization problem is to answer the following question: *Based on the information we know about the system, i.e., the output history learned from observation and the input (action) history, what action should we take at a particular time so that we can obtain the best possible system performance?*

The information history $\boldsymbol{H}_l = \{\boldsymbol{Y}_l, \boldsymbol{A}_{l-1}\}$, with $\boldsymbol{Y}_l = \{Y_0, Y_1, \ldots, Y_l\}$ being the observation history and $\boldsymbol{A}_{l-1} = \{A_0, A_1, \ldots, A_{l-1}\}$ being the action history (with $\boldsymbol{A}_{-1} := \emptyset$), represents all the information available at time l before an action is taken at l, $l = 0, 1, \ldots$. Based on this information, an action can be chosen by following some rules, called a *policy*, denoted as $d_l : A_l = d_l(\boldsymbol{H}_l), A_l \in \mathcal{A}$. (This is called a deterministic policy.) We wish to find a policy that maximizes the system performance. (Since we use reward, rather than cost, as the system performance, optimization is equivalent to maximization here.) Such a policy is called an *optimal policy*. When the number of policies is finite, such optimal policies always exist and may not be unique.

In engineering applications, at the design stage, sample paths can only be obtained by simulation following a system model; and while a system is operating, the paths can also be obtained by direct observation. If learning and optimization is implemented by simulation, then the approach is called a *simulation-based* approach. With simulation, we may even let the system operate under policies that are generally not feasible in a real system. For real systems, performance optimization (or improvement) decisions can be made through learning the system behavior by observing its sample paths recorded while the system is operating without interruption; we call such an approach an *on-line* approach.

1.1.2 Optimal Policies

From the above description, the main element in learning and optimization is the policy. In this subsection, we use some examples to show different types of policies and some main features in searching for optimal policies.

Open-Loop Policies

Example 1.1. Suppose that we run a system for three time instants $l = 0, 1, 2$, and at any time instant we can take one of the two actions, denoted as α_0 and α_1. In this simple example, there are no observations, and we do not know how the actions influence the system evolution at all. Thus, at any time $l = 1, 2$, we only know the action history $\{A_0, \ldots, A_{l-1}\}$, $l = 1, 2$. There are $8 = 2^3$ possible sequences of actions that we can take in the period of $l = 0, 1, 2$, denoted as $\boldsymbol{\alpha}_i$, $i = 0, 1, \ldots, 7$. For each action sequence, there is associated total reward $\eta(\boldsymbol{\alpha}_i)$, $i = 0, 1, \ldots, 7$. We can certainly run the system eight times, each with a different action sequence $\boldsymbol{\alpha}_i$, and obtain all possible values of the system performance. Table 1.1 lists the eight possible action sequences and their corresponding performance. (If the system involves

	Action Sequences	Performance
$\boldsymbol{\alpha}_0$	$\{\alpha_0, \alpha_0, \alpha_0\}$	3
$\boldsymbol{\alpha}_1$	$\{\alpha_0, \alpha_0, \alpha_1\}$	7
$\boldsymbol{\alpha}_2$	$\{\alpha_0, \alpha_1, \alpha_0\}$	8
$\boldsymbol{\alpha}_3$	$\{\alpha_0, \alpha_1, \alpha_1\}$	4
$\boldsymbol{\alpha}_4$	$\{\alpha_1, \alpha_0, \alpha_0\}$	5
$\boldsymbol{\alpha}_5$	$\{\alpha_1, \alpha_0, \alpha_1\}$	10
$\boldsymbol{\alpha}_6$	$\{\alpha_1, \alpha_1, \alpha_0\}$	9
$\boldsymbol{\alpha}_7$	$\{\alpha_1, \alpha_1, \alpha_1\}$	6

Table 1.1. The Action Sequences and the Corresponding Performance in Example 1.1

randomness, we can repeatedly run the system with each action sequence many times and Table 1.1 lists the averages.)

a) (*Open-loop policies*) From the table, we can find the action sequence that yields the maximum performance, which is $\boldsymbol{\alpha}_5 = \{\alpha_1, \alpha_0, \alpha_1\}$. Thus, to get the best performance, we need to take action α_1 at $l = 0$, α_0 at $l = 1$, and then α_1 at $l = 2$. This best action sequence corresponds to the "open-loop" control in control theory. The action sequence is pre-fixed before the system starts running. The structure of an *open-loop policy* is shown in Figure 1.2. With an open-loop policy, the output-and-action history are not used in determining the actions; i.e., the function $A_l = d_l(\boldsymbol{H}_l)$ does not depend on $\boldsymbol{H}_l$, $l = 0, 1, \ldots$, at all. The outputs may be random; the best policy corresponds to the best average performance.

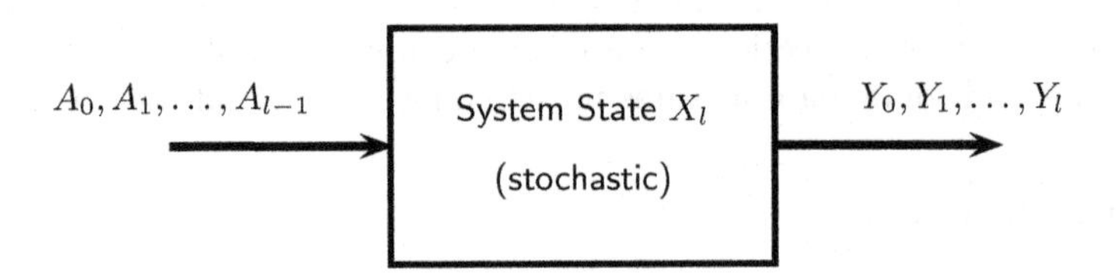

Fig. 1.2. The Structure of an Open-loop Policy

b) (*Action-history-dependent open-loop policies*) In practice, knowing the best action sequence is not enough. Suppose that α_0, instead of α_1, is taken by mistake at $l = 0$. Then, knowing the best action sequence $\boldsymbol{\alpha}_5$ alone does not indicate what next action should be taken in order to get the sub-optimal performance under this circumstance. To answer this question, we need to further examine Table 1.1.

Table 1.1 contains all the information about the optimization problem. From the table, we can find the best action taken at any time given the history of the actions taken up to that time. At $l = 0$, the action history $\boldsymbol{A}_0 = \emptyset$. Because the best performance lies in the lower part of the table, we need to choose $A_0 = d_0(\emptyset) = \alpha_1$. At $l = 1$, we may have two possible histories, $\boldsymbol{A}_1 = \alpha_1$ or α_0, by mistake. Note that $\boldsymbol{\alpha}_2$ is the best action sequence in the upper part of Table 1.1. Therefore, if α_0 is taken at $l = 0$, then we need to take α_1 at $l = 1$; i.e., the best policy at $l = 1$ given $\boldsymbol{A}_1 = \alpha_0$ is $d_1(\alpha_0) = \alpha_1$. Similarly, we have $d_1(\alpha_1) = \alpha_0$. At $l = 2$, we may have four possible action histories $\{\alpha_0, \alpha_0\}$, $\{\alpha_0, \alpha_1\}$, $\{\alpha_1, \alpha_0\}$, and $\{\alpha_1, \alpha_1\}$. Let us start with $\{\alpha_0, \alpha_0\}$. In this case, if we take α_0 at $l = 2$, we obtain the sequence $\boldsymbol{\alpha}_0$; and if instead we take α_1 at $l = 2$, we end up with $\boldsymbol{\alpha}_1$. Because $\boldsymbol{\alpha}_1$ performs better than $\boldsymbol{\alpha}_0$ does, we should take α_1 at $l = 2$. That is, we should set $d_2(\{\alpha_0, \alpha_0\}) = \alpha_1$. Similarly, from the table, we must have $d_2(\{\alpha_0, \alpha_1\}) = \alpha_0$, $d_2(\{\alpha_1, \alpha_0\}) = \alpha_1$, and $d_2(\{\alpha_1, \alpha_1\}) = \alpha_0$. Finally, the optimal policy is $\boldsymbol{d} := \{d_0, d_1, d_2\}$.

The decision on d_0 is based on the comparison: $\max\{\eta(\boldsymbol{\alpha}_i), i = 4, 5, 6, 7\} > \max\{\eta(\boldsymbol{\alpha}_i), i = 0, 1, 2, 3\}$; the decision on d_1 is based on the following two comparisons: $\max\{\eta(\boldsymbol{\alpha}_i), i = 2, 3\} > \max\{\eta(\boldsymbol{\alpha}_i), i = 0, 1\}$ and $\max\{\eta(\boldsymbol{\alpha}_i), i = 4, 5\} > \max\{\eta(\boldsymbol{\alpha}_i), i = 6, 7\}$; and the decision on d_2 is based on four comparisons: $\eta(\boldsymbol{\alpha}_1) > \eta(\boldsymbol{\alpha}_0)$, $\eta(\boldsymbol{\alpha}_2) > \eta(\boldsymbol{\alpha}_3)$, $\eta(\boldsymbol{\alpha}_5) > \eta(\boldsymbol{\alpha}_4)$, and $\eta(\boldsymbol{\alpha}_6) > \eta(\boldsymbol{\alpha}_7)$. There are altogether $2^0 + 2^1 + 2^2 = 7$ comparisons among these maximums.

In summary, in the optimal policy thus obtained, the action at time l, A_l, depends on the action history $\boldsymbol{A}_{l-1} = \{A_0, A_1, \ldots, A_{l-1}\}$; i.e., $A_l = d_l(\boldsymbol{A}_{l-1})$. This structure is shown in Figure 1.3. □

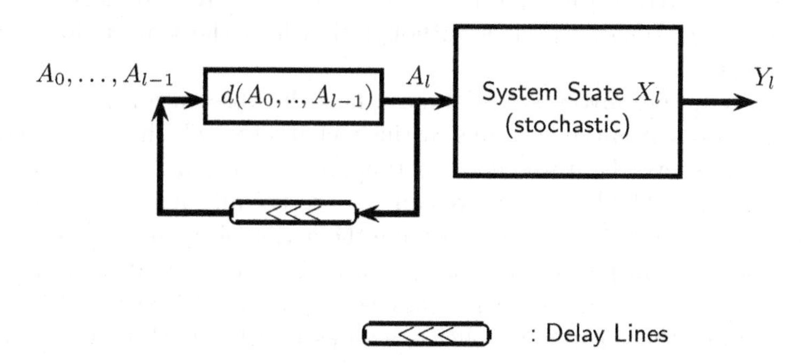

Fig. 1.3. The Optimal Action Depends on the Action History

The approach used in the example is called an *exhaustive search* because to obtain the optimal policy, we need to search through every policy. It is the simplest but the most time-consuming learning-optimization strategy. The example also indicates that, principally, by exhaustive search, we can find the optimal policy for the system performance without any prior knowledge about the structure or the status of the system.

In Example 1.1(a), it is shown that if there is no observation or the system is deterministic, a pre-fixed action sequence $\boldsymbol{\alpha}_5$ is enough for optimization, and the system may not need to do on-line adjustment if no mistake occurs in implementation. However, as shown in Example 1.1(b), on any particular time instant, the best action depends on the action history; this fact leads to the action-history-dependent open-loop policies, which adjust future actions when the past action sequence is not the optimal.

Closed-Loop (Feedback) Policies

If the system is involved with randomness and there are observable outputs, a fixed sequence of actions may not lead to the best performance. In such cases, in addition to the past history of the actions, the action chosen at time l, A_l, should also depend on the observations up to time l. This corresponds to the "closed-loop" or "feedback" control in control theory.

Example 1.2. In Example 1.1, we further assume that at each time instant the system outputs one of the two symbols, denoted as y_0 and y_1. The system's causality works as follows. At time l, $l = 0, 1, 2$, an observation Y_l is made first; after that, an action A_l is taken; then the system evolves to time $l+1$, at that time an observation Y_{l+1} is made; and then an action A_{l+1} is taken; and so on. Finally, at $l = 2$ a total reward (performance) is received after the action A_2 is taken. Now, the history of the system (as far as we can know) at time l is a sequence of observations and actions denoted as $\boldsymbol{H}_l = \{Y_0, A_0, Y_1, A_1, \ldots, Y_{l-1}, A_{l-1}, Y_l\}$, $l = 0, 1, 2$, $A_{-1} = \emptyset$. A history $\{y_0, \alpha_1, y_0, \alpha_0, y_1, \alpha_1\}$ may lead to a different performance from what $\{y_1, \alpha_1, y_1, \alpha_0, y_0, \alpha_1\}$ does, although they have the same action sequence $\{\alpha_1, \alpha_0, \alpha_1\}$.

When a system involves randomness, how the observation Y_l evolves after an action is taken is the system's intrinsic character, which is not controlled by us. In other words, if we choose action A_l, then the nature will determine Y_{l+1}, and so on. The best thing we can do is to use all the information that is available to us up to time l to determine the action $A_l = d_l(\boldsymbol{H}_l)$ such that on average the system performance is the best. Such a policy depending on the observation history is called a *feedback policy* or *closed-loop policy* in control theory. Unlike the open-loop policies, actions in a feedback policy cannot be pre-determined before the system operates.

In this example, $\boldsymbol{H}_0 = Y_0$, which may take two values y_0 and y_1. Thus, at $l = 0$ there are 2^2 sub-policies d_0 that map y_0 and y_1 to α_0 and α_1. At $l = 1$, $\boldsymbol{H}_1 = \{Y_0, A_0, Y_1\}$ may take $2^3 = 8$ different values. Finally, $\boldsymbol{H}_2 = \{Y_0, A_0, Y_1, A_1, Y_2\}$ may take $2^5 = 32$ different values. (How many different policies? see Problem 1.3.)

Table 1.2 gives an example of possible histories of actions and observations and their corresponding performance. To save space, in the table, we assume that for some reasons the observations of the system at time $l = 0$

and $l = 1$ are fixed as y_0 and y_1, respectively. The observation at $l = 2$, however, may take either y_0 or y_1, depending on the randomness involved. We further assume that the probabilities of $Y_2 = y_0$ and $Y_2 = y_1$ are both 0.5, equally. As shown in the table, there are two possible histories corresponding to each action sequence; e.g., if we take action sequence $\{\alpha_0, \alpha_1, \alpha_0\}$, we may in fact have either $\boldsymbol{h}_4 := \{y_0, \alpha_0, y_1, \alpha_1, y_0, \alpha_0\}$ or $\boldsymbol{h}_5 := \{y_0, \alpha_0, y_1, \alpha_1, y_1, \alpha_0\}$ with an equal probability of 0.5, respectively. We find that the average performance for every action sequence is the same as the performance corresponding to the same action sequence in Table 1.1 (e.g., the average performance for $\{\alpha_0, \alpha_0, \alpha_0\}$ is $3 = \frac{2+4}{2}$ and that for $\{\alpha_0, \alpha_0, \alpha_1\}$ is $7 = \frac{8+6}{2}$, etc.).

From the table, it is clear that given the history (up to $l = 1$) $\{y_0, \alpha_1, y_1, \alpha_0\}$, if we observe $Y_2 = y_1$ at $l = 2$, we should take α_0 to receive a reward of 10 (instead of taking α_1 to get 8). However, if we observe $Y_2 = y_0$, we definitely should take action α_1 at $l = 2$ to receive a reward of 12 (instead of taking α_0 to get 0!) Therefore, the pre-fixed action sequence $\{\alpha_1, \alpha_0, \alpha_1\}$ is no longer optimal. With the feedback policy, the average performance given the action sequence $\{\alpha_1, \alpha_0\}$ is $\frac{1}{2}(10 + 12) = 11$, which is larger than 10, the best performance for all the open-loop policies. Thus, *a feedback policy may achieve a better performance than all the open-loop policies.* □

	Action Sequences	Action-Observation Histories	Performance
$\boldsymbol{h}_0$	$\{\alpha_0, \alpha_0, \alpha_0\}$	$\{y_0, \alpha_0, y_1, \alpha_0, y_0, \alpha_0\}$	2
$\boldsymbol{h}_1$	$\{\alpha_0, \alpha_0, \alpha_0\}$	$\{y_0, \alpha_0, y_1, \alpha_0, y_1, \alpha_0\}$	4
$\boldsymbol{h}_2$	$\{\alpha_0, \alpha_0, \alpha_1\}$	$\{y_0, \alpha_0, y_1, \alpha_0, y_0, \alpha_1\}$	8
$\boldsymbol{h}_3$	$\{\alpha_0, \alpha_0, \alpha_1\}$	$\{y_0, \alpha_0, y_1, \alpha_0, y_1, \alpha_1\}$	6
$\boldsymbol{h}_4$	$\{\alpha_0, \alpha_1, \alpha_0\}$	$\{y_0, \alpha_0, y_1, \alpha_1, y_0, \alpha_0\}$	12
$\boldsymbol{h}_5$	$\{\alpha_0, \alpha_1, \alpha_0\}$	$\{y_0, \alpha_0, y_1, \alpha_1, y_1, \alpha_0\}$	4
$\boldsymbol{h}_6$	$\{\alpha_0, \alpha_1, \alpha_1\}$	$\{y_0, \alpha_0, y_1, \alpha_1, y_0, \alpha_1\}$	6
$\boldsymbol{h}_7$	$\{\alpha_0, \alpha_1, \alpha_1\}$	$\{y_0, \alpha_0, y_1, \alpha_1, y_1, \alpha_1\}$	2
$\boldsymbol{h}_8$	$\{\alpha_1, \alpha_0, \alpha_0\}$	$\{y_0, \alpha_1, y_1, \alpha_0, y_0, \alpha_0\}$	0
$\boldsymbol{h}_9$	$\{\alpha_1, \alpha_0, \alpha_0\}$	$\{y_0, \alpha_1, y_1, \alpha_0, y_1, \alpha_0\}$	10
$\boldsymbol{h}_{10}$	$\{\alpha_1, \alpha_0, \alpha_1\}$	$\{y_0, \alpha_1, y_1, \alpha_0, y_0, \alpha_1\}$	12
$\boldsymbol{h}_{11}$	$\{\alpha_1, \alpha_0, \alpha_1\}$	$\{y_0, \alpha_1, y_1, \alpha_0, y_1, \alpha_1\}$	8
$\boldsymbol{h}_{12}$	$\{\alpha_1, \alpha_1, \alpha_0\}$	$\{y_0, \alpha_1, y_1, \alpha_1, y_0, \alpha_0\}$	10
$\boldsymbol{h}_{13}$	$\{\alpha_1, \alpha_1, \alpha_0\}$	$\{y_0, \alpha_1, y_1, \alpha_1, y_1, \alpha_0\}$	8
$\boldsymbol{h}_{14}$	$\{\alpha_1, \alpha_1, \alpha_1\}$	$\{y_0, \alpha_1, y_1, \alpha_1, y_0, \alpha_1\}$	3
$\boldsymbol{h}_{15}$	$\{\alpha_1, \alpha_1, \alpha_1\}$	$\{y_0, \alpha_1, y_1, \alpha_1, y_1, \alpha_1\}$	9

Table 1.2. The Action-Observation Histories and Their Rewards in Example 1.2

The structure of a feedback or a closed-loop policy is shown in Figure 1.4. In both examples, we do not know how the actions control the system's operation.

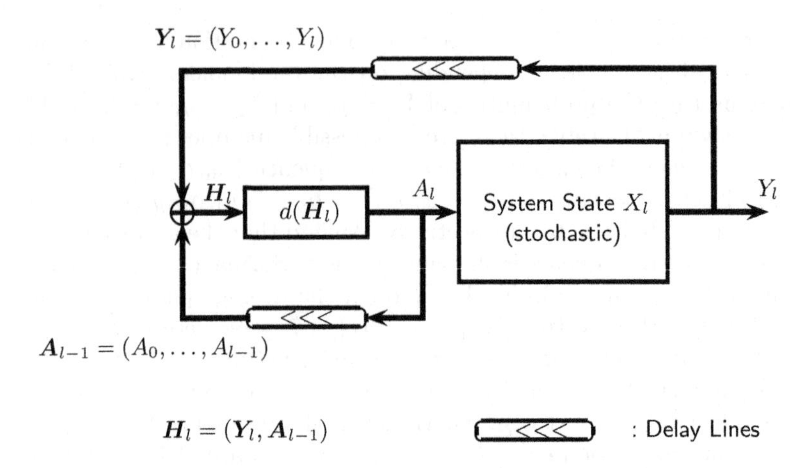

Fig. 1.4. The Optimal Action Depends on the Action-Observation History

Stationary and Randomized Policies

For simplicity, in the above examples, we used very short sample paths to explain the ideas. Practically, the sample paths can be very, even infinitely, long.

If the system is completely observable, then the observation history $\boldsymbol{Y}_l$ is the same as the state history $\boldsymbol{X}_l$ and the policy is $A_l = d_l(\boldsymbol{X}_l, \boldsymbol{A}_{l-1})$. Because of the Markov property of a state process, the current state X_l contains all the information in the system's history in regard to its future behavior. We may expect that in many cases a policy depending on only X_l may do as well as a policy depending on the entire history $\boldsymbol{H}_l = (\boldsymbol{X}_l, \boldsymbol{A}_{l-1})$ for controlling the system's future behavior. Therefore, we may only consider the policies $A_l = d_l(X_l)$, $l = 0, 1, \ldots$.

A policy $A_l = d_l(X_l)$, $X_l \in \mathcal{S}$, $A_l \in \mathcal{A}$, $l = 0, 1, \ldots$, is called a *stationary policy* if it does not depend on time l; such a policy is denoted as $A = d(X)$, $X \in \mathcal{S}$, which is a mapping from the state space $\mathcal{S}$ to the action space $\mathcal{A}$.

The action $d(i)$, $i \in \mathcal{S}$, controls the transition probabilities of state i. With a stationary policy d, the transition probabilities when the state is $i \in \mathcal{S}$ are denoted as $p^{d(i)}(j|i)$, $j \in \mathcal{S}$. The system under policy $d(X)$ is Markov, and the corresponding transition matrix is denoted as $P^d := [p^{d(i)}(j|i)]$. Therefore, a policy d is also referred to as a policy P^d. When performance is involved, the reward function may also depend on the policy d and is denoted as $f^d(i)$, $i \in \mathcal{S}$. We may write it in a vector form $f^d := (f^d(1), \ldots, f^d(S))^T$. With the reward vector, we may refer to (P^d, f^d) as a policy. In addition, we will simply use (P, f) as a generic notation for a policy. In many cases, $f^d(i)$ depends only on $d(i)$, $i \in \mathcal{S}$, and the reward vector is then $f^d = (f(1, d(1)), \ldots, f(S, d(S)))^T$, where the superscript "T" denotes transpose.

Finally, it should be noted that when the action depends on the "state" history, the resulting "state" process is no longer Markov (we put the word "state" in quotation marks because it is not a state in the strict sense). The following example shows that there may exist a policy $d(X)$ depending only on the current state that does as well as a history-dependent policy, in terms of the long-run average performance.

Example 1.3. Consider a process $\widetilde{\boldsymbol{X}} = \{\widetilde{X}_0, \widetilde{X}_1, \ldots\}$, $\widetilde{X}_l \in \mathcal{S}$, $l = 0, 1, \ldots$. Suppose that at any time l, $l = 1, 2, \ldots$, the transition probability is determined by the current state $\widetilde{X}_l$ and the previous state $\widetilde{X}_{l-1}$ and is denoted as $p[k|(i,j)] := p[\widetilde{X}_{l+1} = k|\widetilde{X}_{l-1} = i, \widetilde{X}_l = j]$, $i, j, k \in \mathcal{S}$. This process is not Markovian on the state space $\mathcal{S}$ because the state at $l+1$, $\widetilde{X}_{l+1}$, depends on state at $l-1$, $\widetilde{X}_{l-1}$.

Define $\widetilde{Y}_l := (\widetilde{X}_{l-1}, \widetilde{X}_l)$, $l = 1, 2 \ldots$. Then, $\widetilde{\boldsymbol{Y}} = \{\widetilde{Y}_1, \widetilde{Y}_2, \ldots,\}$ is a Markov chain with transition probabilities

$$\begin{aligned} p[(j,k)|(i,j)] &= p[k|(i,j)], \qquad i,j,k \in \mathcal{S}; \\ p[(k',k)|(i,j)] &= 0, \qquad \text{if } k' \neq j,\ i,j,k \in \mathcal{S}. \end{aligned}$$

Let $\pi(i,j)$, $i, j \in \mathcal{S}$, be the steady-state probabilities of the Markov chain $\widetilde{\boldsymbol{Y}}$. Then, the steady-state probability flow balance equation is

$$\pi(k',k) = \sum_{i,j} \pi(i,j) p[(k',k)|(i,j)], \qquad k, k' \in \mathcal{S}. \tag{1.4}$$

The steady-state probability of $\widetilde{\boldsymbol{X}}$ is

$$\pi(k) = \sum_{k'} \pi(k',k).$$

Summing over $k' \in \mathcal{S}$ on both sides of (1.4), we get

$$\pi(k) = \sum_{i,j} \pi(i,j) p[k|(i,j)], \qquad i,j,k \in \mathcal{S}. \tag{1.5}$$

Let $\pi(i|j)$ be the steady-state conditional probability of $\widetilde{X}_{l-1} = i$ given that $\widetilde{X}_l = j$ (i.e., the conditional probability of $\widetilde{X}_{l-1} = i$ given that $\widetilde{X}_l = j$ in a stationary process $\widetilde{\boldsymbol{X}}$), and let $p(k|j)$ be the steady-state conditional probability of $\widetilde{X}_{l+1} = k$ given $\widetilde{X}_l = j$. We have

$$\pi(i|j) = \frac{\pi(i,j)}{\pi(j)}$$

and

$$p(k|j) = \sum_i \{\pi(i|j) p[k|(i,j)]\}.$$

Thus,

$$\pi(j)p(k|j) = \sum_{i} \{\pi(i,j)p[k|(i,j)]\}.$$

Summing over $j \in \mathcal{S}$ on both sides and using (1.5), we get

$$\pi(k) = \sum_{j} \pi(j)p(k|j). \tag{1.6}$$

Let $\boldsymbol{X}$ be a Markov chain with transition probabilities $p(k|j)$, $j,k \in \mathcal{S}$. From (1.6), $\pi(k)$, the steady-state probability of state k of $\widetilde{\boldsymbol{X}}$ equals the steady-state probability of state k of $\boldsymbol{X}$, $k \in \mathcal{S}$. Therefore, a Markov chain with a history-independent policy $p(k|j)$, $j \in \mathcal{S}$, has the same steady-state probabilities as a non-Markov chain with a history-dependent policy $p[k|(i,j)]$, $i,j,k \in \mathcal{S}$. Finally, for any reward function $f(X_l)$, $l = 0,1,\ldots$, from (1.7), the average performance $\eta = \sum_i \pi(i)f(i)$ is the same for both processes.

Furthermore, from the definition of $\boldsymbol{X}$, we can easily check that the steady-state probabilities of $(X_l = i, X_{l+1} = j)$, $\pi(i,j)$, $i,j \in \mathcal{S}$, are the same for both $\boldsymbol{X}$ and $\widetilde{\boldsymbol{X}}$. Thus, for any reward function $f(X_l, X_{l+1})$, $l = 0,1,\ldots$, the average performance $\eta = \sum_{i,j} \pi(i,j)f(i,j)$ is the same for both processes.

However, as shown in Problem 1.6, if the reward function depends on three consecutive states (X_l, X_{l+1}, X_{l+2}), there may not exist a history-independent policy defined on the same state space that is equivalent to a history-dependent policy. □

Of course, even if an equivalent history-independent policy on the same state space does not exist, we can always construct an equivalent Markov chain by enlarging the state dimension. For example, the process $\widetilde{\boldsymbol{Y}}$ in Example 1.3 is Markov.

A (stationary) *randomized policy* $\nu = d(X)$ assigns a distribution ν over the action space $\mathcal{A}$ for every state $X = i \in \mathcal{S}$; it is a mapping from the state space $\mathcal{S}$ to the space of the distributions over the action space. For example, suppose that $\mathcal{A} = \{\alpha_1, \alpha_2, \ldots, \alpha_M\}$. For any state $i \in \mathcal{S}$, a randomized policy assigns a distribution $\nu = (p_1(i), p_2(i), \ldots, p_M(i))$ on $\mathcal{A}$. When the system state is i, we take action α_k with probability $p_k(i)$, $k = 1,2,\ldots,M$, $i \in \mathcal{S}$ and $\sum_{k=1}^{M} p_k(i) = 1$. A deterministic policy is a special case of a randomized policy ν where $p_k(i) = 1$ for some $k \in \{1,2,\ldots,M\}$, with k depending on i, $i \in \mathcal{S}$.

1.1.3 Fundamental Limitations of Learning and Optimization

Exhaustive Search is Not Feasible

In the last section, we have defined policies. A set of policies constitutes a policy space. To find an optimal policy in a given policy space is a typical search problem. As illustrated in Example 1.1, an optimal policy can be found by exhaustive search; i.e., by comparing the performance of all the policies. However, even for a small problem, the policy space is too large for the exhaustive search approach. For example, for a (small) system with $S = 100$ states and $M = 2$ actions available in each state, the number of stationary policies is $M^S = 2^{100} \approx 10^{30}$! That is, the number of policies increases exponentially with respect to the number of states. Therefore, exhaustive search, which requires computing or comparing the performance of every policy, is not computationally feasible for most practical problems.

Moreover, if there is no additional information about the system structure (such as the Markov model), any optimization scheme is no better than blind searching. This is formulated as the "No Free Lunch Theorem", see [152]. The recently developed *"Ordinal Optimization"* approach deals with the tradeoff between accuracy and efficiency of random search. It proposes a novel and interesting idea of a "soft goal" and opens up a new perspective for optimization. This is beyond the scope of this book and the readers are referred to [80, 93, 123, 139, 140, 149, 150, 151, 152, 175, 181, 184, 266] for details.

Learning and Optimization

To develop efficient algorithms for performance optimization, we need to explore the special features of a system. This process is called *learning.* For dynamic systems, learning may involve observing and analyzing a sample path of a system to obtain necessary information; this is in the normal sense of the word "learning", as it is used in research areas such as *reinforcement learning.* Simulation-based and on-line optimization approaches are based on learning from sample paths. On the other hand, we may also analytically study the behavior of a system under a policy to learn how to improve the system performance. In a wide sense, we shall also call this analytical process "learning".

So far, we have described the optimization problem. We know that exhaustive search, which requires no information about the system dynamics except the performance for every policy, is not computationally feasible for most practical problems. To develop efficient optimization approaches, we need to explore the special feature of a system by learning. Naturally, our next step is to determine what information we need to learn, either from sample paths or analytically, and how we use such information to achieve our goal; i.e., to find an optimal policy, or to improve the system performance, and how we can achieve our goal with as little information about the system structure as possible.

Obviously, the task is complicated and we are facing a vast forest and wish to find a path in it to reach our destination at the top of a peak. It is wise to pause for a short while and take an overview of the forest from the outside to see which directions may possibly lead us to our goal quickly. Indeed, we are constrained by some philosophical and logical facts that significantly limit what we can do. These facts are simple and intuitively obvious, yet they provide general principles that chart the paths in our journey of developing optimization theories and methodologies. Because of the importance as well as the simplicity of these facts, we state them as the "fundamental limitations":

The Fundamental Limitations of Learning and Optimization

A. A system can be run and/or studied under only one policy at a time.
B. By learning from the behavior of a system under one policy, we cannot obtain the performance of other policies, if no structural information of the system is available.
C. We can only compare two policies at a time.

These simple rules describe the boundaries in developing learning and optimization approaches. First of all, if there is no structural information for the system, from the fundamental limitations A and B, we need to observe/analyze every policy to get or to estimate its performance, and from the fundamental limitation C, for M policies we need to make $M-1$ comparisons. This is the exhaustive search method.

Exhaustive Search

Given M policies d_i, $i = 1, 2, \ldots, M$. Let η^{d_i} be the performance of policy d_i, $i = 1, 2, \ldots, M$.

i. Set $\widetilde{d} := d_1$, and $\widetilde{\eta} := \eta^{d_1}$;
ii. For $i := 2$ to M, do
   if $\eta^{d_i} > \widetilde{\eta}$ then set $\widetilde{d} := d_i$ and $\widetilde{\eta} := \eta^{d_i}$.

The algorithm outputs an optimal policy $\widetilde{d} = d^*$. The main operation in the algorithm is the comparison of the performance of two policies, $\eta^{d_i} > \widetilde{\eta}$. It is important to note that to verify this relationship we may not need to obtain the exact values of the performance of these two policies. For example, if the performance of two policies is quite different, then we may need only run a short simulation for each policy to verify this relationship. This may save computation significantly, see the references for ordinal optimization [150, 151, 175, 181, 184, 266].

The fundamental limitation B indicates that if we want to do better than exhaustive search, we need to use the special features of a system. However, we wish to develop approaches that require as little structural information and can be applied to as many systems as possible. The question is *"HOW"*. These fundamental limitations also provide us with some hints.

Performance Gradient

As indicated by the fundamental limitations A and B, if we analyze a system's behavior under one policy, we can hardly know its behavior under other policies. It is natural to believe that if two policies are "close" to each other, then the system under these two policies may behave similarly. If this is the case, when we are analyzing a system under a policy, it might be easier to "predict" the system behavior under a "close" policy and then to calculate its performance than to do the same for a policy that is "far away". In other words, to predict the performance for a "close" policy may require as little knowledge about the system structure as possible.

If a policy space can be characterized by a continuous parameter θ, then two policies are "close" if their corresponding values for θ are close. Such a policy space is called a *continuous policy space*. For example, for Markov systems policies correspond to transition probability matrices. Therefore, two policies can be viewed as "close" if their transition probability matrices are close (item-by-item). In modelling manufacturing or communication networks, policies may be characterized by production rates or transmission rates. Two policies are close if their corresponding rates are close. In randomized policies, the distributions $(p_1, p_2, \ldots, p_M)$ over the action space $\mathcal{A} = \{\alpha_1, \alpha_2, \ldots, \alpha_M\}$ are continuous variables. Two randomized policies are close if their corresponding distributions are close.

Therefore, a reasonable step towards developing efficient and generally applicable approaches is to look at a "neighborhood" of a policy. The neighborhood must be small enough, so that the behavior of the system under the policies in this neighborhood of the policy can be predicted with as little knowledge about the system structure as possible. In mathematical terms, "small enough" is precisely described by the word "infinitesimal". When the performance of the policies in an infinitesimal neighborhood of a policy is known, we can further get the gradient of the performance in the policy space at this policy.

We may summarize the above discussion by the following *Corollary A:*

Corollary A:

With some knowledge about the system structure under different policies, by studying the behavior of a system under one policy, we can determine the performance of the system under the policies in a small neighborhood of this policy; i.e., determine the performance gradient.

The prediction of the performance for other (neighboring) policies while analyzing the system under one policy can be done analytically, if we know the structure (usually based on a model) and the values of its parameters.

However, in many cases, we always start by analyzing a sample path of the system. This is because

1. A sample path clearly illustrates the system dynamics, and sample-path-based analysis stimulates intuitive thinking.
2. In many practical problems, the size of the problem is too large for any analytical solution, or we may have only partial information about the system; for example, in some cases, we may only know the structure of the system but do not know values of its parameters, or in some other cases, we know the values of the parameters, but the system structure is too complicated to model. Sample-path-based algorithms may be implemented easily even with these constraints.

The results obtained by the sample-path-based approach can also be expressed in an analytical form. In Chapter 2, we can see that the sample-path-based approach leads to analytical formulas that may not be easily perceived purely from an analytical point of view.

Performance Differences

The gradient method does not apply to discrete policy spaces. For discrete policy spaces, we need to compare the performance of different policies that may not be close to each other.

The fundamental limitation C implies more than it seems on the surface. It says that all we can do in terms of optimization is based on a simple comparison of two policies. In other words, if we cannot compare two policies, then we have no way to do optimization. Furthermore, we may even emphasize that the performance difference formula contains all the information about what we can do in performance optimization. This simple philosophical point guides the direction of our research in optimization: *We should always start with developing a formula for the difference of the performance measure of any two policies and then to investigate what we can learn from this performance difference formula.* In many cases, it is not difficult to derive such a difference formula for a particular problem, yet the insights provided and the results thus obtained can be remarkable.

The above philosophical point is better illustrated in Chapters 4 and 8. In Chapter 4, we show that the theory of MDPs is based on the performance difference formulas and this point of view provides direct, simple, and intuitively clear proofs for many results. In Chapters 9 and 8, we propose and develop the event-based optimization approach with this simple point of view; the approach further utilizes the system structure to overcome the computational issues and some other difficulties.

How much we can get from the performance difference formula depends on the system structure. So far, the best result is that with some assumptions such as the independent-action assumption in Markov decision processes, by analyzing the system's behavior under one policy, we can find another policy that performs better, if such a policy exists. See Chapters 4 and 8 for details.

We may summarize the above discussion by the following *Corollary B:*

Corollary B:

With some assumptions on the system structure, by studying the behavior of a system under one policy, we can find a policy that performs better, if such a policy exists.

Like many other rules, physical or social, the fundamental limitations of learning and optimization are summaries of many research and explorations. In this sense, these limitations and many statements in this section are hindsight. In addition, they are philosophical remarks rather than precise and accurate scientific statements. They can only be used as guidelines. In this book, we will show that if we pursue these philosophical thoughts, we may obtain many results in learning and optimization. The meaning and significance of these remarks will become clear after reading the remaining chapters. Readers may revisit this section after understanding the contents of other chapters in this book.

1.1.4 A Sensitivity-Based View of Learning and Optimization

In summary, if no information about the system structure is available, we can only do exhaustive, or blind, searches in the policy space. There are different types of policies: those depending on the histories of both input actions and output observations, and those depending only on the current state. If we know something about the structure and the dynamics of the system, we may develop efficient optimization techniques (analytic, simulation, on-line, learning, etc.).

The fundamental limitations of learning and optimization and their corollaries sketch out the directions of developing efficient and widely applicable learning and optimization approaches with as little information about the system structure as possible. There are two feasible directions: First, because we can only learn from one policy at a time, we may at most obtain local (in the neighborhood of a policy) information in the policy space; this leads to performance gradients or derivatives. Second, because we can only compare two policies at a time, we may start with the performance difference formulas of any two policies in developing learning and optimization methods. In short, these directions can be characterized by performance derivatives and performance differences, respectively. We will refer to this as a sensitivity-based view [56].

Over the years, different disciplines have been developed from different perspectives. The main theme of this book is to show *how the main concepts and results in these different disciplines can be introduced and derived through a unified framework from the sensitivity-based view explained above in a clear and intuitive way, and to show how this sensitivity-based view leads to new results and research directions.*

As explained above, the central piece of this unified framework is the two types of performance sensitivity formulas, one for performance differences, and the other for performance derivatives. These two types of sensitivity formulas lead to two types of learning and optimization approaches.

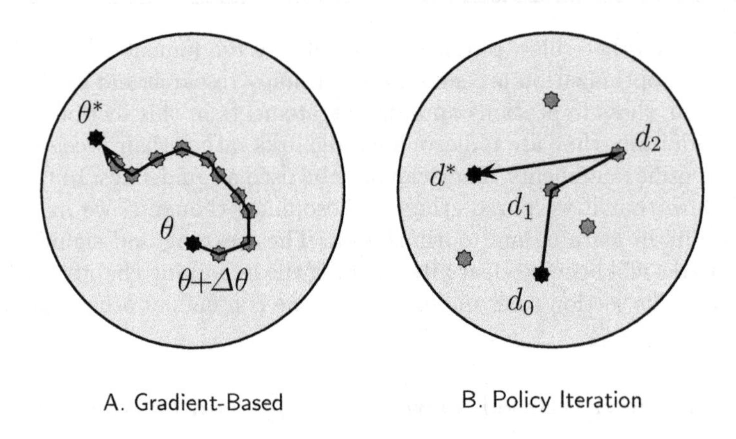

Fig. 1.5. Two Types of Optimization Approaches

The first type of approach is centralized on performance gradients. In this book, we mainly discuss the technique called *perturbation analysis (PA)*. With PA, we can predict the effect of small (infinitesimal) changes in system parameters (policies) on system performance and obtain the performance derivatives with respect to the parameters. We can develop gradient-based optimization approaches using PA (also called the *policy-gradient* method in reinforcement learning). This approach applies to problems where policy spaces are parameterized. The basic idea is shown in Figure 1.5.A. We first set the parameter θ to be any value and determine the performance gradient at θ with PA. Then we change θ slightly along the direction of the gradient and determine the gradient again at the new θ. We repeat this procedure until reaching a point θ^* at which the performance gradient is zero; this is a local optimal point. The performance gradient can be calculated analytically, or estimated from a sample path. When the gradient estimates contain noise, stochastic approximation techniques can be used in the optimization procedure.

The second type of learning and optimization approach is based on the comparison of the system performance measures of two different policies. The approach strongly depends on the system structure. A well-known result in this direction is: When the actions taken in different states are independent, it may be possible to use the information learned by observing or analyzing the system behavior under the current policy to determine a policy under which the performance of the system is better, if such a policy exists (albeit it is not possible to determine the exact value of the performance for the "better" policy). This leads to the *policy iteration* procedure shown in Figure 1.5.B. We start with any policy d_0, learn from its behavior and find a better policy d_1, then learn from d_1 and find a better policy d_2, and so on until the best policy d^* is reached. This approach is based on the performance difference formula and depends heavily on its particular form.

However, it is not always possible to find a better policy using the information obtained only from the current policy. In fact, for many problems especially when the actions in different states are correlated, such a policy-iteration type of approach does not work. It is shown in Chapters 8 and 9 that, under these circumstances, the optimization problem can be modelled by using the event-based policies and we can determine whether policy iteration works for the event-based policies by using the performance difference formulas.

In summary, the two types of approaches illustrated in Figure 1.5 are based on two types of performance sensitivity formulas, one for performance derivatives, and the other for performance differences. In this book, we show that this sensitivity-based view provides a unified framework for different formulations and solutions to different optimization problems with different performance criteria in different research disciplines (Chapters 2 to 6). In addition, we also show that with the sensitivity-based view, new results and approaches can be developed, by following the same ideas illustrated in Figure 1.5 (Chapter 8), for many systems that may not fit the standard formulation in the existing literature. This leads to new research topics including event-based optimization (Chapters 8 and 9).

1.2 Problem Formulations in Different Disciplines

Different disciplines in learning and optimization, such as Markov decision processes (MDPs) in operations research, perturbation analysis (PA) in discrete event dynamic systems (DEDSs), reinforcement learning (RL) in computer science, and identification and adaptive control (I&AC) in control systems, have different formulations of the system structures. These different disciplines also differ in the way they utilize the structural information.

Roughly speaking, both MDPs and RL assume a Markov structure for the systems; PA started with a queueing network-type of structure and was extended to the Markov structure later; in I&AC, the system evolution is described by dynamic (differential or difference) equations. By and large, the MDP literature focuses on analytical solutions, and the parameters are usually assumed to be known to us. RL emphasizes the learning aspect, and algorithms are developed based on sample paths obtained from simulation. PA extracts information from a sample path to answer the what-if type of questions: What is the effect on a system performance if the system parameter or structure changes? This is done by predicting the system behavior after the parameter or structural changes. I&AC is mainly an analytical approach that utilizes special features of the system's dynamic structure described by differential or difference equations.

In this section, we briefly describe the problem formulations for these different areas and introduce a new area, event-based optimization. We try to provide an overview with our sensitivity-based view in each of these areas. The details will be discussed in the remaining chapters of the book.

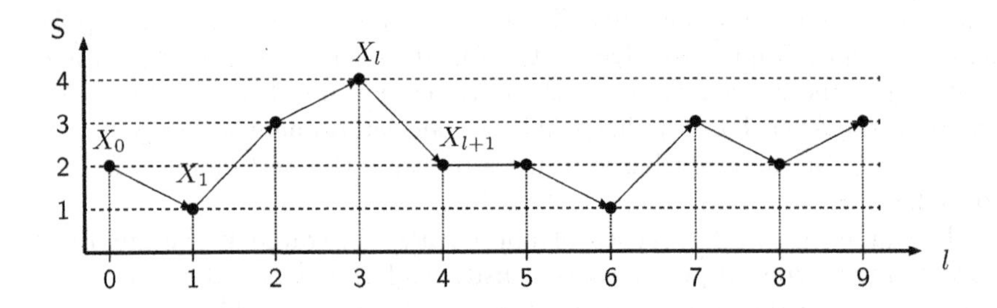

Fig. 1.6. A Sample Path of a Markov Chain

The System Model

In many problems, the behavior of the black box in Figure 1.1 is described by a Markov chain. That is, we assume that under the influence of any possible input sequence, the system's evolution possesses the Markov property. In most problems discussed in this book, we also assume that the state space is finite and is denoted as $\mathcal{S} = \{1, 2, \ldots, S\}$. A sample path of the Markov chain (see Figure 1.6) is denoted as $\boldsymbol{X} = \{X_0, X_1, \ldots\}$, with $X_l \in \mathcal{S}$ being the state at time $l = 0, 1, \ldots$. The evolution (transition) of the system is determined by the transition probabilities $p(j|i)$, $i, j \in \mathcal{S}$, which is the conditional probability of $X_{l+1} = j$ given that $X_l = i$. The matrix $P = [p(j|i)]_{i,j=1}^{S}$ is called the transition probability matrix, which is usually pictorially illustrated by a state-transition diagram shown in Figure 1.7. In the figure, the number near an arrow is the transition probability of the state transition indicated by the arrow.

In most of this book, we study the long-run average performance measure η of a Markov chain defined as

$$\eta(i) = \lim_{L \to \infty} \frac{1}{L} \sum_{l=0}^{L-1} E\left[f(X_l) | X_0 = i\right],$$

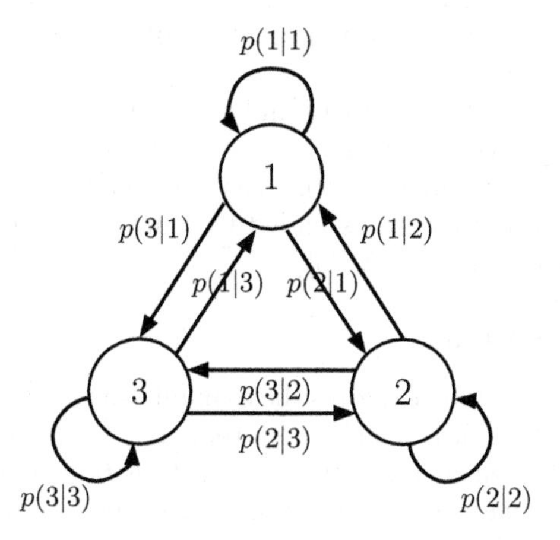

Fig. 1.7. The State-Transition Diagram of a Three-State Markov Chain

where $f(x), x \in \mathcal{S}$, is a mapping $\mathcal{S} \to \mathcal{R}$, called a *performance (or reward) function*, and "E" denotes the expectation in the probability space on which the Markov chain is defined. When the system is ergodic, this becomes

$$\eta = \lim_{L\to\infty} \frac{1}{L} \sum_{l=0}^{L-1} f(X_l) = \sum_{i\in\mathcal{S}} \pi(i) f(i), \qquad \text{w.p.1}, \tag{1.7}$$

which is independent of the initial state X_0. For a more comprehensive review of Markov chains and Markov processes, see Appendix A.

With the Markov model, in Part I of this book (Chapters 2 to 7), we assume that the observation at time l, Y_l, equals X_l, $l = 0, 1, \ldots$; that is, we assume that the system is completely observable. When a stationary policy d is used, we denote the corresponding quantities as P^d, f^d, and η^d, etc.

In some cases such as for queueing systems, we need the continuous-time model. Then, P^d is replaced by the infinitesimal generator or other system parameters, and the summation in (1.7) is replaced by an integration over $[0, \infty)$.

1.2.1 Perturbation Analysis (PA)

PA estimates the performance derivatives with respect to system parameters by analyzing a single sample path of a stochastic dynamic system. This is in line with Corollary A of the fundamental limitations in Section 1.1.3. The most significant contribution of PA is that it testifies to the fact that *a sample path of a dynamic system may contain information not only for the performance of the system under observation, but also for the derivatives of the performance with respect to system parameters.* Because PA emphasizes the dynamic nature of a stochastic system, such a system is also called a *discrete event dynamic system (DEDS) [72, 138]*.

PA was proposed in the late 1970s and early 1980s [72, 107, 112, 142]. The early work on PA focused on queueing systems. Appendix C contains a survey of the main results of queueing theory that are related to this book. Later, the basic principles of PA were extended to Markov systems (both discrete- and continuous-time models) [62, 70].

PA of Queueing Systems

PA of queueing systems fully explores the dynamic nature of queueing systems. The dynamic evolution of a queueing system can be clearly illustrated by a sample path. In PA of queueing systems, a *perturbation* on a sample path is a small delay (*infinitesimal perturbation*) in a customer transition time. A small change in a system parameter, say the mean service time of a server, induces a series of perturbations on a sample path. (E.g., if the mean service time of a server increases by a small amount, then the service time of each customer at this server will increase by a small amount, and the service completion time of each customer will be delayed by a small amount.) The average effect of a perturbation at server i when the system is in state $\boldsymbol{n}$ on a system's performance can be measured by a fundamental quantity called a *perturbation realization factor*, denoted as $c^{(f)}(\boldsymbol{n}, i)$, which satisfies a set of linear equations and can be estimated by observing and analyzing a single sample path of the queueing system with the current parameters. Finally, the effect of a small (infinitesimal) change in a system parameter on the system performance equals the sum of the effects of all the perturbations induced by the parameter change on a sample path. See Section 2.4 for a detailed discussion. These basic principles are illustrated in Figure 1.8.

With these principles, many results have been developed. In particular, it has been shown that in a closed Jackson network, the "normalized" derivative of the system throughput (the number of customers served by all the servers in the network per unit of time) with respect to a mean service rate equals the expected value of the perturbation realization factor, which can be estimated on a single sample path with a very effective algorithm; see Section 2.4.3 and [46, 49, 51]. The approach fits the general framework of learning and optimization: We learn the information from a sample path using the dynamics with the queueing structure to predict the performance of the system under a slightly changed parameter. Optimization can be implemented by stochastic approximation methods together with the performance derivative estimates; Refs. [82, 83, 84, 141] contain some examples of PA-based performance optimization.

PA of Markov Systems

The PA principles illustrated in Figure 1.8 were extended to performance sensitivities with respect to the transition probabilities of Markov systems in the

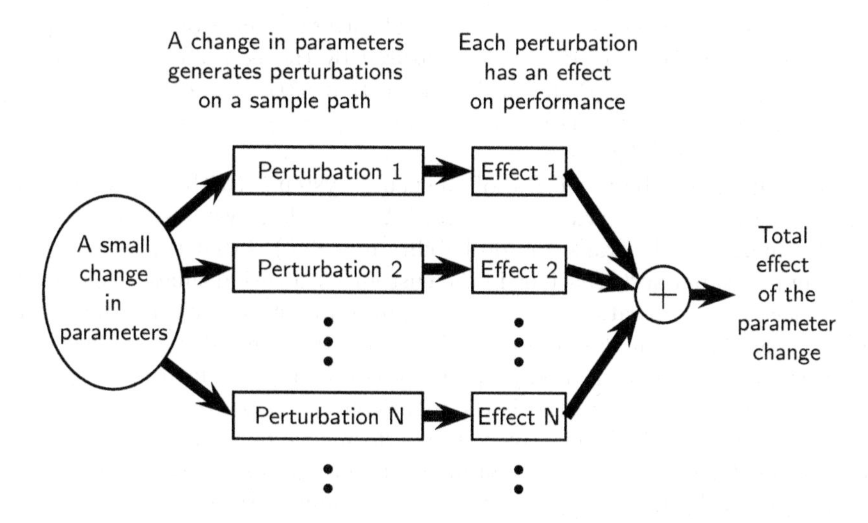

Fig. 1.8. The Basic Principles of Perturbation Analysis

mid-1990s (see [70] for discrete-time Markov chains and [62] for continuous-time Markov processes). In this approach, the behavior of the black box in Figure 1.1 is described by a Markov model with transition probability matrix P and the performance measure η is defined in (1.7). We assume that the states X_l are observable, i.e.; $Y_l = X_l$, $l = 0, 1, \ldots$. In this model, a policy d corresponds to a transition probability matrix denoted as P^d. We wish to get the performance sensitivity around a policy P^d in the policy space by analyzing the system's behavior under this policy P^d.

Let P^h be another policy, and we assume that the performance function f is the same for both policies P^d and P^h, and let $\Delta P = P^h - P^d$. (Sometimes we need to specifically denote it as $(\Delta P)^{d,h}$.) Define $P_\delta = P^d + \delta(\Delta P) = (1-\delta)P^d + \delta P^h$, $0 \leq \delta \leq 1$. P_δ (sometimes we need to specifically denote it as $P_\delta^{d,h}$) is a randomized policy: With policy P_δ, in any state $k \in \mathcal{S}$ the system moves according to $p^{h(k)}(j|k)$, $j \in \mathcal{S}$, with probability δ, and moves according to $p^{d(k)}(j|k)$, $j \in \mathcal{S}$, with probability $1-\delta$. Let π_δ and η_δ be the steady-state probability and the performance measure associated with P_δ. (Sometimes we need to specifically denote π_δ and η_δ as $\pi_\delta^{d,h}$ and $\eta_\delta^{d,h}$.) We have $P_0 = P^d$, $P_1 = P^h$ and $\eta_0 = \eta^d$. The performance derivative at policy P^d along the direction ΔP (from P^d to P^h) is $\frac{d\eta_\delta}{d\delta}|_{\delta=0} = \lim_{\delta\to 0} \frac{\eta_\delta - \eta}{\delta}$. Different P^hs correspond to different directional derivatives in the policy space.

The performance derivatives are obtained by predicting how the system would behave if we slightly perturb the transition probability matrix from P^d to P_δ, $\delta << 1$. Small changes in P^d induce a series of perturbations on a sample path of P^d. A perturbation on a sample path is a "jump" from one state i to another state j, $i, j \in \mathcal{S}$ (i.e., at some time l, the Markov chain with P^d was in state i, $X_l = i$, however, after the transition probabilities change slightly to

P_δ, the system becomes in state $X_l = j$). The average effect of such a jump on the system performance η^d can be measured by the *perturbation realization factor*, denoted as $\gamma^d(i,j)$. It can be shown that $\gamma^d(i,j) = g^d(j) - g^d(i)$, for all $i, j \in \mathcal{S}$, where $g^d(i)$ is called the *performance potential* (or simply the *potential*) of state i [62, 70].

The difference between PA of queueing systems (with respect to the changes in the mean service times) and PA of Markov systems (with respect to the changes in the transition probability matrices) is that the former deals with infinitesimal perturbations in the customer transition times, while in the latter case, a perturbation is a change ("jump") in states, which is finite and has a "big" effect on the system performance. In both cases, the effect of a perturbation can be measured by perturbation realization factors.

The performance potential is the main concept of performance optimization of Markov systems. Intuitively, the performance potential of state i, $g(i)$, of a policy P measures the "potential" contribution of state i to the long-run average reward η in (1.7). It is defined on a sample path of P as

$$g(i) := E\left\{\sum_{l=0}^{\infty}[f(X_l) - \eta]\Big|X_0 = i\right\}. \tag{1.8}$$

From this, we can easily derive

$$\begin{aligned} g(i) &= \text{contribution of the current state } i \\ &\qquad + \text{expected long term "potential" contribution of the next state} \\ &= (f(i) - \eta) + \sum_{j \in \mathcal{S}} p(j|i)g(j). \end{aligned}$$

This can be written in a matrix form called the *Poisson equation*:

$$(I - P)g + \eta e = f, \tag{1.9}$$

where $g = (g(1), \ldots, g(S))^T$ is the potential vector, and $e = (1, 1, \ldots, 1)^T$ is a column vector with all components being ones.

From the definition of $g(i)$ in (1.8), we can see that the effect of a jump from state i to j on the long-run average reward (1.7) can be measured by $\gamma(i,j) = g(j) - g(i)$. Finally, the effect of a small (infinitesimal) change in a Markov chain's transition probability matrix (from P^d to P_δ) on the long-run average reward (1.7) can be decomposed into the sum of the effects of all the single perturbations (jumps on a sample path) induced by the change on a sample path. These PA principles are the same as those illustrated in Figure 1.8. With these principles, we can intuitively derive the formulas for the performance derivative along any direction in the policy space (see Chapter 2):

$$\left.\frac{d\eta_\delta}{d\delta}\right|_{\delta=0} = \pi^d(\Delta P)^{d,h} g^d = \pi^d(P^h - P^d)g^d. \tag{1.10}$$

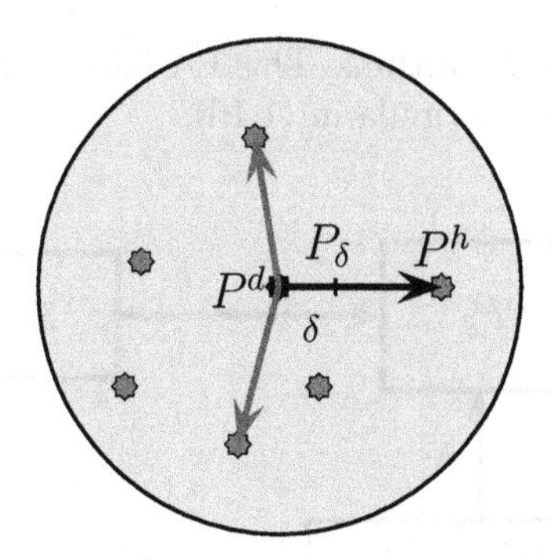

Fig. 1.9. The Directional Derivatives Along Any Direction

(This formula can be also easily derived from the Poisson equation, see Problem 1.8; however, the PA principles provide a clear and intuitive explanation that can be easily extended to other non-standard problems.)

From (1.10), knowing the steady-state probability π^d and the potential g^d of policy P^d, we can obtain the directional derivative $\frac{d\eta_\delta}{d\delta}|_{\delta=0}$ along any direction $(\Delta P)^{d,h}$ pointing to any given policy P^h from P^d. This is illustrated in Figure 1.9. It is explained in the next subsection that the potentials also play a key role in policy iteration.

The extension of (1.10) to the case where the transition matrix depends arbitrarily on any parameter θ (denoted as P_θ with $P_0 = P$) is straightforward. Replacing ΔP in (1.10) with $(\frac{dP_\theta}{d\theta})|_{\theta=0}$, we have

$$\left.\frac{d\eta_\theta}{d\theta}\right|_{\theta=0} = \pi \left.\frac{dP_\theta}{d\theta}\right|_{\theta=0} g.$$

Therefore, without loss of generality, we need only to discuss the linear case (1.10).

In (1.10), the performance derivatives are expressed in terms of potentials, which can be estimated (or "learned") from a sample path of the Markov chain. Optimization can be carried out using the performance derivatives together with stochastic approximation. This is illustrated in Figure 1.5.A. The block diagram of the PA-based optimization is shown in Figure 1.10. The system (the largest block with the "logo" of a Markov chain) generates a sample path $\{X_0, X_1, \ldots, X_L\}$; this sample path is the input to the PA algorithm, which yields an estimate of the performance gradient used in the stochastic approximation algorithms to determine the value of the parameter θ with a better performance. θ determines the next policy P_θ, which gives the transition probabilities $p_\theta(j|k)$, $j, k \in \mathcal{S}$. This process converges to a local optimal point θ^*. The length of the sample path $\{X_0, X_1, \ldots, X_L\}$ may vary in different algorithms; the errors of the gradient estimates are larger if the length is shorter; stochastic approximation techniques have to be used to ensure convergence of the algorithms. This PA-based optimization is based on the performance derivative formula in (1.10).

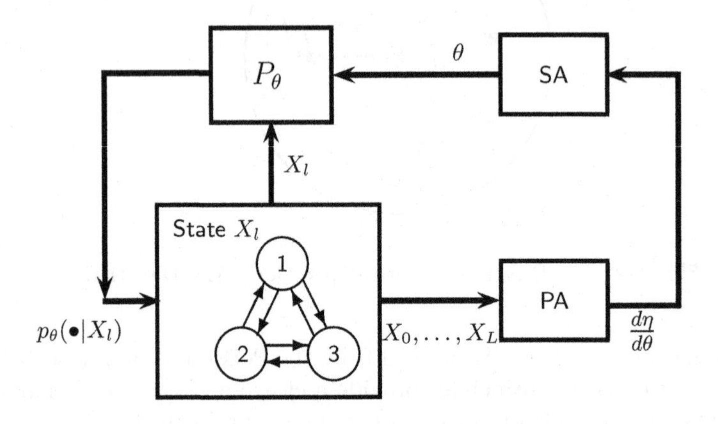

Fig. 1.10. The Block Diagram of PA-Based Optimization

There are a number of advantages of PA: It can estimate performance derivatives along all directions based on a single sample path of a Markov chain; the derivatives can be estimated as a whole without estimating each potential for every state, and thus the "curse of dimensionality" issue disappears (compared with the MDP approach); it can be implemented on line without disturbing the operation of a system; and, furthermore, the approach applies to any policy space or subspace with constraints. The disadvantage of PA-based approaches is common for all gradient-type approaches: It may reach a local optimal point. See Chapter 2 for details of all these points.

Extensions of PA of queueing systems include *finite perturbation analysis (FPA)*[43, 129, 143] and *smoothed perturbation analysis (SPA)*[119]. Recently, the fluid model of queueing systems was introduced into PA, which provided good approximations [74, 75, 210, 211, 231, 252, 262, 263]. In the literature, there are other gradient-estimation approaches, including the *likelihood ratio (LR)* method [44, 115, 116, 117, 118, 130, 176, 177, 178, 179, 205, 217], which is also called the *score function method* [221, 222], and the weak derivative (WD) method [130, 132, 134].

1.2.2 Markov Decision Processes (MDPs)

In MDPs [8, 21, 24, 98, 135, 136, 137, 163, 216], a system's behavior is modelled as a Markov chain $\boldsymbol{X} = \{X_0, X_1, \ldots\}$ with state space $\mathcal{S} = \{1, 2, \ldots, S\}$. In addition to the state space, there is an action space $\mathcal{A}$ consisting of all (finite) available actions. If the system is in state i, $i \in \mathcal{S}$, an action $\alpha \in \mathcal{A}(i)$ can be taken and applied to the system, where $\mathcal{A}(i) \subseteq \mathcal{A}$ is the set of actions that are available in state $i \in \mathcal{S}$, $\mathcal{A} = \cup_{i \in \mathcal{S}} \mathcal{A}(i)$. The action determines the state transition probabilities. When action α is taken in state i, the state transition probabilities are denoted as $p^{\alpha}(j|i)$, $j \in \mathcal{S}$. The (instant) reward that the system receives when it is in state i with action α is $f(i, \alpha)$, $i \in \mathcal{S}$, $\alpha \in \mathcal{A}(i)$. The function f is called a *reward (or performance) function.* A *stationary and deterministic* policy is a mapping from $\mathcal{S}$ to $\mathcal{A}$, denoted as $d : \alpha = d(i)$, that determines the action taken in state i. Therefore, if policy d is adopted, the transition probability matrix is $P^d := [p^{d(i)}(j|i)]_{i,j \in \mathcal{S}}$; and the reward function is $f(i, d(i))$, $i \in \mathcal{S}$, which is a special case of $f^d(i)$, $i \in \mathcal{S}$. We further assume that under all policies, the Markov chains that model the system behavior are ergodic.

The long-run average reward is

$$\eta^d = \lim_{L \to \infty} \left\{ \frac{1}{L} \sum_{l=0}^{L-1} f(X_l, d(X_l)) \right\} = \pi^d f^d, \tag{1.11}$$

which exists with probability 1 (w.p.1) and is independent of the initial state for ergodic chains. The goal of MDPs is to find a policy d^* such that its performance is the best among all policies.

In a more general setting, the performance function may depend on the next state; i.e., it may take the form $f(X_l, X_{l+1}, \alpha)$, $\alpha \in \mathcal{A}(X_l)$. As shown in Problem 1.9, this is equivalent to (1.11), with $f(i, \alpha)$ replaced by $\bar{f}(i, \alpha) = \sum_{j \in \mathcal{S}} [f(i, j, \alpha) p^{\alpha}(j|i)]$.

The above formulation is called *the standard MDPs.* In this formulation, we assume that the state X_l is observable; i.e., $Y_l = X_l$, in Figure 1.1. The Markov model specifies the structure and dynamics inside the black box in the figure. Because of the Markov property, if we know X_l, the past state-and-action histories do not add any information for predicting the future behavior; therefore, we may let the policy depend only on the current state (cf., Problem 1.6).

In the standard MDPs, we assume that all the functions and parameters, e.g., the sets $\mathcal{S}$, $\mathcal{A}(i)$, $i \in \mathcal{S}$, and $p^{\alpha}(j|i)$, $f(i, \alpha)$, $i, j \in \mathcal{S}$, $\alpha \in \mathcal{A}(i)$, etc., are known. Therefore, analysis can be carried out. However, since the policy space is too large (there are $\prod_{i \in \mathcal{S}} |\mathcal{A}(i)|$ policies, with $|\mathcal{A}(i)|$ denoting the numbers of actions in $\mathcal{A}(i)$), it is not possible to solve for the steady-state probabilities of all the policies to make comparisons of their performance. That is, except for very small problems, the exhaustive search is not feasible.

There are two basic approaches to the standard MDPs, *value iteration* and *policy iteration.* Value iteration is basically a numerical approach; the corresponding sample-path-based approach is Q-learning (see the next subsection for Reinforcement Learning). Policy iteration fits well the framework from a sensitivity viewpoint. The basic principle of policy iteration is: *Under some assumptions, by analyzing the behavior of the system under one policy, we can always find another policy under which the system performs better, if such a policy exists.* After a better policy is found, we can analyze this better policy to find another even better policy. Repeating this updating procedure until no better policy exists, we can obtain the best policy (see Figure 1.5.B).

The policy iteration approach can be easily explained from a performance sensitivity view, as shown in Figure 1.5.B. For simplicity, in this chapter, we assume that all the policies share the same performance function $f(i)$, $i \in \mathcal{S}$. As will be explained in detail in Chapter 3, the updating procedure is based on a simple formula for the performance difference:

$$\eta^h - \eta^d = \pi^h (\Delta P) g^d, \tag{1.12}$$

where η^h and η^d are the performance measures corresponding to the transition matrices P^h and P^d, respectively, $\Delta P = P^h - P^d$, π^h is the steady-state probability corresponding to P^h; and g^d is the performance potential vector associated with P^d, which is the same as what appeared in the performance derivative formula (1.10). Equation (1.12) can be easily derived from the Poisson equation (1.9) for P^d (see Problem 1.8). It can be also derived using a sample-path-based argument by first principles, see Chapter 9. The sample-path-based argument provides a clear intuition that can be extended to problems where the Poisson equation may not exist.

The only difference between (1.12) and (1.10) is that π^d in (1.10) is replaced by π^h in (1.12). This difference, however, represents a major difficulty in applying the performance difference formula in learning and optimization. In the performance derivative formula (1.10), both π^d and g^d can be obtained from analyzing the system with transition probability matrix P^d. Thus, given P^d, we can obtain the directional derivatives along any given direction $\Delta P = P^h - P^d$ without analyzing the system under P^h. However, to obtain the performance difference with (1.12), we need to solve for both π^h and g^d. This is the same as a comparison in exhaustive search because we need to analyze both systems in order to compare the performance of the two systems.

Fortunately, all is not lost. The particular factorized form of (1.12) can be utilized. In fact, the updating procedure in policy iteration is based on (1.12) and the following simple fact: $\pi^h > 0$ (i.e., $\pi^h(i) > 0$ for all $i \in \mathcal{S}$) for any P^h. Thus, for any given P^d, if we can find a P^h such that $(\Delta P) g^d = (P^h - P^d) g^d \geq 0$ with at least one positive component, then $\eta^h > \eta^d$. In particular, there is no need to solve for π^h in the procedure. Conventionally, in state i we choose the action that maximizes the ith component of $P^h g^d$ as $h(i)$; i.e., we choose $h(i)$, $i \in \mathcal{S}$, such that

$$\sum_{j=1}^{S} p^{h(i)}(j|i) g^d(j) = \max \left\{ \sum_{j=1}^{S} p^{\alpha}(j|i) g^d(j) : \alpha \in \mathcal{A}(i) \right\}. \tag{1.13}$$

In words, we choose the action such that after the next transition with this action the expected potential is maximized. Let h be the policy determined by (1.13). We have $\eta^h > \eta^d$ if P^d is not the optimal policy; however, h may not be optimal even if (1.13) holds.

The fundamental quantity in (1.12) is the performance potential. From a learning point of view, we need to analyze the behavior of a system under one policy to "learn" its potential of each state in order to determine how to make the system perform better. In fact, potentials can be obtained either by solving the Poisson equation, as in the standard MDPs approach, or by estimation (i.e., learning from sample paths). See Chapter 3 for details. The block diagram of the standard policy-iteration optimization procedure in MDPs is shown in Figure 1.11. In this approach, the Poisson equation is solved analytically to obtain the potentials. The potentials can also be "learned" (estimated) from sample paths, which will be the topic of reinforcement learning discussed in the next subsection.

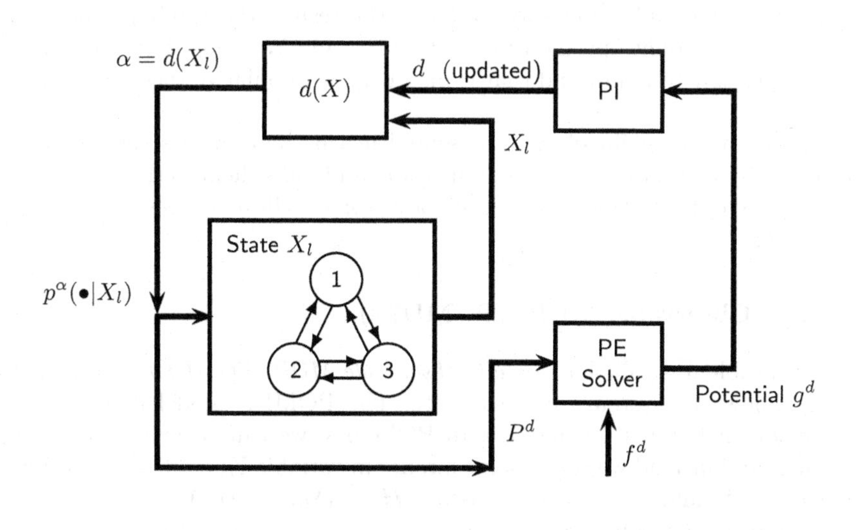

Fig. 1.11. The Block Diagram of Policy-Iteration-Based Optimization

Potential is equivalent to the bias or the relative cost in the MDP literature, up to an additive constant. In this book, we use the word "potential" because of its physical meaning. As explained in Section 1.2.1, roughly speaking, the performance potential of a state i, $i \in \mathcal{S}$, measures the "potential" contribution of the state i to the system performance; the difference between the potentials of two states measures the effect of a jump from one state to the other state on the system performance; and to improve the performance, in any state we should choose an action that leads to the best expected potential after the state transition with this action (i.e., the largest $P^h g^d$ in (1.12)). We will revisit the meaning of this terminology in Chapter 2.

As shown above, with the performance difference equation (1.12), the policy iteration procedure for ergodic chains can be derived clearly, concisely, and intuitively. This sensitivity-based approach also applies to multi-chain Markov systems, systems with absorbing states, and problems with other performance criteria such as discounted performance and even bias and nth-bias optimality etc., (see Chapter 4 for more). Furthermore, no discounting is needed to prove the results for average performance problems. In all these cases, the approach is based on the performance difference formulas, which depend on the performance potentials defined for these different problems. Therefore, the sensitivity-based view provides a unified approach to all these MDP-types of optimization problems; we will see that the approach is surprisingly simple and clear. In particular, the nth-bias optimality formulation and solution [71] provide a simple and direct way to derive the results that are equivalent to the sensitive discount optimality in [216, 248, 249]. We hope that our simple approach is easy to follow and that it may help to popularize these important results.

The sensitivity point of view presented in this book brings these results out naturally and intuitively in a unified way and links them to the rest of the performance optimization world. The above topics will be addressed in detail in Chapter 4.

Partially Observable MDPs (POMDPs)

A MDP problem is called a *partially observable MDP (POMDP)* if we do not exactly know the system's current state. The POMDP model fits well the description in Figures 1.1 and 1.4. In POMDPs, we cannot exactly observe the state, and instead we can observe a random variable Y_l, $l = 0, 1, \ldots$, which depends on X_l and A_{l-1}. Given a history $\boldsymbol{H}_l = \{Y_0, \ldots, Y_l; A_0, \ldots, A_{l-1}\}$, we try to determine a policy $A_l = d_l(\boldsymbol{H}_l)$, $l = 0, 1, \ldots$, to optimize the long-run average performance defined in (1.11).

The essential idea in POMDPs is that knowing $\boldsymbol{H}_l$, we may obtain the conditional probability distribution of X_l, denoted as $\mathcal{P}(\bullet|\boldsymbol{H}_l)$. If we can exactly determine the state X_l by $\boldsymbol{H}_l$, we may do as well as MDPs; otherwise, we do our best by using this conditional distribution of X_l given $\boldsymbol{H}_l$, and a policy is a mapping defined on the space of distributions to the space of actions, i.e., $A_l = d_l[\mathcal{P}(\bullet|\boldsymbol{H}_l)]$.

One difference between POMDPs and MDPs is as follows. In MDPs, the set of actions available in state i are denoted as the set $\mathcal{A}(i)$. An action in $\mathcal{A}(1)$ may not be applied to the system in state 2. Thus, for any state $i \in \mathcal{S}$ and action $\alpha \in \mathcal{A}(i)$, we need only to specify the transition probabilities $p^{\alpha}(\bullet|i)$ for this particular i and α. However, in POMDPs, because we do not exactly know the state, if we apply an action to the system, it may be in any state. Therefore, we only have an action space $\mathcal{A}$, which cannot be further decomposed into subsets $\mathcal{A}(i)$, $i \in \mathcal{S}$; and for any action $\alpha \in \mathcal{A}$, we need to specify the whole transition probability matrix.

POMDPs are much more complicated than MDPs, because even for a simple problem, the space of the conditional probability distributions of X_l may not be finite. In this book, we will not discuss POMDPs in detail. We, however, will make comparison with POMDPs when we discuss the event-based optimization approach.

1.2.3 Reinforcement Learning (RL)

The fundamental model for systems in RL is also the Markov chain. While the MDP is basically an analytical approach, which assumes that all the parameters are known, RL is a simulation-based (or in some cases, on-line) learning approach. In RL, most information is learned from the system's sample paths. The transition probability matrix may not need to be known, or only the parts related to actions need to be known. Simulation can be carried out by following the system structure (e.g., the queueing structure).

There are different approaches in RL, depending on how much we know about the system structure and how much we need to learn from the sample paths. If we know enough information about the transition probabilities to implement policy iteration with potentials, e.g., we know the parts related to the actions (see, e.g., Example 5.1 in Section 5.1), we need only to "learn", or to estimate, the potentials for all the states from a sample path of the system under one policy and then update the policies iteratively. In this sense, any estimation-based or on-line approach for estimating potentials belongs to RL. In this regard, many efficient RL algorithms, such as TD(λ) (see Chapter 6 and many references such as [25, 159, 223, 226, 229, 230, 236, 237, 238, 239, 244, 254]), and approximate approaches, such as *neuro-dynamic programming* [21, 25], have been developed.

If we do not know anything about the transition probabilities, we cannot implement policy iteration even if we know the potentials. In this case, we need to learn the system behavior for all state-action pairs. Basically, in state i, we need to try all the actions in $\mathcal{A}(i)$ in order to get enough information for comparison. Therefore, this type of RL approach (e.g., Q-learning) requires a sample path to visit all the state-action pairs.

In these approaches, we consider a variant of the potential $g^d(i)$, called the *Q-factor* of a state-action pair (i, α), denoted as $Q^d(i, \alpha)$ for any $i \in \mathcal{S}$ and $\alpha \in \mathcal{A}(i)$. $Q^d(i, \alpha)$ is defined as the average potential of state i if action $\alpha \in \mathcal{A}(i)$ (not necessarily $d(i)$) is taken at a particular time and the rest of the Markov chain is run under a policy d:

$$Q^d(i, \alpha) = \sum_{j=1}^{S} p^{\alpha}(j|i) g^d(j) + f(i, \alpha) - \eta^d, \qquad \alpha \in \mathcal{A}(i).$$

With this definition, (1.13) (in which the reward $f(i)$ is assumed to be the same for all actions) becomes

$$Q^d(i, h(i)) = \max\{Q^d(i, \alpha) \ : \alpha \in \mathcal{A}(i)\}.$$

Thus, we may implement policy iteration by choosing the action that leads to the largest $Q^d(i, \alpha)$ in state i as the improved policy h.

Sample-path-based algorithms may be developed to estimate Q-factors. This leads to the Q-factor-based policy iteration, which can be used when the Markov chain's transition probability matrix is unknown. However, it is only possible to estimate the Q-factors for those state-action pairs that appear on a sample path. The main advantage of this approach is that it estimates the combined effect of the transition probabilities $p^\alpha(j|i)$ and the potentials $g^d(j)$ together without estimating these items separately. But the number of state-action pairs increases to $\sum_{i=1}^{S} |\mathcal{A}(i)|$, compared with the number of potentials S.

In the approach, we need a sample path that visits all the state-action pairs. However, with a deterministic policy d, only the state-action pairs $(i, d(i))$, $i \in \mathcal{S}$, are visited. This issue may be resolved by introducing, with a small probability, other actions into the system as follows: in any state i, we apply action $d(i)$ with probability $1 - \epsilon$ and any other action $\alpha \in \mathcal{A}(i)$ randomly with an equal probability $\epsilon/(|\mathcal{A}(i)| - 1)$, $0 < \epsilon << 1$. We denote such a policy as d_ϵ. However, this may cause undesirable disturbance to a system and may not be feasible in real-world systems. Thus, a sample path that visits all the state-action pairs can only be generated by simulation, and not by observing the operation of a real-world system. Such an approach is called *simulation-based*, whereas an approach that is implementable by observing a real system without disturbance is called an *on-line* approach. Simulation-based approaches are expected to yield more general results than the on-line approaches, because the latter can also be implemented by simulation. The two approaches illustrated in Figure 1.5, PA (gradient based) and policy iteration, can be implemented on-line. Both simulation-based and on-line approaches are called *sample-path-based* approaches.

The block diagram of the reinforcement learning optimization procedure is shown in Figure 1.12. This is similar to the MDP case (Figure 1.11) except that the inputs to the RL block are the sample path $\{X_0, X_1, \ldots, X_L\}$ and the reward observed at each step is $f(X_l)$, $l = 0, 1, \ldots, L$; and the outputs are either the potentials or the Q-factors. The RL and PI blocks in the diagram may update their outputs (potentials or Q-factors for RL, and policies for PI) periodically or even at every transition.

Principally, we can obtain the optimal policy by learning from a sample path that visits all the state-action pairs. Indeed, such a sample path contains the information about the system under all possible policies. This can be seen by the simple "cut-and-paste" method. Let d be any policy and $\alpha_i = d(i)$, $i \in \mathcal{S}$. We can construct a sample path $\boldsymbol{X}^d$ of this policy d on $\boldsymbol{X}$ as its "sub"-sample path as follows. First, on $\boldsymbol{X}$, we find the first state i such that the action taken is α_i and we denote it as X_0^d. Suppose that from this pair (i, α_i), $\boldsymbol{X}$ moves to the next state denoted as j. If the action taken at this next state

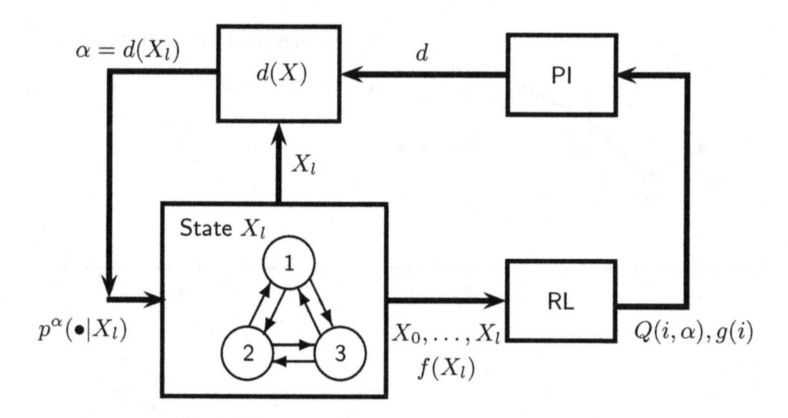

Fig. 1.12. The Block Diagram of Reinforcement Learning

happens to be $\alpha_j = d(j)$, we take this state as X_1^d; otherwise, we look for the earliest state j on $\boldsymbol{X}$ that takes action α_j and we denote it as X_1^d. Repeating this procedure, we can get an "embedded" sample path $\boldsymbol{X}^d$ for policy d on $\boldsymbol{X}$. See Figure 1.13.

Therefore, principally we should be able to determine the optimal policy by analyzing any sample path $\boldsymbol{X}$ that visits all the state-action pairs. The simplest way to do so is to implement policy iteration on the sample path: First, we construct a sub-sample path with an initial policy d_0, $\boldsymbol{X}^{d_0}$ on $\boldsymbol{X}$, estimate its potentials (or Q-factors, with a sample path for the policy $d_{0,\epsilon}$), and find a better policy d_1 by policy iteration. Then, we construct a sub-sample path with d_1, $\boldsymbol{X}^{d_1}$ on $\boldsymbol{X}$, and so on. If the transition probabilities are unknown, they can also be estimated from $\boldsymbol{X}$ since it visits all the state-action pairs.

Of course, this "construction" approach may not be the most efficient. Improvement can be made by applying stochastic approximation techniques. Exploring along this direction leads to various RL algorithms including the *Q-learning* approach in the literature. These algorithms estimate the Q-factors or the potentials for the optimal policy, from a sample path that visits all state-action pairs. They are very aggressive: they update the estimates at every state transition. For more details, see Chapter 6.

In recent years, performance-gradient-based optimization has attracted more and more attention from the RL community. It is called *policy gradient* in RL [2, 17, 18]. The approach follows the same idea of PA: It is based on sample-path-based estimates for performance gradients. There is, however, a distinction between the policy gradient approach and PA: The former focuses on algorithms for estimating the gradients, and the latter (PA), in addition to

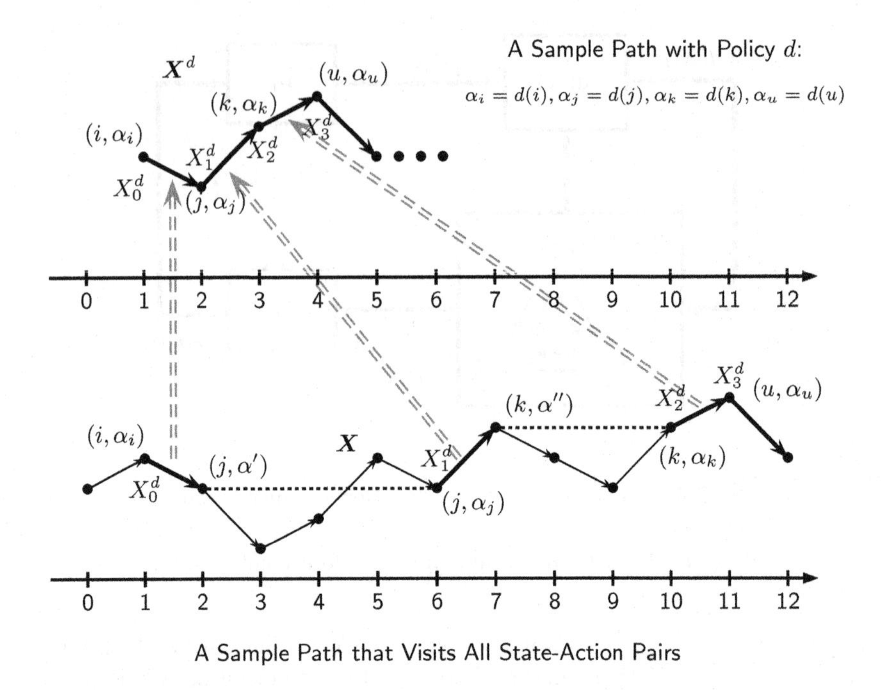

Fig. 1.13. Constructing a Sample Path for Any Policy by Cut-and-Paste

algorithms, emphasizes the construction of gradient formulas using the idea of perturbation.

Finally, we observe that there is another difference between the RL and the MDP formulations: RL assumes that an instant reward $f(i, \alpha)$ can be observed, and the function f may not be known. That is, in observing a sample path, at any time l we indeed know the reward received, without knowing the function f. If the state i is completely observable, as in the standard MDP case, assuming that the reward is observable is the same as assuming that the function f is known, because we can calculate the reward according to $f(i, \alpha)$ if state i is observed and action α is taken. However, in POMDPs, these two assumptions are different: We do not know the reward even if we know f because we do not know exactly what the state is.

In summary, the RL approach focuses on algorithms estimating potentials and its variant Q-factors, or the potentials and Q-factors for optimal policies; and the policy gradient approach focuses on algorithms for performance gradients.

1.2.4 Identification and Adaptive Control (I&AC)

Identification and adaptive control are well-developed areas. In adaptive control theory, system dynamics are modelled by differential or difference equations that determine the system structure. With such a mathematical model, elegant analysis can be carried out, leading to widely deployed adaptive control algorithms. When the system parameters are unknown and/or time varying, they need to be estimated from observations (this is also called system identification), and performance optimization can be achieved by using the adaptive control algorithms with the parameters estimated from observations.

In this book, we cannot review all the results in I&AC. Instead, we show that optimization with such a problem formulation can also be achieved by learning from the performance sensitivity point of view. In particular, we show that under some assumptions, the problem can be solved by the on-line policy iteration approach discussed in Section 1.2.2. The fundamental idea is: A stochastic system under control, although it has its special structure, can be generally modelled as a Markov process, with the control variables viewed as actions.

Example 1.4. (Linear Systems) A (discrete-time) linear stochastic system is modelled as

$$X_{l+1} = AX_l + Bu_l + \xi_l, \qquad l = 0, 1, \ldots, \tag{1.14}$$

where X_l is the system state at time l, which is usually a random vector, u_l is a vector of control variables, ξ_l is a vector of noise, and A and B are matrices with appropriate dimensions. Apparently, $\boldsymbol{X} = \{X_l,\ l = 0, 1, \ldots\}$ is a Markov chain and u_l can be viewed as the actions that determine the transition probabilities of $\boldsymbol{X}$, based on the probability distribution function of ξ_l. The problem is how to choose u_l, $l = 0, 1 \ldots$, such that a performance measure (e.g., the long-run average of a quadratic cost function of X_l and u_l) is minimized.

The one-dimensional case is illustrated in Figure 1.14. At time $l = 0$, the system is in state x_0. If there was no noise, at $l = 1$, the system would move to $Ax_0 + Bu_0$. However, because of the random noise, it moves to x_1 instead, which is a random variable centered at $Ax_0 + Bu_0$ with a distribution determined by ξ_0. The situation at $l = 2$ is similar. Thus, the control variable u_l controls the distribution (or precisely, the center of the distribution) of X_{l+1}. □

In Example 1.4, we assume that X_l, $l = 0, 1, \ldots$, are observable. More generally, we may have an observation equation

$$Y_l = CX_l + D\zeta_l, \qquad l = 0, 1, \ldots, \tag{1.15}$$

where Y_l is an observable vector, ζ_l is a random noise vector, and C and D are two matrices with appropriate dimensions. The system in (1.14) and (1.15) fits well the formulation described by Figure 1.1, in which (1.14) models the dynamics of the black box and (1.15) specifies the relationship between the observations and the states.

Principally, such a problem is amenable to either MDPs, RL, or PA. For example, we can apply policy iteration to find the optimal feedback control

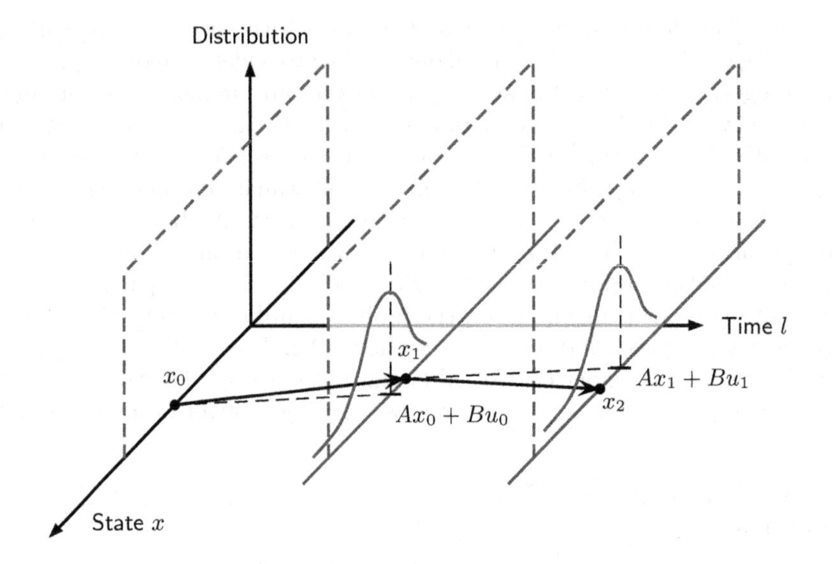

Fig. 1.14. A Control Problem Viewed as an MDP

policy $u_l = d(X_l)$. As discussed in Section 1.2.2, in each iteration when the system is (or is estimated to be) at a particular state, we let the control variables take the values that lead to the largest expected potential at the next state. Under some conditions, such an iteration procedure will converge to an optimal policy. To illustrate and verify this approach, it is shown in Chapter 7 that for the linear stochastic control problem in (1.14) with a quadratic performance reward, we can derive, with policy iteration, the famous Riccati equation for optimal policies.

When the system parameters are unknown, the basic quantities such as potentials have to be learned from a sample path with various RL algorithms. When the system parameters vary with a slow time scale, the policies have to be updated frequently to keep up with the parameter changes. In this sense, the on-line policy iteration, RL, or PA-based optimization are equivalent to system identification and adaptive control.

The on-line policy iteration approach is conceptually quite different from the adaptive control approach. The latter employs dynamic programming, which essentially works backwards: For finite horizon problems, it first finds the optimal solution for the system with k time steps remaining, and then with this optimal solution to the k-step problem, it goes back one step to find the solution for the system with $k + 1$ time steps remaining, $k = 0, 1, \ldots$; and so on. To solve an infinite horizon problem, we let this procedure go on until k approaches to infinity. Therefore, in dynamic programming, we iterate backwards in time, and, at every time step, we find the optimal policy for a finite-step problem. Of course, a backward procedure can only be implemented off-line. On the contrary, policy iteration essentially works forwards: the potentials are estimated along the same direction as time goes forward, and policies are iterated until an optimal one is reached. Another feature is that with policy iteration, we estimate the potentials, and the system parameters may not need to be estimated directly. This corresponds to the direct adaptive control in the literature, where the parameters for the optimal control law, instead of the system, are identified. See Chapter 7 for more details.

One advantage of the on-line or sample-path-based approach is that, from the learning point of view, principally it applies to both linear and non-linear systems in the same way. The system structure affects the transition probability matrix. However, determining the transition probability matrix for different control parameters might be a difficult task. Another difficulty one may encounter is that the theoretical results on convergence, etc., may be limited, because the state and action spaces in such problems are usually continuous, as shown in the linear system example. Since one can always approximate a continuous space with a discrete one by discretization and practical problems are finite in nature, approximate and heuristic methods can be developed. This topic will be discussed in Chapter 7.

1.2.5 Event-Based Optimization and Potential Aggregation

We will show in Chapters 2 and 9 that, with the physical interpretation of potentials, we can use potentials as building blocks to construct the two sensitivity formulas (1.12) and (1.10) by first principles. Furthermore, we can follow the same idea to construct/derive sensitivity formulas for many special and/or non-standard problems by first principles. For example, we can derive sensitivity formulas, similar to (1.12) and (1.10), for two Markov chains that have two different but overlapping state spaces. We can also derive sensitivity formulas for policies that depend on "events" rather than states.

Since the two sensitivity formulas form the basis for learning and optimization, we can develop new learning and optimization approaches based on the sensitivity formulas for these special problems that do not fit the standard MDP formulation. The special structure of a problem can be used in constructing the formulas, and thus the computation can be reduced and/or the policy can meet some special constraints.

One of such approaches is the *event-based optimization*, which can be applied to systems where the actions can be taken only when some events happen. An *event* is defined as a set of state transitions that share some common features. The decision of selecting actions depends on the information associated with the events. Special features of the system structure and logical relations of the system behavior can be captured by events. Potentials can be aggregated using the special features. Many problems can be treated by this event-based approach with properly defined events.

The Main Features of the Event-Based Approach

In addition to the structural information captured by events, which helps to reduce computation by aggregation, compared with MDPs and POMDPs, the event-based approach has some other nice features. We first give a simple example to illustrate the ideas.

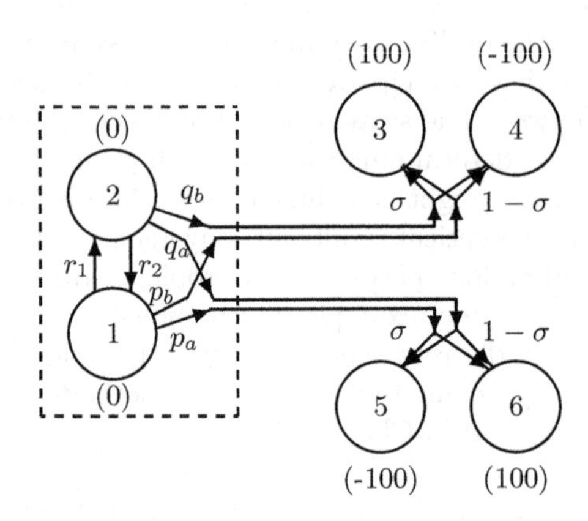

Fig. 1.15. The State Transition Diagram for Example 1.5

Example 1.5. Figure 1.15 illustrates a part of a system's state transition diagram. The transition probabilities are shown by the symbols near the arrows in the figure. First, we assume that when the system is in state 1 or state 2, we only know that the system is in the set $\{1, 2\}$, but do not know exactly which state it is in. This is a POMDP problem, in which we define an aggregated state $0 := \{1, 2\}$, and if $X_l = 1$, or 2, we have $Y_l = 0$, and if $X_l = 3, 4, 5$, or 6, we have $Y_l = X_l$. The observation history is $\boldsymbol{Y} = \{Y_0, Y_1, \ldots\}$. We can take an action to control the value of $\sigma \in [0, 1]$. Figure 1.16 lists the transition probabilities of states 1 and 2 when a particular σ is chosen. The rewards for states 3 and 6 are 100, those for states 4 and 5 are -100, and those for states 1 and 2 are 0.

	1	2	3	4	5	6
1	0	r_1	$p_b\sigma$	$p_b(1-\sigma)$	$p_a\sigma$	$p_a(1-\sigma)$
2	r_2	0	$q_b\sigma$	$q_b(1-\sigma)$	$q_a\sigma$	$q_a(1-\sigma)$

Fig. 1.16. The Transition Probabilities of States 1 and 2 When Action σ is Taken in Example 1.5

In POMDPs, action A_l is specified by a policy depending on the conditional distribution of X_l given $\boldsymbol{H}_l = \{Y_0, A_0, \ldots, Y_{l-1}, A_{l-1}, Y_l\}$. We wish to find a policy $d(\boldsymbol{H}_l)$ that optimizes the long-run average performance defined in (1.11).

The best situation happens when we can exactly determine $X_l = 1$ or $X_l = 2$ by using $\boldsymbol{H}_l$. In this case, the problem becomes a standard MDP. With the MDP formulation, we wish to determine a policy that specifies the actions for states 1 and 2, denoted as $d(1) = \sigma_1$ and $d(2) = \sigma_2$. Note that σ_1 and σ_2 may be different, and the transition probabilities corresponding to state 1 (or 2) are listed in the first (or the second) row of Figure 1.16, with σ replaced by σ_1 (or σ_2).

Let us analyze the structure of the transition diagram. From Figure 1.15, if the system moves from state 1 (or 2) to the two top states, 3 and 4, we need to take the biggest value $\sigma = 1$ to reach state 3 with probability 1 and get a reward of 100; on the contrary, if the system moves from state 1 (or 2) to the two bottom states, 5 and 6, we need to take the smallest value $\sigma = 0$ to reach state 6 with probability 1 and get a reward of 100. Now, suppose that the current state is 1. In the next time instant, the system will move to the top states and to the bottom states with two positive probabilities, p_a and p_b, respectively. As shown above, a big σ is good for the top, but bad for the bottom, and vice versa. When $p_a = p_b = 0.5$, for any σ the average reward at the next step is zero. Therefore, even if we know the states 1 and 2 exactly, the (MDP) optimal policy may not be very good. The above discussion explains the limitation with the state-dependent policies: the optimal performance is not very good even if the history provides the exact information about the current state.

However, the situation improves significantly if we know a bit of information about the state transition. From the structure shown in Figure 1.15, the top four transitions, or the bottom four transitions, have similar properties. This structure can be captured by aggregating these transitions together and defining two events:

$$a := \{\langle 1,5\rangle, \langle 1,6\rangle, \langle 2,5\rangle, \langle 2,6\rangle\},$$

and

$$b := \{\langle 1,3\rangle, \langle 1,4\rangle, \langle 2,3\rangle, \langle 2,4\rangle\},$$

where $\langle i,j\rangle$ denotes a transition from state i to state j, $i,j \in \mathcal{S}$. These two sets of state transitions aggregated into two events are shown as the two ovals, a and b, in Figure 1.17; they are also illustrated by the two black boxes in Figure 1.18.

With this formulation, when event a, or b, occurs, the system is in either state 1 or 2, but we do not know exactly which state the system is in. However, independent of the state, if event a occurs, the system moves to state 5 with probability σ and to state 6 with probability $1-\sigma$; and if event b occurs, the

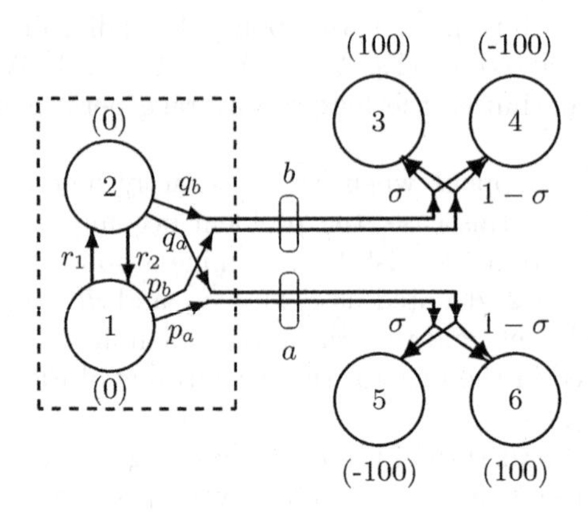

Fig. 1.17. State Transitions Aggregated into Two Events a and b

system moves to state 3 with probability σ and to state 4 with probability $1-\sigma$.

In the event-based approach, we assume that we can observe the events, not the states; i.e., at any time instant l, we can observe whether $\langle X_l, X_{l+1}\rangle \in a$, or $\langle X_l, X_{l+1}\rangle \in b$, occurs. We need to determine an event-based policy that specifies the probabilities chosen for events a and b: $\sigma_a = d(a)$ and $\sigma_b = d(b)$.

Event b *Event a*

	1	2	3	4	5	6
1	0	r_1	$p_b\sigma$	$p_b(1-\sigma)$	$p_a\sigma$	$p_a(1-\sigma)$
2	r_2	0	$q_b\sigma$	$q_b(1-\sigma)$	$q_a\sigma$	$q_a(1-\sigma)$

Fig. 1.18. Events a and b as Sets of State Transitions in Example 1.5

From the reward structure shown in Figure 1.17, we may design a myopic policy: if a occurs, we choose the smallest value, i.e., $\sigma_a = 0$, which leads to state 6 and the reward at the next step is 100; and similarly, if b occurs, we choose the largest value, i.e., $\sigma_b = 1$, and the reward at the next step is also 100. In this example, this myopic event-based policy is better than the optimal MDP policy.

Thus, in this example, the optimal event-based policy is better than the optimal state-based policy; or knowing the event is better than knowing the state. This is because knowing the event implies knowing something about the current transition, which, strictly speaking, belongs to the future. In addition, we can see that a history-independent event-based policy is good enough in this example. □

A real system having the transition probabilities shown in Figure 1.15 is given in Chapter 9, and events a and b have a natural interpretation in the system. Many other real-world systems fit the event-based formulation. For example, in the admission control of a communications system, actions (accept or reject) are taken only when a customer arrives at a network. A customer arrival to a network can be modelled as an event, and the state transition at an arrival instant has the following property: The population of the system increases by one at this transition (if the customer is accepted), or remains the same (if the customer is rejected), and it will not decrease.

In these problems, we can define a policy depending on events rather than on states. We can derive or construct sensitivity formulas similar to (1.12) and (1.10) for any two event-based policies. With these formulas constructed, performance gradients can be estimated on line and in some cases policy iteration can also be developed for performance optimization with event-based policies. The details will be discussed in Chapters 8 and 9.

Because each event may correspond to many states, the technique involves aggregation of the potentials of the states that correspond to the same event. The number of events usually scales to the system size and is much smaller than that of the states, which grows exponentially with the size of the system. Thus, computation can be reduced. In addition, the same action is taken at different states that correspond to the same event; this is not feasible for the standard MDPs, which require the actions at different states be chosen independently.

The event-based approach provides a unified view for solutions to many non-standard problems (some existing and some new), including POMDPs [159, 161, 188], state and time aggregations [4, 67, 101], hierarchical control (hybrid systems) [100, 250], options [15], and singular perturbation analysis [1, 27, 99]. Different events can be defined to capture the special features in these different problems.

In summary, the structural information of a system's transition diagram may not be reflected by the state-dependent policies, and it may be captured by the events defined as sets of state transitions. Knowing the events at a time instant implies knowing something about the future and the event-based policies may perform much better than the state-based ones. The sensitivity-based view also provides natural solutions to the event-based approach and therefore opens a new direction for learning and optimization. These topics will be discussed in detail in Chapters 8 and 9.

1.3 A Map of the Learning and Optimization World

With a sensitivity point of view, the world of learning and optimization can be illustrated by the map shown Figure 1.19. This map summarizes the discussions in this chapter as well as in this book. The central piece of the map is the performance potential. Various RL methods yield sample-path-based estimates for potentials g, or their variant Q-factors, or their values for the optimal policy; the potentials are used as building blocks in constructing the two performance sensitivity formulas; these two formulas form the basis for gradient-based (PA-based) and MDP-type optimization approaches; RL methods can also be developed for directly estimating the performance gradients on sample paths; stochastic approximation techniques can be used to derive efficient optimization algorithms with the sample-path-based gradient estimates, and to derive on-line policy iteration algorithms. Both the gradient-based approaches and policy iteration can be applied to system identification and adaptive control (I&AC) problems, even with non-linear systems. This sensitivity point of view of I&AC may lead to sample-path-based methods in the area.

The map uses the two standard sensitivity formulas as examples. These two formulas can be replaced by any two sensitivity formulas constructed, or derived, for other special or nonstandard problems (possibly with potentials aggregated), and the same logical relationships among different approaches shown in the map still hold. The PA and MDP methods based on these constructed formulas may utilize the special features and thus reduce the computation in learning and optimization. One such special problem is event-based optimization. The sensitivity point of view opens up a new and rich research direction.

This map can be used as a road map for the materials covered in this book.

1.4 Terminology and Notation

One of the difficulties in introducing a number of different research areas in the same book is that it is not easy to unify the terminology and to choose notation that is consistent with those conventionally used in their respective areas. First, just like the translation of languages, there may not exist exact one-to-one correspondences between the concepts in these different areas. For example, both PA in discrete event dynamic system theory and the policy gradient in RL are based on performance gradients, yet they have different emphases: The former mainly focuses on the construction of performance gradients; and the latter, on the development of efficient algorithms. Second, different terminologies represent different perspectives on the same concept and it is these different perspectives that motivate different thinking, which is precisely what we wish to encourage. For a daily-life example, John Smith may be called either Dr. Smith on one occasion, or simply John on others. In our case, the solution to the Poisson equation is called the potentials in PA, which represent the potential contributions of the current states to the performance and have a similar meaning as the potential energy in physics; it

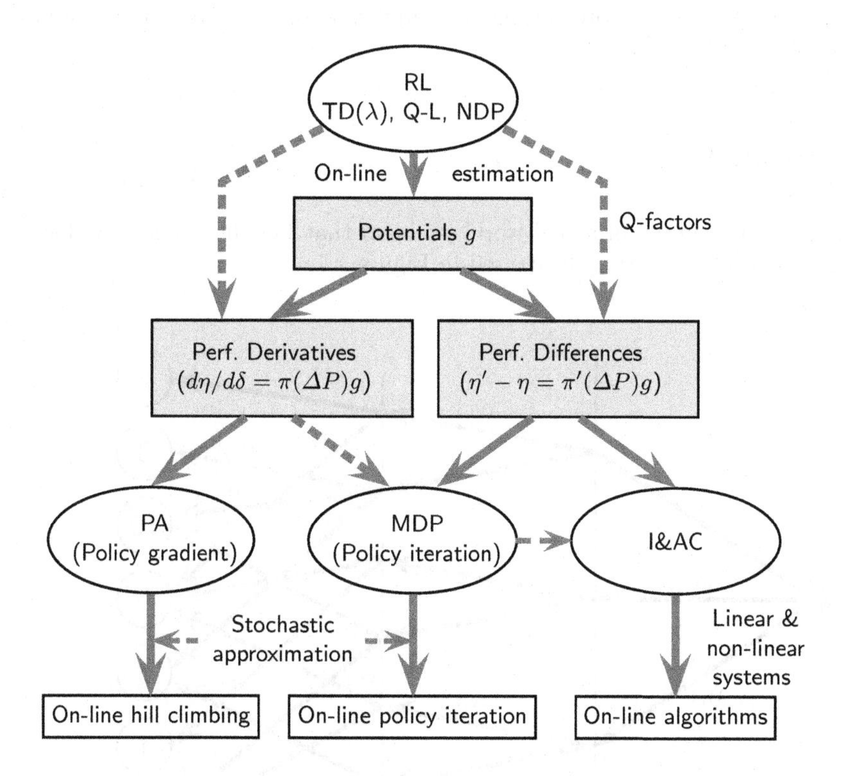

Fig. 1.19. A Map of the Learning and Optimization World (PA: Perturbation Analysis, MDP: Markov Decision Process, RL: Reinforcement Learning, Q-L: Q-Learning, I&AC: Identification and Stochastic Control, NDP: Neuro-dynamic Programming)

is called the biases in MDPs, which emphasize the transient nature, i.e., the deviation of the total average reward (starting from a particular state) from the steady-state value.

In this book, we have successfully unified the terminology and notation, with only a few exceptions: For example, in most parts of the book, we use the terminology "potential" due to its physical interpretations emphasized in the book, but we will use "bias optimality" when discussing the related optimization problems to make it consistent with the other literature.

We also note that some notation used in this book are different from those used in the literature in the related research areas. For example, in the PA literature, a policy is denoted as $\mathcal{L}$ and the transition matrices of two policies are denoted as P and P', but in most of this book, they are denoted as d and P^d and P^h, respectively. These are consistent with the notations in the MDP literature. Also, we use $\gamma(i,j)$, instead of $d(i,j)$ used in the PA literature, to denote the realization factors, because d is already used for a policy.

The notation and abbreviations used in this book are listed at the end of the book, see Page 543.

PROBLEMS

1.1. Give an example of a real-world problem that fits the general model of learning and optimization illustrated in Figure 1.1.

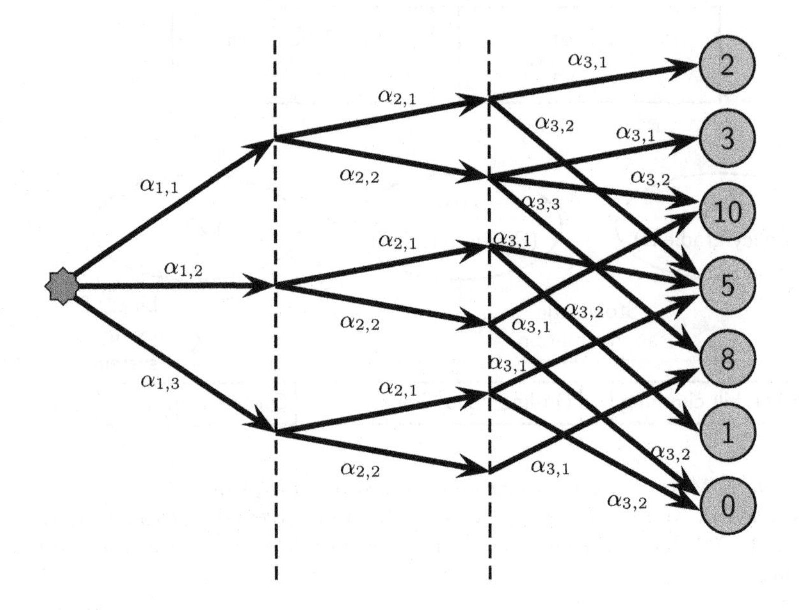

Fig. 1.20. The Travelling Problem

1.2. A person travels from the star point shown in Figure 1.20 to one of the seven destinations indicated as the circles in the figure. The person may receive a reward shown as the number in the corresponding circle when she/he reaches a destination. There are three time steps, $l = 0, 1, 2$, in this problem. The letters $\alpha_{1,1}$, $\alpha_{1,2}$, $\alpha_{1,3}$, and $\alpha_{2,1}, \dots,$ near the arrows represent the actions. Develop an optimal policy for the person to receive the biggest reward. Note that there is more than one optimal policy.

1.3. In Example 1.2, at $l = 0$, there are two possible observations y_0 and y_1. Thus, the number of possible sub-policies $d_0 : \{y_0, y_1\} \rightarrow \{\alpha_0, \alpha_1\}$ is $2^2 = 4$. Next, if we do not follow any policy at $l = 0$, then at $l = 1$, there are eight possible different histories $\{Y_0, A_0, Y_1\}$. In this case, at time $l = 1$ every policy

d_1 needs to specify an action for every one of these eight different action-observation histories. Thus, there are $2^{2^3} = 2^8 = 256$ possible sub-policies d_1 at $l = 1$. However, if we follow any sub-policy at $l = 0$, because $A_0 = d_0(Y_0)$, we have only four (instead of eight) possible different histories for each d_0. Therefore, if we follow any sub-policy at $l = 0$, each sub-policy at $l = 1$ needs to specify actions for these four different action-observation histories. That is, for each sub-policy d_0, there are only 2^4 different sub-policies d_1 at $l = 1$. Thus, there are altogether $2^2 \times 2^4 = 64$ different combined policies $\{d_0, d_1\}$. Convince yourself that the above argument is valid, and continue to calculate how many policies there are for $\boldsymbol{d} = \{d_0, d_1, d_2\}$.

1.4. Prove that the optimal feedback policy based on observations performs better than the optimal open-loop policy on average (cf. Example 1.2).

1.5. Consider an MDP with state space $\mathcal{S} = \{1, 2, \ldots, S\}$. Let the action space be $\mathcal{A} = \{\alpha_1, \alpha_2, \ldots, \alpha_S\}$; suppose that when action α_j is taken in any state i, the system will, with probability 1, move to state j, $j = 1, 2, \ldots, S$.

a. For any $i \in \mathcal{S}$, define a distribution on $\mathcal{A}$ as $\nu_i = \{p(1|i), p(2|i), \ldots, p(S|i)\}$. Let $\nu_i = d(i)$, $i \in \mathcal{S}$, be a randomized policy defined as follows: In any state i, $i \in \mathcal{S}$, action α_j is taken with probability $p(j|i)$, $j \in \mathcal{S}$. What is the Markov chain under this policy $\nu_i = d(i)$, $i \in \mathcal{S}$?
b. Let $\alpha^{(1)}$ and $\alpha^{(2)}$ represent another two actions: if $\alpha^{(k)}$ is taken in state i, then the system moves according to the probability distribution $\nu_i^{(k)} = \{p^{(k)}(1|i), p^{(k)}(2|i), \ldots, p^{(k)}(S|i)\}$, $k = 1, 2$. Let $\nu_i = d(i)$, $i \in \mathcal{S}$, be a randomized policy defined as follows: In any state i, action $\alpha^{(1)}$ is taken with probability p_i, and action $\alpha^{(2)}$ is taken with probability q_i, $p_i + q_i = 1$, $i = 1, 2, \ldots, S$. What is the Markov chain under this policy $\nu_i = d(i)$, $i \in \mathcal{S}$?

1.6. Consider a two-state process $\widetilde{\boldsymbol{X}}$ with history-dependent transition probabilities $p[1|(1,1)] = 0$, $p[0|(1,1)] = 1$; $p[1|(0,0)] = 1$, $p[0|(0,0)] = 0$; $p[0|(1,0)] = 1$, $p[1|(1,0)] = 0$; and $p[1|(0,1)] = 1$, $p[0|(0,1)] = 0$.

a. Draw a sample path of $\widetilde{\boldsymbol{X}}$. What property does it have?
b. Derive the equivalent Markov chain $\boldsymbol{X}$ as shown in Example 1.3.
c. Suppose that the reward function depends on three consecutive states (X_l, X_{l+1}, X_{l+2}) and is defined as $f(1,1,1) = f(0,0,0) = 100$ and $f(i,j,k) = 0$ otherwise. Explain that the steady-state performance measures for both $\widetilde{\boldsymbol{X}}$ and $\boldsymbol{X}$ defined as $\widetilde{\eta} = \sum_{i,j,k} \widetilde{\pi}(i,j,k) f(i,j,k)$ and $\eta = \sum_{i,j,k} \pi(i,j,k) f(i,j,k)$, respectively, are different.

1.7. The exhaustive search algorithm presented in Section 1.1.3 is very "robust". Suppose that because of the estimation error, the relationship $\eta^{d_i} > \widetilde{\eta}$ cannot be accurately verified.

a. If d_M is an optimal policy, then the algorithm outputs a correct optimal policy if only the last comparison is correctly made.

b. Explain that the algorithm outputs the optimal policy as long as the comparisons $\eta^{d_i} > \widetilde{\eta}$ are correctly made when η^{d_i} or $\widetilde{\eta}$ is the optimal performance.
c. Suppose that η^* is the best performance and η^*_- is the next to the best performance, and set $\delta = \eta^* - \eta^*_-$. Then the algorithm outputs the correct optimal policy if the estimation error for the performance is always smaller than $\delta/2$.

1.8. Derive Equation (1.12) by the Poisson equation (1.9), and derive (1.10) by (1.12).

1.9. In the MDP problem, the reward function may depend on the next state; i.e., it may take the form $f(X_l, X_{l+1}, \alpha)$, $\alpha \in \mathcal{A}(X_l)$. Prove that this problem is equivalent to the standard MDP with $f(i, \alpha)$ replaced by $\bar{f}(i, \alpha) = \sum_{j \in \mathcal{S}} [f(i, j) p^{\alpha}(j|i)]$.

1.10. Consider a Markov chain $\{X_0, X_1, \ldots\}$ defined on a finite state space $\mathcal{S}$. In any state $i \in \mathcal{S}$, an action $\alpha \in \mathcal{A}(i)$ can be taken, which determines the transition probability as $p^{\alpha}(j|i)$, $j \in \mathcal{S}$. Now, let us assume that the action chosen at X_l depends on both X_{l-1}, and X_l. Thus, if $X_{l-1} = k$ and $X_l = i$, the action is denoted as $\alpha = d(k, i)$ and the transition probabilities at X_l are $p^{d(k,i)}(j|i)$, $j \in \mathcal{S}$, where $d(k, i)$ is the policy.

a. Prove that this problem is equivalent to a standard MDP with an enlarged state space.
b. Can you find an equivalent standard MDP with state space $\mathcal{S}$?

1.11. Consider the optimization problem of a discrete time M/M/1 queue. When a customer arrives at the server, the number of customers in the system increases by one. The server serves one customer at a time. Other customers have to wait in a queue. When a customer finishes its service, s/he leaves the server, and the number of customers in the system decreases by one. Let X_l be the number of customers in the server at time $l = 0, 1, \ldots$. If $X_l = n$, then the probability that a customer arrives in the lth period (i.e., $X_{l+1} = X_l + 1$) is $a(n)$, and the probability that a customer leaves (i.e., $X_{l+1} = X_l - 1$) is $b(n)$, and X_l stays the same with probability $1 - [a(n) + b(n)]$. If $X_l = 0$, then $b(0) = 0$. The system has a capacity of N; i.e., an arrival customer will be rejected if there are N customers in the system, or equivalently, $a(N) = 0$. Suppose that $a(n)$, $n = 0, 1, \ldots, N-1$, can take M different values: $a_1, a_2, \ldots, a_M \in [0, 1]$. We wish to maximize

$$\eta = \kappa_1 \eta_1 - \kappa_2 \eta_2,$$

where η_1 is the average number of customers accepted to the system, η_2 is the average of $w(X_l)$, with w being a function of the number of customers in the system, and $\kappa_1, \kappa_2 > 0$ are two weighting factors.

Formulate this problem as a standard MDP.

1.12. For an ergodic Markov chain, we have

$$\eta = \lim_{L \to \infty} \frac{1}{L} \sum_{l=0}^{L-1} f(X_l), \qquad \text{w.p.1.}$$

Develop a "learning" algorithm that updates iteratively the estimates of η at every transition of the Markov chain using the reward observed at the transition. That is, find an algorithm

$$\hat{\eta}_l = \kappa_l \hat{\eta}_{l-1} + (1 - \kappa_l) f(X_l),$$

with $\hat{\eta}_{-1} = 0$ and $0 < \kappa_l < 1$, such that $\lim_{l \to \infty} \hat{\eta}_l = \eta$. Determine κ_l for $l = 0, 1, \ldots$.

1.13. Consider a Markov chain run under a deterministic policy $\alpha_i = d(i)$, $i \in \mathcal{S}$. Drive the equation for Q-factors:

$$Q^d(i, \alpha_i) - \sum_{j \in \mathcal{S}} p^{\alpha_i}(j|i) Q^d(j, \alpha_j) + \eta^d = f(i, \alpha_i).$$

1.14. Consider a Markov chain with state space $\mathcal{S}$. In each state $i \in \mathcal{S}$, there are two available actions denoted as $\alpha_{1,i}$ and $\alpha_{2,i}$. Let d be a randomized policy with $d(i) = \nu_i = (p_{1,i}, p_{2,i})$, $p_{1,i}, p_{2,i} > 0$, $p_{1,i} + p_{2.i} = 1$, representing the probabilities of taking actions $\alpha_{1,i}$ and $\alpha_{2,i}$, respectively, $i \in \mathcal{S}$. We also can view ν_i as an action that determines the transition probabilities of state i (see Problem 1.5). Therefore, we have three actions for each state: $\alpha_{1,i}$, $\alpha_{2,i}$, and ν_i, $i \in \mathcal{S}$. Observe a sample path of the system under the randomized policy d. Overall, when the system visits state i, it takes action ν_i. This is equivalent to a system that takes action $\alpha_{1,i}$ sometimes when the system visits state i, and takes action $\alpha_{2,i}$ other times when it visits i, with probabilities $p_{1,i}$ and $p_{2,i}$, respectively. Thus, a sample path of the system under policy d contains the information about $Q^d(i, \alpha_{1,i})$, $Q^d(i, \alpha_{2,i})$, and $Q^d(i, \nu_i)$, $i \in \mathcal{S}$.

a. Prove $Q^d(i, \nu_i) = p_{1,i} Q^d(i, \alpha_{1,i}) + p_{2,i} Q^d(i, \alpha_{2,i})$.
b. If $Q^d(i, \alpha_{1,i}) \geq Q^d(i, \alpha_{2,i})$, then $Q^d(i, \alpha_{1,i}) \geq Q^d(i, \nu_i)$.
c. Prove that for every randomized policy d, there is always a deterministic policy which is at least as good as d.

1.15. Consider a linear control system defined as

$$X_{l+1} = X_l + u_l + \xi_l, \qquad l = 0, 1, \ldots.$$

The state space is the set of integers $\mathcal{S} := \{\ldots, -1, 0, 1, \ldots\}$. The control variable u can take two values -2 and 2. The random noise ξ takes values from the integer set $\{-4, -3, -2, -1, 0, 1, 2, 3, 4\}$ with probabilities $p(\xi = 0) = 0.2$ and $p(\xi = i) = 0.1$ if $i \neq 0$. Describe the system with the MDP formulation.

1.16. Consider an admission control problem of a communications system consisting of three servers. The system is a Jackson type and hence its state can be denoted as $\boldsymbol{n} = (n_1, n_2, n_3)$ with n_i being the number of customers in server i, $i = 1, 2, 3$. Define an event a_{+4} as a customer arriving at the network and finding that there are four customers already in the network. Clearly define this event by a set of state transitions. (Denote the transition from state $\boldsymbol{n}$ to $\boldsymbol{n}'$ as $\langle \boldsymbol{n}, \boldsymbol{n}' \rangle$.)

Part I

Four Disciplines in Learning and Optimization

To climb steep hills requires slow pace at first.

William Shakespeare, English poet and playwright (1564 - 1818)

Don't buy the house; buy the neighborhood.

Russian Proverb

2

Perturbation Analysis

Perturbation analysis (PA) is the core of the gradient-based (or policy gradient) learning and optimization approach. The basic principle of PA is that *the derivative of a system's performance with respect to a parameter of the system can be decomposed into the sum of many small building blocks, each of which measures the effect of a single perturbation on the system's performance, and this effect can be estimated on a sample path of the system.* This decomposition principle applies to the differences in a system's performance with two policies as well and is thus fundamental to other learning and optimization approaches such as the policy iteration approach (see Chapter 4).

Historically, perturbation analysis was first developed for queueing systems and was later extended to Markov systems. Because PA of Markov systems is generally applicable and has a strong connection with other learning and optimization approaches, such as Markov decision processes and reinforcement learning, we first introduce the PA principle to Markov systems. PA of queueing systems will be discussed at the end of this chapter as supplementary material.

There were a number of books published in later 1980's and 1990's on PA of queueing-type systems [51, 72, 107, 112, 142]. The PA principle summarized above was discussed in detail in [45, 51, 141, 142] and extended to Markov systems in [62, 70].

2.1 Perturbation Analysis of Markov Chains

We first discuss PA of discrete-time Markov chains and related topics in this section. PA of continuous-time Markov processes is covered in the next section.

Consider an ergodic (irreducible and aperiodic) Markov chain $\boldsymbol{X} = \{X_l : l \geq 0\}$ on a finite state space $\mathcal{S} = \{1, 2, \dots, S\}$ with a transition probability matrix $P = [p(j|i)]_{i,j=1}^{S}$. Its steady-state probabilities are denoted as a row vector $\pi = (\pi(1), \dots, \pi(S))$ and the reward function is denoted as a (column) vector $f = (f(1), f(2), \dots, f(S))^T$. We have $Pe = e$, where $e = (1, 1, \dots, 1)^T$, and the probability flow balance equation $\pi = \pi P$. We first consider the *long-run average reward* (or, simply, the average reward) as the performance measure, which is defined as follows:

$$\eta = E_\pi(f) = \sum_{i=1}^{S} \pi(i) f(i) = \pi f,$$

where E_π denotes the expectation corresponding to the steady-state probability π on $\mathcal{S}$.

Let P' be another irreducible and aperiodic transition probability matrix on the same state space $\mathcal{S}$. Suppose that P changes to

$$P_\delta = P + \delta \Delta P = \delta P' + (1 - \delta) P, \tag{2.1}$$

with $0 \leq \delta \leq 1$ and $\Delta P = P' - P := [\Delta p(j|i)]$. Since $Pe = P'e = e$, we have $(\Delta P)e = 0$ and $P_\delta e = e$.

P_δ represents a randomized policy, which, at every state transition, implements policy P with probability $1 - \delta$ and policy P' with probability δ. When δ varies from 0 to 1, P_δ fills the line from P to P' in the policy space (Figure 2.1). With randomized policies, we can fill all the policies in the convex set spanned by a set of policies in a policy space. For example, we can fill the triangle with vertices P_0, P_1, and P_2 in the policy space with randomized policies $P(\delta_0, \delta_1, \delta_2) := \delta_0 P_0 + \delta_1 P_1 + \delta_2 P_2$, where $\delta_0 + \delta_1 + \delta_2 = 1$; $P(\delta_0, \delta_1, \delta_2)$ implements policy P_i with probability δ_i, $i = 0, 1, 2$, at every state transition (Figure 2.2).

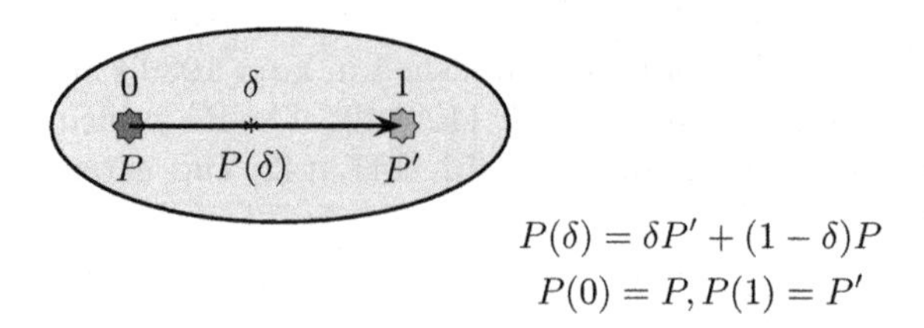

Fig. 2.1. Randomized Policies with Two Base Policies

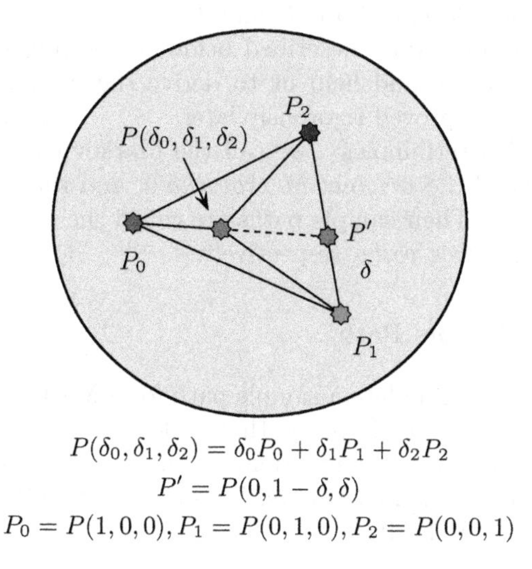

$$P(\delta_0, \delta_1, \delta_2) = \delta_0 P_0 + \delta_1 P_1 + \delta_2 P_2$$
$$P' = P(0, 1-\delta, \delta)$$
$$P_0 = P(1,0,0), P_1 = P(0,1,0), P_2 = P(0,0,1)$$

Fig. 2.2. Randomized Policies with Three Base Policies

For simplicity, we first assume that the Markov chain with transition probability matrix P_δ in (2.1) for all $0 \le \delta \le 1$ has the same reward function f, and we denote it as (P_δ, f). The steady-state probability of transition matrix P_δ is denoted as π_δ and the average reward of the Markov chain (P_δ, f) is denoted as $\eta_\delta = \pi_\delta f$. Then $\eta_0 = \eta = \pi f$ and $\eta_1 = \eta' = \pi' f$. Set $\Delta\eta_\delta = \eta_\delta - \eta$. The derivative of η_δ with respect to δ at $\delta = 0$ is

$$\left.\frac{d\eta_\delta}{d\delta}\right|_{\delta=0} = \lim_{\delta\to 0} \frac{\Delta\eta_\delta}{\delta},$$

which can be viewed as the directional derivative in the policy space along the direction from policy P to policy P' (see Figure 1.9 and Figure 2.1).

The goal of perturbation analysis is to determine the performance derivative $\frac{d\eta_\delta}{d\delta}$ by observing and/or analyzing the behavior of the Markov chain with transition probability matrix P. In particular, we wish to estimate this derivative by observing and analyzing a single sample path of the Markov chain with transition probability matrix P.

2.1.1 Constructing a Perturbed Sample Path

The main idea of PA comes from the fact that given a sample path of the Markov chain with transition probability matrix P, we can construct a sample path of the Markov chain with transition probability matrix P_δ, when δ is small; and this does not require that we rerun or resimulate the Markov chain with P_δ. If δ is small, the additional computation involved is also small. The performance derivative $\frac{d\eta_\delta}{d\delta}$ can be obtained by measurement or analysis once we have the sample paths of both P and P_δ. The above statement as well as the construction procedure described below is not very precise, but they provide a clear intuition and help us to derive the performance derivative formula, which will be proved rigorously later.

Following the PA terminology, we call the Markov chain with transition probability matrix P the *original Markov chain*, and that with P_δ the *perturbed Markov chain*. Their sample paths are called the *original sample paths* and the *perturbed sample paths*, respectively.

Constructing a Sample Path

We first review how to simulate a sample path for a Markov chain with transition probability matrix P. Suppose that at time $l = 0, 1, \ldots$, the Markov chain is in state $X_l = k$. In simulation, the next state after the transition at any time l is determined as follows. We generate a uniformly distributed random variable $\xi_l \in [0, 1)$. If

$$\sum_{k'=1}^{u_{l+1}-1} p(k'|k) \le \xi_l < \sum_{k'=1}^{u_{l+1}} p(k'|k), \qquad u_{l+1} \in \mathcal{S}, \tag{2.2}$$

(with the convention $\sum_{k'=1}^{0} p(k'|k) = 0$), then we set $X_{l+1} = u_{l+1}$. In the case illustrated in Figure 2.3.A, we have $p(i|k) = 0.5$, $p(j|k) = 0.5$, and $p(k'|k) = 0$ for all $k' \neq i, j$. The current state is k. We generate a $[0, 1)$-uniformly distributed random variable ξ. If $0 \le \xi < 0.5$, then the Markov chain moves into state i; otherwise, it moves into state j. Following this process, starting from any initial state X_0, we can construct a sample path for the Markov chain with any transition probability matrix P. Therefore, a sample path of a Markov chain is determined by an initial state X_0 and a sequence of $[0, 1)$-uniformly distributed random variables $\{\xi_0, \xi_1, \ldots\}$. Figure 2.4 illustrates such a sample path $\boldsymbol{X} := \{X_0, X_1, \ldots, X_l, \ldots\}$.

The performance measure η can be estimated from the sample path $\boldsymbol{X}$. In fact, if the Markov chain is ergodic, we have

$$\eta = \lim_{L \to \infty} \frac{1}{L} \sum_{l=0}^{L-1} f(X_l), \qquad \text{w.p.1},$$

where "w.p.1" stands for "with probability 1". Set

$$F_L = \sum_{l=0}^{L-1} f(X_l).$$

Then, we have

$$\eta = \lim_{L \to \infty} \frac{F_L}{L}. \tag{2.3}$$

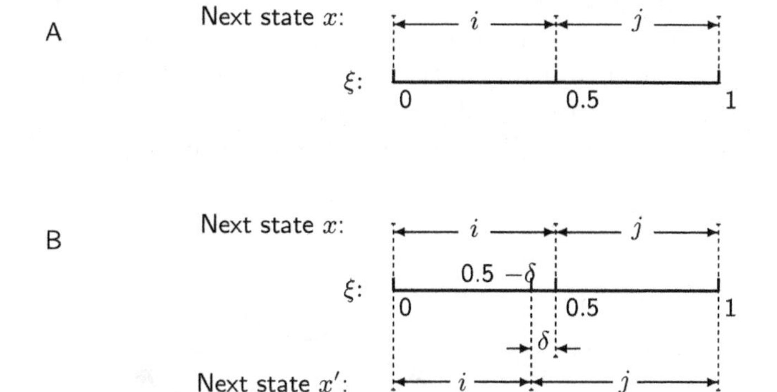

Fig. 2.3. Determining the State Transitions

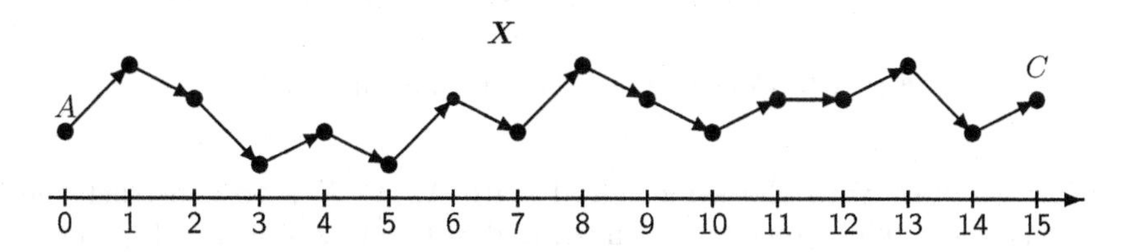

Fig. 2.4. A Sample Path of a Markov Chain

Constructing a Perturbed Sample Path on a Given Original Sample Path

Now, suppose that we are given a sample path of a Markov chain with transition probability matrix P, as shown as $\boldsymbol{X} = \{X_0, X_1, \ldots\}$ in Figure 2.4. It starts with initial state X_0 and is generated according to (2.2) with a sequence of $[0,1)$-uniformly distributed and independent random numbers $\{\xi_0, \xi_1, \ldots, \xi_l, \ldots\}$. We wish to construct a perturbed sample path for the Markov chain with $P_\delta = P + \delta\Delta P$. We denote it as $\boldsymbol{X}_\delta = \{X_{\delta,0}, X_{\delta,1}, \ldots\}$. To this end, we may think as follows.

To save computation, we may try to use the same sequence $\{\xi_0, \xi_1, \ldots, \xi_l, \ldots\}$ to generate the perturbed path. However, we need to use (cf. (2.2))

$$\sum_{k'=1}^{u_{\delta,l+1}-1} [p(k'|k) + \delta\Delta p(k'|k)] \leq \xi_l < \sum_{k'=1}^{u_{\delta,l+1}} [p(k'|k) + \delta\Delta p(k'|k)] \tag{2.4}$$

to determine the state at $X_{\delta,l+1}$; i.e., if (2.4) holds, we set $X_{\delta,l+1} = u_{\delta,l+1}$.

First, we observe that when δ is very small, in most cases we may have $u_{\delta,l+1} = u_{l+1}$, if $X_{\delta,l} = X_l$, $l = 0, 1, \ldots$. For example, let us assume that the transition probabilities of the Markov chain in Figure 2.3 are perturbed to $p_\delta(i|k) = 0.5-\delta$, $p_\delta(j|k) = 0.5+\delta$, and $p_\delta(l|k) = 0$, $l \neq i, j$ (i.e., $\Delta p(i|k) = -1$, $\Delta p(j|k) = 1$, and $\Delta p(l|k) = 0$). In this case, if $X_{\delta,l} = X_l = k$ and the same ξ_l is used to determine the state transition, then $X_{\delta,l+1} \neq X_{l+1}$ if and only if $0.5 - \delta \leq \xi_l < 0.5$, in which case the original Markov chain $\boldsymbol{X}$ moves to $X_{l+1} = i$, but the perturbed one $\boldsymbol{X}_\delta$ moves to $X_{\delta,l+1} = j$. The probability that this discrepancy occurs is δ, which is very small as assumed, see Figure 2.3.B.

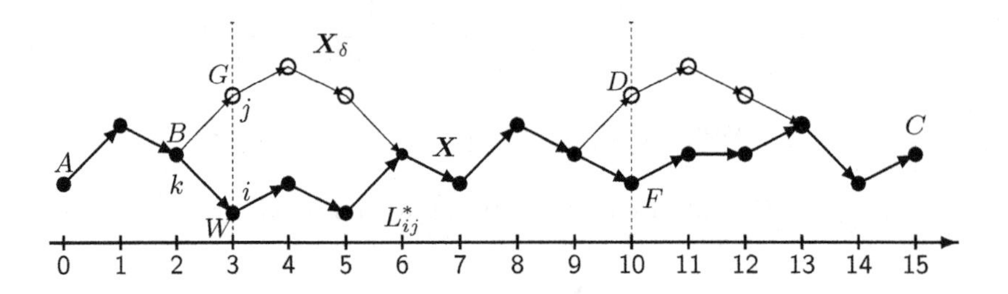

Fig. 2.5. Constructing a Perturbed Sample Path

Now we start with the same initial state $X_0 = X_{\delta,0}$ to construct the perturbed path. This procedure is illustrated in Figure 2.5, in which point A denotes the initial state and path $A-B-W-F-C$ is the given original sample path $\boldsymbol{X}$. As we have explained, starting from the same state the transitions of $\boldsymbol{X}$ and $\boldsymbol{X}_\delta$ differ only with a very small probability. In Figure 2.5, it so happens that with the same random variables ξ_0 and ξ_1, according to (2.2) and (2.4), we have $X_1 = X_{\delta,1}$ and $X_2 = X_{\delta,2}$.

Next, we assume that at $l = 2$, according to (2.2) and (2.4) with the same random variable ξ_2, we determine that $\boldsymbol{X}$ moves to $X_3 = i$ (point W) but $\boldsymbol{X}_\delta$ moves to another state $X_{\delta,3} = j$ (point G). We say that, because of the change of P to P_δ, the system has a *perturbation* (or simply called a "jump") from i to j at $l = 3$. After $l = 3$, the original sample path follows the path $W - F - C$; the perturbed path, however, follows a completely different path starting from point G. For convenience in understanding, let us generate an additional sequence of $[0, 1)$-uniformly and independently distributed random variables $\xi_{\delta,3}, \xi_{\delta,4}, \ldots$, which are also independent of $\xi_3, \xi_4, \ldots$, to construct the perturbed path following (2.4) starting from point G at $l = 3$ until the perturbed path merges with the original one.

Figure 2.5 shows that the perturbed path $\boldsymbol{X}_\delta$ merges with the original one $\boldsymbol{X}$ at $l = L^*_{ij} = 6$. Theoretically, because both sample paths $\boldsymbol{X}$ and $\boldsymbol{X}_\delta$ are ergodic, they will merge in finite steps (i.e., L^*_{ij} is finite) with probability 1. Let $\xi_{\delta,3}, \xi_{\delta,4}$, and $\xi_{\delta,5}$ be the random variables that determine the transitions at $l = 3, 4$, and 5 (or equivalently, the states $X_{\delta,4}, X_{\delta,5}$, and $X_{\delta,6} = X_6$) on $\boldsymbol{X}_\delta$. Then, the original path $\boldsymbol{X}$ from X_0 to X_6 is generated by $\xi_0, \xi_1, \xi_2, \xi_3, \xi_4, \xi_5$, while the perturbed path $\boldsymbol{X}_\delta$ from $X_{\delta,0}$ to $X_{\delta,6}$ is generated by $\xi_0, \xi_1, \xi_2, \xi_{\delta,3}, \xi_{\delta,4}, \xi_{\delta,5}$, with $\xi_{\delta,3}, \xi_{\delta,4}, \xi_{\delta,5}$ independent of ξ_3, ξ_4, ξ_5.

Starting from the merging point $X_6 = X_{\delta,6}$, the situation is the same as at the initial point $X_0 = X_{\delta,0}$. Again, we use the same random variables $\xi_6, \xi_7, \ldots$, to construct the perturbed path until it differs from the original one. In Figure 2.5, it so happens that with the same random variables ξ_6, ξ_7, and ξ_8 according to (2.2) and (2.4) we have $X_7 = X_{\delta,7}$, $X_8 = X_{\delta,8}$ and $X_9 = X_{\delta,9}$. However, there is a perturbation at $l = 10$. In other words, according to (2.2) and (2.4) with the same random variable ξ_9, we determine that $\boldsymbol{X}$ and $\boldsymbol{X}_\delta$ move to two different states X_{10} (point F) and $X_{\delta,10}$ (point D), respectively. After $l = 10$, the situation is the same as at $l = 3$. The two sample paths $\boldsymbol{X}$ and $\boldsymbol{X}_\delta$ follow different paths $D - C$ and $F - C$ until they merge again at $l = 13$. $X_{\delta,11}$, $X_{\delta,12}$, and $X_{\delta,13}$ are generated by random variables $\xi_{\delta,10}$, $\xi_{\delta,11}$, and $\xi_{\delta,12}$, which are independent of ξ_{10}, ξ_{11}, and ξ_{12}. $\boldsymbol{X}_\delta$ and $\boldsymbol{X}$ merge again at $l = 13$. Starting from this merging point $X_{\delta,13} = X_{13}$, once again the situation is the same as at the initial point $X_0 = X_{\delta,0}$.

The above description illustrates how to construct a perturbed sample path $\boldsymbol{X}_\delta$, given an original sample path $\boldsymbol{X}$. At any time instant l, if $X_{\delta,l} = X_l$, then we use the same random variable ξ_l to determine the state transitions (or equivalently X_{l+1} and $X_{\delta,l+1}$) for both $\boldsymbol{X}$ and $\boldsymbol{X}_\delta$ by using (2.2) and (2.4); if it turns out that $X_{\delta,l+1} \neq X_{l+1}$, we say there is a perturbation (jump) at $l+1$. After each jump, $\boldsymbol{X}_\delta$ is completely different from $\boldsymbol{X}$ until they merge together. In these segments in which the two sample paths are different, $\boldsymbol{X}_\delta$ is generated independently of $\boldsymbol{X}$. In Figure 2.5, $\boldsymbol{X}$ and $\boldsymbol{X}_\delta$ are generated by the following sequences of random variables, respectively:

$$\begin{aligned}
\boldsymbol{X} &: \xi_0\ \xi_1\ \xi_2\ \xi_3\ \ \xi_4\ \ \xi_5\ \xi_6\ \xi_7\ \xi_8\ \xi_9\ \xi_{10}\ \ \xi_{11}\ \ \xi_{12}\ \xi_{13}\ \xi_{14}, \\
\boldsymbol{X}_\delta &: \xi_0\ \xi_1\ \xi_2\ \xi_{\delta,3}\ \xi_{\delta,4}\ \xi_{\delta,5}\ \xi_6\ \xi_7\ \xi_8\ \xi_9\ \xi_{\delta,10}\ \xi_{\delta,11}\ \xi_{\delta,12}\ \xi_{13}\ \xi_{14},
\end{aligned}$$

where all the random variables ξ_l, and $\xi_{\delta,l}$ are independent of each other.

Finally, when δ is very small, perturbations rarely happen (see Figure 2.3.B). In this case, in most time instants, the perturbed sample path $\boldsymbol{X}_\delta$ is the same as the original one $\boldsymbol{X}$; i.e., in reality, the lengths of the common segments are much longer than what might be indicated by $X_0 - X_2$, $X_6 - X_9$, and $X_{13} - X_{14}$ in Figure 2.5.

2.1.2 Perturbation Realization Factors and Performance Potentials

To calculate performance derivatives, we need to compare the average rewards of the original and the perturbed Markov chains, η and η_δ, by using the sample paths $\boldsymbol{X}$ and $\boldsymbol{X}_\delta$ constructed above. As shown in Figure 2.5, the difference between $\boldsymbol{X}$ and $\boldsymbol{X}_\delta$ is only reflected in the segments after the perturbations. In other words, the effect of a change of the transition probability matrix from P to P_δ on the system performance can be decomposed into the sum of the effects of the perturbations generated due to the change in P. Therefore, we first need to study the effect of a single perturbation on the system performance. We show that this effect can be measured by a quantity called the *perturbation realization factor.*

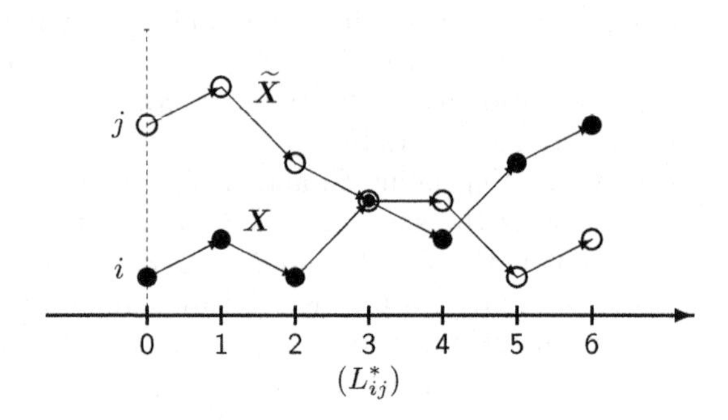

Fig. 2.6. Realization of a Perturbation

Perturbation Realization

Again, we use Figure 2.5 to illustrate the idea. At $l = 3$, the sample path is perturbed from state i to state j. This perturbation will certainly affect the system's behavior and the system's performance. As shown in Figure 2.5, after $l = 3$, the perturbed Markov chain evolves differently from the original chain, until, at $l = L^*_{ij}$, the perturbed path merges with the original one. The effect of the perturbation takes place in the period from $l = 3$ to L^*_{ij}. In PA terminology, we say that the perturbation generated at $l = 3$ is *realized* by the system at $l = L^*_{ij} = 6$.

Strictly speaking, the perturbed path $\boldsymbol{X}_\delta$ follows the perturbed transition probability matrix P_δ. However, because δ is very small and the length from the perturbed point $l = 3$ to the merging point L^*_{ij}, $L^*_{ij} - 3$, is finite (with probability 1), the probability that there is another perturbation in the period from $l = 3$ to L^*_{ij} (i.e., there are two perturbations in the period from $l = 3$ to L^*_{ij}, one at $l = 3$ the other in the period from $l = 4$ to L^*_{ij}) is on the order δ^2. This contributes to the high-order performance derivatives and in the first-order derivatives we may ignore this high-order term. Therefore, to calculate the performance derivatives, as δ approaches zero, we may assume that from $l = 3$ to L^*_{ij} the perturbed path $\boldsymbol{X}_\delta$ is the same as if it follows the original transition probability matrix P.

Thus, to quantify the effect of a single perturbation from i to j, we study two independent Markov chains $\boldsymbol{X} = \{X_l, l \geq 0\}$ and $\widetilde{\boldsymbol{X}} = \{\widetilde{X}_l, l \geq 0\}$ with $X_0 = i$ and $\widetilde{X}_0 = j$, respectively; both of them follow the same transition matrix P (Figure 2.6). Let these two sample paths merge for the first time at L_{ij}^*, i.e.,

$$L_{ij}^* = \min\left\{ l : l \geq 0, \widetilde{X}_l = X_l \middle| \widetilde{X}_0 = j, X_0 = i \right\}.$$

Recall that the performance measure is $\eta \approx \frac{F_L}{L}$ (see (2.3)). Apparently, the average effect of a single perturbation on η is zero, because L_{ij}^* is finite with probability 1. We, therefore, study the effect of a single perturbation on F_L for a large L.

Let E denote the expectation in the probability space spanned by all the sample paths of both $\boldsymbol{X}$ and $\widetilde{\boldsymbol{X}}$. The *perturbation realization factor (PRF)* is defined as [62, 70]:

$$\gamma(i,j) = E\left\{ \sum_{l=0}^{L_{ij}^*-1} \left[f(\widetilde{X}_l) - f(X_l) \right] \middle| \widetilde{X}_0 = j,\ X_0 = i \right\}, \quad i,j = 1, \ldots, S. \tag{2.5}$$

Thus, $\gamma(i,j)$ represents the average effect of a jump from i to j on F_L in (2.3). For convenience, sometimes we may refer to $\gamma(i,j)$ as the effect of a jump on the performance η itself, although this effect is on an "infinitesimal" scale.

By the strong Markov property, the two Markov chains $\boldsymbol{X}$ and $\widetilde{\boldsymbol{X}}$ behave similarly statistically after L_{ij}^*. Thus,

$$\lim_{L \to \infty} E\left\{ \sum_{l=L_{ij}^*}^{L-1} \left[f(\widetilde{X}_l) - f(X_l) \right] \middle| \widetilde{X}_0 = j,\ X_0 = i \right\} = 0.$$

Therefore, (2.5) becomes

$$\begin{aligned} \gamma(i,j) &= \lim_{L \to \infty} E\left\{ \sum_{l=0}^{L-1} \left[f(\widetilde{X}_l) - f(X_l) \right] \middle| \widetilde{X}_0 = j,\ X_0 = i \right\} \\ &= \lim_{L \to \infty} E\left[\widetilde{F}_L - F_L \middle| \widetilde{X}_0 = j,\ X_0 = i, \right], \qquad i,j = 1, \ldots, S. \end{aligned} \tag{2.6}$$

Essentially, the perturbation realization factors use the difference in the sums of the rewards on the perturbed path and the original one to measure the effect of a single perturbation.

The matrix $\Gamma := [\gamma(i,j)]_{i,j=1}^S \in \mathcal{R}^{S \times S}$ is called a *perturbation realization factor (PRF) matrix*. From (2.5), we have

$$\gamma(i,j) = f(j) - f(i) + \sum_{i'=1}^{S}\sum_{j'=1}^{S} E\left\{ \sum_{l=1}^{L^*_{i'j'}-1} \left[f(\widetilde{X}_l) - f(X_l)\right] \middle| \widetilde{X}_1 = j', X_1 = i' \right\}$$
$$\times \mathcal{P}\left(\widetilde{X}_1 = j', X_1 = i' \middle| \widetilde{X}_0 = j, X_0 = i\right)$$
$$= f(j) - f(i) + \sum_{i'=1}^{S}\sum_{j'=1}^{S} p(i'|i)p(j'|j)\gamma(i',j').$$

By writing this in a matrix form, we have the following *PRF equation* [70]

$$\Gamma - P\Gamma P^T = F, \tag{2.7}$$

where $F = ef^T - fe^T$.

If F is a Hermitian matrix, then (2.7) is called the Lyapunov equation in the literature [13, 14, 162, 174]. (A *Hermitian matrix*, also called a self-adjoint matrix, is a square matrix that is equal to its own conjugate transpose. Thus, a real Hermitian matrix is a symmetric matrix.) The PRF equation differs from the Lyapunov equation because F is a skew-symmetric matrix, $F^T = -F$.

Performance Potentials

From (2.6), we have $\gamma(i,i) = 0$ for any $i = 1, \ldots, S$, and $\gamma(i,j) = -\gamma(j,i)$, or $\Gamma^T = -\Gamma$; i.e., Γ is skew-symmetric. In addition, from (2.6), we can easily prove

$$\gamma(i,j) = \gamma(i,k) + \gamma(k,j), \qquad i,j,k = 1,\ldots,S. \tag{2.8}$$

This is the same equation as that for the differences of potential energies in physics. This observation motivates the following analysis: Let us fix any state denoted as $k^* \in \mathcal{S}$. Then, (2.8) becomes

$$\gamma(i,j) = \gamma(i,k^*) + \gamma(k^*,j) = \gamma(k^*,j) - \gamma(k^*,i), \qquad i,j = 1,\ldots,S.$$

Define $g_{k^*}(j) = \gamma(k^*,j)$. Then,

$$\gamma(i,j) = g_{k^*}(j) - g_{k^*}(i), \qquad i,j = 1,\ldots,S. \tag{2.9}$$

For any two states k_1^* and k_2^*, we have

$$g_{k_2^*}(j) - g_{k_1^*}(j) = \gamma(k_2^*, k_1^*), \qquad j = 1,\ldots,S,$$

which does not depend on j. This means that if we choose a different k^*, the resulting $g_{k^*}(j)$'s differ by only the same constant for all $j \in \mathcal{S}$. With this in mind, we omit the subscript k^* and rewrite (2.9) as

$$\gamma(i,j) = g(j) - g(i), \qquad i,j = 1,\dots,S. \tag{2.10}$$

$g(i)$ is called the *performance potential* (or simply the *potential*) of state i, and $g_{k^*}(i)$ denotes a particular version of the potential. (The word "potentials" have been used in the literature in similar contents, e.g., [154, 168].) Just as in physics, different versions of the potentials may differ by a constant. Let $g = (g(1), \dots, g(S))^T$. Then, (2.10) becomes

$$\Gamma = eg^T - ge^T. \tag{2.11}$$

If g is a potential (vector), so is $g + ce$ for any real constant c. For simplicity, we use the same notation g for different versions of the potentials and keep in mind that potential g in different expressions may differ by a constant. A physical interpretation of the performance potentials compared with the potential energy is illustrated in Figure 2.7.

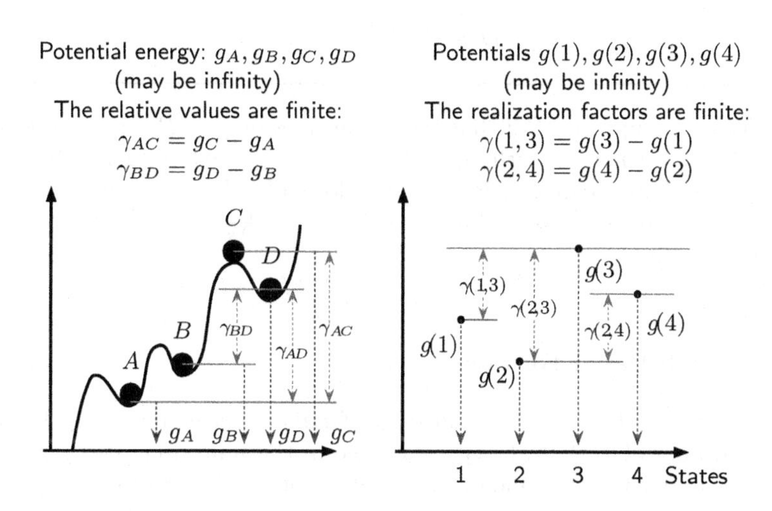

Fig. 2.7. Physical Interpretation of Potential Energy and Performance Potentials

Substituting (2.11) into (2.7), we obtain

$$e[(I-P)g - f]^T = [(I-P)g - f]e^T,$$

i.e., $e[(I-P)g-f]^T$ is a symmetric matrix. Thus, we must have $(I-P)g-f = ce$, where c is a constant. Left-multiplying both sides of this equation by π and using $\pi = \pi P$, we get $c = -\pi f = -\eta$. Finally, we have

$$(I - P)g + \eta e = f. \tag{2.12}$$

This is called the *Poisson equation.* Its solution is unique only up to an additive constant; i.e., if g is a solution to (2.12), then for any constant c, $g + ce$ is also a solution. To write (2.12) for each component, we have

$$g(i) = f(i) - \eta + \sum_{j \in \mathcal{S}} p(j|i)g(j).$$

This equation has a clear interpretation: The long-term contribution of state i to the average performance, $g(i)$, equals its one-step contribution at the current time, $f(i) - \eta$, plus the expected long-term "potential" contribution of the next state. Equation (2.10) shows that the effect of a perturbation from state i to j (the perturbation realization factor $\gamma(i, j)$) equals the difference in the long-term contributions of these two states.

One of the ways to specify a solution to (2.12) is to normalize it by setting $\pi g = \eta$. In this case, (2.12) takes the form

$$(I - P + e\pi)g = f.$$

It is shown in Appendix B.2 that the eigenvalues of $(I - P + e\pi)$ are $\{1, 1 - \lambda_2, \dots, 1 - \lambda_S\}$, where λ_i with $|\lambda_i| < 1$, $i = 2, \dots, S$, are the eigenvalues of the transition probability matrix P [20]. Therefore, $(I - P + e\pi)$ is invertible and the eigenvalues of $(I - P + e\pi)^{-1}$ are $\{1, \frac{1}{1-\lambda_2}, \dots, \frac{1}{1-\lambda_S}\}$, $|\lambda_i| < 1$, $i = 2, \dots, S$. Therefore, we have

$$g = (I - P + e\pi)^{-1} f. \tag{2.13}$$

Sample-Path-Based Formulas

The matrix $(I - P + e\pi)^{-1}$ is called the *fundamental matrix* [202]. Because the eigenvalues of $P - e\pi$, $0, \lambda_2, \dots, \lambda_S$, lie in the unit circle (see Appendix B.2), we can expand the fundamental matrix into a Taylor series:

$$(I - P + e\pi)^{-1} = \sum_{k=0}^{\infty} (P - e\pi)^k = I + \sum_{k=1}^{\infty} (P^k - e\pi). \tag{2.14}$$

Thus, from (2.13), we have

$$g = f + \sum_{k=1}^{\infty} (P^k - e\pi) f.$$

Note that from (A.3), the (i,j)th entry of P^k is $p^{(k)}(j|i) = \mathcal{P}(X_k = j|X_0 = i)$. Then, from (2.14), we have

$$g(i) = \lim_{L\to\infty} \left\{ E\left[\sum_{l=0}^{L-1} f(X_l) \Big| X_0 = i \right] - (L-1)\eta \right\}.$$

We may get a more convenient version of the potentials by adding a constant $-\eta$ to every component of g. Thus, we have another version of g:

$$g = [(I - P + e\pi)^{-1} - e\pi]f, \tag{2.15}$$

for which $\pi g = 0$, and

$$g(i) = \lim_{L\to\infty} E\left\{ \sum_{l=0}^{L-1} [f(X_l) - \eta] \Big| X_0 = i \right\}. \tag{2.16}$$

The Poisson equation (2.12) can be easily derived from (2.16); see Problem 2.5.

From (2.16), we can derive another sample-path-based formula for $\gamma(i,j)$. On a sample path of $\boldsymbol{X}$ starting with $X_0 = j$, define $L(i|j)$ to be the first passage time of $\boldsymbol{X}$ reaching state i; i.e., $L(i|j) = \min\{l : l \geq 0, X_l = i|X_0 = j\}$. Then, we have

$$\gamma(i,j) = E\left\{ \sum_{l=0}^{L(i|j)-1} [f(X_l) - \eta] \Big| X_0 = j \right\}. \tag{2.17}$$

This can be intuitively explained as follows: from (2.16),

$$\begin{aligned}
&\gamma(i,j) = g(j) - g(i) \\
&= \lim_{L\to\infty} \left\{ E\left\{ \sum_{l=0}^{L} [f(X_l) - \eta] \Big| X_0 = j \right\} - E\left\{ \sum_{l=0}^{L} \left[f(\widetilde{X}_l) - \eta\right] \Big| \widetilde{X}_0 = i \right\} \right\} \\
&= \lim_{L\to\infty} \left\{ E\left\{ \left[\sum_{l=0}^{L(i|j)-1} [f(X_l) - \eta] + \sum_{L(i|j)}^{L} [f(X_l) - \eta] \right] \Big| X_0 = j \right\} \right. \\
&\quad \left. - E\left\{ \left[\sum_{l=0}^{L-L(i|j)} \left[f(\widetilde{X}_l) - \eta\right] + \sum_{L-L(i|j)+1}^{L} \left[f(\widetilde{X}_l) - \eta\right] \right] \Big| \widetilde{X}_0 = i \right\} \right\},
\end{aligned} \tag{2.18}$$

where $\{X_l, l \geq 0\}$ and $\{\widetilde{X}_l, l \geq 0\}$ are two independent Markov chains with the same transition probability matrix P. Because of the strong Markov property and $X_{L(i|j)} = i$, the second term equals the third term as long as $L > L(i|j)$. In addition, since $\lim_{l\to\infty} E[f(\widetilde{X}_l)] = \eta$, the last term goes to zero as $L \to \infty$. Thus, (2.18) leads to (2.17). The idea is explained in Figure 2.8: In region (II), both $\boldsymbol{X}$ and $\widetilde{\boldsymbol{X}}$ are statistically identical because $\widetilde{X}_0 = X_{L(i|j)} = i$; the mean of $f(X_l) - \eta$ on $\widetilde{\boldsymbol{X}}$ in region (III) goes to zero as $L \to \infty$. Thus, the only term left in the difference $g(j) - g(i)$ is the summation on $\boldsymbol{X}$ in region (I). For a detailed proof, see [62].

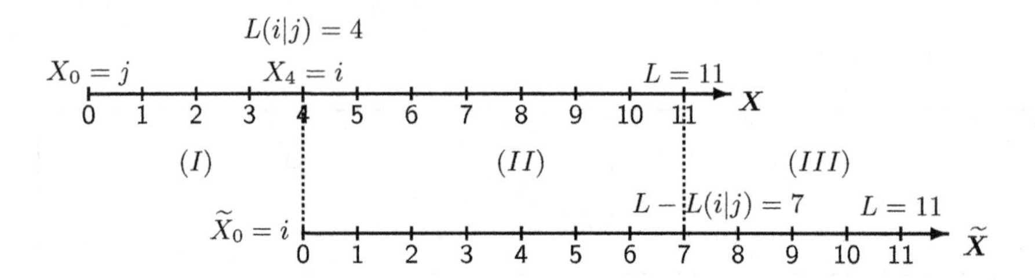

Fig. 2.8. An Explanation for (2.17)

In summary, the perturbation realization factor $\gamma(i,j)$, $i,j \in \mathcal{S}$, measures the "infinitesimal" effect of a perturbation from state i to j on the average reward η (more precisely, it measures the effect on F_L for $L >> 1$). From the physical meaning, the performance potential $g(i)$, $i = 1, \ldots, S$, measures the long-term "potential" contribution of state i to η. Similar to the potential energy in physics, only the differences in the different $g(i)$'s are important for performance sensitivities.

Finally, the reward function can be defined as $f(i,j)$, $i,j \in \mathcal{S}$; i.e, the system gains a reward $f(i,j)$ when it is in state $X_l = i$ and moves to state $X_{l+1} = j$, $l = 0, 1, \ldots$. The average reward is defined as

$$\eta = \lim_{L\to\infty} \frac{1}{L} \sum_{l=0}^{L-1} f(X_l, X_{l+1}).$$

In this case, if we use the average

$$f(i) := \sum_{j=1}^{S} p(j|i) f(i,j)$$

as the reward function, all the results developed in this and the remaining sections for PA of Markov chains hold.

2.1.3 Performance Derivative Formulas

To derive the performance derivative $\frac{d\eta_\delta}{d\delta}$ at policy (P, f) along any direction ΔP, we consider a sample path $\boldsymbol{X}$ with a transition probability matrix P

consisting of L, $L >> 1$, transitions. Among these transitions, on average, there are $L\pi(k)$ transitions at which the system is in state k. Each time when $\boldsymbol{X}$ visits state i after visiting state k, because of the change from P to $P_\delta = P + \delta \Delta P$, the perturbed path $\boldsymbol{X}_\delta$ may have a jump, denoted as from state i to state j (i.e., after visiting k, $\boldsymbol{X}$ moves to i and $\boldsymbol{X}_\delta$ moves to j), as shown in Figure 2.5. For convenience, we allow $i = j$ as a special case. A "real jump" (with $i \neq j$) happens rarely. Denote the probability of a jump from i to j after visiting state k as $p(i, j|k)$. We have

$$p(i, j|k) = p(i|k)p_\delta(k, j|k, i),$$

where $p_\delta(k, j|k, i)$ denotes the conditional probability that $\boldsymbol{X}_\delta$ moves from state k to state j given that $\boldsymbol{X}$ moves from state k to i. By definition, we have $\sum_{j=1}^{S} p_\delta(k, j|k, i) = 1$. Therefore,

$$\sum_{j=1}^{S} p(i, j|k) = p(i|k). \tag{2.19}$$

Similarly,

$$\sum_{i=1}^{S} p(i, j|k) = p_\delta(j|k), \tag{2.20}$$

and $\sum_{i,j=1}^{S} p(i, j|k) = 1$. On average, in the L transitions on the sample path, there are $L\pi(k)p(i, j|k)$ jumps from i to j following the visit to state k. As discussed in Section 2.1.2, each such jump has on average an effect of $\gamma(i, j)$ on F_L.

A real jump happens extremely rarely as $\delta \to 0$. As discussed in Section 2.1.2, the probability that the Markov chain jumps at $l = 3$ and that there is another jump of $\boldsymbol{X}_\delta$ from $l = 4$ to L^*_{ij} (or equivalently, $\boldsymbol{X}_\delta$ would move differently if it followed P from $l = 3$ to L^*_{ij}) is on the order of δ^2; the effect of such a situation can be ignored for performance derivatives. Therefore, we may assume that, from $l = 3$ to L^*_{ij} and in other periods after each jump before merging, $\boldsymbol{X}_\delta$, generated according to P_δ, is the same as following P. Thus, on average, the total effect on F_L due to the change in P to $P_\delta = P + \delta \Delta P$ is

$$\begin{aligned} & E(F_{\delta,L} - F_L) \\ \approx & \sum_{k=1}^{S} \left[\sum_{i,j=1}^{S} L\pi(k)p(i, j|k)\gamma(i, j) \right] \\ = & \sum_{k=1}^{S} \left\{ \sum_{i,j=1}^{S} L\pi(k)p(i, j|k)[g(j) - g(i)] \right\} \\ = & \sum_{k=1}^{S} L\pi(k) \left\{ \sum_{j=1}^{S} \left[g(j) \sum_{i=1}^{S} p(i, j|k) \right] \right. \end{aligned}$$

$$-\sum_{i=1}^{S}\left[g(i)\sum_{j=1}^{S}p(i,j|k)\right]\Bigg\}. \quad (2.21)$$

From (2.19) and (2.20), (2.21) becomes

$$\begin{aligned}E(F_{\delta,L}-F_L) &\approx \sum_{k=1}^{S} L\pi(k)\left\{\left[\sum_{j=1}^{S}p_\delta(j|k)g(j)\right]-\left[\sum_{i=1}^{S}p(i|k)g(i)\right]\right\}\\ &= \sum_{k=1}^{S} L\pi(k)\left\{\sum_{j=1}^{S}[p_\delta(j|k)-p(j|k)]g(j)\right\}\\ &= L\pi(P_\delta - P)g = L\pi(\Delta P)\delta g.\end{aligned}$$

Thus,

$$\eta_\delta - \eta = \lim_{L\to\infty}\frac{1}{L}E(F_{\delta,L}-F_L) \approx \pi(\Delta P)\delta g. \quad (2.22)$$

Finally, letting $\delta \to 0$, we obtain the performance derivative formula

$$\left.\frac{d\eta_\delta}{d\delta}\right|_{\delta=0} = \pi(\Delta P)g. \quad (2.23)$$

Strictly speaking, the approximation in (2.21) is not accurate (the difference of both sides is on the order of $o(L)$, which may not be small for a large L). It is accurate only after both sides of (2.21) are divided by L, resulting in (2.22). Nevertheless, (2.21) provides a good intuition.

From (2.11), we have $\pi\Gamma = g^T - (\pi g)e^T$. Thus, from (2.23), we get

$$\left.\frac{d\eta_\delta}{d\delta}\right|_{\delta=0} = \pi(\Delta P)\Gamma^T\pi^T. \quad (2.24)$$

Note that g, Γ, and π can be estimated on a single sample path of a Markov chain with transition matrix P; thus, given any ΔP, the performance derivative along the direction ΔP can be obtained by (2.23) or (2.24) using the sample path-based estimates of π and g or Γ. Algorithms can be developed for estimating the performance derivative based on a single sample path using (2.23) without estimating each component of g; see Chapter 3.

Finally, (2.23) can be easily derived by using the Poisson equation (2.12). Let P' be the transition probability matrix of another irreducible Markov chain defined on the same state space $\mathcal{S}$, and let η' and π' be its corresponding performance measure and steady-state probability, respectively. Multiplying both sides of (2.12) on the left by π' and using $\pi' e = 1$ and $\pi' = \pi' P'$, we obtain the performance difference formula:

$$\eta' - \eta = \pi'(\Delta P)g. \tag{2.25}$$

Taking P' as $P_\delta = P + \delta(\Delta P)$ and η' as η_δ in (2.25), we get

$$\eta_\delta - \eta = \pi_\delta \delta(\Delta P)g.$$

Letting $\delta \to 0$, we obtain (2.23) (it is easy to see $\lim_{\delta \to 0} \pi_\delta = \pi$). Thus, the performance derivative formula (2.23) follows directly from the Poisson equation (2.12). However, our PA-based reasoning intuitively explains the physical meaning of the realization factors and the potentials. It clearly illustrates the nature of the performance derivatives: They can be constructed by using the potentials as building blocks. More importantly, this PA-based construction approach can be used in constructing performance derivative formulas for other non-standard problems in which the special features of the system can be utilized. New optimization schemes can be developed for such special systems. We discuss these problems in Chapters 8 and 9.

So far, we have assumed that f does not change. Suppose that the reward function associated with P' is f' and, in addition to the change of P to P_δ, f also changes to $f_\delta = f + \delta \Delta f$, $\Delta f = f' - f$. Then, it is easy to obtain the performance derivative formula

$$\left.\frac{d\eta_\delta}{d\delta}\right|_{\delta=0} = \pi[(\Delta P)g + \Delta f]. \tag{2.26}$$

The performance difference formula (2.25) becomes

$$\eta' - \eta = \pi'[(\Delta P)g + \Delta f]. \tag{2.27}$$

The difference between (2.23) (or (2.26)) and (2.25) (or (2.27)) is that π in (2.23) is replaced by π' in (2.25).

With realization factors, we have

$$\left.\frac{d\eta_\delta}{d\delta}\right|_{\delta=0} = \pi\left[(\Delta P)\Gamma^T \pi^T + \Delta f\right] \tag{2.28}$$

and

$$\eta' - \eta = \pi'\left[(\Delta P)\Gamma^T \pi^T + \Delta f\right]. \tag{2.29}$$

Finally, sometimes it may be useful to specifically denote the two policies in performance sensitivity analysis as (P^h, f^h) and (P^d, f^d) (instead of (P', f') and (P, f)). With these notations, (2.26) becomes

$$\left.\frac{d\eta_\delta}{d\delta}\right|_{\delta=0} = \pi^d[(\Delta P)g^d + \Delta f]$$
$$= \pi^d[(P^h - P^d)g^d + (f^h - f^d)],$$

where η_δ is the performance of (P_δ, f_δ), with $P_\delta = P^d + \delta\Delta P$, $\Delta P = P^h - P^d$, $f_\delta = f^d + \delta\Delta f$, and $\Delta f = f^h - f^d$.

2.1.4 Gradients with Discounted Reward Criteria

In this subsection, we show that the idea of performance potentials and the performance derivative formula can be extended to Markov chains with discounted reward criteria.

Consider an ergodic Markov chain $\boldsymbol{X} = \{X_l, l \geq 0\}$ with transition probability matrix P and reward function f. Let β, $0 < \beta \leq 1$, be a discount factor. For $0 < \beta < 1$, we define the discounted reward as a column vector $\eta_\beta = (\eta_\beta(1), \ldots, \eta_\beta(S))^T$ with

$$\eta_\beta(i) := (1-\beta)E\left[\sum_{l=0}^{\infty} \beta^l f(X_l) \middle| X_0 = i\right]. \tag{2.30}$$

The factor $(1-\beta)$ in (2.30) is used to obtain the continuity of η_β at $\beta = 1$. We show that the long-run average reward discussed in the last subsection can be viewed as a special case when $\beta \to 1$ and therefore we denote $\eta_1 := \eta e$. Also, the weighting factors in (2.30) are normalized: $\sum_{l=0}^{\infty}(1-\beta)\beta^l = 1$. In a matrix form, Equation (2.30) is

$$\eta_\beta = (1-\beta)\sum_{l=0}^{\infty} \beta^l P^l f = (1-\beta)(I - \beta P)^{-1} f, \qquad 0 < \beta < 1. \tag{2.31}$$

The second equality in (2.31) holds because for $0 < \beta < 1$, all the eigenvalues of βP are within the unit circle [20]. From (2.39) given below, we know that $\lim_{\beta\uparrow 1} \eta_\beta$ exists and we have

$$\eta_1 := \lim_{\beta\uparrow 1} \eta_\beta = \eta e, \tag{2.32}$$

with $\eta = \pi f$ being the average reward.

β-Potentials

The *discounted Poisson equation* is defined as

$$(I - \beta P + \beta e\pi)g_\beta = f, \qquad 0 < \beta \leq 1. \tag{2.33}$$

g_β is called the *β-potential.* When $\beta = 1$, it is the standard Poisson equation (2.12). Thus, the 1−potential is simply the potential (2.13) and is denoted as $g := g_1$. From (2.33), we have

$$\begin{aligned} g_\beta &= (I - \beta P + \beta e\pi)^{-1} f \\ &= \left[\sum_{l=0}^{\infty} \beta^l (P - e\pi)^l\right] f \\ &= \left\{ I + \left[\sum_{l=1}^{\infty} \beta^l (P^l - e\pi)\right]\right\} f, \qquad 0 < \beta \le 1. \end{aligned} \tag{2.34}$$

The above expansion holds because all the eigenvalues of $P - e\pi$ are in the unit circle. In particular, by setting $\beta = 1$ we obtain (2.14).

It is easy to verify the following equations:

$$\pi(I - \beta P + \beta e\pi)^{-1} = \pi, \tag{2.35}$$

$$(I - \beta P + \beta e\pi)^{-1} e = e,$$

$$(I - \beta P)^{-1} e = \frac{1}{1-\beta} e, \tag{2.36}$$

and

$$(I - \beta P)^{-1} = (I - \beta P + \beta e\pi)^{-1} + \frac{\beta}{1-\beta} e\pi. \tag{2.37}$$

Equation (2.37) is obtained by using (2.36), (2.35), and the following equation

$$(I - \beta P)^{-1}(I - \beta P + \beta e\pi) = I + (I - \beta P)^{-1} \beta e\pi.$$

In addition, we have

$$\lim_{\beta \uparrow 1} g_\beta = g_1,$$

$$\pi g_\beta = \pi f. \tag{2.38}$$

From (2.37), we obtain

$$\lim_{\beta \uparrow 1} (1 - \beta)(I - \beta P)^{-1} = e\pi. \tag{2.39}$$

Performance Sensitivities

Suppose that the transition matrix P and the reward function f change to P' and f', respectively, with P' being another irreducible and aperiodic transition matrix. From (2.31), we have

$$\begin{aligned} \eta'_\beta - \eta_\beta &= (1 - \beta)(f' - f) + \beta(P'\eta'_\beta - P\eta_\beta) \\ &= (1 - \beta)(f' - f) + \beta(P' - P)\eta_\beta + \beta P'(\eta'_\beta - \eta_\beta). \end{aligned}$$

This leads to

$$\eta'_\beta - \eta_\beta = (1-\beta)(I-\beta P')^{-1}(f'-f) + \beta(I-\beta P')^{-1}(P'-P)\eta_\beta. \quad (2.40)$$

From (2.31) and (2.37), we obtain

$$\eta_\beta = (1-\beta)g_\beta + \beta\eta e. \quad (2.41)$$

Substituting this into the right-hand side of (2.40) and noting that $(P'-P)e = 0$, we obtain the *performance difference formula for the discounted reward criterion:*

$$\eta'_\beta - \eta_\beta = (1-\beta)(I-\beta P')^{-1}[(\beta P' g_\beta + f') - (\beta P g_\beta + f)], \qquad 0 < \beta < 1. \quad (2.42)$$

Finally, as a special case, letting $\beta \to 1$ in (2.42) and using (2.39), we obtain the performance difference formula for the long-run average reward (2.27):

$$\eta' - \eta = \pi'[(P'g + f') - (Pg + f)].$$

Now, suppose that P changes to $P_\delta = P + \delta\Delta P$, $\Delta P = P' - P$, and f changes to $f_\delta = f + \delta\Delta f$, $\Delta f = f' - f$, $0 < \delta < 1$. Taking P_δ as the P' in (2.42), we have

$$\eta_{\beta,\delta} - \eta_\beta = (1-\beta)(I-\beta P_\delta)^{-1}[(\beta P_\delta g_\beta + f_\delta) - (\beta P g_\beta + f)], \qquad 0 < \beta < 1. \quad (2.43)$$

Letting $\delta \downarrow 0$, we obtain the *performance derivative formula for the discounted reward criterion:*

$$\left.\frac{d\eta_{\beta,\delta}}{d\delta}\right|_{\delta=0} = (1-\beta)(I-\beta P)^{-1}(\beta\Delta P g_\beta + \Delta f), \qquad 0 < \beta < 1. \quad (2.44)$$

When $\beta \uparrow 1$, this equation reduces to (2.26).

From (2.41) and (2.44), we have

$$\left.\frac{d\eta_{\beta,\delta}}{d\delta}\right|_{\delta=0} = (I-\beta P)^{-1}[\beta\Delta P\eta_\beta + (1-\beta)\Delta f], \qquad 0 < \beta < 1.$$

Similarly, from (2.41) and (2.43), we have

$$\eta_{\beta,\delta} - \eta_\beta = (I - \beta P_\delta)^{-1}[\beta \Delta P \eta_\beta + (1-\beta)\Delta f], \qquad 0 < \beta < 1.$$

All the applications of the performance potentials g_β in optimization depend only on the differences of the components of g_β. In other words, we can replace g_β with $g_\beta + ce$, where c is any constant. These are different versions of the β-potential, and for simplicity, we will use the same notation g_β to denote them. In particular, the performance difference and derivative formulas (2.42) and (2.44) hold when g_β is replaced by $g_\beta + ce$. Therefore, we may add a constant vector $-\eta e$ to (2.34) and obtain a sample-path-based expression for the β-potential (cf., (2.16)) as follows:

$$\begin{aligned} g_\beta(i) &= \lim_{L\to\infty} E\left\{ \sum_{l=0}^{L-1} \beta^l [f(X_l) - \eta] \Big| X_0 = i \right\} \\ &= E\left\{ \sum_{l=0}^{\infty} \beta^l [f(X_l) - \eta] \Big| X_0 = i \right\}, \qquad 0 < \beta \le 1, \end{aligned} \qquad (2.45)$$

in which we have exchanged the order of $\lim_{L\to\infty}$ and "E". Of course, for $0 < \beta < 1$, we can also discard the constant term $(\sum_{l=0}^{\infty} \beta^l)\eta$ and obtain

$$g_\beta(i) = E\left\{ \left[\sum_{l=0}^{\infty} \beta^l f(X_l) \right] \Big| X_0 = i \right\}. \qquad (2.46)$$

In modern Markov theory, (2.46) is called the β-potential of reward function f with $0 < \beta < 1$. (In [87], it is called the α-potential, since α is used as the discount factor there; in this book, we reserve α to denote actions in MDPs.) Therefore, from (2.16), (2.45), and (2.46), the potential for the long-run average reward, $g(i)$, is a natural extension of the β-potential from $0 < \beta < 1$ to $\beta = 1$; a constant η is subtracted from each term in (2.46) to keep the sum finite when extended to $\beta = 1$. This justifies again our terminology of "potential" for g in the long-run average reward case.

It is clear that the β-potential (2.46) is almost the same as the discounted reward (2.30). This explains why the concept of the discounted performance potential is not introduced in many previous works on the optimization of discounted rewards. Nevertheless, this concept puts the approach to the discounted-reward optimization problem in the same framework as the approach to the average-reward problem. This is also true for the policy iteration approach in MDPs, see Chapter 4.

Similar to the average-reward case, we define the (discounted) *PRF matrix* as

$$\Gamma_\beta = e g_\beta^T - g_\beta e^T, \qquad 0 < \beta \le 1.$$

From this equation and (2.38), we have

$$\Gamma_\beta^T \pi^T = g_\beta - \eta e.$$

The (discounted) PRF matrix satisfies the *(discounted) PRF equations:*

$$-\Gamma_\beta + \beta P \Gamma_\beta P^T = -F, \tag{2.47}$$

where $F = ef^T - fe^T$.

This can be easily verified:

$$\begin{aligned}
-\Gamma_\beta + \beta P \Gamma_\beta P^T &= -(e g_\beta^T - g_\beta e^T) + \beta P (e g_\beta^T - g_\beta e^T) P^T \\
&= -(e g_\beta^T - g_\beta e^T) + \beta [e (P g_\beta)^T - P g_\beta e^T] \\
&= -[e (g_\beta - \beta P g_\beta + \beta e \pi g_\beta)^T - (g_\beta - \beta P g_\beta + \beta e \pi g_\beta) e^T] \\
&= -F.
\end{aligned}$$

Equation (2.47) reduces to the standard PRF equation (2.7) when $\beta = 1$.

With the PRF matrix, (2.42) and (2.44) become

$$\eta'_\beta - \eta_\beta = (1-\beta)(I - \beta P')^{-1} [\beta (\Delta P) \Gamma_\beta^T \pi^T + \Delta f], \qquad 0 < \beta < 1,$$

and

$$\frac{d\eta_{\beta,\delta}}{d\delta} = (1-\beta)(I - \beta P)^{-1} [\beta (\Delta P) \Gamma_\beta^T \pi^T + \Delta f], \qquad 0 < \beta < 1.$$

Again, when $\beta \uparrow 1$, these two sensitivity formulas reduce to the average-reward case (2.28) and (2.29).

Figure 2.9 summarizes the results for both the discounted- and average-reward performance sensitivity analysis with a unified view; all the results of the average-reward case can be obtained by setting $\beta \uparrow 1$ from those of the discounted-reward case.

Intuitions

Finally, we offer an intuitive explanation for the discounted reward derivative formula (2.44). For simplicity, we assume that $\Delta f = 0$. Because the discounted reward $\eta_\beta(i)$ depends on the initial state $i \in \mathcal{S}$, we have to consider the transient probabilities on the sample path. Consider a sample path $\boldsymbol{X} = \{X_0, X_1, \ldots\}$ starting from $X_0 = i \in \mathcal{S}$. The conditional probability of $X_l = k$, $l = 1, 2, \ldots$, given that $X_0 = i$ is $\mathcal{P}(X_l = k | X_0 = i) = p^{(l)}(k|i)$ (cf. (A.3)). Let $p_l(u, v|k)$ be the probability that, given $X_l = k$, the system has a jump from

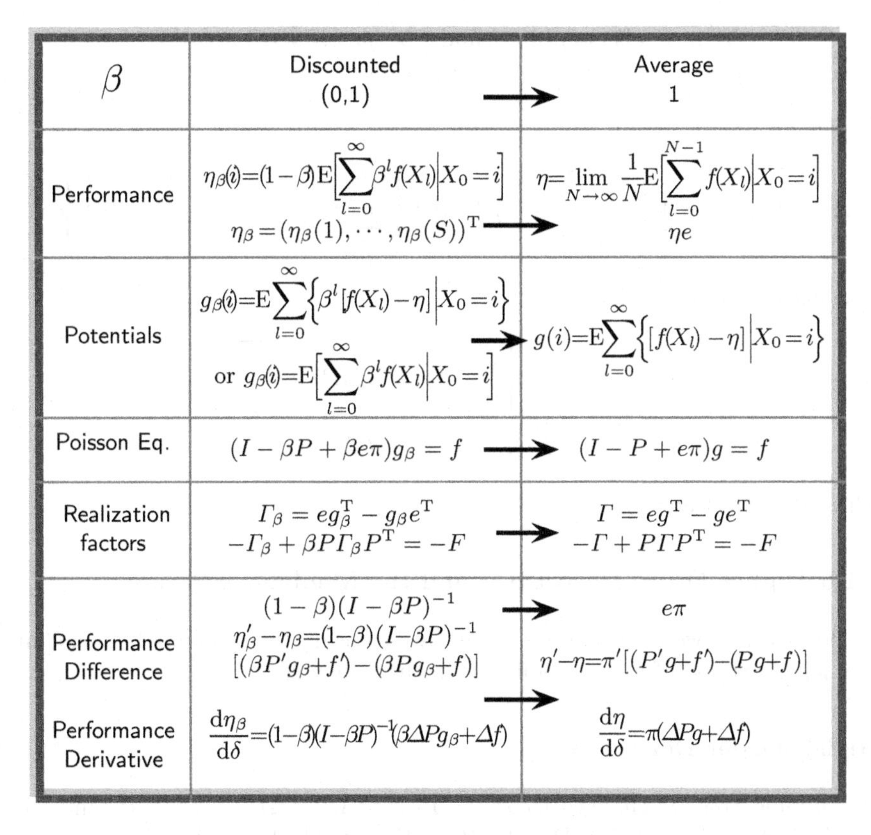

β	Discounted (0,1)	Average 1
Performance	$\eta_\beta(i)=(1-\beta)\mathrm{E}\left[\sum_{l=0}^{\infty}\beta^l f(X_l)\Big\vert X_0=i\right]$ $\eta_\beta=(\eta_\beta(1),\cdots,\eta_\beta(S))^{\mathrm{T}}$	$\eta=\lim_{N\to\infty}\frac{1}{N}\mathrm{E}\left[\sum_{l=0}^{N-1}f(X_l)\Big\vert X_0=i\right]$ ηe
Potentials	$g_\beta(i)=\mathrm{E}\sum_{l=0}^{\infty}\left\{\beta^l[f(X_l)-\eta]\Big\vert X_0=i\right\}$ or $g_\beta(i)=\mathrm{E}\left[\sum_{l=0}^{\infty}\beta^l f(X_l)\Big\vert X_0=i\right]$	$g(i)=\mathrm{E}\sum_{l=0}^{\infty}\left\{[f(X_l)-\eta]\Big\vert X_0=i\right\}$
Poisson Eq.	$(I-\beta P+\beta e\pi)g_\beta=f$	$(I-P+e\pi)g=f$
Realization factors	$\Gamma_\beta=eg_\beta^{\mathrm{T}}-g_\beta e^{\mathrm{T}}$ $-\Gamma_\beta+\beta P\Gamma_\beta P^{\mathrm{T}}=-F$	$\Gamma=eg^{\mathrm{T}}-ge^{\mathrm{T}}$ $-\Gamma+P\Gamma P^{\mathrm{T}}=-F$
	$(1-\beta)(I-\beta P)^{-1}$	$e\pi$
Performance Difference	$\eta'_\beta-\eta_\beta=(1-\beta)(I-\beta P)^{-1}$ $[(\beta P'g_\beta+f')-(\beta Pg_\beta+f)]$	$\eta'-\eta=\pi'[(P'g+f')-(Pg+f)]$
Performance Derivative	$\frac{\mathrm{d}\eta_\beta}{\mathrm{d}\delta}=(1-\beta)(I-\beta P)^{-1}(\beta\Delta Pg_\beta+\Delta f)$	$\frac{\mathrm{d}\eta}{\mathrm{d}\delta}=\pi(\Delta Pg+\Delta f)$

Fig. 2.9. A Comparison of Discounted- and Average-Reward Problems

state u to v at time $l+1$ (i.e., the original system with transition probability matrix P moves from $X_l = k$ to $X_{l+1} = u$, but the perturbed system with $P_\delta = P + \delta(\Delta P)$ moves from $X_l = k$ to $X_{l+1} = v$). The effect of such a jump, measured starting from $l+1$, is $\gamma_\beta(u,v) = g_\beta(v) - g_\beta(u)$. Since the jump happens at time $l+1$, its effect on the discounted reward $\eta_\beta(i)$ in (2.30) is $\beta^{l+1}\gamma_\beta(u,v)$. Therefore, from the physical meaning, we can decompose $\Delta\eta_\beta(i)$ into

$$\begin{aligned}
\Delta\eta_\beta(i) &= (1-\beta)\left[\sum_{l=0}^{\infty}\sum_{k=1}^{S}\sum_{u=1}^{S}\sum_{v=1}^{S}\beta^{l+1}\mathcal{P}(X_l = k|X_0 = i)p_l(u,v|k)\gamma_\beta(u,v)\right] \\
&= (1-\beta)\left\{\sum_{l=0}^{\infty}\sum_{k=1}^{S}\beta^{l+1}p^{(l)}(k|i)\sum_{u=1}^{S}\sum_{v=1}^{S}\{p_l(u,v|k)[g_\beta(v)-g_\beta(u)]\}\right\}.
\end{aligned}$$

Similar to (2.19) and (2.20), we have

$$\sum_{v=1}^{S} p_l(u,v|k) = p(u|k) \quad \text{and} \quad \sum_{u=1}^{S} p_l(u,v|k) = p_\delta(u|k).$$

Thus,

$$\Delta\eta_\beta(i) = (1-\beta)\left\{\sum_{l=0}^{\infty}\sum_{k=1}^{S}\beta^{l+1}p^{(l)}(k|i)\left\{\sum_{j=1}^{S}[p_\delta(j|k)-p(j|k)]g_\beta(j)\right\}\right\}.$$

In a matrix form, this is

$$\begin{aligned}\Delta\eta_\beta &= (1-\beta)\left\{\sum_{l=0}^{\infty}\beta^{l+1}P^l[(\Delta P)\delta g_\beta]\right\}\\ &= (1-\beta)(I-\beta P)^{-1}[\beta(\Delta P)\delta g_\beta],\end{aligned}$$

which directly leads to (2.44).

2.1.5 Higher-Order Derivatives and the MacLaurin Series

In this section, we continue our study by exploring the system's behavior in the neighborhood of a given policy P in the policy space.

Higher-Order Derivatives

We assume that P changes to $P_\delta = P + \delta(\Delta P)$, $\Delta P = P' - P$, and we let $f_\delta \equiv f$, for simplicity. Denote $B = P - I$, which can be viewed as an infinitesimal generator of a Markov process with unit transition rates and a transition probability matrix P for its embedded chain (see Appendix A.2). To study the higher-order derivatives with respect to δ, it is convenient to use short-hand notation defined as

$$\begin{aligned}B^\# &= -(-B+e\pi)^{-1} + e\pi\\ &= -[(I-P+e\pi)^{-1} - e\pi].\end{aligned} \tag{2.48}$$

$B^\#$ is called the *group inverse* of B [202], which satisfies

$$BB^\# = B^\# B = I - e\pi, \tag{2.49}$$

and

$$B^\# e = 0, \qquad \pi B^\# = 0.$$

The term "group" comes from the following fact. For any probability distribution π on state space $\mathcal{S}$, define a set of $S \times S$ matrices

$$\mathcal{B} := \{B :\ \pi B = 0,\ Be = 0\}. \tag{2.50}$$

It is easy to verify that $\mathcal{B}$ is a group (see, e.g., [219] for a definition) with identity element $I - e\pi$ under the operation of matrix multiplication (see Problem 2.11). Equation (2.49) indicates that $B^{\#}$ is indeed the inverse of B in group $\mathcal{B}$.

With the group inverse, the potential in (2.15) becomes

$$g = -B^{\#} f,$$

and the performance derivative formula (2.23) takes the form

$$\left.\frac{d\eta_\delta}{d\delta}\right|_{\delta=0} = \pi(\Delta P)g = \pi[(\Delta P)(-B^{\#})]f.$$

For the irreducible finite Markov chain with transition matrix P_δ, we have

$$\pi_\delta(I - P_\delta) = 0,$$

and $\frac{dP_\delta}{d\delta} = \Delta P$. By taking derivatives on both sides of this equation with respect to δ, we have

$$\frac{d\pi_\delta}{d\delta}(I - P_\delta) = \pi_\delta(\Delta P).$$

Continuously taking derivatives on both sides of the resulting equations, we obtain for any $n \geq 1$,

$$\frac{d^n \pi_\delta}{d\delta^n}(I - P_\delta) = n\frac{d^{n-1}\pi_\delta}{d\delta^{n-1}}(\Delta P).$$

Setting $\delta = 0$ and multiplying both sides of the above equation on the right by $-B^{\#}$ and noting that $BB^{\#} = I - e\pi$ and $\pi e = 1$, we get

$$\left.\frac{d^n \pi_\delta}{d\delta^n}\right|_{\delta=0} = n \left.\frac{d^{n-1}\pi_\delta}{d\delta^{n-1}}\right|_{\delta=0} [(\Delta P)(-B^{\#})].$$

Thus,

$$\left.\frac{d^n \pi_\delta}{d\delta^n}\right|_{\delta=0} = n!\pi[(\Delta P)(-B^{\#})]^n.$$

Finally, for any reward function f, we have

$$\begin{aligned}\left.\frac{d^n \eta_\delta}{d\delta^n}\right|_{\delta=0} &= n!\pi[(\Delta P)(-B^{\#})]^n f \\ &= n!\pi[(\Delta P)(I - P + e\pi)^{-1}]^n f, \qquad n \geq 1.\end{aligned}$$

The MacLaurin Expansion

We note that η_δ is an analytical function of δ. (More precisely, it is a rational function of δ whose denominator and numerator are both polynomials of δ with finite degrees. This can be verified by solving $\pi_\delta(I-P_\delta)=0$ and $\pi_\delta e = 1$.) Thus, η_δ has a MacLaurin expansion at $\delta = 0$:

$$\eta_\delta - \eta = \sum_{n=1}^{\infty} \frac{1}{n!} \left.\frac{d^n \eta_\delta}{d\delta^n}\right|_{\delta=0} \delta^n$$

$$= \pi \left\{ \sum_{n=1}^{\infty} [(\Delta P)(-B^{\#})]^n \delta^n \right\} f, \qquad (2.51)$$

or equivalently,

$$\eta_\delta = \pi \sum_{n=0}^{\infty} \{[(\Delta P)(-B^{\#})]^n f \delta^n\}. \qquad (2.52)$$

Denote the spectrum radius of a matrix W as $\rho(W)$ (i.e., the largest absolute value of the eigenvalues of W). Define

$$r = \frac{1}{\rho[(\Delta P)(-B\#)]} = \frac{1}{\rho[(\Delta P)(I-P+e\pi)^{-1}]}.$$

Then, for $\delta < r$, the eigenvalues of $\delta(\Delta P)B^{\#}$ are all in the unit circle, and the summation in (2.52) converges. Therefore, for $\delta < r$, we have

$$\sum_{n=0}^{\infty} [(\Delta P)(-B^{\#})\delta]^n = [I - \delta(\Delta P)(-B^{\#})]^{-1}$$

$$= [I - \delta(\Delta P)(I - P + e\pi)^{-1}]^{-1}. \qquad (2.53)$$

Next, if we take $f = e_{\cdot i}$, where $e_{\cdot i}$ is a column vector representing the ith column of the identity matrix I, then the corresponding performance is $\pi_\delta e_{\cdot i} = \pi_\delta(i)$, $i \in \mathcal{S}$. Thus, from (2.52), we have

$$\pi_\delta(i) = \pi \sum_{n=0}^{\infty} \{[(\Delta P)(-B^{\#})]^n e_{\cdot i} \delta^n\}.$$

In matrix form, we have

$$\pi_\delta = \pi \sum_{n=0}^{\infty} \{[(\Delta P)(-B^{\#})]^n \delta^n\}, \qquad \delta < r. \qquad (2.54)$$

Thus, from (2.51) we obtain

$$\eta_\delta - \eta = \pi_\delta \delta(\Delta P)(-B^{\#})f.$$

This is consistent with the performance difference formula (2.25). From (2.52), (2.53), and (2.54), we establish a general form:

$$\begin{aligned}
\eta_\delta &= \pi \sum_{k=0}^{n} [\delta(\Delta P)(-B^{\#})]^k f + \pi \sum_{k=n+1}^{\infty} [\delta(\Delta P)(-B^{\#})]^k f \\
&= \pi \sum_{k=0}^{n} [\delta(\Delta P)(-B^{\#})]^k f + \pi_\delta [\delta(\Delta P)(-B^{\#})]^{n+1} f, \\
&\qquad \delta < r, \qquad \text{for any } n \geq 0. \qquad (2.55)
\end{aligned}$$

The last term in (2.55), $\pi_\delta[\delta(\Delta P)(-B^{\#})]^{n+1}f$, is the error in taking the first $(n+1)$ terms in the MacLaurin series as an estimate of η_δ. Equations (2.54) and (2.55) hold for $\delta < r$. If $r > 1$, then we can set $\delta = 1$ in (2.54) and (2.55) and obtain the performance value for $P' = P + \Delta P$ as follows

$$\begin{aligned}
\eta' &= \pi \sum_{k=0}^{\infty} [(\Delta P)(-B^{\#})]^k f \\
&= \pi \sum_{k=0}^{n} [(\Delta P)(-B^{\#})]^k f + \pi' [(\Delta P)(-B^{\#})]^{n+1} f, \qquad \text{for any } n \geq 0.
\end{aligned}$$

The extensions to Markov chains with general state space are in [131, 133].

Numerical Calculations

The saving in computation is significant when we use the MacLaurin series to calculate the performance for many different (ΔP)'s and δ's. There is only one matrix inversion $(I - P + e\pi)^{-1}$ involved. The nth derivative of π_δ at $\delta = 0$, i.e., $n!\pi[(\Delta P)(I - P + e\pi)^{-1}]^n$, can be simply obtained by multiplying the $(n-1)$th derivative, i.e., $(n-1)!\pi[(\Delta P)(I - P + e\pi)^{-1}]^{n-1}$, with the matrix $n(\Delta P)(I - P + e\pi)^{-1}$. For example, π_δ in (2.54) can be calculated simply as follows. First, we set $G_\delta := \delta(\Delta P)(I - P + e\pi)^{-1}$. Then, we

i. solve $\pi = \pi P$ and $\pi e = 1$ to obtain π, calculate G_δ, and set $\pi_\delta := \pi$;
ii. recursively calculate $\pi_\delta := \pi + \pi_\delta G_\delta$, until π_δ reaches a desired precision.

The matrix $(I - P + e\pi)^{-1}$ can be estimated by analyzing a sample path of the Markov chain (see, e.g., Problem 3.18 in Chapter 3). Thus, with a sample-path-based approach, matrix inversion is not even needed. This also implies that, principally, we can obtain the performance of a Markov system with any transition probability matrix P_δ by analyzing a sample path of the Markov system with transition probability matrix P as long as $\delta < r$.

It is not easy to determine the value of r. However, there exist some upper bounds (albeit not tight) for $\rho[\Delta P(-B^{\#})] = \frac{1}{r}$ (or lower bounds for r). From spectrum theory, we have [20]

$$\rho[\Delta P(I - P + e\pi)^{-1}] \le ||\Delta P|| \times ||(I - P + e\pi)^{-1}||,$$

where $||\cdot||$ denotes the norm of a matrix, which is defined as

$$||\Delta P|| = \max_{\text{all } i} \sum_j |\Delta p(j|i)|.$$

Thus, if $||\Delta P||$ is not large, then the spectrum radius of $(\Delta P)(-B^{\#})$ may be small. Let $s^+ = \max_{\text{all } i} \sum_{j:\ \Delta p(i,j)>0} [\Delta p(j|i)]$. If $s^+ \le 0.5$, then $||\Delta P|| \le 1$. Since $\Delta P = P' - P$, at least we have $||\Delta P|| \le 2$.

It is, however, not easy to obtain the norm of the fundamental matrix. As shown in Section 2.1.2, the eigenvalues of $(I - P + e\pi)^{-1}$ are $\{1, \frac{1}{1-\lambda_2}, \ldots, \frac{1}{1-\lambda_S}\}$, with $|\lambda_i| < 1$, $i = 2, \ldots, S$, and 1 being the eigenvalues of P. Thus, we have

$$\rho(I - P + e\pi)^{-1} = \frac{1}{\inf\{1,\ |1 - \lambda_i|, i = 2, \ldots, S\}}.$$

However, there is no direct link between $\rho(I - P + e\pi)^{-1}$ and $\rho\Delta P(I - P + e\pi)^{-1}$.

The worst case happens when P has an eigenvalue that is close to 1. If $P \approx I$ (I is a reducible matrix and hence cannot be chosen as P), then $I - P + e\pi \approx e\pi$, which has an eigenvalue 0. The radius of $(I - P + e\pi)^{-1}$ is thus very large. In this case, the radius of $\Delta P(I - P + e\pi)^{-1}$ may be also very large; i.e., r may be very small.

	1	2	3	4	5
1	0.000	0.300	0.200	0.100	0.400
2	0.500	0.000	0.000	0.300	0.200
3	0.200	0.150	0.000	0.150	0.500
4	0.400	0.200	0.150	0.150	0.100
5	0.250	0.250	0.250	0.250	0.000

Table 2.1. The Matrix P

Example 2.1. For illustrative purposes, we study a Markov chain with five states. The state transition matrix P is listed in Table 2.1; the change of P, ΔP, is listed in Table 2.2; and the reward function f is given in Table 2.3. All the values are arbitrarily chosen with some considerations about generality. Note that ΔP represents some dramatic changes in P, e.g., $p(1|2)$ changes

	1	2	3	4	5
1	0.100	-0.300	0.100	0.000	0.100
2	-0.500	0.000	0.500	0.000	0.000
3	-0.200	0.100	0.000	0.100	0.000
4	-0.100	0.100	-0.050	0.000	0.050
5	-0.250	0.250	0.000	0.000	0.000

Table 2.2. The Matrix ΔP

	1	2	3	4	5
f	10	5	1	15	3
π	0.256	0.192	0.136	0.189	0.228

Table 2.3. f and π

	1	2	3	4	5
1	0.803	0.077	0.048	-0.055	0.127
2	0.180	0.858	-0.094	0.079	-0.023
3	-0.048	-0.026	0.900	-0.021	0.196
4	0.112	0.012	0.008	0.949	-0.082
5	0.006	0.039	0.079	0.049	0.827

Table 2.4. The Matrix $(I - P + e\pi)^{-1}$

	1	2	3	4	5
1	0.022	-0.249	0.131	-0.026	0.122
2	-0.425	-0.052	0.426	0.017	0.034
3	-0.131	0.072	-0.018	0.114	-0.036
4	-0.059	0.081	-0.055	0.017	0.017
5	-0.156	0.195	-0.036	0.033	-0.038

Table 2.5. The Matrix $\Delta P(I - P + e\pi)^{-1}$

from 0.5 to 0, and $p(3|2)$ changes from 0 to 0.5. We calculated the matrices $(I - P + e\pi)^{-1}$ and $\Delta P(I - P + e\pi)^{-1}$, which are listed in Tables 2.4 and 2.5. The eigenvalues of $\Delta P(I - P + e\pi)^{-1}$ are 0.3176, -0.3415, 0, -0.0164 +0.0325i, and -0.0164-0.0325i; all of them are inside the unit circle. Thus, $r > 1$ and the MacLaurin series converges within $\delta \leq 1$. ($\delta > 1$ does not make sense, since for $\delta > 1$, $p_\delta(1|2) < 0$). Table 2.6 lists the coefficients of the first to the tenth terms in the MacLaurin series, i.e., $\pi[\Delta P(I - P + e\pi)^{-1}]^n$, for $n = 1, 2, \ldots, 10$. The coefficients of the terms with orders higher than 10 are all numerically zeros. Table 2.7 lists the performance values of the Markov chains with $P_\delta = P + \delta\Delta P$, $\delta = 0.1, 0.2, \ldots, 0.9, 1$, obtained by using the first n terms of the MacLaurin series, $n = 1, 2, \ldots, 10$. All these values converge

	1	2	3	4	5
1st	-0.14050	-0.00387	0.09417	0.02282	0.02738
2nd	-0.01941	0.04907	-0.02399	0.01562	-0.02129
3th	-0.01577	-0.00232	0.01869	-0.00184	0.00123
4th	-0.00189	0.00547	-0.00334	0.00251	-0.00275
5th	-0.00165	-0.00038	0.00210	-0.00029	0.00022
6th	-0.00017	0.00060	-0.00041	0.00028	-0.00030
7th	-0.00017	-0.00006	0.00024	-0.00004	0.00003
8th	-0.00001	0.00007	-0.00005	0.00003	-0.00003
9th	-0.00002	-0.00001	0.00003	0.00000	0.00000
10th	0.00000	0.00001	-0.00001	0.00000	0.00000

Table 2.6. The Coefficients of the MacLaurin Series of η_δ

to the actual average reward of the corresponding Markov chains. Note that after $n = 6$ the values change very little. The average reward of the original Markov chain ($\delta = 0$) is 7.1647. □

n δ	0.1	0.2	0.3	0.4	0.5	0.6	0.7	0.8	0.9	1.0
1	7.0741	6.9836	6.8930	6.8024	6.7118	6.6212	6.5307	6.4401	6.3495	6.2589
2	7.0761	6.9915	6.9108	6.8340	6.7612	6.6924	6.6275	6.5666	6.5096	6.4566
3	7.0759	6.9901	6.9061	6.8229	6.7394	6.6547	6.5677	6.4773	6.3825	6.2822
4	7.0759	6.9901	6.9063	6.8237	6.7416	6.6592	6.5760	6.4914	6.4051	6.3166
5	7.0759	6.9901	6.9063	6.8235	6.7410	6.6576	6.5726	6.4849	6.3933	6.2967
6	7.0759	6.9901	6.9063	6.8236	6.7410	6.6578	6.5731	6.4860	6.3955	6.3009
7	7.0759	6.9901	6.9063	6.8236	6.7410	6.6578	6.5729	6.4855	6.3944	6.2986
8	7.0759	6.9901	6.9063	6.8236	6.7410	6.6578	6.5729	6.4856	6.3946	6.2991
9	7.0759	6.9901	6.9063	6.8236	6.7410	6.6578	6.5729	6.4855	6.3945	6.2988
10	7.0759	6.9901	6.9063	6.8236	6.7410	6.6578	6.5729	6.4855	6.3946	6.2989

Table 2.7. The Performance Calculated by the MacLaurin Series

The next example shows that r may be less than 1.

Example 2.2. Consider

$$P = \begin{bmatrix} 0.90 & 0.10 \\ 0.15 & 0.85 \end{bmatrix},$$

$$\Delta P = \begin{bmatrix} -0.8 & 0.8 \\ 0.8 & -0.8 \end{bmatrix},$$

and $f = (1, 5)^T$. Then we have $\pi = (0.6, 0.4)$, and

$$\Delta P (I - P + e\pi)^{-1} = \begin{bmatrix} -3.2 & 3.2 \\ 3.2 & -3.2 \end{bmatrix}.$$

Its eigenvalues are 0 and -6.4. Therefore, the MacLaurin series converges only if $\delta < \frac{1}{|-6.4|} = 0.156$. In fact, as n increases, $\pi[\Delta P(I - P + e\pi)^{-1}]^n$ goes to infinity very rapidly. Note that the matrix P in this example is close to I.

The curve in Figure 2.10 shows the performance of the system for $\delta \in [0, 1]$. The five points $(*)$ in the figure show the performance of the system calculated by the MacLaurin series corresponding to $\delta = 0.03, 0.06, 0.09, 0.12$, and 0.15. The first four points are the values given by the first 10 terms of the MacLaurin series, and the fifth point is given by 50 terms. At the first three points ($\delta = 0.03, 0.06, 0.09$), the MacLaurin series almost reaches the true value after the first 10 terms (with an error of less than 0.001). The last point is very close to the convergence range ($\delta = 0.15 \approx r = 0.156$), and as shown in the figure, the MacLaurin series does not converge even after 50 terms. In fact, it does converge after 200 terms. □

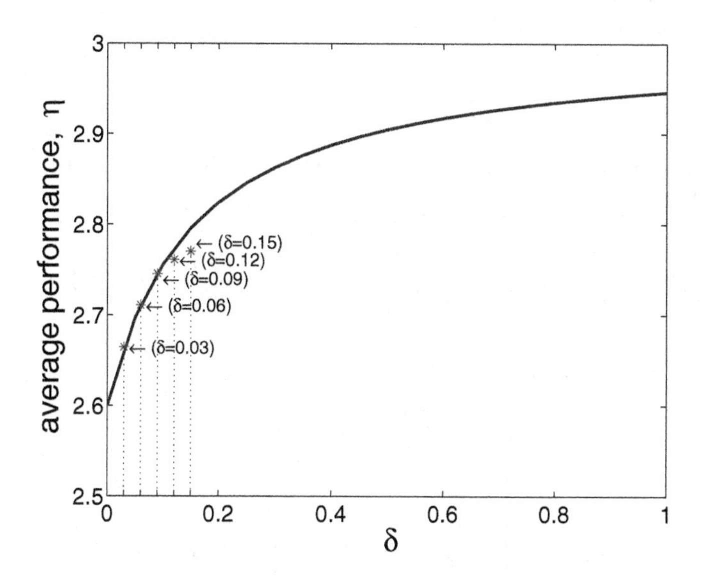

Fig. 2.10. The Performance Compared with MacLaurin Series

Extension to General Function P_θ

Now, let us extend the results to the more general case when the transition probability matrix is a function of θ denoted as P_θ. We assume that the first and all the higher-order derivatives of P_θ with respect to θ exist at $\theta = 0$. Set $P_0 = P$ and $\Delta P_\theta := P_\theta - P$. Let the reward function $f_\theta \equiv f$ for all θ. Let π_θ be the steady-state probability vector of the Markov chain with transition probability matrix P_θ, and η_θ be its corresponding long-run average reward.

We may use (2.52) to get an expansion of η_θ. For any fixed $\theta > 0$, we simply set $\Delta P = \Delta P_\theta$ and $\delta = 1$ in (2.52). Assume that θ is small enough so that $\rho[\Delta P_\theta(-B^{\#})] < 1$ and therefore expansion (2.52) exists. Then, we have

$$\eta_\theta = \pi \sum_{n=0}^{\infty} \{[\Delta P_\theta(-B^{\#})]^n f\}. \tag{2.56}$$

Equation (2.55) becomes

$$\eta_\theta = \pi \sum_{k=0}^{n} [\Delta P_\theta(-B^{\#})]^k f + \pi_\theta [\Delta P_\theta(-B^{\#})]^{n+1} f, \qquad \text{for any } n \geq 0.$$

Note that this expansion is not a MacLaurin series of η_θ in terms of θ. In fact, ΔP_θ has an expansion

$$\Delta P_\theta = \left.\frac{dP_\theta}{d\theta}\right|_{\delta=0} \theta + \frac{1}{2!} \left.\frac{d^2 P_\theta}{d\theta^2}\right|_{\delta=0} \theta^2 + \cdots,$$

where the derivatives are taken at $\theta = 0$. Substituting it into (2.56), we obtain the MacLaurin series of η_θ:

$$\eta_\theta = \pi \left\{ I + \left[\frac{dP_\theta}{d\theta}(-B^{\#})\right] \theta + \left\{ \frac{1}{2!} \frac{d^2 P_\theta}{d\theta^2}(-B^{\#}) + \left[\frac{dP_\theta}{d\theta}(-B^{\#})\right]^2 \right\} \theta^2 + \cdots \right\} f. \tag{2.57}$$

Therefore, we have

$$\frac{d\eta_\theta}{d\theta} = \pi \frac{dP_\theta}{d\theta}(-B^{\#}) f = \pi \frac{dP_\theta}{d\theta} g \tag{2.58}$$

and

$$\frac{d^2\eta_\theta}{d\theta^2} = \pi \left\{ \frac{d^2 P_\theta}{d\theta^2}(-B^{\#}) + 2! \left[\frac{dP_\theta}{d\theta}(-B^{\#})\right]^2 \right\} f.$$

Other higher-order derivatives can be obtained in a similar way.

Benes [19] presented an interesting result on the MacLaurin series of the call blocking probability in terms of the input call intensity λ in a telecommunication network. The results presented in this section are more general and concise and can be applied on-line when the system is running. Other related works are in [29], [120], [153], [158], and [267]. In [120], the MacLaurin series of the moments of the response times in a GI/G/1 queue is derived; the results are extended to inventory systems in [29], [153], [158]; and [267] focuses on the expansion of performance measures in queueing systems.

2.2 Performance Sensitivities of Markov Processes

In this section, we extend the aforementioned sensitivity analysis results to (continuous-time) Markov processes. Consider an irreducible and aperiodic (ergodic) Markov process $\boldsymbol{X} = \{X_t, t \geq 0\}$ with a finite state space $\mathcal{S} = \{1, 2, \ldots, S\}$ and an infinitesimal generator $B = [b(i,j)]$, where $b(i,j) \geq 0$, $i \neq j$, $b(i,i) < 0$. Let π be the steady-state probability (row) vector. We have $\pi e = 1$ and

$$Be = 0, \qquad \pi B = 0.$$

We can construct an embedded Markov chain (discrete-time) that has the same steady-state probability as the Markov process $\boldsymbol{X}$. This is called *uniformization* (see Problem A.8). Thus, the sensitivity analysis of a Markov process can be converted to that of a Markov chain, and then the results in Section 2.1 can be translated to Markov processes. In this section, however, we adopt a direct approach, which provides a clear meaning and intuition.

Perturbation Realization

Let f be a *reward function* on $\mathcal{S}$ and also denote a (column) vector $f = (f(1), \ldots, f(S))^T$. The long-run average performance measure of the Markov process is:

$$\eta = \pi f = \lim_{T \to \infty} \frac{1}{T} E\left[\int_0^T f(X_t)dt\right],$$

which exists for ergodic Markov processes, where E denotes the expectation.

To determine the effect of a perturbation (jump) from state i to state j on the performance η, we study two independent sample paths $\boldsymbol{X}$ and $\widetilde{\boldsymbol{X}}$ with the same infinitesimal generator B, starting from initial states $X_0 = i$ and $\widetilde{X}_0 = j$, respectively. Let E denote the expectation in the probability space spanned by all the sample paths of both $\boldsymbol{X}$ and $\widetilde{\boldsymbol{X}}$. By the ergodicity of $\boldsymbol{X}$ and $\widetilde{\boldsymbol{X}}$, they will merge together with probability 1. Define

$$T^*_{ij} = \inf\left\{t : t \geq 0, X_t = \widetilde{X}_t \middle| X_0 = i, \widetilde{X}_0 = j\right\}.$$

By the strong Markov property, after T^*_{ij}, the two processes $\boldsymbol{X}$ and $\widetilde{\boldsymbol{X}}$ will behave similarly probabilistically. T^*_{ij} is just the coupling time of the two independent Markov processes with different initial states. Readers are referred to [203] for a survey of the relevant results about coupling.

Now, we define the *perturbation realization factor (PRF)* as (cf. (2.5))

$$\gamma(i,j) = E\left\{ \int_0^{T_{ij}^*} [f(\widetilde{X}_t) - f(X_t)]dt \middle| X_0 = i, \widetilde{X}_0 = j \right\}, \qquad i,j \in \mathcal{S}. \tag{2.59}$$

The PRF matrix is $\Gamma := [\gamma(i,j)]$. From the definition, we have

$$\gamma(i,j) = -\gamma(j,i), \qquad i,j \in \mathcal{S},$$

or equivalently, Γ is skew-symmetric:

$$\Gamma^T = -\Gamma.$$

$\gamma(i,j)$ can be written in a more convenient form as shown below. First, for any $T > T_{ij}^*$, we have

$$\int_0^T [f(\widetilde{X}_t) - f(X_t)]dt$$
$$= \int_0^{T_{ij}^*} [f(\widetilde{X}_t) - f(X_t)]dt + \int_{T_{ij}^*}^T [f(\widetilde{X}_t) - f(X_t)]dt.$$

Next, because $\widetilde{X}_{T_{ij}} = X_{T_{ij}}$, by the strong Markov property, we have

$$\lim_{T\to\infty} E\left\{ \int_{T_{ij}^*}^T [f(\widetilde{X}_t) - f(X_t)]dt \middle| X_0 = i, \widetilde{X}_0 = j \right\} = 0.$$

Thus,

$$\lim_{T\to\infty} E\left\{ \int_0^T [f(\widetilde{X}_t) - f(X_t)]dt \middle| X_0 = i, \widetilde{X}_0 = j \right\}$$
$$= E\left\{ \int_0^{T_{ij}^*} [f(\widetilde{X}_t) - f(X_t)]dt \middle| X_0 = i, \widetilde{X}_0 = j \right\},$$

and from (2.59), we have (cf. (2.6))

$$\gamma(i,j) = \lim_{T\to\infty} E\left\{ \left[\int_0^T f(\widetilde{X}_t)dt - \int_0^T f(X_t)dt \right] \middle| X_0 = i, \widetilde{X}_0 = j \right\},$$
$$i,j \in \mathcal{S}. \tag{2.60}$$

A rigorous proof of (2.60) involves proving the exchangeability of the order of $\lim_{T\to\infty}$ and "E", which follows from the dominated convergence theorem and the finiteness of f, see [62]. Equation (2.60) indicates that $\gamma(i,j)$ measures the "infinitesimal" effect of a perturbation ("jump") from state i to state j on the long-run average reward.

In addition to (2.59) and (2.60), we have another formula that is similar to (2.17) for Markov chains. On the sample path of a Markov process $\boldsymbol{X}$ starting with $X_0 = j$, we define its first passage time to state i as $T(i|j) = \inf\{t : t \geq 0, X_t = i | X_0 = j\}$. Then,

$$\gamma(i,j) = E\left\{ \int_0^{T(i|j)} [f(X_t) - \eta]dt \,\middle|\, X_0 = j \right\}. \tag{2.61}$$

An intuitive explanation is similar to Figure 2.8 for (2.17).

For ergodic Markov processes, the PRF matrix Γ satisfies the following *PRF equation:*

$$B\Gamma + \Gamma B^T = -F, \tag{2.62}$$

where $F = ef^T - fe^T$.

Proof. On a Markov process $\boldsymbol{X}$ with $X_0 = i$, we define $p_t(k|i) = \mathcal{P}(X_t = k|X_0 = i)$ and $P_t = [p_t(k|i)]_{i,k\in\mathcal{S}}$. Then, (A.14) gives us

$$P_t = \exp(Bt) = \sum_{n=0}^{\infty} \frac{1}{n!}(Bt)^n, \qquad B^0 = I.$$

It follows that $E[f(X_t)|X_0 = i] = \sum_{k\in\mathcal{S}} p_t(k|i)f(k)$ is the ith entry of $[\exp(Bt)]f$. Let $\widetilde{\boldsymbol{X}}$ be another independent Markov process starting from $\widetilde{X}_0 = j$ and define

$$\begin{aligned}\gamma_T(i,j) &= E\left\{ \int_0^T [f(\widetilde{X}_t) - f(X_t)]dt \,\middle|\, X_0 = i, \widetilde{X}_0 = j \right\} \\ &= \int_0^T \left\{ E[f(\widetilde{X}_t)|\widetilde{X}_0 = j] - E[f(X_t)|X_0 = i] \right\} dt,\end{aligned} \tag{2.63}$$

and $\Gamma_T = [\gamma_T(i,j)]_{i,j=1}^S$. Then, from (2.60),

$$\Gamma = \lim_{T\to\infty} \Gamma_T.$$

The integrand on the right-hand side of (2.63) equals the difference between the jth and ith entries of $[\exp(Bt)]f$. Therefore,

$$\Gamma_T = \int_0^T \left\{ ef^T[\exp(Bt)]^T - [\exp(Bt)]fe^T \right\} dt.$$

Using $Be = 0$ and $[\exp(Bt)]B = B[\exp(Bt)]$, we obtain

$$\begin{aligned} & B\Gamma_T + \Gamma_T B^T \\ &= \int_0^T \{ef^T[\exp(Bt)]^T B^T - B[\exp(Bt)]fe^T\}\, dt \\ &= ef^T \left[\int_0^T [\exp(Bt)]Bdt\right]^T - \left[\int_0^T [\exp(Bt)]Bdt\right] fe^T \\ &= ef^T[\exp(BT) - \exp(0)]^T - [\exp(BT) - \exp(0)]fe^T, \end{aligned} \tag{2.64}$$

where the variable 0 in $\exp(0)$ denotes a matrix whose elements are all zeros. Therefore, $\exp(0) = I$. For ergodic Markov processes, we have $\lim_{T\to\infty} p_T(j|i) = \pi(j)$; thus, $\lim_{T\to\infty}\exp(BT) = \lim_{T\to\infty} P_T = e\pi$. Furthermore, $ef^T(e\pi)^T = (\pi f)ee^T = e\pi fe^T$. Letting $T \to \infty$ in (2.64), we obtain the PRF equation (2.62). □

If F is a Hermitian matrix, then (2.62) is the continuous-time version of the Lyapunov equation [162, 174]. However, the continuous-time PRF equation (2.62) is different from the Lyapunov equation because F here is a skew-symmetric matrix, $F^T = -F$.

Next, it is easy to see that the solution to (2.62) with the form of (2.65), specified below, is unique. Suppose that there are two such solutions to (2.62) denoted as $\Gamma_1 = eg_1^T - g_1e^T$ and $\Gamma_2 = eg_2^T - g_2e^T$. Let $W = \Gamma_1 - \Gamma_2 = ew^T - we^T$, with $w = g_1 - g_2$. Then $BW + WB^T = 0$. Because $Be = 0$, we have $ew^TB^T - Bwe^T = 0$. Multiplying both sides of this equation on the left by the group inverse $B^\#$ and using $B^\#B = I - e\pi$ and $B^\#e = 0$, we have $(I - e\pi)we^T = 0$. Therefore,

$$we^T = e\pi we^T = (\pi w)ee^T,$$

where πw is a constant. From this, we have $W = (we^T)^T - we^T = 0$, i.e., $\Gamma_1 = \Gamma_2$.

Performance Potentials

From (2.60), we have

$$\gamma(i,j) = \gamma(i,k) + \gamma(k,j), \qquad i,j,k \in \mathcal{S}.$$

Similar to the sensitivity analysis of Markov chains, we can define *performance potentials* $g(i)$, $i \in \mathcal{S}$, as follows:

$$\gamma(i,j) = g(j) - g(i), \qquad \text{for all } i,j \in \mathcal{S},$$

or, equivalently,

$$\Gamma = eg^T - ge^T, \tag{2.65}$$

where $g = (g(1), \ldots, g(S))^T$ is called a *potential vector*.

Substituting (2.65) into (2.62), we get $e(Bg + f)^T = (Bg + f)e^T$. Thus, $Bg + f = ce$, with c being a constant. Because $\pi B = 0$, we get $c = \pi f = \eta$. Thus, the performance potentials satisfy the following *Poisson equation*:

$$Bg = -f + \eta e. \tag{2.66}$$

Again, its solution is only up to an additive constant: if g is a solution to (2.66), so is $g + ce$ for any constant c.

For ergodic Markov processes, the group inverse of B is defined as $B^{\#} = (B - e\pi)^{-1} + e\pi$ [202] (cf. (2.48)). We have

$$BB^{\#} = B^{\#}B = I - e\pi.$$

By multiplying both sides of (2.66) on the left by $B^{\#}$, we obtain the general form of its solution

$$g = -B^{\#}f + ce,$$

where $c = \pi g$, which may be any constant. In particular, we can choose a solution that satisfies $c = \pi g = \eta$. In this case, the Poisson equation (2.66) becomes

$$(B - e\pi)g = -f,$$

and its solution is

$$g = -B^{\#}f + \eta e.$$

We may also choose $c = \pi g = 0$. Then,

$$g = -B^{\#}f. \tag{2.67}$$

If we choose $c = \pi g = -\eta$, then

$$g = -(B + e\pi)^{-1}f = -(B^{\#} + e\pi)f = -B^{\#}f - \eta e.$$

For simplicity, we have used the same notation g to denote different versions of the potentials, which may differ by a constant. We need to keep this in mind to avoid possible confusion.

Now, let us develop a sample-path-based explanation for $B^{\#}$ and g. First, we have

$$\int_0^\infty B[\exp(Bt)]dt = \int_0^\infty [\exp(Bt)]Bdt = -(I - e\pi).$$

From this, using $Be = \pi B = 0$, we get

$$B\left\{\int_0^\infty [\exp(Bt) - e\pi]dt\right\} = \left\{\int_0^\infty [\exp(Bt) - e\pi]dt\right\}B = -(I - e\pi). \tag{2.68}$$

Furthermore, we can easily prove that

$$\pi \left\{ \int_0^\infty [\exp(Bt) - e\pi] dt \right\} = \left\{ \int_0^\infty [\exp(Bt) - e\pi] dt \right\} e = 0.$$

By multiplying both sides of (2.68) on the left by $B^{\#}$, we obtain

$$\begin{aligned} B^{\#} &= - \int_0^\infty [\exp(Bt) - e\pi] dt \\ &= - \lim_{T \to \infty} \left[\int_0^T \exp(Bt) dt - Te\pi \right]. \end{aligned} \quad (2.69)$$

From (2.69) and (2.67), and using $P_t = \exp(Bt)$, we get

$$g(i) = \lim_{T \to \infty} E \left\{ \int_0^T [f(X_t) - \eta] dt \middle| X_0 = i \right\}.$$

This is the sample path explanation of the potential $g(i)$ (cf. (2.16)). This is also consistent with (2.60).

In modern Markov theory [87], the α-*potential* of a function f is defined as

$$g^{(f)}(i) = E \left\{ \int_0^\infty [\exp(-\alpha t)] f(X_t) dt \middle| X_0 = i \right\}, \qquad \alpha > 0.$$

Thus, our definition of the potential can be viewed as an extension of the classical α-potential to the case of $\alpha = 0$. To keep the integral finite at $\alpha = 0$, a constant term η is subtracted from the integrand (see (2.46) for the discussion of the discrete-time version).

Performance Derivatives

With the aforementioned results, the performance derivative formulas can be easily derived. Let B and B' be two infinitesimal generators on the same state space $\mathcal{S}$. Suppose that B changes to another infinitesimal generator $B_\delta = [b_\delta(i,j)] = B + \delta \Delta B$, with $\delta > 0$ being a small real number, $\Delta B = B' - B = [\Delta b(i,j)]$. We have $\Delta B e = 0$. Let $\boldsymbol{X}_\delta$ be the Markov process with infinitesimal generator B_δ. We assume that $\boldsymbol{X}_\delta$ is also irreducible. Let π_δ be the vector of the steady-state probabilities of $\boldsymbol{X}_\delta$. The average reward of $\boldsymbol{X}_\delta$ is $\eta_\delta = \eta + \Delta \eta_\delta$. The performance derivative along the direction of ΔB is $\frac{d\eta_\delta}{d\delta}\big|_{\delta=0} = \lim_{\delta \to 0} \frac{\eta_\delta - \eta}{\delta}$. With this notation, we have $\frac{dB_\delta}{d\delta} = \Delta B$.

Taking derivatives of both sides of $\pi_\delta B_\delta = 0$ at $\delta = 0$, we get

$$\left. \frac{d\pi_\delta}{d\delta} \right|_{\delta=0} B = -\pi \frac{dB_\delta}{d\delta} = -\pi(\Delta B).$$

By multiplying both sides of this equation on the right by $B^{\#}$ and using $BB^{\#} = I - e\pi$ and $\frac{d\pi_\delta}{d\delta} e = 0$, we obtain

$$\left.\frac{d\pi_\delta}{d\delta}\right|_{\delta=0} = -\pi(\Delta B)B^{\#}.$$

Therefore,

$$\left.\frac{d\eta_\delta}{d\delta}\right|_{\delta=0} = -\pi(\Delta B)B^{\#} f = \pi(\Delta B)g.$$

Next, by multiplying both sides of (2.62) on the right by π^T and using $\pi B = 0$ and $\pi e = 1$, we have

$$B\Gamma\pi^T = fe^T\pi^T - ef^T\pi^T = (I - e\pi)f.$$

That is, $B\Gamma\pi^T = BB^{\#}f$. By multiplying both sides of this equation on the left by $B^{\#}$, we get $(I - e\pi)\Gamma\pi^T = (I - e\pi)B^{\#}f$. Using $\pi B^{\#} = 0$ and $\pi\Gamma\pi^T = \pi(eg^T - ge^T)\pi^T = 0$, we obtain

$$B^{\#}f = \Gamma\pi^T.$$

This leads to the performance derivative formula in terms of Γ:

$$\left.\frac{d\eta_\delta}{d\delta}\right|_{\delta=0} = -\pi(\Delta B)\Gamma\pi^T.$$

If, in addition to the changes in B, the reward function f also changes to $f_\delta = f + \delta\Delta f$, we have

$$\left.\frac{d\eta_\delta}{d\delta}\right|_{\delta=0} = \pi[(\Delta B)g + \Delta f].$$

The higher-order derivatives can be derived in a way similar to the Markov chains:

$$\left.\frac{d^n\eta_\delta}{d\delta^n}\right|_{\delta=0} = n!\pi\left\{[(\Delta B)(-B^{\#})]^{n-1}[(\Delta B)(-B^{\#})f + \Delta f]\right\}. \quad (2.70)$$

In addition, we have the following MacLaurin expansion:

$$\eta_\delta = \eta + \pi\sum_{k=1}^{n}[\delta(\Delta B)(-B^{\#})]^{k-1}[(\Delta B)(-B^{\#})f + \Delta f]\delta$$
$$+ \pi_\delta[\delta(\Delta B)(-B^{\#})]^n[(\Delta B)(-B^{\#})f + \Delta f]\delta.$$

When $\Delta f = 0$, this becomes

$$\eta_\delta = \pi \sum_{k=0}^{n} [\delta(\Delta B)(-B^\#)]^k f + \pi_\delta [\delta(\Delta B)(-B^\#)]^{n+1} f.$$

Thus, we can use $\pi \sum_{k=0}^{n} [\delta(\Delta B)(-B^\#)]^k f$ to estimate η_δ, and the error in the estimation is $\pi_\delta [\delta(\Delta B)(-B^\#)]^{n+1} f$. All the items in π and $B^\#$ can be estimated on a sample path of the Markov process with infinitesimal generator B, see Problem 3.18.

2.3 Performance Sensitivities of Semi-Markov Processes*

In this section, we extend the above PA results to (continuous-time) semi-Markov processes (SMPs). The previous results on PA of Markov processes become special cases. This section is based on [57], and we only study the long-run average-reward problem (for extensions to the discounted-reward problem, see [57]).

2.3.1 Fundamentals for Semi-Markov Processes*

We study a semi-Markov process $\boldsymbol{X} = \{X_t, t \geq 0\}$ defined on a finite state space $\mathcal{S} = \{1, 2, \ldots, S\}$. Let $T_0, T_1, \ldots, T_l, \ldots$, with $T_0 = 0$, be the transition epoches. The process is right continuous so the state at each transition epoch is the state after the transition. Let $X_l = X_{T_l}, l = 0, 1, 2, \ldots$. Then, $\{X_0, X_1, \ldots\}$ is the embedded Markov chain. The interval $[T_l, T_{l+1})$ is called a *period* and its length is called the *sojourn time* in state X_l.

The Embedded Chain and the Sojourn Time

The semi-Markov kernel [87] is defined as

$$p(j; t|i) := \mathcal{P}(X_{l+1} = j, T_{l+1} - T_l \leq t | X_l = i),$$

which we assume does not depend on l (time-homogenous). Set

$$p(t|i) := \sum_{j \in \mathcal{S}} p(j; t|i) = \mathcal{P}(T_{l+1} - T_l \leq t | X_l = i),$$

$$h(t|i) := 1 - p(t|i),$$

$$p(j|i) := \lim_{t \to \infty} p(j; t|i) = \mathcal{P}(X_{l+1} = j | X_l = i),$$

and

$$p(t|i,j) := \frac{p(j;t|i)}{p(j|i)} = \mathcal{P}(T_{l+1} - T_l \leq t | X_l = i, X_{l+1} = j).$$

Normally, $p(i|i) = 0$, for all $i \in \mathcal{S}$. But, in general, we may allow the process to move from a state to itself at the transition epoches; in such a case, $p(i|i)$ may be nonzero and our results still hold. However, a transition from a state to the same state cannot be determined by observing only the system states of a semi-Markov process.

The matrix $[p(j|i)]$ is the transition probability matrix of the embedded Markov chain. We assume that this matrix is irreducible and aperiodic [20]. Let

$$m(i) = \int_0^\infty s p(ds|i) = E[T_{l+1} - T_l | X_l = i]$$

be the mean of the sojourn time in state i. We also assume that $m(i) < \infty$ for all $i \in \mathcal{S}$. Under these assumptions, the semi-Markov process is irreducible and aperiodic and hence ergodic. Define the *hazard rates* as

$$r(t|i) = \frac{\frac{d}{dt} p(t|i)}{h(t|i)},$$

and

$$r(j;t|i) = \frac{\frac{d}{dt} p(j;t|i)}{h(t|i)}.$$

The latter is the rate at which the process moves from i to j in $[t, t+dt)$ given that the process does not move out from state i in $[0, t)$.

The Equivalent Infinitesimal Generator

Let $p_t(j|i) = \mathcal{P}(X_t = j | X_0 = i)$. By the total probability theorem, we can easily derive

$$p_{t+\Delta t}(j|i) = \sum_{k \in \mathcal{S}} p_t(k|i) \int_0^\infty \widetilde{p}_t(s|k) \{ I_j(k)[1 - r(s|k)\Delta t] + r(j;s|k)\Delta t \} ds, \tag{2.71}$$

where $I_j(k) = 1$ if $k = j$, $I_j(k) = 0$ if $k \neq j$; $\widetilde{p}_t(s|k)ds$ is defined as the probability that, given that the state at time t is k, the process has been in state k for a period of s to $s + ds$. This probability depends on k and therefore may depend on the initial state. Precisely, let l_t be the integer such that $T_{l_t} \leq t < T_{l_t+1}$. Then,

$$\widetilde{p}_t(s|k)ds = \mathcal{P}(s \leq t - T_{l_t} < s + ds | X_t = k). \tag{2.72}$$

It is proved at the end of this subsection that

$$\lim_{t \to \infty} \widetilde{p}_t(s|k) = \frac{h(s|k)}{m(k)}. \tag{2.73}$$

Now, set $\Delta t \to 0$ in (2.71) and we obtain

$$\frac{dp_t(j|i)}{dt} = -\sum_{k \in \mathcal{S}} p_t(k|i) \int_0^\infty \{\widetilde{p}_t(s|k)[I_j(k) r(s|k) - r(j; s|k)]\} ds. \qquad (2.74)$$

Since the semi-Markov process is ergodic, when $t \to \infty$, we have $p_t(j|i) \to \pi(j)$ [87] and $\frac{dp_t(j|i)}{dt} \to 0$, where $\pi(j)$ is the steady-state probability of j. Letting $t \to \infty$ on both sides of (2.74) and using (2.73), we get

$$\begin{aligned}
0 &= -\sum_{k \in \mathcal{S}} \pi(k) \int_0^\infty \frac{1}{m(k)} \left\{ I_j(k) \frac{d}{ds}[p(s|k)] - \frac{d}{ds}[p(j; s|k)] \right\} ds \\
&= -\sum_{k \in \mathcal{S}} \pi(k) \left\{ \frac{1}{m(k)} [I_j(k) - p(j|k)] \right\} \\
&= -\sum_{k \in \mathcal{S}} \pi(k) \{\lambda(k)[I_j(k) - p(j|k)]\},
\end{aligned}$$

where we define

$$\lambda(k) := \frac{1}{m(k)}.$$

Finally, we have

$$\sum_{k \in \mathcal{S}} \pi(k) b(k, j) = 0, \qquad \text{for all } j \in \mathcal{S},$$

where we define

$$b(k, j) = -\lambda(k)[I_j(k) - p(j|k)]. \qquad (2.75)$$

In matrix form, we can write

$$\pi B = 0, \qquad (2.76)$$

where $\pi = (\pi(1), \ldots, \pi(S))$ is the steady-state probability vector and B is a matrix with elements $b(k, j)$. In addition, we can easily verify that

$$Be = 0.$$

Equation (2.76) is exactly the same as the Markov process with B as its infinitesimal generator. Therefore, B in (2.76) is *the equivalent infinitesimal generator* for a semi-Markov process. Note that B depends only on $m(i)$ and $p(j|i)$, $i, j \in \mathcal{S}$. This implies that the steady-state probability is insensitive to the high-order statistics of the sojourn times in any state, and it is independent of whether the sojourn time in state i depends on j, the state it moves into from i.

The Steady-State Probability

We will study the general case where the reward function depends not only on the current state but also on the next state that the semi-Markov process moves into. To this end, for any time $t \in [T_l, T_{l+1})$, we denote $Y_t = X_{l+1}$, and study the process $\{(X_t, Y_t), t \geq 0\}$. Because the process $\{Y_t, t \geq 0\}$ is completely determined by the process $\{X_t, t \geq 0\}$, for notational simplicity, we still denote the process $\{(X_t, Y_t), t \geq 0\}$ as

$$\boldsymbol{X} = \{(X_t, Y_t), t \geq 0\}. \tag{2.77}$$

Let $\pi(i,j)$ be the steady-state probability of $(X_t, Y_t) = (i,j)$ and $\pi(j|i)$ be the steady-state conditional probability of $Y_t = j$ given that $X_t = i$, i.e., $\pi(j|i) = \lim_{t\to\infty} P(Y_t = j|X_t = i)$. (This is different from $\lim_{l\to\infty} P(X_{l+1} = j|X_l = i)$, which is the steady-state conditional probability of the embedded Markov chain.)

Define

$$m(i,j) = \int_0^\infty sp(ds|i,j) = E[T_{l+1} - T_l|X_l = i, X_{l+1} = j].$$

Then, we have

$$m(i) = \sum_{j\in\mathcal{S}} p(j|i)m(i,j) = \int_0^\infty sp(ds|i). \tag{2.78}$$

We can prove (see the end of this subsection)

$$\pi(j|i) = \frac{\int_0^\infty sp(j;ds|i)}{\int_0^\infty sp(ds|i)} = \frac{p(j|i)m(i,j)}{m(i)}. \tag{2.79}$$

Therefore,

$$\pi(i,j) = \pi(j|i)\pi(i) = \pi(i)\frac{p(j|i)m(i,j)}{m(i)}, \tag{2.80}$$

where $\pi(i)$, $i \in \mathcal{S}$, can be obtained from (2.76).

Proofs

A. The Proof of (2.73).

Consider an interval $[0, T_L]$, with $L >> 1$. Let $I_k(x) = 1$ if $x = k$ and $I_k(x) = 0$ if $x \neq k$; and $I(*)$ be an indicator function, i.e., $I(*) = 1$ if the expression in the brackets holds, $I(*) = 0$ otherwise. Let l_t be the integer such that $T_{l_t} \leq t < T_{l_t+1}$. From (2.72), by ergodicity, we have

$$\lim_{t\to\infty} \widetilde{p}_t(s|k)ds = \lim_{T_L\to\infty} \frac{\int_0^{T_L} I(s \leq t - T_{l_t} < s + ds)I_k(X_t)dt}{\int_0^{T_L} I_k(X_t)dt}. \tag{2.81}$$

Let N_k be the number of periods in $[0, T_L]$ in which $X_t = k$. We have

$$\lim_{T_L \to \infty} \frac{1}{N_k} \int_0^{T_L} I_k(X_t)dt = \int_0^\infty sp(ds|k). \tag{2.82}$$

Next, we observe that, for a fixed $s > 0$, $\int_0^{T_L} I(s \le t - T_{l_t}) I_k(X_t)dt$ is the total length of the time period in $[0, T_L]$ in which $s \le t - T_{l_t}$ and $X_t = k$. Furthermore, among the N_k periods, roughly $N_k p(d\tau|k)$ periods terminate with a length of τ to $\tau + d\tau$. For any $s < \tau$, in each of such periods, the length of time in which $s \le t - T_{l_t}$ is $\tau - s$. Thus,

$$\int_0^{T_L} I(s \le t - T_{l_t}) I_k(X_t)dt \approx N_k \int_s^\infty (\tau - s)p(d\tau|k),$$

or

$$\lim_{T_L \to \infty} \frac{1}{N_k} \int_0^{T_L} I(s \le t - T_{l_t}) I_k(X_t)dt = \int_s^\infty (\tau - s)p(d\tau|k).$$

Therefore,

$$\begin{aligned} &\lim_{T_L \to \infty} \frac{1}{N_k} \int_0^{T_L} I(s \le t - T_{l_t} < s + ds) I_k(X_t)dt \\ = &- \lim_{T_L \to \infty} \frac{1}{N_k} \left[\int_0^{T_L} I(s + ds \le t - T_{l_t}) I_k(X_t)dt \right. \\ &\left. - \int_0^{T_L} I(s \le t - T_{l_t}) I_k(X_t)dt \right] \\ = &- \frac{d}{ds} \left[\int_s^\infty (\tau - s)p(d\tau|k) \right] ds = [1 - p(s|k)]ds = h(s|k)ds. \end{aligned} \tag{2.83}$$

From (2.81), (2.82), and (2.83), we get

$$\lim_{t \to \infty} \tilde{p}_t(s|k)ds = \frac{h(s|k)ds}{\int_0^\infty sp(ds|k)}.$$

Therefore, (2.73) holds. □

B. The Proof of (2.79).

Consider a time interval $[0, T_L]$, with $L >> 1$. Let N_i be the number of periods in $[0, T_L]$ in which $X_t = i$. Then,

$$\lim_{T_L \to \infty} \frac{1}{N_i} \int_0^{T_L} I_i(X_t)dt = \int_0^\infty sp(ds|i).$$

Let $I_{i,j}(x, y) = 1$ if $x = i$ and $y = j$, and $I_{i,j}(x, y) = 0$ otherwise. We have

$$\lim_{T_L \to \infty} \frac{1}{N_i} \int_0^{T_L} I_{i,j}(X_t, Y_t)dt = \int_0^\infty sp(j; ds|i).$$

Thus, we have

$$\pi(j|i) = \lim_{T_L \to \infty} \frac{\int_0^{T_L} I_{i,j}(X_t, Y_t)dt}{\int_0^{T_L} I_i(X_t)dt} = \frac{\int_0^\infty sp(j; ds|i)}{\int_0^\infty sp(ds|i)} = \frac{p(j|i)m(i,j)}{m(i)}.$$

Therefore, (2.79) holds. □

2.3.2 Performance Sensitivity Formulas*

Consider a semi-Markov process $\boldsymbol{X} = \{(X_t, Y_t), t \geq 0\}$ (see (2.77)) starting from a *transition epoch* $T_0 = 0$ and an initial state $X_0 = j$. We define the reward function as $f(i,j)$, $i,j \in \mathcal{S}$, where $f : \mathcal{S} \times \mathcal{S} \to \mathcal{R}$. The long-run average reward is

$$\eta = \lim_{T \to \infty} \frac{1}{T} E\left[\int_0^T f(X_t, Y_t)dt \bigg| X_0 = j\right],$$

which does not depend on j because $\boldsymbol{X}$ is ergodic.

The Perturbation Realization Factor

On $\boldsymbol{X}$ with $T_0 = 0$ and $X_0 = j$, denote the instant at which the process moves into state i for the first time as

$$T(i|j) = \inf\{t : t \geq 0,\ X_t = i | X_0 = j\}.$$

Following the same approach as for the PA of Markov processes (2.61), we define the *perturbation realization factors* as (the only difference is that $T_0 = 0$ must be a transition epoch in the semi-Markov case):

$$\gamma(i,j) = E\left\{\int_0^{T(i|j)} [f(X_t, Y_t) - \eta]dt \bigg| X_0 = j\right\}. \tag{2.84}$$

Define $\Gamma = [\gamma(i,j)]_{i,j=1}^S$.

From (2.80) and by ergodicity, we have

$$\eta = \sum_{i,j \in \mathcal{S}} \pi(i,j) f(i,j) = \sum_{i \in \mathcal{S}} \pi(i) f(i) = \pi f,$$

where $f = (f(1), f(2), \ldots, f(S))^T$, and (for simplicity, we use "f" for both $f(i)$ and $f(i,j)$)

$$f(i) = \frac{\sum_{j \in \mathcal{S}} p(j|i) f(i,j) m(i,j)}{m(i)}. \tag{2.85}$$

From (2.84), we have

$$\gamma(i,j) = E\left\{ \int_0^{T_1} [f(X_t,Y_t)-\eta]dt \middle| X_0 = j \right\}$$
$$+ E\left\{ \int_{T_1}^{T(i|j)} [f(X_t,Y_t)-\eta]dt \middle| X_0 = j \right\}$$
$$= \sum_{k\in\mathcal{S}} p(k|j) \left\{ E\left\{ \int_0^{T_1} [f(X_0,Y_0)-\eta]dt \middle| X_0 = j, X_1 = k \right\} \right.$$
$$\left. +E\left\{ \int_{T_1}^{T(i|j)} [f(X_t,Y_t)-\eta]dt \middle| X_0 = j, X_1 = k \right\} \right\}$$
$$= \sum_{k\in\mathcal{S}} p(k|j) \left\{ [f(j,k)-\eta] E[T_1|X_0 = j, X_1 = k] \right.$$
$$\left. +E\left\{ \int_{T_1}^{T(i|k)} [f(X_t,Y_t)-\eta]dt \middle| X_1 = k \right\} \right\}$$
$$= \sum_{k\in\mathcal{S}} p(k|j) \left\{ [f(j,k)-\eta] m(j,k) \right.$$
$$\left. + E\left\{ \int_{T_1}^{T(i|k)} [f(X_t,Y_t)-\eta]dt \middle| X_1 = k \right\} \right\}.$$

From (2.78) and (2.85), the aforementioned equation leads to

$$\gamma(i,j) = m(j)[f(j)-\eta] + \sum_{k\in\mathcal{S}} p(k|j)\gamma(i,k),$$

or, equivalently,

$$-[f(j)-\eta] = \sum_{k\in\mathcal{S}} \{-\lambda(j)[I_j(k) - p(k|j)]\gamma(i,k)\}$$
$$= \sum_{k\in\mathcal{E}} [b(j,k)\gamma(i,k)].$$

In matrix form, this is

$$\Gamma B^T = -(ef^T - \eta ee^T). \tag{2.86}$$

Next, on the process $\boldsymbol{X}$, with $T_0 = 0$ being a transition epoch and $X_0 = j$, for any state $i \in \mathcal{S}$ we define two sequences $u_0, u_1, \ldots$, and $v_0, v_1, \ldots$, as follows:

$$u_0 = T_0 = 0, \tag{2.87}$$
$$v_n = \inf\{t \geq u_n, X_t = i\},$$

and

$$u_{n+1} = \inf\{t \geq v_n, X_t = j\}, \tag{2.88}$$

i.e., v_n is the first time when the process reaches i after u_n, and u_{n+1} is the first time when the process reaches j after v_n, $n = 0, 1, \ldots$. Apparently, $u_0, u_1, \ldots$ are stopping times and X_t is a regenerative process with $\{u_n, n = 0, 1, \ldots\}$ as its associated renewal process. By the theory of regenerative processes [87], we have

$$\eta = \frac{E[\int_{u_0}^{u_1} f(X_t, Y_t)dt]}{E[u_1 - u_0]} = \frac{E[\int_0^{v_0} f(X_t, Y_t)dt] + E[\int_{v_0}^{u_1} f(X_t, Y_t)dt]}{E[v_0] + E[u_1 - v_0]}.$$

Thus,

$$E\left\{\int_0^{v_0} [f(X_t, Y_t) - \eta]dt\right\} + E\left\{\int_{v_0}^{u_1} [f(X_t, Y_t) - \eta]dt\right\} = 0.$$

By the definition of u_0, v_0 and u_1, the above equation is

$$\gamma(i, j) + \gamma(j, i) = 0;$$

therefore, the matrix Γ is skew-symmetric

$$\Gamma^T = -\Gamma.$$

Taking the transpose of (2.86), we get

$$-B\Gamma = -(fe^T - \eta ee^T).$$

From the above equation and (2.86), Γ satisfies the following *PRF equation*

$$B\Gamma + \Gamma B^T = -F,$$

where $F = ef^T - fe^T$.

Performance Potentials

Similar to Equations (2.87) to (2.88), for any three states i, j, k, we define three sequences $u_0, u_1, \ldots$; $v_0, v_1, \ldots$; and $w_0, w_1, \ldots$ as follows. $u_0 = T_0 = 0$, $X_0 = j$, $v_n = \inf\{t \geq u_n, X_t = i\}$, $w_n = \inf\{t \geq v_n, X_t = k\}$, and $u_{n+1} = \inf\{t \geq w_n, X_t = j\}$. By a similar approach, we can prove that

$$\gamma(i, j) + \gamma(j, k) + \gamma(k, i) = 0.$$

In general, we can prove that, for any closed circle $i_1 \to i_2 \to \cdots \to i_n \to i_1$ in the state space, we have

$$\gamma(i_1, i_2) + \gamma(i_2, i_3) + \cdots + \gamma(i_{n-1}, i_n) + \gamma(i_n, i_1) = 0.$$

This is similar to the conservation law of potential energy in physics. Therefore, we can define a performance potential $g(i)$ in any state and write $\gamma(i,j) = g(j) - g(i)$ and

$$\Gamma = eg^T - ge^T, \tag{2.89}$$

where $g = (g(1), \ldots, g(S))^T$. By substituting (2.89) into (2.86), we get the *Poisson equation*:

$$Bg = -f + \eta e.$$

Similar to the Markov process case, we have different versions of g, which differ by a constant vector ce. For example, when $\pi g = \eta$, we have

$$(B - e\pi)g = -f, \tag{2.90}$$

and when $\pi g = 0$, we have

$$g = -B^{\#} f.$$

Finally, a Markov process with transition rates $\lambda(i)$ and transition probabilities $p(j|i)$ can be viewed as a semi-Markov process whose kernel is $p(j;t|i) = p(j|i)\{1 - \exp[-\lambda(i)t]\}$. With this special kernel, we have

$$m(i,j) = m(i) = \frac{1}{\lambda(i)},$$

$$\pi(i,j) = \pi(i)p(j|i),$$

and

$$f(i) = \sum_{j=1}^{S} p(j|i) f(i,j).$$

The results in this section become the same as those in Section 2.2 for Markov processes.

Performance Sensitivity Formulas

We have shown that with properly defined g and B, the Poisson equation and PRF equation hold for potentials and perturbation realization factor matrices, respectively, for semi-Markov processes. Thus, performance sensitivity formulas can be derived in a way similar to Markov processes, and the results are briefly stated here.

First, for two semi-Markov processes with B', η', f' and B, η, f, by multiplying both sides of (2.90) on the left by π' and using $\pi' B' = 0$ and $\pi g = \eta$, we get

$$\begin{aligned}\eta' - \eta &= \pi'[(B' - B)g + (f' - f)] \\ &= \pi'[(B'g + f') - (Bg + f)].\end{aligned} \tag{2.91}$$

As we shall see in Chapter 4, this equation serves as a foundation for semi-Markov decision processes. As shown in Chapter 4, policy iteration for semi-Markov processes can be derived from (2.91).

Next, suppose that B changes to $B_\delta = B + \delta \Delta B$, with $\Delta B = B' - B$, and f changes to $f_\delta = f + \delta \Delta f$, with $\Delta f = f' - f$. We have $\Delta B e = 0$. ΔB can be determined by the changes in the characteristics of the semi-Markov process. For example, if $\lambda(i) = 1/m(i)$ changes to $\lambda(i) + (\Delta\lambda)\delta$, $i = 1, 2, \ldots, S$, $\Delta\lambda > 0$, then, according to (2.75), $b(i,j)$ changes to $b(i,j) - \delta(\Delta\lambda)[I_j(i) - p(j|i)]$; i.e., $\Delta B = -\Delta\lambda(I - P)$, $P = [p(j|i)]$; on the other hand, if P changes to $P + \Delta P$, then $\Delta b(i,j) = \lambda(i)[\Delta P(j|i)]$, $i, j = 1, 2, \ldots, S$. Denote the average reward of the semi-Markov system with B_δ and f_δ as η_δ. We can easily obtain

$$\left.\frac{d\eta_\delta}{d\delta}\right|_{\delta=0} = \pi[-(\Delta B)B^{\#} f + \Delta f]$$
$$= \pi[(\Delta B)\Gamma^T \pi^T + \Delta f].$$

Sample-path-based expressions for g and Γ can be derived. From (2.84), with a similar reasoning as in (2.18), we have

$$\gamma(i,j) = \lim_{T\to\infty} \left\{ E\left\{ \int_0^T [f(\widetilde{X}_t, \widetilde{Y}_t) - \eta] dt \middle| \widetilde{X}_0 = j \right\} \right.$$
$$\left. - E\left\{ \int_0^T [f(X_t, Y_t) - \eta] dt \middle| X_0 = i \right\} \right\},$$

where $\widetilde{\boldsymbol{X}}$ and $\boldsymbol{X}$ have the same kernel; they are independent but start from two different initial states, $\widetilde{X}_0 = j$ and $X_0 = i$, respectively, with $T_0 = \widetilde{T}_0 = 0$ being a transition epoch for both $\widetilde{\boldsymbol{X}}$ and $\boldsymbol{X}$. From this equation, we have

$$g(j) = \lim_{T\to\infty} E\left\{ \int_0^T [f(X_t, Y_t) - \eta] dt \middle| X_0 = j \right\}. \qquad (2.92)$$

This is the same as in the Markov process case, except that the integral starts with a transition epoch. The convergence of the right-hand side of (2.92) can be easily verified by, e.g., using the embedded Markov chain model.

With the equivalent infinitesimal generator, the high-order derivatives are the same as those for the Markov chains (2.70). Again, all the items in π and $B^{\#}$ can be estimated on a sample path of the semi-Markov process with B; see Problem 3.18.

Example 2.3. Consider a communication line (or a switch, a router, etc.) at which packets arrive in a Poisson process with a rate of λ packets per second. The packet length is assumed to have a general probability distribution function $\Phi(x)$; the unit of the length is bit per packet. For each packet, the system manager can choose a transmission rate of θ bits per second. Thus, the transmission time for each packet has a distribution function $\widetilde{\Phi}(\tau) = \mathcal{P}(t \le \tau) = \mathcal{P}(x \le \theta\tau) = \Phi(\theta\tau)$. In a real system, θ takes a discrete value determined by the number of channels; each channel has a fixed amount of bandwidth. Thus, we can view θ as an action and denote the action space as $\{\theta_1, \theta_2, \ldots, \theta_K\}$, with $\theta_k = k\mu$, $k = 1, 2, \ldots, K$, where μ denotes the transmission rate of one channel in bits per second. Of course, in a theoretical study, we can also view θ as a continuous variable.

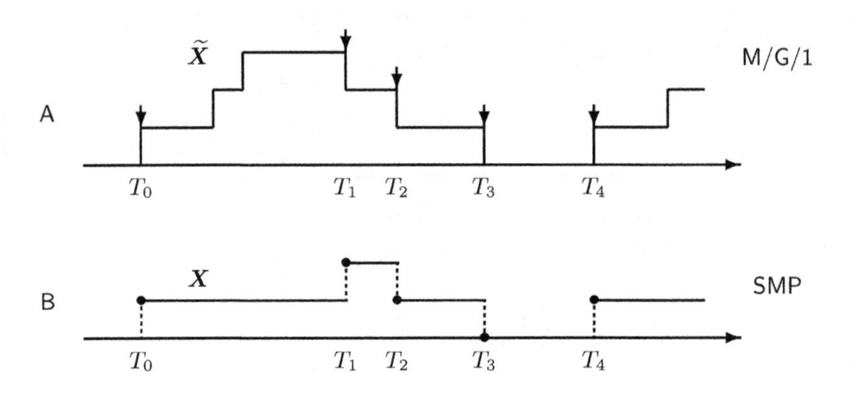

Fig. 2.11. An M/G/1 Queue and the Embedded SMP

The system can be modelled as an M/G/1 queue; the (physical) state at time t is $N(t) = i$ with i being the number of customers (packets) in the queue at time t. Figure 2.11.A illustrates a sample path $\boldsymbol{X} = \{N(t), t \ge 0\}$. For stability, we require that $K\mu > \lambda\bar{x}$, where $\bar{x}$ is the mean length of the packets. The decisions for actions are made at the beginning of the transmission of every packet. Thus, we consider the embedded points consisting of all the service completion times and the arrival times to all the idle periods, denoted as $T_0, T_1, \ldots$. Define $\widetilde{X}_t = N(T_n)$ for $T_n \le t < T_{n+1}$, $n = 0, 1, 2, \ldots$. Then, $\widetilde{\boldsymbol{X}} = \{\widetilde{X}_t, t \ge 0\}$ is a semi-Markov process (SMP). Figure 2.11.B illustrates the embedded SMP corresponding to the sample path in Figure 2.11.A. It is clear that the following equations hold for $\widetilde{\boldsymbol{X}}$:

$$p(1; t|0) = 1 - \exp(-\lambda t),$$

$$p(t|i) = \Phi(\theta t), \qquad i > 0,$$

and

$$p(j; dt|i) = \mathcal{P}(X_{n+1} = j, t \le T_{n+1} - T_n < t + dt | X_n = i)$$

$$= \left[\frac{(\lambda t)^{j-i+1}}{(j-i+1)!} \exp(-\lambda t) \right] \Phi(\theta dt), \qquad i > 0, \ i-1 \leq j,$$

where the term in the braces is the probability that there are $j-i+1$ arrivals in the period of $[0, t)$.

In the optimization problem, the reward (cost) function usually consists of two parts: the holding cost $f_1(i, j)$ and the bandwidth cost $f_2(\theta)$. That is,

$$f_\theta(i, j) = f_1(i, j) + f_2(\theta).$$

It is well known that, if in an interval $[0, t]$, there are k arrivals from a Poisson process, then these k arrivals uniformly distribute over the period (see, e.g., [169]). Thus, it is reasonable to take the average number of customers in $[0, t]$, $(i+j)/2$, as the holding cost, and we may set

$$f_\theta(i, j) = \kappa_1 \frac{i+j}{2} + \kappa_2 \theta, \qquad \kappa_1 + \kappa_2 = 1, \ 0 < \kappa_1, \kappa_2 < 1,$$

where the first term represents the cost for the average waiting time. The problem is now formulated in a semi-Markov framework and the results developed in this section can be applied. □

Finally, many results about SMPs can be obtained by using the embedded Markov chain method (see, e.g., [243]). It is natural to expect that the sensitivity analysis can also be implemented using this approach. However, compared with the embedded-chain-based approach, the approach presented in this section is more direct and concise and hence the results have a clear intuitive interpretation. In addition, with the embedded approach, the expected values (time and cost) on a period $T_{n+1} - T_n$ are used; the sample-path-based approach used here is easier to implement on-line (e.g., see the definition in (2.84)).

The discounted reward with a discount factor $\beta > 0$ for semi-Markov processes is defined as

$$\eta_\beta(i) = \lim_{T \to \infty} E \left[\int_0^T \beta \exp(-\beta t) f(X_t, Y_t) dt \middle| X_0 = i \right], \qquad T_0 = 0. \quad (2.93)$$

Similar to the discrete case in (2.31), the weighting factor in (2.93) is also normalized: $\int_0^\infty \beta \exp(-\beta t) dt = 1$. The performance potential for the discounted reward criterion is

$$g_\beta(i) = \lim_{T \to \infty} E \left\{ \int_0^T \exp(-\beta t)[f(X_t, Y_t) - \eta] dt \middle| X_0 = i \right\}, \qquad i \in \mathcal{S}.$$

The sensitivity analysis of the discounted reward for semi-Markov processes involves an *equivalent Markov process*. We refer readers to [57] for technical details.

2.4 Perturbation Analysis of Queueing Systems

The early works on perturbation analysis (PA) focused on queueing systems. The idea of PA was first proposed in [144] for the buffer allocation problem in a serial production line and was first studied for queueing networks in [141]. The special structure of queueing systems, especially the interactions among different customers or different servers, makes PA a very efficient tool for estimating the performance derivatives with respect to the mean service times based on a single sample path. This section contains an overview of the main results of PA of queueing systems.

The main difference between PA of Markov chains and PA of queueing systems is that in the former, a perturbation is a "jump" on a sample path from one state to another due to parameter changes, while, in the latter, it is a small (infinitesimal) delay in a customer's transition time. Some queueing (such as the Jackson-type) networks can be modelled by Markov processes and therefore the theory and algorithms developed for Markov processes can be applied. However, because of the special features of a queueing system, the performance derivatives with respect to service time changes can be obtained by a much more efficient and more intuitive approach, which applies to non-Markov queueing systems as well.

The dynamic nature of a system's behavior is explored more clearly in PA of queueing systems. Its basic principle can be described as follows: a small increase in the mean service time of a server *generates* a series of small delays, called *perturbations*, in the service completion times of the customers served by that server. Each such perturbation of a customer's service completion time will cause delays in the service completion times of other customers (at the same server or at other servers). In other words, a perturbation will be *propagated* through the system due to the interactions among customers and servers. Thus, a perturbation will affect the system performance through propagation. The average effect of a perturbation on the system performance can be measured by a quantity called the *perturbation realization factor (PRF)*. Finally, the effect of a change in the mean service time of a server equals the sum of the effects of all the perturbations generated on the service completion times of the server due to this change in its mean service time. The above description is precisely captured by *three fundamental rules of PA*:

1. Perturbation generation;
2. Perturbation propagation;
3. Perturbation realization.

These rules will be discussed in subsequent subsections. In PA of Markov chains, the perturbations (jumps) are generated according to (2.2) and (2.4); the perturbation realization is illustrated by Figure 2.6 and measured by (2.5).

However, as we will see, the "propagation" effect in Markov chains is not as explicit as in queueing networks.

Problem Description

Consider a closed Jackson network (cf. Appendix C.2) with M servers and N customers. The service times of every server in the network are independently and exponentially distributed. Let $\bar{s}_i$ be the mean service time of server i, $i = 1, \ldots, M$, and let $q_{i,j}$, $i, j = 1, \ldots, M$, be the routing probabilities. $Q = [q_{i,j}]$ is the routing probability matrix. The system state can be denoted as $\boldsymbol{n} = (n_1, n_2, \ldots, n_M)$, where n_i is the number of customers in server i. For a closed network with M servers and N customers, we have $\sum_{i=1}^{M} n_i = N$. The state space is $\mathcal{S} = \{\text{all } \boldsymbol{n} : \sum_{i=1}^{M} n_i = N\}$. The system state at time t is denoted as $\boldsymbol{N}(t) = (n_1(t), \ldots, n_M(t))$. The system can be modelled by a Markov process $\boldsymbol{X} = \{\boldsymbol{N}(t), t \geq 0\}$. Let T_l, $l = 0, 1, \ldots$, be the lth transition time of $\boldsymbol{X}$, counting the customer transitions at all the servers. Figure 2.12 illustrates a sample path of a three-server five-customer closed queueing network. The vertical dashed arrows signal the customer transitions among servers, and each of the three staircase-like curves indicates the evolution of a server in the network.

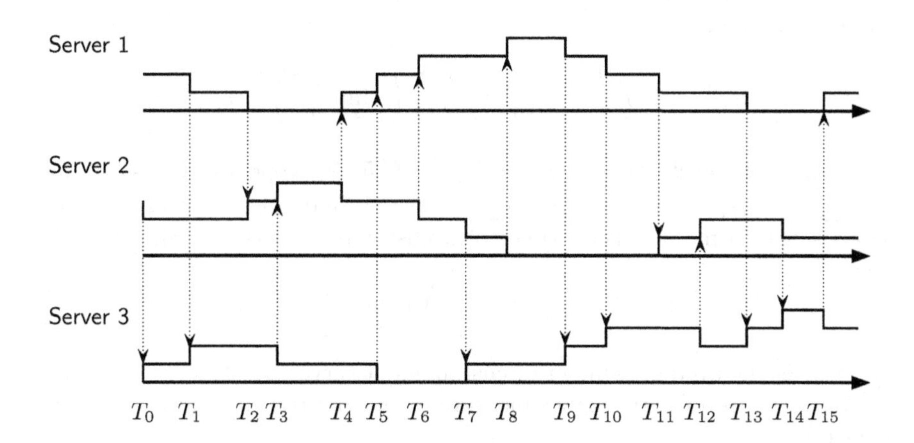

Fig. 2.12. A Sample Path of a Closed Queueing Network with $M = 3$ and $N = 5$

Let $f : \mathcal{S} \to \mathcal{R}$ be a reward (or cost) function. The system performance is defined as the long-run average reward

$$\eta^{(f)} = \lim_{L \to \infty} \frac{1}{L} \int_0^{T_L} f[\boldsymbol{N}(t)] dt, \qquad \text{w.p.1}, \tag{2.94}$$

where T_L is the Lth transition time of the system. In this section, we use the superscript "(f)" to explicitly denote the dependency of a quantity on f for clarity. For closed Jackson networks in which a customer can reach any server in the network while circulating in the network (irreducible networks), the state process $\boldsymbol{N}(t)$ is an ergodic Markov process, and the limit in (2.94) exists with probability 1 and does not depend on the initial state. Set

$$F_L = \int_0^{T_L} f[\boldsymbol{N}(t)]dt.$$

Then, we have

$$\eta^{(f)} = \lim_{L \to \infty} \frac{F_L}{L}.$$

With L being the number of customers' service completions in the period of $[0, T_L]$, the performance measure defined in (2.94) is the *customer average.* These types of performance measures cover a wide range of applications. For example, if $f(\boldsymbol{n}) = I(\boldsymbol{n}) \equiv 1$ for all $\boldsymbol{n} \in \mathcal{S}$, then $F_L = T_L$ and

$$\eta^{(I)} = \lim_{L \to \infty} \frac{T_L}{L} = \frac{1}{\eta}, \tag{2.95}$$

where $\eta = \lim_{L \to \infty} \frac{L}{T_L}$ is the system throughput (the number of service completions per unit of time). If $f(\boldsymbol{n}) = n_i$, then F_L is the area underneath the sample path of server i. Let L_i be the number of service completions at server i in $[0, T_L]$. Then,

$$\eta^{(f)} = \lim_{L \to \infty} \frac{F_L}{L} = \left(\lim_{L \to \infty} \frac{L_i}{L} \right) \left(\lim_{L \to \infty} \frac{F_L}{L_i} \right) = v_i \bar{\tau}_i,$$

where v_i is the visit ratio of server i (see (C.5) in Appendix C), satisfying $v_i = \sum_{j=1}^{M} v_j q_{j,i}$ and normalized to $\sum_{k=1}^{M} v_i = 1$, and $\bar{\tau}_i$ is the mean response time (waiting time + service time) of a customer at server i. Similarly, we have

$$\eta^{(f)} = \lim_{L \to \infty} \frac{F_L}{L} = \left(\lim_{L \to \infty} \frac{T_L}{L} \right) \left(\lim_{L \to \infty} \frac{F_L}{T_L} \right) = \eta^{(I)} \bar{n}_i,$$

where $\bar{n}_i$ is the average number of customers at server i.

Another type of performance measure is the long-run time-average reward defined as

$$\eta_T^{(f)} = \lim_{L \to \infty} \frac{1}{T_L} \int_0^{T_L} f[\boldsymbol{N}(t)]dt,$$

which can be easily converted to customer averages as follows:

$$\eta_T^{(f)} = \eta \eta^{(f)} = \frac{\eta^{(f)}}{\eta^{(I)}}.$$

Now suppose that the mean service time of one of the servers, say server v, changes from $\bar{s}_v$ to $\bar{s}_v + \Delta \bar{s}_v$. We call the closed network with $\bar{s}_i$, $i = 1, 2, \ldots, M$, the original network, and the network with the changed mean service time $\bar{s}_v + \Delta\bar{s}_v$ and $\bar{s}_i$, $i \neq v$, the perturbed network. A sample path of the original network is called an *original sample path*, and a sample path of the perturbed network is called a *perturbed sample path.*

Given a sample path of a network, its average reward $\eta^{(f)}$ can be easily estimated by simple calculation. The goal of PA is to obtain an estimate for the performance derivatives $\frac{d\eta^{(f)}}{d\bar{s}_v}$, $v = 1, 2, \ldots, M$, by observing and analyzing an original sample path. This is shown in Figure 2.13, in which we use θ to denote a generic parameter.

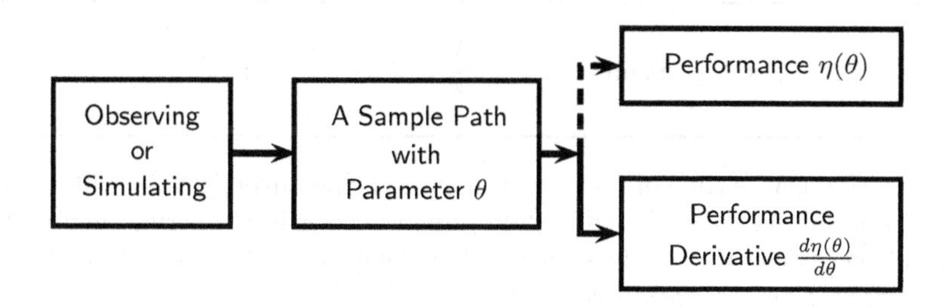

Fig. 2.13. The Goal of Perturbation Analysis

2.4.1 Constructing a Perturbed Sample Path

As in Markov chains, the first step in PA of queueing systems is to construct a perturbed sample path by using an original one.

Suppose that we are given an original sample path with transition times T_l, $l = 0, 1, \ldots$. Let T'_l be the lth transition time on the corresponding perturbed path, $l = 0, 1, \ldots$. Suppose that the lth transition time is a service completion time of server i. Then, $\Delta T_l := T'_l - T_l$ is called the *perturbation of server* i at time T_l; it is also called the perturbation of the customer that completes the service at server i at T_l.

Perturbation Generation

First, we study how the change in the mean service time of a server affects every customer's service time at that server. In general, let $\bar{s}$ be the mean service time of a server with an exponentially distributed service time. Then, the service time of a customer at that server, denoted as s, has the following distribution:

$$\Phi(s) = 1 - \exp\left(-\frac{s}{\bar{s}}\right).$$

In simulation, we use the inverse-transform method to generate the service times (shown in Figure A.2 and reproduced in Figure 2.14). First, we generate a uniformly distributed random number ξ in $[0, 1)$. Then, we set

$$s = \Phi^{-1}(\xi) = -\bar{s}\ \ln(1-\xi). \tag{2.96}$$

It is well known that s in (2.96) is exponentially distributed with mean $\bar{s}$. Assume that the mean service time changes to $\bar{s}+\Delta\bar{s}$ (for the sake of discussion, we may assume that $\Delta\bar{s} > 0$). Then, with the same random variable ξ, the service time in (2.96) changes to

$$s + \Delta s = -(\bar{s} + \Delta\bar{s})\ \ln(1-\xi).$$

Thus, we have

$$\Delta s = -\Delta\bar{s}\ln(1-\xi) = \frac{\Delta\bar{s}}{\bar{s}}s = \kappa s, \qquad \kappa := \frac{\Delta\bar{s}}{\bar{s}}. \tag{2.97}$$

That is, the service time of every customer at the perturbed server will increase by an amount Δs (> 0) shown in (2.97); in other words, the service completion time of every customer at the server will be delayed by Δs (> 0). We call (2.97) the *perturbation generation rule* [142]:

The Perturbation Generation Rule:

At the perturbed server, because of the change in the mean service time $\Delta\bar{s}$, every customer's service completion time obtains a perturbation of Δs, shown in (2.97), on the sample path.

This perturbation obtained during a customer's service period is in addition to the perturbation(s) previously obtained by the server before the customer starts its service.

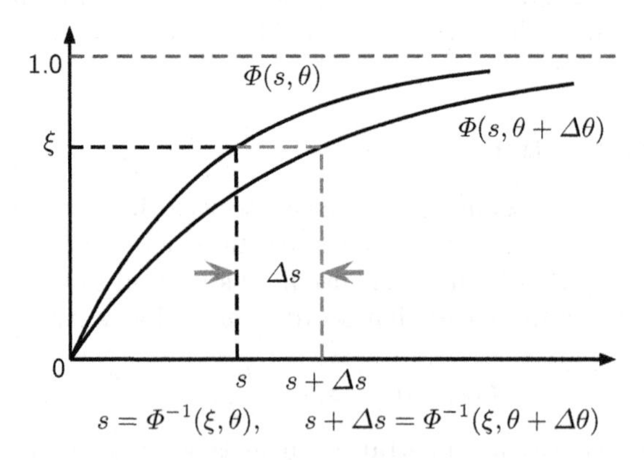

Fig. 2.14. The Perturbation Generation Rule

The inverse-transform method can be used to derive the perturbation generation rule for other service distributions. Let $\Phi(s, \theta)$ be the distribution function of the service times of the customers at a server, which depends on a parameter θ. With the inverse-transform method, we determine a customer's service time by using the inverse function of the distribution function:

$$s = \Phi^{-1}(\xi, \theta) = \sup\{s : \ \Phi(s, \theta) \leq \xi\},$$

where ξ is a uniformly distributed random variable on $[0, 1)$. Suppose that the distribution parameter θ changes to $\theta + \Delta\theta$. Then, the service time of the customer changes to

$$s + \Delta s = \Phi^{-1}(\xi, \theta + \Delta\theta).$$

We have

$$\begin{aligned} \Delta s &= \Phi^{-1}(\xi, \theta + \Delta\theta) - \Phi^{-1}(\xi, \theta) \\ &\approx \left.\frac{\partial \Phi^{-1}(\xi, \theta)}{\partial \theta}\right|_{\xi=\Phi(s,\theta)} \Delta\theta = \left.\frac{\partial s}{\partial \theta}\right|_{\xi=\Phi(s,\theta)} \Delta\theta. \end{aligned} \tag{2.98}$$

Δs is the *perturbation generated* during the service period because of $\Delta\theta$. The same random variable ξ is used for both s and $s + \Delta s$. Pictorially, the perturbation generation rule is illustrated in Figure 2.14.

In practice, calculating the partial derivative $\frac{\partial \Phi^{-1}(\xi, \theta)}{\partial \theta}$ may require a relatively large amount of computation. However, in most applications, such as in communication systems, the packet length distribution, $length = \Phi^{-1}(\xi)$, is fixed, and one can only change the transition rate μ. The service (transition) time is $s = \frac{length}{\mu} = \frac{1}{\mu}\Phi^{-1}(\xi)$. Therefore, for service rate μ, we have

$$\begin{aligned} \Delta s &\approx -\frac{\Delta\mu}{\mu^2}\Phi^{-1}(\xi) = -\frac{\Delta\mu}{\mu}s \\ &= \kappa s, \qquad \kappa = -\frac{\Delta\mu}{\mu}, \end{aligned} \tag{2.99}$$

which is in the same form as (2.97) for the mean service time of the exponential distribution.

Perturbation Propagation

A perturbation of one customer, or one server, will affect the transition times of other customers, or other servers, in the network. Figure 2.15 illustrates the interaction between two servers. Suppose that the first customer in server 1 obtains a perturbation Δ at time T_1; i.e., its service completion time is delayed by Δ. Apparently, the service starting time of the next customer at the same server will be delayed by Δ and its service completion time will also be delayed by the same amount Δ at T_2. In addition, at T_1, because server 2 was idle and was waiting for a customer arriving from server 1, the service starting time of server 2 at T_1 and its completion time at T_3 will also be delayed by Δ. We summarize the above discussion in two *perturbation propagation rules* [142]:

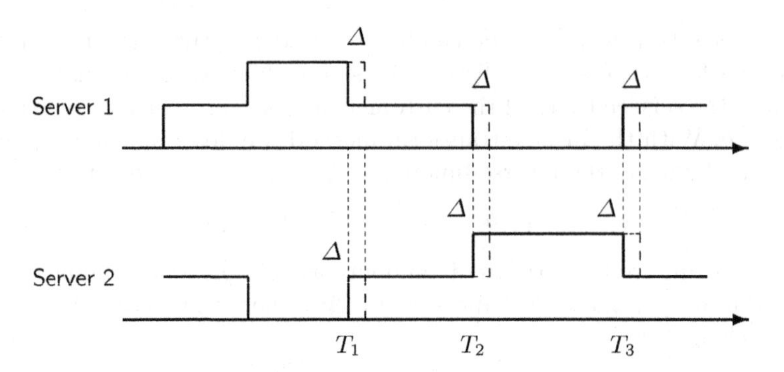

Fig. 2.15. Perturbation Propagation

The Perturbation Propagation Rules:

i. A server keeps its perturbation until it meets an idle period (or the perturbation of a customer's service completion time is propagated to the next customer in the server until the server meets an idle period).
ii. The perturbation of one server will be propagated to another server if a customer at the former moves to the latter and terminates an idle period of the latter server.

The first rule implies that when a server meets an idle period, the server's original perturbation is lost. The second rule implies that after the idle period, the server will acquire a perturbation propagated from another server. That is, after an idle period, a server's perturbation always equals that of the server that terminates the idle period. A special case is illustrated in Figure 2.16, in which server 1 has a perturbation $\Delta_1 = \Delta$ at T_1, but after the idle period, at T_2, the server acquires the perturbation from server 2, which is $\Delta_2 = 0$. Thus, the perturbation Δ_1 of server 1 at T_1 is lost after the idle period at T_2. This explains how a non-perturbed server can be viewed as a server having a perturbation 0 in perturbation propagation.

Note that we assume that the perturbation can be as small as we wish (*infinitesimal perturbation*). Thus, we can always assume that the perturbation is smaller than the length of the idle period. See Section 2.4.4 for more details.

Constructing a Perturbed Path

Now we return to the closed Jackson network with M servers and N customers. Suppose that we are given a sample path of the original system with

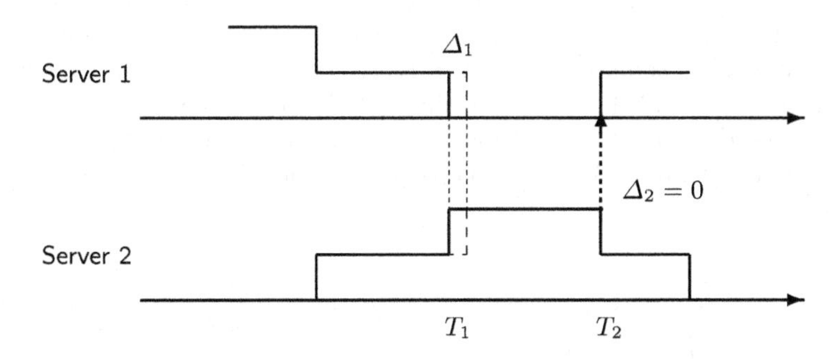

Fig. 2.16. Perturbation Propagation for $\Delta = 0$

mean service times $\bar{s}_i$, $i = 1, 2, \ldots, M$, and one server's (server v) mean service time is perturbed from $\bar{s}_v$ to $\bar{s}_v + \Delta\bar{s}_v$.

From the perturbation generation and propagation rules, we can efficiently determine the perturbations of all the servers and therefore construct a perturbed sample path on an original one without simulating the perturbed system again. We may simply generate a perturbation on the original sample path according to (2.97) whenever a customer completes its service at server v, and then propagate it along the original sample path according to the two propagation rules. Note that we can propagate all the perturbations at a server altogether. This leads to the following simple algorithm for determining the perturbations of all servers on the sample path at any time:

Algorithm 2.1. (Constructing a Perturbed Sample Path)
Given an original sample path for a closed Jackson network:

i. Initialization: Set $\Delta_i := 0$, $i = 1, 2, \ldots, M$;
ii. (Perturbation generation) At the kth service completion time of server v, set $\Delta_v := \Delta_v + s_{v,k}$, $k = 1, 2, \ldots$, $s_{v,k}$ is the service time of the customer;
iii. (Perturbation propagation) After a customer from server i terminates an idle period of server j, set $\Delta_j := \Delta_i$, $i, j = 1, 2, \ldots, M$.

The perturbation of server i is $\kappa\Delta_i$, $i = 1, 2, \ldots, M$.

In the algorithm, Δ_i denotes the (accumulated) perturbation of server i, $i = 1, 2, \ldots, M$. The perturbation of every server is updated whenever it starts a new busy period, and, in addition, the perturbation of the perturbed server is also updated whenever it completes its service to a customer. Because all the perturbations generated and propagated are proportional to $\kappa = \frac{\Delta\bar{s}_v}{\bar{s}_v}$, at any time the perturbation at any server in the network must be proportional to κ. Therefore, for simplicity, in the algorithm, we use $s_{v,k}$ instead of $\kappa s_{v,k}$ as the perturbation generated. Thus, the exact perturbation corresponding to $\Delta \bar{s}_v$ at any server i should be $\kappa \Delta_i$, $i = 1, \ldots, M$. The algorithm determines the perturbations of all the transition times of all servers (i.e., T_l, $l = 1, 2, \ldots$) at the perturbed path. The transition times of the perturbed path equal those of the original path plus the perturbation of the corresponding server.

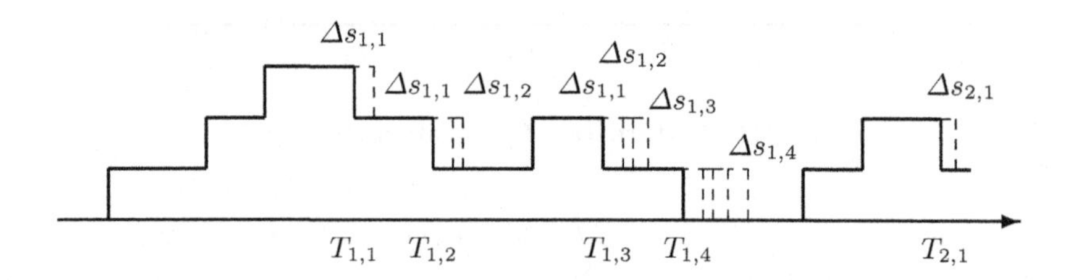

Fig. 2.17. A Perturbed Sample Path for an M/G/1 Queue

Example 2.4. To illustrate perturbation propagation within the same server, we consider a single server queue, in which there is no perturbation propagation among different servers. In such a system, the third step in Algorithm 2.1 is not implemented, and the perturbation of the server is reset to zero at the beginning of every new busy period. Actually, a single server queue is an open network, and the arriving customers can be viewed as from a source that is never perturbed. A sample path of such a single server queue (may be viewed as an M/M/1 or an M/G/1 queue) and its corresponding perturbed path constructed by Algorithm 2.1 are shown in Figure 2.17.

The figure illustrates the first busy period of the sample path, in which there are four customers served by the server. In the kth busy period, the ith customer's service time is denoted as $s_{k,i}$, and its departure time is denoted as $T_{k,i}$, $k, i = 1, 2, \ldots$. At the first customer's departure time $T_{1,1}$, a perturbation $\Delta s_{1,1} = \kappa s_{1,1}$ is generated according to (2.98) or (2.99). This perturbation is propagated to the departure times of the subsequent customers in the same busy period, $T_{1,2}$, $T_{1,3}$, and $T_{1,4}$. At $T_{1,2}$, another perturbation $\Delta s_{1,2} = \kappa s_{1,2}$ is generated; thus, the total perturbation at $T_{1,2}$ is $\Delta s_{1,1} + \Delta s_{1,2}$. This perturbation propagates to $T_{1,3}$ and $T_{1,4}$; and so on. In general, the perturbation of the ith departure time in the kth busy period is

$$\Delta T_{k,i} = \sum_{l=1}^{i} \Delta s_{k,l}, \qquad (2.100)$$

where $\Delta s_{k,l} = \kappa s_{k,l}$ is the perturbation of the lth customer's service time in the kth busy period, generated according to (2.98) or (2.99). □

Figure 2.17 also illustrates a fundamental fact: The simple rules for perturbation propagation hold only if the perturbation accumulated at the end of a busy period is smaller than the length of the idle period following the busy period. For the time being, we may think that we can always choose $\Delta\bar{s}$ or $\Delta\theta$ small enough such that this condition holds. For a rigorous discussion, see Section 2.4.4.

Example 2.5. Suppose that we are given an original sample path of a three-server five-customer closed network shown in Figure 2.12, and server 2's mean service time is perturbed from $\bar{s}_2$ to $\bar{s}_2 + \Delta\bar{s}_2$. We may construct a perturbed sample path by following the perturbation generation and propagation rules, as shown in Figure 2.18. The top figure shows the original path plus the perturbations at all transition instants; and the bottom figure shows the perturbed path thus constructed, in which $T_l' = T_l + \Delta T_l$, with ΔT_l being the perturbation of the transition instant T_l, $l = 0, 1, \ldots, 15$.

There are five perturbations generated, denoted as perturbations Δs_1, Δs_2, Δs_3, Δs_4, and Δs_5 (for simplicity, we omitted the subscript denoting server 2, e.g., we write $\Delta s_{2,1} = \Delta s_1$, etc.) and differentiated by different grays shown in the figure. The five perturbations are induced during the first five customers' service times at the perturbed server, server 2. They are generated according to the perturbation generation rule (2.98).

As shown in the figure, Perturbation Δs_1 obtained at T_4 by server 2 is propagated to server 1 immediately since the customer at server 2 terminates an idle period of server 1 at T_4. This perturbation is also propagated to the subsequent service completion times of server 2, T_6, T_7 and T_8. At T_6, server 2 obtains another perturbation Δs_2 for its second customer, resulting in a total perturbation of $\Delta T_6 = \Delta s_1 + \Delta s_2$. Similarly, we have $\Delta T_7 = \Delta s_1 + \Delta s_2 + \Delta s_3$ and $\Delta T_8 = \Delta s_1 + \Delta s_2 + \Delta s_3 + \Delta s_4$. As shown in the figure, ΔT_7 is propagated to server 3 through an idle period. The perturbation that is propagated to server 1 at T_4, Δs_1, is also propagated to the subsequent customers' service completion times, T_9, T_{10}, T_{11}, and T_{13}, in the same busy period of server 1. Likewise, the perturbation propagated to server 3 at T_7, $\Delta s_1 + \Delta s_2 + \Delta s_3$, is also propagated to the subsequent customers' service completion times, T_{12} and T_{15}, in the same busy period of server 3.

The perturbation that server 2 acquired in the first busy period $\Delta T_8 = \Delta s_1 + \Delta s_2 + \Delta s_3 + \Delta s_4$ is lost after the idle period starting from T_8. Indeed, at the beginning of the next busy period T_{11}, server 2 acquires a perturbation $\Delta T_{11} = \Delta s_1$ through propagation from server 1. There is another perturbation, Δs_5, generated during the service time of the first customer in the second busy period of server 2, resulting in a total perturbation of $\Delta T_{14} = \Delta s_1 + \Delta s_5$ for server 2 at T_{14}. Note that although the arrival time to server 1 at T_8 is delayed by $\Delta s_1 + \Delta s_2 + \Delta s_3 + \Delta s_4$, its effect is temporary: it does not affect any other service completion time at server 1 at all. The same statement holds for the delays in other arrival times except for those arrivals that start a new busy period.

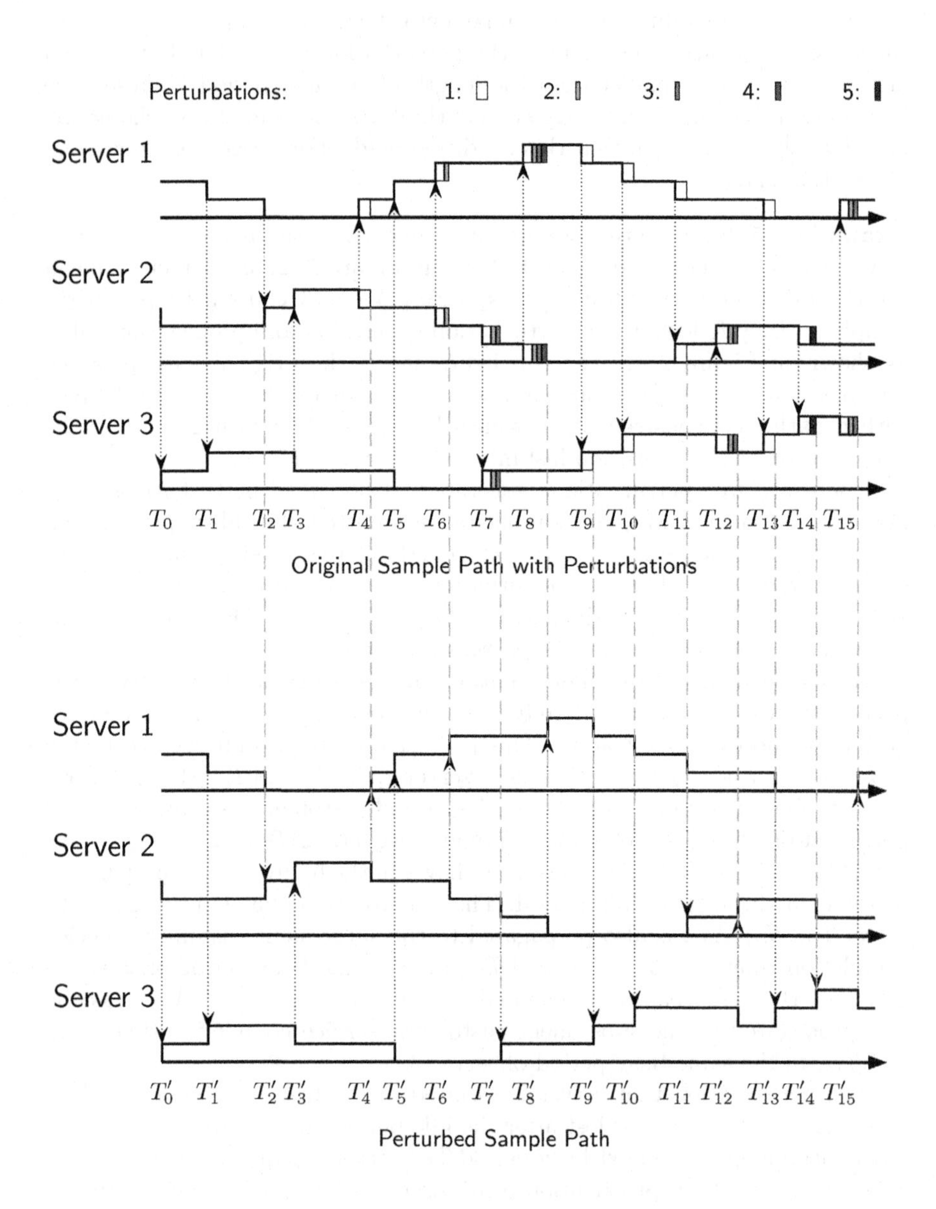

Fig. 2.18. A Perturbed Sample Path of the Network in Figure 2.12

Again, the perturbation propagated to server 1 in the second busy period, Δs_1, is lost after the idle period starting from T_{13}. After that, server 1 acquires a perturbation $\Delta s_1 + \Delta s_2 + \Delta s_3$ from server 3 through propagation at T_{15}.

It is interesting to note that starting from T_7, every server acquires the perturbation Δs_1. We say that Δs_1 is realized at T_7 by the network. In contrast, starting from T_9, no server has the perturbation Δs_4. We say that Δs_4 is lost by the network at T_9 (see Section 2.4.2). □

Calculating the Performance Derivatives

With the perturbed sample path constructed by Algorithm 2.1, the performance of the perturbed system can be calculated. As an example, we consider the system throughput. Recall that T_L is the Lth transition time of a queueing system. Assume that $L >> 1$. Then, the overall system throughput (the number of customers served by all the servers in the network per unit of time) is defined as

$$\eta = \lim_{L \to \infty} \frac{L}{T_L} \approx \frac{L}{T_L}.$$

In the perturbed system with $\bar{s}_v$ changed to $\bar{s}_v + \Delta \bar{s}_v$, it takes $T_L + \Delta T_L$ to finish the L transitions, with $\Delta T_L = \kappa \Delta_u$, $\kappa = \frac{\Delta \bar{s}_v}{\bar{s}_v}$, where u denotes the server for which T_L is the service completion time, and Δ_u is its perturbation at T_L determined by Algorithm 2.1 (in which κ is set to be one). The throughput of the perturbed system is

$$\eta + \Delta \eta \approx \frac{L}{T_L + \Delta T_L} \approx \frac{L}{T_L}(1 - \frac{\Delta T_L}{T_L}) = \eta(1 - \frac{\Delta T_L}{T_L}).$$

Thus, we have

$$\Delta \eta \approx -\eta \frac{\Delta T_L}{T_L},$$

and

$$\frac{\bar{s}_v}{\eta} \frac{\Delta \eta}{\Delta \bar{s}_v} \approx -\frac{\bar{s}_v}{\Delta \bar{s}_v} \frac{\Delta T_L}{T_L} = -\frac{\Delta_u}{T_L}.$$

Therefore, the *elasticity* (or the *normalized derivative*) of η with respect to $\bar{s}_v$ can be estimated on a sample path with PA as follows.

$$\frac{\bar{s}_v}{\eta} \frac{\partial \eta}{\partial \bar{s}_v} \approx -\frac{\Delta_u}{T_L}, \tag{2.101}$$

which does not depend on κ!

To obtain the derivatives of the throughput, in addition to the throughput itself, the algorithm adds only three clauses to the simulation program, one for perturbation generation, one for perturbation propagation (see Algorithm 2.1), and one for calculating the normalized derivative according to (2.101); and it adds only about 5% of computation time [141]. The following example illustrates the accuracy of this algorithm.

Example 2.6. Consider a closed Jackson network with $M = 6$, $N = 12$; the mean service times of the servers are 30, 40, 50, 55, 45, and 35, respectively; and the routing probability matrix is

$$Q = \begin{bmatrix} 0.00 & 0.10 & 0.20 & 0.15 & 0.35 & 0.20 \\ 0.25 & 0.00 & 0.15 & 0.10 & 0.10 & 0.40 \\ 0.35 & 0.15 & 0.00 & 0.25 & 0.25 & 0.00 \\ 0.25 & 0.25 & 0.10 & 0.00 & 0.20 & 0.20 \\ 0.00 & 0.20 & 0.25 & 0.15 & 0.00 & 0.40 \\ 0.40 & 0.30 & 0.00 & 0.15 & 0.15 & 0.00 \end{bmatrix}.$$

We ran a simulation for $L = 500,000$ transitions and applied the PA Algorithm 2.1 to the simulation. The resulting elasticities of the system throughput with respect to each mean service time given by (2.101) and the theoretical values of these elasticities (calculated by (C.19) and (C.15)) are shown in Table 2.8. □

$-\frac{\bar{s}_i}{\eta}\frac{\partial \eta}{\partial \bar{s}_i}$	$i = 1$	2	3	4	5	6
PA estimate	0.0906	0.1374	0.1025	0.2131	0.2736	0.1828
Theoretical	0.0915	0.1403	0.0980	0.2087	0.2812	0.1802

Table 2.8. Elasticities in Example 2.6

Now, we consider the average reward defined with any general reward function f in (2.94):

$$\eta^{(f)} = \lim_{L\to\infty} \frac{1}{L} \int_0^{T_L} f[\boldsymbol{N}(t)]dt = \lim_{L\to\infty} \frac{F_L}{L}, \tag{2.102}$$

where $\boldsymbol{N}(t)$ denotes the state process and

$$F_L = \int_0^{T_L} f[\boldsymbol{N}(t)]dt.$$

The computation of the performance derivative $\frac{\partial \eta^{(f)}}{\partial \bar{s}_i}$ involves more than that of the derivative of the system throughput $\frac{\partial \eta}{\partial \bar{s}_i}$. It depends not only on the final perturbation Δ_u, as shown in (2.101), but also on the perturbations of every transition time. We need to modify Algorithm 2.1 as follows:

Algorithm 2.2. (Calculating the Performance Derivatives)

Given an original sample path for a closed Jackson network:

i. Initialization: Set $\Delta_i := 0$, $i = 1, 2, \ldots, M$, and $\Delta F := 0$;
ii. (Perturbation Generation and Propagation) Same as steps ii and iii in Algorithm 2.1, which determine the perturbations of T_l, ΔT_l, $l = 1, 2, \ldots$;
iii. (Update ΔF) At every transition time T_l, $l = 1, 2 \ldots$, set $\Delta F := \Delta F + [f(\boldsymbol{n}) - f(\boldsymbol{n}')]\Delta T_l$, where $\boldsymbol{n} = \boldsymbol{N}(T_{l-})$ and $\boldsymbol{n}' = \boldsymbol{N}(T_l)$ are the system states before and after the transition, respectively.

Similar to Algorithm 2.1, κ is also set to be one in Algorithm 2.2. Let ΔF_L be the perturbation obtained by the algorithm at T_L. Then, the real perturbation of F_L for the system is $\kappa \Delta F_L$, with $\kappa = \frac{\Delta \bar{s}_v}{\bar{s}_v}$. Thus, when L is sufficiently large, from (2.102), we have $\Delta \eta^{(f)} = \kappa \frac{\Delta F_L}{L}$. From this, we obtain

$$\frac{\bar{s}_v}{\eta^{(I)}} \frac{\partial \eta^{(f)}}{\partial \bar{s}_v} \approx \frac{\bar{s}_v}{\eta^{(I)}} \frac{\Delta \eta^{(f)}}{\Delta \bar{s}_v} = \frac{\Delta F_L}{T_L}.$$

Finally, both Algorithms 2.1 and 2.2 can be implemented on line; i.e., there is no need to store the history of the sample path. Ref. [64] contains some simulation examples for Algorithm 2.2, applied to mean response times.

2.4.2 Perturbation Realization

We derived the PA algorithms for performance derivatives in the previous subsection. In this subsection, we start a more rigorous study of PA.

We first introduce the fundamental concept in PA: the *perturbation realization*. We show that, on average, the final effect of a single perturbation on the system performance (more precisely, on F_L, $L >> 1$ in (2.102)) can be measured by a quantity called the *perturbation realization factor*. Therefore, roughly speaking, the effect of a change in a system's parameter on the performance equals the sum of the realization factors of all the perturbations that are induced by the parameter change. This general principle is the same as in PA of Markov chains. The difference is that a perturbation for a queueing system is a small (infinitesimal) delay in time and that for a Markov chain is a state "jump". Historically, however, this principle was first proposed for PA of queueing systems [45, 49, 50, 51, 113, 141], and was extended later to Markov systems [62, 70].

Perturbation Realization

Consider the M-server closed Jackson network discussed in Section 2.4.1. The performance is defined as (2.94). To study the effect of a single perturbation

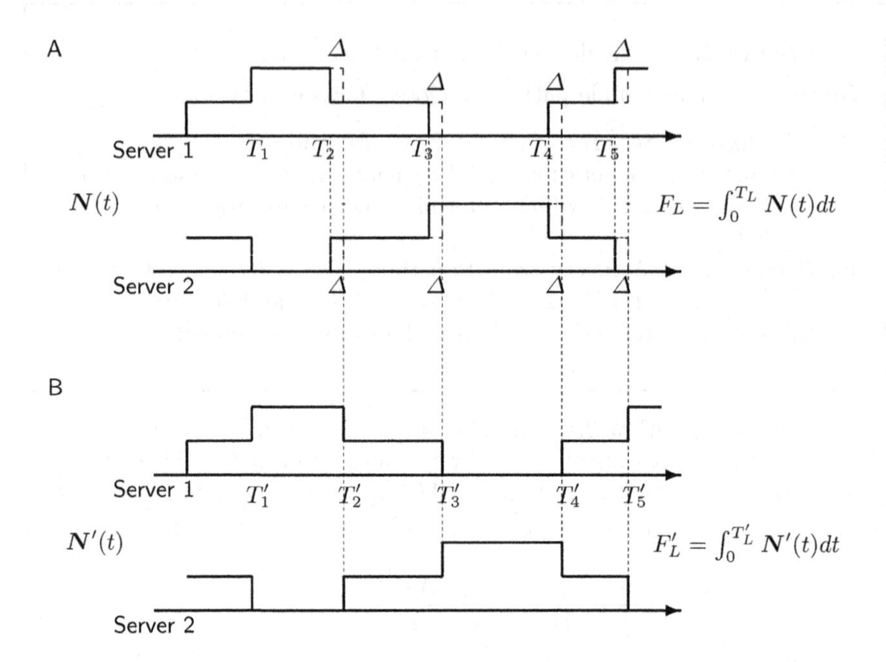

Fig. 2.19. A Sample Path and its Perturbed Counterpart

on the performance $\eta^{(f)}$, we assume that at some time, a perturbation Δ is generated at a server (e.g., in Figure 2.15 a perturbation Δ is generated at server 1 at T_1). As explained in Section 2.4.1, this perturbation will be propagated along a sample path. To study the effect of this single perturbation, we assume that there is no other perturbation generated on the sample path. During propagation, some servers in the network acquire this perturbation (e.g., in Figure 2.15, server 2 obtains a perturbation at T_1); others may lose the perturbation obtained before (e.g., in Figure 2.16, server 1 loses its perturbation at T_2). During propagation, every server has either perturbation Δ or perturbation 0 (no perturbation).

If, through propagation, every server in the network acquires the perturbation Δ, we say that the perturbation is *realized* by the network. After the perturbation is realized, the perturbed sample path is the same as the original one except that the entire sample path is shifted to the right by the amount of Δ. That is, there is an L^*, such that $T'_l = T_l + \Delta$ for all $l \geq L^*$. If, through propagation, every server in the network loses its perturbation (or acquires a perturbation of 0), we say that the perturbation is *lost* by the network. After the perturbation is lost, the perturbed sample path is exactly the same as the original one. That is, there is an L^*, such that $T'_l = T_l$ for all $l \geq L^*$. Apparently, whether a perturbation is realized or lost is random and depends on the sample path.

The solid lines in Figure 2.19.A illustrate a sample path $\boldsymbol{N}(t)$ of a two-server two-customer cyclic queueing network consisting of transition instants T_1 to T_5. A perturbation Δ is generated at server 1 at T_2, which is propagated to server 2 at T_2, and after T_2 all the servers have the same perturbation Δ, and the perturbation is realized by the network. The perturbed sample path corresponding to this perturbation is shown in Figure 2.19.B.

A closed queueing network is called *irreducible* if a customer at any server may visit any other server in the network, either directly, or by going through other servers. That is, for any pair of $i, j \in \{1, 2, \dots, M\}$, there exists a sequence of integers, $k_1, k_2, \dots, k_m \in \{1, 2, \dots, M\}$, such that $q_{i,k_1} q_{k_1,k_2} \cdots q_{k_m,j} > 0$. Such a routing probability matrix Q is also called *irreducible*. The following theorem indicates that a closed irreducible network will eventually "settle down" after being perturbed by a small perturbation.

Theorem 2.1. A perturbation in an irreducible closed Jackson network will either be realized or lost by the network with probability 1.

Proof. Since the network is irreducible, the state process $\boldsymbol{N}(t)$ will visit any state. In particular, with probability 1 every sample path will eventually visit state $(N, 0, \dots, 0)$; i.e, all customers are at server 1. If at that time server 1 has the perturbation, then after that time, all the servers will have the same perturbation; i.e., the perturbation is realized. On the other hand, if at that time server 1 has no perturbation, then after it all the servers in the network will have no perturbation; i.e., the perturbation is lost. □

The probability that a perturbation is realized is called the *perturbation realization probability.* It depends on the system state. The realization probability of a perturbation of server i when the system is in state $\boldsymbol{n}$ is denoted as $c(\boldsymbol{n}, i)$, $\boldsymbol{n} \in \mathcal{S}$, $i = 1, 2, \dots, M$.

Example 2.7. In Figure 2.18, the perturbation generated at T_4, Δs_1, is realized by the network at T_7. The perturbation generated at T_8, Δs_4, is lost at T_{11}. The other three perturbations, Δs_2, Δs_3, and Δs_5, have not been either realized or lost at T_{15}. Whether they will be realized or lost depends on the future evolution of the sample path. □

Perturbation Realization Factors

The effect of a perturbation on the long-run average reward $\eta^{(f)}$ defined in (2.94) can be studied by using the concept of perturbation realization. We first define the *realization factor* of a perturbation Δ of server i in state $\boldsymbol{n}$ for $\eta^{(f)}$ as (cf. (2.6) for realization factors for Markov chains):

$$c^{(f)}(\boldsymbol{n},i) = \lim_{L\to\infty} E\left(\frac{\Delta F_L}{\Delta}\right) = \lim_{L\to\infty} E\left(\frac{F'_L - F_L}{\Delta}\right)$$
$$= \lim_{L\to\infty} E\left\{\frac{1}{\Delta}\left\{\int_0^{T'_L} f[\boldsymbol{N}'(t)]dt - \int_0^{T_L} f[\boldsymbol{N}(t)]dt\right\}\right\}, \quad (2.103)$$

where F'_L is measured on the perturbed path generated by the propagation of this perturbation Δ (see Figure 2.19). It is clear that the realization factor $c^{(f)}(\boldsymbol{n},i)$ measures the average effect of a perturbation at $(\boldsymbol{n},i)$ on F_L in (2.94) as $L\to\infty$.

Recall that if a perturbation is realized, then there is an integer L^*, such that $T'_L = T_L + \Delta$ for all $L \geq L^*$, and if a perturbation is lost, then there is an L^*, such that $T'_L = T_L$ for all $L \geq L^*$. In both cases, there is an L^* (depending on the sample path) such that

$$\int_{T_{L^*}}^{T_L} f[\boldsymbol{N}(t)]dt - \int_{T'_{L^*}}^{T'_L} f[\boldsymbol{N}'(t)]dt = 0,$$

for all $L \geq L^*$ (in Figure 2.19, $L^* = 2$). Therefore, (2.103) becomes

$$c^{(f)}(\boldsymbol{n},i) = E\left\{\frac{1}{\Delta}\left\{\int_0^{T'_{L^*}} f[\boldsymbol{N}'(t)]dt - \int_0^{T_{L^*}} f[\boldsymbol{N}(t)]dt\right\}\right\}. \quad (2.104)$$

Thus, $c^{(f)}(\boldsymbol{n},i)$ defined in (2.103) is finite with probability 1 (cf. (2.5) for Markov chains).

Next, we study the effect of two or more perturbations at different servers. Consider a sample path of a closed network consisting of M servers. Suppose that at time $t = 0$, both server 1 and server 2 obtain a perturbation denoted as Δ_1 and Δ_2, respectively, with the same size $\Delta_1 = \Delta_2 = \Delta$. Let us propagate Δ_1 and Δ_2 separately along the sample path. First, we consider the propagation of Δ_1 at server 1. During the propagation, we use a 0-1 row vector $w_1(t)$ to denote which server has the perturbation at time $t \in [0,\infty)$. Specifically, we define $w_{1,i}(t) = 1$ if server i has the perturbation at time t, $w_{1,i}(t) = 0$ if otherwise, where $w_{1,i}(t)$ is the ith component of $w_1(t)$. Thus, initially the situation is represented by the vector $w_1(0) = (1,0,0,\ldots,0)$. According to the propagation rules, when server i terminates an idle period of server j, server i's perturbation (either 0 or Δ) will be propagated to server j. This is equivalent to simply setting $w_{1,j} := w_{1,i}$ after the propagation.

Similarly, the propagation of the perturbation Δ_2 starts with the vector $w_2(0) = (0,1,0,\ldots,0)$. We combine both vectors $w_1(0)$ and $w_2(0)$ together as an array

$$\begin{bmatrix} 1\,0\,0\,0\ldots 0 \\ 0\,1\,0\,0\ldots 0 \end{bmatrix}. \quad (2.105)$$

Now, let us propagate $\Delta_1(=\Delta)$ and $\Delta_2(=\Delta)$ simultaneously along the same sample path. As explained above, the propagation process is equivalent to copying the ith column of the above array to its jth column when server i terminates an idle period of server j. Thus, it is clear that, during propagation, the columns in the array (2.105) can never be $(1,1)^T$. That is, if we propagate both perturbations Δ_1 and Δ_2 together along the same sample path, any transition time of this sample path can acquire at most one of the perturbations, never both. In other words, if, at any time, a server has a perturbation, then this perturbation is propagated from either Δ_1 or Δ_2. Eventually, the array may reach one of the following three situations:

$$\begin{bmatrix} 0\,0\ldots 0 \\ 0\,0\ldots 0 \end{bmatrix}, \quad \begin{bmatrix} 0\,0\ldots 0 \\ 1\,1\ldots 1 \end{bmatrix}, \quad \begin{bmatrix} 1\,1\ldots 1 \\ 0\,0\ldots 0 \end{bmatrix}.$$

That is, either one of them is realized, or both are lost, on the sample path; but they cannot be both realized. Furthermore, the propagation of one perturbation (say Δ_1) does not interfere (change) the propagation of the other (say Δ_2). That is, each perturbation is propagated along the sample path in the same way as if the other did not exist.

Based on this observation, we have the *superposition of the propagation of perturbations* on a sample path: If we propagate two perturbations of servers i and j, with the same size, simultaneously on a sample path $\boldsymbol{N}(t)$ and obtain a perturbation path $\boldsymbol{N}'(t)$, then we have

$$c^{(f)}(\boldsymbol{n},i)+c^{(f)}(\boldsymbol{n},j)=E\left\{\frac{1}{\Delta}\left\{\int_0^{T'_{L^*}} f[\boldsymbol{N}'(t)]dt-\int_0^{T_{L^*}} f[\boldsymbol{N}(t)]dt\right\}\right\}.$$

The same discussion applies to the propagation of more than two perturbations. Let $V\subseteq\{1,2,\ldots,M\}$. Suppose that at time $t=0$, all the servers in set V obtain the same perturbation $\Delta_i=\Delta$, $i\in V$. We propagate all these perturbations simultaneously on a sample path $\boldsymbol{N}(t)$ and obtain a perturbation path $\boldsymbol{N}'(t)$. Then, we have

$$\sum_{i\in V} c^{(f)}(\boldsymbol{n},i)=E\left\{\frac{1}{\Delta}\left\{\int_0^{T'_{L^*}} f[\boldsymbol{N}'(t)]dt-\int_0^{T_{L^*}} f[\boldsymbol{N}(t)]dt\right\}\right\}. \quad (2.106)$$

Now we are ready to show that $c^{(f)}(\boldsymbol{n},i)$ satisfy the following set of linear equations [43, 51].

1. If $n_i=0$, then $c^{(f)}(\mathbf{n},i)=0$.
2. $\sum_{i=1}^{M} c^{(f)}(\mathbf{n},i)=f(\mathbf{n})$.
3. Let $\mathbf{n}_{-i,+j}=(n_1,\ldots,n_i-1,\ldots,n_j+1,\ldots,n_M)$ be a neighboring state of $\boldsymbol{n}$. Then,

$$\left[\sum_{i=1}^{M}\epsilon(n_i)\mu_i\right]c^{(f)}(\mathbf{n},k)=\sum_{i=1}^{M}\sum_{j=1}^{M}\epsilon(n_i)\mu_i q_{i,j}c^{(f)}(\mathbf{n}_{-i,+j},k)$$
$$+\sum_{j=1}^{M}\mu_k q_{k,j}\left\{[1-\epsilon(n_j)]c^{(f)}(\mathbf{n}_{-k,+j},j)+f(\mathbf{n})-f(\mathbf{n}_{-k,+j})\right\},$$
$$n_k>0,\quad k=1,2,\ldots,M, \qquad (2.107)$$

where $\epsilon(n_j)=0$, if $n_j=0$, and $\epsilon(n_j)=1$, if $n_j>0$.

The above equations can be easily derived. Property 1 is simply a convention: When a server is idle, any perturbation will be lost with probability 1 because after the idle period the server's perturbation is determined by another server that does not have the perturbation. Property 2 is a direct consequence of the superposition of propagation (2.106): Set $V=\{1,2,\ldots,M\}$. By definition, this means that every server has the same perturbation Δ at $T_0=0$, hence $L^*=0$; i.e., $T'_L=T_L+\Delta$ for all $L\geq L^*=0$. In particular, $T_{L^*}=0$ and $T'_{L^*}=\Delta$. Therefore,

$$\begin{aligned}F'_L-F_L&=\int_0^{T'_L}f[\boldsymbol{N}'(t)]dt-\int_0^{T_L}f[\boldsymbol{N}(t)]dt\\&=\left\{\int_0^{T'_{L^*}}f[\boldsymbol{N}'(t)]dt-\int_0^{T_{L^*}}f[\boldsymbol{N}(t)]dt\right\}\\&\quad+\left\{\int_{T'_{L^*}}^{T'_L}f[\boldsymbol{N}'(t)]dt-\int_{T_{L^*}}^{T_L}f[\boldsymbol{N}(t)]dt\right\}\\&=\int_0^{\Delta}f[\boldsymbol{N}'(t)]dt=f(\boldsymbol{n})\Delta.\end{aligned}$$

This leads to the second property. Equation (2.107) can be derived by the theorem of total probability. In (2.107), we assume that server k has a perturbation. $\frac{\epsilon(n_i)\mu_i q_{i,j}}{\sum_{i=1}^{M}\epsilon(n_i)\mu_i}$ is the probability that the next transition is from server i to server j, $i,j=1,2,\ldots,M$. If no idle period is involved in this transition, there is no perturbation propagation and server k keeps the same perturbation after the transition except that the system state changes to $\boldsymbol{n}_{-i,+j}$. This is reflected by the first term on the right-hand side. If there is an idle period at server j (i.e., $1-\epsilon(n_j)=1$), then, in addition to the perturbation in server k, the perturbation will be propagated from server k to server j. This is reflected by the second term on the right-hand side. $f(\mathbf{n})-f(\mathbf{n}_{-k,+j})$ is the effect due to the delay of the transition from server k to server j. Equation (2.107) implies that the effect of a perturbation before a transition equals the weighted sum, by transition probabilities, of the effects of the perturbations after the transition, plus the effect due to the delay of the transition. It has been proved that (2.107) and the equations in Properties 1 and 2 have a unique solution for irreducible closed Jackson networks [51, 113].

From (2.104), if $f(\boldsymbol{n}) = I(\boldsymbol{n}) = 1$ for all $\boldsymbol{n} \in \mathcal{S}$, we have

$$c^{(I)}(\boldsymbol{n}, i) = E\left[\frac{T'_{L^*} - T_{L^*}}{\Delta}\right].$$

From the meaning of the realization probability, we have $E[T'_{L^*} - T_{L^*}] = c(\boldsymbol{n}, i)\Delta$. Thus, $c(\boldsymbol{n}, i) = c^{(I)}(\boldsymbol{n}, i)$. Therefore, the realization probabilities satisfy the following equations:

1. If $n_i = 0$, then $c(\mathbf{n}, i) = 0$.
2. $\sum_{i=1}^{M} c(\mathbf{n}, i) = 1$.
3. If $n_k > 0$, $k = 1, 2, \ldots, M$, then

$$\left[\sum_{i=1}^{M} \epsilon(n_i)\mu_i\right] c(\mathbf{n}, k) = \sum_{i=1}^{M}\sum_{j=1}^{M} \epsilon(n_i)\mu_i q_{i,j} c(\mathbf{n}_{-i,+j}, k)$$
$$+ \sum_{j=1}^{M} \mu_k q_{k,j} \{[1 - \epsilon(n_j)] c(\mathbf{n}_{-k,+j}, j)\}.$$

The following example taken from [51] provides some idea of the numerical values for the realization probabilities.

Example 2.8. Consider a closed Jackson network with $M = 3$, $N = 5$, $\bar{s}_1 = 10$, $\bar{s}_2 = 8$, $\bar{s}_3 = 5$, and routing probability matrix

$$Q = \begin{bmatrix} 0 & 0.5 & 0.5 \\ 0.8 & 0 & 0.2 \\ 0.3 & 0.7 & 0 \end{bmatrix}.$$

The realization probabilities are obtained by solving the set of equations. The results, together with the steady-state probabilities, are listed in Table 2.9.

□

2.4.3 Performance Derivatives

We have now quantified the effect of a single perturbation on the long-run average reward. Next, we will determine the effect of a small change in a mean service time. Suppose that the mean service time of server v changes from $\bar{s}_v$ to $\bar{s}_v + \Delta\bar{s}_v$. Let $s_{v,l}$, $l = 1, 2, \ldots$, be the service time of the lth customer served at server v. Following the perturbation generation rule (2.97), the lth customer's service completion time at server v will gain a perturbation $\Delta_{v,l} = s_{v,l}\frac{\Delta\bar{s}_v}{\bar{s}_v} = \kappa s_{v,l}$, $l = 1, 2, \ldots$. All these perturbations will be propagated along the sample path. To calculate the effect of a small change in the mean

n	π(n)	c(n, 1)	c(n, 2)	c(n, 3)
(5,0,0)	0.19047	1.00000	0.00000	0.00000
(4,0,1)	0.06644	0.90584	0.00000	0.09416
(4,1,0)	0.15061	0.89385	0.10615	0.00000
(3,0,2)	0.02318	0.78826	0.00000	0.21174
(3,1,1)	0.05254	0.77060	0.17336	0.05604
(3,2,0)	0.11908	0.74279	0.25721	0.00000
(2,0,3)	0.00809	0.62556	0.00000	0.37444
(2,1,2)	0.01833	0.60901	0.25029	0.14070
(2,2,1)	0.04154	0.58286	0.37574	0.04141
(2,3,0)	0.09416	0.54528	0.45472	0.00000
(1,0,4)	0.00282	0.38089	0.00000	0.61911
(1,1,3)	0.00639	0.37327	0.34810	0.27863
(1,2,2)	0.01449	0.35728	0.51926	0.12346
(1,3,1)	0.03285	0.33315	0.62079	0.04606
(1,4,0)	0.07445	0.29754	0.70246	0.00000
(0,0,5)	0.00098	0.00000	0.00000	1.00000
(0,1,4)	0.00223	0.00000	0.48951	0.51049
(0,2,3)	0.00505	0.00000	0.71510	0.28490
(0,3,2)	0.01146	0.00000	0.83485	0.16515
(0,4,1)	0.02597	0.00000	0.91819	0.08181
(0,5,0)	0.05887	0.00000	1.00000	0.00000

Table 2.9. A Numerical Example of Realization Probabilities

service time $\bar{s}_v$, we need to add up the effect of all these single perturbations on the system performance.

Let $\pi(\boldsymbol{n})$ be the steady-state probability of state $\boldsymbol{n}$. Consider a time period $[0, T_L]$ with $L >> 1$. The length of the total time when the system is in state $\boldsymbol{n}$ in $[0, T_L]$ is $T_L\pi(\boldsymbol{n})$. The total perturbation generated in this period at server v due to the change $\Delta\bar{s}_v$ in the mean service time is $T_L\pi(\boldsymbol{n})\frac{\Delta\bar{s}_v}{\bar{s}_v}$. Since each perturbation on average has an effect of $c^{(f)}(\boldsymbol{n}, v)$ on F_L, the overall effect on F_L of all the perturbations induced when the system state is $\boldsymbol{n}$ is $[T_L\pi(\boldsymbol{n})\frac{\Delta\bar{s}_v}{\bar{s}_v}]c^{(f)}(\boldsymbol{n}, v)$. Finally, the total effect of the mean service time change, $\Delta\bar{s}_v$, on F_L is

$$\Delta F_L \approx \sum_{\text{all } \boldsymbol{n}} T_L\pi(\boldsymbol{n})\frac{\Delta\bar{s}_v}{\bar{s}_v}c^{(f)}(\boldsymbol{n}, v).$$

From this, we have

$$\frac{\bar{s}_v}{T_L/L}\frac{\Delta F_L/L}{\Delta \bar{s}_v} \approx \sum_{\text{all } \boldsymbol{n}} \pi(\boldsymbol{n}) c^{(f)}(\boldsymbol{n}, v).$$

Letting $L \to \infty$ and then $\Delta \bar{s}_v \to 0$, we obtain the steady-state performance derivative as follows:

$$\frac{\bar{s}_v}{\eta^{(I)}}\frac{\partial \eta^{(f)}}{\partial \bar{s}_v} = \sum_{\text{all } \boldsymbol{n}} \pi(\boldsymbol{n}) c^{(f)}(\boldsymbol{n}, v), \tag{2.108}$$

where $\eta^{(I)} = \lim_{L\to\infty} \frac{T_L}{L} = \frac{1}{\eta}$, see (2.95). Thus, the *normalized* derivative of the average reward (the left-hand side of (2.108)) equals the steady-state expectation of the realization factor. The above discussion provides an intuitive derivation and explanation for (2.108). See (2.116) in the next section for a formal formulation.

Set $f = I$ in (2.108). With $\eta^{(I)} = \frac{1}{\eta}$ and $c(\boldsymbol{n}, v) = c^{(I)}(\boldsymbol{n}, v)$, we can express the "elasticity" (normalized derivative) of the system throughput by using the perturbation realization probabilities:

$$\frac{\bar{s}_v}{\eta}\frac{\partial \eta}{\partial \bar{s}_v} = -\sum_{\text{all } \boldsymbol{n}} \pi(\boldsymbol{n}) c(\boldsymbol{n}, v). \tag{2.109}$$

Summing up both sides over $v = 1, 2, \ldots, M$, we have

$$\sum_{v=1}^{M} \frac{\bar{s}_v}{\eta}\frac{\partial \eta}{\partial \bar{s}_v} = -1. \tag{2.110}$$

Example 2.9. In this example [51], we choose $M = 3$, $N = 8$, $\bar{s}_1 = 5$, $\bar{s}_2 = 10$, and $\bar{s}_3 = 12$. The routing probability matrix is

$$Q = \begin{bmatrix} 0 & 0.5 & 0.5 \\ 0.7 & 0 & 0.3 \\ 0.4 & 0.6 & 0 \end{bmatrix}.$$

The realization probability equations are solved numerically. The elasticities calculated by (2.109) are -0.0365, -0.5133, and -0.4502, which are exactly the same as those calculated by queueing theory formulas. These values also satisfy (2.110). □

As shown in Section 2.4.1, the elasticity of the throughput can be estimated by a very efficient algorithm, Algorithm 2.1, together with equation (2.101). A close examination of the algorithm reveals that it, in fact, estimates the right-hand side of (2.109); i.e., it estimates the total sum as (cf. (2.101)):

$$\sum_{\text{all } \boldsymbol{n}} \pi(\boldsymbol{n}) c(\boldsymbol{n}, v) \approx \frac{\Delta_u}{T_L}.$$

Roughly speaking, $T_L \pi(\boldsymbol{n})$ is proportional to the perturbation generated when the system is in state $\boldsymbol{n}$; we may use $T_L \pi(\boldsymbol{n})$ as the perturbation generated, which corresponds to setting $\kappa = 1$ in Algorithm 2.1. At the end of the simulation, $\Delta_u \approx \sum_{\text{all } \boldsymbol{n}} T_L \pi(\boldsymbol{n}) c(\boldsymbol{n}, v)$ contains all the realized perturbations.

Similarly, with Algorithm 2.2, we, in fact, are estimating the performance derivative by

$$\sum_{\text{all } \boldsymbol{n}} \pi(\boldsymbol{n}) c^{(f)}(\boldsymbol{n}, v) \approx \frac{\Delta F}{T_L}.$$

Again, $T_L \pi(\boldsymbol{n})$ is proportional to the perturbation generated in state $\boldsymbol{n}$, and ΔF reflects the differences in performance realized due to all these perturbations.

Example 2.10. Consider the mean response time $\bar{\tau}$ in an M/G/1 queue in Example 2.4. Let $f = n$ be the number of customers in the server. We have $F_L = \int_0^{T_L} n(t) dt$, and $\bar{\tau} = \lim_{L \to \infty} \frac{F_L}{L}$. (In the definition of $\bar{\tau}$, L should be the number of departures. However, since the number of arrivals roughly equals that of the departures, we may take L be the number of all transitions, including both arrivals and departures, and the normalized derivative will be the same.) Suppose that the arrival rate does not change but the service rate changes. Then, the perturbation generation rule is (2.99), i.e., the perturbations of the service times are proportional to the service times. At a service completion time, the system state changes from n to $n-1$, $n > 0$, so $f(\boldsymbol{n}) - f(\boldsymbol{n}') = 1$ in Algorithm 2.2. The perturbations at the service completion times are calculated in (2.100). Thus, the perturbation of F_L calculated by Algorithm 2.2 is

$$\Delta F = \sum_{k=1}^{K} \sum_{i=1}^{n_k} \sum_{l=1}^{i} s_{k,l},$$

where K is the number of busy periods in the L transitions, and n_k is the number of customers served in the kth busy period. (The real change in F_L should be $\kappa \Delta F_L$.) Finally, we have

$$\frac{\mu}{\eta^{(I)}} \frac{\partial \bar{\tau}}{\partial \mu} \approx -\frac{\Delta F}{T_L} = -\frac{1}{T_L} \sum_{k=1}^{K} \sum_{i=1}^{n_k} \sum_{l=1}^{i} s_{k,l}. \tag{2.111}$$

It can be proved that the right-hand side of (2.111) is indeed a strongly consistent estimate of its left-hand side (see the discussions in [103, 104, 146, 234, 235] and Problem 2.32). □

Comparison of PA of Queueing Systems and PA of Markov Chains

In PA of queueing systems, a small (*"infinitesimal"*, the exact meaning of this word will become clear in the next section) change in a system parameter (such as the mean service time of a server) induces a series of small (infinitesimal) changes of the state transition times on a sample path; each such change is called a perturbation (perturbation generation). These perturbations will be propagated along the sample path and affect the transition times of other state transitions (perturbation propagation). For irreducible networks, the effect of such a small perturbation on a sample path cannot continue forever; eventually, a perturbation will be either realized or lost on any sample path (perturbation realization). The average effect of each perturbation on the system performance can be precisely measured by a quantity called the perturbation realization factor (PRF). The total effect of a small change in a system parameter on the system performance can then be calculated by adding together the average effects of all the perturbations induced by the parameter change. The derivative of the performance with respect to the parameter can then be determined.

In PA of Markov chains, a small change in a system parameter (such as the transition probability matrix) induces a series of changes in the state transitions on a sample path; each such change is a perturbation and is also called a "jump" to intuitively reflect its discrete and finite nature. Thus, in PA of queueing systems a perturbation is an "infinitesimal" change on a sample path; while in PA of Markov chains, it is a finite change on a sample path. Moreover, perturbation propagation is not so distinct in PA of Markov chains, although we may view the Markov system as in a propagation period before the perturbed sample path merges with the original one. The perturbation realization principle and the calculation of performance derivatives for Markov chains are essentially the same as those for queueing systems: a single perturbation (jump) can only affect the system in a finite period (until the perturbed path merges with the original one), and its effect on the system performance can be measured by PRF, and so on.

In general, given a sample path of any system, we may first examine how a parameter change induces perturbations on a sample path (perturbation generation) and then determine how each perturbation affects the system performance (perturbation realization). During this process, we may explore how the system dynamics may help in determining the evolution of perturbations and whether there are simple propagation rules. These PA principles are illustrated in Figure 2.20. Again, this approach is of an intuitive nature and the results obtained need to be rigorously proved (cf. Section 2.4.4).

2.4.4 Remarks on Theoretical Issues*

The previous subsections provided an intuitive explanation for PA. The results have to be theoretically studied in a probability and statistical framework.

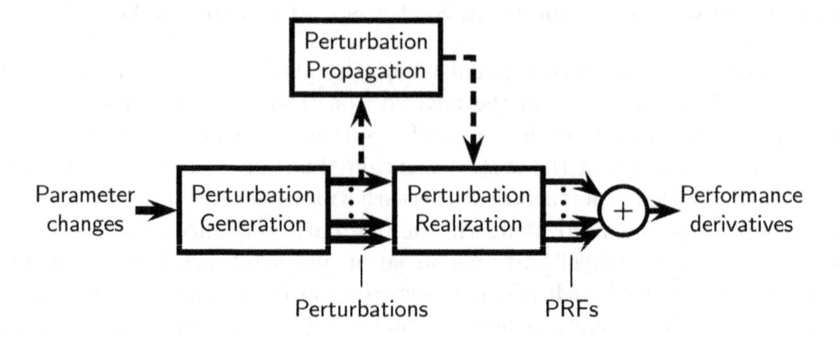

Fig. 2.20. The PA Principles

For example, if we use $-\frac{\Delta_u}{T_L}$ in (2.101) as an estimate for the elasticity of the throughput with respect to a mean service time, then is this estimate strongly consistent as L goes to infinity? Furthermore, is an estimate for performance derivative obtained in a finite sample path an unbiased estimate? These issues were first formulated and studied in [42], and later they were studied for many problems by many authors, and [51] provides a detailed summary of the theory. It is out of the scope of this book to discuss all these issues in detail and we give only a brief review here.

Sample Functions and Sample Derivatives

First let us mathematically describe a sample path of a closed Jackson network obtained by simulation. With the inverse-transform method (2.96), the lth service time at server i, $s_{i,l}$, can be obtained by a uniformly distributed random number $\xi_{i,l} \in [0,1)$ with $s_{i,l} = -\bar{s}_i \ln(1-\xi_{i,l})$. Therefore, all the service times on a sample path depend on a sequence of random numbers $\{\xi_{1,1}, \xi_{1,2}, \ldots; \xi_{2,1}, \xi_{2,2}, \ldots; \ldots; \xi_{M,1}, \xi_{M,2}, \ldots\}$; they are independent and uniformly distributed on $[0,1)$. After the completion of its service, a customer at server i will move to server j, $j = 1, 2, \ldots, M$, with probability $q_{i,j}$. Thus, the next destination of the lth customer at server i can be determined by another uniformly distributed random number $\zeta_{i,l} \in [0,1)$: if $\sum_{k=1}^{j-1} q_{i,k} \le \zeta_{i,l} < \sum_{k=1}^{j} q_{i,k}$ (with the convention $\sum_{k=1}^{0} q_{i,k} = 0$), then this customer moves to server j. Therefore, all the destinations depend on another sequence of independent and uniformly distributed $[0,1)$ random numbers $\{\zeta_{1,1}, \zeta_{1,2}, \ldots; \zeta_{2,1}, \xi_{2,2}, \ldots; \ldots; \zeta_{M,1}, \zeta_{M,2}, \ldots\}$. Finally, let $\xi = \{\xi_{1,1}, \xi_{1,2}, \ldots; \ldots; \xi_{M,1}, \xi_{M,2}, \ldots; \zeta_{1,1}, \zeta_{1,2}, \ldots; \ldots; \zeta_{M,1}, \zeta_{M,2}, \ldots\}$. Then, ξ represents all the randomness involved in the system. Let $\theta = \{\bar{s}_i, q_{i,j}, i, j = 1, 2, \ldots, M\}$ represent all the parameters in the system. With these notations, a sample path of the system is determined by, and therefore is denoted as, (ξ, θ).

For any fixed integer L, F_L in (2.94) and $\eta_L^{(f)} = \frac{F_L}{L}$ are defined on a sample path and therefore are functions of (ξ, θ). We denote them as $F_L(\xi, \theta)$

and $\eta_L^{(f)}(\xi,\theta)$. As we can see from (2.96)-(2.97), in a perturbed sample path, the same sequence of random numbers ξ is used, but the parameters may experience a small change. Thus, a perturbed sample path is in fact $(\xi,\theta+\Delta\theta)$. The perturbed performance is $F_L(\xi,\theta+\Delta\theta)$. Of course, $\Delta\theta$ may be zero for many of its components. In the Jackson network studied in this section, we only choose $\Delta\bar{s}_v \neq 0$ for server v. Since we are concerned with the performance derivatives, for notational simplicity, let us assume that θ is a scalar parameter that changes to $\theta+\Delta\theta$, $\Delta\theta \neq 0$.

The perturbation generation and propagation rules help us to construct the perturbed sample path $(\xi,\theta+\Delta\theta)$ from the original sample path (ξ,θ) for a small $\Delta\theta$ (by Algorithm 2.1), and then to obtain the perturbed performance $F_L(\xi,\theta+\Delta\theta)$ (by Algorithm 2.2). We have

$$\Delta F_L(\xi,\theta) = F_L(\xi,\theta+\Delta\theta) - F_L(\xi,\theta)$$

and

$$\Delta\eta_L^{(f)}(\xi,\theta) = \frac{1}{L}[F_L(\xi,\theta+\Delta\theta) - F_L(\xi,\theta)]. \tag{2.112}$$

For any fixed ξ, $\eta_L^{(f)}(\xi,\theta)$ or $F_L(\xi,\theta)$ is a function of θ. We call it a *sample performance function* [46, 51].

When we apply the propagation rules, we require the perturbation of any server, Δ, to be small enough. In fact, Δ should be smaller than the length of an idle period in order for the perturbation Δ to be propagated through the idle period without changing its size. Figure 2.21 shows the situation when a perturbation is larger than an idle period. Figure 2.21.A illustrates the same sample path as Figure 2.15, except that the perturbation Δ_1 is larger than the length of the idle period $T_2 - T_1$. Figure 2.21.B illustrates the corresponding perturbed path. Indeed, when Δ_1 is larger than $T_2 - T_1$, the idle period in server 1 disappears in the perturbed path and a new idle period appears in server 2. The order of the transition times of server 1 and server 2 changes: $T_2 > T_1$ in the original path, but $T_1' > T_2'$ in the perturbed one. Both servers are delayed by $\Delta_1 - (T_2 - T_1)$ after the idle period. All these facts indicate that the simple propagation rules used in Algorithm 2.1 do not apply.

In fact, Algorithm 2.2 requires a more strict condition: the perturbation of any server in $[0, T_L)$ should be smaller than the shortest sojourn time of the system in any state in $[0, T_L)$. For any finite L and a fixed sample path (ξ,θ), we can always choose (with probability 1) $\Delta\theta$ to be small enough such that this requirement is satisfied (this explains the meaning of infinitesimal). Thus, PA Algorithm 2.2 provides the exact value of $\Delta\eta_L^{(f)}(\xi,\theta)$ in (2.112) if $\Delta\theta$ is small enough. That is, what we obtained from PA is in fact the derivative of a sample performance function, which is called a *sample derivative* [46, 51]:

$$\frac{\partial\eta_L^{(f)}(\xi,\theta)}{\partial\theta} = \lim_{\Delta\theta\to 0}\frac{\eta_L^{(f)}(\xi,\theta+\Delta\theta) - \eta_L^{(f)}(\xi,\theta)}{\Delta\theta}, \qquad \text{for a fixed } \xi.$$

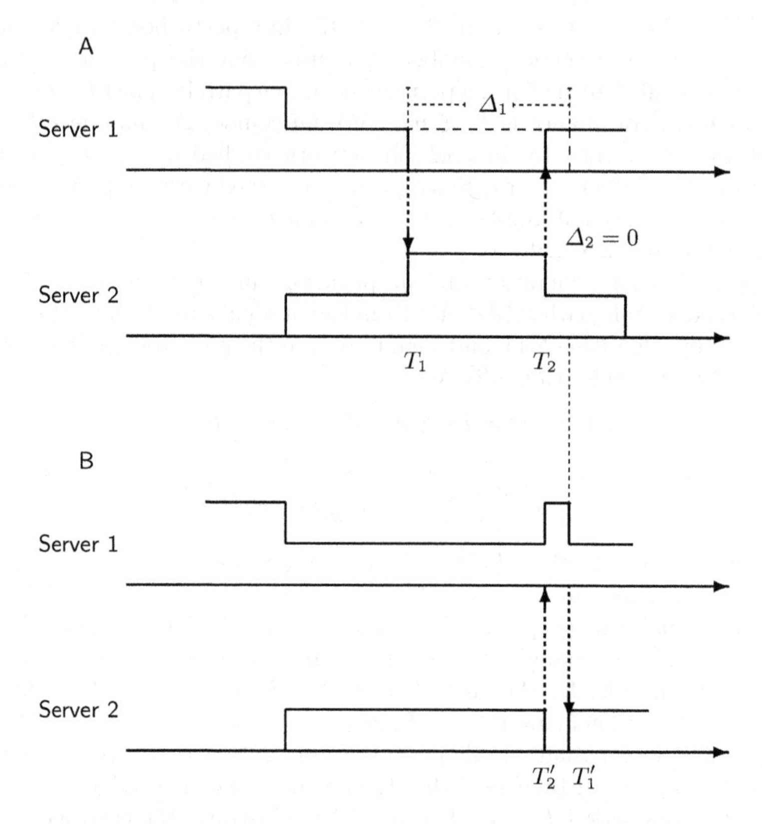

Fig. 2.21. A Large Perturbation Does Not Satisfy the Propagation Rule

Interchangeability

In general, however, we are interested in the derivative of the mean performance $E[\eta_L^{(f)}(\xi,\theta)]$, $\frac{\partial E[\eta_L^{(f)}(\xi,\theta)]}{\partial\theta}$, or the derivative of the steady-state performance $\eta^{(f)}(\theta) = \lim_{L\to\infty}\eta_L^{(f)}(\xi,\theta)$, $\frac{\partial\eta^{(f)}(\theta)}{\partial\theta}$. This raises two questions: Is the sample derivative obtained by PA on a sample path in a finite period $[0,T_L)$ an unbiased estimate? That is, for any $L<\infty$, does

$$E\left\{\frac{\partial}{\partial\theta}[\eta_L^{(f)}(\xi,\theta)]\right\} = \frac{\partial}{\partial\theta}\left\{E[\eta_L^{(f)}(\xi,\theta)]\right\}? \qquad (2.113)$$

Also, is it a strong consistent estimate? That is, does

$$\lim_{L\to\infty}\left\{\frac{\partial}{\partial\theta}[\eta_L^{(f)}(\xi,\theta)]\right\} = \frac{\partial\eta^{(f)}}{\partial\theta} = \frac{\partial}{\partial\theta}\left\{\lim_{L\to\infty}[\eta_L^{(f)}(\xi,\theta)]\right\}? \quad (2.114)$$

In calculus, (2.113) or (2.114) means that the order of the two operators "E" and "$\frac{\partial}{\partial\theta}$", or "$E$" and "$\lim_{L\to\infty}$", is interchangeable. This interchangeability requires some conditions on the sample performance function. The following simple example gives some ideas about why such interchangeability may not hold for some systems.

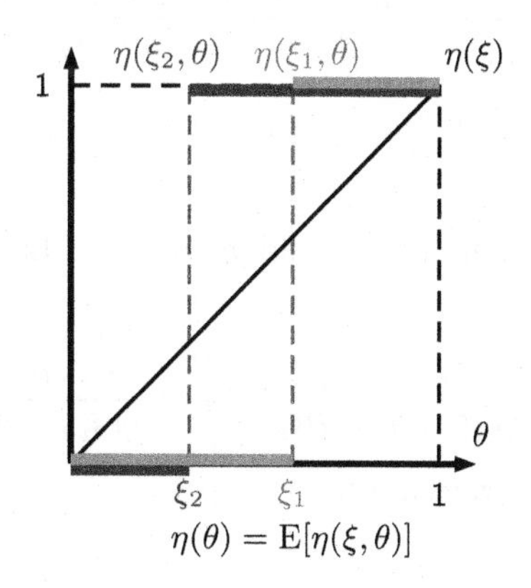

Fig. 2.22. A Sample Function That Does Not Satisfy Interchangeability

Example 2.11. Consider a sample function defined as

$$\eta(\xi,\theta) = \begin{cases} 1, & \text{if } \theta > \xi, \\ 0, & \text{otherwise}, \end{cases} \quad (2.115)$$

where ξ is a uniformly distributed random variable in $[0,1)$. η equals 1 if $\theta \in [\xi, 1)$ and 0 if $\theta \in [0,\xi)$. Two such sample paths corresponding to ξ_1 and ξ_2 are illustrated in Figure 2.22. The mean performance is $\eta(\theta) = E[\eta(\xi,\theta)] = \theta$. The sample derivative is the slope of the sample function $\eta(\xi,\theta)$, which equals 0 with probability 1. Therefore, we have

$$E\left\{\frac{\partial}{\partial\theta}[\eta(\xi,\theta)]\right\} = 0 \neq \frac{\partial}{\partial\theta}\{E[\eta(\xi,\theta)]\} = 1.$$

That is, the interchangeability does not hold for this sample function. □

Fortunately, we can prove that for closed Jackson networks with any finite reward function $f(\boldsymbol{n})$, $\boldsymbol{n} \in \mathcal{S}$, it does hold [46, 51]

$$E\left\{\frac{\partial}{\partial \bar{s}_v}\left[\eta_L^{(f)}(\xi,\bar{s}_v)\right]\middle| X_0=\boldsymbol{n}_0\right\}=\frac{\partial}{\partial \bar{s}_v}E\left\{\left[\eta_L^{(f)}(\xi,\bar{s}_v)\right]\middle| X_0=\boldsymbol{n}_0\right\},$$

where $\boldsymbol{n}_0$ is any initial state. This equation shows that the sample derivative provided by PA, $\frac{\partial}{\partial \bar{s}_v}[\eta_L^{(f)}(\xi,\bar{s}_v)]$, is unbiased for the derivative of the mean (transient) average reward in $[0,T_L)$. In particular, when $f\equiv I$, we have [49]

$$E\left\{\frac{\partial}{\partial \bar{s}_v}[T_L(\xi,\bar{s}_v)]\middle| X_0=\boldsymbol{n}_0\right\}=\frac{\partial}{\partial \bar{s}_v}E\{[T_L(\xi,\bar{s}_v)]|X_0=\boldsymbol{n}_0\}.$$

For long-run average rewards, we also have [51]

$$\lim_{L\to\infty}\left[\frac{\bar{s}_v}{\eta_L^{(I)}(\xi,\bar{s}_v)}\frac{\partial\eta_L^{(f)}(\xi,\bar{s}_v)}{\partial\bar{s}_v}\right]=\frac{\bar{s}_v}{\eta^{(I)}(\bar{s}_v)}\frac{\partial\eta^{(f)}(\bar{s}_v)}{\partial\bar{s}_v}$$
$$=\sum_{\text{all }\boldsymbol{n}}\pi(\boldsymbol{n})c^{(f)}(\boldsymbol{n},v),\qquad \text{w.p.1},\tag{2.116}$$

where $\eta^{(f)}(\bar{s}_v)=\lim_{L\to\infty}\eta_L^{(f)}(\xi,\bar{s}_v)$ and $\eta^{(I)}(\bar{s}_v)=\lim_{L\to\infty}\eta_L^{(I)}(\xi,\bar{s}_v)$. In particular, we have

$$\lim_{L\to\infty}\left[\frac{\bar{s}_v}{\eta_L(\xi,\bar{s}_v)}\frac{\partial\eta_L(\xi,\bar{s}_v)}{\partial\bar{s}_v}\right]=\frac{\bar{s}_v}{\eta(\bar{s}_v)}\frac{\partial\eta(\bar{s}_v)}{\partial\bar{s}_v}$$
$$=-\sum_{\text{all }\boldsymbol{n}}\pi(\boldsymbol{n})c(\boldsymbol{n},v),\qquad \text{w.p.1},\tag{2.117}$$

where $\eta(\bar{s}_v)=\lim_{L\to\infty}\eta_L(\xi,\bar{s}_v)$, and $\eta_L(\xi,\bar{s}_v)=\frac{L}{T_L(\xi,\bar{s}_v)}$. That is, the normalized sample derivatives provided by PA are strongly consistent estimates of the normalized derivatives of the steady-state performance.

However, the nice properties of unbiasedness and strong consistency do not always hold. As illustrated in Example 2.11, the interchangeability may not hold if the sample functions are discontinuous. Roughly speaking, the interchangeability in (2.113) requires that the sample performance functions be "smooth" enough.

The sample derivatives of the performance with respect to the changes in routing probabilities $q_{i,j}$, $i,j=1,2,\ldots,M$, have discontinuities similar to those in Example 2.11. To demonstrate the idea, we consider a closed network and assume that its service time distributions do not change. A sample path of such a network is determined by the random variables $\zeta:=\{\zeta_{1,1},\zeta_{1,2},\ldots;\zeta_{2,1},\xi_{2,2},\ldots;\ldots;\zeta_{M,1},\zeta_{M,2},\ldots\}$, and therefore we may denote a sample path as $(\zeta,q_{i,j},i,j=1,2,\ldots,M)$. For the sake of discussion, we assume that $q_{1,2}$ and $q_{1,3}$ change to $q'_{1,2}=q_{1,2}+\delta$ and $q'_{1,3}=q_{1,3}-\delta$, respectively. As we know, in simulation, the customer transition is determined as follows: we first divide the interval $[0,1]$ into M small segments, each with length $q_{1,i}$, $i=1,2,\ldots,M$ (see Figure 2.23 for $M=3$). If $\zeta_{1,l}$ falls in the kth segment, then the lth customer at server 1 moves to server k. When $q_{1,2}$

and $q_{1,3}$ change to $q'_{1,2}$ and $q'_{1,3}$, respectively, the only change happens when $\zeta_{1,l}$ falls in the small segment with length δ (in the middle of the period $[0,1]$ shown in Figure 2.23). In this case, the customer moves to server 3 in the original sample path but to server 2 in the perturbed path. It is clear that for a fixed realization of ζ and a finite L, there is always (with probability 1) a δ_0 that is small enough such that no $\zeta_{1,l}$, $l = 1, 2, \ldots, L$, falls in that small segment. Therefore, the two sample paths $(\zeta, q_{i,j})$ and $(\zeta, q'_{i,j})$ with $\delta < \delta_0$ are exactly the same in $[0, T_L]$. This implies that the sample function $F_L(\zeta, q_{i,j})$ is a piecewise constant function of $q_{i,j}$. As shown in Example 2.11, the interchangeability in (2.113) does not hold for the derivative of performance $F_L(\zeta, q_{i,j})$ with respect to δ (or the changes in $q_{1,2}$ and $q_{1,3}$).

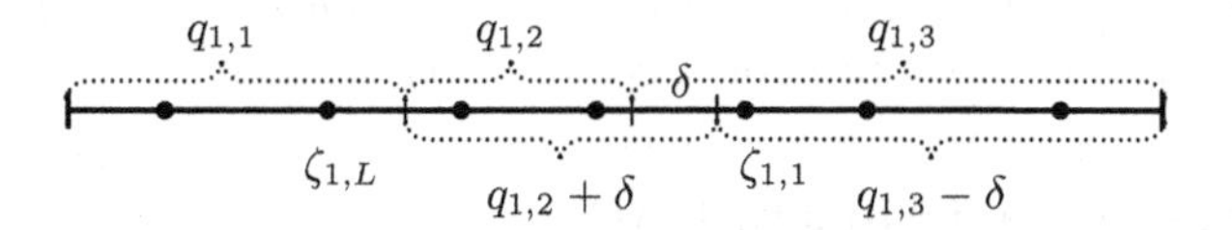

Fig. 2.23. Determine the Customer Transitions

The PA of queueing systems introduced in this section is based on sample derivatives. This approach requires interchangeability, which may not hold if the sample function is not continuous. The discontinuity of a sample function can be explained from a sample path point of view. Essentially, if a small change in a parameter may cause a big change in a sample path, the sample function may be discontinuous. In the case with the routing probabilities, a small change in $q_{1,2}$ (or $q_{1,3}$) may cause a big change in a customer's destination (from server 2 to server 3). Such a big change also occurs when two transitions exchange their order of occurrence, leading to two different states. This sample-path-based explanation gives us an intuitive feeling about whether the discontinuity may exist (see [42] and [126] for more details).

Other examples where the interchangeability does not hold include queueing networks with multi-class customers or with blocking due to finite buffer sizes. They can also be explained by the intuitive explanation described above (see [43] and [126] for more discussion).

For the same reason, the sample performance functions for Markov systems with respect to the transition probability matrix are also piecewise linear, and the sample-derivative is therefore zero and the approach discussed in this section does not apply. However, as shown in Section 2.1, the basic principle of perturbation generation and perturbation realization can be extended to Markov systems. The derivative obtained by using realization factors for Markov systems is not a sample derivative.

Similar results regarding the sample functions and sample derivatives for systems with continuous state spaces are presented in [47], and a comparison of the dynamics of the continuous and discrete event systems is given in [48].

2.5 Other Methods*

Much effort was expended in the 1980's to overcome the difficulty caused by the discontinuity of the sample functions for some systems. Different approaches were developed; these approaches work well for some special problems. Among them are the *smoothed perturbation analysis (SPA)* [105, 107, 114, 119], the *finite perturbation analysis (FPA)* [143], the *rare perturbation analysis (RPA)* [33, 34, 36, 37]. There are also other related works [73, 94, 102, 106, 108, 128, 147, 156, 185, 214, 232, 241, 245, 247, 251]. These topics have been widely discussed in previous books [51, 72, 107, 112, 142], and therefore we will not discuss them in this book.

In this section, we will briefly review some other methods of performance sensitivity analysis. They are the *stochastic fluid model*, the *weak derivative method*, and the *likelihood ratio* or *score function method.*

The Stochastic Fluid Model (SFM)

The stochastic fluid model (SFM) has been recently adopted to model complex, discrete-event dynamic systems such as communication networks, and perturbation analysis has been proposed in SFM as a means for sensitivity analysis. The essential idea of this method is to use a continuous flow to approximately model the packet transmission in a network. Since in communication a data or voice packet consists of small units called bits, SFM is particularly suitable for communication systems.

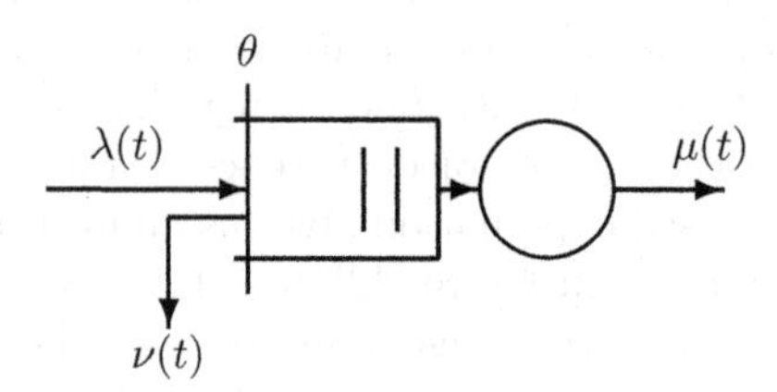

Fig. 2.24. The Stochastic Fluid Model for a Single Queue with Buffer Size θ

Figure 2.24 illustrates a stochastic fluid model for a single queue. The inflow rate and the processing rate at time t are denoted as $\lambda(t)$ and $\mu(t)$ (units/second), respectively; and we use θ (units) to denote the size of the buffer. When the buffer is full, the incoming fluid will overflow, and we denote its rate at $\nu(t)$. Let $x_\theta(t)$ be the volume of the fluid in the buffer. Apparently, the system dynamic can be modelled by

$$\frac{dx_\theta(t)}{dt} = \begin{cases} 0, & \text{if } x_\theta(t) = 0, \text{ and } \lambda(t) \le \mu(t), \\ 0, & \text{if } x_\theta(t) = \theta \text{ and } \lambda(t) > \mu(t), \\ \lambda(t) - \mu(t), & \text{otherwise.} \end{cases}$$

Sample-path-based analysis can be applied to such a SFM to obtain an estimate for the performance derivative. The approach is more suitable (although approximate) for estimating the gradients of packet loss probability with respect to the buffer size. It can be shown that such estimates are unbiased for the derivatives of the performance obtained with the SFM model. Such problems are usually difficult to handle with the standard PA of queueing systems. For more details and applications, see [74, 75, 189, 210, 211, 231, 252, 262, 263].

The Likelihood Ratio (Score Function) Method

Another performance derivative estimation method is called the *likelihood ratio* method, [44, 115, 116, 117, 118, 130, 176, 177, 178, 179, 205, 217], also called the *score function* method [221, 222].

To illustrate the main idea, let us consider a D/M/1 queue in which the inter-arrival time is a fixed number $D > 0$ and the service times are independent and exponentially distributed with mean $\bar{s}$. Let $s_1, s_2, \ldots, s_L$ be the sequence of customers' service times. Then, a sample path of the system can be represented by, and therefore denoted as, a vector $s := (s_1, s_2, \ldots, s_L)$ (instead of in the form of (ξ, θ)). The performance defined on this sample path is denoted as $\eta(s)$. Let $\Phi(s, \theta)$ be the distribution function of s, where $\theta = \bar{s}$ denotes the system parameter. (To help our understanding, we may view s as a scalar variable, otherwise, $\Phi(s, \theta)$ is the joint distribution of $s_1, \ldots, s_L$.) The mean performance is

$$\eta_\theta = E[\eta(s)] = \int_{-\infty}^{\infty} \eta(s) d\Phi(s, \theta). \tag{2.118}$$

Our goal is to estimate the derivative $\frac{d\eta_\theta}{d\theta}$.

In PA, we set $\xi := \Phi(s, \theta)$ to be a $[0, 1)$ uniformly distributed random variable. Then, we have $s = \Phi^{-1}(\xi, \theta)$, and, for notational convenience, we denote it as $s = \Phi^{-1}(\xi, \theta) := s(\xi, \theta)$. Thus, we have

$$\eta_\theta = \int_0^1 \eta[s(\xi, \theta)] d\xi.$$

As explained in Section 2.4.4, for any fixed $\xi \in [0, 1)$, $\eta[s(\xi, \theta)]$ is called a sample performance function. In PA, we use the sample derivative $\frac{d}{d\theta}\eta[s(\xi, \theta)]$ as an estimate of $\frac{d\eta_\theta}{d\theta}$. The issue is whether or not this estimate is unbiased, i.e., whether or not (cf. (2.113))

$$\frac{d\eta_\theta}{dt} = \frac{d}{d\theta}\left\{\int_0^1 \eta[s(\xi, \theta)] d\xi\right\} = \int_0^1 \frac{d}{d\theta}\{\eta[s(\xi, \theta)]\} d\xi?$$

In the sample derivative $\frac{d}{d\theta}\eta[s(\xi, \theta)]$, the same random variable ξ is used for both $\eta[s(\xi, \theta)]$ and $\eta[s(\xi, \theta + \Delta\theta)]$. Thus, PA is also called it a *common random number (CRN)* method. It is known that using the common random number leads to the smallest variance in estimating the difference between two random variables (See Problem A.4). Therefore, the sample derivative usually has a small variance.

The rationale of the likelihood ratios method is as follows: Suppose that the probability density function of $\Phi(s,\theta)$ exists and denote it as $\phi(s,\theta) = \frac{d}{ds}\Phi(s,\theta)$. Then, (2.118) becomes

$$\eta(\theta) = \int_{-\infty}^{\infty} \eta(s)\phi(s,\theta)ds,$$

and we have, assuming that the two operators $\int$ and $\frac{d}{d\theta}$ can change their order,

$$\begin{aligned} \frac{d\eta_\theta}{d\theta} &= \int_{-\infty}^{\infty} \eta(s)\frac{d\phi(s,\theta)}{d\theta}ds \\ &= \int_{-\infty}^{\infty} \eta(s)\frac{d\phi(s,\theta)}{d\theta}ds \\ &= \int_{-\infty}^{\infty} \eta(s)\frac{d\ln[\phi(s,\theta)]}{d\theta}d\Phi(s,\theta) \\ &= E\left\{\eta(s)\frac{d\ln[\phi(s,\theta)]}{d\theta}\right\}. \end{aligned} \quad (2.119)$$

This indicates that we may use

$$\eta(s)\frac{d\ln[\phi(s,\theta)]}{d\theta} \quad (2.120)$$

as an unbiased estimate of the performance derivative $\frac{d\eta_\theta}{d\theta}$. In (2.120), we have

$$\frac{d\ln[\phi(s,\theta)]}{d\theta} = \frac{1}{\phi(s,\theta)}\frac{d[\phi(s,\theta)]}{d\theta}.$$

Observe that

$$\eta(s)\frac{d\ln[\phi(s,\theta)]}{d\theta} = \lim_{\Delta\theta\to 0}\frac{1}{\Delta\theta}\left\{\eta[s(\xi,\theta)]\frac{\phi(s,\theta+\Delta\theta)}{\phi(s,\theta)} - \eta[s(\xi,\theta)]\right\}. \quad (2.121)$$

Therefore, in the LR estimate (2.120), we in fact use

$$\eta[s(\xi,\theta)]\frac{\phi(s,\theta+\Delta\theta)}{\phi(s,\theta)}$$

in the place of $\eta[s(\xi,\theta+\Delta\theta)]$. The reason is that if the system parameter changes from θ to $\theta+\Delta\theta$, the same sample path s, and therefore the same sample performance value $\eta(s)$, will still be observed, but with a different probability that is adjusted by the likelihood ratio

$$\frac{\phi(s, \theta + \Delta\theta)}{\phi(s, \theta)}$$

(see [44] for more discussion). Therefore, this approach is called the *likelihood ratio (LR)*, or the *score function (SF)* method.

From (2.121), the LR estimate essentially uses the same sample path s as a possible realization of the system behavior under parameters $\theta + \Delta\theta$ and adjusts the probability of this sample path; hence, the LR method is also called the *common realization (CR)* method.

LR only requires the interchangeability of $\int$ and $\frac{d}{d\theta}$ to hold for the probability density function, which is usually smoother than the sample performance function (the s in $\eta(s)$ in (2.119) is fixed). Thus, an LR estimate is unbiased more often than a PA estimate is. However, the variance may be too large to be applicable [44]. Variance reduction techniques based on regenerative periods have been developed.

The Weak Derivative Method

In the weak derivative method [130, 132, 134], the derivative of the probability density function is expressed by the difference between two properly chosen probability density functions, and the performance derivative is then expressed by the difference between two expected values. For example, in (2.119), if we have $c(\theta) > 0$ and two density functions $\phi_1(s, \theta)$ and $\phi_2(s, \theta)$ such that

$$\frac{d\phi(s, \theta)}{d\theta} = c(\theta)[\phi_1(s, \theta) - \phi_2(s, \theta)]. \tag{2.122}$$

Then,

$$\frac{d\eta(\theta)}{d\theta} = c(\theta)\left[\int_{-\infty}^{\infty} \eta(s)\phi_1(s, \theta)d\theta - \int_{-\infty}^{\infty} \eta(s)\phi_2(s, \theta)d\theta\right],$$

which is the difference between the mean performance of two sample paths, one with probability density function $\phi_1(s, \theta)$ and the other with $\phi_2(s, \theta)$. The triple $(c(\theta), \phi_1(s, \theta), \phi_2(s, \theta))$ is called a *weak derivative* of $\phi(s, \theta)$. Obviously, it is not unique.

The same principle applies to the performance derivatives of Markov chains. Consider two Markov chains defined on the same state space $\mathcal{S} = \{1, 2, \ldots, S\}$ with two ergodic transition probability matrices P and P' and the same reward function f. Let $\Delta P = P' - P$, $P_\delta = P + \delta\Delta P$. We start with (2.23). From (2.13) and (2.14), the directional derivative along ΔP is

$$\frac{d\eta_\delta}{d\delta} = \pi\Delta P \sum_{l=0}^{\infty}(P^l - e\pi)f. \tag{2.123}$$

Corresponding to (2.122), we have

$$\frac{dP_\delta}{d\delta} = \Delta P = C(P^+ - P^-), \tag{2.124}$$

where P^+, P^-, and C are defined as follows: C is a diagonal matrix with nonzero diagonal components $c(i)$, $i = 1, 2, \ldots, S$,

$$c(i) = \sum_{j=1}^{S} \max\{\Delta p(j|i), 0\},$$

$\Delta p(j|i) = p'(j|i) - p(j|i)$, $i, j = 1, 2, \ldots, S$, and

$$p^+(j|i) = \begin{cases} \frac{1}{c(i)} \max\{\Delta p(j|i), 0\}, & \text{if } c(i) > 0, \\ 0, & \text{if } c(i) = 0; \end{cases}$$

$$p^-(j|i) = \begin{cases} \frac{1}{c(i)} \max\{-\Delta p(j|i), 0\}, & \text{if } c(i) > 0, \\ 0, & \text{if } c(i) = 0. \end{cases}$$

When $c(i) \neq 0$, the ith rows of P^+ and P^- are transition probability vectors; and when $c(i) = 0$, the ith rows of P^+ and P^- are zero. The triple (C, P^+, P^-) is called a weak derivative of P_δ. The decomposition of (2.124) is not unique, and there may be other weak derivatives of P_δ.

From (2.124) and $(\Delta P)e = 0$, the derivative (2.123) becomes

$$\begin{aligned} \frac{d\eta_\delta}{d\delta} &= \pi \Delta P \sum_{l=0}^{\infty} P^l f \\ &= \pi C(P^+ - P^-) \sum_{l=0}^{\infty} P^l f \\ &= \sum_{i=1}^{S} \pi(i) c(i) \sum_{l=0}^{\infty} (p_i^+ P^l f - p_i^- P^l f), \end{aligned} \tag{2.125}$$

where p_i^+ and p_i^- denote the ith rows of P^+ and P^-, respectively.

There is a sample-path-based interpretation of (2.125). Let $\boldsymbol{X}^+ = \{X_l^+, l = 0, 1, \ldots\}$ be a Markov chain obtained as follows: Suppose that $X_0^+ = i$ is the initial state, and the first transition from X_0^+ to X_1^+ follows transition probability vector p_i^+, and the rest of the transitions at $l = 1, 2, \ldots$ follow transition probability matrix P. Let $\boldsymbol{X}^-$ be a similar Markov chain except that the first transition from X_0^- to X_1^- follows p_i^-, with $X_0^- = i$. From (2.125), we have

$$\frac{d\eta_\delta}{d\delta} = \sum_{i=1}^{S} \pi(i) c(i) \sum_{l=0}^{\infty} E\{[f(X_l^+) - f(X_l^-)] | X_0^+ = X_0^- = i\}. \tag{2.126}$$

Therefore, the performance derivative can be expressed via the difference between two expectations on two different Markov chains $\boldsymbol{X}^+$ and $\boldsymbol{X}^-$. Furthermore, by the strong Markov property, the infinite sum $\sum_{l=0}^{\infty}$ can be replaced by a finite one $\sum_{l=0}^{L_{+,-}}$; at $L_{+,-}$, the two sample paths $\boldsymbol{X}^+$ and $\boldsymbol{X}^-$ merge together.

The form of (2.126) resembles the performance realization factors. In fact, from (2.126) we can easily derive (see [130])

$$\frac{d\eta_\delta}{d\delta} = \sum_{i=1}^{S} \pi(i)c(i) \left[\sum_{j_1,j_2=1}^{S} \gamma(j_2,j_1)p^+(j_1|i)p^-(j_2|i) \right]. \tag{2.127}$$

PROBLEMS

2.1. In Figure 2.2, the three points P_0, P_1, and P_2 represent three policies. Every point P in the triangle with these three points as vertices represents a randomized policy denoted as $P(\delta_1, \delta_1, \delta_2) = \delta_0 P_0 + \delta_1 P_1 + \delta_2 P_2$, $\delta_0 + \delta_1 + \delta_2 = 1$, with $P_0 = P(1,0,0)$, $P_1 = P(0,1,0)$, and $P_2 = P(0,0,1)$.

a. Determine the values of δ_0, δ_1, and δ_2 by the lengths of the segments shown in the figure.
b. Along the line from P_0 to P_1, we have the randomized policies $P_\delta = (1-\delta)P_0 + \delta P_1$, $0 < \delta < 1$, and we can obtain the directional derivative in this direction, denoted as $\frac{d\eta_\delta}{d\delta}|_{P_0-P_1}$. Similarly, we can obtain the directional derivative in the direction from P_0 to P_2, denoted as $\frac{d\eta_\delta}{d\delta}|_{P_0-P_2}$. What is the directional derivative from P_0 to P? Express it in terms of $\frac{d\eta_\delta}{d\delta}|_{P_0-P_1}$ and $\frac{d\eta_\delta}{d\delta}|_{P_0-P_2}$. (*Hint: Along this direction, δ_1/δ_2 is fixed.*)

2.2. (Random walk) A random walker moves among five positions $i = 1,2,3,4,5$. At position $i = 2,3,4$, s/he moves to positions $i-1$ and $i+1$ with an equal probability $p(i-1|i) = p(i+1|i) = 0.5$; at the boundary positions $i = 1$ and $i = 5$, s/he bounces back with probability 1 $p(4|5) = p(2|1) = 1$. We are given a sequence of 20 $[0,1)$-uniformly and independently distributed random variables as follows:

0.740, 0.605, 0.234, 0.342, 0.629, 0.965, 0.364, 0.230, 0.599, 0.079,
0.782, 0.219, 0.475, 0.051, 0.596, 0.850, 0.865, 0.434, 0.617, 0.969.

a. With this sequence, construct a sample path $\boldsymbol{X}$ of the random walk from X_0 to X_{20} according to (2.2). Set $X_0 = 3$.
b. Suppose that the perturbed transition probabilities are $p'(i-1|i) = 0.3$, $p'(i+1|i) = 0.7$, $i = 2,3,4$, and $p'(4|5) = p'(2|1) = 1$. Set $p_\delta(j|i) = p(j|i) + \delta[p'(j|i) - p(j|i)]$. By using the original sample path obtained in (a), construct a perturbed sample path $\boldsymbol{X}_\delta$, $\delta = 1$, following Figure 2.5. Use the following $[0,1)$-uniformly and independently distributed random variables when $\boldsymbol{X}_\delta$ is different from $\boldsymbol{X}$ (use the lth number to determine the lth transition of $\boldsymbol{X}_\delta$, if $X_{\delta,l} \neq X_l$):

0.173, 0.086, 0.393, 0.804, 0.011, 0.233, 0.934, 0.230, 0.786, 0.410,
0.119, 0.634, 0.862, 0.418, 0.601, 0.118, 0.626, 0.835, 0.361, 0.336.

c. Repeat b) for $\delta = 0.7, 0.5, 0.3, 0.2, 0.1$.
d. Observe the trend of the perturbed paths $\boldsymbol{X}_\delta$. In particular, when δ is small, most likely the perturbed parts from the jumping point to the merging point are the same as if they follow the original transition probabilities $p(j|i)$, $i, j = 1, 2, \ldots, \mathcal{S}$.

2.3. Let $\boldsymbol{X}$ and $\widetilde{\boldsymbol{X}}$ be two independent ergodic Markov chains with the same transition probability matrix P on the same state space $\mathcal{S}$. Define $\boldsymbol{Y} = (\boldsymbol{X}, \widetilde{\boldsymbol{X}})$.

a. Prove that $\boldsymbol{Y}$ is ergodic.
b. Express L^*_{ij} in Figure 2.6 in terms of the Markov chain $\boldsymbol{Y}$.

2.4. Consider a three-state Markov chain with

$$P = \begin{bmatrix} 0 & 0.5 & 0.5 \\ 0.1 & 0.6 & 0.3 \\ 0.7 & 0.1 & 0.2 \end{bmatrix}, \qquad f = \begin{bmatrix} 10 \\ 5 \\ 8 \end{bmatrix}.$$

a. Solve the Poisson equation (2.12) $(I - P)g + \eta e = f$ for g and η (by, e.g., setting $g(0) = 0$).
b. Solve $\pi = \pi P$ and $\pi e = 1$ for π first. Then, solve $(I - P + e\pi)g = f$ for g.
c. Compare both methods in a) and b).

2.5. For an ergodic Markov chain $\boldsymbol{X} = \{X_l, l = 0, 1, \ldots\}$, derive the Poisson equation using

$$g(i) = \lim_{L \to \infty} \sum_{l=0}^{L-1} E\{[f(X_l) - \eta] | X_0 = i\}.$$

2.6. The Poisson equation for the perturbed Markov chain is

$$(I - P_\delta)g_\delta + \eta_\delta e = f_\delta,$$

where $P_\delta = P + \delta \Delta P$ and $f_\delta = f + \delta \Delta f$. Derive the performance derivative formula (2.26) from the above equation.

2.7. Prove the following results:

a. If $f = ce$ with c being a constant, then $g = ce$ is a constant vector.
b. If $p(j|i) = p_j$ for all $i \in \mathcal{S}$; i.e., every row in the transition probability matrix is the same, then $g = f$.
c. If $p(j|i) = p(i|j)$, for all $i, j \in \mathcal{S}$; i.e., the transition probability matrix P is symmetric, then $\sum_{i=1}^{S} g(i) = \sum_{i=1}^{S} f(i)$.

2.8. Prove $e\frac{d\eta_\delta}{d\delta} = \lim_{\beta\uparrow 1} \frac{d\eta_{\beta,\delta}}{d\delta}$. In other words,

$$\frac{d}{d\delta}\left(\lim_{\beta\uparrow 1} \eta_{\beta,\delta}\right) = \lim_{\beta\uparrow 1} \frac{d\eta_{\beta,\delta}}{d\delta}.$$

2.9. Assume that P changes to $P_\delta = P + \delta(\Delta P)$, $\Delta Pe = 0$, and $f_\delta \equiv f$. Derive the second-order derivative of the discounted reward $\eta_{\beta,\delta}$ with respect to δ, $\frac{d^2\eta_{\beta,\delta}}{d\delta^2}$.

2.10. In Example 2.2, we have

$$G_1 := \Delta P(I - P + e\pi)^{-1} = \begin{bmatrix} -3.2 & 3.2 \\ 3.2 & -3.2 \end{bmatrix}.$$

a. Find the eigenvalues and eigenvectors of G_1.
b. Verify that

$$\begin{bmatrix} -3.2 & 3.2 \\ 3.2 & -3.2 \end{bmatrix} = \begin{bmatrix} 1 & 1 \\ 1 & -1 \end{bmatrix} \begin{bmatrix} 0 & 0 \\ 0 & -6.4 \end{bmatrix} \begin{bmatrix} 1 & 1 \\ 1 & -1 \end{bmatrix}^{-1}.$$

c. Prove that

$$G_1^n = \begin{bmatrix} 1 & 1 \\ 1 & -1 \end{bmatrix} \begin{bmatrix} 0 & 0 \\ 0 & (-6.4)^n \end{bmatrix} \begin{bmatrix} 1 & 1 \\ 1 & -1 \end{bmatrix}^{-1},$$

and

$$\pi_\delta = \pi \sum_{n=0}^{\infty} G_\delta^n = \pi \sum_{n=0}^{\infty} (\delta G_1)^n$$
$$= \pi \begin{bmatrix} 1 & 1 \\ 1 & -1 \end{bmatrix} \begin{bmatrix} 0 & 0 \\ 0 & \sum_{n=0}^{\infty}(-6.4\delta)^n \end{bmatrix} \begin{bmatrix} 1 & 1 \\ 1 & -1 \end{bmatrix}^{-1}.$$

d. Determine the convergence region of π_δ. Extend the discussion to more general case.

2.11. A group is a nonempty set G, together with a binary operation on G, denoted as juxtaposition ab, $a, b \in G$, and $ab \in G$, with the following properties: (i) *(Associativity)* $(ab)c = a(bc)$, for all $a, b, c \in G$; (ii) *(Identity)* There exists an element $e \in G$ for which $ea = ae = a$ for all $a \in G$; and (iii) *(Inverse)* For each $a \in G$, there is an element denoted a^{-1}, for which $aa^{-1} = a^{-1}a = e$, [220].

a. Verify that the set of matrices defined in (2.50) with matrix multiplication as the juxtaposition satisfies the above properties.
b. In Example 2.2, we have

$$B = P - I = \begin{bmatrix} -0.10 & 0.10 \\ 0.15 & -0.15 \end{bmatrix}.$$

What is its group inverse? Is the inverse an infinitesimal generator?

2.12. Assume that the MacLaurin series of P_δ exists in $[0, \delta]$. Equation (2.57) can be derived directly by the following procedure: Taking the derivatives of the both sides of $\pi_\delta(I - P_\delta) = 0$ n times, we can obtain $\frac{d^n \pi}{d\delta^n}$ at $\delta = 0$. Then, we can construct the MacLaurin series of π. Work out the details of this approach and derive the MacLaurin series of η_δ at $\delta = 0$.

2.13. Prove the continuous version of the PRF equation (2.62) from its discrete version (2.7) by setting $B = P - I$, and vice versa.

2.14. Consider a Markov chain $\boldsymbol{X}$ with transition probabilities $p(j|i)$, $i, j \in \mathcal{S}$, and reward function f. For any $0 < p < 1$, we define an equivalent Markov chain $\boldsymbol{X}'$ with transition probabilities $p'(j|i) = (1-p)p(j|i)$, $j \neq i$, and $p'(i|i) = p + (1-p)p(i|i)$, $i \in \mathcal{S}$. Set $f' = f$. Prove that $\eta' = \eta$ and $g' = \frac{g}{1-p}$.

2.15. Consider a Markov process $\boldsymbol{X}$ with transition rates $\lambda(i)$, and transition probabilities $p(j|i)$, $i, j \in \mathcal{S}$, and reward function f. For any $\lambda > \lambda(i)$, $i \in \mathcal{S}$, we define an equivalent Markov process $\boldsymbol{X}'$ with transition rates $\lambda'(i) \equiv \lambda$, and transition probabilities $p'(j|i) = \frac{\lambda(i)}{\lambda} p(j|i)$, $j \neq i$, and $p'(i|i) = [1 - \frac{\lambda(i)}{\lambda}] + \frac{\lambda(i)}{\lambda} p(i|i)$. Set $f' = f$.

a. Prove that $\eta' = \eta$ and $g' = g$.
b. Let the discrete-time Markov chain embedded at the transition epoches of $\boldsymbol{X}'$ as $\boldsymbol{X}^\dagger$. Find the steady-state probability $\pi^\dagger$ and the potential $g^\dagger$ of $\boldsymbol{X}^\dagger$.
c. Suppose that $1 = \lambda > \lambda(i)$, $i \in \mathcal{S}$, prove that $g^\dagger = g$.
d. For any $\kappa > 0$, we define a Markov process $\widetilde{\boldsymbol{X}}$ with transition rates $\widetilde{\lambda}(i) = \kappa\lambda(i)$, $i \in \mathcal{S}$, transition probabilities $\widetilde{p}(j|i) = p(j|i)$, $i, j \in \mathcal{S}$, and reward function $\widetilde{f} = f$. Prove that $\widetilde{\pi} = \pi$ and $\widetilde{g} = \frac{g}{\kappa}$.
e. Given any Markov process $\boldsymbol{X}$, can you find a Markov chain that has the same steady-state probability π and potential g as $\boldsymbol{X}$? (*Hint: use the results in b)-d).*)

2.16.* For semi-Markov processes with the discounted reward defined in (2.93), set $\eta_\beta := (\eta_\beta(1), \ldots, \eta_\beta(S))^T$ and $g_\beta := (g_\beta(1), \ldots, g_\beta(S))^T$. Prove that (cf. [57])

$$\lim_{\beta \downarrow 0} g_\beta = g,$$

$$\lim_{\beta \downarrow 0} \eta_\beta = \eta e,$$

and

$$\eta_\beta = \beta g_\beta + \eta e.$$

2.17. Consider a two-server cyclic Jackson queueing network with service rates μ and λ for servers 1 and 2, respectively. There are N customers in the network. The system's state $\boldsymbol{n} = n$ is the number of customers at server 1. The state process is Markov. Let the performance be the average response time of the customers at server 1, denoted as $\bar{\tau}$. Calculate the performance potentials $g(i)$, $i = 1, 2, \ldots, S$, and the average response time $\bar{\tau}$, and derive the derivative of $\bar{\tau}$ with respect to λ and μ.

2.18. The two-server N-customer cyclic Jackson queueing network studied in Problem 2.17 is equivalent to an $M/M/1/N$ queue with arrival rate λ, service rate μ, and a finite buffer size N. (When the number of customers in the queue is $n = N$, an arriving customer is simply lost.)

a. Suppose that the arrival rate only changes when $n = 0$; i.e., when $n = 0$, λ changes to $\lambda + \Delta\lambda$, and when $n > 0$, λ remains unchanged. What is the derivative of the average response time $\bar{\tau}$ with respect to this change?
b. Suppose that the arrival rate only changes when $n = n^*$, with $0 < n^* < N$. What is the derivative of $\bar{\tau}$ with respect to this change?
c. Suppose that the arrival rate only changes when $n = N$. What is the derivative of $\bar{\tau}$ with respect to this change? (You may view the $M/M/1/N$ queue as the two-server cyclic queue again to verify your result.)

2.19. Consider a Markov chain with one closed recurrent state set $\mathcal{S}_1$ and one transient state set $\mathcal{S}_2$ (a uni-chain). Let the transition probability matrix be

$$P = \begin{bmatrix} P_1 & 0 \\ P_{21} & P_{22} \end{bmatrix},$$

with P_1 corresponding to $\mathcal{S}_1$ and P_{21}, P_{22} corresponding to $\mathcal{S}_2$, and 0 being a matrix with all zero components. Denote the potential vector as $g = (g_1^T, g_2^T)^T$ with $g_1 = (g(1), \ldots, g(S_1))^T$ and $g_2 = (g(S_1 + 1), \ldots, g(S))^T$, $S_1 = |\mathcal{S}_1|$, $S_2 = |\mathcal{S}_2|$, $S_1 + S_2 = S$.

Derive an equation for g_1 and express g_2 in terms of g_1 and P_{21}, P_{22}.

2.20. Consider a Markov chain with transition probability matrix

$$P = \begin{bmatrix} B & b \\ 0 & 1 \end{bmatrix},$$

where B is an $(S-1) \times (S-1)$ irreducible matrix, $b > 0$ is an $(S-1)$ dimensional column vector, 0 represents an $(S-1)$-dimensional row vector whose components are all zero. The last state S is an absorbing state. Clearly, the long-run average reward for this Markov chain is $\eta = f(S)$, independent of B, b, and the initial state. Thus, the long-run average reward does not reflect the transient behavior. Now, we set $f(S) = 0$. Define

$$g(i) = E\left[\sum_{l=0}^{\infty} f(X_l) \middle| X_0 = i\right].$$

Let $L_{i,S} = \min\{l : l \geq 0, X_l = S | X_0 = i\}$ be the first passage time from i to S. Then,

$$g(i) = E\left[\sum_{l=0}^{L_{i,S}-1} f(X_l)\middle| X_0 = i\right].$$

a. Derive an equation for $g = (g(1), \ldots, g(S))^T$.
b. Derive an equation for the average first passage times $E[L_{i,S}]$, $i \in \mathcal{S}$.

2.21.* (This problem helps in understanding the difference between the discounted reward criteria for both the discrete-time and continuous-time models.) Consider a Markov chain $\boldsymbol{X}$ with transition probability matrix $P = [p(j|i)]_{i,j=1}^S$ and reward function $f(i)$, $i = 1, 2, \ldots, S$. For simplicity, we assume that $p(i|i) = 0$ for all $i = 1, 2, \ldots, S$. Let $\widetilde{\boldsymbol{X}}$ be a Markov chain with reward function $\widetilde{f}(i) = f(i)$, $i = 1, 2, \ldots, S$, and transition probability matrix $\widetilde{P}$ defined as $\widetilde{p}(i|i) = q$, $0 < q < 1$, and $\widetilde{p}(j|i) = (1-q)p(j|i)$, $j \neq i$, $i, j = 1, 2, \ldots, S$.

a. Prove that $\widetilde{\boldsymbol{X}}$ is equivalent to $\boldsymbol{X}$ in the sense that they have the same steady-state probabilities: $\widetilde{\pi}(i) = \pi(i)$ for all $i = 1, 2, \ldots, S$.
b. The discounted reward of $\boldsymbol{X}$ is defined as (2.30):

$$\eta_\beta(i) = (1-\beta)E\left[\sum_{l=0}^{\infty} \beta^l f(X_l)\middle| X_0 = i\right],$$

where $0 < \beta < 1$ is a discount factor. Similarly, the discounted reward of $\widetilde{\boldsymbol{X}}$ is defined with a discount factor $0 < \widetilde{\beta} < 1$ as

$$\widetilde{\eta}_{\widetilde{\beta}}(i) = (1-\widetilde{\beta})E\left[\sum_{l=0}^{\infty} \widetilde{\beta}^l f(\widetilde{X}_l)\middle| \widetilde{X}_0 = i\right].$$

Find a value for $\widetilde{\beta}$ such that $\widetilde{\eta}_{\widetilde{\beta}}(i) = \eta_\beta(i)$ for all $i = 1, 2, \ldots, S$.
c. Let $\Delta > 0$ be a positive number. Consider a continuous-time (non-Markov) process $\hat{\boldsymbol{X}} := \{\hat{X}_t, t \in [0, \infty)\}$, where $\hat{X}_t = X_l$ if $l\Delta \leq t < (l+1)\Delta$, $l = 0, 1, \ldots$, with $\boldsymbol{X} = \{X_l, l = 0, 1, \ldots\}$ being the Markov chain considered in a). The discounted reward of $\hat{\boldsymbol{X}}$ is defined by an exponential weighting factor (cf. (2.93)):

$$\eta_\alpha(i) = \lim_{T\to\infty} E\left[\int_0^T \alpha \exp(-\alpha t) f(\hat{X}_t) dt\middle| X_0 = i\right], \qquad T_0 = 0.$$

What is the equivalent β such that $\eta_\beta(i) = \eta_\alpha(i)$ for all $i = 1, 2, \ldots, S$?
d. Repeat c) for continuous-time process $\hat{\boldsymbol{X}} := \{\hat{X}_t, t \in [0, \infty)\}$, with $\hat{X}_t = \widetilde{X}_l$ if $l\Delta \leq t < (l+1)\Delta$, $l = 0, 1, \ldots$.
e. What about in d) when we let $\Delta \to 0$ while keeping $\frac{1-q}{\Delta} = \lambda$ (where λ is a constant)?

(Hint: If $\boldsymbol{X} = \{X_0 = i_0, X_1 = i_1, \ldots\}$*, then we have* $\widetilde{\boldsymbol{X}} = \{\widetilde{X}_0 = \widetilde{X}_1 = \cdots = \widetilde{X}_{n_0-1} = i_0, \widetilde{X}_{n_0} = \widetilde{X}_{n_0+1} = \cdots = \widetilde{X}_{n_0+n_1-1} = i_1, \ldots\}$*, where* n_l *is the number of consecutive visits to state* i_l*,* $l = 0, 1, \ldots$*. Note that* n_l *is geometrically distributed with parameter* q*. Therefore,*

$$\widetilde{\eta}_{\widetilde{\beta}}(i) = (1-\widetilde{\beta})E[(1+\widetilde{\beta}+\cdots+\widetilde{\beta}^{n_0-1})f(i_0)+(\widetilde{\beta}^{n_0}+\cdots+\widetilde{\beta}^{n_0+n_1-1})f(i_1)+\cdots].$$

We conclude that $\widetilde{\eta}_{\widetilde{\beta}}(i) = \eta_\beta(i)$ *if* $\beta = \frac{(1-q)\widetilde{\beta}}{1-q\widetilde{\beta}}$*.)*

2.22. Prove that the random variable s generated according to (2.96) is indeed exponentially distributed.

2.23. Develop a PA algorithm to determine a perturbed sample path for an open Jackson network consisting of M servers, with mean service time $\bar{s}_i$, $i = 1, 2, \ldots, M$. The customers arrive in a Poisson process with mean inter-arrival time $a = \frac{1}{\lambda}$. Both a and $\bar{s}_i$, $i = 1, 2, \ldots, M$, may be perturbed.

2.24. Suppose that at some time the perturbations of the servers in a closed network are $\Delta_1, \Delta_2, \ldots, \Delta_M$ determined by Algorithm 2.1. What is the perturbation that has been realized by the network at that time? As we know, if a perturbation is realized, then the future perturbed sample path looks the same as the original one except that it is shifted to the right by an amount equal to the perturbation. Can we use this fact to simplify the calculation in Algorithm 2.2?

2.25. Using the 0-1 vector array (2.105), discuss the situation of the propagation of M perturbations with the same size, each at one server, along a sample path. Prove that $\sum_{i=1}^{M} c(\boldsymbol{n}, i) = 1$.

2.26. We further study the propagations of two equal perturbations $\Delta_1 = \Delta$ at server 1 and $\Delta_2 = \Delta$ at server 2 simultaneously on the same sample path. Consider the array in (2.105). Set $w(t) = w_1(t) + w_2(t)$.

a. What is the meaning of $w(t)$?
b. What does it mean when $w(t) = (1, 1, \ldots, 1)$ or $w(t) = (0, 0, \ldots, 0)$?
c. How does $w(t)$ evolve?

2.27. In addition to (2.94), we may define the system performance as the long-run time average

$$\eta_T^{(f)} = \lim_{L\to\infty} \frac{1}{T_L} \int_0^{T_L} f[\boldsymbol{N}(t)]dt.$$

We have $\eta_T^{(f)} = \frac{\eta^{(f)}}{\eta^{(I)}}$.

a. Derive the derivative of $\eta_T^{(f)}$ with respect to $\bar{s}_i$, $i = 1, 2, \ldots, M$.

b. Define the reward function f corresponding to the steady-state probability $\pi(\boldsymbol{n})$, with $\boldsymbol{n}$ being any state, and derive $\frac{d\pi(\boldsymbol{n})}{d\bar{s}_i}$, $i = 1, 2, \ldots, M$.

2.28.* Prove that, in a closed Jackson network, the sample function $T_L(\xi, \bar{s}_v)$ (with ξ fixed) is a piecewise linear function of $\bar{s}_v$, $v = 1, 2, \ldots, M$ (see [46]).

2.29. Consider a closed Jackson network in which $\mu_i q_{i,j} = \mu_j q_{j,i}$, $i, j = 1, 2, \ldots, M$. Prove that

$$c(\boldsymbol{n}, k) = \frac{n_k}{N}, \qquad k = 1, 2, \ldots, M;$$

and

$$\frac{\bar{s}_k}{\eta}\frac{\partial \eta}{\partial \bar{s}_k} = -\frac{1}{M},$$

where $k = 1, 2, \ldots, M$, denotes any server in the network.

2.30.* *(This problem requires a good knowledge of queueing theory)* Consider an M/M/1 queue with arrival rate λ and service rate μ. The system state is simply the number of customers in the queue; i.e., $\boldsymbol{n} = n$. The performance measure is the average response time $\tau = \lim_{L\to\infty} \frac{1}{L}\int_0^{T_L} n(t)dt$. Thus, $f(n) = n$. For the M/M/1 queue, there is a source sending customers to the queue with rate λ. Denote the source as server 0, and the server as server 1. Server 0 can be viewed as always having infinitely many customers.

a. Prove that the realization factors $c^{(f)}(n, 0)$ and $c^{(f)}(n, 1)$, $n = 0, 1, \ldots$, satisfy the following equations:

$$c^{(f)}(0, 0) = 0, \; c^{(f)}(0, 1) = 0,$$

$$c^{(f)}(n, 0) + c^{(f)}(n, 1) = n, \qquad n \geq 0,$$

$$(\lambda + \mu)c^{(f)}(n, 0) = \mu c^{(f)}(n-1, 0) + \lambda c^{(f)}(n+1, 0) - \lambda, \qquad n > 0,$$

and

$$(\lambda + \mu)c^{(f)}(n, 1) = \lambda c^{(f)}(n+1, 1) + \mu c^{(f)}(n-1, 1) + \mu, \qquad n > 0.$$

b. To solve for $c^{(f)}(n, i)$, $i = 0, 1$, we need a boundary condition. Using the physical meaning of perturbation realization, prove that $c^{(f)}(1, 1)$ equals the average number of customers served in a busy period of the M/M/1 queue; i.e. (see, e.g., [169]),

$$c^{(f)}(1, 1) = \frac{\mu}{\mu - \lambda} = \frac{1}{1-\rho}, \qquad \rho = \frac{\lambda}{\mu}.$$

c. Prove

$$c^{(f)}(n, 1) = \frac{n}{1-\rho},$$

and

$$c^{(f)}(n, 0) = -\frac{n\rho}{1-\rho}.$$

d. By the same argument as in closed networks, explain and derive

$$\frac{\mu}{\eta^{(I)}}\frac{d\tau}{d\mu} = -\frac{\lambda\mu}{(\mu-\lambda)^2} = -\frac{\rho}{(1-\rho)^2},$$

and

$$\frac{\lambda}{\eta^{(I)}}\frac{d\tau}{d\lambda} = \frac{\lambda^2}{(\mu-\lambda)^2} = \frac{\rho^2}{(1-\rho)^2}.$$

2.31. The head-processing time of a packet in a communication system, or the machine tool set-up time in manufacturing, is usually a fixed amount of time. Consider a two-server cyclic queueing network in which the service times of the two servers are exponentially distributed with mean $\bar{s}_1$ and $\bar{s}_2$, respectively. Suppose that every service time of server 1 increases by a fixed amount of time Δ. Derive the derivative of performance $\eta^{(f)}$ with respect to Δ using performance realization factors $c^{(f)}(\boldsymbol{n}, 1)$.

2.32. Prove that Algorithm 2.2 yields a strongly consistent estimate for the derivative of the average response time in an M/G/1 queue; i.e., in (2.111) we have

$$\frac{\mu}{\eta^{(I)}}\frac{\partial\bar{\tau}}{\partial\mu} = -\lim_{K\to\infty}\frac{1}{T_L}\sum_{k=1}^{K}\sum_{i=1}^{n_k}\sum_{l=1}^{i} s_{k,l}, \qquad \text{w.p.1.}$$

2.33. Consider a closed Jackson network with M servers and N customers. The throughput of server i is $\eta_i = \breve{\eta} v_i$ where $\breve{\eta}$ is the "un-normalized system throughput":

$$\breve{\eta} = \frac{G_M(N-1)}{G_M(N)},$$

where v_i is server i's visiting ratio: The solution to

$$v_i = \sum_{j=1}^{M} q_{j,i} v_j, \qquad j = 1, 2, \ldots, M,$$

and (see (C.16) in Appendix C)

$$G_m(n) = \sum_{n_1+\ldots+n_M=n}\prod_{i=1}^{m} x_i^{n_i},$$

where $x_i = v_i\bar{s}_i$, $i = 1, 2\ldots, M$. We have

$$dx_i = dv_i\bar{s}_i + v_i d\bar{s}_i. \tag{2.128}$$

Now, we consider the derivative of $\breve{\eta}$ with respect to the routing probability matrix $Q = [q_{i,j}]_{i,j=1}^{M}$. It is clear that $\breve{\eta}$ depends on the routing probabilities only through x_i, $i = 1, 2, \ldots, M$. Suppose that v_i changes to $v_i + dv_i$, $i = 1, 2, \ldots, M$. From (2.128), we observe that in terms of the changes in x_i, dx_i, $i = 1, 2, \ldots, M$, this is equivalent to setting $dv_i = 0$ and $d\bar{s}_i = \bar{s}_i\frac{dv_i}{v_i}$ for all $i = 1, 2, \ldots, M$.

a. Explain that, for closed Jackson networks, the derivative of the average reward $\sum_{\text{all } \boldsymbol{n}} \pi(\boldsymbol{n}) f(\boldsymbol{n})$ with respect to the changes in routing probabilities can be obtained through the derivatives of the average reward with respect to the mean service times.
b. Derive the performance derivative formula $\frac{d\eta_i}{dQ}$, by using performance realization factors $c^{(f)}(\boldsymbol{n}, i)$, $i = 1, 2, \ldots$.

2.34.* Consider the same two-server cyclic Jackson queueing network studied in Problem 2.17. Let $\eta_T^{(f)} = \lim_{L\to\infty} \frac{\int_0^{T_L} f(n(t))dt}{T_L}$ denote the time-average performance, where $n(t)$ is the number of customers at time t at server 1, and L denotes the number of transitions. The performance function is $f(n) = n$. Let us assume that the arrival rate λ, or the service rate μ, changes only when the state is n.

a. Derive $\frac{d\eta_T^{(f)}}{d\lambda}$ and $\frac{d\eta_T^{(f)}}{d\mu}$ in terms of the realization factors $c^{(f)}(n, 1), c^{(f)}(n, 2)$ and realization probability $c(n, 1), c(n, 2)$.
b. Express $\frac{d\eta_T^{(f)}}{d\lambda}$ and $\frac{d\eta_T^{(f)}}{d\mu}$ in terms of the performance potentials $g(n)$.
c. Compare both results in a) and b) and derive a relation between the realization factors and the potentials. Give an intuitive explanation for this relation. (cf. [260])

2.35. In weak derivative expression (2.125), we may choose $P^+ = P'$ and $P^- = P$.

a. Derive (2.126) and express its meaning based on sample paths.
b. Derive (2.127).

2.36. Derive (2.23) from (2.127).

2.37. Consider a (continuous-time) Markov process with transition rates $\lambda(i)$ and transition probabilities $p(j|i)$, $i, j = 1, 2, \ldots, S$. Suppose that the transition probability matrix $P := [p(j|i)]_{i,j\in\mathcal{S}}$ changes to $P + \delta\Delta P$ and the transition rates $\lambda(i)$, $i = 1, 2, \ldots, S$ remain unchanged. Let η be the average reward with reward function f. Derive the performance derivative formula for $\frac{d\eta_\delta}{d\delta}$ using the construction approach illustrated in Section 2.1.3.

You can observe a lot just by watching.

Berra's Law - Yogi Berra,
US baseball player, coach,
and manager (1925 -)

3

Learning and Optimization with Perturbation Analysis

As shown in Chapter 2, performance derivatives for Markov systems depend heavily on performance potentials. In this chapter, we first discuss the numerical methods and sample-path-based algorithms for estimating performance potentials, and we then derive the sample-path-based algorithms for estimating performance derivatives. In performance optimization, the process of estimating the potentials and performance derivatives from a sample path is called *learning*.

Policy gradients (PG) in reinforcement learning (RL) is almost a synonym for perturbation analysis (PA) in discrete event dynamic systems (DEDS). However, because the terms PG and PA are used by researchers in two different disciplines, there is a different emphasis on different aspects of the analysis. With PA in DEDS, we construct sensitivity formulas by exploring the system's dynamic nature and develop sample-path-based and on-line estimation algorithms for performance derivatives; while with PG in RL we emphasize the algorithmic features of gradient estimation algorithms, such as their efficiency and recursiveness. Therefore, this chapter is closely related to Chapter 6 on reinforcement learning. We will introduce performance gradient algorithms from a sample-path-based perspective and leave the algorithmic features, especially those related to the stochastic approximation approach, to Chapter 6 (see Figure 3.1).

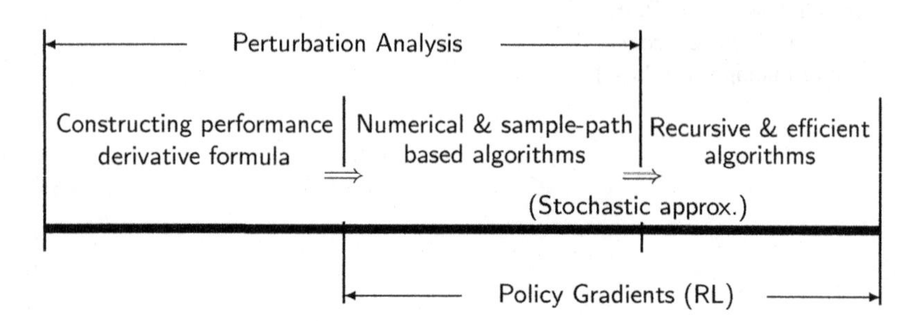

Fig. 3.1. Perturbation Analysis vs. Policy Gradients

3.1 The Potentials

We first study the potentials for ergodic Markov chains (discrete time), and the results can be extended to ergodic Markov processes (continuous time) naturally.

3.1.1 Numerical Methods

With $\pi g = \eta$

The first numerical method depends on the equation for performance potentials ((2.13) and (2.14)):

$$\begin{aligned} g &= (I - P + e\pi)^{-1} f \\ &= \sum_{k=0}^{\infty} [(P - e\pi)^k] f = \Big\{ I + \sum_{k=1}^{\infty} (P^k - e\pi) \Big\} f. \end{aligned}$$

Thus, g can be calculated iteratively by setting:

$$g_0 = f, \quad g_k = f + (P - e\pi) g_{k-1}, \qquad k \geq 0, \tag{3.1}$$

and $g = \lim_{k \to \infty} g_k$. This method requires solving for π first.

With the Realization Factors

Alternatively, we can solve the PRF equation (2.7)

$$\Gamma - P \Gamma P^T = F, \tag{3.2}$$

which does not contain π. Again, its solution is also only up to an additive constant; i.e, if Γ is a solution to (3.2), so is $\Gamma + cee^T$ for any constant c. In addition to (3.2), the PRF matrix $\Gamma = eg^T - ge^T$ also satisfies $\pi\Gamma\pi^T = 0$ or simply $e^T\Gamma e = 0$.

From (3.2), we have

$$\begin{aligned}\Gamma &= P\Gamma P^T + F \\ &= P(P\Gamma P^T + F)P^T + F = P^2\Gamma(P^2)^T + PFP^T + F \\ &= P^k\Gamma(P^k)^T + P^{k-1}F(P^{k-1})^T + \cdots + PFP^T + F.\end{aligned}$$

Since

$$\lim_{k\to\infty} P^k\Gamma(P^k)^T = e\pi\Gamma\pi^T e^T = 0,$$

we have

$$\Gamma = \sum_{k=0}^{\infty} P^k F(P^k)^T,$$

with $P^0 = I$. Therefore, we have the following iterative algorithm

$$\Gamma_0 = F, \quad \Gamma_k = P\Gamma_{k-1}P^T + F, \qquad k \geq 1, \tag{3.3}$$

and $\lim_{k\to\infty}\Gamma_k = \Gamma$. While this algorithm (3.3) does not require solving for π, it has two matrix multiplications in each iteration.

With $g(S) = 0$

Note that in the Poisson equation $(I-P)g+\eta e = f$, the same term η appears in every row. Using this feature, we may develop another numerical algorithm as follows. First, denote the Sth row of P as p_{S*}. Define

$$P_- = P - ep_{S*}.$$

The last row of P_- is zero. Let

$$f_- = [f(1) - f(S), \ldots, f(S-1) - f(S), 0]^T.$$

Subtracting the last row of the Poisson equation from all the rows, and by setting $g(S) = 0$, we get

$$g = P_- g + f_-. \tag{3.4}$$

From this, we can write

$$g = \lim_{L\to\infty}\Big(\sum_{l=0}^{L} P_-^l\Big) f_-. \tag{3.5}$$

Note that because $f_-(S) = 0$ and the last row of P_- is zero, from (3.4) or (3.5) we indeed have $g(S) = 0$. This is consistent with the fact that the potential vector g is unique only up to an additive constant vector, and the g in (3.5) represents one form of the potential vector.

Let $\{1, \lambda_1, \ldots, \lambda_{S-1}\}$ be the set of the eigenvalues of P (see Lemma B.1 in Appendix B). First, we assume that all the eigenvalues are simple. For ergodic chains, we have $|\lambda_i| < 1$ for $i = 1, 2, \ldots, S-1$ [20]. Let $x \neq 0$ be an eigenvector corresponding to one of the eigenvalues, denoted as $\lambda \neq 1$; i.e., $Px = \lambda x$. If $\lambda = 0$, then $Px = 0$ and it is easy to verify that $P_- x = 0$ and $x \neq ce$, with $c \neq 0$ being any constant. That is, $\lambda = 0$ is also an eigenvalue of P_- with eigenvector $x \neq ce$.

Now, we assume that $\lambda \neq 0$. Define $x' = x - \frac{1}{\lambda}(p_{S*}x)e$. Then, we can verify that $x' \neq 0$ and

$$\begin{aligned} P_- x' &= (P - ep_{S*})\Big[x - \frac{1}{\lambda}(p_{S*}x)e\Big] \\ &= \lambda\Big[x - \frac{1}{\lambda}(p_{S*}x)e\Big] = \lambda x', \end{aligned} \tag{3.6}$$

i.e., λ is an eigenvalue of P_- with eigenvector x'. In addition, $P_- e = 0$, i.e., 0 is an eigenvalue of P_-. Therefore, the eigenvalues of $P_- = P - ep_{S*}$ are $\{0, \lambda_1, \ldots, \lambda_{S-1}\}$, with all $|\lambda_i| < 1$, $i = 1, \ldots, S-1$, which are the same as the eigenvalues of $P - e\pi$. One of λ_i, $i = 1, \ldots, S-1$ may be zero (note that we assumed that λ_i, $i = 1, 2, \ldots, S-1$, are different). Therefore, the limit in (3.5) converges at the same rate as (or as fast as) the rate of $\lim_{k\to\infty}(P - e\pi)^k = 0$, or the rate of $\lim_{k\to\infty} P^k = e\pi$.

When there are multiple eigenvalues, we need to examine the multiplicities of the eigenvalues of both P and P_-. First, we assume that $\lambda = 0$ is an eigenvalue of P with $m_0 \geq 0$ multiplicity. We note that for any $x \neq 0$ if $Px = 0$ or $Px = e$ (i.e., $x = e$), then $P_- x = 0$. This means that the space spanned by the eigenvectors of P corresponding to both $\lambda = 0$ and $\lambda = 1$ is a subspace of the space spanned by the eigenvectors of P_- corresponding to $\lambda = 0$.

On the other hand, if $x \neq 0$ and $P_- x = 0$, then either $Px = 0$ or $Px = e$. This can be proved as follows: Because $P_- = P - ep_{S*}$, we have $Px = e(p_{S*}x)$. If $p_{S*}x = 0$, then we have $Px = 0$. If $p_{S*}x \neq 0$, then, without loss of generality, we may assume that $p_{S*}x = 1$. Thus, $Px = e$. This means that the space spanned by the eigenvectors of P_- corresponding to $\lambda = 0$ is a subspace of the space spanned by the eigenvectors of P corresponding to both $\lambda = 0$ and $\lambda = 1$.

Finally, the space spanned by the eigenvectors of P_- corresponding to $\lambda = 0$ is the same as the space spanned by the eigenvectors of P corresponding to both $\lambda = 0$ and $\lambda = 1$; and the multiplicity of $\lambda = 0$ for P_- is $m_0 + 1$.

Let $\lambda \neq 0, 1$ be one of the eigenvalues of P with multiplicity m and x_k, $k = 1, \ldots, m$, be the corresponding linearly independent eigenvectors. As shown in (3.6), $x'_k = x_k - \frac{1}{\lambda}(p_{S*}x_k)e$, $k = 1, \ldots, m$, are eigenvectors of P_-.

We wish to prove that x'_k, $k = 1, \ldots, m$, are linearly independent. Suppose that the opposite is true, i.e., there is a set of real numbers c_k, $k = 1, \ldots, m$, not all of them are zeros, such that $\mathbf{v} = \sum_{k=1}^{m} c_k x'_k = 0$. Set $\mathbf{u} = \sum_{k=1}^{m} c_k x_k$. Because $P_- e = 0$, we have $P_- x'_k = P_- x_k$, $k = 1, 2, \ldots, m$, and

$$P_- \mathbf{u} = P_- \left(\sum_{k=1}^{m} c_k x_k \right) = 0.$$

Thus, $\mathbf{u} \neq 0$ is an eigenvalue of P_- for $\lambda = 0$. Note that $\mathbf{u}$, being a vector spanned by x_k, $k = 1, \ldots, m$, which are eigenvalues of P corresponding to eigenvalue $\lambda \neq 0, 1$, is linearly independent of the eigenvectors of P corresponding to $\lambda = 0$ and 1. Thus, $\mathbf{u}$ adds one to the multiplicity of $\lambda = 0$ for P_-. This implies that the multiplicity of $\lambda = 0$ for P_- is larger than $m_0 + 1$, which is impossible. Therefore, x'_k, $k = 1, \ldots, m$, are linearly independent, and the multiplicity of λ for P_- is the same as that for P.

In summary, we conclude that the eigenvalues of P_- are $\{0, \lambda_1, \ldots, \lambda_{S-1}\}$, with all $|\lambda_i| < 1$, $i = 1, \ldots, S-1$, being the same as those of P. The multiplicity of $\lambda_i \neq 0$, $i = 1, \ldots, S-1$, for P_- are the same as that for P, and the multiplicity of 0 or P_- is $m_0 + 1$.

From (3.5), we have the following iterative algorithm:

$$g_0 = f_-, \quad g_k = f_- + P_- g_{k-1}, \qquad k \geq 1, \tag{3.7}$$

and $g = \lim_{k \to \infty} g_k$.

The above three numerical algorithms have about the same convergence rate (determined by the eigenvalues of P), which is the same as the rate in computing the steady-state probability π using $\lim_{k \to \infty} P^k = e\pi$. The algorithm in (3.7) does not require solving for π, and only one matrix multiplication is needed in each iteration.

In queueing systems, the perturbation realization factors satisfy the set of linear equations (2.107). They can be solved numerically by any standard method for linear equations, and an example is shown in Table 2.9. Further results exploring the special features of these linear equations have not yet been developed in the literature.

3.1.2 Learning Potentials from Sample Paths

The sample-path-based learning algorithms can be derived from (2.16)

$$g(i) = \lim_{L \to \infty} E \left\{ \sum_{l=0}^{L-1} [f(X_l) - \eta] \Big| X_0 = i \right\}, \tag{3.8}$$

and (2.17)

$$\gamma(i,j) = E\left\{ \sum_{l=0}^{L(i|j)-1} [f(X_l) - \eta] \Big| X_0 = j \right\}, \tag{3.9}$$

where $L(i|j) = \min\{l \geq 0 : X_l = i | X_0 = j\}$; or from (2.5) and (2.6),

$$\begin{aligned} \gamma(i,j) &= g(j) - g(i) \\ &= \lim_{L \to \infty} E\left\{ \sum_{l=0}^{L-1} \left[f(\widetilde{X}_l) - f(X_l) \right] \Big| \widetilde{X}_0 = j,\ X_0 = i \right\} \quad (3.10) \\ &= E\left\{ \sum_{l=0}^{L^*_{ij}-1} \left[f(\widetilde{X}_l) - f(X_l) \right] \Big| \widetilde{X}_0 = j,\ X_0 = i \right\}, \quad i,j = 1,\ldots,S; \quad (3.11) \end{aligned}$$

at L^*_{ij}, the two sample paths $\widetilde{\boldsymbol{X}}$ and $\boldsymbol{X}$ merge together for the first time.

Algorithms for *g*

From (3.8), we have the following approximation for $g(i)$,

$$g_L(i) = E\left[\sum_{l=0}^{L-1} f(X_l) \Big| X_0 = i \right] - L\eta, \tag{3.12}$$

with $\lim_{L \to \infty} g_L(i) = g(i)$. The average reward η can be estimated from a sample path by

$$\eta_L = \frac{1}{L} \sum_{l=0}^{L-1} f(X_l), \tag{3.13}$$

with $\eta = \lim_{L \to \infty} \eta_L$, with probability 1. However, because potentials are valid only up to an additive constant, we may ignore the constant $L\eta$ in (3.12) and use its first term as an estimate,

$$g_L(i) = E\left\{ \sum_{l=0}^{L-1} f(X_l) \Big| X_0 = i \right\}. \tag{3.14}$$

With (3.14), the potential g can be estimated on a sample path in a way similar to the estimation of η in (3.13). Let $I_i(x) = 1$ if $x = i$ and $I_i(x) = 0$ if $x \neq i$. Define

$$g_{L,N}(i) = \frac{\sum_{n=0}^{N-L+1} \left\{ I_i(X_n) \left[\sum_{l=0}^{L-1} f(X_{n+l}) \right] \right\}}{\sum_{n=0}^{N-L+1} I_i(X_n)}, \tag{3.15}$$

in which $\sum_{n=0}^{N-L+1} I_i(X_n)$ is the number of visits to state i of the Markov chain in the period of $[0, N-L+1]$. After each such visit, we add up $f(X_n)$ for L transitions, and $g_{L,N}$ is the average of these sums. We have

$$\lim_{N\to\infty} g_{L,N}(i) = g_L(i), \qquad \text{w.p.1.} \tag{3.16}$$

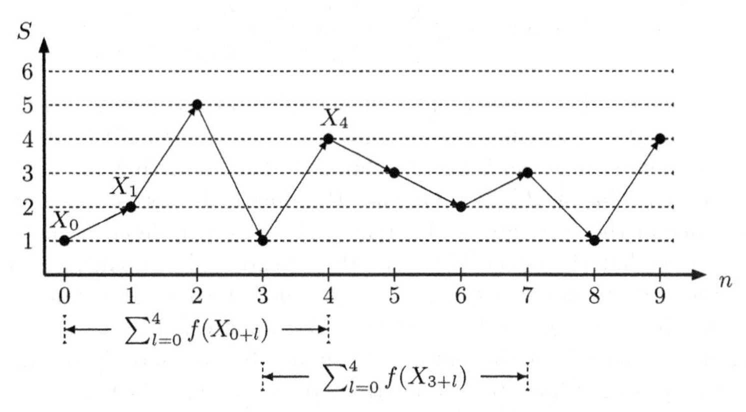

Fig. 3.2. Items in (3.15) Are Not Independent

The proof of (3.16) is not straightforward, since the items $\sum_{l=0}^{L-1} f(X_{n+l})$ for different n may not be independent. For example, given a particular sample path, say $\{1, 2, 5, 1, 4, 3, 2, 3, 1, 6, \ldots\}$ as shown in Figure 3.2, with $L = 5$, the two periods starting from $X_0 = 1$ and $X_3 = 1$ overlap. Both items $\sum_{l=0}^{4} f(X_{0+l}) = f(1) + f(2) + f(5) + f(1) + f(4)$ and $\sum_{l=0}^{4} f(X_{3+l}) = f(1)+f(4)+f(3)+f(2)+f(3)$ contain the same term $f(1)+f(4)$. Therefore, the standard law of large numbers does not apply in this case. The proof of (3.16) is based on a fundamental theorem on ergodicity (see [32]; we state its version on a finite state space $\mathcal{S}$):

The Fundamental Ergodicity Theorem:

Let $\boldsymbol{X} = \{X_n, n \geq 0\}$ be an ergodic Markov chain on state space $\mathcal{S}$; $\phi(x_1, x_2, \ldots)$, $x_i \in \mathcal{S}$, $i = 1, 2\ldots$, be a function on $\mathcal{S}^\infty$. Then the process $\boldsymbol{Z} = \{Z_n, n \geq 0\}$ with $Z_n = \phi(X_n, X_{n+1}, \ldots)$ is an ergodic Markov chain. In particular, we have

$$\lim_{N\to\infty} \frac{1}{N} \sum_{n=0}^{N-1} \phi(X_n, X_{n+1}, \ldots) = E[\phi(X_n, X_{n+1}, \ldots)], \qquad \text{w.p.1,} \tag{3.17}$$

where "E" denotes the steady-state expectation of the Markov chain $\boldsymbol{Z}$, and the right-hand side of (3.17) does not depend on n.

Since this theorem is very useful in proving the convergence results related to sample-path-based algorithms, we will refer to it as the *Fundamental Ergodicity Theorem*. In our case, we define $Z_n = I_i(X_n)[\sum_{l=0}^{L-1} f(X_{n+l})]$; then, $\{Z_n, n \geq 0\}$ is ergodic. From (3.15), we have

$$g_{L,N}(i) = \frac{\frac{1}{N-L+2}\sum_{n=0}^{N-L+1} Z_n}{\frac{1}{N-L+2}\sum_{n=0}^{N-L+1} I_i(X_n)}.$$

By the fundamental ergodicity theorem, the numerator converges to $E(Z_n) = \pi(i)g_L(i)$ and the denominator converges to $\pi(i)$. Thus, (3.16) holds.

One remaining problem is how to choose L. It is clear that the larger L is, the smaller the bias of (3.15) is. On the other hand, the larger L is, the larger the variance of the estimate is. Therefore, there is a tradeoff in choosing L. We first note that the effect of potentials depends only on their differences, i.e., on the realization factors $\gamma(i,j) = g(j) - g(i)$. Ideally, to estimate $\gamma(i,j)$, the length should be the first passage time from state j to state i (see (3.9)). Therefore, the length of the period, L, should be comparable to the mean of the first passage times from one state to the others. On the other hand, from (3.8), L should be large enough so that $E[f(X_l)]$ is close to η when $l > L$. Because the l-step state transition probability $\mathcal{P}(X_l|X_0)$ converges exponentially fast to the steady-state probability, we may expect that L can be chosen as a small number. The following simulation example provides some empirical evidence.

Example 3.1. We simulated a Markov chain with ten states. The state transition matrix is arbitrarily chosen as

$$P = \begin{bmatrix} 0.20 & 0.00 & 0.05 & 0.10 & 0.15 & 0.15 & 0.05 & 0.05 & 0.05 & 0.20 \\ 0.30 & 0.00 & 0.00 & 0.20 & 0.10 & 0.15 & 0.15 & 0.05 & 0.05 & 0.00 \\ 0.00 & 0.15 & 0.05 & 0.30 & 0.00 & 0.05 & 0.20 & 0.20 & 0.05 & 0.00 \\ 0.05 & 0.10 & 0.25 & 0.00 & 0.30 & 0.00 & 0.05 & 0.20 & 0.05 & 0.000 \\ 0.00 & 0.20 & 0.15 & 0.00 & 0.15 & 0.00 & 0.15 & 0.25 & 0.00 & 0.100 \\ 0.00 & 0.10 & 0.30 & 0.00 & 0.20 & 0.10 & 0.10 & 0.00 & 0.15 & 0.050 \\ 0.10 & 0.10 & 0.10 & 0.10 & 0.10 & 0.10 & 0.10 & 0.10 & 0.10 & 0.100 \\ 0.00 & 0.20 & 0.00 & 0.20 & 0.00 & 0.20 & 0.00 & 0.20 & 0.00 & 0.200 \\ 0.05 & 0.15 & 0.25 & 0.00 & 0.15 & 0.15 & 0.15 & 0.00 & 0.00 & 0.100 \\ 0.15 & 0.05 & 0.00 & 0.20 & 0.15 & 0.10 & 0.20 & 0.10 & 0.05 & 0.000 \end{bmatrix},$$

and the reward function is

$$f = [10, 5, 1, 15, 3, 0, 7, 20, 2, 18]^T.$$

Table 3.1 lists the theoretical and estimated values of the potentials g, in the form of (3.15) and normalized to $\pi g = 0$, estimated with $L = 5$ on a sample path with length $N = 100,000$. The means and standard deviations (SD) are the results of ten simulations. These results indicate that, for a ten-state Markov chain, $L = 5$ yields very accurate estimates for g. □

i	1	2	3	4	5	6	7	8	9	10
Theoretic	1.865	-4.025	-5.121	6.268	-3.259	-13.553	-1.997	14.098	-10.033	9.614
Mean	1.845	-4.056	-5.132	6.243	-3.266	-13.520	-1.893	14.162	-9.902	9.654
SD	0.098	0.088	0.163	0.140	0.140	0.187	0.116	0.110	0.185	0.160

Table 3.1. The Potentials in Example 3.1 with 100,000 Transitions and $L = 5$

Algorithms for Γ

Next, we derive a sample-path-based algorithm from (3.9). On a sample path of $\boldsymbol{X} = \{X_l, l \geq 0\}$ with $X_0 = i$, for each pair of states j and i, we define two sequences of epochs $\{l_k(j)\}$ and $\{l_k(i)\}$ as follows:

$$\begin{aligned} l_0(i) &= 0, \\ l_k(j) &= \text{the epoch that } \{X_l\} \text{ first visits state } j \text{ after } l_{k-1}(i), \quad k \geq 1, \\ l_k(i) &= \text{the epoch that } \{X_l\} \text{ first visits state } i \text{ after } l_k(j), \quad k \geq 1. \quad (3.18) \end{aligned}$$

Note that $\{l_k(j)\}$ and $\{l_k(i)\}$ are well defined on a sample path; see Figure 3.3.

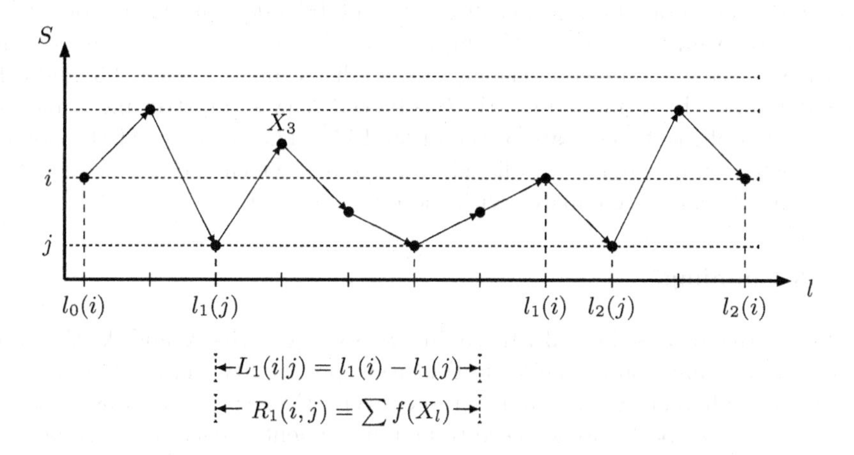

Fig. 3.3. Estimating $\gamma(i, j)$

Now, define $L_k(i|j) = l_k(i) - l_k(j)$ and

$$R_k(i, j) = \sum_{l=l_k(j)}^{l_k(i)-1} f(X_l).$$

The Markov property ensures that the $L_k(i|j)$ and $R_k(i,j)$, $k = 0, 1, \ldots$, are identically and independently distributed (i.i.d), respectively. By the law of large numbers, we have

$$\lim_{N\to\infty} \frac{1}{N} \sum_{k=1}^{N} L_k(i|j) = E[L(i|j)], \qquad \text{w.p.1},$$

and

$$\lim_{N\to\infty} \frac{1}{N} \sum_{k=1}^{N} R_k(i,j) = E\left[\sum_{l=0}^{L(i|j)-1} f(X_l) \Big| X_0 = j \right], \qquad \text{w.p.1.}$$

Therefore,

$$\lim_{N\to\infty} \frac{1}{N} \left\{ \sum_{k=1}^{N} [R_k(i,j) - L_k(i|j)\eta] \right\} = \gamma(i,j), \qquad \text{w.p.1}, \qquad (3.19)$$

where η can be estimated on the sample path using (3.13). Potentials can be obtained by using any row of Γ. We may also use $g = (\pi\Gamma)^T$, which may lead to more accurate estimates because it employs all the rows of Γ.

Example 3.2. We consider the same Markov chain as in Example 3.1. We did ten simulation runs, and each consists of 100,000 state transitions. The theoretical values as well as the means and the standard deviations of the estimated realization factors using (3.19) are listed in Tables 3.2, 3.3, and 3.4, respectively. The estimated matrix Γ is indeed skew-symmetric and standard deviations of most items are of the order 10^{-2}. The statistics of the potentials based on $g = (\pi\Gamma)^T$ are listed in Table 3.5, which shows much smaller standard deviations compared with those in Table 3.1. □

3.1.3 Coupling*

The algorithms based on (3.10) require two sample paths $\boldsymbol{X}$ and $\widetilde{\boldsymbol{X}}$; they are independent and follow the same transition probability matrix P but start from two different states i and j, respectively. However, to estimate $\gamma(i,j)$, the two sample paths do not need to be independent of each other. In fact, it is well known that introducing co-relation between the random samples of two random variables may reduce the variance in estimating the difference of their mean values [257] (also see Problem 3.7). For example, we may use the same sequence of random variables $\{\xi_0, \xi_1, \ldots\}$ to simulate the two sample paths $\boldsymbol{X}$ and $\widetilde{\boldsymbol{X}}$ to obtain estimates for $\gamma(i,j)$ in (3.11). Introducing co-relation between the two sample paths $\boldsymbol{X}$ and $\widetilde{\boldsymbol{X}}$ is called the *coupling approach* in simulation (see, [212]). In the following, we will study this coupling issue in greater detail.

	1	2	3	4	5	6	7	8	9	10
1	0.000	-5.890	-6.987	4.403	-5.124	-15.418	-3.863	12.232	-11.899	7.749
2	5.890	0.000	-1.097	10.293	0.766	-9.528	2.028	18.122	-6.009	13.639
3	6.987	1.097	0.000	11.389	1.863	-8.430	3.124	19.219	-4.912	14.735
4	-4.403	-10.293	-11.389	0.000	-9.527	-19.820	-8.265	7.830	-16.302	3.346
5	5.124	-0.766	-1.863	9.527	0.000	-10.294	1.260	17.356	-6.775	12.873
6	15.418	9.528	8.430	19.820	10.294	0.000	11.556	27.650	3.519	23.167
7	3.863	-2.028	-3.124	8.265	-1.261	-11.556	0.000	16.095	-8.036	11.611
8	-12.232	-18.122	-19.219	-7.830	-17.356	-27.650	-16.095	0.000	-24.130	-4.484
9	11.899	6.009	4.912	16.302	6.775	-3.519	8.036	24.130	0.000	19.647
10	-7.749	-13.639	-14.735	-3.346	-12.873	-23.167	-11.610	4.484	-19.647	0.000

Table 3.2. The Theoretical Values of the Realization Factors in Example 3.2

	1	2	3	4	5	6	7	8	9	10
1	0.000	-5.801	-6.983	4.336	-5.106	-15.377	-3.827	12.286	-11.727	7.756
2	5.800	0.000	-1.012	10.294	0.780	-9.474	2.097	18.204	-5.898	13.682
3	6.983	1.012	0.000	11.381	1.838	-8.416	3.146	19.217	-4.912	14.769
4	-4.336	-10.294	-11.381	0.000	-9.492	-19.690	-8.143	7.876	-16.217	3.442
5	5.105	-0.782	-1.838	9.491	0.000	-10.223	1.390	17.408	-6.582	12.918
6	15.376	9.472	8.414	19.689	10.221	0.000	11.684	27.629	3.647	23.214
7	3.827	-2.098	-3.147	8.142	-1.391	-11.687	0.000	16.014	-7.999	11.629
8	-12.285	-18.204	-19.218	-7.875	-17.409	-27.630	-16.014	0.000	-24.069	-4.491
9	11.726	5.895	4.910	16.214	6.579	-3.653	7.997	24.067	0.000	19.709
10	-7.755	-13.683	-14.768	-3.440	-12.920	-23..216	-11.629	4.493	-19.713	0.000

Table 3.3. The Mean Realization Factors in Example 3.2

Define a composed Markov chain $\widehat{\boldsymbol{X}} := \{(X_l, \widetilde{X}_l), l = 0, 1, \ldots\}$; its state space is

$$\widehat{\mathcal{S}} = \mathcal{S} \times \mathcal{S} = \{(1,1), (1,2), \ldots, (1,S), \\ (2,1), (2,2), \ldots, (2,S), \ldots, (S,1), \ldots, (S,S)\},$$

and its transition probabilities are

$$\widehat{p}[(i',j')|(i,j)] := \mathcal{P}\left(X_{l+1} = i', \widetilde{X}_{l+1} = j' \middle| X_l = i, \widetilde{X}_l = j \right), \\ i, i', j, j' \in \mathcal{S},$$

which equal

	1	2	3	4	5	6	7	8	9	10
1	0.000	0.017	0.060	0.043	0.052	0.017	0.047	0.077	0.144	0.063
2	0.017	0.000	0.020	0.027	0.011	0.025	0.035	0.025	0.109	0.040
3	0.060	0.020	0.000	0.029	0.049	0.054	0.025	0.024	0.065	0.076
4	0.043	0.028	0.029	0.000	0.037	0.028	0.025	0.041	0.116	0.100
5	0.052	0.011	0.050	0.038	0.000	0.021	0.037	0.045	0.041	0.026
6	0.017	0.024	0.054	0.027	0.020	0.000	0.025	0.041	0.037	0.070
7	0.047	0.036	0.025	0.025	0.037	0.025	0.000	0.039	0.059	0.032
8	0.077	0.024	0.024	0.041	0.044	0.041	0.039	0.000	0.128	0.064
9	0.146	0.110	0.065	0.117	0.042	0.039	0.059	0.130	0.000	0.102
10	0.064	0.040	0.076	0.101	0.025	0.068	0.031	0.065	0.097	0.000

Table 3.4. The Standard Deviations of the Realization Factors in Example 3.2

	1	2	3	4	5	6	7	8	9	10
Theoretic	1.865	-4.025	-5.121	6.268	-3.259	-13.553	-1.997	14.098	-10.033	9.614
Mean	1.859	-4.039	-5.092	6.237	-3.273	-13.517	-1.912	14.132	-9.932	9.671
SD	0.0122	0.0074	0.0105	0.0127	0.0111	0.0038	0.0148	0.0146	0.0405	0.0206

Table 3.5. The Potentials Based on the Realization Factors in Example 3.2

$$\widehat{p}[(i',j')|(i,j)] := \mathcal{P}\left(X_{l+1} = i' \middle| X_l = i, \widetilde{X}_l = j\right)$$
$$\times \mathcal{P}\left(\widetilde{X}_{l+1} = j' \middle| X_l = i, \widetilde{X}_l = j, X_{l+1} = i'\right).$$

The transition probability matrix of $\widehat{\boldsymbol{X}}$ is denoted as

$$\widehat{P} = \left[\widehat{p}[(i',j')|(i,j)]\right]_{(i,j),(i,j')\in\widehat{\mathcal{S}}}.$$

To simplify the notation, we denote

$$p_j(i'|i) := \mathcal{P}\left(X_{l+1} = i' \middle| X_l = i, \widetilde{X}_l = j\right),$$

which is the conditional transition probability distribution of $\boldsymbol{X}$ from state $X = i$ when the Markov chain $\widetilde{\boldsymbol{X}}$ is in state $\widetilde{X} = j$; and

$$\widetilde{p}_{i'|i}(j'|j) := \mathcal{P}\left(\widetilde{X}_{l+1} = j' \middle| \widetilde{X}_l = j, X_l = i, X_{l+1} = i'\right),$$

which is the conditional transition probability of the Markov chain $\widetilde{\boldsymbol{X}}$ moving from state j to state j', given that the Markov chain $\boldsymbol{X}$ moves from state i to state i'. Thus,

$$\widehat{p}[(i',j')|(i,j)] = p_j(i'|i)\widetilde{p}_{i'|i}(j'|j), \qquad i,i',j,j' \in \mathcal{S}. \tag{3.20}$$

With similar definitions, we have

$$\widehat{p}[(i',j')|(i,j)] = \widetilde{p}_i(j'|j)p_{j'|j}(i'|i), \qquad i,i',j,j' \in \mathcal{S}. \tag{3.21}$$

Summing up both sides of (3.20) and (3.21) over $i' \in \mathcal{S}$, we have

$$\widetilde{p}_i(j'|j) = \sum_{i' \in \mathcal{S}} p_j(i'|i)\widetilde{p}_{i'|i}(j'|j), \qquad j' \in \mathcal{S}.$$

Summing up both sides of (3.20) and (3.21) over $j' \in \mathcal{S}$, we have

$$p_j(i'|i) = \sum_{j' \in \mathcal{S}} \widetilde{p}_i(j'|j)p_{j'|j}(i'|i), \qquad j' \in \mathcal{S}.$$

If $\boldsymbol{X}$ and $\widetilde{\boldsymbol{X}}$ are independent, then $\widetilde{p}_{i'|i}(j'|j) = \widetilde{p}_i(j'|j) = p(j'|j)$ and $p_{j'|j}(i'|i) = p_j(i'|i) = p(i'|i)$, for all $i,i',j,j' \in \mathcal{S}$.

Now, let the reward function of $\widehat{\boldsymbol{X}}$ be $\hat{f}(i,j) = f(j) - f(i)$, the corresponding performance potentials of $\widehat{\boldsymbol{X}}$ be $\widehat{g}(i,j)$, and the steady-state probability distribution of $\widehat{\boldsymbol{X}}$ be $\widehat{\pi}(i,j)$, $i,j \in \mathcal{S}$. We have the Poisson equation for $\widehat{\boldsymbol{X}}$ (assuming $\widehat{P}$ is irreducible):

$$(I - \widehat{P})\widehat{g} + \widehat{\eta}e = \widehat{f}, \tag{3.22}$$

where $\widehat{\eta} = \widehat{\pi}\widehat{f}$ is the average reward.

Equation (3.22) holds for $\widehat{\boldsymbol{X}} = (\boldsymbol{X}, \widetilde{\boldsymbol{X}})$. In our case, both $\boldsymbol{X}$ and $\widetilde{\boldsymbol{X}}$ have the same transition probability matrix P. Thus, their steady-state probabilities are equal; i.e., $\pi(i) = \widetilde{\pi}(i)$, $i \in \mathcal{S}$. Thus, we have $\widehat{\eta} = \widehat{\pi}\widehat{f} = \sum_{i,j \in \mathcal{S}} \widehat{\pi}(i,j)\widehat{f}(i,j) = 0$, and (3.22) becomes

$$(I - \widehat{P})\widehat{g} = \widehat{f}.$$

In addition, although the transitions of $\boldsymbol{X}$ and $\widetilde{\boldsymbol{X}}$ are coupled, the transition of each of $\boldsymbol{X}$ and $\widetilde{\boldsymbol{X}}$ at any time must follow the transition probability matrix P. Precisely, we may require that

$$\widetilde{p}_i(j'|j) = p(j'|j), \qquad i \in \mathcal{S}, \tag{3.23}$$

and that

$$p_j(i'|i) = p(i'|i), \qquad j \in \mathcal{S}. \tag{3.24}$$

Under these conditions the coupling is reflected by the conditional transition probabilities $p_{j'|j}(i'|i)$ and $\widetilde{p}_{i'|i}(j'|j)$. Next, we show that, under these conditions, $\widehat{g}(i,j) = g(j) - g(i)$, $i,j \in \mathcal{S}$ is indeed a solution to (3.22).

To facilitate the matrix manipulation, we need to introduce some notation. Let $A = [a(i,j)]$ be an $m \times n$ matrix and $B = [b(i',j')]$ be an $m' \times n'$ matrix.

The *Kronecker product* of A and B is defined as the $(mm') \times (nn')$ matrix denoted as

$$A \otimes B = \begin{bmatrix} a(1,1)B & \dots & a(1,n)B \\ \vdots & \ddots & \vdots \\ a(m,1)B & \dots & a(m,n)B \end{bmatrix}.$$

For clarity, we use e_m to denote an m-dimensional column vector with all components being one.

With this notation, we can verify that

$$\hat{f} = (e_S \otimes f) - (f \otimes e_S).$$

Conditions (3.23) and (3.24) are equivalent to

$$\widehat{P}(I \otimes e_S) = P \otimes e_S, \tag{3.25}$$

and

$$\widehat{P}(e_S \otimes I) = e_S \otimes P. \tag{3.26}$$

We can easily derive that, for any matrix A and vector g, if Ag is well defined, then $(A \otimes e)g = (Ag) \otimes e$ and $(e \otimes A)g = e \otimes (Ag)$, for an e with any dimension.

Finally, from (3.25) and (3.26), we have

$$\begin{aligned} & (I - \widehat{P})(e_S \otimes I - I \otimes e_S)g \\ = & (e_S \otimes I - I \otimes e_S)g - [e_S \otimes (Pg) - (Pg) \otimes e_S] \\ = & e_S \otimes [(I - P)g] - [(I - P)g] \otimes e_S \\ = & e_S \otimes [(I - P + e_S\pi)g] - [(I - P + e_S\pi)g] \otimes e_S \\ = & e_S \otimes f - f \otimes e_S \\ = & \widehat{f}. \end{aligned}$$

Thus, under conditions (3.23) and (3.24),

$$\widehat{g} = e_S \otimes g - g \otimes e_S$$

is indeed one of the solutions to (3.22). That is, $\widehat{g}(i,j) = g(j) - g(i) = \gamma(i,j)$, $i,j \in \mathcal{S}$, are the realization factors of $\boldsymbol{X}$. We have

$$e_{S^2}^T \widehat{g} = 0, \quad \text{and} \quad e_{S^2}^T \widehat{f} = 0.$$

Equation (3.22) is the *perturbation realization factor (PRF) equation with coupled sample paths.*

Now, we discuss the numerical method for solving (3.22). Let ν be any S^2 dimensional row vector such that $\nu e_{S^2} = 1$, and $\nu \widehat{g} = 0$. For example, we can take $\nu = \frac{1}{S^2} e_{S^2}^T$. We can write the PRF equation (3.22) as

$$(I - \widehat{P} + e_{S^2}\nu)\widehat{g} = \hat{f}.$$

We can prove (see Problem 3.2) that the eigenvalues of $\widehat{P} - e_{S^2}\nu$ are all in the unit circle. Thus, we have the following expansion:

$$\begin{aligned} \widehat{g} &= (I - \widehat{P} + e_{S^2}\nu)^{-1}\widehat{f} \\ &= \sum_{l=0}^{\infty}(\widehat{P} - e_{S^2}\nu)^l \widehat{f}. \end{aligned} \tag{3.27}$$

Let λ be one of the eigenvalues of P and x be its corresponding eigenvector. Define $\widehat{x} = x \otimes e_S$. It is easy to verify that λ is the eigenvalue of $\widehat{P}$ with eigenvector $\widehat{x}$. Therefore, all the eigenvalues of P are the eigenvalues of $\widehat{P}$ (which may have other eigenvalues). Thus, the convergence rate of (3.27) cannot be better than (3.1) or (3.3). Therefore, the coupling approach cannot improve the convergence rate of the numerical algorithms for calculating Γ ($\gamma(i,j) = \widehat{g}(i,j)$).

The coupling method is generally used in simulation to reduce the variance of the estimates for the difference of the mean of two different random variables. Relative references include [31, 33, 91, 92, 115, 127, 177, 179, 212, 213]. Applying this approach to estimate $\gamma(i,j) = g(j) - g(i)$ with two coupled sample paths still requires further research and we will not discuss the details in this book. Problems 3.9 and 3.10 provide a brief introduction to this variance-reduction simulation approach.

3.2 Performance Derivatives

3.2.1 Estimating through Potentials

The performance potentials obtained by numerical methods or by learning from sample paths can be used to calculate the performance derivatives using the performance derivative formula (2.23):

$$\frac{d\eta_\delta}{d\delta} = \pi(\Delta P)g. \tag{3.28}$$

We first give a few numerical examples.

Example 3.3. We consider a Markov chain with the same transition probability matrix P and reward function f as those in Examples 3.1 and 3.2. To study the derivatives of the average reward, we arbitrarily choose the direction of P as

$$\Delta P = \begin{bmatrix} -0.010 & 0.000 & 0.005 & 0.005 & -0.010 & 0.010 & 0.010 & 0.005 & 0.005 & -0.020 \\ -0.010 & 0.000 & 0.000 & 0.015 & 0.005 & 0.005 & -0.005 & -0.005 & -0.005 & 0.000 \\ 0.000 & 0.010 & 0.010 & 0.010 & 0.000 & -0.010 & 0.000 & -0.010 & -0.010 & 0.000 \\ 0.005 & -0.020 & 0.005 & 0.000 & 0.005 & 0.000 & 0.010 & -0.010 & 0.005 & 0.000 \\ 0.000 & 0.010 & -0.010 & 0.000 & 0.010 & 0.000 & -0.010 & 0.010 & 0.000 & -0.010 \\ 0.000 & 0.010 & -0.010 & 0.000 & -0.020 & 0.005 & 0.005 & 0.000 & 0.005 & 0.005 \\ 0.010 & -0.010 & 0.010 & -0.010 & 0.010 & -0.010 & 0.010 & -0.010 & 0.010 & -0.010 \\ 0.000 & 0.010 & 0.000 & -0.010 & 0.000 & 0.010 & 0.000 & -0.005 & 0.000 & -0.005 \\ 0.010 & -0.010 & -0.020 & 0.000 & 0.010 & 0.010 & 0.010 & 0.000 & 0.000 & -0.010 \\ 0.010 & -0.010 & 0.000 & 0.010 & -0.010 & -0.010 & 0.010 & -0.010 & 0.010 & 0.000 \end{bmatrix}.$$

We use the sample-path-based estimates of potentials in the form of (3.15) to compute the derivatives. To study the effect of L, we choose $L = 1, 2, 3, 5, 10, 15, 20$. For each value of L, we do two sets of simulation, with each set having ten runs. Each simulation run contains 100,000 state transitions in the first set and 1,000,000 transitions in the second set. The theoretical value of the derivative is -0.1176. The means and standard deviations of the derivatives calculated by (3.28) using the sample-path-based potential estimates in these two sets of simulations are listed in Tables 3.6 and 3.7.

L	1	2	3	5	10	15	20	Theoretic
Mean	-0.0979	-0.1224	-0.1162	-0.1172	-0.1180	-0.1183	-0.1176	-0.1176
SD	0.00045	0.00059	0.00070	0.00151	0.00186	0.00261	0.00216	-

Table 3.6. The Performance Derivatives in Example 3.3 with 100,000 Transitions

L	1	2	3	5	10	15	20	Theoretic
Mean	-0.0989	-0.1229	-0.1167	-0.1176	-0.1178	-0.1176	-0.1174	-0.1176
SD	0.00009	0.00015	0.00016	0.00025	0.00026	0.00047	0.00059	-

Table 3.7. The Performance Derivatives in Example 3.3 with 1,000,000 Transitions

These tables show that the estimate is quite accurate even when L is as small as 2 or 3. The standard deviation is acceptable even if L is 20. Thus, the results are not so sensitive to the value of L. It is interesting to note that even if we choose $L = 1$ in this case, the error is only about 17%. $L = 1$ means using the reward function to approximate the potentials, i.e., assuming that $g \approx f$. This corresponds to the "myopic" view in optimization: When the system jumps to state i, we just use the one step reward $f(i)$ to represent its effect on the long-run performance. □

3.2.2 Learning Directly

One disadvantage of the approach in Section 3.2.1 is that it requires us to estimate the potentials for all the states. This is sometimes difficult for a number of reasons: The number of states may be too large; some states may be visited very rarely; and for systems with special structures (e.g. queueing networks), it may not be convenient even to list out all the states. In this subsection, we show that the performance derivatives can be estimated directly from sample paths without estimating each individual potential.

An analogue is the estimation of the performance measure itself. There are two ways to do the estimation: We may estimate all $\pi(i)$ first and then use $\eta = \pi f$ to calculate the performance, or we may estimate η directly by

$$\eta = \lim_{L \to \infty} \frac{1}{L} \sum_{l=0}^{L-1} f(X_l), \qquad \text{w.p.1.} \tag{3.29}$$

This direct estimation balances the accuracy of $\pi(i)$ and the frequency of the visits to i: If i is not visited often, then $\pi(i)$ may not be accurately estimated; meanwhile, its effect on η is also small. We wish to develop equations similar to (3.29) for the derivatives of average rewards.

A Basic Formula and a General Algorithm

We first present a basic formula for the direct estimation of the derivatives of average rewards. This formula is the foundation of the sample-path-based algorithms. With this formula, a general algorithm for derivatives can be developed; many other algorithms can be viewed as special cases of this general algorithm [61].

Consider a stationary Markov chain $\boldsymbol{X} = (X_0, X_1, \ldots)$. (This implies that the initial probability distribution is the steady-state distribution π.) Let E denote the expectation on the probability space generated by $\boldsymbol{X}$. Because it is impossible for a sample path with transition matrix P to contain information about $\Delta P = P' - P$, we need to use a standard technique in simulation called *importance sampling*. First, we make a standard assumption in importance sampling: For any $i, j \in \mathcal{S}$, if $\Delta p(j|i) \neq 0$, then $p(j|i) \neq 0$. This assumption allows us to analyze the effect of $\Delta p(j|i)$ based on the information observed when the system moves from state i to state j on $\boldsymbol{X}$. If the assumption does not hold, we may have $p'(j|i) > 0$ while $p(j|i) = 0$ for some $i, j \in \mathcal{S}$. In this case, a sample path of $\boldsymbol{X}$ does not contain any transition from i to j, and we may need to observe two or more transitions (see Problem 3.11).

First, we have (2.23)

$$\begin{aligned} \frac{d\eta_\delta}{d\delta} &= \pi \Delta P g = \sum_{i \in \mathcal{S}} \sum_{j \in \mathcal{S}} [\pi(i) \Delta p(j|i) g(j)] \\ &= \sum_{i \in \mathcal{S}} \sum_{j \in \mathcal{S}} \left\{ \pi(i) p(j|i) \left[\frac{\Delta p(j|i)}{p(j|i)} g(j) \right] \right\}. \end{aligned}$$

For a stationary Markov chain $\boldsymbol{X} = \{X_l, l = 0, 1, \ldots\}$, this is

$$\frac{d\eta_\delta}{d\delta} = E \left\{ \frac{\Delta p(X_{l+1}|X_l)}{p(X_{l+1}|X_l)} g(X_{l+1}) \right\}, \tag{3.30}$$

which does not depend on l. Next, let $\hat{g}(X_{l+1}, X_{l+2}, \ldots)$ be an unbiased estimate of $g(X_{l+1})$; i.e., let

$$g(i) = E\{\hat{g}(X_{l+1}, X_{l+2}, \ldots) | X_{l+1} = i\}, \qquad i \in \mathcal{S}. \tag{3.31}$$

With (3.31), we have

$$
\begin{aligned}
& E\left\{\frac{\Delta p(X_{l+1}|X_l)}{p(X_{l+1}|X_l)}\hat{g}(X_{l+1}, X_{l+2}, \ldots)\right\} \\
= & E\left\{E\left[\frac{\Delta p(X_{l+1}|X_l)}{p(X_{l+1}|X_l)}\hat{g}(X_{l+1}, X_{l+2}, \ldots)\Big| X_l, X_{l+1}\right]\right\} \\
= & E\left\{\frac{\Delta p(X_{l+1}|X_l)}{p(X_{l+1}|X_l)} E\left[\hat{g}(X_{l+1}, X_{l+2}, \ldots)\Big| X_l, X_{l+1}\right]\right\} \\
= & E\left\{\frac{\Delta p(X_{l+1}|X_l)}{p(X_{l+1}|X_l)} g(X_{l+1})\right\}.
\end{aligned}
$$

Therefore, we have the following *basic formula*:

$$
\frac{d\eta_\delta}{d\delta} = E\left\{\frac{\Delta p(X_{l+1}|X_l)}{p(X_{l+1}|X_l)}\hat{g}(X_{l+1}, X_{l+2}, \ldots)\right\}. \tag{3.32}
$$

With this formula, we can develop a general algorithm for estimating derivatives. In fact, for an ergodic Markov chain $\boldsymbol{X} = \{X_0, X_1, \ldots\}$, we have

$$
\frac{d\eta_\delta}{d\delta} = \lim_{N\to\infty} \frac{1}{N} \sum_{n=0}^{N-1} \left\{\left[\frac{\Delta p(X_{n+1}|X_n)}{p(X_{n+1}|X_n)}\right] \hat{g}(X_{n+1}, X_{n+2}, \ldots)\right\}, \quad \text{w.p.1}, \tag{3.33}
$$

where $\hat{g}(X_{n+1}, X_{n+2}, \ldots)$ is any function satisfying (3.31).

The proof of (3.33) follows directly from the fundamental ergodicity theorem (3.17) by simply defining

$$
\phi(X_n, X_{n+1}, \ldots) = \frac{\Delta p(X_{n+1}|X_n)}{p(X_{n+1}|X_n)}\hat{g}(X_{n+1}, X_{n+2}, \ldots).
$$

Specific Algorithms

With different estimates or approximations of the potentials, (3.33) leads to a few specific approximate algorithms for the derivatives of average rewards.

Algorithm 3.1. *(Approximation by truncation)*
With (3.14), we have

$$
g(i) \approx g_L(i) = E\left\{\sum_{l=0}^{L-1} f(X_l)\Big| X_0 = i\right\}.
$$

Therefore, from (3.31), we may choose

$$\hat{g}(X_{n+1}, X_{n+2}, \cdots) \approx \sum_{l=0}^{L-1} f(X_{n+l+1}).$$

Using this $\hat{g}$ in (3.32) and (3.33), we get

$$\begin{aligned}\frac{d\eta_\delta}{d\delta} &\approx E\left\{\frac{\Delta p(X_{n+1}|X_n)}{p(X_{n+1}|X_n)}\left[\sum_{l=0}^{L-1} f(X_{n+l+1})\right]\right\} \\ &= \lim_{N\to\infty}\frac{1}{N}\left\{\sum_{n=0}^{N-1}\left[\frac{\Delta p(X_{n+1}|X_n)}{p(X_{n+1}|X_n)}\right]\left[\sum_{l=0}^{L-1} f(X_{n+l+1})\right]\right\}, \qquad \text{w.p.1.}\end{aligned} \tag{3.34}$$

This is equivalent to

$$\frac{d\eta_\delta}{d\delta} \approx \lim_{N\to\infty}\frac{1}{N}\sum_{n=0}^{N-1}\left\{f(X_{n+L})\sum_{l=0}^{L-1}\left[\frac{\Delta p(X_{n+l+1}|X_{n+l})}{p(X_{n+l+1}|X_{n+l})}\right]\right\}, \qquad \text{w.p.1.} \tag{3.35}$$

This algorithm and similar ones for Markov processes and queueing networks are presented in [69].

Example 3.4. We repeat the simulation for the same Markov chain as in Example 3.3 and apply (3.35) to estimate the derivative of the average reward. We perform ten simulation runs for each value of L and the results are listed in Table 3.8. The table shows that for $L = 2$ to 15, (3.35) yields very accurate estimates. When L increases further from 20, the estimate becomes inaccurate because the variance becomes larger. □

By the ergodicity of the Markov chain, (3.35) can be written as

$$\frac{d\eta_\delta}{d\delta} \approx E\left\{f(X_{n+L})\sum_{l=0}^{L-1}\left[\frac{\Delta p(X_{n+l+1}|X_{n+l})}{p(X_{n+l+1}|X_{n+l})}\right]\right\}, \tag{3.36}$$

where "E" denotes the steady-state expectation. Define

$$\rho_L(i) = E\left\{\sum_{l=0}^{L-1}\left[\frac{\Delta p(X_{l+1}|X_l)}{p(X_{l+1}|X_l)}\right]\middle| X_L = i\right\}.$$

Then, (3.36) becomes

L	1	2	3	5	10	15
Mean	-0.0973	-0.1221	-0.1157	-0.1163	-0.1151	-0.1137
SD	0.0033	0.0067	0.0104	0.0162	0.0305	0.00443

L	25	50	75	100	200	Theoretic
Mean	-0.1098	-0.1035	-0.0933	-0.0797	-0.0434	-0.1176
SD	0.0760	0.1522	0.2300	0.3086	0.6351	-

Table 3.8. The Performance Derivatives in Example 3.4 with 100,000 Transitions

$$\frac{d\eta_\delta}{d\delta} \approx \sum_{i \in \mathcal{S}} \pi(i) f(i) \rho_L(i).$$

Define

$$\rho(i) = \lim_{L \to \infty} \rho_L(i), \qquad i \in \mathcal{S}. \tag{3.37}$$

Then, by the above derivation, we have

$$\frac{d\eta_\delta}{d\delta} = \sum_{i \in \mathcal{S}} \pi(i) f(i) \rho(i). \tag{3.38}$$

Equation (3.38) and the convergence of (3.37) can be rigorously proved; see Problem 3.15.

Algorithm 3.2. *(Approximation by discounting)*

Because $\lim_{\beta \uparrow 1} g_\beta = g$, potential g can be approximated by the β-potential g_β in (2.45):

$$g_\beta(i) = E\left\{ \sum_{l=0}^{\infty} \beta^l [f(X_l) - \eta] \Big| X_0 = i \right\},$$

with $0 < \beta < 1$ being a discount factor. Ignoring the constant term, we have the approximation of the potential as follows:

$$g_\beta(i) = E\left\{ \sum_{l=0}^{\infty} \beta^l f(X_l) \Big| X_0 = i \right\}.$$

Therefore, we can choose

$$\hat{g}(X_{n+1}, X_{n+2}, \ldots) \approx \sum_{l=0}^{\infty} \beta^l f(X_{n+l+1}).$$

Using this as the $\hat{g}$ in (3.33), we get

$$\frac{d\eta_\delta}{d\delta} \approx \lim_{N\to\infty} \frac{1}{N} \left\{ \sum_{n=0}^{N-1} \left[\frac{\Delta p(X_{n+1}|X_n)}{p(X_{n+1}|X_n)} \right] \left[\sum_{l=0}^{\infty} \beta^l f(X_{n+l+1}) \right] \right\}, \qquad \text{w.p.1}$$

$$= E\left\{ \frac{\Delta p(X_{n+1}|X_n)}{p(X_{n+1}|X_n)} \left[\sum_{l=0}^{\infty} \beta^l f(X_{n+l+1}) \right] \right\}. \tag{3.39}$$

The right-hand side of (3.39) equals the sum of the two terms:

$$\lim_{N\to\infty} \frac{1}{N} \sum_{n=0}^{N-1} \left\{ \frac{\Delta p(X_{n+1}|X_n)}{p(X_{n+1}|X_n)} \left[\sum_{l=0}^{N-n-1} \beta^l f(X_{n+l+1}) + \sum_{l=N-n}^{\infty} \beta^l f(X_{n+l+1}) \right] \right\}. \tag{3.40}$$

For the second term, we have

$$\left| \frac{1}{N} \sum_{n=0}^{N-1} \left\{ \frac{\Delta p(X_{n+1}|X_n)}{p(X_{n+1}|X_n)} \left[\sum_{l=N-n}^{\infty} \beta^l f(X_{n+l+1}) \right] \right\} \right|$$
$$\leq \max_{i,j\in\mathcal{S}} \left| \frac{\Delta p(j|i)}{p(j|i)} \right| \max_{i\in\mathcal{S}} |f(i)| \left\{ \frac{1}{N} \sum_{n=0}^{N-1} \sum_{l=N-n}^{\infty} \beta^l \right\}$$
$$\to 0, \qquad \text{as } N \to \infty.$$

Therefore, the second term in (3.40) is zero, and (3.39) becomes

$$\frac{d\eta_\delta}{d\delta} \approx \lim_{N\to\infty} \frac{1}{N} \sum_{n=0}^{N-1} \left\{ \frac{\Delta p(X_{n+1}|X_n)}{p(X_{n+1}|X_n)} \left[\sum_{l=0}^{N-n-1} \beta^l f(X_{n+l+1}) \right] \right\}, \qquad \text{w.p.1.}$$

We exchange the order of the above two finite sums and obtain

$$\frac{d\eta_\delta}{d\delta} \approx \lim_{N\to\infty} \frac{1}{N} \sum_{n=1}^{N} \left\{ f(X_n) \sum_{l=0}^{n-1} \left[\beta^{n-l-1} \frac{\Delta p(X_{l+1}|X_l)}{p(X_{l+1}|X_l)} \right] \right\}, \qquad \text{w.p.1.} \tag{3.41}$$

This is the policy-gradient algorithm developed in [17, 18].

We can calculate $z_n := \sum_{l=0}^{n-1} \left[\beta^{n-l-1} \frac{\Delta p(X_{l+1}|X_l)}{p(X_{l+1}|X_l)} \right]$ recursively: set $z_0 = 0$ and

$$z_{k+1} = \beta z_k + \frac{\Delta p(X_{k+1}|X_k)}{p(X_{k+1}|X_k)}, \qquad k \geq 0.$$

On the other hand, to calculate $\sum_{l=0}^{L-1} \left[\frac{\Delta p(X_{n+l+1}|X_{n+1})}{p(X_{n+l+1}|X_{n+1})} \right]$ in Algorithm 3.1, we have to store L values.

Finally, the discount factor approximation can also be used to reduce the variance in estimating the performance gradients [198].

Example 3.5. We repeat the simulation for the same Markov system as in Examples 3.3 and 3.4 and apply (3.41) to estimate the performance derivative. We perform ten simulation runs for each value of the discount factor β, and the means and standard deviations of the estimates are listed in Table 3.9. The table shows that for $\beta = 0.8$ to 0.9, the algorithm in (3.41) yields very accurate estimates. Of course, when β increases, g_β increases and goes closer to g. However, when β increases, the variance of the estimate also increases. This explains why the estimation error becomes larger for $\beta > 0.9$. Thus, when we choose the value of β, we need to balance both bias and variance. The table shows that $\beta = 0.8$ to 0.9 are the best choices. □

β	0.80	0.85	0.90	0.95	0.97	Theoretic
Mean	-0.113	-0.114	-0.114	-0.111	-0.125	-0.1176
SD	0.016	0.021	0.031	0.061	0.065	-

Table 3.9. The Performance Derivatives in Example 3.5 with 100,000 Transitions

Algorithm 3.3. *(Based on perturbation realization factors)*

It is sometimes easier and more accurate to estimate the potentials via perturbation realization factors $\gamma(i,j) = g(j) - g(i)$, $i, j, \in \mathcal{S}$. This is based on (2.17)

$$\gamma(i,j) = E\left\{ \sum_{l=0}^{L(i|j)-1} [f(X_l) - \eta] \Big| X_0 = j \right\}.$$

To develop a direct algorithm for derivatives of average rewards, we first use the above equation to obtain $\hat{g}$. To this end, we choose any regenerative state i^* as a reference point and set $g(i^*) = 0$. Then, for any state $i \in \mathcal{S}$, we have

$$g(i) = g(i) - g(i^*) = \gamma(i^*, i).$$

For convenience, we set $X_0 = i^*$ and define $u_0 = 0$, and we let $u_{k+1} = \min\{n : n > u_k, X_n = i^*\}$ be the sequence of regenerative points. For any time instant $n \geq 0$, we define an integer $m(n)$ such that $u_{m(n)} \leq n < u_{m(n)+1}$. This implies that $u_{m(n)} = n$ when $i = i^*$. From (2.17), we have

$$g(i) = \gamma(i^*, i) = E\left\{ \sum_{l=n}^{u_{m(n)+1}-1} \Big[f(X_l) - \eta\Big] \Big| X_n = i \right\}.$$

Choosing $\hat{g}(X_{n+1}, \ldots) = \sum_{l=n+1}^{u_{m(n+1)+1}-1} [f(X_l) - \eta]$ in (3.32) and (3.33), we have

$$\frac{d\eta_\delta}{d\delta} = E\left\{\left[\frac{\Delta p(X_{n+1}|X_n)}{p(X_{n+1}|X_n)}\right]\sum_{l=n+1}^{u_{m(n+1)+1}-1}[f(X_l)-\eta]\right\}$$

$$= \lim_{N\to\infty}\frac{1}{N}\sum_{n=0}^{N-1}\left\{\left[\frac{\Delta p(X_{n+1}|X_n)}{p(X_{n+1}|X_n)}\right]\sum_{l=n+1}^{u_{m(n+1)+1}-1}[f(X_l)-\eta]\right\}, \quad \text{w.p.1} \tag{3.42}$$

$$= \lim_{N\to\infty}\frac{1}{N}\sum_{n=1}^{N}\left\{[f(X_n)-\eta]\sum_{l=u_{m(n)}}^{n}\frac{\Delta p(X_l|X_{l-1})}{p(X_l|X_{l-1})}\right\}, \quad \text{w.p.1}, \tag{3.43}$$

where $u_{m(n+1)+1}$ is the first time after X_{n+1} that the Markov chain reaches state i^*.

Next, define

$$\hat{w}_{n+1} = \sum_{l=n+1}^{u_{m(n+1)+1}-1}[f(X_l)-\eta].$$

By the regenerative property, from (3.42) we have

$$\frac{d\eta_\delta}{d\delta} = \frac{E\left\{\sum_{k=u_m}^{u_{m+1}-1}\left(\frac{\Delta p(X_{k+1}|X_k)}{p(X_{k+1}|X_k)}\hat{w}_{k+1}\right)\right\}}{E[u_{m+1}-u_m]}$$
$$= E\left(\frac{\Delta p(X_{k+1}|X_k)}{p(X_{k+1}|X_k)}\hat{w}_{k+1}\right). \tag{3.44}$$

Define

$$\hat{r}_n = \sum_{l=u_{m(n)}}^{n}\frac{\Delta p(X_l|X_{l-1})}{p(X_l|X_{l-1})}.$$

Therefore, (3.43) takes the following form

$$\frac{d\eta_\delta}{d\delta} = \frac{E\left\{\sum_{k=u_m}^{u_{m+1}-1}[f(X_k)-\eta]\hat{r}_k\right\}}{E[u_{m+1}-u_m]}$$
$$= E\left\{[f(X_k)-\eta]\hat{r}_k\right\}. \tag{3.45}$$

The optimization scheme proposed in [197] is essentially a result of combining the above algorithms with stochastic approximation techniques. See Section 6.3.1 for additional discussion.

Example 3.6. We repeat the simulation for the same Markov system as in Examples 3.3, 3.4, and 3.5. We perform ten simulation runs and apply (3.43) to estimate the performance derivatives. The mean is -0.1191 and the standard deviation is 0.0075. □

Algorithm 3.4. *(Parameterized policy spaces)*

Now, we consider a parameterized space of transition probability matrices denoted as $P_\theta = [p_\theta(j|i)]$, $i, j \in \mathcal{S}$, where θ is a continuous parameter and $\frac{d}{d\theta}\{p_\theta(j|i)\}$ exists for all $i, j \in \mathcal{S}$. We assume that the Markov chains under all transition probability matrices are ergodic. The corresponding steady-state probabilities and average rewards are denoted as π_θ and $\eta_\theta = \pi_\theta f$. For simplicity, we assume that the reward function f is the same for all P_θ. (The extension to f_θ depending on θ is straightforward.)

Algorithms for the derivatives of average rewards can be developed by replacing $\Delta p(j|i)$ in (3.32) and (3.33) with $\frac{d}{d\theta}\{p_\theta(j|i)\}$. For example, the basic formula (3.32) becomes

$$\frac{d\eta_\theta}{d\theta} = E\left\{ \frac{\frac{d}{d\theta}p_\theta(X_{n+1}|X_n)}{p(X_{n+1}|X_n)} \hat{g}(X_{n+1}, X_{n+2}, \ldots) \right\},$$

where $\hat{g}(X_{n+1}, X_{n+2}, \ldots)$ is an unbiased estimate of $g(i)$, given $X_{n+1} = i$. The specific algorithms (3.34) and (3.39) become

$$\frac{d\eta_\theta}{d\theta} = E\left\{ \frac{\frac{d}{d\theta}p_\theta(X_{n+1}|X_n)}{p(X_{n+1}|X_n)} \left[\sum_{l=0}^{L-1} f(X_{n+l+1}) \right] \right\}, \tag{3.46}$$

$$\frac{d\eta_\theta}{d\theta} = E\left\{ \frac{\frac{d}{d\theta}p_\theta(X_{n+1}|X_n)}{p(X_{n+1}|X_n)} \left[\sum_{l=0}^{\infty} \beta^l f(X_{n+l+1}) \right] \right\}. \tag{3.47}$$

From (3.44), we have

$$\begin{aligned} \frac{d\eta_\theta}{d\theta} &= E\left\{ \frac{\frac{d}{d\theta}p_\theta(X_{n+1}|X_n)}{p(X_{n+1}|X_n)} \left[\sum_{l=n+1}^{u_{m(n+1)+1}-1} [f(X_l) - \eta] \right] \right\} \\ &= \frac{E\left\{ \sum_{n=u_m}^{u_{m+1}-1} \left[\frac{\frac{d}{d\theta}p(X_{n+1}|X_n)}{p(X_{n+1}|X_n)} \left(\sum_{l=n+1}^{u_{m(n+1)+1}-1} [f(X_l) - \eta] \right) \right] \right\}}{E[u_{m+1} - u_m]}. \end{aligned} \tag{3.48}$$

From(3.45), we have

$$\begin{aligned} \frac{d\eta_\theta}{d\theta} &= E\left\{ [f(X_n) - \eta] \sum_{l=u_{m(n)}}^{n} \frac{\frac{d}{d\theta}p_\theta(X_l|X_{l-1})}{p(X_l|X_{l-1})} \right\} \\ &= \frac{E\left\{ \sum_{n=u_m}^{u_{m+1}-1} \left[[f(X_n) - \eta] \sum_{l=u_{m(n)}}^{n} \frac{\frac{d}{d\theta}p_\theta(X_l|X_{l-1})}{p(X_l|X_{l-1})} \right] \right\}}{E[u_{m+1} - u_m]}. \end{aligned} \tag{3.49}$$

Other equations similar to (3.35) and (3.41) can be developed.

Example 3.7. The above estimation algorithms are applied to the partially-observable Markov decision processes (POMDPs) in [17, 18]. (This example can be better understood after reading the materials in Chapter 4 about the Markov decision processes.)

In this example, we use the following simple parameterized model. In addition to the state space $\mathcal{S}$, there is a finite action space denoted as $\mathcal{A}$ and a finite observation space denoted as $\mathcal{Y}$. Each $\alpha \in \mathcal{A}$ determines a transition probability matrix $P^\alpha = [p^\alpha(j|i)]$. When the Markov chain is in state $i \in \mathcal{S}$, an observation $y \in \mathcal{Y}$ is obtained according to a probability distribution $\nu_i(y)$. For any observation y, we may choose a randomized policy $\mu_y(\alpha)$, which is a probability distribution over the action space $\mathcal{A}$. It is assumed that the distribution depends on a parameter θ and therefore is denoted as $\mu_y(\theta, \alpha)$. When $y \in \mathcal{Y}$ is observed, with policy $\mu_y(\theta, \alpha)$, we take action $\alpha \in \mathcal{A}$ with probability $\mu_y(\theta, \alpha)$. Furthermore, we assume that P^α does not depend on θ.

Given an observation distribution $\nu_i(y)$ and a randomized policy $\mu_y(\theta, \alpha)$, the corresponding transition probability is

$$p_\theta(j|i) = \sum_{\alpha, y} \{\nu_i(y)\mu_y(\theta, \alpha)p^\alpha(j|i)\}.$$

Therefore,

$$\frac{d}{d\theta}p_\theta(j|i) = \sum_{\alpha, y} \left[\nu_i(y)p^\alpha(j|i)\frac{d}{d\theta}\mu_y(\theta, \alpha)\right]. \tag{3.50}$$

We further assume that although the state X_n, $n = 0, 1, \ldots$, is not completely observable, the cost at any time $f(X_n)$ is known (e.g., by observation, or it depends only on the action). Then, the algorithms in (3.46) and (3.47) can be used with (3.50). If, in addition, there is a state i^*, which is irreducible for all policies, then the algorithm in (3.49) can be used. □

Finally, all the above algorithms are expressed in sample-path-based averages. Stochastic approximation based recursive algorithms can be developed based on these average-type algorithms. We will study these topics in Chapter 6.

Performance Derivatives for Queueing Systems

A direct learning algorithm for performance derivatives of queueing networks has been presented as Algorithm 2.2 in Section 2.4.1. An algorithm for the derivatives of the mean response time with respect to service rate in an M/G/1 queue is given in Example 2.10 in Section 2.4.3.

As explained in Section 2.4.3, Algorithm 2.2 directly estimates the performance derivative via $\sum_{\text{all } \boldsymbol{n}} \pi(\boldsymbol{n})c^{(f)}(\boldsymbol{n}, v)$ (see (2.108)) without estimating every perturbation realization factor $c^{(f)}(\boldsymbol{n}, v)$ and every steady-state probability $\pi(\boldsymbol{n})$ separately. The same explanation applies to the algorithm in Example 2.10 in Section 2.4.3.

It is interesting to note the difference in the process of developing the PA theory for both queueing systems and Markov systems. For queueing systems, the performance derivative estimation algorithms were developed first, and the concept of the perturbation realization factor and the performance derivative formula were developed later to provide a theoretical background for the algorithms. For Markov systems, the concept of performance potentials and performance derivatives were developed first, and the sample-path-based algorithms, both for potentials and for derivatives directly, were proposed later, by using the formulas.

The algorithms for estimating $c^{(f)}(\boldsymbol{n}, v)$ in queueing systems should be easy to develop; however, there has not been much effort in this direction, perhaps because there have not been many applications with $c^{(f)}(\boldsymbol{n}, v)$ alone so far. On the other hand, as we will see in Chapter 4, the estimated potentials can also be used in policy iteration optimization of Markov systems. In a recent study, the relation between the realization factors $c^{(f)}(\boldsymbol{n}, v)$ (with a queueing model) and the potentials $g(\boldsymbol{n})$ (with a Markov model) is established, and policy-iteration-type algorithms are developed for (customer-average) performance optimization of queueing systems based on $c^{(f)}(\boldsymbol{n}, v)$; see [260]. In such algorithms, the realization factors $c^{(f)}(\boldsymbol{n}, v)$ or their aggregations need to be calculated or estimated on sample paths.

3.3 Optimization with PA

3.3.1 Gradient Methods and Stochastic Approximation

The PA gradient estimates can be used to implement sample-path-based performance optimization. When the sample path is long enough, the estimates are very accurate and we can simply use them in any gradient-based optimization procedure [22, 23, 85] for deterministic systems. If the sample path is short, then the gradient estimates contain stochastic errors, and stochastic approximation techniques have to be used in developing optimization algorithms.

As shown in Figure 3.1, we will leave the stochastic approximation-based recursive algorithms to Chapter 6, in which we first introduce the related material in stochastic approximation in some detail. In this section, we discuss some fundamental methods in performance optimization with accurate estimates of the performance gradients.

Gradient Methods and the Robbins-Monro Algorithm

In general, we consider the optimization of a performance function $\eta(\theta) : \mathcal{D} \rightarrow \mathcal{R}$, where $\mathcal{R} = (-\infty, \infty)$ and $\mathcal{D} \subseteq \mathcal{R}^M$ is a convex M-dimensional parameter subset. Denote the performance gradients at any point $\theta \in \mathcal{D}$ as $\frac{d\eta(\theta)}{d\theta} := (\frac{\partial\eta(\theta)}{\partial\theta(1)}, \ldots, \frac{\partial\eta(\theta)}{\partial\theta(M)})^T$, where $\theta(i)$, $i = 1, 2, \ldots, M$, is the ith component of θ. Let θ^* be a local optimal point of $\eta(\theta)$ in $\mathcal{D}$. We have $\frac{d\eta(\theta^*)}{d\theta} = 0$. We want to find out a local optimal point. This is a constrained optimization problem.

Suppose that the performance gradients $\frac{d\eta(\theta)}{d\theta}$ can be accurately estimated. We may find θ^* iteratively by using any gradient-based method (see, e.g., Chapter 2 of [23]). We start with an initial point $\theta_0 \in \mathcal{D}$. At the kth iteration, we run the system with parameter θ_k, $k = 0, 1, \ldots$, and apply PA on a long sample path to estimate the performance gradients at θ_k, $\frac{d\eta(\theta_k)}{d\theta}$. Set $h_k := \frac{d\eta(\theta_k)}{d\theta}$. In the simplest gradient method, the parameter θ is updated according to

$$\theta_{k+1} = \Pi_{\mathcal{D}}\left(\theta_k + \kappa_k h_k\right), \tag{3.51}$$

where $\Pi_{\mathcal{D}}$ denotes a projection onto $\mathcal{D}$, and $\kappa_k > 0$ is called a *step size.*

It can be shown that under some conditions on $\eta(\theta)$, θ_k converges to a local optimal point θ^* as $k \to \infty$, when κ_k is a small positive constant (e.g., in Example 6.1 in Chapter 6, $\theta_k \to \theta^*$ if $0 < \kappa_k = \kappa < 1$). Under some other conditions on $\eta(\theta)$, the convergence of θ_k requires $\kappa_k \to 0$ and $\sum_{k=0}^{\infty} \kappa_k = \infty$. The convergence of the algorithm (3.51) may be slow, and other methods such as Newton's method and Armijo's rule etc. can be used to improve the convergence rate. The detailed analysis of the gradient algorithms is beyond the scope of this book and can be found in, e.g., [23].

Because of the stochastic nature of the system, the gradient estimate obtained from any sample path with a finite length contains stochastic errors. We denote such a noisy (usually unbiased) estimate as

$$\hat{h}_k := \frac{\widehat{d\eta(\theta_k)}}{d\theta}.$$

The problem becomes to find the zeros of a function $\frac{d\eta(\theta)}{d\theta}$, which cannot be measured accurately. This is a topic in stochastic approximation (SA). With SA, we may simply replace the accurate value of the gradient in (3.51) with its estimate (cf. (6.6) and (6.7) in Chapter 6):

$$\theta_{k+1} = \Pi_{\mathcal{D}}\left(\theta_k + \kappa_k \hat{h}_k\right). \tag{3.52}$$

It is well known that with a properly chosen sequence of κ_k (in general $\sum_{k=1}^{\infty} \kappa_k = \infty$ and $\sum_{k=1}^{\infty} \kappa_k^2 < \infty$, e.g., $\kappa_k = \frac{1}{k}$; these conditions are more strict than those for the deterministic case (3.51)) and under some conditions for the noise in the gradient estimates $\hat{h}_k$ and for the performance function $\eta(\theta)$, we are guaranteed to obtain a sequence of θ_k that converges almost surely (with probability 1) to a local optimal point θ^*, as $k \to \infty$. Equation (3.52) corresponds to the *Robbins-Monro* algorithm in finding a zero point for the performance gradient; it will be discussed in greater detail in Chapter 6.

Sample-Path-Based Implementation

In sample-path-based implementation, the gradient estimation error depends on the length of the sample path. Therefore, the convergence of the optimization algorithm relies on the coordination among the lengths of sample paths in every iteration and the step sizes.

As discussed, there are two ways to implement optimization algorithms with PA. First, we can run a Markov, or a queueing, system under one set of parameters for a relatively long period to obtain an accurate gradient estimate and then update the parameters according to (3.51). When the estimation error is small, we hope that this standard gradient-based method for performance optimization of deterministic systems works well.

Second, when the sample paths are short, we need to use the stochastic approximation based algorithm (3.52). It is well known that the standard step size sequence (e.g., $\kappa_k = \frac{1}{k}$) makes the algorithm very slow, so some ad hoc methods are usually used in practice to speed up the convergence.

Both (3.51) and (3.52) take the same form and the difference is only on the choice of step sizes. On the other hand, there are always stochastic errors even when we run a relatively long sample path. Therefore, ad hoc methods are also useful even when we apply the deterministic version (3.51).

One of the ad hoc methods works as follows. When the sample path of the kth iteration is not long enough, we do not use the gradient estimate obtained when the system is under parameters θ_k in the kth iteration in (3.51). Instead, we may use a weighted sum of the current estimate under θ_k and the previous estimates under θ_{k-1}, θ_{k-2}, etc. as the gradient estimate. This may maintain the accuracy of the estimate since the step size is usually very small, (i.e., θ_k, θ_{k-1}, θ_{k-2} are very close) and, therefore, it may avoid instability caused by the large deviation of the gradient estimates due to the short length of each iteration and therefore it may speed up the convergence process.

There is a trade-off between the lengths of the sample paths and the number of iterations in reaching the optimal point. When the lengths are longer, fewer iterations may be required; and when the lengths are shorter, more iterations may be required. There are not much work in stochastic approximation dealing with the convergence speeds of the algorithms. Therefore, it is not clear which method, with long lengths or short ones, is faster (in terms of the number of transitions). Figure 3.4 illustrates the two optimization approaches with PA-based gradient estimates.

3.3.2 Optimization with Long Sample Paths

To illustrate the optimization approach with long sample paths, we consider the optimization of the system throughput η (see (2.95)) with respect to the mean service times, $\bar{s}_i$, $i = 1, 2, \dots, M$, in a closed Jackson network (Section C.2). We assume that the mean service times must meet a constraint: The total mean service time is a constant, i.e., $\sum_{i=1}^{M} \bar{s}_i = const$, where "*const*" denotes

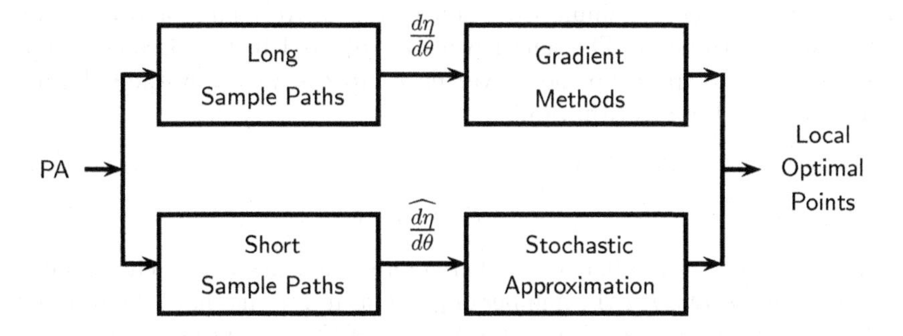

Fig. 3.4. Two Optimization Approaches with PA-Based Gradient Estimates

a constant. This constraint defines the region $\mathcal{D}$ in $\mathcal{R}^M$. The performance gradient

$$\frac{d\eta}{d\bar{s}} := \left[\frac{\partial\eta}{\partial\bar{s}_1}, \ldots, \frac{\partial\eta}{\partial\bar{s}_M}\right]^T$$

can be obtained by PA with a sample path of the queueing system. Let $\bar{s}_{i;k}$ be server i's mean service time at the kth iteration. We update the mean service times as follows:

$$\bar{s}_{i;k+1} = \bar{s}_{i;k} + \kappa_k \left\{\frac{\partial\eta}{\partial\bar{s}_i} - \frac{1}{M}\sum_{j=1}^{M}\frac{\partial\eta}{\partial\bar{s}_j}\right\}_{\bar{s}_i=\bar{s}_{i;k},\ i=1,\ldots,M} . \tag{3.53}$$

It can be easily verified that $\sum_{i=1}^{M}\bar{s}_{i;k} = const$, $k = 1, 2, \ldots$, as long as the initial values satisfy $\sum_{i=1}^{M}\bar{s}_{i;0} = const$.

Next, we provide a numerical example to show how the optimization approach works in practice. Some ad hoc modifications are added in the example to speed up the optimization process.

Example 3.8. Consider a closed Jackson network with $M = 3$ servers and $N = 5$ customers; let the routing matrix be

$$Q = \begin{bmatrix} 0 & 0.3 & 0.7 \\ 0.6 & 0 & 0.4 \\ 0.5 & 0.5 & 0 \end{bmatrix}.$$

The mean service times satisfy the constraint $\sum_{i=1}^{3}\bar{s}_i = 100$. We wish to maximize the system throughput.

We start with arbitrarily chosen initial values $\bar{s}_{1;0} = 80$, $\bar{s}_{2;0} = 10$, and $\bar{s}_{3;0} = 10$. We run the system with these initial values for 1,000 transitions and apply the PA algorithm to obtain an estimate of the performance gradient. Then, we follow (3.53) to update the mean service times. The initial length of 1,000 transitions is relatively short in estimating the gradients, because it is expected that at the beginning the gradient is relatively large and therefore is easy to be estimated. The length will be adjusted in the parameter updating process. To speed up the convergence process, we apply the following modifications to the algorithm (3.53):

1. We choose the step size as

$$\kappa_k = a_1 \times a_2^r + b,$$

where $1 > \{a_1, a_2\} > 0$, and $b > 0$ are more or less arbitrarily chosen positive numbers, and r is the number of previous iterations that have resulted in degradation, rather than improvement, of the system performance. In this example, we choose $a_1 = 0.2$, $a_2 = 0.2$, and $b = 0.01$.
To speed up the process, we use an exponential decreasing step size rather than an inverse-proportional one. In addition, we reduce the step size only when the performance degrades, indicating that the update in the last iteration might be too large. Finally, we add a positive constant b to set up a lower bound for the step size. Theoretically, such a step size may not guarantee the convergence of the algorithm, but it may reach close enough to the optimal point.
2. If the performance degrades (or r increases), we quadruple the length of simulation in the next iteration to obtain a more accurate estimate of the performance gradient.
3. At each iteration, we update the gradient estimate by a weighted sum of the current estimate and the previous one as follows.

$$\left\{\frac{\bar{s}_i}{\eta}\frac{\partial \eta}{\partial \bar{s}_i}\right\}_{k+1} = w_1 \left\{\frac{\bar{s}_i}{\eta}\frac{\partial \eta}{\partial \bar{s}_i}\right\}_k + w_2 \left\{\frac{\bar{s}_i}{\eta}\frac{\partial \eta}{\partial \bar{s}_i}\right\}_{\text{the }(k+1)\text{th run}}, \quad (3.54)$$

where $w_1 = \frac{cL_k}{L_{k+1}+cL_k}$, $w_2 = \frac{L_{k+1}}{L_{k+1}+cL_k}$, and $c < 1$; L_k and L_{k+1} are the lengths of the kth and $(k+1)$th iterations, respectively. In (3.54), $\left\{\frac{\partial \eta}{\partial \bar{s}_i}\right\}_{k+1}$ is the value used in (3.53) to update the mean service times, and $\left\{\frac{\partial \eta}{\partial \bar{s}_i}\right\}_{\text{the }(k+1)\text{th run}}$ is the estimate obtained in the $(k+1)$th run.

After 36 iterations, the algorithm reaches a near-optimal point as

$$(\bar{s}_1, \bar{s}_2, \bar{s}_3) = (30.61,\ 39.69,\ 29.70),$$

with a throughput of 0.06512, as compared with the optimal value obtained by analytical formulas

$$(\bar{s}_1, \bar{s}_2, \bar{s}_3) = (30.58,\ 39.94,\ 29.49),$$

with the optimal throughput of 0.06513. □

In stochastic approximation based approaches with recursive algorithms, the system parameters can be updated within a short period or even at every transition. These topics are discussed in Chapter 6.

3.3.3 Applications

There have been hundreds of papers in the area of PA and its applications in the literature, and it is impossible to review all of them in this book. By and large, the applications cover a wide range of subjects such as capacity planning, inventory problems, resource allocation, flow control, bandwidth provisioning, traffic shaping, pricing, and stability and reliability analysis, in many areas including communications, networking, manufacturing, and logistics. References include [10, 35, 38, 74, 95, 144, 145, 158, 164, 180, 186, 187, 196, 199, 200, 204, 211, 210, 215, 224, 225, 233, 240, 258, 261, 263].

PROBLEMS

3.1. Study the potential with $g(S) = 0$:

a. Prove that the solution to (3.4) satisfies $p_{S*}g = \eta - f(S)$.
b. Derive (3.4) from the Poisson equation $(I - P)g + \eta e = f$ with the normalization condition $p_{S*}g = \eta - f(S)$.

3.2. Let P be an $S \times S$ ergodic stochastic transition matrix and ν be an S-dimensional (row) vector with $\nu e = 1$. Set $P_{-\nu} = P - e\nu$.

a. Suppose that there is a potential g such that $\nu g = \eta$. Prove that $g = P_{-\nu}g + f$.
b. Prove that the eigenvalues of $P - e\nu$ are 0 and λ_i, $i = 1, \ldots, S-1$, where λ_i, with $|\lambda_i| < 1$, $i = 1, \ldots, S-1$, are the eigenvalues of P.
c. Develop an iterative algorithm similar to (3.7).
d. For any vector ν with $\nu e = 1$, we can develop the algorithm in c) without presetting $\nu g = \eta$. Prove that the potential obtained by the algorithm indeed satisfies $\nu g = \eta$.
e. Prove that the algorithm (3.4)-(3.7) is a special case of the above algorithm. Verify that $p_{S*}g = \eta$.

3.3. For any vector ν with $\nu e = 1$,

a. Prove that $g = (I - P + e\nu)^{-1}f$ is a potential vector with normalization condition $\nu g = \eta$.
b. Can you derive a sample-path-based algorithm similar to (2.16) based on a)?

3.4. Consider

$$P = \begin{bmatrix} 0 & 0.5 & 0.5 \\ 0.7 & 0 & 0.3 \\ 0.4 & 0.6 & 0 \end{bmatrix}, \qquad f = \begin{bmatrix} 10 \\ 2 \\ 7 \end{bmatrix}.$$

a. Calculate the potential vector using algorithm (3.1).
b. Calculate the potential vector using algorithm (3.3).

c. Calculate the potential vector using algorithm (3.7).
d. Calculate the potential vector using the algorithm proposed in Problem 3.2.

Observe the convergence speeds and compare them with that of $\lim_{k\to\infty} P^k = e\pi$.

3.5. Suppose that a Markov chain starts from state i and that we use the consecutive visits to the state i as the regenerative points (cf. (3.18)). That is, we set

$$l_0 = 0, \quad \text{with } X_0 = i,$$
$$l_k = \text{the epoch that } \{X_l\} \text{ first visits state } i \text{ after } l_{k-1}, \qquad k \geq 1.$$

Then we denote the first visit epoch to state j in the kth regenerative period as $l_{j,k}$; i.e., $l_{j,k} = \min\{l_{k-1} < l \leq l_k : X_l = j\}$. We note that in some periods, such a point may not exist. Can we use the average of the sum $\sum_{l=l_{k-1}}^{l_{j,k}-1} f(X_l)$ as the estimate of $\gamma(j,i)$? If not, why not?

3.6. Let $p(1|1) = 0.5$, $p(2|1) = 0.2$, and $p(3|1) = 0.3$; and $p(1|2) = 0.3$, $p(2|2) = 0.5$, and $p(3|2) = 0.2$. Suppose that $X = 1$ and $\widetilde{X} = 2$ and that we use the same uniformly distributed random variable $\xi \in [0,1)$ to determine the transitions from both $X = 1$ and $\widetilde{X} = 2$, according to (2.2). In this case, what are the conditional transition probabilities $\widetilde{p}_{1|1}(*|2)$, $\widetilde{p}_{2|1}(*|2)$, and $\widetilde{p}_{3|1}(*|2)$?

3.7. Let X and Y be two random variables with probability distributions $\Phi(x)$ and $\Psi(y)$, respectively. Their means are denoted as $\bar{x} = E(X)$ and $\bar{y} = E(Y)$. We wish to estimate $\bar{x} - \bar{y} = E(X - Y)$ by simulation. We generate random variables X and Y using the inverse transformation method. Thus, we have $X = \Phi^{-1}(\xi_1)$ and $Y = \Psi^{-1}(\xi_2)$, where ξ_1 and ξ_2 are two uniformly distributed random variables in $[0,1)$. Prove that if we choose $\xi_1 = \xi_2$, then the variance $Var[X - Y]$ is the smallest among all possible pairs of ξ_1 and ξ_2.

3.8. In the coupling approach, prove the following statements:

a. Let $\widehat{\pi}$ be the S^2 dimensional steady-state probability (row) vector of $\widehat{P}$, i.e., $\widehat{\pi}\widehat{P} = \widehat{\pi}$, and π be the steady-state probability vector of P, i.e., $\pi = \pi P$. Then $\widehat{\pi}(e_S \otimes I) = \widehat{\pi}(I \otimes e_S) = \pi$, and $\widehat{\pi}\widehat{g} = \widehat{\pi}\widehat{f} = 0$.
b. Equation (3.22) can take the form

$$(I - \widehat{P} + e_{S^2}\widehat{\pi})\widehat{g} = \widehat{f},$$

with $\widehat{\pi}\widehat{g} = 0$. Therefore, we have

$$\widehat{g} = \sum_{l=0}^{\infty} \widehat{P}^l \widehat{f}.$$

3.9. To illustrate the coupling approach used in simulation for speeding up the estimation of $\gamma(i,j)$, let us consider a simple Markov chain with transition probability matrix

$$P = \begin{bmatrix} 0.2 & 0.3 & 0.5 \\ 0.2 & 0.3 & 0.5 \\ 0.2 & 0.3 & 0.5 \end{bmatrix}.$$

a. Suppose that we generate two independent Markov chains with initial states $X_0 = 1$ and $X'_0 = 2$, respectively. What is the average length from $l = 0$ to L^*_{12}, $E(L^*_{12})$?
b. If we use the same $[0,1)$ uniformly distributed random variable ξ to determine the state transitions for both Markov chains, what is $E(L^*_{12})$?
c. Answer the questions in a) and b), if

$$P = \begin{bmatrix} 0.2 & 0.4 & 0.4 \\ 0.4 & 0.2 & 0.4 \\ 0.4 & 0.4 & 0.2 \end{bmatrix}.$$

3.10. The realization factor $\gamma(i,j)$ can be obtained by simulating two sample paths initiated with i and j, respectively, up to its merging point $L_{i,j}$:

$$\gamma(i,j) = E\left\{ \sum_{l=0}^{L_{i,j}-1} [f(X'_l) - f(X_l)] \middle| X_0 = i, X'_0 = j \right\}.$$

If the two sample paths are independent, as shown in the text, we can obtain the perturbation realization factor equation. However, in simulation, we may use coupling to reduce the variance in estimating the difference of the mean values of two random variables ($\gamma(i,j) = g(j) - g(i)$). In our case, we wish to let the two sample paths, initiated with i and j, merge as early as possible.

To this end, in simulation we can force the two sample paths $\boldsymbol{X}$ and $\boldsymbol{X}'$ with two initial states i and j, respectively, to merge as fast as possible. We may use the same random variable to determine the state transitions in the two paths. For example, if $p(k|i) = 0.3$ and $p(k|j) = 0.2$, instead of using two independent random numbers in $[0,1)$ to determine the state transitions for $X_0 = i$ and $X'_0 = j$, respectively, we generate one uniformly distributed random number $\xi \in [0,1)$, and if $\xi \in [0, 0.2)$, we let both $X_1 = X'_1 = k$.

We use an example to show this coupling method: Let $p(1|2) = 0.5$, $p(2|2) = 0.3$, $p(3|2) = 0.2$, and $p(1|3) = 0.2$, $p(2|3) = 0.7$, $p(3|3) = 0.1$. The largest probabilities for the two paths starting from $X_0 = 2$ and $X'_0 = 3$ to merge at $X_1 = X'_1 = 1$ is $\min\{p(1|2), p(1|3)\} = 0.2$, to merge at $X_1 = X'_1 = 2$ is $\min\{p(2|2), p(2|3)\} = 0.3$, and to merge at $X_1 = X'_1 = 3$ is $\min\{p(3|2), p(3|3)\} = 0.1$. Thus, the largest probability that the two sample paths merge at $X_1 = X'_1$ with the coupling technique is $0.2+0.3+0.1 = 0.6$. We simulate the two sample paths in two steps. In the first step, we generate a uniformly distributed random variable $\xi \in [0,1)$. If $\xi \in [0, 0.2)$, we set $X_1 = X'_1 = 1$; if $\xi \in [0.2, 0.5)$, we set $X_1 = X'_1 = 2$; if $\xi \in [0.5, 0.6)$, we set $X_1 = X'_1 = 3$. If $\xi \in [0.6, 1)$, we go to the second step: using two other independent random numbers to determine the transitions for the two sample paths.

Continue with the above reasoning and mathematically formulate it. Work on $\gamma(i, S)$ for all states $i \in \mathcal{S}$ and derive the following equation

$$g(i) - g(S) = f(i) - f(S) + \sum_{j=1}^{S} [p(j|i) - p(j|S)] g(j), \qquad i \in \mathcal{S}.$$

Prove that this equation is the same as (3.4).

3.11. One of the restrictions of the basic formula (3.32) is that it requires $p(j|i) > 0$ if $\Delta p(j|i) > 0$ for all $i, j \in \mathcal{S}$. This condition can be relaxed. For example, we may assume that whenever $\Delta p(j|i) > 0$, there exists a state, denoted as $k_{i,j}$, such that $p(k_{i,j}|i)p(j|k_{i,j}) > 0$. Under this assumption, we have

$$\frac{d\eta_\delta}{d\delta} = \sum_{i \in \mathcal{S}} \sum_{j \in \mathcal{S}} \left\{ \pi(i) \left[p(k_{i,j}|i)p(j|k_{i,j}) \frac{\Delta p(j|i)}{p(k_{i,j}|i)p(j|k_{i,j})} g(j) \right] \right\}.$$

Furthermore, we have

$$\frac{d\eta_\delta}{d\delta} = \sum_{i \in \mathcal{S}} \sum_{j \in \mathcal{S}} \left\{ \pi(i) \left[\sum_{k \in \mathcal{S}} p(k|i)p(j|k) \right] \left[\frac{\Delta p(j|i)}{\sum_{k \in \mathcal{S}} p(k|i)p(j|k)} g(j) \right] \right\}.$$

a. Continue the analysis and develop the direct learning algorithms for the performance derivatives.
b. Compared with (3.32), what are the disadvantages of this "improved" approach, if any?
c. Extend this analysis to the more general case of irreducible Markov chains.

3.12. In the gradient estimate (3.34), we have ignored the constant term η in the expression of g. A more accurate estimate should be

$$\frac{d\eta_\delta}{d\delta} \approx \lim_{N \to \infty} \frac{1}{N} \left\{ \sum_{n=0}^{N-1} \left[\frac{\Delta p(X_{n+1}|X_n)}{p(X_{n+1}|X_n)} \sum_{l=0}^{L-1} [f(X_{n+l+1}) - \eta] \right] \right\}, \qquad \text{w.p.1.}$$

Prove that

$$\frac{d\eta_\delta}{d\delta} \approx \lim_{N \to \infty} \frac{1}{N} \left\{ \sum_{n=0}^{N-1} \left[\frac{\Delta p(X_{n+1}|X_n)}{p(X_{n+1}|X_n)} \sum_{l=0}^{L-1} f(X_{n+l+1}) \right] \right\}, \qquad \text{w.p.1,}$$

and discuss the estimation error caused by ignoring the term $L\eta$ in the estimate.

3.13. Discuss the error in the gradient estimate (3.41) caused by ignoring the second term of (3.40) for a finite N. You may set $f \equiv 1$.

3.14. Let η_r be the performance of a Markov chain with transition probability matrix P_r defined as $p_r(i|i) = r$ for all $i \in \mathcal{S}$ and $p_r(j|i) = (1-r)q_{i,j}$, $j \neq i$, $i, j \in \mathcal{S}$, with $\sum_{j\in\mathcal{S}} q_{i,j} = 1$ for all $i \in \mathcal{S}$. Prove $\frac{d\eta_r}{dr} = 0$ for all $0 < r < 1$ using performance derivative formula (3.30).

3.15. In Algorithm 3.1, prove that the following equation holds

$$\lim_{L\to\infty} \left\{ \sum_{l=0}^{L-1} P^l (\Delta P) P^{L-l-1} \right\} = e\pi(\Delta P)(I - P + e\pi)^{-1}.$$

In addition, prove that, at the steady state, we have

$$\begin{aligned} \pi(i)\rho_L(i) &= E\left\{ \sum_{l=0}^{L-1} \frac{\Delta p(X_{l+1}|X_l)}{p(X_{l+1}|X_l)} I_i(X_L) \right\} \\ &= \pi \left\{ \sum_{l=0}^{L-1} P^l (\Delta P) P^{L-l-1} \right\} e_{\cdot i}, \end{aligned}$$

where $e_{\cdot i}$ is the ith column vector of the identity matrix I. Equation (3.38) and the convergence of (3.37) follow directly from these two equations.

3.16. In Problem 3.15, we set $G_L = \sum_{l=0}^{L-1} P^l (\Delta P) P^{L-l-1}$. Prove that

$$G_{L+1} = PG_L + G_L P - PG_{L-1}P,$$

with $G_0 = 0$, $G_1 = \Delta P$. Set $G = \lim_{L\to\infty} G_L$. Explain the meaning of G. Finally, letting $L \to \infty$ on both sides of the above equation, we obtain $G = PG + GP - PGP$. Is this equation useful in any sense?

3.17. Write a computer simulation program

a. to estimate potentials by using (3.15) and (3.19);
b. to estimate the performance derivatives by using (3.35), (3.41), and (3.43).

3.18. The group inverse (2.48) $B^{\#} = -[(I - P + e\pi)^{-1} - e\pi]$ (for ergodic chains) plays an important role in performance sensitivity analysis. Let $b^{\#}(i,j)$ be the (i,j)th component of $B^{\#}$. Consider a Markov chain starting from state $i \in \mathcal{S}$. Let $N_{ij}^{(L)}$ be the expected number of times that the Markov chain visits state $j \in \mathcal{S}$ in the first L stages. Prove (cf. [168]) that

$$\lim_{L\to\infty} \left(N_{ji}^{(L)} - N_{ki}^{(L)} \right) = b^{\#}(k,i) - b^{\#}(j,i).$$

3.19. Given a direction defined by ΔP, is it possible to estimate the second-order derivative $\frac{d^2\eta_\delta}{d\delta^2}$ using a sample path of the Markov chain with transition probability matrix P (cf. Section 2.1.5)? How about the second-order performance derivatives of any given reward function $f(\theta)$?

3.20. Consider a continuous-time Markov process with transition rates $\lambda(i)$ and transition probabilities $p(j|i)$, $i,j = 1,2,\ldots,S$. Suppose that the transition probability matrix $P := [p(j|i)]_{i,j\in\mathcal{S}}$ changes to $P + \delta\Delta P$, and the transition rates $\lambda(i)$, $i = 1,2,\ldots,S$, remain unchanged. Let η be the average reward with reward function f. Develop a direct learning algorithm for $\frac{d\eta_\delta}{d\delta}$.

3.21. Consider a closed Jackson network consisting of M servers and N customers with mean service times $\bar{s}_i$, $i = 1,2,\ldots,S$, and routing probabilities $q_{i,j}$, $i,j = 1,2\ldots,M$. Let

$$\eta_T^{(f)} = \lim_{L\to\infty} \frac{1}{T_L} \int_0^{T_L} f(\boldsymbol{N}(t))dt$$

be the time-average performance. Suppose that the routing probabilities change to $q_{i,j} + \delta\Delta q_{i,j}$, $i,j = 1,2\ldots,M$. Develop a direct learning algorithm for the derivative of the time-average reward using performance potentials. Use the intuition explained in Section 2.1.3 to develop the performance derivative formula.

One of the principal objects of theoretical research in my department of knowledge is to find the point of view from which the subject appears in its greatest simplicity.

Josiah Willard Gibbs
American Scientist
(1839 - 1903)

4
Markov Decision Processes

In Chapter 2, we introduced the basic principles of PA and derived the performance derivative formulas for queueing networks and Markov and semi-Markov systems with these principles. In Chapter 3, we developed sample-path-based (on-line learning) algorithms for estimating the performance derivatives and sample-path-based optimization schemes. In this chapter, we will show that *the performance sensitivity based view leads to a unified approach to both PA and Markov decision processes (MDPs).* In MDPs, since the policy space is discrete, performance derivatives do not make sense, and correspondingly we consider the difference in performance measures under two policies. We show that the policy-iteration type of optimization scheme can be derived intuitively and straightforwardly with the performance difference formulas. This approach applies to MDPs with different performance criteria, including the long-run average reward, the discounted reward, and the bias. We will also introduce the nth-bias optimization theory, which is a complete extension of the theory for average reward and bias optimality. We show that the performance difference formulas provide an easy and intuitive way for developing the theory.

In general, we use the word "performance" to refer to any performance criteria, including the long-run average reward, the discounted reward, the bias, and the nth bias, etc. Therefore, "performance difference formula" may refer to any among the "average-reward difference formula", "bias difference formula", etc, which are used when we discuss a specific performance criterion.

In Section 4.1, we discuss the policy iteration approach for optimization of long-run average reward, bias, and discounted reward of ergodic systems. We use this simple case to show the basic ideas of the approach. In Section 4.2, we extend these results to multi-chain models. In Section 4.3, we introduce and study the nth-bias optimization problem for multi-chain models, which is an extension of the bias optimization, and this problem also provides a complete spectrum for the family of MDP optimization problems. Just like with ergodic systems, the main results in both Sections 4.2 and 4.3 can be derived almost directly from the performance difference formulas. The main difficulty involved for the multi-chain case (which explains why this sensitivity-based approach was proposed only recently) is that the performance difference formulas contain two terms. We give a simple example to illustrate the structural relations of these two terms in the difference formulas that allow us to overcome this main difficulty. The presentation in this chapter is based on recent papers [55, 63, 71].

MDPs and Policies

We first describe the discrete-time MDP model used in this chapter. In an MDP, at any time l, $l = 0, 1, 2, \dots$, the system is in a state $X_l \in \mathcal{S}$, where $\mathcal{S} = \{1, 2, \dots, S\}$ is a finite state space. In addition to the state space, there is an action space $\mathcal{A}$. We assume that the number of available actions is finite. If the system is in state i, $i \in \mathcal{S}$, we can take (independently from the actions taken in other states) any action $\alpha \in \mathcal{A}(i) \subseteq \mathcal{A}$ and apply it to the system, where $\mathcal{A}(i)$ is the set of actions that are available in state $i \in \mathcal{S}$, $\mathcal{A} = \cup_{i \in \mathcal{S}} \mathcal{A}(i)$. The action determines the transition probabilities of the system as well as the reward received. If action α is applied to the system when the state is i, the state transition probabilities are denoted as $p^{\alpha}(j|i)$, $j \in \mathcal{S}$, and the reward that the system receives is denoted as $f(i, \alpha)$.

A (*stationary and deterministic*) *policy* is a mapping from $\mathcal{S}$ to $\mathcal{A}$, denoted as d (or h, or other letters specified), with $d(i) \in \mathcal{A}(i)$, that determines the action taken in state i, $i \in \mathcal{S}$. We use

$$\mathcal{D} = \times_{i \in \mathcal{S}} \mathcal{A}(i)$$

to denote the space of all possible (stationary and deterministic) policies, where "$\times$" is called a *Cartesian product*, which is a direct product of sets. Specifically, the Cartesian product of two sets X and Y, denoted $X \times Y$, is the set of all possible ordered pairs (x, y), where $x \in X$ and $y \in Y$. The definition can be generalized to the n-ary Cartesian product over n sets $X_1, \dots, X_n$: $\times_{i=1}^{n} X_i = \{(x_1, x_2, \dots, x_n) : \text{for all } x_1 \in X_1, \dots, x_n \in X_n\}$.

Sometimes we also write d as a vector $d = (d(1), d(2), \dots, d(S))$. We use superscript d to indicate that the quantities are associated with policy $d \in \mathcal{D}$. Therefore, if policy d is adopted, the state transition probability matrix is $P^d = [p^{d(i)}(j|i)]_{i,j=1}^{S}$, and its ith row is denoted as a vector $p^d(\bullet|i)$. The reward vector is denoted as $f^d = (f(1, d(1)), f(2, d(2)), \dots, f(S, d(S))^T$, and the steady-state probability vector of a Markov chain under policy d is denoted as $\pi^d = (\pi^d(1), \pi^d(2), \dots, \pi^d(S))$. For convenience, we also call a quantity associated with P^d (and f^d) a quantity of policy d. For example, we may call π^d the steady-state probability vector of policy d, and g^d the potential vector of d.

Let $|\mathcal{A}(i)|$ be the number of actions in $\mathcal{A}(i)$, $i = 1, 2, \ldots, S$. Then, the number of policies in $\mathcal{D}$ is $\prod_{i=1}^{S} |\mathcal{A}(i)|$. For example, if $S = 100$ and $|\mathcal{A}(i)| = 2$ for all i, then there are altogether $2^{100} \approx 10^{30}$ policies! Even for such a small problem, an exhaustive search in the policy space, which requires us to solve for the steady-state probabilities for every policy, is not feasible for performance optimization.[1]

Because the Markov model is widely applicable to many systems, the MDP formulation for optimization encompasses a wide range of applications in many areas, including inventory management, manufacturing, communication networks, computer systems, transportation, financial engineering, and control systems. We refer readers to other textbooks for examples of such applications (e.g., see [21, 25, 216]). Problems 4.1, 4.2, 4.3, and 4.4 also provide some examples for applications.

Since a policy corresponds to a state transition probability matrix and a reward function, we sometimes refer to a pair (P^d, f^d) as a policy and denote it as $(P^d, f^d) \in \mathcal{D}$. When the reward function f^d does not play a role, e.g., it is the same for all policies, we simply write a policy as $P^d \in \mathcal{D}$. To simplify the notation, when we are discussing only one policy, we sometimes omit the superscript d and denote it as $(P, f) \in \mathcal{D}$.

4.1 Ergodic Chains

We first study the case where the Markov chains under all policies are ergodic and share the same state space $\mathcal{S}$. In this case, the long-run average reward (also called the "gain" for short) under policy d is

$$\eta^d = \lim_{L \to \infty} \left\{ \frac{1}{L} \sum_{l=0}^{L-1} f[X_l, d(X_l)] \right\} = \pi^d f^d,$$

in which the limit exists with probability 1 and is independent of the initial state.

A policy $\widehat{d}$ is called a (*gain*) *optimal policy* if

$$\eta^{\widehat{d}} \geq \eta^d, \quad \text{for all } d \in \mathcal{D},$$

[1] A recently developed theory, *Ordinal Optimization*, shows that the search space can be significantly reduced if we are willing to relax the optimization criterion from searching for a best policy to searching for a "good enough" policy, whose performance is within, say, the top 10% of all policies' performance, see e.g., [150, 151, 175, 181, 184, 266]. This creates a new research area that is different from the subjects discussed in this book, in which we deal only with optimal policies.

and its gain $\eta^{\widehat{d}}$ is called the *optimal gain*. We denote the optimal gain as η^*, i.e., $\eta^* = \max_{d \in \mathcal{D}} \eta^d$, and define

$$\mathcal{D}_0 := \left\{ d \in \mathcal{D} \, : \, \eta^d = \eta^* \right\}$$

to be the set of all gain-optimal policies. The goal of optimization is to find an optimal policy $\widehat{d} \in \mathcal{D}_0$. Sometimes we also use a shorthand notation "arg" and denote the set of all optimal policies as follows

$$\mathcal{D}_0 = \arg \left\{ \max_{d \in \mathcal{D}} \eta^d \right\} := \left\{ d \in \mathcal{D} \, : \, \eta^d = \eta^* \right\}.$$

Thus, the goal of optimization is to find a policy

$$\widehat{d} \in \arg \left\{ \max_{d \in \mathcal{D}} \eta^d \right\}.$$

As we have noted, except for very small systems, an exhaustive search in $\mathcal{D}$ for an optimal policy is not feasible. We need to develop efficient algorithms, and policy iteration is one such algorithm.

4.1.1 Policy Iteration

As shown in Figure 1.5, the fundamental idea of policy iteration is that, by observing and/or analyzing the behavior of a system under a policy, we may find another policy that performs better, if such a policy exists. This fact can be easily seen from the performance (average-reward) difference formula (2.27), which we rewrite as follows.

The Average-Reward Difference Formula

Let η^h and η^d be the long-run average rewards corresponding to two policies h and d, respectively, π^h be the steady-state probability vector of h, and g^d be the vector of *performance potentials* of d. Then, from (2.27), we have

The Average-Reward Difference Formula for Ergodic Chains:

$$\eta^h - \eta^d = \pi^h \left[(f^h + P^h g^d) - (f^d + P^d g^d) \right]. \tag{4.1}$$

In the equation, g^d and η^d are the solution to the Poisson equation $(I - P^d)g^d + \eta^d e = f^d$.

For two S-dimensional vectors u and v, we define $u = v$ if $u(i) = v(i)$ for all $i \in \mathcal{S}$; $u \leq v$ if $u(i) \leq v(i)$ for all $i \in \mathcal{S}$; $u < v$ if $u(i) < v(i)$ for all $i \in \mathcal{S}$; and $u \preceq v$ if $u(i) < v(i)$ for at least one i, and $u(j) = v(j)$ for other components. The relation $\leq$ includes $=$, $\preceq$, and $<$. Similar definitions are used for the relations $>$, $\succeq$, and $\geq$, and for matrices.

Next, we note that $\pi^h > 0$ for any h with an ergodic transition probability matrix. This simple fact plays a fundamental role in the development of the optimization theory and we emphasize it as follows.

A Fundamental Fact:

$$\pi^h > 0 \text{ for any ergodic } P^h.$$

The average-reward difference formula (4.1) and the fundamental fact lead immediately to the following result.

Comparison Lemma:

$$\text{If } f^h + P^h g^d \succeq f^d + P^d g^d, \text{ then } \eta^h > \eta^d. \tag{4.2}$$

$$\text{If } f^h + P^h g^d \preceq (\geq, \leq) f^d + P^d g^d, \text{ then } \eta^h < (\geq, \leq) \eta^d. \tag{4.3}$$

With the Comparison Lemma (4.2) and (4.3), we can easily prove the following result.

Optimality Condition:

A policy $\widehat{d}$ is gain optimal if and only if

$$f^{\widehat{d}} + P^{\widehat{d}} g^{\widehat{d}} \geq f^d + P^d g^{\widehat{d}}, \qquad \text{for all } d \in \mathcal{D}. \tag{4.4}$$

Proof of the Optimality Condition: The proof follows almost directly from the Comparison Lemma (4.2) and (4.3). First, if (4.4) holds, then $(f^d + P^d g^{\widehat{d}}) \leq (f^{\widehat{d}} + P^{\widehat{d}} g^{\widehat{d}})$ for all $d \in \mathcal{D}$. Setting P^d, f^d as P^h and f^h, and $P^{\widehat{d}}$ and $f^{\widehat{d}}$ as P^d, f^d, respectively, in the Comparison Lemma (4.3), we have $\eta^d \leq \eta^{\widehat{d}}$, for all $d \in \mathcal{D}$; i.e., $\widehat{d}$ is gain optimal.

Next, we prove the "only if" part: Let $\widehat{d}$ be a gain-optimal policy. We need to prove that (4.4) holds. Suppose that this is not true. Then, there must exist one policy, denoted as d', such that (4.4) does not hold. That is, there must be at least one state, denoted as i, such that

$$f(i, \widehat{d}(i)) + \sum_{j=1}^{S} \left\{ p^{\widehat{d}(i)}(j|i) g^{\widehat{d}}(j) \right\} < f(i, d'(i)) + \sum_{j=1}^{S} \left\{ p^{d'(i)}(j|i) g^{\widehat{d}}(j) \right\}.$$

We create a policy $\widetilde{d}$ by setting $\widetilde{d}(i) = d'(i)$, and $\widetilde{d}(j) = \widehat{d}(j)$ for all $j \neq i$. We have

$$f^{\widetilde{d}} + P^{\widetilde{d}} g^{\widehat{d}} \succeq f^{\widehat{d}} + P^{\widehat{d}} g^{\widehat{d}}.$$

("$>$" holds for the ith component, and "$=$" holds for all others.) Thus, by the Comparison Lemma (4.2), we have $\eta^{\widetilde{d}} > \eta^{\widehat{d}}$. This contradicts the fact that $\widehat{d}$ is an optimal policy. □

In the proof, we construct a policy that violates the optimal-policy assumption. This method will be used again in proving similar necessary conditions for other general cases.

It is important to note that by the Comparison Lemma (4.2) and (4.3), only the potentials of one policy are needed to compare the average rewards of two policies in the particular situation; and by the Optimality Condition (4.4), only the potentials of the current policy $\widehat{d}$ are needed to check whether this policy is optimal. Also note that, when we constructed the policy $\widetilde{d}$ in the proof of the lemma, we used (implicitly) the assumption of MDPs: We can choose actions independently in different states. This is called the *independent-action assumption.*

The Policy Iteration (PI) Algorithm

From the Comparison Lemma (4.2), for any given policy d, we can find a "better" policy h, if such a policy exists, by using the potential g^d of d. From this, it is natural to propose the following *policy iteration algorithm.*

Algorithm 4.1. A Policy Iteration Algorithm for a Gain-Optimal Policy:

1. Guess an initial policy d_0, set $k = 0$.
2. (Policy evaluation) Obtain the potential g^{d_k} by solving the Poisson equation $(I - P^{d_k})g^{d_k} + \eta^{d_k} e = f^{d_k}$, or by estimation on a sample path of the system under policy d_k (see Chapter 3).
3. (Policy improvement) Choose

$$d_{k+1} \in \arg\left\{\max_{d \in \mathcal{D}} \left[f^d + P^d g^{d_k}\right]\right\}, \tag{4.5}$$

   component-wisely (i.e., to determine an action for each state). If in state i, action $d_k(i)$ attains the maximum, then set $d_{k+1}(i) = d_k(i)$.
4. If $d_{k+1} = d_k$, stop; otherwise, set $k := k + 1$ and go to step 2.

Note that the ith component of $f^d + P^d g^{d_k}$ depends only on $d(i)$, $i \in \mathcal{S}$, and therefore (4.5) is equivalent to

$$d_{k+1}(i) \in \arg\left\{ \max_{\alpha \in \mathcal{A}(i)} \left[f(i,\alpha) + \sum_{j \in \mathcal{S}} p^{\alpha}(j|i) g^{d_k}(j) \right] \right\}, \qquad \text{for all } i \in \mathcal{S}.$$

From step 3 of the algorithm, if $d_{k+1} \neq d_k$, then

$$f^{d_{k+1}} + P^{d_{k+1}} g^{d_k} \succeq f^{d_k} + P^{d_k} g^{d_k}, \tag{4.6}$$

and by the Comparison Lemma (4.2), we have $\eta^{d_{k+1}} > \eta^{d_k}$. That is, the average reward increases at each iteration before it stops. Because the number of policies is finite, the iteration procedure has to stop after a finite number of iterations. When it stops at step k, we set $\widehat{d} := d_k = d_{k+1}$. Equation (4.5) becomes

$$\widehat{d} \in \arg\left\{ \max_{d} \left[f^d + P^d g^{\widehat{d}} \right] \right\},$$

or

$$f^{\widehat{d}} + P^{\widehat{d}} g^{\widehat{d}} \geq f^d + P^d g^{\widehat{d}}$$

for all $d \in \mathcal{D}$. Thus, by the Optimality Condition (4.4), $\widehat{d}$ is the gain-optimal policy. The above discussion proves that *the policy iteration (PI) algorithm stops at a gain-optimal policy after a finite number of iterations.*

In step 3, according to (4.5), in every state i the PI algorithm chooses the action that leads to the largest expected "potential" reward in one transition (the current reward $f(i,\alpha)$ plus the expected "potential" rewards of the states after the transition $\sum_{j \in \mathcal{S}} p^{\alpha}(j|i) g^{d_k}(j)$). Figure 4.1 illustrates the situation in a particular state 3 in an MDP with four states. Such a policy (generated according to (4.5)) is called a *greedy* (or *myopic*) policy since it only looks one transition ahead.

At each iteration, we first need to solve a Poisson equation for g^{d_k} in step 2 (it is not necessary to solve for π^{d^k}, see Section 3.1.1); then, we need to do $\sum_{i=1}^{S} |\mathcal{A}(i)|$ comparisons, according to (4.5), in step 3. When the action space $\mathcal{A}$ is small, the computational complexity of the algorithm depends on the number of iterations required for the algorithm to stop. In general, the number of iterations required is very small (for example, we randomly generated many transition probabilities for a ten-state MDP problem, with 10 actions in each state. There are 10^{10} policies in this problem. In most cases, it only takes two iterations to reach the best policy). But there is no general theory for the speed of convergence. When the action space is large, the computation of the policy iteration is still very complicated; to further reduce the computational complexity is an on-going research topic, see [77, 78, 157].

It is not necessary to choose the greedy policy in step 3. In fact, step 3 can be very flexible. The PI algorithm converges to the optimal policy as long as at every iteration the average reward increases if the current policy is not optimal. Therefore, in step 3 we only need to choose a policy d_{k+1} such that (4.6) holds. There are many such choices; for example, we may choose a policy d_{k+1} such that the ">" sign holds for only one state in (4.6), and the actions for other states are the same for both d_k and d_{k+1}. Moreover, because we do not solve for $\pi^{d_{k+1}}$, we really do not know whether the average reward of the greedy policy is better than any other policy satisfying (4.6) with equalities for many states. We simply hope that the greedy policy is usually not a bad choice. (Of course, except for the last iteration, the greedy policy is not the best choice, since it is not an optimal policy.)

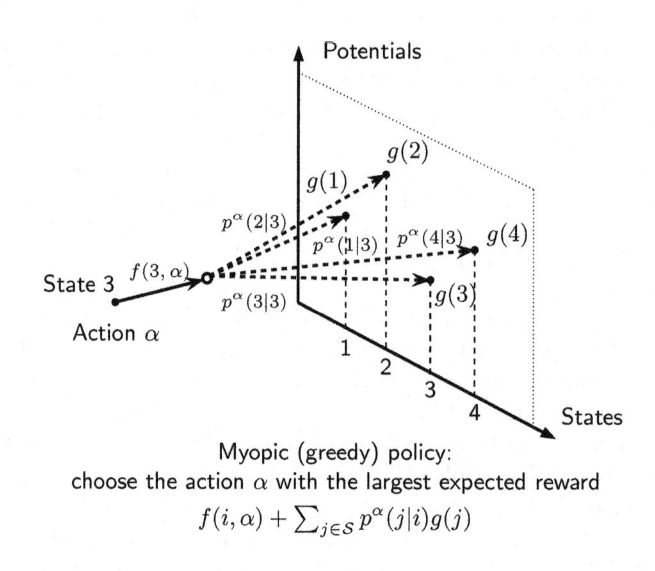

Fig. 4.1. The Expected "Potential" Reward from State $i = 3$

From the Poisson equation, the Optimality Condition (4.4) is equivalent to Bellman's *optimality equation* in the literature.

Optimality Equation:

A policy $\widehat{d}$ is gain optimal if and only if

$$\eta^{\widehat{d}}e + g^{\widehat{d}} = \max_{d \in \mathcal{D}} \left\{ f^d + P^d g^{\widehat{d}} \right\}. \qquad (4.7)$$

This is the necessary and sufficient condition for optimal policies.

The aforementioned simple analysis shows that policy iteration follows directly from the average-reward difference formula. In particular, it employs the property that one can find a better policy by using only the potentials of the current policy. This is based on the special form of the average-reward difference formula and the fundamental fact that $\pi^h > 0$ for all $h \in \mathcal{D}$. In addition, this requires the independent-action assumption, i.e., the actions at different states can be chosen independently. The approach is based on the comparison of the average rewards of two policies; when no better policy can be found, the iteration reaches the maximal reward. Bellman's optimality condition (4.7) is not directly used in verifying optimality.

Relation to PA

From the performance derivative formula (2.26), the derivative of the average reward along the direction from d_k to any policy $d \in \mathcal{D}$ is

$$\left.\frac{d\eta^{d_\delta}}{d\delta}\right|_{\delta=0} = \pi^{d_k}\left[(f^d + P^d g^{d_k}) - (f^{d_k} + P^{d_k} g^{d_k})\right],$$

where η^{d_δ} is the average reward of the randomized policy d_δ, which implements policy d with probability δ and implements policy d_k with probability $1-\delta$. For the Markov chain under policy d_δ, we have $P^{d_\delta} = P^{d_k} + \delta(P^d - P^{d_k})$ and $f^{d_\delta} = f^{d_k} + \delta(f^d - f^{d_k})$. From (4.5), the term $(f^d + P^d g^{d_k}) - (f^{d_k} + P^{d_k} g^{d_k})$ takes the maximal value (component-wisely) along the direction pointing to the greedy policy d_{k+1}. Thus, the performance derivative $\frac{d\eta^{d_\delta}}{d\delta}$ also reaches its maximum along the direction from d_k to d_{k+1}. That is,

PA vs. Policy Iteration:
At each iteration, the policy iteration Algorithm 4.1 in fact chooses the next policy along the steepest direction in the policy space.

In other words, the greedy policy at each iteration is along the steepest direction. *A policy is optimal if and only if at this policy the performance derivatives along the directions to all other policies are non-positive.* This point is interesting: In the discrete policy space $\mathcal{D} = \times_{i \in \mathcal{S}} \mathcal{A}(i)$, a local optimal policy is also a global optimal policy.

Summary

In summary, the policy-iteration-based optimization theory follows naturally from the average-reward difference formula. With the fundamental fact that $\pi^h > 0$, we obtain the Comparison Lemma, which, together with the independent-action assumption, leads to the optimality equation and the policy iteration optimization algorithm. This logical structure applies to general cases, such as bias optimality, multi-chain models, and the more general case of the nth-bias optimality, as well. We summarize this structure in Figure 4.2. The policy iteration algorithm stops at a policy that satisfies the optimality equation and, therefore, the policy iteration approach provides a constructive proof for the existence of the solution to the optimality equation.

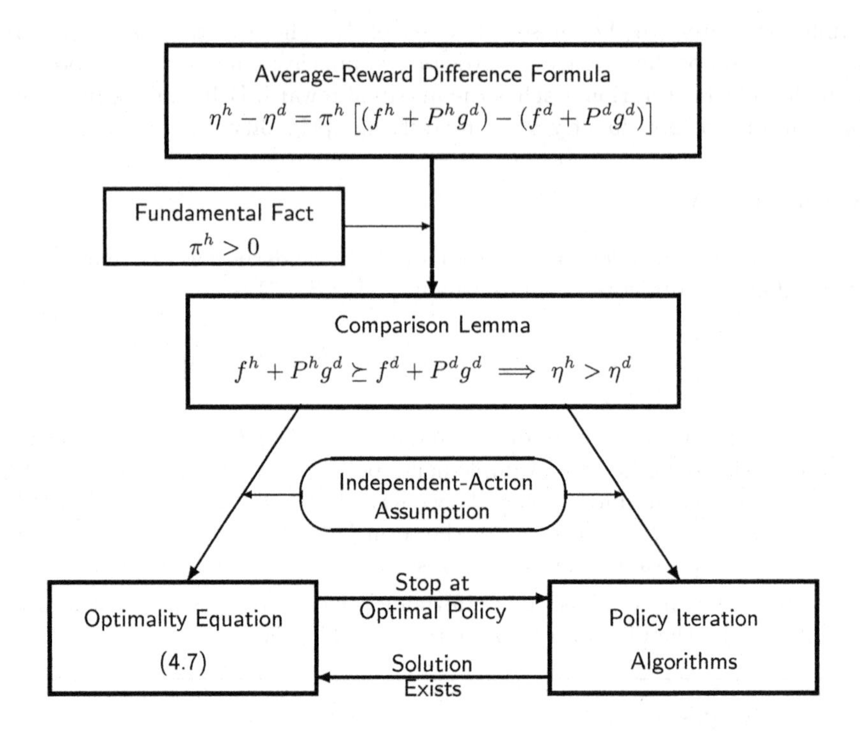

Fig. 4.2. The Logical Relationships Among the Results

4.1.2 Bias Optimality

While the policy iteration algorithm presented in the last subsection leads to a policy that maximizes the average (or steady-state) reward, it ignores the system's transient behavior. In this subsection, we show that the approach based on the comparison of any two policies can be applied to obtain policy iteration algorithms that optimize not only the steady-state performance but also the transient performance (bias). We first introduce the concept and formulate the problem (see [182, 183] for discussions and references on the bias optimality problem.)

The Bias

We discuss any ergodic Markov chain with transition probability matrix P and reward function f. The transient performance starting from any state $i \in \mathcal{S}$ can be measured by a quantity called *bias*, defined as:

$$g(i) = \lim_{L \to \infty} \sum_{l=0}^{L-1} \{E[f(X_l) - \eta] \,|X_0 = i\}. \tag{4.8}$$

The bias has a physical interpretation: It measures the sum of the deviations of the expected reward at every step from its steady-state mean starting from an initial state; it is represented by the dashed area in Figure 4.3.

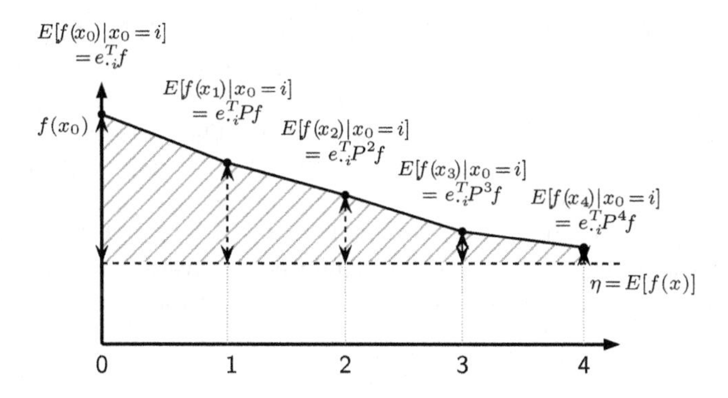

Fig. 4.3. The Meaning of the Bias

Clearly, the bias in (4.8) is a particular version of the potential defined in (2.16). Thus, it satisfies the Poisson equation (see Problem 2.5)

$$(I - P)g + \eta e = f, \tag{4.9}$$

and the normalization condition $\pi g = 0$. We also have (2.15)

$$g = \left[(I - P + e\pi)^{-1} - e\pi\right] f. \tag{4.10}$$

Because of its clear physical meaning, we call this particular version of the potential the "bias" of the Markov chain. It should be kept in mind that we use the same notation g for both the potential and the bias; when it denotes the potential, it is only up to an additive constant, and when it denotes the bias, it is uniquely determined by $\pi g = 0$.

Bias-Optimal Policies

The policy iteration algorithm in Section 4.1.1 leads to a gain-optimal policy in $\mathcal{D}_0$. The bias optimality problem is to find a policy in $\mathcal{D}_0$, $\widehat{d}$, that has the largest bias for all states $i \in \mathcal{S}$ among all the policies in $\mathcal{D}_0$:

$$g^{\widehat{d}} \geq g^d, \qquad \text{for all } d \in \mathcal{D}_0,$$

or

$$\widehat{d} \in \arg\left\{\max_{d \in \mathcal{D}_0} g^d, \text{ all components}\right\}.$$

Such a policy is called a *bias-optimal policy.* In the definition, $\widehat{d}$ maximizes all components of g^d, $d \in \mathcal{D}_0$, and we will see that such a bias-optimal policy indeed exists.

The following example helps us to have an intuition about the bias-optimality problem.

Example 4.1. Consider a two-state MDP with $\mathcal{S} = \{1, 2\}$. In each state, there are two actions: $\mathcal{A}(1) = \{\alpha_1, \alpha_2\}$ and $\mathcal{A}(2) = \{\beta_1, \beta_2\}$, with transition probabilities $p^{\alpha_1}(\bullet|1) = (0.5, 0.5)$, $p^{\alpha_2}(\bullet|1) = (0.75, 0.25)$, $p^{\beta_1}(\bullet|2) = (0.5, 0.5)$, and $p^{\beta_2}(\bullet|2) = (0.75, 0.25)$, respectively. The corresponding reward functions are $f(1, \alpha_1) = 1$, $f(1, \alpha_2) = 0.5$, $f(2, \beta_1) = -1$, and $f(2, \beta_2) = -1.5$. Therefore, we have four polices $d_1 = (\alpha_1, \beta_1)$, $d_2 = (\alpha_1, \beta_2)$, $d_3 = (\alpha_2, \beta_1)$, and $d_4 = (\alpha_2, \beta_2)$. Denote $P_k = P^{d_k}$ and $f_k = f^{d_k}$, $k = 1, 2, 3, 4$. By calculating the corresponding steady-state probabilities π_k, average rewards η_k, and biases g_k, $k = 1, 2, 3, 4$, we have the results listed in Table 4.1.

k	P_k	f_k	π_k	η_k	Bias g_k
d_1	$\begin{bmatrix} 0.5 & 0.5 \\ 0.5, & 0.5 \end{bmatrix}$	$\begin{bmatrix} 1 \\ -1 \end{bmatrix}$	$[0.5, 0.5]$	0	$\begin{bmatrix} 1 \\ -1 \end{bmatrix}$
d_2	$\begin{bmatrix} 0.5 & 0.5 \\ 0.75 & 0.25 \end{bmatrix}$	$\begin{bmatrix} 1 \\ -1.5 \end{bmatrix}$	$[0.6, 0.4]$	0	$\begin{bmatrix} 0.8 \\ -1.2 \end{bmatrix}$
d_3	$\begin{bmatrix} 0.75 & 0.25 \\ 0.5 & 0.5 \end{bmatrix}$	$\begin{bmatrix} 0.5 \\ -1 \end{bmatrix}$	$[\frac{2}{3}, \frac{1}{3}]$	0	$\begin{bmatrix} \frac{2}{3} \\ -\frac{4}{3} \end{bmatrix}$
d_4	$\begin{bmatrix} 0.75 & 0.25 \\ 0.75 & 0.25 \end{bmatrix}$	$\begin{bmatrix} 0.5 \\ -1.5 \end{bmatrix}$	$[0.75, 0.25]$	0	$\begin{bmatrix} 0.5 \\ -1.5 \end{bmatrix}$

Table 4.1. Biases in Example 4.1

From the table, we note that all the policies have the same average reward $\eta_k = 0$, $k = 1, 2, 3, 4$. Thus, all of them are gain-optimal policies. Their biases g_k, $k = 1, 2, 3, 4$, are different. We observe that $g_k = g_{k'} + c_{kk'}e$, with some constants $c_{kk'}$, $k, k' = 1, 2, 3, 4$. (For a general formula, see (4.13).) Therefore, for any particular k, g_k satisfies all four Poisson equations for d_1, d_2, d_3, and d_4. In other words, any particular g_k can be viewed as the potential of all four policies. However, each g_k is the bias only for policy d_k, which satisfies $\pi_k g_k = 0$.

Policy d_1 has the largest biases in both states. Because of the normalization condition $\pi_k g_k = 0$ for all k, π_k must satisfy some constraints when both $g_k(1)$ and $g_k(2)$ become larger. For example, because $g_1(1) > g_2(1)$ and $g_1(2) > g_2(2)$, we must have $\pi_1(1) < \pi_2(1)$ and $\pi_1(2) > \pi_2(2)$ to maintain the condition $\pi_1 g_1 = \pi_2 g_2 = 0$. That is, larger biases are achieved by redistribution of the steady-state probabilities among the states. As shown in Figure 4.4, we have $g_1 > g_2 > g_3 > g_4$, component-wisely. State 1 has a larger bias than state 2 has for all policies. Thus, we must have $\pi_1(1) < \pi_2(1) < \pi_3(1) < \pi_4(1)$. □

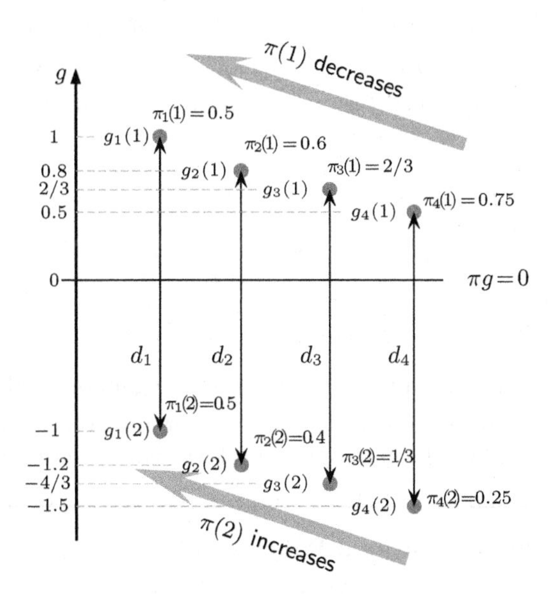

Fig. 4.4. The Biases of the Gain-Optimal Policies in Example 4.1

The first step in our approach to finding a bias-optimal policy is to derive a formula that compares the biases of any two policies in $\mathcal{D}_0$. Before doing so, we need to specify the set $\mathcal{D}_0$.

The Set of Gain-Optimal Policies $\mathcal{D}_0$

From now on, we will use superscript "$*$" to denote the optimal value; e.g., η^* is the optimal gain and g^* is the optimal bias. We denote a gain-optimal policy as $\widehat{d}$. Its average reward $\eta^{\widehat{d}}$ equals the optimal gain η^*, but its bias $g^{\widehat{d}}$ may or may not equal the optimal bias g^*.

The set of all gain-optimal policies, $\mathcal{D}_0$, can be determined by the following lemma.

Lemma 4.1. *Let $\widehat{d}$ be a gain-optimal policy, i.e., $\eta^{\widehat{d}} = \eta^*$. Any other policy $d \in \mathcal{D}$ is gain optimal if and only if*

$$f^d + P^d g^{\widehat{d}} = f^{\widehat{d}} + P^{\widehat{d}} g^{\widehat{d}}. \tag{4.11}$$

Proof. The "if" part follows directly from the difference formula (4.1); now, we prove the "only if" part. Suppose that both d and $\widehat{d}$ are gain optimal. We have $\eta^d = \eta^{\widehat{d}}$. Thus, by (4.1), we have $\pi^d \left[(f^d + P^d g^{\widehat{d}}) - (f^{\widehat{d}} + P^{\widehat{d}} g^{\widehat{d}})\right] = 0$. Suppose that (4.11) does not hold; then, because $\pi^d > 0$, there must exist a state, denoted as i, such that $\left[f(i, d(i)) + (P^d g^{\widehat{d}})(i)\right] - \left[f(i, \widehat{d}(i)) + (P^{\widehat{d}} g^{\widehat{d}})(i)\right] > 0$. We then construct a policy $\widetilde{d}$ by setting $\widetilde{d}(k) = \widehat{d}(k), k \neq i$, and $\widetilde{d}(i) = d(i)$. Thus, we have $p^{\widetilde{d}(k)}(j|k) = p^{\widehat{d}(k)}(j|k)$, $k \neq i$, $j \in \mathcal{S}$ and $p^{\widetilde{d}(i)}(j|i) = p^{d(i)}(j|i)$, $j \in \mathcal{S}$; and $f(k, \widetilde{d}(k)) = f(k, \widehat{d}(k)), k \neq i$, and $f(i, \widetilde{d}(i)) = f(i, d(i))$. With this construction, we have

$$\begin{aligned}\eta^{\widetilde{d}} - \eta^{\widehat{d}} &= \pi^{\widetilde{d}} \left[(f^{\widetilde{d}} + P^{\widetilde{d}} g^{\widehat{d}}) - (f^{\widehat{d}} + P^{\widehat{d}} g^{\widehat{d}})\right] \\ &= \pi^{\widetilde{d}}(i) \left\{ \left[(f(i, \widetilde{d}(i)) + (P^{\widetilde{d}} g^{\widehat{d}})(i)\right] - \left[f(i, \widehat{d}(i)) + (P^{\widehat{d}} g^{\widehat{d}})(i)\right] \right\} > 0.\end{aligned}$$

This is impossible because $\eta^{\widehat{d}}$ is the maximal reward. □

The lemma provides a way to determine the gain-optimal set $\mathcal{D}_0$: Choose any gain-optimal policy $\widehat{d}$, calculate its bias (or any potential) $g^{\widehat{d}}$ (which may not be optimal), then the gain-optimal set $\mathcal{D}_0$ can be determined as follows.

The Set of Gain-Optimal Policies $\mathcal{D}_0$:

$$\mathcal{D}_0 := \times_{i \in \mathcal{S}} \widehat{\mathcal{A}}_1(i), \tag{4.12}$$

where

$$\begin{aligned}\widehat{\mathcal{A}}_1(i) = \Bigg\{ \alpha \in \mathcal{A}(i) : f(i, \alpha) + \sum_{j=1}^{S} p^{\alpha}(j|i) g^{\widehat{d}}(i) \\ = f(i, \widehat{d}(i)) + \sum_{j=1}^{S} p^{\widehat{d}(i)}(j|i) g^{\widehat{d}}(i) \Bigg\}.\end{aligned}$$

For any state $i \in \mathcal{S}$, we can choose any actions $\alpha \in \widehat{\mathcal{A}}_1(i)$ to form a gain-optimal policy. From Lemma 4.1, (4.11) holds for any two gain-optimal policies. Thus, $\mathcal{D}_0$ does not depend on the choice of the initial policy $\widehat{d} \in \mathcal{D}_0$ (cf. Problem 4.11).

We will see that Lemma 4.1 may not hold for the multi-chain case.

The Bias Difference Formula

Consider any two gain-optimal policies h and d in $\mathcal{D}_0$, with biases g^h and g^d, respectively. From the Poisson equation, we have $g^h = (f^h + P^h g^h) - \eta^h e$ and

$g^d = (f^d + P^d g^d) - \eta^d e$. Thus, using Lemma 4.1 and noting $\eta^h = \eta^d (= \eta^*)$, we have

$$g^h - g^d = (f^h + P^h g^h) - (f^d + P^d g^d) = P^h (g^h - g^d).$$

Iteratively applying this equation, we have

$$g^h - g^d = P^h (g^h - g^d) = (P^h)^2 (g^h - g^d) = \cdots = (P^h)^k (g^h - g^d).$$

Letting $k \to \infty$, for ergodic chains, we have $(P^h)^k \to e\pi^h$. Because $\pi^h g^h = 0$, we have the following equation for biases of the gain-optimal policies:

$$g^h - g^d = e\pi^h (g^h - g^d) = -(\pi^h g^d) e. \tag{4.13}$$

Similarly, $g^d - g^h = -(\pi^d g^h) e$ and we have $\pi^h g^d = -\pi^d g^h$.

Example 4.2. From Table 4.1, Equation (4.13) holds for all policies. For example,

$$g_1 - g_2 = \begin{bmatrix} 0.2 \\ 0.2 \end{bmatrix} = (\pi_2 g_1) e = -(\pi_1 g_2) e.$$

This equation also holds for any other pairs of biases, see Figure 4.4. □

Equation (4.13) shows that the difference in the biases of any two gain-optimal policies $h, d \in \mathcal{D}_0$ in any state is a constant $-\pi^h g^d$. This explains that a bias-optimal policy maximizing the biases in all the states indeed exists. If we pick up any policy $d \in \mathcal{D}_0$, then the bias optimality problem becomes to find a policy $h \in \mathcal{D}_0$ that maximizes $-\pi^h g^d$, or equivalently, minimizes $\pi^h g^d$. This insight is significant: the bias optimization problem is almost the same as the gain-optimization problem; in the latter, we optimize $\eta^h = \pi^h f$ (assuming that f is the same for all policies), and in the former, we optimize $-\pi^h g^d$. The difference is that f is replaced by $-g^d$, which is the bias of any particular gain-optimal policy. This motivates our further study.

In the gain-optimization problem, we need the potential associated with the reward function f. Now, in the bias-optimization problem, we need the potential associated with $(-g^d)$. If we replace f with $-g^d$ in the Poisson equation (4.9), the corresponding average reward η should be $\pi^d (-g)^d = 0$. Thus, we obtain the following Poisson equation with $-g^d$, the bias of policy d, as the reward function:

$$(I - P^d) w^d = -g^d.$$

This is the Poisson equation for biases. In general, for a policy with transition probability matrix P and bias g, it takes the form

Poisson Equation for Biases:

$$(I - P) w = -g. \tag{4.14}$$

This equation specifies the vector w, which is the negative value of the *potential of the potential* g, or called the *bias-potential.* The solution to (4.14) is not unique. Again, if w is a solution, then so is $w + ce$ for any constant c. Note that to maximize the bias g^h, $h \in \mathcal{D}_0$, we need to maximize $\pi^h(-g^d)$ for a fixed $d \in \mathcal{D}_0$. This explains why the "−" sign appears in (4.14).

From (4.13) and (4.14), the bias difference formula for two gain-optimal policies $d, h \in \mathcal{D}_0$ is

The Bias Difference Formula for Two Gain-Optimal Policies:

$$g^h - g^d = \left\{\pi^h(P^h - P^d)w^d\right\} e. \tag{4.15}$$

In (4.15), w^d is the bias-potential of policy d. Comparing this formula with the average-reward difference formula (4.1), we can obtain the following results for two gain-optimal policies h and d.

Comparison Lemma:

$$\text{If } P^h w^d \succeq P^d w^d, \text{ then } g^h > g^d. \tag{4.16}$$

$$\text{If } P^h w^d \preceq (\geq, \leq) P^d w^d, \text{ then } g^h < (\geq, \leq)\ g^d. \tag{4.17}$$

With this Comparison Lemma, we can easily prove the following result.

Bias Optimality Condition:

A gain-optimal policy $\widehat{d} \in \mathcal{D}_0$ is bias optimal if and only if

$$P^{\widehat{d}} w^{\widehat{d}} \geq P^d w^{\widehat{d}}, \qquad \text{for all } d \in \mathcal{D}_0. \tag{4.18}$$

The Bias Optimality Condition (4.18) is equivalent to Bellman's optimality equation for bias-optimal policies:

Bias Optimality Equation:

A gain-optimal policy $\widehat{d} \in \mathcal{D}_0$ is bias optimal if and only if

$$g^{\widehat{d}} + w^{\widehat{d}} = \max_{d \in \mathcal{D}_0} \left\{P^d w^{\widehat{d}}\right\}. \tag{4.19}$$

In (4.19), $\mathcal{D}_0$ is defined by (4.12). Equation (4.18) together with (4.4), or equivalently (4.7) and (4.19), are the (necessary and sufficient) bias optimality conditions.

The Bias Potential

The bias-potential w can be obtained either by solving the Poisson equation (4.14) or by analyzing a sample path. We may normalize it by setting $\pi w = 0$ and (4.14) becomes

$$(I - P + e\pi)w = -g,$$

where g is a bias with $\pi g = 0$. Therefore,

$$\begin{aligned} w &= -(I - P + e\pi)^{-1} g \\ &= -\sum_{l=0}^{\infty} (P - e\pi)^l g = -\sum_{l=0}^{\infty} P^l g \\ &= -\sum_{l=0}^{\infty} \left\{ P^l \sum_{k=0}^{\infty} P^k (f - \eta e) \right\} \\ &= -\sum_{l=0}^{\infty} \{ (l+1) P^l (f - \eta e) \} . \end{aligned}$$

Thus, we have

$$w(i) = -\sum_{l=0}^{\infty} \{ (l+1) E \left[f(X_l) - \eta | X_0 = i \right] \} . \tag{4.20}$$

This $w(i)$ can be estimated on a sample path of the Markov chain under (P, f). The bias-potential (4.20) has a similar form as the bias (4.8), except that a weighting factor $-(l+1)$ is added to the deviation $E[f(X_l) - \eta]$ at time l, $l = 0, 1, 2, \ldots$. Again, like the performance potentials, only the difference between the components of w is important in sensitivity analysis and in comparison of biases. Thus, we may ignore the constant $-(\pi f)e$ in (4.10) and obtain

$$w = -\left[(I - P + e\pi)^{-1} \right]^2 f. \tag{4.21}$$

The Policy Iteration Algorithm for a Bias-Optimal Policy

From the Comparison Lemma (4.16) and (4.17), we can derive *the policy iteration algorithm for a bias-optimal policy*:

Algorithm 4.2. A Policy Iteration Algorithm for a Bias-Optimal Policy:

1. Select any gain-optimal policy d_0, which may be the one obtained from the policy iteration algorithm 4.1 for gain-optimal policies; set $k = 0$.
2. Determine $\mathcal{D}_0$ by (4.12).
3. Obtain the bias g^{d_k}, by solving (4.9), and the bias-potential w^{d_k}, by solving (4.14) or by using (4.21). They can also be estimated on a sample path of the system under policy d_k with (4.8) and (4.20), respectively.
4. Choose
$$d_{k+1} \in \arg\left\{\max_{d \in \mathcal{D}_0}\left[P^d w^{d_k}\right]\right\},$$
component-wisely (i.e., to determine an action for each state). If in a state i, action $d_k(i)$ attains the maximum, then set $d_{k+1}(i) = d_k(i)$.
5. If $d_{k+1} = d_k$, stop; otherwise, set $k := k+1$ and go to step 3.

Example 4.3. In Example 4.1, we can easily verify that $(I - P_1 + e\pi_1) = I$ and, therefore, that $w_1 = -g_1 = (-1, 1)^T$. Thus, from $[p^{\alpha_2}(\bullet|1) - p^{\alpha_1}(\bullet|1)]\, w_1 = -0.5 < 0$ and $\left[p^{\beta_2}(\bullet|2) - p^{\beta_1}(\bullet|2)\right] w_1 = -0.5 < 0$, we conclude that $d_1 = (\alpha_1, \beta_1)$ is a bias-optimal policy. □

We can also derive a policy iteration algorithm that starts from any policy in $\mathcal{D}$ and reaches a bias-optimal policy without first determining the policy subspace $\mathcal{D}_0$. This algorithm uses both the average-reward and bias difference formulas (4.1) and (4.15) at each iteration. This is stated as the *second policy iteration algorithm for a bias-optimal policy*:

Algorithm 4.3. A Second Policy Iteration Algorithm for a Bias-Optimal Policy:

1. Select any policy $d_0 \in \mathcal{D}$, and set $k = 0$.
2. Obtain the bias g^{d_k} by solving (4.9) and bias-potential w^{d_k} by solving (4.14) or by (4.21), or by estimation.
3. Set (component-wisely)
$$\widetilde{\mathcal{D}} := \arg\left\{\max_{d \in \mathcal{D}}\left[f^d + P^d g_k^d\right]\right\},$$
and choose

$$d_{k+1} = \arg\left\{ \max_{d \in \widetilde{\mathcal{D}}} \left[P^d w^{d_k} \right] \right\}.$$

If in a state i, action $d_k(i)$ attains the maximum, then set $d_{k+1}(i) = d_k(i)$.

4. If $d_{k+1} = d_k$, stop; otherwise, set $k := k + 1$ and go to step 2.

Although we try to "improve" both the average reward and the bias at every iteration in this algorithm, we are not sure if this is indeed true. Because (4.15) holds only when $\eta^h = \eta^d$, we can only assert that the bias increases after the average reward reaches its maximum. Therefore, we are not sure whether this algorithm is "faster" than Algorithm 4.2. The algorithm converges because at least the average reward improves at every iteration before it reaches the set $\mathcal{D}_0$, and after that the bias improves. Of course, the computation at each iteration increases.

We have shown that a bias-optimal policy can be obtained by policy iteration algorithms derived from the bias difference formula (4.15). The development of the results follows the same logic as that for the gain-optimality problem shown in Figure 4.2.

4.1.3 MDPs with Discounted Rewards

In this section, we show that the same sensitivity-based approach applies to MDPs with the discounted reward criterion. For any policy (P, f), the discounted reward is defined as a column vector $\eta_\beta = (\eta_\beta(1), \ldots, \eta_\beta(S))^T$ with (cf. (2.30))

$$\eta_\beta(i) = (1-\beta) E\left\{ \sum_{l=0}^{\infty} \beta^l f(X_l) \Big| X_0 = i \right\}, \qquad 0 < \beta < 1, \tag{4.22}$$

where $\{X_l, l = 0, 1, \ldots\}$ is a sample path of the Markov chain under (P, f). The factor $(1-\beta)$ in (4.22) is used to obtain the continuity of η_β at $\beta = 1$; we have (cf. (2.32)) $\eta_1 := \lim_{\beta \uparrow 1} \eta_\beta = \eta e$ with η being the average reward. Thus, (4.22) covers both the average reward case with $\beta = 1$ and the discounted reward case with $0 < \beta < 1$. For any particular policy d, (4.22) becomes

$$\eta_\beta^d(i) = (1-\beta) E\left\{ \sum_{l=0}^{\infty} \beta^l f(X_l, d(X_l)) \Bigg| X_0 = i \right\}, \qquad 0 < \beta < 1.$$

The optimization problem is to find a policy $\widehat{d} \in \mathcal{D}$ such that its discounted reward in all states is the maximum among all the policies in $\mathcal{D}$:

$$\eta_\beta^{\widehat{d}} \geq \eta_\beta^d, \qquad \text{for all } d \in \mathcal{D},$$

or

$$\widehat{d} \in \arg\left\{\max_{d\in\mathcal{D}} \eta_\beta^d, \text{ all components}\right\}.$$

We will see that such an optimal policy maximizing $\eta_\beta(i)$ for all $i \in \mathcal{S}$ indeed exists.

The discounted-reward difference formula is (cf. (2.42)) as follows.

The Discounted-Reward Difference Formula:

$$\eta_\beta^h - \eta_\beta^d = (1-\beta)(I-\beta P^h)^{-1}\left[(f^h + \beta P^h g_\beta^d) - (f^d + \beta P^d g_\beta^d)\right], \quad 0 < \beta < 1. \tag{4.23}$$

When $\beta \uparrow 1$, we have $(1-\beta)(I-\beta P^h)^{-1} \to e\pi^h$ (cf. (2.39)). Thus, for $\beta = 1$, we replace $(1-\beta)(I-\beta P^h)^{-1}$ by $e\pi^h$ (cf. (2.39)), and (4.23) reduces to (4.1).

In (4.23), g_β is the β-potential, a solution to the discounted Poisson equation

$$(I - \beta P + \beta e\pi)g_\beta = f, \qquad 0 < \beta \leq 1. \tag{4.24}$$

When $\beta = 1$, (4.24) becomes the standard Poisson equation.

A Fundamental Fact:

$$(I - \beta P^h)^{-1} > I,\ 0 < \beta < 1, \text{ for any ergodic } P^h.$$

This fundamental fact follows from the simple fact that $(I - \beta P^h)^{-1} = I + \beta P^h + \beta^2 (P^h)^2 + \cdots$. The discounted reward difference formula (4.23) and the fundamental fact lead immediately to the following result.

Comparison Lemma:

If $f^h + \beta P^h g_\beta^d \succeq f^d + \beta P^d g_\beta^d$, $0 < \beta \leq 1$, then $\eta_\beta^h \succeq \eta_\beta^d$.

With this Comparison Lemma, we can easily prove the following result.

Optimality Condition:

A policy $\widehat{d}$ is optimal if and only if

$$f^{\widehat{d}} + \beta P^{\widehat{d}} g_\beta^{\widehat{d}} \geq f^d + \beta P^d g_\beta^{\widehat{d}}, \qquad \text{for all } d \in \mathcal{D}.$$

When $\beta = 1$, this condition becomes the same as that for the average-reward optimal policies.

With this optimality condition, we can develop the following *policy iteration algorithm for a discounted-reward optimal policy with* $0 < \beta \le 1$. It covers the average-reward case as a special case with $\beta = 1$.

Algorithm 4.4. A Policy Iteration Algorithm for a Discounted-Reward Optimal Policy:

1. Select an initial policy $d_0 \in \mathcal{D}$, and set $k = 0$.
2. Obtain the potential $g_\beta^{d_k}$ by solving the discounted Poisson equation $(I - \beta P^{d_k} + \beta e \pi^{d_k}) g_\beta^{d_k} = f^{d_k}$, or by estimation on a sample path of the system under policy d_k (cf. (2.46)).
3. Choose
$$d_{k+1} \in \arg\left\{\max_{d \in \mathcal{D}} \left[f^d + \beta P^d g_\beta^{d_k}\right]\right\}, \tag{4.25}$$
component-wisely. If in a state i, action $d_k(i)$ attains the maximum, then set $d_{k+1}(i) = d_k(i)$.
4. If $d_{k+1} = d_k$, stop; otherwise, set $k := k + 1$ and go to step 2.

Again, the development of the above results follows the same logic as shown in Figure 4.2. The performance difference formula-based approach provides a unified framework for developing policy iteration algorithms for both average and discounted reward criteria, as well as for the bias optimality problem. In particular, with this approach, all the results for the average-reward MDPs can be derived independently of the discounted-reward case.

4.2 Multi-Chains

In this section, we extend the results in Section 4.1 to multi-chains. (A multi-chain contains more than one closed subsets of recurrent states, see Appendix B.1.) We show that, in the multi-chain case, the policy iteration optimization algorithms can also be derived almost directly from the performance difference formulas.

The Multi-Chain Markov Model

Consider a multi-chain Markov chain $\{X_l, l = 0, 1, \ldots\}$ defined on a finite state space $\mathcal{S} = \{1, 2, \ldots, S\}$. Let $P = [p(j|i)]_{i,j=1}^S$ be the transition probability matrix and $f(i)$, $i \in \mathcal{S}$, be the reward function. We have $Pe = e$, with $e = (1, \ldots, 1)^T$. The long-run average reward is defined as a vector η, with components

$$\eta(i) = \lim_{L\to\infty} \frac{1}{L} E\left\{\sum_{l=0}^{L-1} f(X_l)\middle| X_0 = i\right\},$$

which depend on the initial state i, $i \in \mathcal{S}$, where E denotes the expectation corresponding to the probability space generated by all the sample paths with transition probability P. For simplicity, we discuss only the aperiodic case. There is no loss of generality because, as shown in Problem B.4, for any periodic Markov chain, we can always construct an equivalent aperiodic Markov chain. Let[2] $\eta = (\eta(1), \ldots, \eta(S))^T$. In matrix form, we have

$$\eta = \lim_{L\to\infty} \frac{1}{L}\left\{\sum_{l=0}^{L-1} P^l f\right\} = P^* f, \tag{4.26}$$

where $f = (f(1), \ldots, f(S))^T$, and P^* is the Cesaro limit

$$P^* = \lim_{L\to\infty} \frac{1}{L} \sum_{l=0}^{L-1} P^l, \tag{4.27}$$

which exists and represents the steady-state probabilities of the Markov chain, see (B.11) or (4.31). With (4.27), we can easily prove that $P^*e = e$ and

$$PP^* = P^*P = P^*P^* = P^*. \tag{4.28}$$

From (4.26) and (4.28), we get

$$P\eta = P^*\eta = \eta. \tag{4.29}$$

More results on P and P^* for multi-chains can be found in Appendix B.

Given a policy $d \in \mathcal{D}$, the corresponding average reward is denoted as a vector $\eta^d = (P^d)^* f^d$, $P^d = [p^{d(i)}(j|i)]_{i,j=1}^S$ and $f^d = (f(1, d(1)), \ldots, f(S, d(S)))^T$, with $\eta^d(i)$ being the long-run average reward starting from initial state i, $i = 1, 2, \ldots, S$.

A policy $\widehat{d}$ is said to be *gain (average-reward) optimal* if

$$\eta^{\widehat{d}} \geq \eta^d, \qquad \text{for all } d \in \mathcal{D},$$

and its gain $\eta^{\widehat{d}}$ is called the *optimal gain*. Let η^* denote the optimal gain, $\eta^* = \max_{d\in\mathcal{D}} \eta^d$, and define

$$\mathcal{D}_0 := \left\{d \in \mathcal{D} : \eta^d = \eta^*\right\}$$

as the set of all gain-optimal policies. A gain-optimal policy has the largest average reward in every state, and we will see that such policies indeed exist. We wish to develop an efficient algorithm to find a gain-optimal policy $\widehat{d} \in \mathcal{D}_0$.

[2] Please do not be confused with the ergodic case, where η denotes a scaler.

Recall that in deriving the results for ergodic chains, the fundamental fact that $\pi^h > 0$ for any $h \in \mathcal{D}$ is used. The corresponding simple fact for multi-chains is stated below, which will be used often in this chapter. Let u be an S-dimensional vector.

A Fundamental Fact:

If $u \geq 0$ (or $u \leq 0$) and $(P^h)^* u = 0$,

then $u(i) = 0$ for all recurrent states i of P^h. (4.30)

Proof. Following the canonical form of $(P^h)^*$ in (B.11), we partition the vector u as $u = (u_1^T, \ldots, u_m^T, u_{m+1}^T)^T$, with $u_1, \ldots, u_m$ corresponding to the m different classes of recurrent states of P^h. We have

$$(P^h)^* u = \begin{bmatrix} e\pi_1^h & 0 & 0 & \cdots & \cdot & 0 \\ 0 & e\pi_2^h & 0 & \cdots & \cdot & 0 \\ \cdot & \cdot & \cdot & \cdots & \cdot & \cdot \\ 0 & 0 & 0 & \cdots & e\pi_m^h & 0 \\ w_1\pi_1^h & w_2\pi_2^h & w_3\pi_3^h & \cdots & w_m\pi_m^h & 0 \end{bmatrix} \begin{bmatrix} u_1 \\ u_2 \\ \cdots \\ u_m \\ u_{m+1} \end{bmatrix} = \begin{bmatrix} (\pi_1^h u_1)e \\ (\pi_2^h u_2)e \\ \cdots \\ (\pi_m^h u_m)e \\ \sum_{i=1}^m (\pi_i^h u_i) w_i \end{bmatrix}. \quad (4.31)$$

Because $\pi_i^h > 0$ for all $i = 1, 2, \ldots, m$, we must have $\left(u_1^T, \ldots, u_m^T\right)^T = 0$, if $(P^h)^* u = 0$ and $u \geq 0$. □

Note that $u(i)$ may also be zero for transient states. For ergodic chains, we have $(P^h)^* = e\pi^h$. The lemma follows directly from $\pi^h > 0$.

4.2.1 Policy Iteration

Performance Potentials

The potential $g = (g(1), \ldots, g(S))^T$ is defined by the Poisson equation

$$(I - P)g + \eta = f. \quad (4.32)$$

If g satisfies (4.32), then so does $g + u$, for any vector u satisfying $(I - P)u = 0$. For example, we can choose $u = ce$, or $u = c\eta$ (see (4.29)), with c being any constant. Therefore, there are different versions of potentials, each may differ by an additive vector u. We will call all of them potentials and use the same notation g.

A potential satisfying $P^*g = 0$ is called a *bias*. For any given potential g', if we set $g = g' + u$ and $u = -P^*g'$, then g is a bias. (4.32) becomes

$$(I - P + P^*)g = f - \eta.$$

From (B.12) (or Theorem A.7 of [216]), the matrix $(I - P + P^*)$ is nonsingular. We can easily prove that

$$(I - P + P^*)^{-1}P^* = P^*. \tag{4.33}$$

Thus, for a bias, we have

$$\begin{aligned} g &= (I - P + P^*)^{-1}(f - \eta) = \sum_{l=0}^{\infty}(P - P^*)^l(f - \eta) \\ &= \sum_{l=1}^{\infty}(P^l - P^*)(f - \eta) + f - \eta \\ &= \sum_{l=0}^{\infty} P^l(f - \eta) - \sum_{l=1}^{\infty} P^*(f - \eta) \\ &= \sum_{l=0}^{\infty} P^l(f - \eta). \end{aligned}$$

That is,

$$g(i) = \sum_{l=0}^{\infty} E\left\{[f(X_l) - \eta(X_l)]|X_0 = i\right\}. \tag{4.34}$$

Because $P\eta = \eta$, we have $P^l\eta = \eta$. That is, $E[\eta(X_l)|X_0 = i] = \eta(i)$. Thus,

$$g(i) = \sum_{l=0}^{\infty} E\left\{[f(X_l) - \eta(i)]|X_0 = i\right\}. \tag{4.35}$$

From (4.34) and (4.35), sample-path-based learning algorithms can be developed to estimate g, see Section 3.1.2.

The Average-Reward Difference Formula

The sensitivity-based optimization approach starts with the performance difference formula. Let η^h and η^d be the average rewards of the two policies $h, d \in \mathcal{D}$, respectively, and $(P^h)^*$ be the Cesaro limit of P^h. We have

The Average-Reward Difference Formula for Multi-Chains:

$$\eta^h - \eta^d = (P^h)^* \left[(f^h + P^h g^d) - (f^d + P^d g^d)\right] + [(P^h)^* - I]\eta^d. \tag{4.36}$$

Proof. By (4.28) and (4.32), we have

$$\begin{aligned}
\eta^h - \eta^d &= (P^h)^* f^h - \eta^d \\
&= (P^h)^* f^h + (P^h)^* g^d - (P^h)^* g^d - (P^h)^* \eta^d + (P^h)^* \eta^d - \eta^d \\
&= (P^h)^* \left(f^h + P^h g^d - g^d - \eta^d\right) + [(P^h)^* - I]\eta^d \\
&= (P^h)^* \left[(f^h + P^h g^d) - (f^d + P^d g^d)\right] + [(P^h)^* - I]\eta^d.
\end{aligned}$$

This is (4.36). □

In policy iteration, given the current policy d, we wish to find another policy h, which has a better average reward $\eta^h > \eta^d$, without evaluating the system under policy h. We have done so for ergodic chains with the average reward difference formula (4.1), using the fact that $\pi^h > 0$. However, in the multi-chain case, π^h becomes $(P^h)^*$ and there are two terms on the right-hand side of the average-reward difference formula (4.36). This causes a major difficulty in extending the results from ergodic chains to the multi-chain case. Fortunately, as the following example indicates, these two terms can be "decoupled". To simplify the notation, we set $v = (f^h + P^h g^d) - (f^d + P^d g^d)$ and $u = [(P^h)^* - I]\eta^d$ and rewrite (4.36) as

$$\eta^h - \eta^d = (P^h)^* v + u. \tag{4.37}$$

Example 4.4. Let $\mathcal{S} = \{1, 2, 3, 4, 5\}$, $f^d = (5, 2, 1, 3, 1)^T$, $f^h = (4, 1, 1, 2, 0)^T$, and

$$P^d = \begin{bmatrix} 0.5 & 0.5 & 0 & 0 & 0 \\ 0.4 & 0.6 & 0 & 0 & 0 \\ 0 & 0 & 0.2 & 0.8 & 0 \\ 0 & 0 & 0.7 & 0.3 & 0 \\ 0.1 & 0.2 & 0.2 & 0.3 & 0.2 \end{bmatrix}, \quad P^h = \begin{bmatrix} 0.9 & 0.1 & 0 & 0 & 0 \\ 0.8 & 0.2 & 0 & 0 & 0 \\ 0.2 & 0.4 & 0.1 & 0.2 & 0.1 \\ 0.2 & 0.1 & 0.2 & 0.3 & 0.2 \\ 0.3 & 0.1 & 0.2 & 0.1 & 0.3 \end{bmatrix}.$$

After some calculations for g^d, u, v and $(P^h)^*$, we can write the performance difference (4.37) as follows

$$\begin{aligned}
&\eta^h - \eta^d = (P^h)^* v + u \\
&= \begin{bmatrix} 0.8889 & 0.1111 & 0.0000 & 0.0000 & 0.0000 \\ 0.8889 & 0.1111 & 0.0000 & 0.0000 & 0.0000 \\ 0.8889 & 0.1111 & 0.0000 & 0.0000 & 0.0000 \\ 0.8889 & 0.1111 & 0.0000 & 0.0000 & 0.0000 \\ 0.8889 & 0.1111 & 0.0000 & 0.0000 & 0.0000 \end{bmatrix} \begin{bmatrix} 0.3333 \\ 0.3333 \\ -0.2650 \\ -0.5534 \\ -7.7077 \end{bmatrix} + \begin{bmatrix} 0.0000 \\ 0.0000 \\ 2.0832 \\ 2.0832 \\ 1.3019 \end{bmatrix}.
\end{aligned}$$

Observe the following structure: The components in the second term, $u(i)$, for recurrent states of P^h ($i = 1$ and 2) are zeros; the components in $v(i)$ in the first term for recurrent states of P^h are all positive, and the columns of $(P^h)^*$ for transient states ($i = 3, 4,$ and 5) are all zeros. This "decouples" the effect of the two terms: Because $u(1) = u(2) = 0$, the components of $\eta^h - \eta^d$ for the recurrent states, $\eta^h(1) - \eta^d(1)$ and $\eta^h(2) - \eta^d(2)$, are determined by the first term in the average-reward difference formula, $(P^h)^* v$, and furthermore, by $v(1)$ and $v(2)$; and the components of $\eta^h - \eta^d$ for the transient states, $\eta^h(3) - \eta^d(3)$, $\eta^h(4) - \eta^d(4)$, and $\eta^h(5) - \eta^d(5)$, take additional contributions from the second terms, $u(3)$, $u(4)$, and $u(5)$, in the performance difference formula. The negative values in the components of v for the transient states, $v(3)$, $v(4)$, and $v(5)$, do not play a role. □

Comparison of Two Policies

The above example provides an important insight: It is possible to compare the average rewards of two policies by using the structures illustrated in the above example. We state this result in a general form as a Comparison Lemma. We will see that many optimization problems about multi-chains fit this general form.

Suppose that the performance difference formula takes the form

$$\eta^h - \eta^d = (P^h)^* v + Gu, \tag{4.38}$$

where u, v are S-dimensional vectors and G is any $S \times S$ matrix satisfying $Gu \geq 0$, as specified below.

Comparison Lemma (General Form): (4.39)

If (a) $u \geq 0$, $(P^h)^* u = 0$, $Gu \geq 0$, and
(b) $v(i) \geq 0$ when $u(i) = 0$ for $i \in \mathcal{S}$,

$\Longrightarrow$ then $\eta^h \geq \eta^d$.

The lemma also holds if all the signs " $\geq$ " are changed to " $\leq$ ".

Proof. By the Fundamental Fact (4.30) and (a), $u(i) = 0$ for all recurrent states $i \in \mathcal{S}$ of P^h. Thus, by (b), $v(i) \geq 0$ for all recurrent states $i \in \mathcal{S}$ of P^h. Then by the canonical form of $(P^h)^*$ (see (B.11), or (4.31)), we have $(P^h)^* v \geq 0$. Thus, $\eta^h \geq \eta^d$ because $Gu \geq 0$. □

In the lemma, the form Gu on the right-hand side is not crucial. In fact, the lemma holds if we replace (4.38) with

$$\eta^h - \eta^d = (P^h)^* v + z, \tag{4.40}$$

and replace the condition $Gu \geq 0$ with $z \geq 0$.

If $v(i) > 0$ whenever $u(i) = 0$, $i \in \mathcal{S}$, then $\eta^h > \eta^d$. If $v(i) > 0$ for at least one recurrent state of P^h and $v(i) = 0$ for others, we have $\eta^h \succeq \eta^d$. However, if $u(i) = 0$, we do not know whether i is recurrent or transient. Therefore, if when $u(i) = 0$, $v(i) > 0$ for some states and $v(i) = 0$ for others, we are not sure if $\eta^h \succeq \eta^d$.

For ergodic chains, $(P^h)^* = e\pi^h$, and Condition (a) in the Comparison Lemma (General Form (4.39)) implies $u = 0$. Thus, Condition (b) implies $v \geq 0$. The lemma becomes the Comparison Lemma for ergodic chains (4.2) and (4.3) in Section 4.1.

We will see that the difference formulas for average rewards, biases, and nth biases all have the general form of (4.38). Therefore, the Comparison Lemma (4.39) plays an important role; it allows us to find a policy h that is "better" (in terms of average reward, bias, or nth biases) than d without calculating $(P^h)^*$. This is very crucial to the optimization theory since policy iteration algorithms are based on it. We use Figure 4.5 to clearly illustrate the idea.

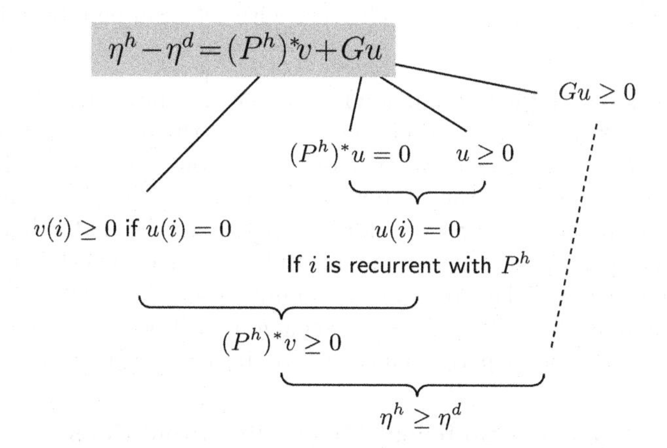

Fig. 4.5. The Fundamental Fact in Optimization with a Difference Formula

We first apply this lemma to the average reward case. Let h and d be two policies.

Comparison Lemma for the Average Reward of Multi-Chains:

$$\text{If} \quad \begin{array}{l} (a) \ P^h \eta^d \geq \eta^d, \text{ and} \\ (b) \ f^h(i) + (P^h g^d)(i) \geq f^d(i) + (P^d g^d)(i) \\ \qquad \textit{when } (P^h \eta^d)(i) = \eta^d(i) \text{ for } i \in \mathcal{S}, \end{array} \tag{4.41}$$

$$\Longrightarrow \text{ then } \eta^h \geq \eta^d.$$

The lemma also holds if all the signs " $\geq$ " are changed to " $\leq$ ".

Proof. In the average reward difference formula (4.36), we set $u = P^h \eta^d - \eta^d \geq 0$, $v = (f^h + P^h g^d) - (f^d + P^d g^d)$, and $G = \left[I - P^h + (P^h)^*\right]^{-1}$. We have $(P^h)^* u = 0$ and thus

$$Gu = \sum_{l=0}^{\infty} (P^h)^l u = [(P^h)^* - I]\eta^d.$$

Then, the difference formula (4.36) takes the form in (4.38). Furthermore, from $P^h \eta^d \geq \eta^d$ we have $(P^h)^{l+1}\eta^d \geq (P^h)^l \eta^d \geq \cdots \geq \eta^d$, or $(P^h)^l u \geq 0$. Thus, $Gu \geq 0$. In addition, (*b*) in this lemma implies (*b*) in the Comparison Lemma (General Form (4.39)). Therefore, the lemma follows directly from the Comparison Lemma (4.39). The case with "$\leq$" can be proved in a similar way. □

Compared with the Comparison Lemma for ergodic chains (4.2), we cannot prove $\eta^h \succeq \eta^d$ in this lemma, because $u(i) = 0$ (therefore $v(i) > 0$) may also hold for some transient states (for more discussion, see Problem 4.16).

In the ergodic case, the property we used in making performance comparisons is the fact that $\pi^h > 0$ for any policy. In the multi-chain case, the corresponding property is the Fundamental Fact (4.30), which is based on the structure of $(P^h)^*$. The Comparison Lemma (4.41) deals with the two terms in the difference formulas. The same technique will be used again in deriving many results for other performance criteria in multi-chain MDPs.

The Necessary and Sufficient Optimality Conditions

From the Comparison Lemma (4.41), we can easily derive the sufficient optimality conditions.

Sufficient Optimality Conditions:

A policy $\widehat{d}$ is gain optimal if

$$\eta^{\widehat{d}}(i) = \max_{\alpha \in \mathcal{A}(i)} \left\{ \sum_{j \in \mathcal{S}} p^\alpha(j|i) \eta^{\widehat{d}}(j) \right\}, \qquad \text{for all } i \in \mathcal{S}, \tag{4.42}$$

$$\eta^{\widehat{d}}(i) + g^{\widehat{d}}(i) = \max_{\alpha \in \widehat{\mathcal{A}}_0(i)} \left\{ f(i, \alpha) + \sum_{j \in \mathcal{S}} p^\alpha(j|i) g^{\widehat{d}}(j) \right\}, \tag{4.43}$$

for all $i \in \mathcal{S}$,

where $\widehat{\mathcal{A}}_0(i) := \left\{ \alpha \in \mathcal{A}(i) : \sum_{j \in \mathcal{S}} p^\alpha(j|i) \eta^{\widehat{d}}(j) = \eta^{\widehat{d}}(i) \right\}$, $i \in \mathcal{S}$.

If the sufficient optimality conditions (4.42) and (4.43) hold, then $\eta^{\widehat{d}}$ is the optimal average reward; i.e., $\eta^{\widehat{d}} = \eta^*$, but the $g^{\widehat{d}}$ in (4.43) is the performance potential, not necessarily the bias; it may not be the optimal bias even after being normalized by $\pi^{\widehat{d}} g^{\widehat{d}} = 0$.

Proof. This is a direct consequence of the Comparison Lemma (4.41) with the "$\leq$ " sign. Let $d \in \mathcal{D}$ be any other policy. Then, (4.42) implies that $P^d \eta^{\widehat{d}} \leq \eta^{\widehat{d}}$; and (4.43) together with the Poisson equation (4.32) means that $f(i, d(i)) + (P^d g^{\widehat{d}})(i) \leq f(i, \widehat{d}(i)) + (P^{\widehat{d}} g^{\widehat{d}})(i)$, whenever $(P^d \eta^{\widehat{d}})(i) = \eta^{\widehat{d}}(i)$. Now, set d and $\widehat{d}$ to be the h and d, respectively, in the Comparison Lemma (4.41). Then, it follows directly (by the Comparison Lemma with relation $\leq$) that $\eta^d \leq \eta^{\widehat{d}}$. □

By $P^{\widehat{d}} \eta^{\widehat{d}} = \eta^{\widehat{d}}$ and the Poisson equation for $\widehat{d}$, the optimality equations also take the form:

$$\sum_{j \in \mathcal{S}} p^{\widehat{d}}(j|i) \eta^{\widehat{d}}(j) = \max_{\alpha \in \mathcal{A}(i)} \left\{ \sum_{j \in \mathcal{S}} p^{\alpha}(j|i) \eta^{\widehat{d}}(j) \right\}, \qquad \text{for all } i \in \mathcal{S},$$

and

$$f(i, \widehat{d}(i)) + \sum_{j \in \mathcal{S}} p^{\widehat{d}}(j|i) g^{\widehat{d}}(j) = \max_{\alpha \in \widehat{\mathcal{A}}_0(i)} \left\{ f(i, \alpha) + \sum_{j \in \mathcal{S}} p^{\alpha}(j|i) g^{\widehat{d}}(j) \right\}, \quad \text{for all } i \in \mathcal{S}.$$

We note that these optimality equations are only the sufficient conditions. This is different from the ergodic chains, for which Bellman's optimality equation is both necessary and sufficient. However, we have

A Necessary Optimality Condition:

If a policy $\widehat{d}$ is gain optimal with $\eta^{\widehat{d}} = \eta^*$, then the optimality equation (4.42) holds.

Proof. Suppose that the result does not hold. Then, there exist an action α and a state k such that

$$\eta^{\widehat{d}}(k) < \sum_{j \in \mathcal{S}} p^{\alpha}(j|k) \eta^{\widehat{d}}(j).$$

Based on this, we can construct another policy $\widetilde{d}$ by setting $\widetilde{d}(k) = \alpha$ and $\widetilde{d}(i) = \widehat{d}(i)$ for all $i \neq k$. Then, we have $(P^{\widetilde{d}} \eta^{\widehat{d}})(k) > \eta^{\widehat{d}}(k)$ and $(P^{\widetilde{d}} \eta^{\widehat{d}})(i) = \eta^{\widehat{d}}(i)$ for $i \neq k$. Thus,

$$P^{\widetilde{d}} \eta^{\widehat{d}} - \eta^{\widehat{d}} \succeq 0. \qquad (4.44)$$

Therefore, $(P^{\widetilde{d}})^l \eta^{\widehat{d}} \geq \eta^{\widehat{d}}$ for any $l \geq 1$. Thus, $(P^{\widetilde{d}})^* \eta^{\widehat{d}} \geq \eta^{\widehat{d}}$. Because $(P^{\widetilde{d}})^*(P^{\widetilde{d}}\eta^{\widehat{d}} - \eta^{\widehat{d}}) = 0$, by the Fundamental Fact (4.30), we have $P^{\widetilde{d}}\eta^{\widehat{d}}(i) = \eta^{\widehat{d}}(i)$ for any recurrent state i under policy $\widetilde{d}$. Thus, the particular state k must be a transient state under policy $\widetilde{d}$. By the construction of $\widetilde{d}$, we have $(P^{\widetilde{d}})^* \left[(f^{\widetilde{d}} + P^{\widetilde{d}} g^{\widehat{d}}) - (f^{\widehat{d}} + P^{\widehat{d}} g^{\widehat{d}})\right] = 0$. (The only non-zero component of the vector in the bracket corresponds to a transient state.) Finally, from the average reward difference formula (4.36), we have

$$\eta^{\widetilde{d}} - \eta^{\widehat{d}} = \left[(P^{\widetilde{d}})^* - I\right] \eta^{\widehat{d}} \geq 0.$$

If $\eta^{\widetilde{d}} = \eta^{\widehat{d}}$, then $P^{\widetilde{d}}\eta^{\widehat{d}} = P^{\widetilde{d}}\eta^{\widetilde{d}} = \eta^{\widetilde{d}} = \eta^{\widehat{d}}$. This conflicts with (4.44). Thus, we have $\eta^{\widetilde{d}} \succeq \eta^{\widehat{d}}$. This is impossible because $\eta^{\widehat{d}}$ is the optimal gain. $\square$

Example 4.5. This example shows that (4.43) is not a necessary condition for gain-optimal policies. We consider an MDP with three states $\mathcal{S} = \{1, 2, 3\}$ and three action sets $\mathcal{A}(1) = \{\alpha_1\}$, $\mathcal{A}(2) = \{\alpha_2\}$, and $\mathcal{A}(3) = \{\alpha_{31}, \alpha_{32}\}$. The reward functions are $f(1, \alpha_1) = 2$, $f(2, \alpha_2) = 4$, $f(3, \alpha_{31}) = 5$, and $f(3, \alpha_{32}) = 8$. There are two policies $d = (\alpha_1, \alpha_2, \alpha_{31})$ and $h = (\alpha_1, \alpha_2, \alpha_{32})$. Their transition probability matrices are

$$P^d = \begin{bmatrix} 0.2 & 0.8 & 0 \\ 0.8 & 0.2 & 0 \\ 0.2 & 0.3 & 0.5 \end{bmatrix}, \quad P^h = \begin{bmatrix} 0.2 & 0.8 & 0 \\ 0.8 & 0.2 & 0 \\ 0.1 & 0.1 & 0.8 \end{bmatrix}.$$

Thus, we have

$$(P^d)^* = \begin{bmatrix} 0.5 & 0.5 & 0 \\ 0.5 & 0.5 & 0 \\ 0.5 & 0.5 & 0 \end{bmatrix}, \quad (P^h)^* = \begin{bmatrix} 0.5 & 0.5 & 0 \\ 0.5 & 0.5 & 0 \\ 0.5 & 0.5 & 0 \end{bmatrix}.$$

From $f^d = (2, 4, 5)^T$, we obtain

$$\eta^d = [3, 3, 3]^T, \quad g^d = \left[-\frac{5}{8}, \frac{5}{8}, \frac{33}{8}\right]^T.$$

From $f^h = (2, 4, 8)^T$, we obtain

$$\eta^h = [3, 3, 3]^T, \quad g^h = \left[-\frac{5}{8}, \frac{5}{8}, 25\right]^T.$$

Since $\eta^d = \eta^h$, both policies d and h are optimal, and therefore both satisfy the first optimality equation (4.42). Furthermore, we have

$$f^d + P^d g^d = \eta^d + g^d = \left[\frac{19}{8}, \frac{29}{8}, \frac{57}{8}\right]^T$$

and

$$f^h + P^h g^d = \left[\frac{19}{8}, \frac{29}{8}, \frac{113}{10}\right]^T.$$

Thus, we have

$$f^h + P^h g^d \succeq g^d + g^d.$$

Therefore, policy d does not satisfy the second optimality equation (4.43). □

The Policy Iteration Algorithm

Next, for any non-optimal policy, we can always construct a "better" policy by using the Comparison Lemma (4.41). This can be formally described as follows. Given any policy $d \in \mathcal{D}$, for any $i \in \mathcal{S}$ and $\alpha \in \mathcal{A}(i)$, let (cf. the Q-factor defined as (6.28) in Section 6.2.2)

$$Q^d(i,\alpha) := f(i,\alpha) + \sum_{j\in\mathcal{S}} p^\alpha(j|i) g^d(j) \tag{4.45}$$

and

$$\mathcal{A}(i,d) := \left\{ \alpha \in \mathcal{A}(i): \begin{array}{l} \sum_{j\in\mathcal{S}} p^\alpha(j|i)\eta^d(j) > \eta^d(i); \\ \text{or } Q^d(i,\alpha) > Q^d(i,d(i)) \\ \text{when } \sum_{j\in\mathcal{S}} p^\alpha(j|i)\eta^d(j) = \eta^d(i) \end{array} \right\}. \tag{4.46}$$

We then define an improvement policy $d' \in \mathcal{D}$ (depending on d) as follows:

$$d'(i) \in \mathcal{A}(i,d) \text{ if } \mathcal{A}(i,d) \neq \emptyset, \text{ and } d'(i) = d(i) \text{ if } \mathcal{A}(i,d) = \emptyset. \tag{4.47}$$

Note that such a policy may not be unique, since there may be more than one action in $\mathcal{A}(i,d)$ for some state $i \in \mathcal{S}$. In such cases, we usually choose

$$d'(i) \in \arg\left\{ \max_{\alpha\in\mathcal{A}(i)} \left[\sum_{j\in\mathcal{S}} p^\alpha(j|i)\eta^d(j)\right]\right\}, \tag{4.48}$$

if $\max_{a\in\mathcal{A}(i)} \left[\sum_{j\in\mathcal{S}} p^\alpha(j|i)\eta^d(j)\right] > \eta^d(i)$; otherwise, we set

$$\widetilde{\mathcal{A}}(i,d) := \arg\left\{ \max_{\alpha\in\mathcal{A}(i)} \left[\sum_{j\in\mathcal{S}} p^\alpha(j|i)\eta^d(j)\right]\right\}$$

and choose

$$d'(i) \in \arg\left\{ \max_{\alpha\in\widetilde{\mathcal{A}}(i,d)} \left[Q^d(i,\alpha)\right]\right\}. \tag{4.49}$$

Another point to note is that we do not need to choose $d'(i) \in \mathcal{A}(i,d)$ for all i such that $\mathcal{A}(i,d) \neq \emptyset$ in (4.47). We may only choose $d'(i) \in \mathcal{A}(i,d)$ for at least one state i and set $d'(j) = d(j)$ for all other states. The results in Lemmas 4.2 and 4.3 and the convergence of the policy iteration algorithm, etc., still hold (cf. the discussion on page 189 for ergodic chains).

By the construction of d' and the Comparison Lemma (4.41), we have

Lemma 4.2. *For any policy $d \in \mathcal{D}$, let d' be a policy constructed by (4.45) to (4.47). We have $\eta^{d'} \geq \eta^d$.*

Proof. For any state $i \in \mathcal{S}$, if $\mathcal{A}(i,d) = \emptyset$, then $d'(i) = d(i)$ and we have $p^{d'(i)}(j|i) = p^{d(i)}(j|i)$ and $\sum_{j\in\mathcal{S}} p^{d'(i)}(j|i)\eta^d(j) = \sum_{j\in\mathcal{S}} p^{d(i)}(j|i)\eta^d(j) = \eta^d(i)$. Next, if $\mathcal{A}(i,d) \neq \emptyset$, from the construction by (4.46), we have $\sum_{j\in\mathcal{S}} p^{d'(i)}(j|i)\eta^d(j) \geq \eta^d(i)$. Thus, Condition ($a$) in the Comparison Lemma (4.41) holds. In addition, if $\sum_{j\in\mathcal{S}} p^{d'(i)}(j|i)\eta^d(j) = \eta^d(i)$, then either $Q^d(i,\alpha) = Q^d(i,d(i))$ when $\mathcal{A}(i,d) = \emptyset$, or $Q^d(i,\alpha) > Q^d(i,d(i))$ when $\mathcal{A}(i,d) \neq \emptyset$. That is, Condition ($b$) in (4.41) holds. Thus, $\eta^{d'} \geq \eta^d$. □

Now, we propose the (standard) *policy iteration algorithm for multi-chain MDPs* as follows:

Algorithm 4.5. A Policy Iteration Algorithm for Multi-Chains:

1. Select an arbitrary policy $d_0 \in \mathcal{D}$, and set $k = 0$.
2. (Policy evaluation) Obtain (by (4.32) or (4.34)) g^{d_k} and η^{d_k}.
3. (Policy improvement) Set $d = d_k$ in (4.46) and (4.47); construct a policy d' according to (4.46) and (4.47), and set $d_{k+1} = d'$.
4. If $d_{k+1} = d_k$, then stop and d_k is optimal (as shown below). Otherwise, set $k := k+1$ and go to step 2.

If there is more than one d' in step 3, we may choose d_{k+1} according to (4.48) and (4.49). If the Markov chain is ergodic, the average reward η^d is a constant vector and we have $\sum_{j\in\mathcal{S}} p^\alpha(j|i)\eta^d(j) = \eta^d(j)$ for any policy $\alpha \in \mathcal{A}(i)$. Thus, $\widetilde{\mathcal{A}}(i,d) = \mathcal{A}(i)$ in (4.49). The PI algorithm becomes the same as that in Section 4.1.1 for the ergodic chains.

The Convergence of the Policy Iteration Algorithm

The convergence of the policy iteration algorithm for ergodic chains in Section 4.1.1 is straightforward because we have the strict relation $\eta_{k+1} > \eta_k$ at each iteration before the algorithm stops. However, in the multi-chain case, we only have $\eta_{k+1} \geq \eta_k$, as shown in Lemma 4.2. Is it possible for the policy iteration process to go in cycles with $\eta_{k+1} = \eta_k$ forever without reaching the optimal value?

To show that this is not possible, we need the following result about the difference of the biases, which is of its own merits in the study of bias optimality in the next section. First, we define the *bias potential* (or the *potential of the potential*) of a policy (P,f), w, as the vector satisfying

$$(I-P)w = -g,$$

where g is the bias of (P, f) (cf. (4.14) for ergodic Markov chains). The solution to this equation is only up to an additive vector, i.e., if w is a bias-potential, then $w + u$ is also a bias-potential for any u satisfying $(I - P)u = 0$. The physical meaning of w will be discussed in the next section.

Consider two policies $h, d \in \mathcal{D}$. Let g^h and g^d be their biases, respectively, w^d be the bias-potential of d, and $(P^h)^*$ be the Cesaro limit of P^h. If $\eta^h = \eta^d$, we have

$$g^h - g^d = (P^h)^*(P^h - P^d)w^d + \sum_{n=0}^{\infty}(P^h)^n \left[(f^h + P^h g^d) - (f^d + P^d g^d)\right]. \tag{4.50}$$

This equation will be proved in the next section. Now, we use this difference formula (4.50) to prove the following lemma.

Lemma 4.3. *For any policy $d \in \mathcal{D}$, let d' be the policy constructed by (4.45) to (4.47). If $\eta^{d'} = \eta^d$ and $d' \neq d$, then $g^{d'} \succeq g^d$.*

Proof. Set $u := (f^{d'} + P^{d'} g^d) - (f^d + P^d g^d)$. Then $u(i) = Q^d(i, d'(i)) - Q^d(i, d(i))$, $i \in \mathcal{S}$. Since $\eta^{d'} = \eta^d$, we have

$$P^{d'}\eta^d = P^{d'}\eta^{d'} = \eta^{d'} = \eta^d, \tag{4.51}$$

and $(P^{d'})^*\eta^d = \eta^d$. Using $(P^{d'})^* P^{d'} = (P^{d'})^*$ and $(P^{d'})^* f^{d'} = \eta^{d'}$, we get

$$\begin{aligned}(P^{d'})^* u &= (P^{d'})^* \left[(f^{d'} + P^{d'} g^d) - (f^d + P^d g^d)\right] \\ &= (P^{d'})^* \left[(f^{d'} + P^{d'} g^d) - (\eta^d + g^d)\right] \\ &= (P^{d'})^* \left(f^{d'} - \eta^d\right) = 0. \end{aligned} \tag{4.52}$$

From (4.51), for the pair of d' and d, $\sum_{j \in \mathcal{S}} p^{d'(i)}(j|i)\eta^d(j) = \eta^d(i)$ in (4.46) holds for all states $i \in \mathcal{S}$. Because d' is constructed by (4.46) and $d' \neq d$, we must have $u \succeq 0$. Then, from (4.52) and by the Fundamental Fact (4.30), $u(i) = 0$ for all recurrent states i of $P^{d'}$. From (4.51) and by the construction in (4.46), we know that for all recurrent states i of $P^{d'}$, $\mathcal{A}(i, d)$ are empty; thus, $d'(i) = d(i)$ for all recurrent states i of $P^{d'}$. Therefore, the rows for all recurrent states of $P^{d'} - P^d$ are all zeros. Because all the columns corresponding to transient states of $P^{d'^*}$ are zeros, we have

$$(P^{d'})^* \left(P^{d'} - P^d\right) = 0.$$

From (4.50), we have

$$\begin{aligned} g^{d'} - g^d &= (P^{d'})^*(P^{d'} - P^d)w^d + \sum_{n=0}^{\infty}(P^{d'})^n \left[(f^{d'} + P^{d'} g^d) - (f^d + P^d g^d)\right] \\ &= \sum_{n=0}^{\infty} \left[(P^{d'})^n u\right]. \end{aligned}$$

Because $P^{d'} \geq 0$ and $u \succeq 0$, we have $(P^{d'})^n u \geq 0$ for $n > 0$. Finally, we have $g^{d'} - g^d \geq (P^{d'})^0 u = u \succeq 0$. □

From the proof, if $\eta^{d'} = \eta^d$, then we must have $P^{d'}\eta^d = \eta^d$ in (4.46). By Lemma 4.2, at each iteration, the average reward either increases or stays the same, and by Lemma 4.3, if the average reward stays the same at an iteration, the bias must increase. These two lemmas guarantee that the policies do not cycle in the policy iteration procedure.

Anti-Cycling in PI:

At each iteration, the average reward either increases or stays the same; If the average reward stays the same, the bias increases.

The anti-cycling property guarantees that the policy iteration algorithm stops in a finite number of iterations. Let $d_0, d_1, \ldots, d_k \ldots$ be the sequence of policies generated by the algorithm. By the anti-cycling property, as k increases, η^{d_k} either increases or stays the same, and when η^{d_k} stays the same, the bias g^{d_k} increases. Thus, any two policies in the sequence of d_k, $k = 0, 1, \ldots$, either have different average rewards or have different biases. Therefore, every policy in the iteration sequence is different. Since the number of policies is finite, the iteration must stop after a finite number of iterations.

Suppose that the PI algorithm stops at a policy denoted as $\widehat{d}$. Then $\widehat{d}$ must satisfy the sufficient optimality equations (4.42) and (4.43), because otherwise for some i the set $\mathcal{A}(i, \widehat{d})$ in (4.46) is non-empty and we can find the next improved policy in the policy iteration. Thus, policy $\widehat{d}$ is gain optimal. Therefore, *the PI Algorithm stops at a gain-optimal policy in a finite number of iterations.* Because the optimal policy satisfies the sufficient optimality equations, this also proves the existence of the solution to the optimality equations (4.42) and (4.43).

4.2.2 Bias Optimality

In this section, we discuss the bias optimality problem. The bias g^d of a policy d is the potential (defined by the Poisson equation $(I - P^d)g^d + \eta^d = f^d$ (4.32)) that satisfies

$$(P^d)^* g^d = 0.$$

In a bias optimality problem, we search for a gain-optimal policy that has the largest bias.

Recall that $\mathcal{D}_0 \subseteq \mathcal{D}$ is the set of all gain-optimal policies. A policy $\widehat{d} \in \mathcal{D}_0$ is said to be *bias optimal* if

$$g^{\widehat{d}} \geq g^d, \qquad \text{for all } d \in \mathcal{D}_0,$$

and its bias $g^{\widehat{d}}$ is called the *optimal bias*. Let g^* denote the optimal bias, i.e., $g^* := \max_{d \in \mathcal{D}_0} g^d$. We define

$$\mathcal{D}_1 := \left\{ d \in \mathcal{D}_0 : g^d = g^* \right\}$$

to be the set of all bias-optimal policies. We wish to develop an efficient algorithm to find a bias-optimal policy $\widehat{d} \in \mathcal{D}_1$.

Just like in the ergodic case, bias represents the transient behavior (see the sample-path-based expression (4.34)). Because the columns of $(P^d)^*$ corresponding to the transient states are all zeros, the condition $(P^d)^* g^d = 0$ does not put any restrictions on the biases of the transient states. As shown in the following simple example, the bias of a transient state may take almost any large value, even if the steady-state performance is fixed. An optimal policy needs to optimize the biases of all states.

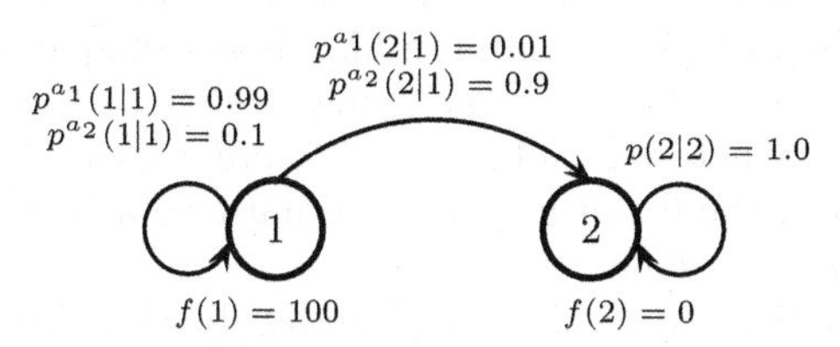

Fig. 4.6. The Multi-Chain MDP in Example 4.6

Example 4.6. Consider the multi-chain MDP problem shown in Figure 4.6. We have $\mathcal{A}(1) = \{\alpha_1, \alpha_2\}$, and $\mathcal{A}(2)$ contains only one action. Thus, there are two policies corresponding to α_1 and α_2, respectively; and for simplicity, these two policies are also denoted as α_1 and α_2. Their transition probabilities are shown in the figure. The reward function is the same for the two policies: $f(1,\alpha_1) = f(1,\alpha_2) = 100$ and $f(2) = 0$. It is clear that both policies have the same long-run average reward $\eta^{\alpha_1} = \eta^{\alpha_2} = 0$. However, the biases in the transient state 1 are quite different: they are $g^{\alpha_1}(1) = 10000$ and $g^{\alpha_2}(1) = 1000/9$. For both policies, we have $g^{\alpha_1}(2) = g^{\alpha_2}(2) = 0$, and

$$(P^{\alpha_1})^* = (P^{\alpha_2})^* = \begin{bmatrix} 0 & 1 \\ 0 & 1 \end{bmatrix}.$$

Both biases satisfy the condition $(P^{\alpha_1})^* g^{\alpha_1} = (P^{\alpha_2})^* g^{\alpha_2} = 0$.

In the ergodic case discussed in Example 4.1, the bias in one state can be very large, if the steady-state probability in this state is very small (due to the normalization condition $\pi g = 0$). In the multi-chain case, the steady-state probabilities for transient states are zero, and the biases of these states may be arbitrarily large, even when the long-run average reward of the Markov chain remains the same. □

By now, we are familiar with the sensitivity-based optimization approach. To study the bias optimality, we first derive the bias difference formula for multi-chains. Recall that in the average-reward difference formula, we need the performance potentials. Similarly, in the bias difference formula, we need to define the potentials of the bias, or the potentials of the potentials.

The Bias Difference Formula

The bias-potential, w, of a policy (P, f) is defined as

$$(I - P)w = -g, \tag{4.53}$$

where g is the bias satisfying $(I - P)g + \eta = f$ and $P^*g = 0$.

Equation (4.53) is almost the same as the Poisson equation (4.32) except that $f - \eta$ is replaced by $-g$. Because $P^*g = 0$, with $-g$ as the reward function, the steady-state performance (corresponding to η) should be $P^*(-g) = 0$. Therefore, (4.53) can be viewed as the Poisson equation by using $-g$ as the reward function. Again, the solution to (4.53) is not unique: if w is a solution to (4.53), then so is $w + u$ for any u satisfying $(I - P)u = 0$. If a bias-potential w satisfies $P^*w = 0$, then it is called a *second order bias* (or simply a second bias). In addition, in the definition (4.53), g must be a bias, not a potential, because $P^*g = -P^*(I - P)w = 0$.

Consider two policies $h, d \in \mathcal{D}$. Let g^h and g^d be their biases, respectively, w^d be the bias-potential of d, and $(P^h)^*$ be the Cesaro limit of P^h. We have the following formula.

The Bias Difference Formula:

If $\eta^h = \eta^d$, then

$$g^h - g^d = (P^h)^*(P^h - P^d)w^d + \sum_{n=0}^{\infty} \{(P^h)^n [(f^h + P^h g^d) - (f^d + P^d g^d)]\}. \tag{4.54}$$

Proof. From the Poisson equation (4.32) and $\eta^h = \eta^d$, we have

$$\begin{aligned} g^h - g^d &= (f^h + P^h g^h - \eta^h) - (f^d + P^d g^d - \eta^d) \\ &= (f^h + P^h g^d) - (f^d + P^d g^d) + P^h(g^h - g^d). \end{aligned} \tag{4.55}$$

For policy d, (4.53) is

$$(I - P^d)w^d = -g^d.$$

Left-multiplying $(P^h)^*$ on both sides of this equation, we get $(P^h)^*g^d = -(P^h)^*(P^h - P^d)w^d$. Thus, from (4.55) and $(P^h)^*g^h = 0$, we have

$$\begin{aligned} &[I - P^h + (P^h)^*] (g^h - g^d) \\ &= (f^h + P^h g^d) - (f^d + P^d g^d) + (P^h)^*(g^h - g^d) \\ &= (f^h + P^h g^d) - (f^d + P^d g^d) - (P^h)^* g^d \\ &= (f^h + P^h g^d) - (f^d + P^d g^d) + (P^h)^*(P^h - P^d)w^d. \end{aligned}$$

From (4.33), we have

$$\begin{aligned}
& g^h - g^d \\
&= (P^h)^*(P^h - P^d)w^d + \left[I - P^h + (P^h)^*\right]^{-1} \left[(f^h + P^h g^d) - (f^d + P^d g^d)\right] \\
&= (P^h)^*(P^h - P^d)w^d + \sum_{n=0}^{\infty} \left\{ \left[P^h - (P^h)^*\right]^n \left[(f^h + P^h g^d) - (f^d + P^d g^d)\right] \right\} \\
&= (P^h)^*(P^h - P^d)w^d \\
&\quad + \sum_{n=0}^{\infty} \left\{ \left[(P^h)^n - (P^h)^*\right] \left[(f^h + P^h g^d) - (f^d + P^d g^d)\right] \right\}. \qquad (4.56)
\end{aligned}$$

Next, from (4.32), (4.28) and (4.26), we have

$$\begin{aligned}
& (P^h)^* \left[(f^h + P^h g^d) - (f^d + P^d g^d)\right] \\
&= (P^h)^* \left[(f^h + P^h g^d) - (\eta^d + g^d)\right] \\
&= (P^h)^* f^h - (P^h)^* \eta^d = (P^h)^* f^h - (P^h)^* \eta^h = 0.
\end{aligned}$$

The lemma then follows directly from (4.56). □

We note that any solution to (4.53) (called a *version* of the bias-potential) for policy d can be used as the bias-potential w^d in the bias difference formula (4.54). The g^h and g^d on both sides are the biases. We observe that (4.54) takes the same form as (4.38). Thus, the following Comparison Lemma follows directly.

Suppose that $\widehat{d} \in \mathcal{D}_0$ is a gain-optimal policy with average reward $\eta^{\widehat{d}} = \eta^*$, bias $g^{\widehat{d}}$, and bias-potential $w^{\widehat{d}}$. Let $h \in \mathcal{D}$ be any other policy.

Comparison Lemma for Biases of Gain-Optimal Policies:

If

(a) $P^h \eta^{\widehat{d}} = \eta^{\widehat{d}}$,

(b) $f^h + P^h g^{\widehat{d}} \geq f^{\widehat{d}} + P^{\widehat{d}} g^{\widehat{d}}$, and

(c) $(P^h w^{\widehat{d}})(i) \geq (P^{\widehat{d}} w^{\widehat{d}})(i)$ when $f^h(i) + (P^h g^{\widehat{d}})(i) = f^{\widehat{d}}(i) + (P^{\widehat{d}} g^{\widehat{d}})(i)$ for some $i \in \mathcal{S}$,

$\Longrightarrow$ then $\eta^h = \eta^{\widehat{d}} = \eta^*$ and $g^h \geq g^{\widehat{d}}$. (4.57)

In addition, if Condition (a) is replaced by the following Condition (a')

(a') policy h is gain optimal,

then $g^h \leq g^{\widehat{d}}$ if we change all the signs " $\geq$ " in (b) and (c) to " $\leq$ ".

Proof. Setting $d = \widehat{d}$ in the average-reward difference formula (4.36) and using Conditions (a) and (b), we get $(P^h)^* \eta^{\widehat{d}} = \eta^{\widehat{d}}$ and $\eta^h \geq \eta^{\widehat{d}}$. Because $\eta^{\widehat{d}}$ is the optimal average reward, we have $\eta^h = \eta^{\widehat{d}}$. Thus, we can use the bias difference formula (4.54). Again, we set $d = \widehat{d}$ in (4.54). Next, set $u := (f^h + P^h g^{\widehat{d}}) - (f^{\widehat{d}} + P^{\widehat{d}} g^{\widehat{d}}) = (f^h + P^h g^{\widehat{d}}) - (\eta^{\widehat{d}} + g^{\widehat{d}})$ and $v = (P^h - P^{\widehat{d}}) w^{\widehat{d}}$. Then, (4.54) takes the form

$$g^h - g^{\widehat{d}} = (P^h)^* v + z,$$

which is the same as (4.40). Condition (b) implies $u \geq 0$. Given that $(P^h)^* f^h = \eta^h = \eta^{\widehat{d}}$, we can easily verify that $(P^h)^* u = 0$. We also have $z = \sum_{n=0}^{\infty} \{(P^h)^n u\} \geq (P^h)^0 u \geq u \geq 0$. Thus, Condition ($a$) in the Comparison Lemma (General Form (4.39)) holds. Condition (c) is the same as Condition (b) in the Comparison Lemma (4.39). Thus, it follows from the Comparison Lemma (4.39) that $g^h \geq g^{\widehat{d}}$.

When the signs "$\geq$" change to "$\leq$", we cannot prove that h is a gain-optimal policy from Conditions (a) and (b). Note that Condition (a') implies Condition (a). With Condition (a'), the other half of the lemma can be proved in a similar way. □

It is interesting to compare this Comparison Lemma (4.57) with Lemma 4.3, which essentially claims that if h is constructed from $\widehat{d}$ by using (4.46), (4.47), $P^h \eta^{\widehat{d}} = \eta^{\widehat{d}} = \eta^h$, and $f^h + P^h g^{\widehat{d}} \geq f^{\widehat{d}} + P^{\widehat{d}} g^{\widehat{d}}$, then $g^h \geq g^{\widehat{d}}$. If, in addition, $f^h + P^h g^{\widehat{d}} \succeq f^{\widehat{d}} + P^{\widehat{d}} g^{\widehat{d}}$, then $g^h \succeq g^{\widehat{d}}$.

To get a specific solution to the Poisson equation (4.53), we may set

$$P^* w = 0 \tag{4.58}$$

as the normalization condition. Thus, we can rewrite (4.53) as

$$(I - P + P^*) w = -g.$$

Therefore,

$$\begin{aligned} w &= -(I - P + P^*)^{-1} g = -\sum_{l=0}^{\infty} (P - P^*)^l g \\ &= -\sum_{l=0}^{\infty} P^l g = -\sum_{l=0}^{\infty} P^l \sum_{k=0}^{\infty} P^k (f - \eta) \\ &= -\sum_{l=0}^{\infty} (l+1) P^l (f - \eta). \end{aligned}$$

Thus,

$$w(i) = -\sum_{l=0}^{\infty} \{(l+1) E[f(X_l) - \eta(i) | X_0 = i]\}. \tag{4.59}$$

This is the same as (4.20). To distinguish, we will refer to the bias-potential with condition (4.58) as the *second order bias*, or simply the *second bias*.

Any version of the bias-potential differs from the second bias by a vector u satisfying $(I-P)u = 0$. With (4.59), the second bias can be estimated on a single sample path without knowing the transition probability matrix P.

The Sufficient and Necessary Conditions for Bias Optimality

With the Comparison Lemma (4.57), we can easily derive the sufficient bias optimality conditions. First, we need a lemma.

Lemma 4.4. *Let η^* be the optimal gain.*
(a) For any policy $d \in \mathcal{D}$, we have $P^d\eta^* \leq \eta^*$.
(b) For any policy $d \in \mathcal{D}$, if $P^d\eta^* \preceq \eta^*$, then $\eta^d \preceq \eta^*$.
(c) For any gain-optimal policy $\widehat{d} \in \mathcal{D}_0$, we have $P^{\widehat{d}}\eta^* = (P^{\widehat{d}})^*\eta^* = \eta^*$.

Proof. (*a*) This is the necessary optimality equation (4.42) for the optimal gain.

(*b*) From (4.29) and $\eta^d \leq \eta^*$, we have $\eta^d = P^d\eta^d \leq P^d\eta^* \preceq \eta^*$.

(*c*) The gain of $\widehat{d}$ is $\eta^{\widehat{d}} = \eta^*$. Thus, $P^{\widehat{d}}\eta^* = \eta^*$. From this, $(P^{\widehat{d}})^n\eta^* = \eta^*$ for any integer n. Then, $(P^{\widehat{d}})^*\eta^* = \eta^*$ holds by noting $(P^{\widehat{d}})^* = \lim_{N\to\infty}\frac{1}{N}\sum_{n=0}^{N-1}(P^{\widehat{d}})^n$. □

A policy $\widehat{d} \in \mathcal{D}$ is a bias-optimal policy if the following conditions hold.

Sufficient Optimality Conditions:
A policy $\widehat{d} \in \mathcal{D}$ is bias optimal if

$$\eta^{\widehat{d}}(i) = \max_{\alpha\in\mathcal{A}(i)}\left\{\sum_{j\in\mathcal{S}} p^\alpha(j|i)\eta^{\widehat{d}}(j)\right\}, \quad \text{for all } i \in \mathcal{S}, \tag{4.60}$$

$$\eta^{\widehat{d}}(i) + g^{\widehat{d}}(i) = \max_{\alpha\in\widehat{\mathcal{A}}_0(i)}\left\{f(i,\alpha) + \sum_{j\in\mathcal{S}} p^\alpha(j|i)g^{\widehat{d}}(j)\right\}, \quad \text{for all } i \in \mathcal{S}, \tag{4.61}$$

$$g^{\widehat{d}}(i) + w^{\widehat{d}}(i) = \max_{\alpha\in\widehat{\mathcal{A}}_1(i)}\left\{\sum_{j\in\mathcal{S}} p^\alpha(j|i)w^{\widehat{d}}(j)\right\}, \quad \text{for all } i \in \mathcal{S}, \tag{4.62}$$

where

$$\widehat{\mathcal{A}}_0(i) := \left\{\alpha \in \mathcal{A}(i) : \sum_{j\in\mathcal{S}} p^\alpha(j|i)\eta^{\widehat{d}}(j) = \eta^{\widehat{d}}(i)\right\}, \tag{4.63}$$

$$\widehat{\mathcal{A}}_1(i) := \left\{\alpha \in \widehat{\mathcal{A}}_0(i) : \eta^{\widehat{d}}(i) + g^{\widehat{d}}(i) = f(i,\alpha) + \sum_{j\in\mathcal{S}} p^\alpha(j|i)g^{\widehat{d}}(j)\right\}. \tag{4.64}$$

If these sufficient conditions hold, then $\eta^{\widehat{d}}$ is the optimal gain, i.e., $\eta^{\widehat{d}} = \eta^*$, and $g^{\widehat{d}}$ is the optimal bias, i.e., $g^{\widehat{d}} = g^*$. However, $w^{\widehat{d}}$ is the bias-potential, which is only determined up to an additive vector u satisfying $(I - P^{\widehat{d}})u = 0$. It may not be the optimal second order bias even after being normalized by $(P^{\widehat{d}})^* w^{\widehat{d}} = 0$.

Proof. First, $\widehat{d}$ is a gain-optimal policy because both $\eta^{\widehat{d}}$ and $g^{\widehat{d}}$ satisfy the gain-optimality equations (4.60) and (4.61) (the same as (4.42) and (4.43)). Next, we prove that $g^{\widehat{d}} \geq g^d$ for any gain-optimal policy $d \in \mathcal{D}_0$. We set d and $\widehat{d}$ to be h and d, respectively, in the Comparison Lemma (4.57). From Lemma 4.4 (c), we have $P^d \eta^{\widehat{d}} = \eta^{\widehat{d}} = \eta^d = \eta^*$. From (4.61), we have $(f^d + P^d g^{\widehat{d}}) \leq (f^{\widehat{d}} + P^{\widehat{d}} g^{\widehat{d}})$. From (4.62), if $(f^d + P^d g^{\widehat{d}})(i) = (f^{\widehat{d}} + P^{\widehat{d}} g^{\widehat{d}})(i)$ then $(P^d w^{\widehat{d}})(i) \leq (P^{\widehat{d}} w^{\widehat{d}})(i)$. Thus, the three conditions $(a'), (b)$, and (c) in the Comparison Lemma (4.57) hold with the "$\leq$" sign. Thus, it follows that $g^d \leq g^{\widehat{d}}$. □

The three equations (4.60), (4.61), and (4.62) are the sufficient conditions for bias-optimal policies; they are not necessary. While a gain-optimal policy satisfies the first optimality equation (4.60) (or (4.42)), we can prove that any bias-optimal policy must satisfy the first two optimality equations (4.60) and (4.61).

Necessary Optimality Conditions:

> If a policy $\widehat{d}$ is bias optimal with $\eta^{\widehat{d}} = \eta^*$ and $g^{\widehat{d}} = g^*$, then the optimality equations (4.60) and (4.61) hold.

Proof. Since $\eta^{\widehat{d}}$ is the optimal gain, it must satisfy the Necessary Optimality Condition for gain-optimal policies (4.42) in Section 4.2.1, which is the same as (4.60). Now, we assume that (4.61) does not hold. That is, there exist a state $k \in \mathcal{S}$ and an action $\alpha \in \mathcal{A}(k)$ such that

$$\eta^{\widehat{d}}(k) = \sum_{j \in \mathcal{S}} p^{\alpha}(j|k) \eta^{\widehat{d}}(j),$$

and

$$f(k, \alpha) + \sum_{j \in \mathcal{S}} p^{\alpha}(j|k) g^{\widehat{d}}(j) > f(k, \widehat{d}(k)) + \sum_{j \in \mathcal{S}} p^{\widehat{d}(k)}(j|k) g^{\widehat{d}}(j).$$

Based on this, we can construct another policy $\widetilde{d}$ by setting $\widetilde{d}(k) = \alpha$ and $\widetilde{d}(i) = \widehat{d}(i)$, for all $i \neq k$. By the construction of $\widetilde{d}$, we have $P^{\widetilde{d}} \eta^{\widehat{d}} = \eta^{\widehat{d}}$ and therefore, $(P^{\widetilde{d}})^* \eta^{\widehat{d}} = \eta^{\widehat{d}}$. Next, set $h = \widetilde{d}$ and $d = \widehat{d}$ in the difference formula (4.36), and we get $\eta^{\widetilde{d}} \geq \eta^{\widehat{d}}$. However, $\eta^{\widehat{d}}$ is the optimal gain, so $\eta^{\widetilde{d}} = \eta^{\widehat{d}}$.

Finally, by the construction of $\widetilde{d}$ and following the same arguments as in Lemma 4.3 to $\widetilde{d}$ and $\widehat{d}$, we obtain $g^{\widetilde{d}} \succeq g^{\widehat{d}}$. This contradicts the fact that $g^{\widehat{d}}$ is the optimal bias. Thus, (4.61) must hold. □

The $\widehat{\mathcal{A}}_0(i)$ in (4.63) depends on $\eta^{\widehat{d}}$, and the $\widehat{\mathcal{A}}_1(i)$ in (4.64) depends on $\eta^{\widehat{d}}$ and $g^{\widehat{d}}$. If they satisfy the sufficient conditions (4.60), (4.61), and (4.62), then $\eta^{\widehat{d}} = \eta^*$ and $g^{\widehat{d}} = g^*$ are the optimal gain and optimal bias, respectively. To be more specific, we define

$$\mathcal{A}_0^*(i) := \left\{ \alpha \in \mathcal{A}(i) : \sum_{j \in \mathcal{S}} p^\alpha(j|i)\eta^*(j) = \eta^*(i) \right\}, \tag{4.65}$$

and

$$\mathcal{A}_1^*(i) := \left\{ \alpha \in \mathcal{A}_0^*(i) : \eta^*(i) + g^*(i) = f(i,\alpha) + \sum_{j \in \mathcal{S}} p^\alpha(j|i)g^*(j) \right\}. \tag{4.66}$$

We further define

$$\mathcal{D}_k^* = \times_{i \in \mathcal{S}} \mathcal{A}_k^*(i), \qquad k = 0, 1, \tag{4.67}$$

to be the sets of policies whose components are actions in $\mathcal{A}_k^*(i)$, $k = 1, 2$, respectively. Recall that $\mathcal{D}_1 \subseteq \mathcal{D}_0$ is the set of all bias-optimal policies. Obviously $\mathcal{D}_0 \subseteq \mathcal{D}_0^*$. By the average-reward difference formula (4.36), we have $\mathcal{D}_1^* \subseteq \mathcal{D}_0$. Now assume that $d \in \mathcal{D}_1$. Then, for average reward, we have $\eta^d = \eta^*$, and for bias, we have $g^d = g^*$. Thus, $P^d\eta^* = P^d\eta^d = \eta^d = \eta^*$, and from the Poisson equation for policy d, we have $f^d + P^d g^* = g^* + \eta^*$. Therefore, $d \in D_1^*$, which implies $\mathcal{D}_1 \subseteq \mathcal{D}_1^*$. Therefore, we have (Figure 4.7)

$$\mathcal{D}_1 \subseteq \mathcal{D}_1^* \subseteq \mathcal{D}_0 \subseteq \mathcal{D}_0^* \subseteq \mathcal{D}. \tag{4.68}$$

For ergodic chains, from (4.12), we have $\mathcal{D}_0 = \mathcal{D}_1^*$.

Policy Iteration for Bias Optimality

Following the same procedure as that for the gain-optimal policy, by the Comparison Lemma (4.57), from any gain-optimal policy we can construct another gain-optimal policy whose bias is larger (or at least equal, see the discussion below on anti-cycling) if such a policy exists. Specifically, given a policy $d \in \mathcal{D}_0$, for any state $i \in \mathcal{S}$ and $\alpha \in \mathcal{A}_0^*(i)$, let (cf. (4.45) and (4.46))

$$Q^d(i,\alpha) := f(i,\alpha) + \sum_{j \in \mathcal{S}} p^\alpha(j|i)g^d(j),$$

and

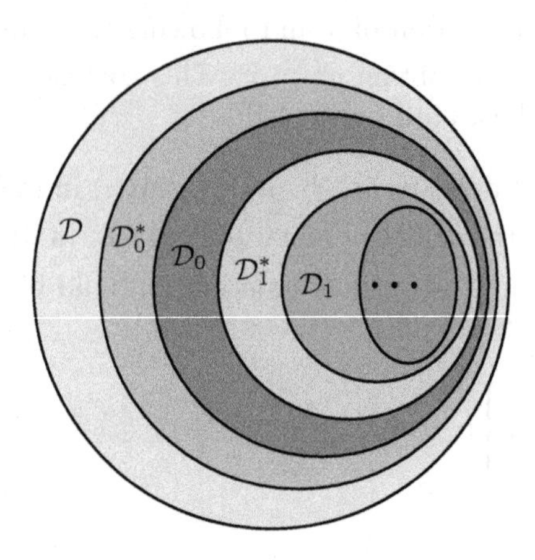

Fig. 4.7. The Relations Among the Policy Sets

$$\mathcal{A}_1(i,d) := \left\{ \alpha \in \mathcal{A}_0^*(i) : \begin{array}{l} Q^d(i,\alpha) > Q^d(i,d(i)); \\ \text{or } \sum_{j\in\mathcal{S}} p^\alpha(j|i) w^d(j) \\ \quad > \sum_{j\in\mathcal{S}} p^{d(i)}(j|i) w^d(j) \\ \text{when } Q^d(i,\alpha) = Q^d(i,d(i)) \end{array} \right\}. \tag{4.69}$$

We then define an improvement policy $d' \in \mathcal{D}$ (depending on d) as follows:

$$d'(i) \in \mathcal{A}_1(i,d) \text{ if } \mathcal{A}_1(i,d) \neq \emptyset, \text{ and } d'(i) = d(i) \text{ if } \mathcal{A}_1(i,d) = \emptyset. \tag{4.70}$$

Note that such a policy may not be unique, since there may be more than one action in $\mathcal{A}_1(i,d)$ for some state $i \in \mathcal{S}$.

Again, there is no need to choose $d'(i) \in \mathcal{A}_1(i,d)$ for all i such that $\mathcal{A}_1(i,d) \neq \emptyset$; we may choose $d'(i) \in \mathcal{A}_1(i,d)$ for at least one state, and set $d'(k) = d(k)$ for other states (cf. the discussion on page 213 for the average reward case).

Because there is no easily verifiable condition for $\mathcal{D}_0$, when a gain-optimal policy $d \in \mathcal{D}_0$ is given, we have to search for the bias-optimal policy in a slightly larger set $\mathcal{D}_0^*$ (cf. Figure 4.7), which can be determined by (4.65). This is why we set $\alpha \in \mathcal{A}_0^*(i)$ in (4.69). Because of this, we need to make sure that the new policy constructed, d', stays in $\mathcal{D}_0$, see Lemma 4.5 (a) below.

Lemma 4.5. *For any given $d \in \mathcal{D}_0$, let d' be defined as in (4.70). Then,*
(a) $\eta^{d'} = \eta^d$.
(b) $g^{d'} \geq g^d$.
(c) If $g^{d'} = g^d$ *and* $d' \neq d$*, then for second bias we have* $w^{d'} \succeq w^d$.

Proof. (a) By construction (4.69), if $d'(i) \neq d(i)$, then $d'(i) \in \mathcal{A}_0^*(i)$, for all $i \in \mathcal{S}$. By definition of $\mathcal{A}_0^*(i)$, because $\eta^d = \eta^*$, we have $P^{d'}\eta^d = \eta^d$. Again by

(4.69), we have $Q^d(i, d'(i)) \geq Q^d(i, d(i))$ for all $i \in \mathcal{S}$. Thus, from the average-reward difference formula (4.36), we have $\eta^{d'} \geq \eta^d$. Therefore, we must have $\eta^{d'} = \eta^d = \eta^*$. That is, $d' \in \mathcal{D}_0$. (Note that the policy iteration procedure for the average reward may or may not stop at the given optimal policy d. If it does, then because $P^{d'}\eta^d = \eta^d$, we can conclude that $Q^d(i, d'(i)) = Q^d(i, d(i))$ for all $i \in \mathcal{S}$.)

(*b*) We take policy d as policy $\widehat{d}$ and policy d' as policy h in the Comparison Lemma (4.57). Then, by the construction in (4.69) and (4.70), Conditions (*a*), (*b*), and (*c*) in the Comparison Lemma hold. Thus, we have $g^{d'} \geq g^d$.

(*c*) Since $g^{d'} = g^d$, we must have $Q^d(i, d'(i)) = Q^d(i, d(i))$ for all $i \in \mathcal{S}$. Otherwise, according to Lemma 4.3, we would have $g^{d'} \succeq g^d$. Therefore, by (4.69) and $d' \neq d$, we have

$$P^{d'} w^d \succeq P^d w^d.$$

Next, from the bias difference formula (4.54), we have

$$g^{d'} - g^d = (P^{d'})^*(P^{d'} - P^d)w^d = 0,$$

which implies that $(P^{d'} w^d)(i) = (P^d w^d)(i)$ for all recurrent states i under policy d'. Therefore, from (4.69) and (4.70), we have $\mathcal{A}_1(i, d) = \emptyset$ and $d'(i) = d(i)$ for all recurrent states i under policy d'.

Similar to the proof of Lemma 4.3, we need a difference formula for the second bias to proceed. To this end, for a second bias w (with $P^* w = 0$) of any policy (P, f), we define the potential of the second bias, z, as

$$(I - P)z = -w.$$

For any two policies h and d with $g^h = g^d$, just the same as (4.54), we can derive (see (4.84) with $n = 1$):

$$w^h - w^d = \sum_{n=0}^{\infty} \left[(P^h)^n (P^h - P^d) w^d\right] + (P^h)^*(P^h - P^d) z^d. \tag{4.71}$$

In our case, we set $h = d'$ in the above equation. Because $d'(i) = d(i)$ for all recurrent states i under policy d', we get $(P^{d'})^*(P^{d'} - P^d) = 0$. Finally, from (4.71), we have

$$w^{d'} - w^d = \sum_{n=0}^{\infty} \left[(P^{d'})^n (P^{d'} - P^d) w^d\right] \geq (P^{d'} - P^d) w^d \succeq 0.$$

This completes the proof. □

With Lemma 4.5, we can state the (standard) *policy iteration algorithm for bias-optimal policies* as follows:

Algorithm 4.6. A Policy Iteration Algorithm for Bias-Optimal Policies:

1. Select an arbitrary gain-optimal policy $d_0 \in \mathcal{D}_0$, and set $k = 0$. Determine the sets $\mathcal{A}_0^*(i)$ for all $i \in \mathcal{S}$ by (4.65).
2. (Policy evaluation) Obtain bias g^{d_k}, by solving (4.32) with $(P^{d_k})^* g^{d_k} = 0$, and bias-potential w^{d_k}, by solving (4.53) (w^{d_k} may not be unique).
3. (Policy improvement) Set $d = d_k$ in (4.69) and (4.70), construct a policy d' accordingly, and set $d_{k+1} = d'$.
4. If $d_{k+1} = d_k$, stop; otherwise, increase k by 1 and return to step 2.

Similar to the policy iteration algorithm for the gain-optimal policies, there may be more than one d' in (4.69) and (4.70) in step 3. In such cases, we usually choose the action with the largest $Q^d(i,\alpha)$ (when it is larger than $Q^d(i,d(i))$), or with the largest $\sum_{j\in\mathcal{S}} p^\alpha(j|i)w^d(j)$ (when the largest $Q^d(i,\alpha)$ equals $Q^d(i,d(i))$) (cf. (4.48) and (4.49)). That is, we choose

$$d'(i) \in \arg\left\{\max_{\alpha\in\mathcal{A}_0^*(i)} \left[Q^d(i,\alpha)\right]\right\},$$

if $\max_{\alpha\in\mathcal{A}_0^*(i)} \left[Q^d(i,\alpha)\right] > Q^d(i,d(i))$; otherwise, set $\widetilde{\mathcal{A}}_0(i,d) := \arg\left\{\max_{\alpha\in\mathcal{A}_0^*(i)} \left[Q^d(i,\alpha)\right]\right\}$ and choose

$$d'(i) \in \arg\left\{\max_{\alpha\in\widetilde{\mathcal{A}}_0(i,d)} \sum_{j\in\mathcal{S}} p^\alpha(j|i)w^d(j)\right\}.$$

Clearly, Lemma 4.5 plays the same role as Lemma 4.3 in the anti-cycling property of the policy iteration algorithm. By Lemma 4.5, any two policies in the sequence of d_k, $k = 0, 1, \ldots$, either have different biases or have different second biases. That is, every policy in the iteration sequence is different. Since the number of policies is finite, the iteration must stop after a finite number of iterations. Suppose that it stops at a policy denoted as $\widehat{d}$. Then $\widehat{d}$ must satisfy the sufficient optimality conditions (4.60), (4.61), and (4.62), because otherwise for some i the set $\mathcal{A}_1(i,\widehat{d})$ in (4.69) is non-empty and we can find the next improved policy in the policy iteration. Therefore, policy $\widehat{d}$ is bias optimal. We conclude that *the policy iteration algorithm stops at a bias-optimal policy in a finite number of iterations.*

4.2.3 MDPs with Discounted Rewards

The principles and approaches in MDPs with the discounted reward criterion are the same as those in the average-reward MDPs. We will only briefly state the results. The discounted reward for multi-chains is defined in the same way as that for ergodic chains in (4.22) and (2.31). Let $\{X_l, l = 0, 1, \ldots\}$ be a sample path of the Markov chain under policy (P, f). We set

$$\eta_\beta(i) = (1-\beta)E\left\{\sum_{l=0}^{\infty} \beta^l f(X_l) \Big| X_0 = i\right\}, \qquad i \in \mathcal{S}.$$

Then we have (cf. (2.31)),

$$\eta_\beta = (1-\beta)\sum_{l=0}^{\infty} \beta^l P^l f = (1-\beta)(I-\beta P)^{-1} f, \qquad 0 < \beta < 1. \quad (4.72)$$

The discounted β-potential is the solution to the discounted Poisson equation (4.24) or (2.33):

$$(I - \beta P + \beta P^*) g_\beta = f, \qquad 0 < \beta \leq 1.$$

We have (cf. (2.39))

$$\lim_{\beta \uparrow 1} (1-\beta)(I-\beta P)^{-1} = P^*,$$

and $\lim_{\beta\uparrow 1} \eta_\beta = \eta$, with η being the vector of the long-run average reward. For any two policies h and d, we have the discounted-reward difference formula as follows.

The Discounted-Reward Difference Formula for Multi-Chains:

$$\eta_\beta^h - \eta_\beta^d = (I-\beta P^h)^{-1}\left\{[(1-\beta)f^h + \beta P^h \eta_\beta^d] - [(1-\beta)f^d + \beta P^d \eta_\beta^d]\right\}, \qquad 0 < \beta < 1. \quad (4.73)$$

We also have

$$\eta_\beta^h - \eta_\beta^d = (1-\beta)(I-\beta P^h)^{-1}\left[(f^h + \beta P^h g_\beta^d) - (f^d + \beta P^d g_\beta^d)\right] + \beta^2 (I-\beta P^h)^{-1}(P^h - I)\eta^d, \qquad 0 < \beta < 1. \quad (4.74)$$

The relation between η_β, g_β, and η is (similar to (2.41) for ergodic chains):

$$\eta_\beta = (1-\beta) g_\beta + \beta \eta.$$

When the system is ergodic, η is a constant vector; the second term on the right-hand side of (4.74) is zero, and (4.74) reduces to (4.23).

With the fundamental fact that $(I-\beta P^h)^{-1} \succeq I$ for $0 < \beta < 1$, we have

Comparison Lemma:
If $(1-\beta)f^h + \beta P^h \eta_\beta^d \succeq (1-\beta)f^d + \beta P^d \eta_\beta^d$, $0 < \beta < 1$, then $\eta_\beta^h \succeq \eta_\beta^d$. (4.75)

The standard policy iteration algorithm for MDPs with the discounted reward follows easily from the Comparison Lemma (4.75). Given a policy $d \in \mathcal{D}$, for any state $i \in \mathcal{S}$ and $\alpha \in \mathcal{A}(i)$, and $0 < \beta < 1$, we let

$$Q_\beta^d(i,\alpha) := (1-\beta)f(i,\alpha) + \beta \sum_{j \in S} p^\alpha(j|i)\eta_\beta^d(j),$$

and define

$$\mathcal{A}_\beta(i,d) := \left\{\alpha \in \mathcal{A}(i) :\ Q_\beta^d(i,\alpha) > Q_\beta^d(i,d(i))\right\}. \tag{4.76}$$

We then define a policy $d' \in \mathcal{D}$ (depending on d) as follows:

$$d'(i) \in \mathcal{A}_\beta(i,d) \text{ if } \mathcal{A}_\beta(i,d) \neq \emptyset, \text{ and } d'(i) = d(i) \text{ if } \mathcal{A}_\beta(i,d) = \emptyset. \tag{4.77}$$

From the Comparison Lemma (4.75), for any given $d \in \mathcal{D}$, let d' be defined as in (4.77), if $d' \neq d$, then $\eta_\beta^{d'} \succeq \eta_\beta^d$. We can develop the standard *policy iteration algorithm for MDPs with discounted rewards:*

Algorithm 4.7. A Policy Iteration Algorithm for an Optimal Policy with a Discounted Reward:

1. Select an arbitrary policy $d_0 \in \mathcal{D}$, and set $k = 0$.
2. (Policy evaluation) Obtain $\eta_\beta^{d_k}$ by using (4.72).
3. (Policy improvement) Set $d = d_k$ in (4.76) and construct a policy d' according to (4.77), and set $d_{k+1} = d'$.
4. If $d_{k+1} = d_k$, then stop and d_k is optimal. Otherwise, increase k by 1 and return to step 2.

4.3 The nth-Bias Optimization*

Among the policies that optimize the bias g (or equivalently, the potential with $P^*g = 0$), there are policies that optimize the second bias w (the bias of the bias, with $P^*w = 0$). This process can go on; that is, among the policies that optimize both the bias and second bias, we can find a policy that optimizes the third bias (the bias of the second bias) and so on. In this section, we propose the concept of *nth bias*, or in a general form the *nth potential*, $n = 1, 2, \ldots,$ and develop a unified theory for the optimization of the nth biases, which include the gain optimality and bias optimality as special cases. We also show that, as n increases, the nth-bias optimal policies eventually reach the set of Blackwell policies (see Problem 4.30). The approach is again based on difference formulas for the nth biases and is essentially the same as that for the gain and bias optimality.

4.3.1 nth-Bias Difference Formulas*

nth Potentials and nth Biases

In this formulation, the average reward (gain) is also called the *0th bias*. Thus, for a policy (P, f), the 0th bias is a vector g_0 with components

$$g_0(i) := \eta(i) = \lim_{L\to\infty} \frac{1}{L} E\Big\{ \sum_{l=0}^{L-1} f(X_l) \Big| X_0 = i \Big\}, \qquad i \in \mathcal{S},$$

where $\{X_l, l = 0, 1, \ldots\}$ is a sample path of the Markov chain with P. As we know,

$$g_0 = P^* f, \qquad P^* g_0 = g_0.$$

The potential g is also called the *first potential* and is denoted as $g_1 := g$, which is the solution to the Poisson equation

$$(I - P)g_1 + g_0 = f.$$

The first potential has different versions that may differ with an additive vector u satisfying $(I - P)u = 0$. The bias, also called the first bias, is a special version of the first potential with normalization condition $P^* g_1 = 0$. Thus, the bias satisfies

$$(I - P + P^*)g_1 = f - g_0.$$

The ith component of the bias is

$$g_1(i) := g(i) = \sum_{l=0}^{\infty} E\left[f(X_l) - \eta(i) | X_0 = i \right].$$

The *nth potential* g_n, $n > 1$, of policy (P, f) is defined as the solution to the Poisson equation of the $(n - 1)$th bias g_{n-1}:

$$(I - P)g_n = -g_{n-1}, \qquad n > 1. \tag{4.78}$$

This definition implies that $P^* g_{n-1} = 0$. (Thus, the g_{n-1} in (4.78) must be a bias, not a potential.) Again, the nth potential is unique up to an additive vector u. If g_n is a solution to (4.78), then for any vector u satisfying $(I-P)u = 0$, $g_n + u$ is also a solution to (4.78). The nth potential can also be defined by the $(n - 1)$th potential, see Problem 4.20.

The *nth bias* g_n, $n > 1$, of policy (P, f) is an nth potential satisfying the normalization condition

$$P^* g_n = 0.$$

Thus, from (4.78), the nth bias satisfies

$$(I - P + P^*)g_n = -g_{n-1}, \qquad n > 1.$$

From this and $P^* g_{n-1} = 0$, for the nth bias g_n we have

$$g_n = -\left[g_{n-1} + P g_{n-1} + P^2 g_{n-1} + \cdots\right], \qquad n > 1, \tag{4.79}$$

where g_{n-1} is the $(n-1)$th bias. Therefore, the ith component of g_n is

$$g_n(i) = -\sum_{l=0}^{\infty} E[g_{n-1}(X_l)|X_0 = i], \qquad n > 1.$$

Generally, we have

$$\begin{aligned} g_n &= -(I - P + P^*)^{-1} g_{n-1} \\ &= (-1)^{n-1}(I - P + P^*)^{-n}(f - \eta). \end{aligned} \tag{4.80}$$

Furthermore, from (4.79), we get

$$\begin{aligned} g_{n+1} &= (-1)^n \sum_{k=0}^{\infty} \frac{\prod_{m=1}^{n}(k+m)}{n!} (P^k f - \eta) \\ &= (-1)^n \sum_{k=0}^{\infty} \binom{n+k}{n} (P^k f - \eta), \qquad n > 0. \end{aligned} \tag{4.81}$$

Then,

$$g_{n+1}(i) = (-1)^n \sum_{k=0}^{\infty} \frac{\prod_{m=1}^{n}(k+m)}{n!} E\{[f(X_k) - \eta(i)]|X_0 = i\}, \qquad n > 0.$$

This can be used to develop sample-path-based estimates of g_n.

Optimal nth Biases

The nth potential, $n \geq 0$, associated with a policy $d \in \mathcal{D}$ (with (P^d, f^d)) is denoted as g_n^d. Without confusion, we will use the same notation g_n^d to denote the nth bias of policy $d \in \mathcal{D}$. Obviously, we can only study optimization of biases, not potentials, because the latter may differ with an additive vector.

A policy $\widehat{d}$ is said to be *gain optimal* and its gain $g_0^{\widehat{d}}$ is called the *optimal gain*, if

$$g_0^{\widehat{d}} \geq g_0^d, \qquad \text{for all } d \in \mathcal{D}.$$

Let g_0^* be the optimal gain, and

$$\mathcal{D}_0 := \left\{d \in \mathcal{D} : g_0^d = g_0^*\right\}$$

be the set of all gain-optimal policies. The *nth-bias optimal policies* for $n > 0$ are defined recursively. We only care about the nth-bias optimal policies in the set of $(n-1)$th-bias optimal policies. A policy $\widehat{d}$ is said to be *nth-bias optimal*, $n > 0$, if $\widehat{d} \in \mathcal{D}_{n-1}$ and

$$g_n^{\widehat{d}} \geq g_n^d, \qquad \text{for all } d \in \mathcal{D}_{n-1}, \quad n > 0.$$

Let g_n^* be the optimal nth bias, and

$$\mathcal{D}_n := \left\{ d \in \mathcal{D}_{n-1} : g_n^d = g_n^* \right\}, \qquad n > 0 \tag{4.82}$$

be the set of all nth-bias optimal policies, $n > 0$. From the definition, we can see that if $\widehat{d}$ is an nth-bias optimal policy, then $\widehat{d}$ is also a kth-bias optimal policy, where $0 \leq k < n$; i.e., $g_k^{\widehat{d}} = g_k^*$, for $k = 0, 1, \ldots, n$; but the higher order biases of $\widehat{d}$ may not be optimal. For example, $g_{n+1}^{\widehat{d}}$ may not take the optimal value if $\widehat{d}$ is an nth-bias optimal policy (i.e., $g_{n+1}^{\widehat{d}} \neq g_{n+1}^*$).

nth-Bias Difference Formulas

To develop an nth-bias optimization theory, we first derive the difference formulas for the nth biases. Let g_n^h and g_n^d, $n = 0, 1, \ldots$, be the nth biases of the two policies $h, d \in \mathcal{D}$, respectively.

The nth-Bias Difference Formula:

(a) $g_0^h - g_0^d = (P^h)^* \left[(f^h + P^h g_1^d) - (f^d + P^d g_1^d) \right] + \left[(P^h)^* - I \right] g_0^d;$

(b) if $g_0^h = g_0^d$, then

$$g_1^h - g_1^d = (P^h)^*(P^h - P^d) g_2^d + \sum_{k=0}^{\infty} \left\{ (P^h)^k \left[(f^h + P^h g_1^d) - (f^d + P^d g_1^d) \right] \right\};$$

$$\tag{4.83}$$

(c) if $g_n^h = g_n^d$ for a particular $n \geq 1$, then

$$g_{n+1}^h - g_{n+1}^d = (P^h)^*(P^h - P^d) g_{n+2}^d + \sum_{k=0}^{\infty} \left\{ (P^h)^k (P^h - P^d) g_{n+1}^d \right\}.$$

$$\tag{4.84}$$

Proof. (a) and (b) are the same as (4.36) and (4.54), respectively. Now, we prove (c). From (4.78) and $g_n^h = g_n^d$, we have

$$\begin{aligned} g_{n+1}^h - g_{n+1}^d &= (-g_n^h + P^h g_{n+1}^h) - (-g_n^d + P^d g_{n+1}^d) \\ &= P^h g_{n+1}^h - P^d g_{n+1}^d \\ &= (P^h - P^d) g_{n+1}^d + P^h (g_{n+1}^h - g_{n+1}^d). \end{aligned}$$

Thus,

$$(I - P^h)(g_{n+1}^h - g_{n+1}^d) = (P^h - P^d)g_{n+1}^d. \tag{4.85}$$

From (4.78), we have $g_{n+1}^d = (P^d - I)g_{n+2}^d$. Left-multiplying $(P^h)^*$ on both sides of this equation, we obtain

$$(P^h)^* g_{n+1}^d = (P^h)^*(P^d - I)g_{n+2}^d = (P^h)^*(P^d - P^h)g_{n+2}^d. \tag{4.86}$$

Because $(P^h)^* g_{n+1}^h = 0$ and from (4.85) and (4.86), we have

$$\begin{aligned}
&[I - P^h + (P^h)^*](g_{n+1}^h - g_{n+1}^d) \\
&= (I - P^h)(g_{n+1}^h - g_{n+1}^d) - (P^h)^* g_{n+1}^d \\
&= (I - P^h)(g_{n+1}^h - g_{n+1}^d) + (P^h)^*(P^h - P^d)g_{n+2}^d \\
&= (P^h - P^d)g_{n+1}^d + (P^h)^*(P^h - P^d)g_{n+2}^d.
\end{aligned}$$

Noting $[I - P^h + (P^h)^*]^{-1}(P^h)^* = (P^h)^*$, we have

$$\begin{aligned}
&g_{n+1}^h - g_{n+1}^d \\
&= \left[I - P^h + (P^h)^*\right]^{-1} (P^h - P^d)g_{n+1}^d + (P^h)^*(P^h - P^d)g_{n+2}^d \\
&= \sum_{k=0}^{\infty} \left[P^h - (P^h)^*\right]^k (P^h - P^d)g_{n+1}^d + (P^h)^*(P^h - P^d)g_{n+2}^d. \quad (4.87)
\end{aligned}$$

Again, from (4.78) and $g_n^h = g_n^d$, we obtain

$$\begin{aligned}
&(P^h)^*(P^h - P^d)g_{n+1}^d \\
&= (P^h)^*(P^h g_{n+1}^d - g_{n+1}^d - g_n^d) \\
&= -(P^h)^* g_n^d = -(P^h)^* g_n^h = 0.
\end{aligned}$$

Then (c) follows directly from (4.87). □

Finally, we note that $(P^h)^*(P^h - P^d)g_{n+2}^d = (P^h)^*(I - P^d)g_{n+2}^d$. Therefore, for any u satisfying $(I - P^d)u = 0$, we have $(P^h)^*(P^h - P^d)g_{n+2}^d = (P^h)^*(P^h - P^d)(g_{n+2}^d + cu)$. Thus, we can use potentials (instead of biases) for the g_{n+2}^d in (4.84) (for the same reason, we can use potentials for g_2^d and g_1^d on the right-hand sides of (b) and (a)).

4.3.2 Optimality Equations*

From the nth bias difference formulas, we can easily derive the following Comparison Lemma, which is the basis for the nth-bias optimality equations and the policy iteration algorithms.

Suppose that $\widehat{d}$ is an nth-bias optimal policy, $n \geq 1$, with optimal biases $g_k^{\widehat{d}} = g_k^*$, $k = 0, 1, \ldots, n$. Let h be any $(n-1)$th-bias optimal policy.

Comparison Lemma for $(n+1)$th Biases of Two nth-Bias Optimal Policies:

$$\text{If}\quad \begin{array}{l} (a)\ \ f^h + P^h g_1^{\widehat{d}} = f^{\widehat{d}} + P^{\widehat{d}} g_1^{\widehat{d}},\ \text{if } n=1;\ \ P^h g_n^{\widehat{d}} = P^{\widehat{d}} g_n^{\widehat{d}},\ \text{if } n>1, \\ (b)\ \ P^h g_{n+1}^{\widehat{d}} \geq P^{\widehat{d}} g_{n+1}^{\widehat{d}},\ \text{and} \\ (c)\ \ (P^h g_{n+2}^{\widehat{d}})(i) \geq (P^{\widehat{d}} g_{n+2}^{\widehat{d}})(i)\ \text{when}\ (P^h g_{n+1}^{\widehat{d}})(i) = (P^{\widehat{d}} g_{n+1}^{\widehat{d}})(i) \\ \quad\ \ \text{for some } i \in \mathcal{S}, \end{array}$$

$$\Longrightarrow \quad \text{then } h \text{ is } n\text{th-bias optimal, i.e., } g_n^h = g_n^*,\ \text{and } g_{n+1}^h \geq g_{n+1}^{\widehat{d}}. \tag{4.88}$$

In addition, if Condition (a) is replaced by the following Condition (a')

(a') policy h is nth-bias optimal,

then $g_{n+1}^h \leq g_{n+1}^{\widehat{d}}$ if we change all the signs " $\geq$ " in (b) and (c) to " $\leq$ ".

Proof. Because both h and $\widehat{d}$ are $(n-1)$th-bias optimal, we have $g_{n-1}^h = g_{n-1}^{\widehat{d}} = g_{n-1}^*$. Thus, we have the bias difference formulas

$$g_1^h - g_1^{\widehat{d}} = \sum_{k=0}^{\infty} (P^h)^k \left[(f^h + P^h g_1^{\widehat{d}}) - (f^{\widehat{d}} + P^{\widehat{d}} g_1^{\widehat{d}}) \right] + (P^h)^* (P^h - P^{\widehat{d}}) g_2^{\widehat{d}},$$

and

$$g_n^h - g_n^{\widehat{d}} = \sum_{k=0}^{\infty} (P^h)^k (P^h - P^{\widehat{d}}) g_n^{\widehat{d}} + (P^h)^* (P^h - P^{\widehat{d}}) g_{n+1}^{\widehat{d}}, \qquad \text{if } n > 1.$$

From Condition (a), the first sum on the right-hand sides of these two equations is zero. Then, from Condition (b), we have $g_n^h \geq g_n^{\widehat{d}}$. Because $g_n^{\widehat{d}}$ is the optimal nth bias, we must have $g_n^h = g_n^{\widehat{d}} = g_n^*$, and, furthermore, from the aforementioned two equations, we must have

$$(P^h)^* (P^h - P^{\widehat{d}}) g_{n+1}^{\widehat{d}} = 0, \qquad n \geq 1. \tag{4.89}$$

The $(n+1)$th-bias difference formula is

$$\begin{aligned} g_{n+1}^h - g_{n+1}^{\widehat{d}} &= (P^h)^* (P^h - P^{\widehat{d}}) g_{n+2}^{\widehat{d}} + \sum_{k=0}^{\infty} \left[(P^h)^k (P^h - P^{\widehat{d}}) g_{n+1}^{\widehat{d}} \right] \\ &= (P^h)^* v + z, \qquad n \geq 1, \end{aligned}$$

where $u := (P^h - P^{\widehat{d}}) g_{n+1}^{\widehat{d}}$, $v := (P^h - P^{\widehat{d}}) g_{n+2}^{\widehat{d}}$, and $z := \sum_{k=0}^{\infty} \left[(P^h)^k u \right]$. From (4.89), we have $(P^h)^* u = 0$, $n \geq 1$. From Condition (b), we have $u \geq 0$ and then $z \geq 0$. Thus, Condition (a) in the Comparison Lemma (General

Form) (4.39) holds. Condition (c) here implies Condition (b) in the Comparison Lemma (4.39). Finally, it follows that $g_{n+1}^{h} \geq g_{n+1}^{\widehat{d}}$, $n \geq 1$.

When the signs "$\geq$" change to "$\leq$", we cannot prove that h is an nth-bias optimal policy from Conditions (a) and (b). Note that Condition (a') implies Condition (a) (cf. (4.61) for $n = 1$ and (4.78) for $n > 1$). With Condition (a'), the other half of the lemma can be proved in a similar way. □

This Comparison Lemma (4.88) tells us that given an nth-bias optimal policy, how we can find another policy in the space of the $(n-1)$th-bias optimal policies that is nth-bias optimal and has a larger (or at least equal) $(n+1)$th bias, $n > 0$. A similar case we have seen in Section 4.2.2 is that if we are given a gain-optimal policy, we can find another policy in the policy space that is gain optimal and has a large (or equal) bias, which is based on the Comparison Lemma for biases (4.57). We can continue this improvement procedure until it reaches a policy whereby no improvement can be made in the $(n+1)$th bias by following this procedure. This policy satisfies the following sufficient optimality equations.

Sufficient Optimality Equations

Suppose that the following $n+2$ optimality equations hold for policy $\widehat{d}$. Then, $\widehat{d}$ is nth-bias optimal, i.e., $g_k^{\widehat{d}} = g_k^*$ is the optimal kth bias, for $k = 0, 1, \ldots, n$.

Sufficient Optimality Conditions:

A policy $\widehat{d}$ is nth-bias optimal, $n \geq 0$, if, for all $i \in \mathcal{S}$, we have

$$g_0^{\widehat{d}}(i) = \max_{\alpha \in \mathcal{A}(i)} \left\{ \sum_{j \in \mathcal{S}} p^{\alpha}(j|i) g_0^{\widehat{d}}(j) \right\}. \tag{4.90}$$

$$g_0^{\widehat{d}}(i) + g_1^{\widehat{d}}(i) = \max_{\alpha \in \mathcal{A}_0(g_0^{\widehat{d}})(i)} \left\{ f(i,\alpha) + \sum_{j \in \mathcal{S}} p^{\alpha}(j|i) g_1^{\widehat{d}}(j) \right\}. \tag{4.91}$$

$$g_k^{\widehat{d}}(i) + g_{k+1}^{\widehat{d}}(i) = \max_{\alpha \in \mathcal{A}_k(g_0^{\widehat{d}}, g_1^{\widehat{d}}, \ldots, g_k^{\widehat{d}})(i)} \left\{ \sum_{j \in \mathcal{S}} p^{\alpha}(j|i) g_{k+1}^{\widehat{d}}(j) \right\},$$
$$k = 1, 2, \ldots, n, \tag{4.92}$$

where, for each $i \in \mathcal{S}$,

$$\mathcal{A}_0(g_0^{\widehat{d}})(i) := \arg \max_{\alpha \in \mathcal{A}(i)} \left\{ \sum_{j \in \mathcal{S}} p^{\alpha}(j|i) g_0^{\widehat{d}}(j) \right\}$$
$$= \left\{ \alpha \in \mathcal{A}(i) : \sum_{j \in \mathcal{S}} p^{\alpha}(j|i) g_0^{\widehat{d}}(j) = g_0^{\widehat{d}}(i) \right\},$$

$$\mathcal{A}_1(\widehat{g_0^d}, \widehat{g_1^d})(i) := \arg \max_{\alpha \in \mathcal{A}_0(\widehat{g_0^d})(i)} \left\{ f(i,\alpha) + \sum_{j \in \mathcal{S}} p^\alpha(j|i) \widehat{g_1^d}(j) \right\}$$

$$= \left\{ \alpha \in \mathcal{A}_0(\widehat{g_0^d})(i) : f(i,\alpha) + \sum_{j \in \mathcal{S}} p^\alpha(j|i) \widehat{g_1^d}(j) = \widehat{g_0^d}(i) + \widehat{g_1^d}(i) \right\},$$

$$\mathcal{A}_k(\widehat{g_0^d}, \dots, \widehat{g_k^d})(i) := \arg \max_{\alpha \in \mathcal{A}_{k-1}(\widehat{g_0^d}, \dots, \widehat{g_{k-1}^d})(i)} \left\{ \sum_{j \in \mathcal{S}} p^\alpha(j|i) \widehat{g_k^d}(j) \right\}$$

$$= \left\{ \alpha \in \mathcal{A}_{k-1}(\widehat{g_0^d}, \dots, \widehat{g_{k-1}^d})(i) : \sum_{j \in \mathcal{S}} p^\alpha(j|i) \widehat{g_k^d}(j) = \widehat{g_{k-1}^d}(i) + \widehat{g_k^d}(i) \right\},$$
$$k = 2, \dots, n.$$

In these equations, the sets $\mathcal{A}_0(\widehat{g_0^d})(i)$'s are defined only for these $(\widehat{g_0^d})$'s that satisfy (4.90); that is, if for a policy $\widehat{d}$, (4.90) does not hold, then the definition makes no sense. Similarly, the sets $\mathcal{A}_1(\widehat{g_0^d}, \widehat{g_1^d})(i)$ are defined only for the $\widehat{g_0^d}$ and $\widehat{g_1^d}$ that satisfy (4.90) and (4.91); and so on, and the sets $\mathcal{A}_n(\widehat{g_0^d}, \dots, \widehat{g_n^d})(i)$ are defined only for those $\widehat{g_0^d}, \dots, \widehat{g_n^d}$ that satisfy Equations (4.90), (4.91), and those in (4.92) up to the equation with $k = n - 1$. We define the policy set as

$$\mathcal{D}_k(\widehat{g_0^d}, \dots, \widehat{g_k^d}) := \times_{i \in \mathcal{S}} \mathcal{A}_k(\widehat{g_0^d}, \dots, \widehat{g_k^d})(i), \quad k = 0, 1, \dots, n.$$

For simplicity, we may set $\widehat{\mathcal{A}}_k(i) := \mathcal{A}_k(\widehat{g_0^d}, \dots, \widehat{g_k^d})(i)$. This is what used in (4.61) and (4.62).

We refer to (4.90)-(4.92) as the *nth-bias optimality equations* and refer to (4.90) as the first equation, to (4.91) as the second equation, and to (4.92) with index k, $k \geq 1$, as the $(k+2)$th equation, etc. We observe that

$$\mathcal{D}_0(\widehat{g_0^d}) = \left\{ \text{all } d \in \mathcal{D} : P^d \widehat{g_0^d} = \widehat{g_0^d} \right\},$$

$$\mathcal{D}_1(\widehat{g_0^d}, \widehat{g_1^d}) = \left\{ \text{all } d \in \mathcal{D} : P^d \widehat{g_0^d} = \widehat{g_0^d}, f^d + P^d \widehat{g_1^d} = \widehat{g_0^d} + \widehat{g_1^d} \right\},$$

and

$$\mathcal{D}_k(\widehat{g_0^d}, \widehat{g_1^d}, \dots, \widehat{g_k^d})$$
$$= \left\{ \text{all } d \in \mathcal{D} : P^d \widehat{g_0^d} = \widehat{g_0^d}, f^d + P^d \widehat{g_1^d} = \widehat{g_0^d} + \widehat{g_1^d}, \right.$$
$$\left. P^d \widehat{g_{l+1}^d} = \widehat{g_l^d} + \widehat{g_{l+1}^d},\ l = 1, 2, \dots, k-1 \right\}, \qquad k \geq 2. \quad (4.93)$$

We have

$$\mathcal{D}_k(g_0^{\widehat{d}}, g_1^{\widehat{d}}, \ldots, g_k^{\widehat{d}}) \subseteq \mathcal{D}_{k-1}(g_0^{\widehat{d}}, g_1^{\widehat{d}}, \ldots, g_{k-1}^{\widehat{d}}), \qquad k \geq 1.$$

Note that $\mathcal{D}_k(g_0^{\widehat{d}}, g_1^{\widehat{d}}, \ldots, g_k^{\widehat{d}})$ is different from the set of kth-bias optimal policies $\mathcal{D}_k$. By definition, for any policy d, if $d(i) \in \mathcal{A}_k(g_0^{\widehat{d}}, \ldots, g_k^{\widehat{d}})(i)$ for all $i \in \mathcal{S}$, we say that $d \in \mathcal{D}_k(g_0^{\widehat{d}}, \ldots, g_k^{\widehat{d}})$. Now, we can rewrite (4.90)-(4.92) in a matrix form.

$$\max_{d \in \mathcal{D}} \left\{ (P^d - I) g_0^{\widehat{d}} \right\} = 0,$$

$$\max_{d \in \mathcal{D}_0(g_0^{\widehat{d}})} \left\{ f^d - g_0^{\widehat{d}} + (P^d - I) g_1^{\widehat{d}} \right\} = 0,$$

$$\max_{d \in \mathcal{D}_k(g_0^{\widehat{d}}, \ldots, g_k^{\widehat{d}})} \left\{ -g_k^{\widehat{d}} + (P^d - I) g_{k+1}^{\widehat{d}} \right\} = 0, \qquad k = 1, 2, \ldots, n.$$

Obviously, if $d \in \mathcal{D}_n(g_0^{\widehat{d}}, g_1^{\widehat{d}}, \ldots, g_n^{\widehat{d}})$, then $d \in \mathcal{D}_k(g_0^{\widehat{d}}, g_1^{\widehat{d}}, \ldots, g_k^{\widehat{d}})$, for $k = 1, 2, \ldots, n$.

Let us first understand the meanings of these mathematical equations. The first equation (4.90) says that in every state $i \in \mathcal{S}$, we have $g_0^{\widehat{d}}(i) \geq \sum_{j \in \mathcal{S}} p^\alpha(j|i) g_0^{\widehat{d}}(j)$ for all actions $\alpha \in \mathcal{A}(i)$. If this is the case, then all the actions that make the equality hold form the set $\mathcal{A}_0(g_0^{\widehat{d}})(i)$. Furthermore, $\mathcal{A}_0(g_0^{\widehat{d}})(i)$ is not empty because $P^{\widehat{d}} \eta^{\widehat{d}} = \eta^{\widehat{d}}$, so $\widehat{d}(i) \in \mathcal{A}_0(g_0^{\widehat{d}})(i)$. The second equation (4.91) says that for all the actions for which the above equality holds (i.e., $\alpha \in \mathcal{A}_0(g_0^{\widehat{d}})(i)$), we have $g_0^{\widehat{d}}(i) + g_1^{\widehat{d}}(i) \geq \left\{ f(i, \alpha) + \sum_{j \in \mathcal{S}} p^\alpha(j|i) g_1^{\widehat{d}}(j) \right\}$. If this is the case, then all the actions in $\mathcal{A}_0(g_0^{\widehat{d}})(i)$ that make the equality hold form the set $\mathcal{A}_1(g_0^{\widehat{d}}, g_1^{\widehat{d}})(i)$. Furthermore, from the Poisson equation for $\widehat{d}$, we have $\widehat{d}(i) \in \mathcal{A}_1(g_0^{\widehat{d}}, g_1^{\widehat{d}})(i)$; so the set is non-empty. This statement goes on until the last (the $(n+2)$th) equation. The set $\mathcal{A}_k(g_0^{\widehat{d}}, g_1^{\widehat{d}}, \ldots, g_k^{\widehat{d}})(i)$ shrinks as k increases. If at some point $\mathcal{A}_k(g_0^{\widehat{d}}, g_1^{\widehat{d}}, \ldots, g_k^{\widehat{d}})(i)$, $k \leq n$, contains only one action denoted as α_0, then we must have $g_{k-1}^{\widehat{d}}(i) + g_k^{\widehat{d}}(i) = \sum_{j \in \mathcal{S}} p^{\alpha_0}(j|i) g_k^{\widehat{d}}(j)$. Furthermore, we have $\mathcal{A}_l(g_0^{\widehat{d}}, g_0^{\widehat{d}}, \ldots, g_l^{\widehat{d}})(i) = \{\alpha_0\}$ and $g_l^{\widehat{d}}(i) + g_{l+1}^{\widehat{d}}(i) = \sum_{j \in \mathcal{S}} p^{\alpha_0}(j|i) g_{l+1}^{\widehat{d}}(j)$ for all $k \leq l \leq n$. Obviously, we have $\alpha_0 = \widehat{d}(i)$, because $\widehat{d}(i) \in \mathcal{A}_k(g_0^{\widehat{d}}, g_1^{\widehat{d}}, \ldots, g_k^{\widehat{d}})(i)$, $k = 1, 2, \ldots, n$, if (4.90)-(4.92) holds.

According to the Sufficient Optimality Conditions, if $g_0^{\widehat{d}}, \ldots, g_n^{\widehat{d}}, g_{n+1}^{\widehat{d}}$ satisfy (4.90)-(4.92), then $g_0^{\widehat{d}}, \ldots, g_n^{\widehat{d}}$ are the optimal kth biases, $g_k^{\widehat{d}} = g_k^*$, $k = 0, 1, \ldots, n$, respectively. However, $g_{n+1}^{\widehat{d}}$ may take different values that are not optimal. (cf. the discussion after the Sufficient Conditions for Bias Optimality in Section 4.2.2.)

To prove the Sufficient Optimality Conditions, we need a result from the Necessary Optimality Conditions stated later. Basically, by definition, in order to get an nth-bias optimal policy, we need to determine the space of all the $(n-1)$th-bias optimal policies. The Necessary Optimality Conditions tell us that the space of all the $(n-1)$th-bias optimal policies is contained in $\mathcal{D}_{n-1}(g_0^{\widehat{d}}, \ldots, g_{n-1}^{\widehat{d}})$. The following proof may be read after getting familiar with the Necessary Optimality Conditions.

Proof of the Sufficient Optimality Conditions. We have proved the cases of $n = 0$ and $n = 1$ in Sections 4.2.1 and 4.2.2, respectively. Now we prove the general case of $n > 1$ by induction. Assume that the theorem holds for the case of n; that is, if the vectors $g_0^{\widehat{d}}, \ldots, g_{n+1}^{\widehat{d}}$, $n \geq 1$, satisfy the first $(n+2)$ equations, then $g_k^{\widehat{d}}$ is the optimal kth bias, $k = 0, 1, \ldots, n$. We wish to prove that the theorem holds for the case of $(n+1)$; that is, if the vectors $g_0^{\widehat{d}}, \ldots, g_{n+2}^{\widehat{d}}$, $n \geq 1$, satisfy the first $(n+3)$ equations, then $g_k^{\widehat{d}}$ is the optimal kth bias, $k = 0, 1, \ldots, n+1$.

To achieve this, we need only to prove that $g_{n+1}^{\widehat{d}} \geq g_{n+1}^{d}$, where g_{n+1}^{d} is the $(n+1)$th bias of any nth-bias optimal policy $d \in \mathcal{D}_n$, $n \geq 1$. By the assumption made through induction, $g_n^{\widehat{d}} = g_n^*$ is the optimal nth bias. Because $d \in \mathcal{D}_n$, from the Necessary Optimality Conditions stated just after this proof, we have $d \in \mathcal{D}_n(g_0^{\widehat{d}}, \ldots, g_n^{\widehat{d}})$. Therefore,

$$f^d + P^d g_1^{\widehat{d}} = f^{\widehat{d}} + P^{\widehat{d}} g_1^{\widehat{d}}, \qquad \text{if } n = 1,$$

or

$$P^d g_n^{\widehat{d}} = P^{\widehat{d}} g_n^{\widehat{d}}, \qquad \text{if } n > 1.$$

Now, we set d and $\widehat{d}$ to be h and $\widehat{d}$ in the Comparison Lemma (4.88). Because $d \in \mathcal{D}_n$, Condition (a') in the lemma holds. In addition, the $(n+2)$th equation, together with $d \in \mathcal{D}_n(g_0^{\widehat{d}}, \ldots, g_n^{\widehat{d}})$, indicates that Condition (b) in the lemma, with the "$\leq$" sign, holds, and the $(n+3)$th equation implies Condition (c) in the lemma, with the "$\leq$" sign, holds. Finally, it follows from the lemma that $g_{n+1}^{\widehat{d}} \geq g_{n+1}^{d}$. □

Necessary Optimality Equations

Necessary Optimality Conditions:
If a policy $\widehat{d}$ is nth-bias optimal with biases $g_k^{\widehat{d}} = g_k^*$, $k = 0, 1, \ldots, n$, then $g_k^{\widehat{d}}$, $k = 0, 1, \ldots, n$, satisfy the first $(n+1)$ optimality equations (4.90)-(4.92).

Proof. The two cases with $n = 0$ and $n = 1$ are proved in Sections 4.2.1 and 4.2.2, respectively. Now, we prove the general case with $n > 1$ by induction. That is, we assume that all the nth-bias optimal policies satisfy the first $(n+1)$ optimality equations. Then, we prove that any $(n+1)$th-bias optimal policy must satisfy the first $(n+2)$ optimality equations, $n \geq 1$.

Because an $(n+1)$th-bias optimal policy is also nth-bias optimal, the first $n+1$ equations hold by assumption. We only need to check the $(n+2)$th equation. In other words, we need to show that for an $(n+1)$th-bias optimal policy $\widehat{d}$, $n \geq 1$, with kth bias $g_k^{\widehat{d}} = g_k^*$, $k = 0, 1, \ldots, n+1$, we have

$$P^d g_{n+1}^{\widehat{d}} \leq g_n^{\widehat{d}} + g_{n+1}^{\widehat{d}}, \qquad \text{for all } d \in \mathcal{D}_n(g_0^{\widehat{d}}, g_1^{\widehat{d}}, \ldots, g_n^{\widehat{d}}). \tag{4.94}$$

Assume that (4.94) does not hold. Then, there must exist a policy denoted as $d' \in \mathcal{D}_n(g_0^{\widehat{d}}, g_1^{\widehat{d}}, \ldots, g_n^{\widehat{d}})$ and a state $i \in \mathcal{S}$ such that

$$(P^{d'} g_{n+1}^{\widehat{d}})(i) > g_n^{\widehat{d}}(i) + g_{n+1}^{\widehat{d}}(i) = (P^{\widehat{d}} g_{n+1}^{\widehat{d}})(i). \tag{4.95}$$

Based on this, we can construct another policy $\widetilde{d}$ by setting $\widetilde{d}(i) = d'(i)$ and $\widetilde{d}(j) = \widehat{d}(j)$, for all $j \neq i$. Because $d' \in \mathcal{D}_n(g_0^{\widehat{d}}, g_1^{\widehat{d}}, \ldots, g_n^{\widehat{d}})$, we have $(P^{d'} g_k^{\widehat{d}})(i) = g_k^{\widehat{d}}(i) + g_{k-1}^{\widehat{d}}(i) = (P^{\widehat{d}} g_k^{\widehat{d}})(i)$, for $k = 2, \ldots, n$; and $(f^{d'} + P^{d'} g_1^{\widehat{d}})(i) = g_0^{\widehat{d}}(i) + g_1^{\widehat{d}}(i) = (f^{\widehat{d}} + P^{\widehat{d}} g_1^{\widehat{d}})(i)$. Therefore, we have $(P^{\widetilde{d}} g_k^{\widehat{d}})(i) = (P^{\widehat{d}} g_k^{\widehat{d}})(i)$ for $k = 2, \ldots, n$, and $(f^{\widetilde{d}} + P^{\widetilde{d}} g_1^{\widehat{d}})(i) = (f^{\widehat{d}} + P^{\widehat{d}} g_1^{\widehat{d}})(i)$. This means that the ith component of the vector $\left[(P^{\widetilde{d}} - P^{\widehat{d}}) g_k^{\widehat{d}}\right](i) = 0$ for $k = 2, \ldots, n$, and $\left[(f^{\widetilde{d}} + P^{\widetilde{d}} g_1^{\widehat{d}}) - (f^{\widehat{d}} + P^{\widehat{d}} g_1^{\widehat{d}})\right](i) = 0$. In addition, by construction, $P^{\widetilde{d}}$ and $P^{\widehat{d}}$, and $f^{\widetilde{d}}$ and $f^{\widehat{d}}$, are the same except for their ith rows. Thus,

$$(P^{\widetilde{d}} - P^{\widehat{d}}) g_k^{\widehat{d}} = 0, \qquad \text{for } k = 2, \ldots, n, \tag{4.96}$$

and

$$(f^{\widetilde{d}} + P^{\widetilde{d}} g_1^{\widehat{d}}) - (f^{\widehat{d}} + P^{\widehat{d}} g_1^{\widehat{d}}) = 0. \tag{4.97}$$

If $n = 1$, we only have (4.97).

For $n = 1$, $d' \in \mathcal{D}_1(g_0^{\widehat{d}}, g_1^{\widehat{d}})$, i.e., it satisfies the first two optimization equations and, by the sufficient conditions for gain-optimal policies, d' is a gain-optimal policy and $\eta^{d'} = \eta^{\widehat{d}} = \eta^*$. Thus, we can apply the bias difference formula and obtain

$$\begin{aligned} g_1^{\widetilde{d}} - g_1^{\widehat{d}} &= (P^{\widetilde{d}})^* (P^{\widetilde{d}} - P^{\widehat{d}}) g_2^{\widehat{d}} + \sum_{k=0}^{\infty} (P^{\widetilde{d}})^k \left[(f^{\widetilde{d}} + P^{\widetilde{d}} g_1^{\widehat{d}}) - (f^{\widehat{d}} + P^{\widehat{d}} g_1^{\widehat{d}})\right] \\ &= (P^{\widetilde{d}})^* (P^{\widetilde{d}} - P^{\widehat{d}}) g_2^{\widehat{d}}. \end{aligned} \tag{4.98}$$

For $n > 1$, we have (4.96), and from (4.98), we conclude that $g_1^{\widetilde{d}} = g_1^{\widehat{d}}$ and therefore we may use the kth-bias difference formula (4.84) for $k > 1$

$$g_k^{\widetilde{d}} - g_k^{\widehat{d}} = (P^{\widetilde{d}})^*(P^{\widetilde{d}} - P^{\widehat{d}})g_{k+1}^{\widehat{d}} + \sum_{l=0}^{\infty}(P^{\widetilde{d}})^l(P^{\widetilde{d}} - P^{\widehat{d}})g_k^{\widehat{d}}, \qquad k = 2, \ldots, n.$$

Setting $k = 2$ and by (4.96), we may get $g_2^{\widetilde{d}} = g_2^{\widehat{d}}$. Repeating this process for $k = 3, \ldots, n-1$, we may obtain $g_k^{\widetilde{d}} = g_k^{\widehat{d}}$, $k = 2, \ldots, n-1$, and finally, we have

$$g_n^{\widetilde{d}} - g_n^{\widehat{d}} = (P^{\widetilde{d}})^*(P^{\widetilde{d}} - P^{\widehat{d}})g_{n+1}^{\widehat{d}}, \qquad n > 1, \tag{4.99}$$

which has the same form as (4.98). Thus, we need to discuss only (4.99) with $n \geq 1$.

Next, by construction, all the components of $(P^{\widetilde{d}} - P^{\widehat{d}})g_{n+1}^{\widehat{d}}$ except its ith component are zeros for $n \geq 1$. Thus,

$$g_n^{\widetilde{d}}(j) - g_n^{\widehat{d}}(j) = \left\{ \left[(P^{\widetilde{d}} - P^{\widehat{d}})g_{n+1}^{\widehat{d}} \right](i) \right\} (P^{\widetilde{d}})^*(i|j).$$

From (4.95), $\left[(P^{\widetilde{d}} - P^{\widehat{d}})g_{n+1}^{\widehat{d}} \right](i) > 0$. Thus, we have $(P^{\widetilde{d}} - P^{\widehat{d}})g_{n+1}^{\widehat{d}} \succeq 0$. Also, by (4.99), we must have $(P^{\widetilde{d}})^*(i|j) = 0$ for all $j \in \mathcal{S}$, because $g_n^{\widehat{d}}$ is nth-bias optimal. This means that i is a transient state under policy $P^{\widetilde{d}}$. Therefore, from the structure of $(P^{\widetilde{d}})^*$, we conclude that $(P^{\widetilde{d}})^*(P^{\widetilde{d}} - P^{\widehat{d}}) = 0$. Finally, from the bias difference equation (4.84), for $n \geq 1$, we have

$$\begin{aligned} & g_{n+1}^{\widetilde{d}} - g_{n+1}^{\widehat{d}} \\ &= \sum_{l=0}^{\infty}(P^{\widetilde{d}})^l(P^{\widetilde{d}} - P^{\widehat{d}})g_{n+1}^{\widehat{d}} + (P^{\widetilde{d}})^*(P^{\widetilde{d}} - P^{\widehat{d}})g_{n+2}^{\widehat{d}} \\ &= \sum_{l=0}^{\infty}(P^{\widetilde{d}})^l(P^{\widetilde{d}} - P^{\widehat{d}})g_{n+1}^{\widehat{d}} \\ &\geq (P^{\widetilde{d}} - P^{\widehat{d}})g_{n+1}^{\widehat{d}} \\ &\succeq 0. \end{aligned}$$

This is impossible because $g_{n+1}^{\widehat{d}}$ is an $(n+1)$th-bias optimal policy. □

To be more precise, when $g_0^{\widehat{d}}, \ldots, g_n^{\widehat{d}}$ are optimal biases, we write $\mathcal{D}_n(g_0^{\widehat{d}}, \ldots, g_n^{\widehat{d}})$ as $\mathcal{D}_n(g_0^*, \ldots, g_n^*)$. We may further simplify the notation by setting $\mathcal{D}_n^* := \mathcal{D}_n(g_0^*, \ldots, g_n^*)$. That is, we define

$$\mathcal{D}_0^* = \mathcal{D}_0(g_0^*) = \left\{ \text{all } d \in \mathcal{D} : P^d g_0^* = g_0^* \right\}, \tag{4.100}$$

$$\begin{aligned} \mathcal{D}_1^* &= \mathcal{D}_1(g_0^*, g_1^*) \\ &= \left\{ \text{all } d \in \mathcal{D} : P^d g_0^* = g_0^*, f^d + P^d g_1^* = g_0^* + g_1^* \right\}, \end{aligned} \tag{4.101}$$

$$\begin{aligned} \mathcal{D}_n^* &= \mathcal{D}_n(g_0^*, \ldots, g_n^*) \\ &= \left\{ \text{all } d \in \mathcal{D} : P^d g_0^* = g_0^*, f^d + P^d g_1^* = g_0^* + g_1^*, \right. \\ &\qquad \left. P^d g_{l+1}^* = g_l^* + g_{l+1}^*, \ l = 1, \ldots, n-1 \right\}, \qquad n \geq 2, \end{aligned} \tag{4.102}$$

in which (4.100) and (4.101) are the same as (4.65), (4.66), and (4.67).

One obvious corollary is that an nth-bias optimal policy $\widehat{d}$ reaches the maximum in the first $(n+1)$ optimality equations. In other words, if $\widehat{d}$ is nth-bias optimal, then $\widehat{d} \in \mathcal{D}_0(g_0^*)$, ..., $\widehat{d} \in \mathcal{D}_n(g_0^*, \dots, g_n^*)$.

One implication of the Necessary Optimality Conditions is as follows. If we are given an $(n-1)$th-bias optimal policy with kth biases g_k^*, $k = 0, 1, \dots, n-1$, and we need to search for an nth-bias optimal policy among all the $(n-1)$th-bias optimal polices, we can search among the policies in the set $\mathcal{D}_{n-1}(g_0^*, \dots, g_{n-1}^*)$. This set may be a bit larger than the space of all the $(n-1)$th-bias optimal policies. For example, in case of the bias optimality, we need to search in $\mathcal{D}_0(g_0^*)$, see (4.69).

From the nth-bias difference formulas and the Poisson equations, it is easy to prove (following the same steps as for (4.68) and see Figure 4.7) that

$$\mathcal{D}_{n+1}(g_0^*, \dots, g_{n+1}^*) \subseteq \mathcal{D}_n \subseteq \mathcal{D}_n(g_0^*, \dots, g_n^*), \qquad n \geq 0. \tag{4.103}$$

This formula holds for every individual component $i \in \mathcal{S}$.

4.3.3 Policy Iteration*

We can devise a policy iteration algorithm for the nth-bias optimal policy by following the same procedure as for the gain and bias optimization problems. By the Comparison Lemma (4.88), from any $(n-1)$th-bias optimal policy, we can construct another $(n-1)$th-bias optimal policy that has a larger (or at least equal) nth bias if such a policy exists. By (4.103), we need to search in the set $\mathcal{D}_{n-1}(g_0^*, \dots, g_{n-1}^*)$.

For a given $(n-1)$th-bias optimal policy $d \in \mathcal{D}_{n-1}$ with kth biases g_k^d, $k = 0, 1, \dots, n+1$, $n > 1$, and $g_k^d = g_k^*$, $k = 0, 1, \dots, n-1$, being the optimal kth bias, we first define

$$\mathcal{A}_n(i,d) := \left\{ \begin{array}{l} \alpha \in \mathcal{A}_{n-1}(g_0^*, g_1^*, \dots, g_{n-1}^*)(i): \\ \sum_{j \in \mathcal{S}} p^\alpha(j|i) g_n^d(j) > \sum_{j \in \mathcal{S}} p^{d(i)}(j|i) g_n^d(j); \text{ or} \\ \sum_{j \in \mathcal{S}} p^\alpha(j|i) g_{n+1}^d(j) > \sum_{j \in \mathcal{S}} p^{d(i)}(j|i) g_{n+1}^d(j) \\ \text{when } \sum_{j \in \mathcal{S}} p^\alpha(j|i) g_n^d(j) = \sum_{j \in \mathcal{S}} p^{d(i)}(j|i) g_n^d(j) \end{array} \right\}. \tag{4.104}$$

We then define an improvement policy d' (depending on d) as follows:

$$d'(i) \in \mathcal{A}_n(i,d) \text{ if } \mathcal{A}_n(i,d) \neq \emptyset, \text{ and } d'(i) = d(i) \text{ if } \mathcal{A}_n(i,d) = \emptyset. \tag{4.105}$$

Note that such a policy may not be unique, since there may be more than one action in $\mathcal{A}_n(i,d)$ for some state $i \in \mathcal{S}$.

Again, there is no need to choose $d'(i) \in \mathcal{A}_n(i,d)$ for all i such that $\mathcal{A}_n(i,d) \neq \emptyset$; we may choose $d'(i) \in \mathcal{A}_n(i,d)$ for at least one state, and set $d'(k) = d(k)$ for other states (cf. the discussion on page 224 for the bias optimality case).

Lemma 4.6. *For any given $(n-1)$th-bias optimal policy $d \in \mathcal{D}_{n-1}$, $n > 1$, let d' be defined as in (4.105). Then, we have*

(a) $g_{n-1}^{d'} = g_{n-1}^{d}$,

(b) $g_n^{d'} \geq g_n^d$, *and*

(c) if $g_n^{d'} = g_n^d$ *and* $d' \neq d$, *then* $g_{n+1}^{d'} \succeq g_{n+1}^d$.

Proof. The proof is similar to that of Lemma 4.5 for the case of $n = 1$.
(a) By construction, if $d' \neq d$ then $d'(i) \in \mathcal{A}_{n-1}(g_0^d, g_1^d, \ldots, g_{n-1}^d)(i)$ for all $i \in \mathcal{S}$, $g_k^d = g_k^*$, $k = 0, 1, \ldots, n-1$. Thus, from (4.93), we have $f^{d'} + P^{d'} g_1^d = f^d + P^d g_1^d$ for $n = 2$, and $P^{d'} g_{n-1}^d = g_{n-2}^d + g_{n-1}^d = P^d g_{n-1}^d$ for $n > 2$. Then, the second term on the right-hand side of the two bias difference equations in (4.83) for $n - 1 = 1$ and in (4.84) for $n - 1 \geq 2$ are zeros. Thus, the difference equations become

$$g_{n-1}^{d'} - g_{n-1}^d = (P^{d'})^*(P^{d'} - P^d) g_n^d,$$

for both $n = 2$ and $n > 2$. Again, by construction (4.104), we have $(P^{d'} - P^d) g_n^d \geq 0$. Therefore, $g_{n-1}^{d'} - g_{n-1}^d \geq 0$. Since d is $(n-1)$th-bias optimal, we conclude that d' is also an $(n-1)$th-bias optimal policy.

(b) We first set d to be the policy $\widehat{d}$ and d' to be the h in the Comparison Lemma (4.88). Consider the lemma for the case of $\widehat{d}$ being an $(n-1)$th-bias optimal. By the constructions in (4.104) and (4.105), Conditions (a), (b), and (c) in the Comparison Lemma hold. Thus, it follows that $g_n^{d'} \geq g_n^d$.

(c) If $g_n^{d'} = g_n^d$, $n \geq 2$, we have

$$P^{d'} g_n^d = P^{d'} g_n^{d'} = g_{n-1}^{d'} + g_n^{d'} = g_{n-1}^d + g_n^d = P^d g_n^d.$$

Then $P^{d'} g_{n+1}^d \succeq P^d g_{n+1}^d$ holds by (4.104), (4.105), and $d' \neq d$. From (4.84), we have

$$g_n^{d'} - g_n^d = \sum_{k=0}^{\infty} (P^{d'})^k (P^{d'} - P^d) g_n^d + (P^{d'})^*(P^{d'} - P^d) g_{n+1}^d.$$

From this equation, we must have $(P^{d'})^*(P^{d'} - P^d) g_{n+1}^d = g_n^{d'} - g_n^d = 0$. From the Fundamental Fact (4.30), we have $(P^{d'} g_{n+1}^d)(i) = (P^d g_{n+1}^d)(i)$ for all recurrent states i under policy d'. From (4.104) and (4.105),

$$d'(i) = d(i), \qquad \text{for all recurrent states } i \text{ under policy } d'. \qquad (4.106)$$

From (4.84), we have

$$g_{n+1}^{d'} - g_{n+1}^d = (P^{d'})^*(P^{d'} - P^d) g_{n+2}^d + \sum_{k=0}^{\infty} (P^{d'})^k (P^{d'} - P^d) g_{n+1}^d.$$

By (4.106) and the structure of $(P^{d'})^*$, we know that $(P^{d'})^*(P^{d'} - P^d) = 0$. Then, $g_{n+1}^{d'} - g_{n+1}^d = \sum_{k=0}^{\infty} (P^{d'})^k (P^{d'} - P^d) g_{n+1}^d \geq (P^{d'} - P^d) g_{n+1}^d \succeq 0$. □

Lemma 4.6 leads to the following *nth-bias optimality policy iteration algorithm*, $n \geq 2$:

Algorithm 4.8. A Policy Iteration Algorithm for nth-Bias Optimal Policies:

1. Set $k = 0$ and select an arbitrary $(n-1)$th-bias optimal policy $d_0 \in \mathcal{D}_{n-1}$, which may be obtained from policy iteration for the $(n-1)$th-bias optimal policies.
2. (Policy evaluation) Obtain $g_n^{d_k}$ and $g_{n+1}^{d_k}$ by solving
$$-g_{n-1}^* + (P^{d_k} - I)g_n = 0,$$
$$-g_n + (P^{d_k} - I)g_{n+1} = 0,$$
subject to $(P^{d_k})^* g_n = 0$.
3. (Policy improvement) Set $d = d_k$ in (4.104) and (4.105); construct a policy d' accordingly, and set $d_{k+1} = d'$. (Determine D_{n-1}^* by (4.102).)
4. If $d_{k+1} = d_k$, stop and set $\widehat{d} = d_k$ and $g_n^{\widehat{d}} = g_n^{d_k}$ is the optimal nth bias g_n^*; otherwise, increase k by 1 and return to step 2.

There may be more than one d' in (4.104) and (4.105) in step 3. In such cases, we usually choose the action with the largest $\sum_{j \in \mathcal{S}} p^\alpha(j|i) g_n^d(j)$ (when it is larger than $\sum_{j \in \mathcal{S}} p^{d(i)}(j|i) g_n^d(j)$), or with the largest $\sum_{j \in \mathcal{S}} p^\alpha(j|i) g_{n+1}^d(j)$ (when the largest $\sum_{j \in \mathcal{S}} p^\alpha(j|i) g_n^d(j)$ equals $\sum_{j \in \mathcal{S}} p^{d(i)}(j|i) g_n^d(j)$). That is, we choose

$$d'(i) \in \arg\left\{ \max_{\alpha \in \mathcal{A}_{n-1}^*(i)} \left[\sum_{j \in \mathcal{S}} p^\alpha(j|i) g_n^d(j) \right] \right\},$$

where $\mathcal{A}_{n-1}^*(i) := \mathcal{A}_{n-1}(g_0^*, \ldots, g_{n-1}^*)(i)$, if

$$\max_{\alpha \in \mathcal{A}_{n-1}^*(i)} \left[\sum_{j \in \mathcal{S}} p^\alpha(j|i) g_n^d(j) \right] > \sum_{j \in \mathcal{S}} p^{d(i)}(j|i) g_n^d(j);$$

otherwise, set

$$\widetilde{\mathcal{A}}_n(i, d) := \arg\left\{ \max_{\alpha \in \mathcal{A}_{n-1}^*(i)} \left[\sum_{j \in \mathcal{S}} p^\alpha(j|i) g_n^d(j) \right] \right\},$$

and choose

$$d'(i) \in \arg\left\{ \max_{\alpha \in \widetilde{\mathcal{A}}_n(i,d)} \left[\sum_{j \in \mathcal{S}} p^\alpha(j|i) g_{n+1}^d(j) \right] \right\}.$$

Again, Lemma 4.6 can be used to compare the nth biases of two $(n-1)$th-bias optimal policies and to prove the anti-cycling property in the policy iteration procedure. By Lemma 4.6 (a), all d_k, $k = 0, 1, \ldots$, produced by the algorithm are $(n-1)$th-bias optimal. By Lemma 4.6 (b), we have $g_n^{d_{k+1}} \geq g_n^{d_k}$. That is, as k increases, the nth bias $g_n^{d_k}$ either increases or stays the same. Furthermore, by Lemma 4.6 (c), when $g_n^{d_k}$ stays the same, $g_{n+1}^{d_k}$ increases. Thus, any two policies in the sequence of d_k, $k = 0, 1, \ldots$, either have different nth biases g_n, or have different $(n+1)th$ biases g_{n+1}. That is, every policy in the iteration sequence is different. Since the number of policies is finite, the iteration must stop after a finite number of iterations. Suppose that it stops at a policy denoted as $\widehat{d}$. Then $\widehat{d}$ must satisfy the first $(n+2)$ bias optimality equations because, otherwise, for some i the set $\mathcal{A}_n(i, d)$ in (4.104) is non-empty and we can find the next improved policy in the policy iteration. Thus, by the Sufficient Optimality Conditions (4.90)-(4.92), policy $\widehat{d}$ is nth-bias optimal. In summary, *the policy iteration algorithm stops at an nth-bias optimal policy in a finite number of iterations.*

The above discussion also shows, by construction through policy iteration, that the solution to the sufficient bias optimality equations exists.

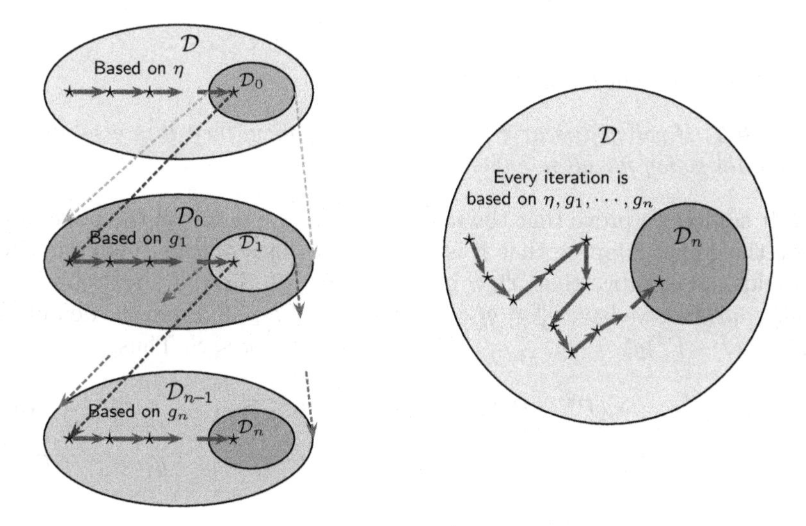

Fig. 4.8. Two Types of Policy Iteration for nth-Bias Optimal Policies

The optimization procedure of the nth bias with the policy iteration algorithms discussed above consists of n phases. Each phase is based on two optimality equations for g_m and g_{m+1} and reaches an optimal mth bias, $m = 0, 1, \ldots, n$. We first find a gain-optimal policy, then determine the set $\mathcal{D}_0^*$, then search the set $\mathcal{D}_0^*$ and find a bias-optimal policy, and then determine the set $\mathcal{D}_1^*$, then search the set $\mathcal{D}_1^*$ and find a second bias-optimal policy, and so on. This procedure is illustrated on the left-hand side of Figure 4.8. In addition to this algorithm, we can also develop another algorithm which works roughly as follows: at each iteration k, we choose an action that maximizes (myopically) all the expected mth biases, $f(i,\alpha)+\sum_{j\in\mathcal{S}}p^{\alpha}(j|i)g_1^{d_k}(j)$ for $m=1$ and $\sum_{j\in\mathcal{S}}p^{\alpha}(j|i)g_m^{d_k}(j)$, for $m=2,\ldots,n$. This is illustrated on the right side of Figure 4.8. We leave the readers to work out the details. Hopefully, this algorithm may take fewer iterations to reach an nth-bias optimal policy, but each iteration involves more computation. (cf. the second policy iteration algorithm for a bias-optimal policy in Section 4.1.2.)

4.3.4 nth-Bias Optimal Policy Spaces*

From the definition of the nth-bias optimal policy space (4.82), we have

$$\mathcal{D}_n \subseteq \mathcal{D}_{n-1}, \qquad n \geq 1.$$

For finite-state and finite-action Markov decision chains, the shrinking of the optimal policy spaces cannot go forever. In fact, we can prove that for a finite state space $\mathcal{S}=\{1,2,\ldots,S\}$, we have

$$\mathcal{D}_S=\mathcal{D}_{S+1}=\mathcal{D}_{S+2}=\cdots.$$

This fact is equivalent to the following lemma.

Lemma 4.7. *If policy d is an Sth-bias optimal policy, then it is also an nth-bias optimal policy for all $n\geq 0$.*

Proof. It suffices to prove that the nth biases g_n, $n\geq 0$, for all the policies in $\mathcal{D}_S$ are the same. Suppose that h is another Sth-bias optimal policy in $\mathcal{D}_S$. We would like to prove that $g_n^h=g_n^d$, for all $n\geq 0$. Since d and h are both Sth-bias optimal, we have $g_k^h=g_k^d=g_k^*$, for all $0\leq k\leq S$. From the definition in (4.78), $(I-P^d)g_k^*=-g_{k-1}^*=(I-P^h)g_k^*$, for $1<k\leq S$. Thus,

$$(P^h-P^d)g_k^*=0, \qquad 1<k\leq S. \tag{4.107}$$

For $k=1$, by the Poisson equation we have $f^h+P^hg_1^*=\eta^*+g_1^*=f^d+P^dg_1^*$. Thus,

$$f^h-f^d+(P^h-P^d)g_1^*=0.$$

For $k=0$, we have $g_0^*=\eta^*$ and $P^dg_0^*=P^d\eta^*=\eta^*=P^hg_0^*$. Thus,

$$(P^h-P^d)g_0^*=0. \tag{4.108}$$

From (4.80), we have $g_k^*=(-1)^{k-1}\left[I-P^d+(P^d)^*\right]^{-k}(f^d-g_0^*)$ for $1\leq k\leq S$. Combining this expression with (4.107), we obtain

$$(P^h-P^d)\left[I-P^d+(P^d)^*\right]^{-k}(f^d-g_0^*)=0, \qquad 1<k\leq S. \tag{4.109}$$

From (4.108) and (4.109), the vectors g_0^* and $\left[I-P^d+(P^d)^*\right]^{-k}(f^d-g_0^*)$, $1<k\leq S$, all lie in the null space of P^h-P^d. Because $P^h\neq P^d$, the rank of

$(P^h - P^d)$ must be at least 1. Hence, the dimension of the null space of $P^h - P^d$ is at most $S - 1$. Thus, the S vectors g_0^* and $\left[I - P^d + (P^d)^*\right]^{-k}(f^d - g_0^*)$, $k = 2, \ldots, S$, are linearly dependent. That is, there is a nonnegative integer $i < S$ such that $\left[I - P^d + (P^d)^*\right]^{-(i+1)}(f^d - g_0^*)$ is a linear combination of g_0^* and $\left[I - P^d + (P^d)^*\right]^{-2}(f^d - g_0^*)$, $\left[I - P^d + (P^d)^*\right]^{-3}(f^d - g_0^*), \ldots,$ $\left[I - P^d + (P^d)^*\right]^{-i}(f^d - g_0^*)$.

We now show by induction that for all $n \geq i+1$, $\left[I - P^d + (P^d)^*\right]^{-n}(f^d - g_0^*)$ is a linear combination of g_0^* and $\left[I - P^d + (P^d)^*\right]^{-2}(f^d - g_0^*)$, $\left[I - P^d + (P^d)^*\right]^{-3}(f^d - g_0^*), \ldots,$ $\left[I - P^d + (P^d)^*\right]^{-i}(f^d - g_0^*)$. Suppose that this statement holds for a particular $n \geq i+1$, i.e.,

$$\left[I - P^d + (P^d)^*\right]^{-n}(f^d - g_0^*) = \kappa_1 g_0^* + \sum_{j=2}^{i} \kappa_j \left[I - P^d + (P^d)^*\right]^{-j}(f^d - g_0^*),$$

where κ_j, $j = 1, 2, \ldots, i$, are real numbers. Pre-multiplying this equation by $\left[I - P^d + (P^d)^*\right]^{-1}$, we have

$$\begin{aligned}
&\left[I - P^d + (P^d)^*\right]^{-(n+1)}(f^d - g_0^*) \\
&= \kappa_1 \left[I - P^d + (P^d)^*\right]^{-1} g_0^* + \sum_{j=2}^{i} \kappa_j \left[I - P^d + (P^d)^*\right]^{-(j+1)}(f^d - g_0^*) \\
&= \kappa_1 \left[I - P^d + (P^d)^*\right]^{-1} (P^d)^* f^d + \sum_{j=2}^{i} \kappa_j \left[I - P^d + (P^d)^*\right]^{-(j+1)}(f^d - g_0^*) \\
&= \kappa_1 g_0^* + \sum_{j=2}^{i} \kappa_j \left[I - P^d + (P^d)^*\right]^{-(j+1)}(f^d - g_0^*).
\end{aligned}$$

Since $\left[I - P^d + (P^d)^*\right]^{-(i+1)}(f^d - g_0^*)$ is a linear combination of g_0^* and $\left[I - P^d + (P^d)^*\right]^{-2}(f^d - g_0^*)$, $\left[I - P^d + (P^d)^*\right]^{-3}(f^d - g_0^*), \ldots,$ $\left[I - P^d + (P^d)^*\right]^{-i}(f^d - g_0^*)$, so is $\left[I - P^d + (P^d)^*\right]^{-(n+1)}(f^d - g_0^*)$. Therefore, $\left[I - P^d + (P^d)^*\right]^{-(n+1)}(f^d - g_0^*)$, for any $n \geq i+1$, lies in the null space of $(P^h - P^d)$, i.e., $(P^h - P^d)g_n^d = 0$ for all $n \geq S$.

Finally, from the nth-bias difference equation in (4.84) and by induction on n, we can prove

$$g_n^h - g_n^d = (P^h)^*(P^h - P^d)g_{n+1}^d + \left[I - P^h + (P^h)^*\right]^{-1}(P^h - P^d)g_n^d = 0,$$

for all $n \geq S$. That is, the nth biases of the policies in $\mathcal{D}_S$ are all the same for all $n \geq 0$. □

Finally, every policy in $\mathcal{D}_S$ is a Blackwell policy (see Problem 4.30). The relationship between nth-biases and the n-discount optimality is discussed in Problem 4.29.

PROBLEMS

4.1. Consider a discrete-time M/M/1 queue. The system state at time $l \geq 0$ is denoted as $X_l = n$, $l = 0, 1, \ldots$, with n being the number of customers in the server. The arrival rate is reflected by the transition probabilities $p(X_{l+1} = n+1|X_l = n) = r$, $0 < r < 1$, $n = 0, 1, \ldots$, and $l = 0, 1, \ldots$. The service rate depends on the number of customers in the server and is reflected by $p(X_{l+1} = n-1|X_l = n) = \mu_n$, $0 < \mu_n < 1 - r$, $n = 1, 2, \ldots$. When the system is in state n and with service rate μ_n, the cost is $\alpha n + \beta \mu_n$, in which αn represents the cost for waiting time, and $\beta \mu_n$ represents the cost for the service. We wish to minimize the average cost by choosing the right service rates μ_n, $n = 1, 2, \ldots$, among all the available choices. Model this problem as a Markov decision process.

4.2. A retailer orders N pieces of a merchandize every evening based on the stock left on that day. The every day's demand on the merchandize can be described by an integer random variable with distribution p_n, $n = 0, 1, \ldots$. The retailer earns c_1 dollars for every piece sold, and s/he suffers a penalty of c_2 dollars for each piece left in every evening. The retailer wishes to make the right order to maximize his/her earnings in a long term. Model the problem as an MDP.

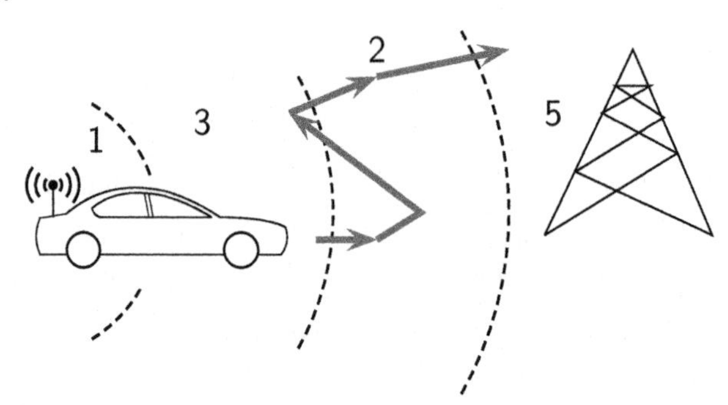

Fig. 4.9. A Wireless Communication System

4.3. A mobile phone user travels through different regions shown in Figure 4.9; each region is characterized into one of the M classes according to the transmission condition in the region. In a region with a "bad" condition, the transmission of signals requires a high power or the bit error rate is high. Therefore, in a "bad" region, the mobile phone user may prefer to delay the transition, by transmitting fewer bits, until s/he reaches a better region. On the other hand, to ensure a reasonable quality of service, the transmission can not be postponed for too long. In a class i region, $i = 1, 2, \ldots, M$, if the mobile phone has n bits in its buffer, the user may choose different level of powers, denoted as $d(i, n)$, $i = 1, 2, \ldots, M$, and $n = 0, 1, \ldots$. Time is discrete and is denoted as $l = 1, 2, \ldots$. When the mobile phone is in a class i region and there are n bits in its buffer, if power $d(i, n)$ is used, then the number of the correctly transmitted bit in the time slot, k, has a distribution $q_k^{d(i,n)}$, $k = 0, 1, \ldots, n$, $\sum_{k=0}^{n} q_k^{d(i,n)} = 1$. When the user is in a class i region in one time slot, s/he will travel to a class j region in the next time slot with probability $p_{i,j}$, $i, j = 1, 2, \ldots, M$. In each time slot, the user generates r bits with a probability of p_r, $\sum_{r=0}^{\infty} p_r = 1$. The cost function is $f(i, n) = \alpha n + \beta_i d(i, n)$, where β_i is the cost per unit of power in a class i region and α represents a weighting factor between the cost of power and the queue length. Model the problem as a discrete MDP.

4.4. Consider a closed network consisting of M single-server stations and N customers. Let n_i be the number of customers in server i, $i = 1, 2, \ldots, M$, and $\boldsymbol{n} := (n_1, n_2, \ldots, n_M)$. The service rate of server i, $i = 1, 2, \ldots, M$, depends on the system state $\boldsymbol{n}$ and is denoted as $\mu_{i,\boldsymbol{n}}$. That is, if at time $t \in [0, \infty)$ the system is in state $\boldsymbol{n}$, then server i completes its service to its customer in $[t, t + \Delta t)$ with probability $\mu_{i,\boldsymbol{n}} \Delta t$. After a customer completes its service at server i, the customer will move to server j with probability $q_{i,j}$, $i, j = 1, 2, \ldots, M$. We may control the service rates $\mu_{i,\boldsymbol{n}}$, $i = 1, 2, \ldots, M$, $\boldsymbol{n} \in \mathcal{S} := \left\{(n_1, \ldots, n_M) : \sum_{k=1}^{M} n_k = N\right\}$, to optimize a properly defined average reward η. We assume that the reward function f is independent of $\mu_{i,\boldsymbol{n}}$.

a. Model the problem as a Markov decision process.
b. Suppose that the service rate of server i, $i = 1, 2, \ldots, M$, depends only on the number of customers in server i, n_i, and is denoted as μ_{i,n_i}, and we may control the load-dependent service rates μ_{i,n_i}, $n_i = 1, 2, \ldots, N$, $i = 1, 2, \ldots, M$, to optimize an average reward. Can we model this problem as a standard MDP? Why?

4.5. Derive the average-reward difference formula for continuous-time ergodic Markov processes with a finite state space and a finite number of actions, and derive a policy iteration algorithm from it.

4.6. Derive the bias difference formula for continuous-time ergodic Markov processes with a finite state space and a finite number of actions, and derive a policy iteration algorithm from it.

4.7. Policy iteration requires the actions in different states be chosen independently. Consider the following optimization problem. The state space consists of $2S$ states denoted as (i, j), $i = 1, 2, \ldots, S$, $j = 1, 2$. The same action has to be taken when the system is in state $(i,1)$ or $(i,2)$ for the same i, $i = 1, 2, \ldots$. Thus, if action α is taken at both $(i,1)$ and $(i,2)$, then the transition probabilities from both state $(i,1)$ and $(i,2)$, $p^\alpha(\bullet|(i,1))$ and $p^\alpha(\bullet|(i,2))$, are determined simultaneously.

a. Explain why the standard policy iteration algorithm does not apply to this problem.
b. Let $\pi(i) := \pi(i,1) + \pi(i,2)$ be the steady-state marginal distribution and $\pi(j|i) = \frac{\pi(i,j)}{\pi(i)}$ be the steady-state conditional probabilities, $i = 1, 2, \ldots, S$, $j = 1, 2$. In this problem, a policy determines an action based on the first component of the state, i. Consider any two policies $h = h(i)$ and $d = d(i)$. We assume that these conditional probabilities are the same for all policies. Thus, $\pi^d(j|i) = \pi^h(j|i)$ for all $i = 1, 2, \ldots, S$ and $j = 1, 2$. Now we have the average-reward difference formula:

$$\eta^h - \eta^d = \sum_{i=1}^{S} \pi^h(i) \left\{ \sum_{j=1}^{2} \pi^d(j|i) \left\{ \left[f^h(i,j) + \sum_{i'=1}^{S} \sum_{j'=1}^{2} p^h[(i',j')|(i,j)] g^d(i',j') \right] - \left[f^d(i,j) + \sum_{i'=1}^{S} \sum_{j'=1}^{2} p^d[(i',j')|(i,j)] g^d(i',j') \right] \right\} \right\}.$$

The $\pi^d(j|i)$ and $g^d(i,j)$ in the big bracket do not depend on P^h. Derive a policy iteration optimization algorithm for the "aggregated" state i.
c. Can we derive a sample-path-based optimization algorithm for the problem in b)?

4.8. Are the following statements true?

a. When the average-reward policy iteration algorithm stops at a policy $\widehat{d}$, the directional performance derivative from $\widehat{d}$ to any other policy in $\mathcal{D}$ is non-positive.
b. If $\widehat{d}$ is a gain-optimal policy, then another policy d is gain optimal, if the directional performance derivative from $\widehat{d}$ to d is zero.
c. If $\widehat{d}$ is a gain-optimal policy, then the directional performance derivative from $\widehat{d}$ to any other gain-optimal policy d is zero.
d. The bias-optimal policy has the largest bias in the policy space $\mathcal{D}$.
e. The difference in the biases of any two ergodic policies is a constant vector (i.e., all its components are equal).

4.9. Let $\widehat{d}$ and d be two ergodic gain-optimal policies in Lemma 4.1. We define a randomized policy d_δ by setting $P^{d_\delta} = P^d + \delta(P^{\widehat{d}} - P^d)$, $f^{d_\delta} = f^d + \delta(f^{\widehat{d}} - f^d)$.

a. Let η^{d_δ} be the average reward of d_δ. Prove $\eta^{d_\delta} = \eta^*$.

b. Derive a directional bias-derivative equation from d to $\widehat{d}$, denoted as $\frac{dg^{d_\delta}}{d\delta}$.
c. When the bias policy iteration algorithm stops at a policy $\widehat{h}$, what are the directional derivatives of the bias from this policy to other policies in $\mathcal{D}_0$?
d. Calculate the bias derivatives between various policies in Example 4.1.

4.10. In Section 4.1.1, we proved that at an optimal policy the average-reward derivatives along the directions to all other policies are non-positive.

a. Suppose $\frac{d\eta^{d_\delta}}{d\delta} > 0$ at a policy d along a direction defined by d_δ: $P^{d_\delta} := P^d + \delta \Delta P$, $f^\delta := f^d + \delta \Delta f$, with $\Delta P = P^h - P^d$, $\Delta f = f^h - f^d$. Can we claim $\eta^h > \eta^d$? If not, give a counter example. If yes, what would this imply in terms of policy iteration?
b. Prove that a policy $d \in \mathcal{D}$ is average-reward optimal if and only if at this policy the performance derivatives along the directions to all other policies are non-positive.

4.11. Suppose that $\widehat{d}$ is the gain-optimal policy with potential $g^{\widehat{d}}$ in Lemma 4.1. Then for any policy $d \in \mathcal{D}_0$, we have $f^d + P^d g^{\widehat{d}} = f^{\widehat{d}} + P^{\widehat{d}} g^{\widehat{d}}$. From this, prove that for any other policy $d' \in \mathcal{D}_0$, we have $f^d + P^d g^{d'} = f^{d'} + P^{d'} g^{d'}$, for all $d \in \mathcal{D}_0$.

4.12. Prove that the second policy iteration algorithm for bias optimality in Section 4.1.2 converges to a bias-optimal policy in a finite number of iterations.

4.13. Calculate the bias-potential w in Example 4.1 for policy d_2 and then find the bias-optimal policy by policy iteration.

4.14. Consider a two-state Markov chain. There are two actions in state 1, corresponding to transition probabilities (0.5, 0.5) and (0.25, 0.75) and rewards 1 and 1.5, respectively; and there are three actions in state 2, corresponding to transition probabilities (0.5, 0.5), (0.25, 0.75), and (0.75, 0.25) and rewards −1, −0.5, and −1.5, respectively. Apply policy iteration to obtain the set of gain-optimal policies and a bias-optimal policy.

4.15. For multi-chains, prove

a. There are more than one solution to $(I - P)u = 0$.
b. The Poisson equation $(I - P)g + \eta = f$ and the normalization condition $P^* g = 0$ uniquely determine the bias of the Markov chain.

4.16. Suppose that d and h are the two policies satisfying Conditions (a) and (b) in the Comparison Lemma (4.41). Prove

a. If in addition to (a) and (b), we have $v(i) = [f^h(i) + (P^h g^d)(i)] - [f^d(i) + (P^d g^d)(i)] > 0$ for some recurrent state i of P^h, then $\eta^h > \eta^d$.
b. If in addition to (a) and (b), we have $P^h \eta^d \neq \eta^d$, then $\eta^h > \eta^d$.

4.17. Find both the gain- and bias-optimal policies using policy iteration for the multi-chain MDP in Example 4.6.

4.18. Consider a Markov chain studied in Problem 2.20 with transition probability matrix

$$P = \begin{bmatrix} B & b \\ 0 & 1 \end{bmatrix},$$

where B is an $(S-1) \times (S-1)$ irreducible matrix, $b > 0$ is an $(S-1)$ dimensional column vector, 0 represents an $(S-1)$ dimensional row vector whose all components are zeros. The last state S is an absorbing state. Set $f(S) = 0$. Clearly, the long-run average reward for this Markov chain is $\eta = 0$. The total reward obtained before reaching the absorbing state, $E\{\sum_{l=0}^{\infty} f(X_l)|X_0 = i\}$, can be viewed as the bias for the problem:

$$g(i) = E\left\{ \sum_{l=0}^{\infty} f(X_l) \middle| X_0 = i \right\}.$$

The Poisson equation for $g = (g(1), \ldots, g(S))^T$ has been derived in Problem 2.20.

a. Derive the bias difference formula for any two policies h and d.
b. Derive a policy iteration algorithm for the bias-optimal policy.

This problem indicates that optimization of the total reward of Markov chains with absorbing states can be solved by the policy iteration for bias-optimal policies.

4.19. For the MDPs with discounted-reward criterion,

a. Prove the discounted-reward difference formulas (4.73) and (4.74).
b. Prove that in (4.77), if $d' \neq d$, then $\eta_\beta^{d'} \succeq \eta_\beta^d$.
c. Prove the convergence of the policy iteration algorithm.

4.20. In (4.53), the bias-potential w is defined as the potential of the bias g satisfying $P^* g = 0$. We can also define a potential of potential by using the potentials g, which is only up to an additive vector u satisfying $(I - P)u = 0$, as follows:

$$(I - P)w - P^* g = -g.$$

a. Prove that the potential of potential defined in this way is the same as the bias-potential defined in (4.53).
b. Define the nth potential by using the $(n-1)$th potential g_{n-1}, and prove that this definition is the same as (4.78).

4.21. Derive a general bias difference formula for $g^h - g^d$, when $\eta^h \neq \eta^d$, for ergodic chains. Discuss whether we can use this equation to derive policy iteration algorithms.

4.22. This problem helps to understand the bias optimality. First, if $\widehat{d}$ and its gain and potential (not necessary bias) $\eta^{\widehat{d}}$ and $g^{\widehat{d}}$ satisfy (4.60) and (4.61), then $\eta^{\widehat{d}} = \eta^*$ is the optimal gain (and $g^{\widehat{d}}$ may not be optimal), and

$$\widehat{\mathcal{A}}_0(i) = \mathcal{A}_0^*(i) := \left\{ \alpha \in \mathcal{A}(i) : \sum_{j \in \mathcal{S}} p^\alpha(j|i)\eta^*(j) = \eta^*(i) \right\},$$

and

$$\widehat{\mathcal{A}}_1(i) := \left\{ \alpha \in \widehat{\mathcal{A}}_0(i) : \eta^*(i) + g^{\widehat{d}}(i) = f(i,\alpha) + \sum_{j \in \mathcal{S}} p^\alpha(j|i) g^{\widehat{d}}(j) \right\}.$$

Now let $d \in \times_{i \in \mathcal{S}} \widehat{\mathcal{A}}_1(i)$. Then by definition we have

$$P^d \eta^* = \eta^*,$$
$$f^d + P^d g^{\widehat{d}} = \eta^* + g^{\widehat{d}}.$$

a. Let g^d be the potential of d. Prove $\eta^d = \eta^*$ and $g^d = g^{\widehat{d}} + u$ with $(I - P^d)u = 0$.
b. Let g^d and $g^{\widehat{d}}$ be the biases of d and $\widehat{d}$, respectively. Prove $g^d - g^{\widehat{d}} = -(P^d)^* g^{\widehat{d}}$.
c. From b), the bias can be improved by optimizing $(P^d)^*(-g^{\widehat{d}})$ (cf. (4.13) for the ergodic case). Can we develop a policy iteration algorithm for bias optimality by using this property? What, if any, are the problems with this approach?

4.23. Prove $(I - P)(I - P + P^*)^{-n}\eta = 0$, and therefore from (4.80) $g_n = (-1)^{-1}(I - P + P^*)^{-1} f$ is a solution to (4.78) with $P^* g_n = (-1)^{n-1}\eta$.

4.24. Derive (4.81) recursively.

4.25. Suppose that a sequence of vectors $g_0^{\widehat{d}}, g_1^{\widehat{d}}, \ldots, g_n^{\widehat{d}}$, and $g_{n+1}^{\widehat{d}}$ satisfies the optimality equations (4.90)-(4.92). Find a policy that has $g_0^{\widehat{d}}, g_1^{\widehat{d}}, \ldots, g_n^{\widehat{d}}$, and $g_{n+1}^{\widehat{d}}$ as its kth biases $k = 0, 1, \ldots, n+1$, respectively.

By the sufficient optimality equations (4.90)-(4.92), $g_k^{\widehat{d}}$ is the optimal kth biases, $k = 0, 1, \ldots, n$, respectively. Therefore, in the sufficient optimality conditions (4.90)-(4.92), we may replace the sentence "A policy $\widehat{d}$ is nth-bias optimal if ..." by "If a sequence of vectors $g_0^{\widehat{d}}, g_1^{\widehat{d}}, \ldots, g_n^{\widehat{d}}$, and $g_{n+1}^{\widehat{d}}$ satisfies (4.90)-(4.92), then $g_k^{\widehat{d}}$ is the optimal kth bias, for $k = 0, 1, \ldots, n$."

4.26. Develop a policy iteration algorithm that "myopically" maximizes the expected mth biases, $m = 1, \ldots, n$, of the actions at each iteration, as illustrated on the right-hand side of Figure 4.8. Prove its convergence.

4.27. A slightly weak version of Lemma 4.7 can be easily established by the well-known Cayley-Hamilton theorem [155]: *For any $n \times n$ matrix A, define its characteristic polynomial as $r(s) = det(sI - A)$. Then we have $r(A) = 0$.* Use the Cayley-Hamilton theorem to prove that if policy d is an $(S+1)$th-bias optimal policy, then it is also an nth-bias optimal policy for all $n \geq 0$. (*Hint: set $A = [I - P^d + (P^d)^*]^{-1}$ in the Cayley-Hamilton theorem.*)

4.28. Let $d, h \in \mathcal{D}$ be two policies.

a. Prove that the following expansion holds for any $N \geq 1$:

$$\eta^h - \eta^d = f^h - f^d + \sum_{k=1}^{N} (P^h - I)^{k-1}(P^h - P^d) g_k^d + (P^h - I)^N (g_N^h - g_N^d).$$

b. Give the conditions under which $(P^h - I)^N (g_N^h - g_N^d)$ converges to zero as $N \to \infty$.
c. What do a) and b) indicate?

4.29. The results presented in this chapter are strongly related to the sensitive discount optimality (n-discount optimality and Blackwell optimality), see [194, 216, 248, 249]. For any Markov chain with transition probability matrix P and reward function f, the discounted reward is defined as (cf. (4.72)):

$$v_\beta(i) := E\Big\{ \sum_{l=0}^{\infty} \beta^l f(X_l) \Big| X_0 = i \Big\}, \qquad 0 < \beta < 1.$$

Denote $v_\beta = (v_\beta(1), \ldots, v_\beta(S))^T$. Set $\beta = (1 + \rho)^{-1}$, or $\rho = (1 - \beta)/\beta$. $0 < \beta < 1$ implies $\rho > 0$. Let ρ_0 be the non-zero eigenvalue of $I - P$ with the smallest absolute value. We have the Laurent series expansion:

$$v_\beta = (1 + \rho) \sum_{n=-1}^{\infty} \rho^n y_n, \qquad 0 < \rho < \rho_0,$$

where $y_{-1} = P^* f$ and $y_n = (-1)^n H_P^{n+1} f$, $n = 0, 1, \ldots$, $H_P = (I - P + P^*)^{-1}(I - P^*)$.

a. Explain the meaning of ρ. (*Hint: inflation rate.*)
b. Prove the Laurent series expansion (cf. Theorem 8.2.3 of [216], but there exists a simpler and direct proof).
c. Prove $y_n = g_{n+1}$ be the $(n+1)$th bias of (P, f), $n = -1, 0, 1, \ldots$. Thus, we have

$$v_\beta = (1 + \rho) \sum_{n=0}^{\infty} \rho^{n-1} g_n, \qquad 0 < \rho < \rho_0.$$

4.30. A policy $d_b \in \mathcal{D}$ is called a (stationary and deterministic) Blackwell policy if there exists a β^*, $0 \leq \beta^* < 1$, such that

$$v_\beta^{d_b} \geq v_\beta^d, \qquad \text{for all } d \in \mathcal{D} \text{ and all } \beta \in [\beta^*, 1).$$

a. Prove that if $d \in \mathcal{D}_S$, with S being the number of states, then d is a Blackwell optimal policy.
b. Prove $d_b \in \mathcal{D}_n$ for all $n \geq 0$.

It is a mistake to try to look too far ahead. The chain of destiny can only be grasped one link at a time.

Sir Winston Churchill
British politician
(1874 - 1965)

5

Sample-Path-Based Policy Iteration

In Chapter 3, we showed that potentials and performance gradients can be estimated with a sample path of a Markov chain, and the estimated potentials and gradients can be used in gradient-based performance optimization of Markov systems. In this chapter, we show that we can use sample-path-based potential estimates in policy iteration to find optimal policies. We focus on the average-reward optimality criterion and ergodic Markov chains. The main idea is as follows. At each iteration k with policy d_k, instead of solving the Poisson equation for potential g^{d_k}, we use its sample-path-based estimate $\bar{g}^{d_k}$ as an approximation in the policy improvement step to determine an improved policy. This leads to sample-path-based policy iteration algorithms.

This approach has several advantages. For example, it does not require solving a large number of linear equations and/or knowing the exact form/value of the transition probability matrix (see Section 5.1). These advantages make the approach practically useful, because for many real engineering systems such as communication networks or manufacturing systems the state spaces are too large and the transition probability matrices may not be entirely known due to unknown parameters and/or to the complexity of the system's structure. However, because the estimates may contain errors, a sample-path-based policy iteration algorithm may not converge, or if it does, it may not converge to an optimal policy. In this chapter, we propose some sample-path-based policy iteration algorithms and provide some conditions that ensure the convergence (either in probability, or with probability 1) of these algorithms to optimal policies.

Similar to the PA-based optimization in Section 6.3.1, there are two ways to implement sample-path-based policy iteration. We may first run the system long enough under one policy at every iteration to get accurate estimates of the potentials and then use them to update the policy, or we may run the system for a short period to get noisy estimates of the potentials, especially at the beginning of the policy iteration, and then gradually improve the estimates as we approach an optimal policy. These topics are discussed in Sections 5.2 and 5.3, respectively.

This chapter complements Chapter 3. Sample-path-based perturbation analysis applies to optimization problems with continuous parameters, while sample-path-based policy iteration applies to optimization problems in discrete policy spaces. This chapter is mainly based on [54], [88], and [97].

5.1 Motivation

We first use a well-designed example to show the advantages of the sample-path-based policy iteration approach.

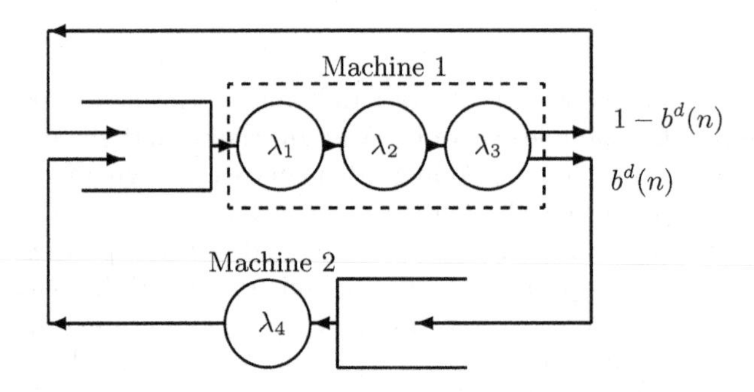

Fig. 5.1. A Two-Machine Manufacturing System

An Illustrative Example

Example 5.1. A manufacturing system consists of two machines and N pieces of works, which are circulating between the two machines, as shown in Figure 5.1. Each work piece has to undertake three consecutive operations at machine 1; thus, machine 1 is illustrated by three circles in the figure, each for one operation. The service times at these three operations are exponentially distributed with rates λ_1, λ_2, and λ_3, respectively. Machine 2 has only one operation with an exponentially distributed service time with rate λ_4. A work piece, after the completion of its service at machine 1, goes to machine 2 with probability $b^d(n)$ and feeds back to machine 1 with probability $1 - b^d(n)$. The superscript "d" represents a policy with $d \in \mathcal{D}$. For any $d \in \mathcal{D}$

and $n = 1, 2, \ldots, N$, $b^d(n) \in [0,1]$. The system can be modelled as a Markov process with its state denoted as (n,i), $0 \le n \le N$, where n is the number of pieces at machine 1 and $i = 1,2,3$ denotes the operation that the piece at machine 1 is undertaking. When $n = 0$, we simply denote the state as 0. To apply the results for discrete-time Markov chains, we study the Markov chain embedded at the transition epochs. We assume that the cost function f does not depend on the actions.

The transition probabilities of the embedded Markov chain are

$$\begin{aligned}
p\left[(n,1),(n+1,1)\right] &= \frac{\lambda_4}{\lambda_1+\lambda_4}, \\
p\left[(n,1),(n,2)\right] &= \frac{\lambda_1}{\lambda_1+\lambda_4}, \\
p\left[(n,2),(n+1,2)\right] &= \frac{\lambda_4}{\lambda_2+\lambda_4}, \\
p\left[(n,2),(n,3)\right] &= \frac{\lambda_2}{\lambda_2+\lambda_4}, \\
p\left[(n,3),(n+1,3)\right] &= \frac{\lambda_4}{\lambda_3+\lambda_4}, \\
p^d\left[(n,3),(n-1,1)\right] &= \frac{\lambda_3}{\lambda_3+\lambda_4} b^d(n), \\
p^d\left[(n,3),(n,1)\right] &= \frac{\lambda_3}{\lambda_3+\lambda_4}\left[1-b^d(n)\right],
\end{aligned}$$

for $0 < n < N$; and

$$\begin{aligned}
p\left[0,(1,1)\right] &= 1, \\
p\left[(N,1),(N,2)\right] &= p\left[(N,2),(N,3)\right] = 1, \\
p^d\left[(N,3),(N,1)\right] &= 1 - b^d(N), \\
p^d\left[(N,3),(N-1,1)\right] &= b^d(N).
\end{aligned}$$

The other transition probabilities are zeros.

We can see that (4.5) and (4.6) in step 3 of the policy iteration algorithm in Chapter 4 can be simplified. The transitions from states $(n,1)$ and $(n,2)$ do not depend on actions. The comparison of actions in the policy improvement step for state $(n,3)$, $0 < n < N$, becomes (recall that the cost function does not depend on actions):

$$\begin{aligned}
&\frac{1}{\lambda_3+\lambda_4}\left\{\lambda_4 g^d(n+1,3) + \lambda_3 b^d(n) g^d(n-1,1) + \lambda_3\left[1-b^d(n)\right] g^d(n,1)\right\} \\
\ge{}& \frac{1}{\lambda_3+\lambda_4}\left\{\lambda_4 g^d(n+1,3) + \lambda_3 b^{d'}(n) g^d(n-1,1) + \lambda_3\left[1-b^{d'}(n)\right] g^d(n,1)\right\},
\end{aligned}$$

for all $d' \in \mathcal{D}$. This is equivalent to

$$\left[b^d(n) - b^{d'}(n)\right] g^d(n-1,1) - \left[b^d(n) - b^{d'}(n)\right] g^d(n,1) \ge 0. \qquad (5.1)$$

The system parameters, λ_1, λ_2, λ_3, and λ_4, do not appear in (5.1). □

In the above example, the service rates govern the evolution of the system, which runs automatically. The control action can affect only some of the system transitions. The transition probabilities corresponding to the uncontrolled transitions (e.g., the transition from $(n, 3)$ to $(n + 1, 3)$) are the same under all policies; they cancel each other in the comparison equation and hence do not appear in the final form. If we can estimate the (g^d)'s on a sample path, then we can implement policy iteration without knowing these transition probabilities, or the corresponding service rates.

Many practical systems have the same features as the above example. Indeed, in many systems, control can be exercised only in a very limited region (e.g., admission control can be applied only at the access points of a high-speed communications network); the remaining parts of such systems simply evolve through their own natures. In Example 5.1, the dashed box in Figure 5.1 can also be viewed as a machine whose service time has an Erlangian distribution. In such cases, the transitions between the different stages are not controllable. This type of service distribution and the more general forms, such as Coxian distributions and phase-type distributions, are very common in practical systems.

The Advantages of the Sample-Path-Based Approach

In summary, Example 5.1 and the above discussion illustrate that the sample-path-based approach has the following advantages.

1. Given a sample path, policy iteration can be implemented *without knowing the whole transition matrix*; only those items related to control actions, $b^d(n)$, have to be known (we do not even need to know the related transition probabilities, e.g., we only need to know $b^d(n)$, not $\frac{\lambda_3}{\lambda_3+\lambda_4}b^d(n)$). In particular, we do not need to estimate all the system parameters λ_i, $i = 1, 2, 3, 4$. Matrix inversion is not required.
2. The approach *saves memory space* required for implementing MDP. In general, only the S potentials, not the $S \times S$ transition matrix, have to be stored. This can be further reduced when there are some states that cannot be reached by "controllable" states, in which actions can be applied. As shown in (5.1), in Example 5.1 only $g^d(n, 1)$, $n = 0, 1, \ldots, N$, have to be estimated and stored; $g^d(n, 2)$ and $g^d(n, 3)$, $n = 0, 1, \ldots, N$, do not even need to be estimated.
3. In the standard computational approach, all the potentials are obtained together through a matrix inversion; thus, obtaining the potential of one state involves the same effort as obtaining the potentials of all the states. In the sample-path-based approach *potentials can be estimated one by one.* This feature makes the policy iteration procedure much more flexible.

a) The computational efforts and memory space of each iteration may be further reduced at the cost of the convergence rate. The idea is, if a state space is too large, at each iteration we may estimate the potentials of only a subset of the state space and update the actions that control the system moving to the states in this subset. For instance, in Example 5.1, we may set $0 = n_0 < n_1 < n_2 < \cdots < n_{k-1} < n_k = N$. Then, in the ith iteration, $i = 1, 2, \ldots, k$, we may estimate $g^d(n, 1)$ only for $n_{i-1} \leq n \leq n_i$, $i = 1, 2, \ldots, k$. Then, by (5.1), we may update the actions in $b^d(n)$ for $n = n_{i-1} + 1, \ldots, n_i$. Of course, it may need more iterations to reach the optimal policy; however, at each iteration, the computation and the memory requirement may be reduced to fit the capacity of the available computing equipment. This feature may be important for on-line optimization using specially designed hardware which may have limited capacity (e.g., in sensor networks). In high speed communications networks, the effect of a slow convergence rate in terms of iterations may be compensated by the fast speed in the system evolution.
b) For many practical systems, we may have some *a priori* knowledge about which states are more important than others. Then, we can estimate only the potentials of the states that are needed for updating the actions on these important states. This may reduce the computation and memory at the cost of the best performance achieved.
c) For large systems for which matrix inversion is not feasible (even if the matrix is completely known), we may simulate the system by using its particular structure (e.g., the queueing structure) and apply the above two methods to reach the optimal solution with more iterations or to obtain a near optimal solution.
d) Distributed optimization may be possible. For example, suppose that we have a communications network consisting of K nodes, which can be modelled as a closed queueing network of K single-server stations, with each server representing one node. Then, the routing decision can be made at each individual node with only the relevant potentials being estimated. This approach depends on state aggregation to further reduce the number of potentials to be estimated (see Chapters 8 and 9 for more discussion). This is an important research direction and more work needs to be done.

The convergence property of sample-path-based policy iteration depends on the errors of the potential estimates, which depend on the length of the sample paths used in the estimation. The remaining sections in this chapter are devoted to the study of the convergence issue.

5.2 Convergence Properties

We can use any algorithm in Section 3.1.2 to estimate potentials. The study in this section is based on (3.19), which expresses the potentials as the average of independent samples, each observed in one regenerative period defined in (3.18). Now let us write it in a slightly different form. First, we choose a reference state $i^* \in \mathcal{S}$. For convenience, we assume that $X_0 = i^*$. Define

$$l_0(i^*) = 0$$
$$l_k(i^*) = \min\{l: \ X_l = i^*, \ l > l_{k-1}(i^*)\}, \qquad k \geq 1.$$

The instants $l_0(i^*), l_1(i^*), \ldots, l_k(i^*), \ldots$, are regenerative points of the Markov chain $\boldsymbol{X} = \{X_l, l = 0, 1, \ldots\}$, and the sample path between $l_k(i^*)$ and $l_{k+1}(i^*)$ is the kth regenerative period. Next, we define $l_k(j) = \min\{l: l > l_k(i^*), X_l = j\}$, $k = 0, 1, \ldots$, and $\chi_k(j) = 1$ if $l_k(j) < l_{k+1}(i^*)$, and $\chi_k(j) = 0$ otherwise. $\chi_k(j)$ indicates whether the system visits state j in the kth regenerative period. The definition is notationally different from but essentially the same as (3.18): The Markov chain may not visit a given state j in a regenerative period.

Consider N regenerative periods. If $\chi_k(j) = 1$, we define

$$V_k(i^*, j) := \sum_{l=l_k(j)}^{l_{k+1}(i^*)-1} [f(X_l) - \bar{\eta}_N],$$

where $\bar{\eta}_N$ is the estimated performance based on N regenerative periods:

$$\bar{\eta}_N := \frac{\sum_{k=0}^{N-1} \left[\sum_{l=l_k(i^*)}^{l_{k+1}(i^*)-1} f(X_l)\right]}{\sum_{k=0}^{N-1} [l_{k+1}(i^*) - l_k(i^*)]} = \frac{1}{l_N(i^*)} \sum_{l=0}^{l_N(i^*)-1} f(X_l). \tag{5.2}$$

$V_k(i^*, j)$ is undefined if $\chi_k(j) = 0$. Let

$$N(j) := \sum_{k=0}^{N-1} \chi_k(j). \tag{5.3}$$

Because of the ergodicity, we have $\lim_{N\to\infty} N(j) = \infty$.

Now, we set $g(i^*) = 0$. Then, the estimated potential of state j, $j \neq i^*$, using N regenerative periods, is

$$\bar{g}_N(j) = \frac{1}{N(j)} \left\{ \sum_{k=0}^{N-1} \chi_k(j) V_k(i^*, j) \right\} \approx \gamma(i^*, j) = g(j), \tag{5.4}$$

if $N(j) > 0$. $\bar{g}_N(j)$ is undefined if $N(j) = 0$.

5.2.1 Convergence of Potential Estimates

By the law of large numbers [26, 28], we have

$$\lim_{N\to\infty} \bar{\eta}_N = \eta, \qquad \text{w.p.1.} \tag{5.5}$$

Lemma 5.1. *As the number of regenerative periods $N \to \infty$, the sample-path-based potential estimate $\bar{g}_N(j)$ in (5.4) converges to its true value $g(j)$ with probability 1.*

Proof. First, let

$$\begin{aligned} \widetilde{V}_k(i^*, j) &:= \sum_{l=l_k(j)}^{l_{k+1}(i^*)-1} [f(X_l) - \eta] \\ &= V_k(i^*, j) - \sum_{l=l_k(j)}^{l_{k+1}(i^*)-1} (\eta - \bar{\eta}_N) \\ &= V_k(i^*, j) - \{l_{k+1}(i^*) - l_k(j)\} (\eta - \bar{\eta}_N). \end{aligned} \tag{5.6}$$

Then, we have

$$\begin{aligned} \bar{g}_N(j) = \frac{1}{N(j)} &\left\{ \sum_{k=0}^{N-1} \chi_k(j) \widetilde{V}_k(i^*, j) \right\} \\ &+ (\eta - \bar{\eta}_N) \left\{ \frac{1}{N(j)} \left\{ \sum_{k=0}^{N-1} \chi_k(j) \left[l_{k+1}(i^*) - l_k(j) \right] \right\} \right\}. \end{aligned}$$

By the law of large numbers [26, 28], we have

$$\lim_{N\to\infty} \frac{1}{N(j)} \left\{ \sum_{k=0}^{N-1} \chi_k(j) \widetilde{V}_k(i^*, j) \right\} = g(j), \qquad \text{w.p.1,} \tag{5.7}$$

and

$$\lim_{N\to\infty} \frac{1}{N(j)} \left\{ \sum_{k=0}^{N-1} \chi_k(j) \left[l_{k+1}(i^*) - l_k(j) \right] \right\} = E\left[l_{k+1}(i^*) - l_k(j) \right], \qquad \text{w.p.1,} \tag{5.8}$$

which is the average first-passage time from state j to state i^*. Because $E\left[l_{k+1}(i^*) - l_k(j)\right] < \infty$ and from (5.5), (5.7), and (5.8), we have

$$\lim_{N\to\infty} \bar{g}_N(j) = g(j), \qquad j \in \mathcal{S}, \quad \text{w.p.1.} \tag{5.9}$$

This completes the proof. □

We note that convergence with probability 1 implies convergence in probability (see Appendix A.1). Thus, as $N \to \infty$, $\bar{g}_N(j)$ in (5.4) also converges to $g(j)$ in probability, i.e., for any $\delta > 0$ and $1 > \epsilon > 0$, there is an integer $N_{\delta,\epsilon}$ such that when $N > N_{\delta,\epsilon}$ we have

$$\mathcal{P}(|\bar{g}_N(j) - g(j)| > \delta) < \epsilon. \tag{5.10}$$

5.2.2 Sample Paths with a Fixed Number of Regenerative Periods

In this subsection, we study the case in which the number of regenerative periods N used in estimating the potentials in each iteration is fixed. We will see that because of the estimation error in $\bar{g}_N^d$, instead of using the maximum value of $f^d + P^d g^{d_k}$ in the policy-improvement step (4.5) in policy iteration Algorithm 4.1, it is more appropriate to use a small region for the expected potentials (cf. $\phi(g)$ and $\psi(g)$ defined in (5.11) and (5.13)).

First, to simplify the notation, for any S-dimensional vector g, we define

$$\phi(g) := \arg\left\{\max_{d\in\mathcal{D}}(f^d + P^d g)\right\} \subseteq \mathcal{D}. \tag{5.11}$$

Precisely, $\phi(g) := \times_{i=1}^S \phi_i(g)$, with

$$\phi_i(g) = \left\{\alpha \in \mathcal{A}(i) : \ f(i,\alpha) + \sum_{j=1}^S p^\alpha(j|i) g(j) \right.$$
$$\left. = \max_{\alpha' \in \mathcal{A}(i)} \left[f(i,\alpha') + \sum_{j=1}^S p^{\alpha'}(j|i) g(j) \right] \right\}.$$

With this notation, the optimality equation for ergodic chains (4.7) becomes:

$$\widehat{d} \in \phi(g^{\widehat{d}}).$$

The set of optimal policies is

$$\mathcal{D}_0 := \left\{ d \in \mathcal{D} \ : \ d \in \phi(g^d) \right\}.$$

In addition, for any S-dimensional vector g and a small positive number $\nu > 0$, we set[1]

$$U_\nu(g) := \left[\max_{d\in\mathcal{D}}(f^d + P^d g) - \nu e, \ \max_{d\in\mathcal{D}}(f^d + P^d g) \right]. \tag{5.12}$$

Similar to (5.11), we define

$$\psi(g) := \left\{ d : f^d + P^d g \in U_\nu(g) \right\} \tag{5.13}$$

as the set of improved policies. Precisely, we have $\psi(g) = \times_{i=1}^S \psi_i(g)$, with

[1] For any two S-dimensional vectors a and b with $a < b$, we use $[a,b]$ to denote an S-dimensional array of intervals $[a,b] := ([a(1),b(1)],[a(2),b(2)],\ldots,[a(S),b(S)])$. An S-dimensional vector $c \in [a,b]$ means that $c(i) \in [a(i),b(i)]$ for all $i = 1,2,\ldots,S$.

$$\psi_i(g) = \Bigg\{ \alpha \in \mathcal{A}(i) : \; f(i,\alpha) + \sum_{j=1}^{S} p^{\alpha}(j|i) g(j) \in$$

$$\left[\max_{\alpha' \in \mathcal{A}(i)} \left\{ f(i,\alpha') + \sum_{j=1}^{S} p^{\alpha'}(j|i) g(j) \right\} - \nu, \; \max_{\alpha' \in \mathcal{A}(i)} \left\{ f(i,\alpha') + \sum_{j=1}^{S} p^{\alpha'}(j|i) g(j) \right\} \right] \Bigg\},$$

where the large square bracket denotes an interval. We certainly have

$$\phi(g) \subseteq \psi(g)$$

for any $\nu > 0$.

The Algorithm

The *sample-path-based policy iteration algorithm with a fixed N* works as follows.

Algorithm 5.1. A Sample-Path-Based Policy Iteration Algorithm With a Fixed N:

1. Choose an integer $N > 0$, a real number $\nu > 0$, and an initial policy d_0; Set $k = 0$.
2. Observe the system under policy d_k to obtain a sample path with N regenerative periods. Estimate the potentials using (5.4). Denote the estimates as $\bar{g}_N^{d_k}$. (Set $\bar{g}_N^{d_k}(j) = \bar{g}_N^{d_{k-1}}(j)$ if $N_k(j) = 0$, where $N_k(j)$ is the $N(j)$ in (5.3) in the kth iteration, with $\bar{g}^{d_{-1}} = 0$).
3. Choose any policy
$$d_{k+1} \in \psi(\bar{g}_N^{d_k}), \qquad (5.14)$$
component-wisely. If at a state i, action $d_k(i)$ is in the set (5.14), then set $d_{k+1}(i) = d_k(i)$.
4. If $d_{k+1} = d_k$, then stop; otherwise, set $k := k+1$ and go to step 2.

There may be multiple policies in the set on the right-hand side of (5.14). If $d_k(i) \in \psi_i(\bar{g}_N^{d_k})$, then we choose $d_{k+1}(i) = d_k(i)$; otherwise, we may choose randomly in $\psi_i(\bar{g}_N^{d_k})$. We will see that if we choose $d_{k+1} \in \phi(\bar{g}_N^{d_k})$ in (5.14), then it will have some problems in setting the stopping criterion in step 4.

The Effect of the Estimation Errors

Because of the errors in estimating potentials, two issues need to be addressed for sample-path-based policy iteration algorithms. The first one is that, at each iteration, the "true" performance may not necessarily improve and the stopping criterion may not be met; thus, we have to study if the algorithm ever stops. The second issue is that, if it does stop, whether it stops at a "true" optimal policy.

The answers to these two questions depend on the following property.

For a set of finite real numbers $\mathcal{C} := \{c_1, c_2, \ldots, c_M\}$, define the distance of c_i and c_j as $\rho_{c_i c_j} \equiv \rho_{ij} := |c_i - c_j|$, $i, j = 1, 2, \ldots, M$ and set $\delta := \min\{\rho_{ij}, c_i \neq c_j, \ i, j = 1, \ldots, M\}$. If we know that two numbers, $x \in \mathcal{C}$ and $y \in \mathcal{C}$, satisfy $\rho_{xy} = |x - y| < \delta$, then they must be the same, i.e., $x = y$.

In an MDP with a finite number of policies, the average reward takes only a finite number of different values. Define

$$\sigma = \frac{1}{2} \min_{d, d' \in \mathcal{D}} \left\{ \left| \eta^d - \eta^{d'} \right| : \ \eta^d \neq \eta^{d'} \right\} \tag{5.15}$$

to be the minimum "distance" between any two policies. We have $\sigma > 0$ (if the average rewards of all policies are not the same). Therefore, if the absolute value of the difference in the average rewards of two policies in $\mathcal{D}$ is less than σ, then either the average rewards of these two policies are the same, or they are simply the same policy. Thus, if the estimation error is small enough, this error can be adjusted and it will not affect the outcome of the policy iteration. This fact is formally stated in Lemma 5.2 below.

At each iteration, let g^d be the true potential vector under the current policy d (we omitted the subscript k in d_k), $\bar{g}^d$ be its estimate, and η^d be the corresponding (true) average reward. Denote the error in the potential estimate as a vector $r := \bar{g}^d - g^d$. Let $h \in \psi(\bar{g}^d)$ be an (improved) policy that reaches the neighborhood of the maximum in (5.12) by using the estimate $\bar{g}^d$ as g, and let π^h and η^h be the (true) steady-state probability and the average reward of h, respectively. The policy h depends on the estimate $\bar{g}^d$.

Lemma 5.2. *We choose $\nu = \sigma/2$ in the sample-path-based policy iteration Algorithm 5.1. Suppose that the Markov chain under every policy is ergodic with a finite state space, and the number of policies is finite. Then, the following holds.*

(a) If the algorithm does not stop at an iteration and $|r| < (\sigma/2)e$,[2] then $\eta^h \geq \eta^d$; i.e., at this iteration, the performance does not decrease.
(b) If the algorithm stops at an iteration and $|r| < (\sigma/2)e$, then it stops at a (true) optimal policy.

[2] For an S-dimensional vector r, we define $|r| = (|r(1)|, |r(2)|, \ldots, |r(S)|)^T$.

Proof. (*a*) From the average-reward difference formula $\eta^h - \eta^d = \pi^h\left[(P^h - P^d)g^d + (f^h - f^d)\right]$, we have

$$\eta^h - \eta^d = \pi^h\left[(P^h - P^d)\bar{g}^d + (f^h - f^d) + (P^d - P^h)r\right]. \tag{5.16}$$

Because the iteration procedure does not stop at this iteration, according to (5.14), we have $h \in \psi(\bar{g}^d)$ and, therefore,

$$(P^h - P^d)\bar{g}^d + (f^h - f^d) + \nu e \geq 0.$$

Thus, $\pi^h\left[(P^h - P^d)\bar{g}^d + (f^h - f^d)\right] \geq -\nu$, then, from (5.16),

$$\eta^h - \eta^d \geq \pi^h(P^d - P^h)r - \nu.$$

However,

$$\begin{aligned}\left|\pi^h(P^d - P^h)r - \nu\right| &\leq \left|\pi^h P^d r\right| + \left|\pi^h P^h r\right| + |\nu| \\ &< \frac{\sigma}{2} + \frac{\sigma}{2} + \frac{\sigma}{2} < \min_{d,d' \in \mathcal{D}}\left\{\left|\eta^d - \eta^{d'}\right| : \eta^d \neq \eta^{d'}\right\}.\end{aligned}$$

Therefore, $\eta^h - \eta^d > -\min_{d,d' \in \mathcal{D}}\left\{\left|\eta^d - \eta^{d'}\right| : \eta^d \neq \eta^{d'}\right\}$. This is only possible if $\eta^h \geq \eta^d$.

(*b*) Suppose that the algorithm stops at an iteration, and the policy at this iteration is denoted as $\widehat{d}$. Let $\bar{g}^{\widehat{d}}$ be the estimate of its potential. Then, from (5.14), for any policy $d \in \mathcal{D}$, we have

$$(P^d - P^{\widehat{d}})\bar{g}^{\widehat{d}} + (f^d - f^{\widehat{d}}) \leq \nu e.$$

Then

$$\begin{aligned}\eta^d - \eta^{\widehat{d}} &= \pi^d\left[(P^d - P^{\widehat{d}})\bar{g}^{\widehat{d}} + (f^d - f^{\widehat{d}}) + (P^{\widehat{d}} - P^d)r\right] \\ &\leq \pi^d(P^{\widehat{d}} - P^d)r + \nu.\end{aligned}$$

Thus, $\eta^d - \eta^{\widehat{d}} \leq (3\sigma)/2$, and hence $\eta^d \leq \eta^{\widehat{d}}$, for all policies $d \in \mathcal{D}$. That is, $\eta^{\widehat{d}}$ is the true optimal average reward. □

The next lemma shows that if the estimation error $|r| = |\bar{g} - g|$ is small enough, the policy iteration using the potential estimate $\bar{g}$ can be viewed as if the true potential g is used. First, we define

$$\begin{aligned}\kappa = \frac{1}{2}\min_{\text{all } d,h,h' \in \mathcal{D}}&\left\{\left|(f^h + P^h g^d)(i) - (f^{h'} + P^{h'} g^d)(i)\right| : \quad \text{all } i \in \mathcal{S},\right. \\ &\left.\text{with } \left[(f^h + P^h g^d)(i) - (f^{h'} + P^{h'} g^d)(i)\right] \neq 0\right\}.\end{aligned}$$

Because there is only a finite number of policies and the state space is finite, we have $\kappa > 0$.

Lemma 5.3. *We choose $\nu = \kappa/2$ in the sample-path-based policy iteration Algorithm 5.1. Suppose that the Markov chain under every policy is ergodic with a finite state space, and the number of policies is finite. If $|r| = |\bar{g}^d - g^d| < (\kappa/2)e$, where g^d and $\bar{g}^d$ are the potential of policy $d \in \mathcal{D}$ and its estimate, then*

$$\psi(\bar{g}^d) \subseteq \phi(g^d).$$

Proof. Let $h \in \phi(g^d)$ and $h' \in \psi(\bar{g}^d)$. By the definition of $\phi(g)$ in (5.11), we have $f^h + P^h g^d \geq f^{h'} + P^{h'} g^d$. By the definition of $\psi(g)$ in (5.13), we have $f^{h'} + P^{h'} \bar{g}^d \geq f^h + P^h \bar{g}^d - \nu e$. From this equation, we have

$$f^{h'} + P^{h'} g^d + (P^{h'} - P^h)(\bar{g}^d - g^d) \geq f^h + P^h g^d - \nu e.$$

Therefore,

$$(f^h + P^h g^d) - (f^{h'} + P^{h'} g^d) \leq (P^{h'} - P^h)(\bar{g}^d - g^d) + \nu e.$$

This, together with $f^h + P^h g^d \geq f^{h'} + P^{h'} g^d$, leads to

$$\left|(f^h + P^h g^d) - (f^{h'} + P^{h'} g^d)\right| \leq \left|(P^{h'} - P^h)(\bar{g}^d - g^d)\right| + \nu e. \tag{5.17}$$

From (5.17), if $|r| = |\bar{g}^d - g^d| < (\kappa/2)e$ and $\nu = \kappa/2$, then $|(f^h + P^h g^d) - (f^{h'} + P^{h'} g^d)| < (2\kappa)e$. By the definition of κ, we must have $f^h + P^h g^d = f^{h'} + P^{h'} g^d$. In other words, $h' \in \phi(g^d)$. Thus, $\psi(\bar{g}^d) \subseteq \phi(g^d)$. □

Note that $\psi(\bar{g}^d)$ may be smaller than $\phi(g^d)$. The implication of this lemma is as follows. Suppose that, at every iteration, the estimation error is $|r| < (\kappa/2)e$. If the sample-path-based algorithm does not stop at an iteration, then the improved policy picked up by using the estimated potentials with (5.14) is one of the policies that may be chosen by the standard policy iteration with the exact potentials. If the sample-path-based iteration stops at a policy $\widehat{d}$, then we have $\widehat{d} \in \psi(\bar{g}^{\widehat{d}})$ and by Lemma 5.3, we have $\widehat{d} \in \phi(g^{\widehat{d}})$; i.e, it will stop if the true potentials are used.

However, because of the random error in the estimates, we do not know if $\widehat{d} \in \psi(\bar{g}^{\widehat{d}})$, although $\widehat{d} \in \phi(g^{\widehat{d}})$; i.e., we do not know if the algorithm will stop even if it reaches an optimal policy. We may determine its probability. Suppose that $d_k = \widehat{d}$ is an optimal policy. Then, according to (5.13), the probability that the algorithm stops at this iteration is

$$p_0 := \mathcal{P}\left\{ f^{\widehat{d}} + P^{\widehat{d}} \bar{g}^{\widehat{d}} \geq \max_{d \in \mathcal{D}} \left[f^d + P^d \bar{g}^{\widehat{d}} \right] - \nu e \right\},$$

where

$$\max_{d \in \mathcal{D}} \left[f^d + P^d \bar{g}^{\widehat{d}} \right]$$
$$= \max_{d \in \mathcal{D}} \left\{ f^d + P^d g^{\widehat{d}} + P^d \left[\bar{g}^{\widehat{d}} - g^{\widehat{d}} \right] \right\}$$

$$\leq \max_{d \in \mathcal{D}} \left\{ f^d + P^d g^{\widehat{d}} \right\} + \max_{d \in \mathcal{D}} \left\{ P^d \left[\bar{g}^{\widehat{d}} - g^{\widehat{d}} \right] \right\}$$
$$= \left[f^{\widehat{d}} + P^{\widehat{d}} g^{\widehat{d}} \right] + \max_{d \in \mathcal{D}} \left\{ P^d \left[\bar{g}^{\widehat{d}} - g^{\widehat{d}} \right] \right\}.$$

Thus,

$$p_0 \geq \mathcal{P} \left\{ P^{\widehat{d}} \left[\bar{g}^{\widehat{d}} - g^{\widehat{d}} \right] \geq \max_{d \in \mathcal{D}} P^d \left[\bar{g}^{\widehat{d}} - g^{\widehat{d}} \right] - \nu e \right\}. \tag{5.18}$$

We wish to find out under what condition this probability is positive. Let $\widehat{r} = \bar{g}^{\widehat{d}} - g^{\widehat{d}}$. Suppose that $|\widehat{r}| < (\nu/2)e$. Then, we have $\max_{d \in \mathcal{D}} \left\{ P^d \left[\bar{g}^{\widehat{d}} - g^{\widehat{d}} \right] \right\} < (\nu/2)e$ and $\max_{d \in \mathcal{D}} \left\{ P^d \left[\bar{g}^{\widehat{d}} - g^{\widehat{d}} \right] - \nu e \right\} < -(\nu/2)e$. On the other hand, we have $\left| P^{\widehat{d}} \left[\bar{g}^{\widehat{d}} - g^{\widehat{d}} \right] \right| < (\nu/2)e$. Thus, we have

$$P^{\widehat{d}} \left[\bar{g}^{\widehat{d}} - g^{\widehat{d}} \right] > -(\nu/2)e > \max_{d \in \mathcal{D}} \left\{ P^d \left[\bar{g}^{\widehat{d}} - g^{\widehat{d}} \right] - \nu e \right\}.$$

Therefore, from (5.18) we have

$$p_0 \geq \mathcal{P} \left[|\widehat{r}| < (\nu/2)e \right]. \tag{5.19}$$

Convergence Property

As shown in (5.9), as the length of a sample path in each iteration $N \to \infty$, the estimate in each iteration converges with probability 1 to the exact value of g. Thus, by Lemmas 5.2 and 5.3, we can show that the sample-path-based policy iteration stops with probability 1 if N is large enough, and it stops at the optimal policy in probability as N goes to infinity.

Theorem 5.1. Convergence Property with Fixed Lengths

We choose $\nu = \min\{\sigma/2, \kappa/2\}$ in the sample-path-based policy iteration Algorithm 5.1. Suppose that the Markov chain under every policy is ergodic with a finite state space, and the number of policies is finite. Then, the following holds.

(*a*) When the length of the sample path N is large enough, the sample-path-based policy iteration (Algorithm 5.1) stops with probability 1.

(*b*) Let η^* be the true optimal average reward and η^*_N be the average reward of the "optimal" policy given by the sample-path-based policy iteration (Algorithm 5.1) with N regenerative periods in each iteration. Then,

$$\lim_{N \to \infty} \mathcal{P}(\eta^*_N = \eta^*) = 1.$$

Proof. (a) can be proved by Lemma 5.3 and (5.19). Because there are only a finite number of states and a finite number of policies, from (5.10), for any $1 > \epsilon > 0$, there is an $N_{\frac{\nu}{2},\epsilon}$ such that if $N > N_{\frac{\nu}{2},\epsilon}$ then

$$\mathcal{P}\left\{\left|\bar{g}_N^d(j) - g^d(j)\right| > \frac{\nu}{2}\right\} < \epsilon$$

holds for all $j \in \mathcal{S}$ and all $d \in \mathcal{D}$. Thus, for this $\nu > 0$, if N is large enough (meaning $N > N_{\frac{\nu}{2},\epsilon}$), we have $|\bar{g}_N^d - g^d| < (\nu/2)e \leq (\kappa/2)e$ with probability $p > 1-\epsilon > 0$ for all $d \in \mathcal{D}$. Therefore, from Lemma 5.3, we have $\psi(\bar{g}_N^d) \subseteq \phi(g^d)$ with probability $p > 0$ for all $d \in \mathcal{D}$.

Therefore, if N is large enough, then at each iteration with probability $p > 0$, the sample-path-based policy iteration Algorithm 5.1 produces a correct and improved policy d_{k+1} in its step 3, which may be chosen by the standard policy iteration algorithm using the true potentials, and the average reward improves if the algorithm does not stop at the iteration.

Suppose that there are K different values of the average rewards η corresponding to all the policies in $\mathcal{D}$. If we have K consecutive iterations, and, in each of them, the sample-path-based algorithm produces a correct policy (i.e.,with a better performance), then the sample-path-based policy iteration process must reach the set of optimal policies $\mathcal{D}_0$. Now, we group every $K+1$ iterations together in the policy iteration sequence: The first group consists of the first $K+1$ iterations, the second group consists of the second $K+1$ iterations, and so on. As discussed above, the probability that the sample-path-based algorithm produces a correct policy in every iteration in the first K iterations in the same group is larger than $p^K > 0$. Thus, the probability that the policy at the $(K+1)$th iteration is an optimal policy, denoted as $\widehat{d}$, is larger than $p^K > 0$. Once the algorithm reaches an optimal policy, we may apply (5.19). That is, under the condition that the algorithm reaches an optimal policy, the probability that the algorithm stops at the $(K+1)$th iteration is $p_0 > 0$. Therefore, the policy iteration algorithm does not stop at any group is less than $q = 1 - p^K p_0 < 1$. Thus, the probability that policy iteration does not stop at the first L groups is less than q^L, which goes to zero as $L \to \infty$. That is, the probability that policy iteration never stops is zero if N is large enough.

For (b), note that η_N^* is a random variable depending on the sample path. We need to prove that for any $\epsilon > 0$, there is an integer $N_\epsilon > 0$ such that if $N > N_\epsilon$ then

$$\mathcal{P}(\eta_N^* \neq \eta^*) < \epsilon. \tag{5.20}$$

Recall that, in Lemma 5.2 and (5.15), we have

$$\sigma = \frac{1}{2}\min_{d,d' \in \mathcal{D}}\left\{\left|\eta^d - \eta^{d'}\right| : \eta^d \neq \eta^{d'}\right\}.$$

Because there is only a finite number of policies, from (5.10), there is an N_ϵ such that if $N > N_\epsilon$ then the probability that the potential-estimation error

$|r| > (\sigma/2)e$ for all policies is less than ϵ. Then, (5.20) follows directly from Lemma 5.2(b). □

Some comments are in order.

1. Because $\phi(g) \subseteq \psi(g)$, we may choose $d_{k+1} \in \phi(g^{d_k})$ (i.e., set $\nu = 0$ in (5.12)) in step 3 of Algorithm 5.1 to replace (5.14). If we do so, the average reward does not decrease at each iteration if the estimation error is small enough. However, we will meet a problem for choosing a stopping criterion: The condition $d_k \in \phi(\bar{g}^{d_k})$ may not hold even if $d_k = \widehat{d}$ is an optimal policy. This can be explained as follows. Suppose that there are two optimal policies $\widehat{d}, d' \in \mathcal{D}_0$. Then, we have $f^{\widehat{d}} + P^{\widehat{d}} g^{\widehat{d}} = f^{d'} + P^{d'} g^{\widehat{d}}$. Because of the error in $\bar{g}^{\widehat{d}}$, it is entirely possible that $f^{d'} + P^{d'} \bar{g}^{\widehat{d}} \succeq f^{\widehat{d}} + P^{\widehat{d}} \bar{g}^{\widehat{d}}$, and thus $\widehat{d} \notin \phi(\bar{g}^{\widehat{d}})$. That means that, if we choose $d_{k+1} \in \phi(\bar{g}^{d_k})$, the algorithm may not stop even if it reaches an optimal policy.
2. Because the probability of the estimation error $r = \bar{g}^d - g^d$ may be widely distributed, it is clear that for any fixed N, the probability that the error of a potential estimate is larger than any $\delta > 0$ is positive. Thus, no matter how large N is, the probability that the fixed-length sample-path-based policy iteration does not stop at the true optimal policy is positive. This means that any algorithm with a fixed N cannot converge to the true optimal with probability 1.
3. If we use a sequence of increasing numbers of regenerative periods, N_1, $N_2, \ldots, N_{k+1} > N_k$, in the iteration, we may face the problem that, at some iterations, the algorithm stops at a false optimal policy because the improved policy is the same as the original one ($i.e.$, $d_k \in \psi(\bar{g}_{N_k}^{d_k})$). This probability may be large at the beginning of the iteration procedure when N_k is small. Therefore, if we use a sequence of increasing integers N_k, we should let the iteration continue even if we obtain the same policy in some iterations, i.e., even if $d_{k+1} = d_k$. In the next subsection, we will prove that under some conditions for the sequence of N_k, the policy iteration, if we let it continue even if $d_{k+1} = d_k$, converges to the true optimal policy either in probability or with probability 1.

5.2.3 Sample Paths with Increasing Lengths

The Algorithm

As discussed at the end of the last subsection, in order to converge with probability 1 to an optimal policy, the policy iteration algorithm with an increasing number of regenerative periods in each iteration should never stop. Because we do not need to set a stopping criterion, we may use $d_{k+1} \in \phi(\bar{g}_{N_k}^{d_k})$ in the policy improvement step.

The algorithm is stated as follows.

Algorithm 5.2 A Sample-Path-Based Policy Iteration Algorithm with Increasing Lengths:

1. Choose a sequence of integers, $N_0, N_1, \ldots$, with $N_{k+1} \geq N_k$, $k = 0, 1, 2, \ldots$, $\lim_{k\to\infty} N_k = \infty$. Set $k = 0$. Choose an initial policy d_0.
2. Observe the system with d_k for N_k regenerative periods. Estimate the potentials using (5.4). Denote the estimates as $\bar{g}_{N_k}^{d_k}$.
3. Choose
$$d_{k+1} \in \phi(\bar{g}_{N_k}^{d_k}) = \arg\left\{\max_{d\in\mathcal{D}}\left[f^d + P^d \bar{g}_{N_k}^{d_k}\right]\right\},$$
component-wisely. (If there is more than one policy in $\phi(\bar{g}_{N_k}^{d_k})$, we may randomly choose one of them.)
4. Set $k := k + 1$; go to step 2.

No stopping criterion is used in the algorithm because it never stops. Thus, in step 3, there is no requirement to set $d_{k+1}(i) = d_k(i)$ whenever possible (as Algorithm 5.1 does). One implication of this change is that after the algorithm reaches an optimal policy, it may oscillate among different optimal policies even if the accurate values of the potentials are used.

The General Conditions for Convergence

The algorithm produces a sequence of policies denoted as $d_0, d_1, \ldots, d_k, \ldots$, and we now study its convergence property. We first study the probability of a wrong decision because of the errors in the potential estimates. We define

$$q(N, d) = \mathcal{P}\left[\phi(\bar{g}_N^d) \subseteq \phi(g^d)\right].$$

This is the probability that the estimated potential will definitely lead to the right choice of the improved policy. Indeed, if at the kth iteration $\phi(\bar{g}_{N_k}^{d_k}) \subseteq \phi(g^{d_k})$, then the improved policy based on the estimated potential is one policy that could be chosen if the true potential were used. We denote it as $d_{k+1} \in \phi(\bar{g}_{N_k}^{d_k}) \subseteq \phi(g^{d_k})$. Thus, we have

$$q(N_k, d_k) \leq \mathcal{P}\left[d_{k+1} \in \phi(g^{d_k}) | d_k\right]. \tag{5.21}$$

We need the following lemma.

Lemma 5.4. Convergence of Products of Infinite Many Numbers: If $\sum_{k=0}^{\infty}(1 - y_k) < \infty$ and $0 \leq y_k \leq 1$ for all k, then $\lim_{n\to\infty} \prod_{k\geq n} y_k = 1$.

Proof. Set $x_k = 1 - y_k$, $k = 0, 1, \ldots$. Then $0 \le x_k \le 1$ and $\sum_{k=0}^{\infty} x_k < \infty$. Because $x_k \ge 0$, then $\sum_{k=0}^{n} x_k$ is nondecreasing, and it must converge to a finite number as $n \to \infty$. Thus,

$$\lim_{n\to\infty} \sum_{k\ge n} x_k = 0. \tag{5.22}$$

Next, for any $0 \le x < 1$, we have the MacLaurin series

$$\ln(1-x) = -x\left(1 + \frac{x}{2} + \frac{x^2}{3} + \cdots\right).$$

If $0 \le x < \frac{1}{2}$, then $1 \le 1 + \frac{x}{2} + \frac{x^2}{3} + \cdots \le 1 + x + x^2 + \cdots < 1 + \frac{1}{2} + \frac{1}{2^2} + \cdots = 2$, and

$$-2x < \ln(1-x) \le -x. \tag{5.23}$$

From (5.22), we can assume that $x_n < \frac{1}{2}$ if n is large enough. Therefore, it follows from (5.23) that if n is large enough, we have

$$-2\sum_{k\ge n} x_k \le \sum_{k\ge n} \{\ln(1-x_k)\} \le -\sum_{k\ge n} x_k.$$

From (5.22), we get $\lim_{n\to\infty} \sum_{k\ge n} \{\ln(1-x_k)\} = 0$. The lemma then follows from $\prod_{k\ge n} y_k = \prod_{k\ge n}(1-x_k) = \exp\left\{\sum_{k\ge n} [\ln(1-x_k)]\right\}$. □

We are now ready to give sufficient conditions for the sample-path-based policy iteration algorithm to reach the set of optimal policies and remain there indefinitely with probability 1 (the proof here follows [88] with some modifications).

Theorem 5.2. Convergence Property with Increasing Lengths

Consider the sample-path-based policy iteration Algorithm 5.2 starting from an initial policy d_0. If the sample paths in different iterations are independently generated and

$$\sum_{k=0}^{\infty} (1 - q_k) < \infty, \tag{5.24}$$

where $q_k := \min_{d\in\mathcal{D}} q(N_k, d)$, then there exists an almost surely finite random integer L such that

$$\mathcal{P}(d_k \in \mathcal{D}_0, \text{for all } k \ge L) = 1.$$

Proof. Denote the underlying probability space as Ω. Any point $\omega \in \Omega$ represents all the sample paths (with policies $d_0, d_1, \ldots$, and lengths $N_0, N_1, \ldots$)

generated in one run of the policy iteration with the initial policy d_0. Every variable or quantity observed in a policy iteration run depends on ω; e.g., we may denote the policy used in its kth iteration as $d_k = d_k(\omega)$. Define

$$L(\omega) = \min\{l : \ d_k \in \mathcal{D}_0 \text{ for all } k \geq l\},$$

provided that the set of integers $\{l : d_k \in \mathcal{D}_0, \text{for all } k \geq l\}$, which depends on ω, is non-null ($\neq \emptyset$). To simplify the notation, we denote

$$\begin{aligned} &\{(l : \ d_k \in \mathcal{D}_0 \text{ for all } k \geq l) \neq \emptyset\} \\ := &\{\omega \in \Omega : (l : \ d_k \in \mathcal{D}_0 \text{ for all } k \geq l) \neq \emptyset\} \subseteq \Omega, \end{aligned}$$

and similar expressions will be used. It suffices to prove

$$\mathcal{P}\{(l : \ d_k \in \mathcal{D}_0 \text{ for all } k \geq l) \neq \emptyset\} = 1,$$

or

$$\mathcal{P}\{\exists\, l : d_k \in \mathcal{D}_0 \text{ for all } k \geq l\} = 1.$$

For any integer $n \geq 0$, define $A_n := \{d_k \in \mathcal{D}_0, \text{ for all } k \geq n\} \subseteq \Omega$. Then, we have $A_n \subseteq A_{n+1}$, $n \geq 0$, and $\{\exists\, l : d_k \in \mathcal{D}_0 \text{ for all } k \geq l\} = \cup_{n \geq 0} A_n$. Hence,

$$\mathcal{P}\{\exists\, l : d_k \in \mathcal{D}_0 \text{ for all } k \geq l\} = \mathcal{P}(\cup_{n \geq 0} A_n) = \lim_{n \to \infty} \mathcal{P}(A_n).$$

Let $K < \infty$ be the number of all policies in $\mathcal{D}$. As proved in Section 4.1.1, in policy iteration with accurate potentials, policies do not repeat in the iteration procedure before it reaches an optimal policy, and once it reaches $\mathcal{D}_0$, it stays there forever. Thus, if $d_{k+1} \in \phi(g^{d_k})$ for consecutive K iterations, then the policy iteration must reach $\mathcal{D}_0$. Therefore, if $d_{k+1} \in \phi(g^{d_k})$ for all $k \geq n - K$, $n \geq K$, then we have $d_k \in \mathcal{D}_0$ for all $k \geq n$. Thus,

$$\left\{d_{k+1} \in \phi(g^{d_k}) \text{ for all } k \geq n - K\right\} \subseteq A_n.$$

Therefore,

$$\mathcal{P}(A_n) \geq \mathcal{P}\left\{d_{k+1} \in \phi(g^{d_k}) \text{ for all } k \geq n - K\right\}.$$

Next, given any sequence of policies $d_0, d_1, \ldots$, the potential estimates at different iterations are independently generated. Note, however, that d_{k+1} depends on d_k, $k = 0, 1, \ldots$. For any d_{n-K}, we have

$$\begin{aligned} &\mathcal{P}\left\{d_{k+1} \in \phi(g^{d_k}) \text{ for all } k \geq n - K | d_{n-K}\right\} \\ = &\Bigg\{ \sum_{d_{n-K+1} \in \phi(g^{d_{n-K}})} \left\{\mathcal{P}\left[d_{k+1} \in \phi(g^{d_k}) \text{ for all } k \geq n - K + 1 | d_{n-K+1}\right]\right. \\ &\left.\mathcal{P}\left[d_{n-K+1} | d_{n-K+1} \in \phi(g^{d_{n-K}})\right]\right\} \Bigg\} \mathcal{P}\left[d_{n-K+1} \in \phi(g^{d_{n-K}}) | d_{n-K}\right], \end{aligned} \quad (5.25)$$

where $\mathcal{P}\left[d_{n-K+1}|d_{n-K+1} \in \phi(g^{d_{n-K}})\right]$ is the conditional probability of d_{n-K+1} given that $d_{n-K+1} \in \phi(g^{d_{n-K}})$. In addition, we have

$$\begin{aligned}
&\mathcal{P}\left\{d_{k+1} \in \phi(g^{d_k}) \text{ for all } k \geq n-K+1|d_{n-K+1}\right\} \\
&= \left\{ \sum_{d_{n-K+2} \in \phi(g^{d_{n-K+1}})} \left\{ \mathcal{P}\left[d_{k+1} \in \phi(g^{d_k}) \text{ for all } k \geq n-K+2|d_{n-K+2}\right] \right. \right. \\
&\qquad \left. \left. \times \mathcal{P}\left[d_{n-K+2}|d_{n-K+2} \in \phi(g^{d_{n-K+1}})\right] \right\} \right\} \\
&\quad \times \mathcal{P}\left\{d_{n-K+2} \in \phi(g^{d_{n-K+1}})|d_{n-K+1}\right\}
\end{aligned}$$

Continuing this process, we obtain

$$\begin{aligned}
&\mathcal{P}\left\{d_{k+1} \in \phi(g^{d_k}) \text{ for all } k \geq n-K|d_{n-K}\right\} \\
= &\left\{ \prod_{k=n-K}^{\infty} \mathcal{P}\left\{d_{k+1} \in \phi(g^{d_k})|d_k\right\} \right\} \\
&\times \left\{ \sum_{d_{k+1} \in \phi(g^{d_k}),\ k \geq n-K} \prod_{k=n-K}^{\infty} \mathcal{P}\left\{d_{k+1}|d_{k+1} \in \phi(g^{d_k})\right\} \right\}. \qquad (5.26)
\end{aligned}$$

From (5.21), we have $\mathcal{P}\left\{d_{k+1} \in \phi(g^{d_k})|d_k\right\} \geq q_k$. Also, we have

$$\sum_{d_{k+1} \in \phi(g^{d_k}),\ k=n-K,\ldots} \left\{ \prod_{k=n-K}^{\infty} \mathcal{P}\left\{d_{k+1}|d_{k+1} \in \phi(g^{d_k})\right\} \right\} = 1.$$

Finally, from (5.25) and (5.26), we get, for any d_{n-K}, that

$$\mathcal{P}\left\{d_{k+1} \in \phi(g^{d_k}) \text{ for all } k \geq n-K|d_{n-K}\right\} \geq \prod_{k \geq n-K} q_k.$$

Thus, with any initial policy d_0, we have

$$\mathcal{P}\left\{d_{k+1} \in \phi(g^{d_k}) \text{ for all } k \geq n-K\right\} \geq \prod_{k \geq n-K} q_k.$$

By (5.24) and Lemma 5.4, we have $\lim_{n\to\infty} \prod_{k \geq n-K} q_k = 1$. Thus, $\lim_{n\to\infty} \mathcal{P}(A_n) = 1$ and the theorem holds. □

Theorem 5.2 means that the sample-path-based policy iteration algorithm converges with probability 1 to the set of optimal policies if condition (5.24) holds. We will see that, to meet this condition, the length of the sample path N_k must increase fast enough. However, we have a weaker result under a weaker condition.

Theorem 5.3. *If the sample paths in different iterations are independently generated, and* $\lim_{k\to\infty} q_k = 1$, *where* $q_k := \min_{d\in\mathcal{D}} q(N_k, d)$, *then*

$$\lim_{n\to\infty} \mathcal{P}(d_n \in \mathcal{D}_0) = 1.$$

Proof. From the proof of Theorem 5.2, if $d_{k+1} \in \phi(d_k)$ for $k = n-K, n-K+1, \ldots, n-1$, then $d_n \in \mathcal{D}_0$. Thus, we have

$$\mathcal{P}(d_n \in \mathcal{D}_0) \geq \prod_{k=n-K}^{n-1} q_k.$$

The theorem follows directly from $\lim_{k\to\infty} q_k = 1$. □

More Specific Conditions

The conditions in Theorems 5.2 and 5.3 are not very easy to verify directly, so we need some further work. First, we observe that, as discussed in the last subsection, because the policy space is finite, small errors in potential estimates can be corrected. This leads to the following lemma. To simplify the notation, for any S-dimensional vector v, we define $||v|| = \max_{i\in\mathcal{S}} |v(i)|$.

Lemma 5.5. *There exists a* $\delta > 0$ *such that if*

$$\sum_{k=0}^{\infty} \max_{d\in\mathcal{D}} \mathcal{P}(||\bar{g}^d_{N_k} - g^d|| > \delta) < \infty,$$

then condition (5.24) holds.

Proof. By Lemma 5.3 and $\phi(\bar{g}^d_N) \subseteq \psi(\bar{g}^d_N)$, for any policy d, there is a $\delta^d > 0$, such that if $||\bar{g}^d_N - g^d|| \leq \delta^d$, then $\phi(\bar{g}^d_N) \subseteq \phi(g^d)$. Set $\delta := \min_{d\in\mathcal{D}} \delta^d > 0$.

$$\begin{aligned} q(N,d) &= \mathcal{P}\left[\phi(\bar{g}^d_N) \subseteq \phi(g^d)\right] \\ &\geq \mathcal{P}(||\bar{g}^d_N - g^d|| \leq \delta^d) \geq \mathcal{P}(||\bar{g}^d_N - g^d|| \leq \delta). \end{aligned}$$

Thus, $1 - q(N,d) \leq 1 - \mathcal{P}(||\bar{g}^d_N - g^d|| \leq \delta) = \mathcal{P}(||\bar{g}^d_N - g^d|| > \delta)$, and

$$\begin{aligned} 1 - \min_{d\in\mathcal{D}} q(N,d) &= \max_{d\in\mathcal{D}} \{1 - q(N,d)\} \\ &\leq \max_{d\in\mathcal{D}} \mathcal{P}(||\bar{g}^d_N - g^d|| > \delta). \end{aligned}$$

From this, we have

$$1 - q_k \leq \max_{d\in\mathcal{D}} \mathcal{P}(||\bar{g}^d_{N_k} - g^d|| > \delta).$$

Condition (5.24) now follows directly. □

Note that we not only proved the lemma, but also found the δ required in the lemma. The next lemma follows immediately.

Lemma 5.6. *Suppose that $\bar{g}_{N_k}$ is an unbiased estimate of g. If*

$$E\left[\bar{g}^d_{N_k}(j) - g^d(j)\right]^2 \le \frac{c^d}{N_k}, \qquad \text{for all } d \in \mathcal{D} \text{ and } j \in \mathcal{S},$$

$c^d > 0$, and

$$\sum_{k=1}^{\infty} \frac{1}{N_k} < \infty,$$

then condition (5.24) holds.

Proof. By Chebychev's inequality, for any $\delta > 0$, we have

$$\begin{aligned} &\mathcal{P}(\|\bar{g}^d_{N_k} - g^d\| > \delta) = \mathcal{P}\left[\cup_{j\in\mathcal{S}}\left\{|\bar{g}^d_{N_k}(j) - g^d(j)| > \delta\right\}\right] \\ &\le \sum_{j\in\mathcal{S}} \mathcal{P}(|\bar{g}^d_{N_k}(j) - g^d(j)| > \delta) \le \sum_{j\in\mathcal{S}} \frac{E\left[\bar{g}^d_{N_k}(j) - g^d(j)\right]^2}{\delta^2} \\ &\le \frac{c^d S}{N_k \delta^2}. \end{aligned}$$

Since $\mathcal{D}$ is finite, we may set $c = \max_{d\in\mathcal{D}} c^d < \infty$. Therefore, for any $\delta > 0$ we have

$$\max_{d\in\mathcal{D}} \mathcal{P}(\|\bar{g}^d_{N_k} - g^d\| > \delta) < \frac{cS}{N_k \delta^2}.$$

Now, let us choose δ as the one that satisfies Lemma 5.5. Then, condition (5.24) holds. □

Note that the conditions in this lemma can be changed to $E\left[\bar{g}^d_{N_k}(j) - g^d(j)\right]^2 \le c^d \kappa(N_k)$ for all $j \in \mathcal{S}$ and $d \in \mathcal{D}$, $c^d > 0$, where $\kappa(N)$ is a non-negative function of N, and $\sum_{k=1}^{\infty} \kappa(N_k) < \infty$.

Convergence of the Algorithm with Estimate (5.4)

We now study the policy iteration algorithms that are based on a particular estimate (5.4). We first note that for any finite N, $\bar{g}_N$ in (5.4) is biased because $E\left[\bar{\eta}_N\right] \ne \eta$. To get some insight, we first simplify the problem by using the unbiased potential estimate

$$\widetilde{g}_N(j) = \frac{1}{N}\left\{\sum_{k=1}^{N} \widetilde{V}_k(i^*, j)\right\}, \tag{5.27}$$

with $\widetilde{V}_k(i^*, j)$ defined in (5.6). To simplify the discussion, we assume that $\chi_k(j) = 1$ for every regenerative period. We want to apply Lemma 5.6. The first condition can be easily verified as follows. Because all $\widetilde{V}_k$, $k = 0, 1, \ldots, N$, are independent and $E(\widetilde{V}_k) = g^d$, we have, for any policy d,

$$E\left[\widetilde{g}_N^d(j) - g^d(j)\right]^2 = E\left[\frac{1}{N}\sum_{k=1}^{N}\widetilde{V}_k^d(i^*,j) - g^d(j)\right]^2$$

$$= \frac{1}{N^2}E\left\{\sum_{k=1}^{N}\left[\widetilde{V}_k^d(i^*,j) - g^d(j)\right]\right\}^2 = \frac{1}{N}\left\{E\left[\widetilde{V}_k^d(i^*,j)\right]^2 - \left[g^d(j)\right]^2\right\}. \quad (5.28)$$

Next, because

$$\left|\widetilde{V}_k^d(i^*,j)\right| \le \max_{i\in\mathcal{S}}\left|f(i,d(i)) - \eta^d\right|\left[l_k^d(i^*) - l_{k-1}^d(i^*)\right]$$

and

$$E\left[l_k^d(i^*) - l_{k-1}^d(i^*)\right]^2 < \infty$$

for finite ergodic chains, we have

$$E\left[\widetilde{V}_k^d(i^*,j)\right]^2 < \infty.$$

Thus, the first condition in Lemma 5.6 holds. Therefore, from Theorem 5.2 and Lemma 5.6, the sample-path-based policy iteration Algorithm 5.2 with potential estimate (5.27) converges with probability 1 to the optimal policy if $\sum_{k=1}^{\infty}\frac{1}{N_k} < \infty$.

Next, we consider the biased estimate (5.4). Lemma 5.6 cannot be applied, and we need to use Lemma 5.5. First, we study the bias of the potential estimate. Set $\Delta_N(j) := \left|E\left[\bar{g}_N^d(j)\right] - g^d(j)\right|$ and $\Delta_N := \max_{j\in\mathcal{S}}\Delta_N(j)$. Because the regenerative periods are independent, we have (from (5.27)):

$$E\left[\bar{g}_N^d(j)\right] = \frac{1}{N}\sum_{k=1}^{N}E\left\{\sum_{l=l_k^d(j)}^{l_{k+1}^d(i^*)-1}\left[f(X_l,d(X_l)) - \bar{\eta}_N^d\right]\right\}$$

$$= \frac{1}{N}\sum_{k=1}^{N}\left\{E\left[\sum_{l=l_k^d(j)}^{l_{k+1}^d(i^*)-1}\left[f(X_l,d(X_l)) - \eta^d\right]\right]\right.$$

$$\left. + E\left[\sum_{l=l_k^d(j)}^{l_{k+1}^d(i^*)-1}\left[\eta^d - \bar{\eta}_N^d\right]\right]\right\}$$

$$= g^d(j) + \frac{1}{N}\sum_{k=1}^{N}E\left\{\sum_{l=l_k^d(j)}^{l_{k+1}^d(i^*)-1}\left[\eta^d - \bar{\eta}_N^d\right]\right\}.$$

Thus,

$$\Delta_N(j) = \frac{1}{N}\sum_{k=1}^{N}E\left\{\sum_{l=l_k^d(j)}^{l_{k+1}^d(i^*)-1}\left[\eta^d - \bar{\eta}_N^d\right]\right\} = E\left\{\sum_{l=l_k^d(j)}^{l_{k+1}^d(i^*)-1}\left[\eta^d - \bar{\eta}_N^d\right]\right\},$$

for any fixed k. Because there are a finite number of states and actions, we have $|f(i,d(i))| < R < \infty$ for some $R > 0$, all $i \in \mathcal{S}$, and all $d \in \mathcal{D}$. Therefore, from (5.2), we have $\bar{\eta}_N^d < R$. Thus,

$$\left| \sum_{l=l_k^d(j)}^{l_{k+1}^d(i^*)-1} \left[\eta^d - \bar{\eta}_N^d\right] \right| < (\eta^d + R)\left[l_{k+1}^d(i^*) - l_k^d(i^*)\right],$$

with $E\left[l_{k+1}^d(i^*) - l_k^d(i^*)\right] < \infty$. Thus, by applying the Lebesgue dominated convergence theorem [28], we have

$$\lim_{N\to\infty} \Delta_N(j) = E\left\{ \lim_{N\to\infty} \sum_{l=l_k^d(j)}^{l_{k+1}^d(i^*)-1} \left[\eta^d - \bar{\eta}_N^d\right] \right\} = 0,$$

and $\lim_{N\to\infty} \Delta_N = 0$. Therefore, for the $\delta > 0$ specified in Lemma 5.5, there is an integer $N_0 > 0$ such that $0 < \Delta_N < \delta/2$ for all $N > N_0$.

Now, assume that $N > N_0$. We have

$$\begin{aligned} &\left|\bar{g}_N^d(j) - g^d(j)\right| \\ &= \left|\bar{g}_N^d(j) - E\left[\bar{g}_N^d(j)\right] + E\left[\bar{g}_N^d(j)\right] - g^d(j)\right| \le \left|\bar{g}_N^d(j) - E\left[\bar{g}_N^d(j)\right]\right| + \Delta_N. \end{aligned}$$

Therefore, if $\left|\bar{g}_N^d(j) - g^d(j)\right| > \delta$, $\delta > 0$, then $\left|\bar{g}_N^d(j) - E\left[\bar{g}_N^d(j)\right]\right| > \delta - \Delta_N > \delta/2$. Thus,

$$\mathcal{P}\left\{\left|\bar{g}_N^d(j) - g^d(j)\right| > \delta\right\} \le \mathcal{P}\left\{\left|\bar{g}_N^d(j) - E\left[\bar{g}_N^d(j)\right]\right| > \delta/2\right\}.$$

Then, by Chebychev's inequality, we get

$$\mathcal{P}\left\{\left|\bar{g}_N^d(j) - E\left[\bar{g}_N^d(j)\right]\right| > \delta/2\right\} \le \frac{E\left\{\bar{g}_N^d(j) - E\left[\bar{g}_N^d(j)\right]\right\}^2}{(\delta/2)^2}.$$

Similar to (5.28), we have

$$E\left\{\bar{g}_N^d(j) - E\left[\bar{g}_N^d(j)\right]\right\}^2 = \frac{1}{N}\left\{E\left[V_k^d(i^*,j)\right]^2 - \left[E(\bar{g}_N^d(j))\right]^2\right\},$$

where $V_k^d(i^*,j) = \sum_{l=l_k^d(j)}^{l_k^d(i^*)-1}\left[f(X_l,d(X_l)) - \bar{\eta}_N^d\right]$. It is easy to verify that $E\left[V_k^d(i^*,j)\right]^2 < \infty$. From the above three equations, we can obtain

$$\mathcal{P}\left\{\left|\bar{g}_N^d(j) - g^d(j)\right| > \delta\right\} \le \frac{4c^d(j)}{N\delta^2},$$

for some $c^d(j) > 0$, and

$$\mathcal{P}\left\{||\bar{g}_N^d - g^d|| > \delta\right\} \le \frac{4c^d}{N\delta^2},$$

for some $c^d > 0$. Therefore,

$$\begin{aligned}
&\sum_{k=0}^{\infty} \max_{d \in \mathcal{D}} \mathcal{P}\left\{||\bar{g}_{N_k}^d - g^d|| > \delta\right\} \\
&= \sum_{k=0}^{N_0} \max_{d \in \mathcal{D}} \mathcal{P}\left\{||\bar{g}_{N_k}^d - g^d|| > \delta\right\} + \sum_{k=N_0+1}^{\infty} \max_{d \in \mathcal{D}} \mathcal{P}\left\{||\bar{g}_{N_k}^d - g^d|| > \delta\right\} \\
&\leq \sum_{k=0}^{N_0} \max_{d \in \mathcal{D}} \mathcal{P}\left\{||\bar{g}_{N_k}^d - g^d|| > \delta\right\} + \frac{4 \max_{d \in \mathcal{D}} c^d}{\delta^2} \sum_{k=N_0+1}^{\infty} \frac{1}{N_k},
\end{aligned}$$

in which the first term is finite. Thus, Lemma 5.5 holds if $\sum_{k=1}^{\infty} \frac{1}{N_k} < \infty$.

In the above analysis, we have assumed that every regenerative period visits state j. This may not be true for all $j \in \mathcal{S}$, especially for those states that are not visited often at steady state. To make sure that we can apply Lemmas 5.5 or 5.6, we may need to extend the length of the kth iteration N_k to a larger number N_k' such that, in the iteration, the number of regenerative periods that visit state j is larger than the N_k required by the algorithms, i.e., $N_k(j) = \sum_{l=1}^{N_k'} \chi_k(j) \geq N_k$, for all $j \in \mathcal{S}$. N_k' may be too large if some states are rarely visited. However, such states are usually not so "important", and, furthermore, the results in Lemmas 5.5 or 5.6 for true optimal policies may be a bit conservative. Further research in this direction is needed.

The results show that for the sample-path-based policy iteration to converge to the optimal policy with probability 1, the lengths of the sample paths in the iterations have to increase fast enough. In addition, in the algorithms with increasing lengths, it is difficult to determine the stopping criteria. At any iteration, it is always possible to have an estimate with a large error that leads to a wrong policy. We cannot be absolutely sure if the obtained policy is optimal even if the iteration stays at the same policy for a few (any finite number of) iterations. On the other hand, if the length is long enough, we may guarantee that the probability of the iteration stopping at a wrong policy is less that any given small positive number, by using the stopping criterion $d_{k+1} = d_k$.

The algorithm updates the policy (or the actions for all states) at the end of each iteration. Therefore, the required computation may be overwhelming at the end of every iteration. This may require a powerful machine for real-time applications, and the computation power may be wasted in the middle of every iteration. To overcome this disadvantage, we may determine the action for a state only when this state is visited during the iteration. More specifically, we may implement step 2 in Algorithms 5.1 and 5.2 at the end of each iteration and implement step 3 in these two algorithms for state i when this state is visited in the next iteration period. In this way, the computation is distributed to all the state transition instants. See [97] for more discussion.

Figure 5.2 illustrates the difference between the fixed-length and increasing-length algorithms. In the fixed-length algorithm, for any fixed length N_k, the

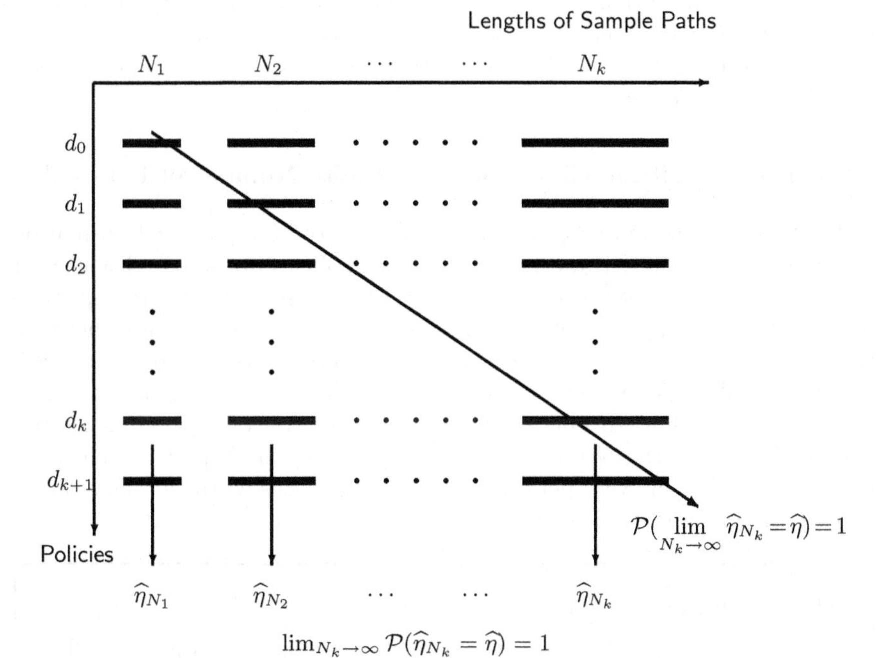

Fig. 5.2. Comparison of the Fixed- and Increasing-Length Policy Iteration Algorithms

algorithm stops at a near-optimal performance $\widehat{\eta}_{N_k}$, which converges to the optimal performance $\widehat{\eta}$ in probability as the length of the regenerative period N_k goes to infinity. In the increasing-length algorithm, the policy iteration goes in the diagonal direction in the figure and converges to the set of optimal policies with probability 1. However, it is difficulty to design stopping criteria for the increasing-length policy iteration algorithm.

Most results in this subsection appeared in [88].

5.3 "Fast" Algorithms*

In the algorithms presented in the above two sections, the potentials are estimated and policies are updated every iteration consisting of N regenerative periods, with N being a relatively large integer. In these algorithms, the potentials are estimated separately in each iteration. The estimates are relatively accurate with large N's. In this section, we explore the possibility of updating the potential estimates as well as the policies in every regenerative period, or after a few regenerative periods, in policy-iteration based performance optimization. The length of a regenerative period is not long enough for applying the algorithms in Sections 5.2.2 and 5.2.3, and therefore the information in the previous regenerative periods need to be used together with that in the current regenerative period to obtain an estimate, and stochastic approximation techniques may be employed.

5.3.1 The Algorithm That Stops in a Finite Number of Periods*

In the "fast" algorithm proposed in this subsection, the potential estimation is also based on (5.4). However, because the policies are updated whenever the system visits a reference state i^* in the algorithm, the different periods between the consecutive visits to i^* may be under different policies; therefore they are not identically distributed and hence are no longer "regenerative". We simply call them "periods". The kth period is denoted as Y_k, $k = 1, 2, \ldots$.

In the algorithm, to obtain an accurate estimate of the potentials, we start with running the system under an initial policy for N periods. Then, we update the policy in every period. The algorithm stops when the same policy is used for N consecutive periods.

Algorithm 5.3 Updating Policies in Every Period:

1. Choose an integer N; set $c := 0$ and $k := 0$; choose an initial policy d_0.
2. Observe the system under policy d_0 for N periods, and get an estimate $\bar{g}^{d_0}$ by applying (5.4) to these N periods.
3. Determine the next policy d_{k+1} by applying (5.14) with $\bar{g}^{d_k}$ as the estimated potentials.
4. If $d_{k+1} = d_k$, set $c := c + 1$; otherwise, set $c := 0$. If $c = N$, then exit; otherwise, go to the next step.
5. Change the policy to d_{k+1}, set $k := k + 1$, observe the system for one period with policy d_k, and update $\bar{g}^{d_k}$ by applying (5.4) to the latest N consecutive periods. Go to step 3.

In the Markov chain generated by the above algorithm, the initial policy d_0 is used in the first N periods, $Y_1, \ldots, Y_N$. $\bar{g}^{d_0}$ is estimated using these N periods and then d_1 is determined by $\bar{g}^{d_0}$. Policy d_1 is then used in the $(N+1)$th period, Y_{N+1}. In general, d_k and its corresponding transition matrix P^{d_k}, which is used in the $(N+k)$th period Y_{N+k}, are determined by $\bar{g}^{d_{k-1}}$, which is estimated using the kth period, Y_k, to the $(N+k-1)$th period, Y_{N+k-1}, $k = 1, 2, \ldots$. This is illustrated in Figure 5.3.

Strictly speaking, in this algorithm, $\bar{g}^{d_k}$, $k \geq 1$, may not be the potential vector corresponding to policy d_k, which is only used in the last period. With this in mind, we will keep the same notation with superscript d_k to denote the estimated potential, since no confusion will be caused.

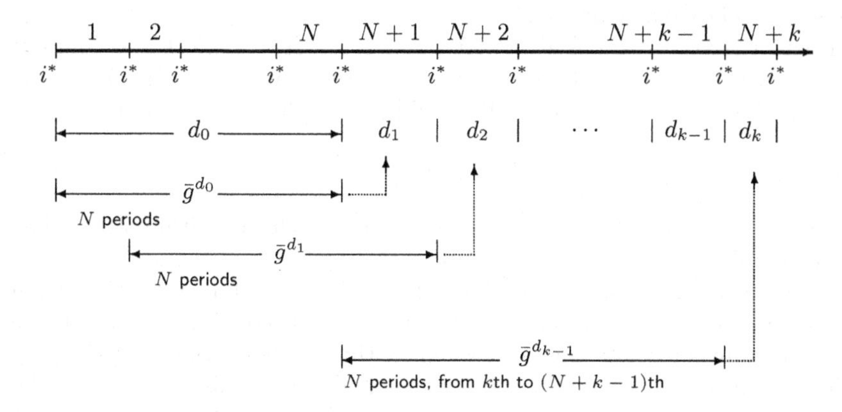

Fig. 5.3. The "Fast" Algorithm 5.3

The rationale behind the algorithm is as follows. If d_{k+1} is "close" to d_k, then the previous data under d_k can be used in obtaining $\bar{g}^{d_{k+1}}$. If d_{k+1} is not "close" to d_k, then the data collected in one period under d_{k+1} would not make a big impact on $\bar{g}^{d_{k+1}}$, which is estimated on N periods, i.e., we may have $\bar{g}^{d_{k+1}} \approx \bar{g}^{d_k}$. Therefore, most likely we would have $d_{k+2} = d_{k+1}$. Thus, the potential estimates would be more accurate for $\bar{g}^{d_{k+2}}$ in the next period, since two periods under the same policy ($d_{k+1} = d_{k+2}$) have been used. The policy gets updated when enough data under this policy $d_{k+1}(= d_{k+2})$ is collected. This also roughly explains that the algorithm might be "fast" because it wastes no periods to collect data that are more than needed to update the policies.

The Policy Reached When the Algorithm Stops

Lemma 5.7. *Suppose that Algorithm 5.3 stops at $\widehat{d}_N$; let η_N^* be the corresponding average reward. For any $\epsilon > 0$, there is an integer $N_\epsilon > 0$ such that if $N > N_\epsilon$, then $\mathcal{P}(\eta_N^* \neq \eta^*) < \epsilon$, where η^* is the true optimal average reward.*

Proof. When the algorithm stops at the last period, denoted as Y_{K+N}, the policies used in Y_K to Y_{K+N} are the same, i.e., $d_K = d_{K+1} = \cdots = d_{K+N} :=$ d. The potential estimated from the N periods Y_K to Y_{K+N-1}, $\bar{g}_N^d$, are based on the same policy d. By Algorithm 5.3, $\bar{g}_N^d$ leads to the same improved policy $d \in \psi(\bar{g}_N^d)$, which is used in Y_{K+N}. The theorem then follows directly from Theorem 5.1(b). □

The lemma claims that if the algorithm stops, then it stops at the true optimal policy with a large probability, if N is large enough. However, it does not indicate whether the algorithm will stop.

Does the Algorithm Stops?

Define the kth period as $Y_k := \{X_{l_k(i^*)+1}, \ldots, \ X_{l_{k+1}(i^*)}\}, k > 0$, and put N consecutive periods together as an augmented state $Z_k := (Y_k, Y_{k+1}, \ldots, Y_{N+k-1})$. Let $\mathcal{Z}$ be the space of all possible Z_k's. Then, we can write

$$d_k = \varphi(Z_k), \tag{5.29}$$

where φ is a mapping from $\mathcal{Z}$ to the policy space. The algorithm stops when the same policy is used for N consecutive basic periods.

From (5.29), the augmented chain $\boldsymbol{Z} = \{Z_1, Z_2, \ldots\}$ is a Markov chain defined on state space $\mathcal{Z}$. However, it may not be irreducible. In fact, if the algorithm converges to a policy (e.g., an optimal policy), then as time goes to infinity, $\boldsymbol{Z}$ tends to stay in the states generated by this policy (e.g., the optimal policy) and the other states may not be reached. Therefore, some conditions on $\boldsymbol{Z}$ may be required for the algorithm to stop. Let us study the issue formally. First, we have a lemma.

Lemma 5.8. *If $Z_{K+N} = Z_K$, then Algorithm 5.3 stops at the end of Z_{K+N}.*

Proof. $Z_{K+N} = Z_K$ means that $Y_{K+l} = Y_{K+N+l}$, $l = 0, \ldots, N-1$ (see Figure 5.4). Let the policies used in the $(K+N+l)$th period be d_{K+l}, $l = 0, \ldots, N-1$. Note that d_{K+1} depends on $Z_{K+1} = (Y_{K+1}, Y_{K+2}, \ldots, Y_{K+N-1}, Y_{K+N})$, which is the same as $Z_K = (Y_K, Y_{K+1}, \ldots, Y_{K+N-1})$ (because $Y_K = Y_{K+N}$), regardless of the order. Therefore, $d_K = \varphi(Z_K) = d_{K+1}$. In the same way, we can prove $d_{K+N} = d_{K+N-1} = \cdots = d_{K+N-2} = \cdots = d_K$. That is, the $N+1$ consecutive policies are the same. Thus, $c = N$ in the algorithm and hence it stops. □

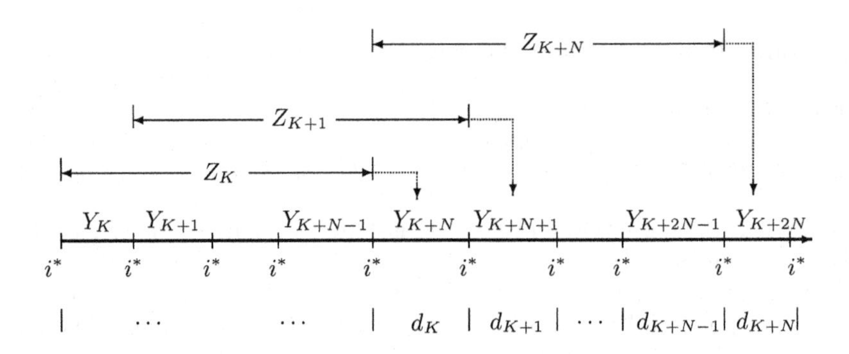

Fig. 5.4. The Periods in Lemma 5.8

Under many conditions, we may find $Z_{K+N} = Z_K$ on a sample path. These conditions require that the transition probability matrices used in different periods have some similarity. For instance, if the transition probability matrix used in the Kth period is completely different from that in the $(K+N)$th period, then a Y_K that is the same as Y_{K+N} may not exist.

To study the structure of a transition probability matrix P, we define a graph G consisting of S nodes. In the graph, two nodes i and j, $i, j \in \mathcal{S}$, are connected by an arrow from i to j if and only if $p(j|i) > 0$. A *loop* in G is a sequence of arrows starting from one node and ending at the same node. Let G^d be the graph corresponding to the transition probability matrix P^d, $d \in \mathcal{D}$.

The Common Loop Condition:
All the graphs G^d, $d \in \mathcal{D}$, have a *common loop.*

Any state lying on the common loop can be picked up as the reference state i^* in generating regenerative periods. Denote the common loop as $i^*, i_1, \ldots, i_m, i^*$. The common loop condition means that there is a period consisting of the sequence of states, $i^*, i_1, \ldots, i_m, i^*$, that can be generated with a positive probability by any policy in $\mathcal{D}$. We call this a *common period.*

Many policies satisfy this condition. For example, if for each $i \in \mathcal{S}$, we have a state $j \neq i$, such that $p^\alpha(j|i) > 0$ for all $\alpha \in \mathcal{A}(i)$, then the set of policies in $\mathcal{D}$ satisfies the common loop condition. Let us find one of the common loops under this condition. We start from any state i. Suppose that j_1 is the state such that $p^\alpha(j_1|i) > 0$ for all actions $\alpha \in \mathcal{A}(i)$, and j_2 is the state such that $p^\alpha(j_2|j_1) > 0$ for all actions $\alpha \in \mathcal{A}(j_1)$, and so on. In this way, we may obtain a sequence of states j_1, j_2, ... Since there are only a finite number of states, there must be two states denoted as j_{k_1} and j_{k_2}, with $k_1 \leq k_2$, such that $j_{k_1} = j_{k_2}$. We then have a common loop $j_{k_1} \to j_{k_1+1} \to \cdots \to j_{k_2} = j_{k_1}$.

Here is an example in which the two graphs do not have a common loop: G^{d_1}: $1 \to 2 \to 3 \to 4 \to 3$ and $4 \to 1$; G^{d_2}: $1 \to 2 \to 3 \to 1$ and $1 \to 4 \to 1$. In this example, the transition in state 3 is completely different for G^{d_1} and G^{d_2}: the system goes to 4 in G^{d_1} and to 1 in G^{d_2}. Therefore, the same path cannot be generated with P^{d_1} and P^{d_2} after the system reaches state 3.

Lemma 5.9. *Under the common loop condition, for any finite integer $N > 0$, Algorithm 5.3 stops with probability 1.*

Proof. Let us choose any state i^* in the common loop as the reference state. Because the number of policies is finite, the probability that any period is a common period is at least $p > 0$. Now, we divide the sample path into many intervals, each consisting of $2N$ periods. Consider a very special interval in which all the $2N$ periods are the same as the common period. The probability that an interval is such a special interval is larger than $p^{2N} > 0$. Therefore, the probability that in the first k intervals there is no such special interval is less than $(1-p^{2N})^k$, where $1 - p^{2N} < 1$. As $k \to \infty$, this probability goes to zero. That is, the probability that on a sample path the special interval never appears is zero. Because the special interval is a special case of $Z_{K+N} = Z_K$, the lemma follows directly from Lemma 5.8. □

Remark

It should be noted that although we only proved that under the common loop condition the algorithm stops with probability 1, it does not mean that the algorithm only stops when the special situation in the proof of Lemma 5.9 holds. In fact, in most cases, the algorithm stops before $\boldsymbol{Z}$ reaches such a special situation. To prove "stop with probability 1", we only need to find any special case that may stop the algorithm and prove such a case occurs with probability 1. It is true that, under this special case, the general property, e.g., Lemma 5.7, may not hold; however, it does hold in general because when the algorithm stops, the special case usually does not occur. See Problem 5.14 for more understanding.

5.3.2 With Stochastic Approximation*

In Algorithm 5.3, the potentials are estimated by using a fixed N number of periods (albeit possibly under different policies). This is similar to what is discussed in Section 5.2.2. In this subsection, we propose an algorithm (Algorithm 5.4) based on the stochastic approximation technique. In the algorithm, the potentials are estimated recursively at each period. This subsection parallels Section 6.3.1.

Algorithm 5.4 A Sample-Path-Based Algorithm with Stochastic Approximation:

1. Choose an initial policy d_0, and set $k = 0$ and $\bar{g}^{d_{-1}} = 0$. Choose an $\epsilon \in (0, 1/2)$ and a $C > 0$.
2. Observe the system under policy d_k for one period. For all $j \in \mathcal{S}$, calculate

$$V_k^{d_k}(i^*, j) = \begin{cases} \sum_{l=l_k^{d_k}(j)}^{l_{k+1}^{d_k}(i^*)-1} \left[f(X_l, d_k(X_l)) - \breve{\eta}_k^{d_k} \right], & \text{if } \chi_k(j) = 1, \\ \bar{g}^{d_{k-1}}(j), & \text{if } \chi_k(j) = 0, \end{cases}$$

where $\chi_k(j) = 1$ if the period contains j, and $\chi_k(j) = 0$ otherwise, and

$$\breve{\eta}_k^{d_k} = \frac{\sum_{l=l_k^{d_k}(i^*)}^{l_{k+1}^{d_k}(i^*)-1} f(X_l, d_k(X_l))}{l_{k+1}^{d_k}(i^*) - l_k^{d_k}(i^*)}.$$

Update the potential estimates as follows:

$$\bar{g}^{d_k}(j) = \bar{g}^{d_{k-1}}(j) + \frac{1}{k+1}\left[V_k^{d_k}(i^*, j) - \bar{g}^{d_{k-1}}(j)\right]. \qquad (5.30)$$

3. For every $i \in \mathcal{S}$, set

$$\beta(i) \in \arg\left\{\max_{\alpha\in\mathcal{A}(i)}\left[\sum_{j=1}^{S} p^{\alpha}(j|i)\bar{g}^{d_k}(j) + f(i,\alpha)\right]\right\}.$$

If there exists an i' such that

$$\sum_{j=1}^{S} p^{\beta(i')}(j|i')\bar{g}^{d_k}(j) + f(i', \beta(i'))$$
$$\geq \sum_{j=1}^{S} p^{d_k(i')}(j|i')\bar{g}^{d_k}(j) + f(i', d_k(i')) + \frac{C}{(k+1)^{1/2-\epsilon}}, \qquad (5.31)$$

then set $d_{k+1}(i) = \beta(i)$ for all i; otherwise, let $d_{k+1}(i) = d_k(i)$ for all i.

4. Set $k := k+1$ and go to step 2.

In step 2, $V_k^{d_k}(i^*, j)$ is the new information obtained in the kth period for potential $g^{d_k}(j)$. This information is used in (5.30) to update the estimate, in a way similar to stochastic approximation. If j does not appear in the period, no new information about $g^{d_k}(j)$ can be obtained in this period, and $\bar{g}^{d_{k-1}}(j)$ is used again.

Because the policies are updated often, the potential estimates may not be so accurate, especially at the beginning of the iteration procedure. This may cause the algorithm to be unstable. To avoid unnecessary oscillation between policies due to estimation errors, we add a threshold $\frac{C}{(k+1)^{1/2-\epsilon}}$ in (5.31) in step 3. The policy is not updated unless the difference in the comparison inequality exceeds a threshold. The value of the threshold gradually goes to zero as the policy approaches to the optimal one. The rate of the threshold approaching zero is controlled by ϵ. With this carefully designed updating scheme with a threshold, the algorithm converges to the optimal policy with probability 1 (a slightly different algorithm is proposed in [97], and its convergence is proved there).

It should be mentioned that there are many ways to propose such "fast" algorithms. The two proposed in Sections 5.3.1 and 5.3.2 just serve as examples. For such algorithms, the convergence speed is not known even if the convergence is proved. That is, we are not sure if they are really "faster" than the other sample-path-based algorithms, and in what sense they may be faster.

PROBLEMS

5.1. Repeat Example 5.1 by using the continuous-time Markov model.

5.2. A machine produces M different products, denoted as $1, 2, \ldots, M$. To process product i, the machine has to perform N_i different operations, denoted as $(i, 1), (i, 2), \ldots, (i, N_i)$. We use a discrete time model. At each time l, $l = 0, 1, \ldots$, the machine can only process one product and perform one operation. If at time instant l the machine is producing product i and is at operation (i, j), $j \neq N_i$, then at time instant $l + 1$ the machine will take operation (i, j') with probability $p_i(j'|j)$, $i = 1, 2, \ldots, M$, $j = 1, \ldots, N_i - 1$, and $j' = 1, \ldots, N_i$. If the machine is at operation (i, N_i), then it will pick up a new product i' and start to process it at operation $(i', 1)$ at the next time instant with probability $p^{\alpha}(i'|i)$, $i, i' = 1, 2, \ldots, M$, where $\alpha \in \mathcal{A}(i)$ represents an action. The operation $(i, 1)$ is called an *entrance operation* and (i, N_i) is called an *exit operation*. The system can be modelled as a Markov chain with state space $\mathcal{S} := \{(i, j) : i = 1, 2, \ldots, M, j = 1, \ldots, N_i\}$. Let f be the properly defined reward function. Derive the policy iteration condition (similar to (5.1) in Example 5.1) for this problem and show that with the sample-path-based approach we do not need to estimate the potentials for all the states.

5.3. In Problem 4.1, prove that if we use the sample-path-based approach, then we do not need to know the value of r.

5.4. As discussed in Section 5.1, to save memory and computation at each iteration, we may partition the state space $\mathcal{S} = \{1, 2, \ldots, S\}$ into N subsets and at each iteration we may only update the actions for the states in one of the subsets. In the extreme case, at each iteration, we may update the action for only one state. That is, at the first iteration, we update $d(1)$; at the second iteration, we update $d(2), \ldots$, and at the Sth iteration, we update $d(S)$. Then, at the $(S + 1)$th iteration, we update $d(1)$ again, and so on in a round robin manner. In such an iteration procedure, we cannot stop if there is no improvement in the performance at some iteration. We let the iteration algorithm stop after the performance does not improve in S consecutive iterations.

a. Formally state this policy iteration algorithm.
b. Prove that the algorithm stops after a finite number of iterations.
c. Prove that the algorithm stops at a gain-optimal policy.
d. Extend this algorithm to the general case where $\mathcal{S}$ is partitioned into N subsets.

5.5. To illustrate the idea behind Lemma 5.2, we consider the following simple problem. There are N different balls with identical appearance but different weights, denoted as $m_1, m_2, \ldots, m_N$, respectively, $m_i \neq m_j$, $i \neq j$. These weights are known to us. You have a scale in your hand that is inaccurate with a maximal absolute error of $r > 0$. Under what condition will you accurately identify these balls using this scale?

5.6. Suppose that when the sample-path-based policy iteration algorithm 5.2 stops, the estimation error of the potentials satisfies $|r| = |\bar{g} - g| < \delta/2$, where $\delta > 0$ is any positive number. Let $\bar{\eta}$ be the optimal average reward thus obtained. Prove

$$|\bar{\eta} - \eta^*| < \delta,$$

where η^* is the true optimal average reward.

5.7. If we use

$$\frac{\sum_{n=0}^{N-L+1} \left\{ I_i(X_n) \left[\sum_{l=0}^{L-1} f(X_{n+l}) - \eta \right] \right\}}{\sum_{n=0}^{N-L+1} I_i(X_n)}$$

to estimate the potentials, then the estimates are biased.

a. Convince yourself that the results in Section 5.2 still hold, and
b. Revise the proofs in Section 5.2 for the sample-path-based policy iteration with the above potential estimates.

5.8. With the sample-path-based policy iteration Algorithm 5.1, suppose that the Markov chain is ergodic with a finite state space under all policies, and the number of policies is finite. Let $|r| = |\bar{g}^d - g^d| < (\kappa/2)e$, where g^d and $\bar{g}^d$ are the potential of policy d and its estimate. Following the same argument as that in Lemma 5.3, prove that

$$\phi(\bar{g}^d) \subseteq \phi(g^d).$$

5.9. In Problem 5.8, we proved that $\phi(\bar{g}^d) \subseteq \phi(g^d)$.

a. On the surface, it looks like the same method as that in Lemma 5.3 can be used to prove $\phi(g^d) \subseteq \phi(\bar{g}^d)$. Give it a try.
b. If you cannot prove the result in a), explain why; if you feel that you did prove it, determine what is wrong in your proof.
c. Suppose that $h, h' \in \phi(g^d)$, and thus $f^h + P^h g^d = f^{h'} + P^{h'} g^d$. Because of the error in $\bar{g}^d$, we may have $f^h + P^h \bar{g}^d \neq f^{h'} + P^{h'} \bar{g}^d$. Therefore, one of them cannot be in $\phi(\bar{g}^d)$. Give an example to show that no matter how small the error $r = g^d - \bar{g}^d$ is, this fact is true.

5.10. Are the following statements true? Please explain the reasons for your answers:

a. Suppose that we use $d_{k+1} \in \phi(\bar{g}_N^{d_k})$ to replace (5.14) in step 3 of Algorithm 5.1 (i.e., set $\nu = 0$ in (5.12)). Then, the algorithm may not stop even if $\phi(\bar{g}_N^{d_k}) \subseteq \phi(g^{d_k})$ for $K' > K$ consecutive iterations $k = n, n+1, \ldots, n+K-1$, where K is the number of policies in $\mathcal{D}$.
b. Algorithm 5.2 may not always stay in $\mathcal{D}_0$ even after $\phi(\bar{g}_{N_k}^{d_k}) = \phi(g^{d_k})$ for K consecutive iterations, where K is any large integer.
c. Statement b) above is true even if we add the following sentence to step 3 of Algorithm 5.2: "If at a state i, action $d_k(i)$ attains the maximum, then set $d_{k+1}(i) = d_k(i)$."

5.11. Can you propose any stopping criteria for the sample-path-based algorithms to stop at an optimal policy in a finite number of iterations with probability 1?

5.12. In Lemma 5.4, $\sum_{k=0}^{\infty}(1 - y_k) < \infty$ implies $\lim_{k\to\infty} y_k = 1$, which, however, is not enough for $\lim_{n\to\infty} \prod_{k\geq n} y_k = 1$. For the latter to hold, y_k has to approach 1 fast enough.

a. For $y_k = 1 - \frac{1}{k}$, $k = 1, 2, \ldots$, we have $\lim_{k\to\infty} y_k = 1$. What is $\lim_{n\to\infty} \prod_{k\geq n} y_k$?
b. Verify the lemma for $y_k = 1 - \frac{1}{k^2}$, $k = 1, 2, \ldots$. What is $\lim_{n\to\infty} \prod_{k\geq n} y_k$?
c. For a sequence y_k, $0 \leq y_k \leq 1$, $k = 1, 2, \ldots$, if $\sum_{k=0}^{\infty}(1 - y_k) < \infty$ we have $\sum_{k=0}^{\infty}(1 - y_k^c) < \infty$ for any $c < 1$ and we can apply this lemma. How about $c > 1$?

5.13.* Write a simulation program for the "fast" Algorithm 5.3. Run it for a simple example with, say, $S = 3$, and each $\mathcal{A}(i)$, $i \in \mathcal{S}$, containing three to five actions. Record the sequence of d_k, $k = 0, 1, 2, \ldots$, and observe its behavior, e.g., how it changes from one policy to another one. Run it a few times with different N's.

5.14.* This problem is designed to help you to understand the remark on the proofs in Section 5.3.1. Consider an ergodic Markov chain $\boldsymbol{X} = \{X_0, X_1, \ldots, X_l, \ldots\}$ with state space $\mathcal{S}$ and reward function $f(i)$, $i \in \mathcal{S}$. Let $i^* \in \mathcal{S}$ be a special state. We repeat the following game: Every time we run the Markov chain, we let it stop when $X_l = X_{l+2} = i^*$; and when it stops, we receive a total reward of $f(X_{l+1})$.

a. We may prove that the Markov chain stops with probability 1 under the special condition $p(i^*|i^*) \neq 0$.
b. Suppose that the Markov chain stops with probability 1. Then, the expected total reward we receive is $\bar{r} = \sum_{k\in\mathcal{S}} p(k|i^*) f(k)$.

Obviously, $p(i^*|i^*) \neq 0$ is not a necessary condition, and this special condition does not change the expected total reward $\bar{r}$ in part b).

5.15.* If we implement Algorithm 5.3 for a few reference states $i^*, j^*, k^*, \ldots$ in parallel on the same sample path, then we can update the policy whenever the system reaches one of these states. In the extreme case, if we implement the algorithm using every state as the reference state separately on the same sample path, we may update the policy at every state transition on the sample path.

We need to study the convergence of such algorithms. Consider, for example, the case where we have two reference states i^* and j^*. Whenever we meet states i^* or j^*, we will update the policy. Therefore, if in a period starting from one i^* to the next i^*, the sample path visits state j^*, then the policy used in this period before visiting j^* is different from that used after the visit. Does this cause a major problem in the convergence of the algorithm? How about the algorithm in which we use all states as reference states?

Get your facts first, and then you can distort them as much as you please.

Mark Twain, American writer (1835 - 1910)

6 Reinforcement Learning

Reinforcement learning (RL) is one of the most active research areas in machine learning and artificial intelligence. The goal of RL, as for many other related areas, is to determine an optimal policy to obtain the best performance. However, the word "learning" clearly indicates the focus of this research area on the computational and exploratory aspects. As defined in [238], *"Reinforcement learning is defined not by characterizing learning methods, but by characterizing a learning problem. Any method that is well suited to solving that problem, we consider to be a reinforcement learning method."* Thus, in a broad sense, any topics discussed in this book, including perturbation analysis and Markov decision processes, belong to RL. We, however, will emphasize the computational and algorithmic aspects of RL in this book. With this view, the relationship between RL and other optimization approaches is described in Figure 1.19 in Chapter 1. In this sense, with RL, we mainly develop efficient algorithms for estimating (on line or off line) the performance potentials, or their variant Q-factors, of a given policy, based on which a better policy can be identified; we develop algorithms for estimating the potentials and Q-factors of an optimal policy, based on which an optimal policy can be found; we develop algorithms for estimating performance gradients, based on which gradient-based stochastic optimization procedures can be developed; or we develop efficient algorithms for finding a local optimal point at which the performance gradient is zero. These topics will be discussed in various sections in this chapter.

There is a large number of books and papers in the literature on RL (e.g., [15, 188, 236, 238, 239, 244, 254, 256]). It is not the purpose of this book to introduce all the aspects of this area. Rather, we will introduce the main ideas and principles with a focus on the computational aspect and the algorithmic nature, by following the relationship between RL and other optimization approaches, as described in Figure 1.19. Many computational issues have been discussed in the previous chapters when the various optimization methods were discussed. The contents in these chapters, in particular those in Chapters 3 and 5, fit within the scope of RL and should certainly be considered as a part of RL. The algorithms developed in these chapters and sections are mainly based on Monte Carlo simulations. These algorithms follow almost directly from the analytical formulas, such as those for potentials, expressed in a form of a mean value (see e.g., (2.16)).

In this chapter, we will focus on computational algorithms other than the relatively straightforward Monte Carlo approach. Those include temporal-difference (TD) learning, Q-learning and their variants, the value-iteration algorithm for MDPs, as well as the TD method for performance gradients and PA-based optimization. One of the important features of these approaches is that they are all related to and can be derived from stochastic approximation techniques. Therefore, stochastic approximation theory is a powerful tool in RL; it not only provides insights in developing computational learning algorithms, but also provides guidelines to the proof of convergence results. We will first briefly introduce the main ideas of stochastic approximation, and then we will use this theory to derive various algorithms. We will focus on explaining how we can obtain these algorithms from the principles of stochastic approximation; the rigorous proofs of the convergence of the algorithms are beyond the scope of this book. (Some proofs do not even exist in the current literature.)

As in the other chapters of this book, we will mainly discuss the optimization problem in relation to ergodic Markov chains with long-run average reward as the performance criteria. However, most works in the RL literature address the relatively easier topics with discounted performance criteria; and rigorous proofs for some results presented in this chapter (such as R-learning, a variant of Q-learning for the average reward problems) may still be lacking. In addition, research on RL algorithms for PA-based performance optimization started just in recent years. There are many open problems in these areas.

In summary, compared with the other books on RL, this chapter has the following main features:

1. The algorithms are stated in the framework of stochastic approximation;
2. It focuses on average reward problems; and
3. It contains results on PA-based performance optimization.

6.1 Stochastic Approximation

In this section, we briefly review some very basic concepts of stochastic approximation [79, 173, 191]. These concepts help the understanding of recursive algorithms developed in PA, RL, and other stochastic learning and optimization theories.

6.1.1 Finding the Zeros of a Function Recursively

The fundamental idea of stochastic approximation can be explained by the recursive algorithms for finding the zeros of a function $f(x)$, $x \in (-\infty, \infty)$ (or the roots of the equation $f(x) = 0$). In optimization, if we want to find a local minimum of a performance function $\eta(x)$, we need to find a point at which the derivative of $\eta(x)$, denoted as $f(x) := \frac{d\eta(x)}{dx}$, is zero.

$f(x)$ is Known

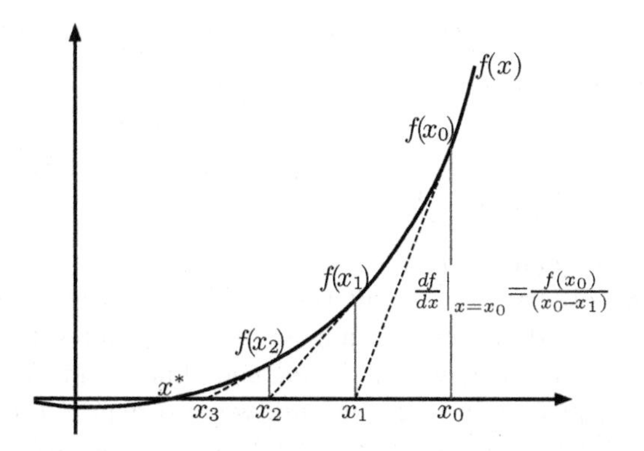

Fig. 6.1. The Recursive Algorithm in the Gradient Method

As illustrated in Figure 6.1, if the function $f(x)$ is known and is smooth enough, we can use the gradient method to find a root of $f(x)$, denoted as x^*, by a recursive algorithm:

$$x_{k+1} = x_k - \frac{1}{(\frac{df(x)}{dx})_{x=x_k}} f(x_k), \qquad k = 0, 1, 2, \ldots, \tag{6.1}$$

where x_{k+1} is the crossing point of the tangent line of the function $f(x)$ at $x = x_k$ with the x-axis. It can be proved that under some conditions on $f(x)$, starting from any x_0, we have $\lim_{k\to\infty} x_k = x^*$ and $f(x^*) = 0$. For example, if $f(x)$ is a convex and increasing function (i.e., $\frac{df(x)}{dx} > 0$ and $\frac{d^2 f(x)}{dx^2} > 0$), then such convergence can be easily proved as follows. Suppose that $f(x_0) > 0$. Since $(\frac{df(x)}{dx})_{x=x_0} > 0$, we have $x_0 > x_1$. Because $f(x)$ is increasing, we have $f(x_0) > f(x_1)$. By the same argument, we have $x_k > x_{k+1}$ and $f(x_k) > f(x_{k+1})$, $k = 1, 2, \ldots$. Suppose that the root x^* exists. Because the function is convex, the curve of $f(x)$ always lies on the same side of the tangent lines. Therefore, $x_k > x^*$ and $f(x_k) > f(x^*) = 0$. Thus we have two decreasing sequences

$$\begin{aligned} x_0 > x_1 > \cdots > x_k > x_{k+1} > \cdots > x^* \qquad &\text{and} \\ f(x_0) > f(x_1) > \cdots > f(x_k) > f(x_{k+1}) > \cdots > f(x^*) = 0. \end{aligned} \tag{6.2}$$

The decreasing sequence $x_0, x_1, \ldots, x_k, \ldots,$ converges to a point denoted as $\hat{x}$. Because $(\frac{df}{dx})_{x=\hat{x}} > 0$, we must have $f(\hat{x}) = 0$. Otherwise the sequence cannot stop at $\hat{x}$. This can be precisely proved. Suppose that $f(\hat{x}) > 0$. We define $\epsilon = \frac{1}{(\frac{df}{dx})_{x=\hat{x}}} f(\hat{x}) > 0$. Since $\lim_{k\to\infty} x_k = \hat{x}$ and

$$\lim_{k\to\infty} \frac{1}{(\frac{df}{dx})_{x=x_k}} f(x_k) = \frac{1}{(\frac{df}{dx})_{x=\hat{x}}} f(\hat{x}) = \epsilon > 0,$$

there must be a point denoted as $x_{\tilde{k}}$ such that $x_{\tilde{k}} - \hat{x} < \epsilon/2$ and $\frac{1}{(\frac{df}{dx})_{x=x_{\tilde{k}}}} f(x_{\tilde{k}})$ $> \epsilon/2$. Thus, we must have

$$x_{\tilde{k}+1} = x_{\tilde{k}} - \frac{1}{(\frac{df}{dx})_{x=x_{\tilde{k}}}} f(x_{\tilde{k}}) < x_{\tilde{k}} - \frac{\epsilon}{2} < \hat{x}.$$

This is impossible. Thus, $f(\hat{x}) = 0$; i.e., $\hat{x} = x^*$.

We have two observations about the proof of the convergence: First, the convexity of $f(x)$ and the use of tangent lines guarantee that the curve of $f(x)$ is always on the one side of the tangent lines, and this results in the two decreasing sequences in (6.2). Second, we also require the function to be increasing, i.e., $\frac{df(x)}{dx} > 0$ for $x > x^*$, which is used in the last part of the proof (cf. Problem 6.1).

$f(x)$ is Unknown

However, if $f(x)$ is unknown but can be precisely observed or measured, then we can obtain $f(x_k)$ by measurement but $(\frac{df}{dx})_{x=x_k}$ is not available. The above gradient-based algorithm cannot be utilized. All is not lost, however. Let us observe the nature of the algorithm in (6.1). The convergence of $\lim_{k\to\infty} x_k = x^*$ depends on the two sequences in (6.2), and it does not depend on the exact values of the derivatives. Thus, we may hope that the algorithm still works if we replace $1/(\frac{df(x)}{dx})_{x=x_k}$ with some real number κ_k that has the same sign as $1/(\frac{df(x)}{dx})_{x=x_k}$. That is, we may hope that the following algorithm works under some conditions for the function $f(x)$ and the sequence κ_k, $k = 0, 1, 2, \ldots$:

$$x_{k+1} = x_k - \kappa_k f(x_k), \qquad \kappa_k > 0, \quad k = 0, 1, 2, \ldots, \tag{6.3}$$

for increasing functions (with $\left.\frac{df(x)}{dx}\right|_{x=x_k} > 0$), and

$$x_{k+1} = x_k + \kappa_k f(x_k), \qquad \kappa_k > 0, \quad k = 0, 1, 2, \ldots, \tag{6.4}$$

for decreasing functions (with $\left.\frac{df(x)}{dx}\right|_{x=x_k} < 0$). κ_k, $k = 0, 1, 2, \ldots$, are called *step sizes.*

Example 6.1. To illustrate the idea, we consider a simple example. Let $f(x) = x - b$. We have $x^* = b$ and $\frac{df(x)}{dx} = 1$ for all x. Starting from any initial point x_0, the algorithm in (6.1) reaches x^* in one step: $x_1 = x_0 - 1 \times f(x_0) = x_0 - 1 \times (x_0 - b) = x^*$. Now let us use (6.3) with $\kappa_k = \kappa$, for all $k = 0, 1, \ldots$, with $1 > \kappa > 0$. We have

$$\begin{aligned} x_k &= x_{k-1} - \kappa(x_{k-1} - b) \\ &= (1-\kappa)x_{k-1} + \kappa b \\ &= (1-\kappa)^k x_0 + \left[1 + (1-\kappa) + (1-\kappa)^2 + \cdots + (1-\kappa)^{k-1}\right] \kappa b \\ &= (1-\kappa)^k x_0 + \left[1 - (1-\kappa)^k\right] b. \end{aligned}$$

Obviously, we have $\lim_{k\to\infty} x_k = b = x^*$. □

In this example, when $1 > \kappa > 0$, the two observations made at the end of the subsection for $f(x)$ being known remain the same: The curve $f(x)$ is on the same side of the "tangent" line, and the decreasing rate is $\kappa > 0$. These properties may not be necessary, though. As shown in Figure 6.2, Algorithm (6.3) may be faster than Algorithm (6.1), if we properly choose the step sizes κ_k, $k = 0, 1, \ldots$. Note that in the right-hand graph for Algorithm (6.3), at the beginning of the recursive procedure, the "tangent" lines intersect the curve $f(x)$.

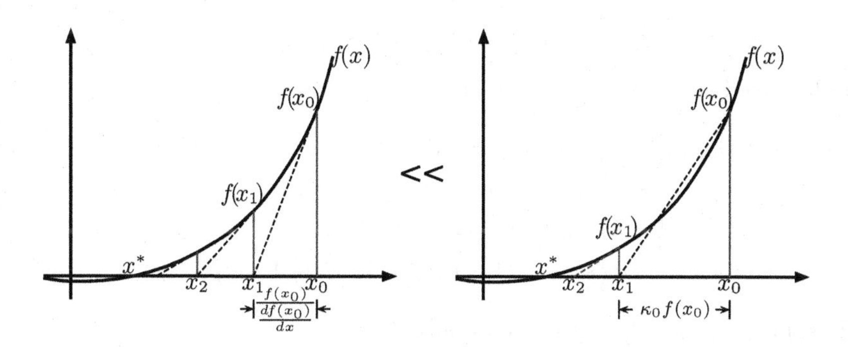

Fig. 6.2. Algorithm (6.3) (Right) May be Faster Than (6.1) (Left)

Observations with Random Noise

In many applications, the observation of the performance $f(x)$ contains random noise. Assume that the noise is additive. Then for any x_k, $k = 0, 1, \ldots$, we have an error-corrupted observation

$$y_k = f(x_k) + \epsilon_k, \tag{6.5}$$

where ϵ_k denotes the observation error at step k. We naturally hope that under some conditions, the recursive algorithm (6.3), or (6.4), works (meaning $\lim_{k\to\infty} x_k = x^*$, w.p.1) with $f(x_k)$ replaced with y_k, $k = 0, 1, \ldots$. Thus, we have the following (*Robbins-Monro*) algorithm:

For increasing functions,

$$x_{k+1} = x_k - \kappa_k y_k, \qquad \kappa_k > 0, \qquad k = 0, 1, 2, \ldots, \tag{6.6}$$

and for decreasing functions,

$$x_{k+1} = x_k + \kappa_k y_k, \qquad \kappa_k > 0, \qquad k = 0, 1, 2, \ldots. \tag{6.7}$$

First, we note that for problems with noisy observations, conditions have to be imposed on the function $f(x)$, the step sizes κ_k, $k = 0, 1, \ldots$, as well as the statistical properties of the noise. In general, the condition on the step sizes with noisy observations must be more strict than that with exact observations.

Example 6.2. We continue the simple problem considered in Example 6.1. Suppose that the observation contains noise and that we have

$$y_k = f(x_k) + \epsilon_k = (x_k - b) + \epsilon_k, \qquad k = 0, 1, \ldots.$$

With a fixed step size $\kappa_k = \kappa$, we have

$$\begin{aligned} x_k &= x_{k-1} - \kappa(x_{k-1} - b + \epsilon_{k-1}) \\ &= (1-\kappa)x_{k-1} + \kappa b - \kappa\epsilon_{k-1} \\ &= (1-\kappa)^k x_0 + \left[1 - (1-\kappa)^k\right] b \\ &\qquad - \kappa\left\{\epsilon_{k-1} + (1-\kappa)\epsilon_{k-2} + \cdots + (1-\kappa)^{k-1}\epsilon_0\right\}. \end{aligned}$$

Thus, $\lim_{k\to\infty} x_k = b$, w.p.1, if and only if

$$\lim_{k\to\infty} \left\{\epsilon_{k-1} + (1-\kappa)\epsilon_{k-2} + \cdots + (1-\kappa)^{k-1}\epsilon_0\right\} = 0, \qquad \text{w.p.1.} \tag{6.8}$$

However, (6.8) may not always hold because if κ is too large then the discount factor $(1-\kappa)$ may be too small and hence the earlier random noise $(1-\kappa)\epsilon_{k-2} + \cdots + (1-\kappa)^{k-1}\epsilon_0$ cannot cancel out the effect of the current noise ϵ_{k-1}.

If we take $\kappa_k = \frac{1}{k+1}$, $k = 0, 1, \ldots$, then we have

$$\begin{aligned} x_k &= x_{k-1} - \frac{1}{k}(x_{k-1} - b + \epsilon_{k-1}) \\ &= \frac{k-1}{k}x_{k-1} + \frac{1}{k}b - \frac{1}{k}\epsilon_{k-1} \\ &= \frac{1}{k}x_1 + \frac{k-1}{k}b - \frac{1}{k}\sum_{i=1}^{k-1}\epsilon_i. \end{aligned}$$

From this equation, it is clear that $\lim_{k\to\infty} x_k = b$, w.p.1, if and only if

$$\lim_{k\to\infty}\left\{\frac{1}{k}\sum_{i=1}^{k-1}\epsilon_i\right\} = 0, \qquad \text{w.p.1.} \tag{6.9}$$

Equation (6.9) is satisfied if ϵ_k, $k = 1, 2, \ldots$, are i.i.d random variables with $E(\epsilon_k) = 0$. □

Example 6.2 shows that, with random noise, if the step sizes κ_k, $k = 0, 1, \ldots$, are too large (do not decrease fast enough) then the effect of the random noise at subsequent steps cannot be cancelled out. Common conditions for the step sizes are:

$$\kappa_k > 0, \qquad \sum_{k=0}^{\infty}\kappa_k = \infty, \qquad \sum_{k=0}^{\infty}\kappa_k^2 < \infty. \tag{6.10}$$

The condition may be weakened under some more strict conditions on the noise:

$$\kappa_k > 0, \qquad \lim_{k\to\infty}\kappa_k = 0, \qquad \sum_{k=0}^{\infty}\kappa_k = \infty. \tag{6.11}$$

Note that the condition $\sum_{k=0}^{\infty}\kappa_k^2 < \infty$ in (6.10) implies $\lim_{k\to\infty}\kappa_k = 0$. Both $\sum_{k=0}^{\infty}\kappa_k^2 < \infty$ and $\lim_{k\to\infty}\kappa_k = 0$ means that the step sizes should not be too large. This can be understood by (6.8) in Example 6.2: If the discounting $(1-\kappa)$ is too small, the noise at subsequent steps cannot be averaged out.

On the other hand, $\sum_{k=0}^{\infty}\kappa_k = \infty$ implies that the step sizes should not decrease too fast. This is true even when there is no noise. Consider the simplest function $f(x) = x - b$ in Example 6.1. With the variant step sizes $\kappa_k > 0$, $k = 0, 1, \ldots$, we have $x_k = x_{k-1} - \kappa_{k-1}(x_{k-1} - b)$. Thus,

$$x_{k-1} - x_k = \kappa_{k-1}(x_{k-1} - b) < \kappa_{k-1}(x_0 - b).$$

Adding up the above inequality for k, $k-1$, down to 0, we obtain

$$x_0 - x_k < \left(\sum_{i=0}^{k-1} \kappa_i \right) (x_0 - b).$$

Now, suppose that $\sum_{k=0}^{\infty} \kappa_k < 1$. Then, letting $k \to \infty$ in the above inequality, we have

$$x_0 - \lim_{k \to \infty} x_k < x_0 - b.$$

Therefore, $\lim_{k\to\infty} x_k > b = x^*$. That is, with the step sizes $\sum_{k=0}^{\infty} \kappa_k < 1$ the algorithm cannot converge to the root of the function. This is shown in Figure 6.3. (As shown in the figure, $b < \cdots < x_{k+1} < x_k$, so $\lim_{k\to\infty} x_k$ exists.) In general, if $\sum_{k=0}^{\infty} \kappa_k < \infty$, it is not guaranteed that the recursive algorithm will converge to the root of $f(x)$.

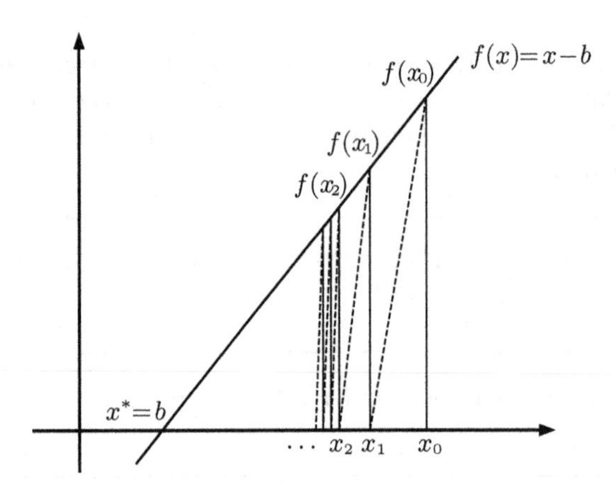

Fig. 6.3. x_k Does Not Converge to x^* When κ_k Are Too Small

Algorithm (6.6) or (6.7) is called the *Robbins-Monro (RM) algorithm.* It has been proved that with the step sizes (6.11) or (6.10), and under a set of conditions (different for (6.11) or (6.10)) on function $f(x)$ and on the statistical properties of the observation noise, the RM algorithm converges to a zero of $f(x)$ with probability 1. These conditions and the proof of the convergence of the algorithm are very technical and are beyond the scope of this book. Readers may refer to [79, 173, 191] etc. for details.

Finally, it should be mentioned that the RM algorithm applies to multi-variable functions $f(x)$, $\Re^n \to \Re$, as well. When the algorithm is applied to finding a local minimum of a function $f(x)$, or equivalently a zero point of its gradient, (6.6), or (6.7), simply suggests moving along the direction of the gradient.

6.1.2 Estimating Mean Values

Stochastic approximation approaches are often used to develop recursive algorithms in estimating mean values of random variables. Let w be a random variable and $x = E(w)$ be its mean. Define $f(x) = x - E(w)$. The problem of estimating $E(w)$ becomes to find the zero of the function $f(x)$. In this problem, the function is unknown because we do not know $E(w)$, and the observation of $f(x)$ contains errors because the observation of $E(w)$ does.

We start with any initial guess of x_0. At step k, $k = 0, 1, \ldots$, we make an observation of w, denoted as w_k, and use it as an estimate of $E(w)$. (Apparently, w_k usually does not equal $E(w)$ and we may write $w_k = E(w) - \epsilon_k$, with ϵ_k denoting the estimation error.) Therefore, the observation y_k at step k in (6.5) is $y_k = x_k - w_k$, $k = 0, 1, \ldots$. Applying the RM algorithm (6.6), we obtain (note that $f(x) = x - E(w)$ is an increasing function of x):

$$\begin{aligned} x_{k+1} &= x_k - \kappa_k y_k \\ &= x_k - \kappa_k (x_k - w_k) \\ &= (1 - \kappa_k) x_k + \kappa_k w_k, \qquad 0 < \kappa_k < 1. \end{aligned} \tag{6.12}$$

When $\kappa_k = \frac{1}{k+1}$, the algorithm in (6.12) yields the average of the observation sequence:

$$x_{k+1} = \frac{k}{k+1} x_k + \frac{1}{k+1} w_k = \frac{1}{k+1} \{w_k + w_{k-1} + \cdots + w_0\},$$

which, indeed, converges to the mean $E(w)$ with probability 1. The RM algorithm tells us that we may give the observations $w_0, w_1, \ldots, w_k, \ldots$ weights that are different from $\frac{1}{k+1}$, and the weighted sum still converges to the mean value.

Summary of the Chapter

The RL approaches to be introduced in this chapter are closely related to the stochastic approximation (SA) principles discussed above. The content of the reminder of the chapter is summarized as follows:

1. Using the SA approach in estimating the mean values, we develop TD algorithms to estimate the potentials, Q-factors, and performance derivatives.
2. Using the SA approach in finding the zeros, we develop TD algorithms to find the zeros of the performance derivatives, i.e., the local optimal points.
3. Using the SA approach in finding the zeros, we develop TD algorithms to find the zeros of the optimality equations, i.e., the global optimal points.

We also make comparisons of the different approaches.

6.2 Temporal Difference Methods

6.2.1 TD Methods for Potentials

The Algorithm TD(0)

The temporal difference (TD) methods applied to estimating the performance potentials can be explained clearly by the stochastic approximation approach; in particular, it can be viewed as a special case of (6.12) for estimating a mean value. From (2.16), we have

$$g(i) = E\left\{\sum_{l=0}^{\infty}[f(X_l) - \eta]\middle|X_0 = i\right\}. \tag{6.13}$$

Now, consider the lth transition on a sample path $\boldsymbol{X} = \{X_0, \ldots, X_l, X_{l+1}, \ldots\}$. Denote $X_l = i$. We have

$$\begin{aligned} g(i) &= E\left\{\sum_{k=0}^{\infty}[f(X_{l+k}) - \eta]\middle|X_l = i\right\} \\ &= [f(i) - \eta] + E\left\{E\left[\sum_{k=1}^{\infty}[f(X_{l+k}) - \eta]\middle|X_{l+1}\right]\middle|X_l = i\right\} \\ &= [f(i) - \eta] + E\left[g(X_{l+1})|X_l = i\right]. \end{aligned} \tag{6.14}$$

From this, when $X_l = i$, we have

$$g(i) = E\left\{[f(X_l) - \eta] + g(X_{l+1})\,|X_l = i\right\}.$$

Therefore, when $X_l = i$, we can use $[f(X_l) - \eta] + g(X_{l+1})$ as an estimate of $g(i)$. Thus, by the stochastic approximation algorithm (6.12), at time l we can update $g(X_l)$ as follows:

$$\begin{aligned} g(X_l) &:= g(X_l) - \kappa_l\left\{g(X_l) - [f(X_l) - \eta + g(X_{l+1})]\right\} \\ &= g(X_l) + \kappa_l\delta_l, \end{aligned} \tag{6.15}$$

in which we have defined the *temporal difference (TD)* as

$$\delta_l = [f(X_l) - \eta + g(X_{l+1}) - g(X_l)], \qquad l = 0, 1, \ldots, \tag{6.16}$$

which reflects the possible stochastic error observed at time l.

It is easy to develop algorithms for estimating η. For example, we can use the standard algorithm (6.12):

$$\eta_{l+1} = \eta_l - \kappa_{l+1}\left[\eta_l - f(X_{l+1})\right] \tag{6.17}$$

with $\kappa_l = \frac{1}{l+1}$ and $\eta_0 = f(X_0)$. (Later we will explain that, in fact, η does not need to be estimated.) Replacing the η in (6.15) with η_l in (6.17), we obtain an algorithm for estimating the performance potentials:

$$\begin{cases} g(X_l) := g(X_l) - \kappa_l \{g(X_l) - [f(X_l) - \eta_l + g(X_{l+1})]\}, \\ \eta_{l+1} := \eta_l - \kappa_{l+1} [\eta_l - f(X_{l+1})], \end{cases} \quad l = 0, 1, \dots,$$

in which η_l is the estimate of η at time l, and $g(X_l)$ is the estimate of the potential of state X_l at time l.

This algorithm is called a *TD*(0) *algorithm*. The algorithm works for any initial values of η and g. The procedure of (6.15) and (6.16) is illustrated in Figure 6.4, in which the sign $:\Longrightarrow$ indicates "replaced with".

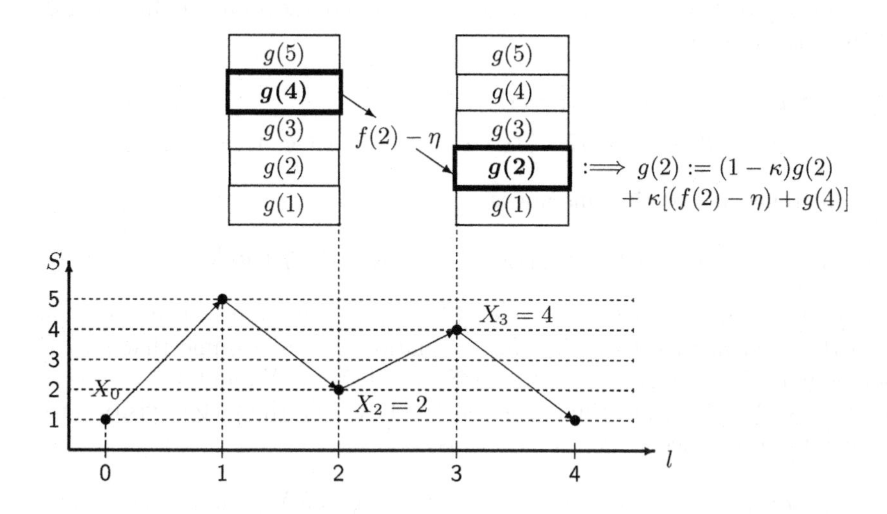

Fig. 6.4. TD(0) for Potentials $g(i)$'s

Comparison with the Monte Carlo Method

We are not going to prove the convergence of the TD(0) algorithm, which requires a deeper knowledge of stochastic approximation. Next, we will discuss some features of this algorithm and compare it with the Monte Carlo method (e.g., (3.15)). We do this by considering two examples.

Example 6.3. In this example, we study a modified version of the TD(0) algorithm in which the potentials are updated every N transitions by using the temporal difference obtained in these N transitions. The standard TD(0) algorithm becomes a special case with $N = 1$.

The sample path is divided into periods, each consisting of N transitions. The TD(0) algorithm is modified in two aspects:

1. The step sizes for the time instants in every period $[(k-1)N, kN-1]$ are set to be the same; i.e., set $\kappa_l \equiv \kappa'_k$, $l \in [(k-1)N, kN-1]$, $k = 1, 2, \ldots$, (e.g., $\kappa'_k = \frac{1}{k}$).
2. At the beginning of each period $[(k-1)N, kN-1]$, we "freeze" the estimates of the potentials $g(X_{l+1})$, $X_{l+1} \in \mathcal{S}$, and use them in the temporal difference δ_l in the entire period. In other words, while updating the values of $g(X_l)$'s, we use the potential estimates obtained at time instant $(k-1)N-1$ for the potentials g's in all the δ_l's in the entire period of $l \in [(k-1)N, kN-1]$.

Consider the kth period denoted as $\{X_{(k-1)N}, X_{(k-1)N+1}, \ldots, X_{(k-1)N+l}, \ldots, X_{kN-1}\}$, $l = 0, 1, \ldots, N-1$. By the first modification, in this period, (6.15) takes the form

$$\begin{aligned} g(X_{(k-1)N+l}) &:= g(X_{(k-1)N+l}) \\ &- \kappa'_k \left\{ g(X_{(k-1)N+l}) - \left[f(X_{(k-1)N+l}) - \eta + g(X_{(k-1)N+l+1}) \right] \right\}. \end{aligned}$$

The difference at this time instant is

$$\Delta g(X_{(k-1)N+l}) = \kappa'_k \left\{ g(X_{(k-1)N+l}) - \left[f(X_{(k-1)N+l}) - \eta + g(X_{(k-1)N+l+1}) \right] \right\}.$$

By the second modification, we need to "freeze" the values of the function g on the right-hand side of the above equation. Let us assume that on the sample path $X_{(k-1)N+l_u} = i$, with $i \in \mathcal{S}$, $u = 1, 2, \ldots, N_i$, and $0 \le l_1 < l_2 < \cdots < l_{N_i} < N-1$; i.e., at time epoches $(k-1)N + l_u$, the system visits state i. Then the difference at $(k-1)N + l_u$ is

$$\begin{aligned} \{\Delta g(i)\}_{\text{at } (k-1)N+l_u} &= \kappa'_k \left\{ g(i) - \left[f(i) - \eta + g(X_{(k-1)N+l_u+1}) \right] \right\}, \\ u &= 1, 2, \ldots, N_i. \end{aligned}$$

Now, suppose that among the states $X_{(k-1)N+l_u+1}$, $u = 1, 2, \ldots, N_i$, there are $N_{i,j}$ times the system visits state j, $j \in \mathcal{S}$, and $\sum_{j \in \mathcal{S}} N_{i,j} = N_i$. Adding the above N_i equations together we obtain the difference in the entire period of $[(k-1)N, kN-1]$:

$$\begin{aligned} \Delta g(i) &= \sum_{u=1}^{N_i} \{\Delta g(i)\}_{\text{at } (k-1)N+l_u} \\ &= \kappa'_k N_i \left\{ g(i) - \left[[f(i) - \eta] + \sum_{j \in \mathcal{S}} \frac{N_{i,j}}{N_i} g(j) \right] \right\}. \end{aligned}$$

Denoting $\kappa'_{k,i} = \kappa'_k N_i$ and noting that $\frac{N_{i,j}}{N_i} \approx p(j|i)$, we can write

$$\Delta g(i) = \kappa'_{k,i} \left\{ g(i) - \left[[f(i) - \eta] + \sum_{j \in \mathcal{S}} p(j|i) g(j) \right] \right\}. \tag{6.18}$$

This is the temporal difference from the Poisson equation $g - [(f - \eta) - Pg] = 0$. Therefore, this modified TD(0) algorithm in fact is a sample-path-based version of the stochastic approximation algorithm for finding the zeros of the Poisson equation . □

When $N = 1$, the above algorithm becomes the TD(0) method. Thus, TD(0) can be viewed as an "aggressive" version of the sample-path-based algorithm for *finding the zeros of the Poisson equation.* In this aggressive version, $p(j|i)$ in (6.18) is also estimated together with $g(j)$ in the stochastic approximation.

Furthermore, as shown in Figure 6.2, TD(0) provides more flexibility because we are allowed to use different step sizes κ_l, $l = 1, 2, \ldots$, which may improve the convergence rate. In addition, as shown in the next example, with TD(0) we may update the potential of one state, $g(i)$, by using the potential of another state, $g(j)$, $j \neq i$. This is called *"bootstrapping"* in [238]. Therefore, in some cases *TD(0) estimates may be more accurate than the Monte-Carlo estimates.*

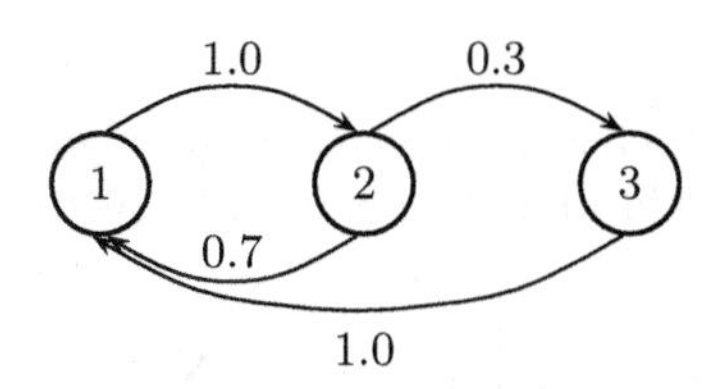

Fig. 6.5. The Transition Diagram for Example 6.4

Example 6.4. Consider a three-state Markov chain illustrated in Figure 6.5, with the numbers on the arcs indicating the state transition probabilities. Suppose that we get a sample path as listed in Table 6.1. Note that in the period from $l = 1$ to $l = 6$, the Markov chain visits states 1 and 2, but it does not visit state 3. Let us start from any initial value of $g(i)$, (say $g(i) = 0$) for all $i = 1, 2, 3$. In the period from $l = 1$ to $l = 6$, we may update the values of $g(1)$ and $g(2)$, with the TD(0) method. Therefore, at $l = 7$, with (6.15) $g(3)$ is assigned a value of $g(3) := \kappa [f(3) - \eta + g(1)]$. Because $g(1)$ has already been updated three times at $l = 2, 4$, and 6, we may get a more accurate estimate of $g(3)$ at $l = 7$ than the Monte Carlo method, which collects only one point for $g(3)$ at $l = 7$. In this sense, TD(0) may be more accurate than the Monte Carlo method, especially when the sample path is short. □

l	1 2 3 4 5 6 7 8 9
X_l	1 2 1 2 1 2 3 1 2

Table 6.1. A Sample Path for Example 6.4

Finally, we note that the solution to the Poisson equation is only up to an additive constant. This feature is kept in the TD method. For example, suppose that starting with any initial guess $g_0(i)$, for all $i \in \mathcal{S}$, we obtained a TD(0) estimate $g(i)$, for all $i \in \mathcal{S}$, by using (6.15). Then if we start with an initial guess $g_0(i)+c$, with c being a constant for all $i \in \mathcal{S}$, we will get a TD(0) estimate $g(i) + c$ for all $i \in \mathcal{S}$. In this sense, the TD method only estimates the differences in the potentials. On the other hand, the Monte Carlo estimate (3.15) tries to estimate a particular solution to the Poisson equation (based on $g = \{I + \sum_{k=1}^{\infty}(P^k - e\pi)\} f$).

More on TD(0)

As explained in Example 6.3, the TD(0) algorithm can be viewed as a stochastic approximation algorithm for finding a solution to the Poisson equation. In fact, the algorithm can be directly derived (explained) from the Poisson equation (2.12):

$$g = Pg + f - \eta e. \tag{6.19}$$

First, we can apply the stochastic approximation algorithm (6.6) to obtain an iterative numerical algorithm for the solution to (6.19):

$$g(i) := g(i) - \kappa_l \left\{ g(i) - \left[[f(i) - \eta] + \sum_{j \in \mathcal{S}} p(j|i) g(j) \right] \right\}. \tag{6.20}$$

This is the same as (6.18). Next, given $X_l = i$, $l = 0, 1, \ldots$, on a sample path, we have $E\left[g(X_{l+1})|X_l = i\right] = \sum_{j \in \mathcal{S}} p(j|i)g(j)$. That is, we can use $g(X_{l+1})$ as a sample path (very noisy) estimate of $\sum_{j \in \mathcal{S}} p(j|i)g(j)$. Replacing this item in (6.20) by $g(X_{l+1})$ yields the TD(0) algorithm (6.15).

As discussed above, the TD(0) algorithm converges to a solution to the Poisson equation, which is only up to an additive constant. Thus, the constant η in (6.15) does not really matter. We may therefore try to remove it and obtain

$$g(X_l) := g(X_l) - \kappa_l \left\{ g(X_l) - [f(X_l) + g(X_{l+1})] \right\}. \tag{6.21}$$

However, the problem with (6.21) is that the values of the $g(i)$'s thus obtained may go to infinity as l increases. (Just as in physics, the potential energy may go to infinity if we change the reference point at each iteration.) We have two ways to fix this problem.

The first method is very simple; it is of a numerical or algorithmic nature. We fix a state $i^* \in \mathcal{S}$ and choose a large number $G > 0$ and add the following step in the algorithm (6.21):

$$\text{If } \min_{i \in \mathcal{S}} |g(i)| > G, \text{ set } g(i) := g(i) - g(i^*) \qquad \text{for all } i \in \mathcal{S}. \tag{6.22}$$

We call this a *G-adjustment* step. After this step, the differences between the potentials of different states remain the same, but the potential of state i^* is brought to zero.

Setting up a stopping criterion for this algorithm requires some consideration: Obviously, we cannot stop the algorithm based on the absolute values of the potentials, and we need to use $\gamma(i^*, i) = g(i) - g(i^*)$. The potentials are only comparable at the iterations right after a G-adjustment step, at which $g(i^*) = 0$. Therefore, we may stop the algorithm after a G-adjustment step. If, for example, we observe that after a few such G-adjustment steps, the change in the norm of the potential vector is smaller than a pre-determined positive number, we may stop the algorithm.

As shown in Example 6.5, the convergence of the G-adjustment method is slow. In addition, there is an issue relating to how to choose the value of G. There is not much work on this method and we present it here simply because its conceptual simplicity.

In the second method, we choose a state i^* and set $g(i^*) = \eta$. For this particular potential vector, the Poisson equation becomes

$$g(i) = \sum_{j=1}^{S} p(j|i) [g(j) - g(i^*)] + f(i). \tag{6.23}$$

With this equation, we can derive the TD(0) algorithm as

$$g(X_l) := g(X_l) - \kappa_l \{g(X_l) - [f(X_l) + g(X_{l+1}) - g(i^*)]\}. \tag{6.24}$$

Compared with (6.15), algorithm (6.24) does not require us to estimate η. We can expect that under some conditions, $g(i^*)$ produced by this algorithm indeed converges to η. This can be explained as follows. Under some standard stochastic approximation conditions, $g(i)$, $i \in \mathcal{S}$, converges to the solution to (6.23), which is

$$g = P [g - g(i^*)e] + f.$$

Pre-multiplying both sides of this equation by π, we get $g(i^*) = \pi f = \eta$. A similar method will be used with Q-factors and Q-learning as discussed later.

Example 6.5. Consider a Markov chain with transition probability matrix

$$P = \begin{bmatrix} \epsilon & 1-\epsilon \\ 1-\epsilon & \epsilon \end{bmatrix},$$

where $0 < \epsilon << 1$. The Markov chain is ergodic for any positive number ϵ. We assume that ϵ is very small, therefore, the Markov chain behaves very like a periodic chain that visits states 0 and 1 alternately. Thus, a sample path looks similar to $1, 0, 1, 0, 1, 0, \ldots$. Let $f(1) = 1$ and $f(0) = 0$. We have $\pi(1) = \pi(0) = 0.5$ and $\eta = \pi f = 0.5$. Solving the Poisson equation, we get $g(1) - g(0) \approx 0.5$.

Now we apply different sample-path-based algorithms to estimate the potentials $g(0)$ and $g(1)$. Consider an approximate sample path $1, 0, 1, 0, 1, 0, \ldots$. The rewards received at every transition on this sample path are listed in the following Table 6.2:

l	0 1 2 3 4 5 6 7 8 9 ...
$f(X_l)$	1 0 1 0 1 0 1 0 1 0 ...

Table 6.2. The Reward Sequence for Example 6.5

Given this sequence of $f(X_l)$, it is easy to estimate η by either taking the average or using the TD(0) method. Therefore, we assume that η is known.

1. (*Monte Carlo*) When ϵ is very small, the convergence of $E\left[f(X_l)|X_0=i\right] \to \eta$, $i = 0, 1$ is very slow. Therefore, the convergence of

$$E\left\{\sum_{l=0}^{L-1} [f(X_l) - \eta] \middle| X_0 = i\right\} \to g(i)$$

is also very slow. Thus, the Monte Carlo algorithm (3.15) is very slow and requires a large L. On the other hand, the convergence of (2.17) $E\left\{\sum_{l=0}^{L(i|j)-1} [f(X_l) - \eta] \middle| X_0 = j\right\} \to \gamma(i,j)$ is very fast, and thus the corresponding algorithm (3.19) is also very fast.

2. (*TD(0) with (6.15) and (6.16)*) First, we rewrite (6.15) and (6.16) as follows:

$$g(X_l) - g(X_{l+1}) := (1 - \kappa_l)\left[g(X_l) - g(X_{l+1})\right] + \kappa_l\left[f(X_l) - \eta\right], \quad l = 0, 1, \ldots. \tag{6.25}$$

The step sizes are taken as $\kappa_l = \frac{1}{l+1}$, $l = 0, 1, \ldots$. With initial values $g(0) = g(1) = 0$, we apply (6.25) to the sequence in Table 6.2 and obtain

$$\text{at } l = 0: \quad g(1) - g(0) = (1-1)(0-0) + [f(1) - \eta] = [f(1) - \eta] = 0.5,$$

$$\text{at } l=1: \quad g(0)-g(1) = (1-\frac{1}{2})\{-[f(1)-\eta]\} + \frac{1}{2}[f(0)-\eta]$$
$$= -0.5,$$

and

$$\text{at } l=2: \quad g(1)-g(0) = (1-\frac{1}{3})\frac{1}{2}\{[f(1)-\eta]-[f(0)-\eta]\}+\frac{1}{3}[f(1)-\eta]$$
$$= 0.5, \cdots\cdots$$

We can see that the estimated value of $g(1)-g(0)$ approaches the correct value 0.5 very fast. This is the same as the perturbation realization factor based Monte Carlo estimate (3.19) and is much better than (3.15).

3. (*TD*(0) *with* $i^*=0$, $g(0)=\eta$) Set the initial values $g(1)=g(0)=0$. With $i^*=0$, (6.24) becomes

$$\begin{aligned} &\text{If } X_l=1, X_{l+1}=0: \quad g(1) := (1-\kappa_l)g(1)+\kappa_l f(1), \\ &\text{If } X_l=0, X_{l+1}=1: \quad g(0) := (1-\kappa_l)g(0)+\kappa_l [f(0)+g(1)-g(0)]. \end{aligned}$$

Applying the above equation to the sequence in Table 6.2, we obtain

$$\begin{aligned} &\text{at } l=0: \quad g(1) = (1-1)g(1)+f(1) = 1, \\ &\text{at } l=1: \quad g(0) = (1-\frac{1}{2})0+\frac{1}{2}[0+1-0] = 0.5, \\ &\text{at } l=2: \quad g(1) = (1-\frac{1}{3})g(1)+\frac{1}{3}f(1) = 1, \\ &\text{at } l=3: \quad g(0) = (1-\frac{1}{4})g(0)+\frac{1}{4}[0+1-0.5] = 0.5. \end{aligned}$$

Indeed, the algorithm converges very fast to $g(0)=0.5=\eta$ and $g(1)-g(0)=0.5$.

4. (*TD*(0) *with G-adjustment*) In this problem, (6.21) becomes

$$\begin{aligned} &\text{If } X_l=1, X_{l+1}=0: \quad g(1) := (1-\kappa_l)g(1)+\kappa_l [f(1)+g(0)], \\ &\text{If } X_l=0, X_{l+1}=1: \quad g(0) := (1-\kappa_l)g(0)+\kappa_l [f(0)+g(1)]. \end{aligned}$$

Applying the above equation to the sequence in Table 6.2 with initial values $g(1)=g(0)=0$, we obtain

$$\begin{aligned} &\text{at } l=0: \quad g(1) = (1-1)g(1)+[f(1)+g(0)] = 1, \\ &\text{at } l=1: \quad g(0) = (1-\frac{1}{2})0+\frac{1}{2}[f(0)+g(1)] = 0.5, \\ &\text{at } l=2: \quad g(1) = (1-\frac{1}{3})g(1)+\frac{1}{3}[f(1)+g(0)] = \frac{7}{6} = 1.167, \\ &\text{at } l=3: \quad g(0) = (1-\frac{1}{4})g(0)+\frac{1}{4}[f(0)+g(1)] = \frac{2}{3} = 0.667, \\ &\text{at } l=4: \quad g(1) = (1-\frac{1}{5})g(1)+\frac{1}{5}[f(1)+g(0)] = \frac{19}{15} = 1.267, \end{aligned}$$

$$\begin{aligned}
\text{at } l = 5: \quad & g(0) = (1 - \frac{1}{6})g(0) + \frac{1}{6}[f(0) + g(1)] = \frac{23}{30} = 0.767, \\
\text{at } l = 6: \quad & g(1) = (1 - \frac{1}{7})g(1) + \frac{1}{7}[f(1) + g(0)] = \frac{281}{210} = 1.338, \\
\text{at } l = 7: \quad & g(0) = (1 - \frac{1}{8})g(0) + \frac{1}{8}[f(0) + g(1)] = \frac{88}{105} = 0.838, \\
\text{at } l = 8: \quad & g(1) = (1 - \frac{1}{9})g(1) + \frac{1}{9}[f(1) + g(0)] = 1.393.
\end{aligned}$$

The estimated potentials increase, and the resulting sequence of $g(1) - g(0) = 1,\ 0.5, 0.667, 0.5, 0.6, 0.5, 0.571, 0.5, 0.555, \ldots$ indeed tends to converge to 0.5. Suppose that we set $G = \frac{2}{3}$, and set $i^* = 1$. Then we adjust the estimates as $g(1) = 0$ and $g(0) = -0.5$ at $l = 5$. Continuing the process, we get

$$\begin{aligned}
\text{at } l = 6: \quad & g(1) = (1 - \frac{1}{7})g(1) + \frac{1}{7}[f(1) + g(0)] = \frac{1}{14} = 0.0714, \\
\text{at } l = 7: \quad & g(0) = (1 - \frac{1}{8})g(0) + \frac{1}{8}[f(0) + g(1)] = -0.4286, \\
\text{at } l = 8: \quad & g(1) = (1 - \frac{1}{9})g(1) + \frac{1}{9}[f(1) + g(0)] = 0.1270.
\end{aligned}$$

The estimated values of $g(1) - g(0)$ are 0.5 and 0.556, the same as if we do not implement the G-adjustment step, which, indeed, prevents the estimates from going to infinity. However, as we can see, the price we pay for not estimating η with the G-adjustment is that it converges slowly. □

K-Step TD and TD(λ)

There are many variants and extensions of the TD method. First, given a Markov chain $\boldsymbol{X} = \{X_0, \ldots, X_l, X_{l+1}, \ldots\}$, similar to (6.14), at the lth transition with $X_l = i$ and for any $K > 1$, we have

$$\begin{aligned}
g(i) &= E\left\{ \sum_{k=0}^{K-1} [f(X_{l+k}) - \eta] + \sum_{k=K}^{\infty} [f(X_{l+k}) - \eta] \middle| X_l = i \right\} \\
&= E\left\{ \sum_{k=0}^{K-1} [f(X_{l+k}) - \eta] \middle| X_l = i \right\} \\
&\quad + E\left\{ E\left\{ \sum_{k=K}^{\infty} [f(X_{l+k}) - \eta] \middle| X_{l+1}, \ldots, X_{l+K} \right\} \middle| X_l = i \right\} \\
&= E\left\{ \sum_{k=0}^{K-1} [f(X_{l+k}) - \eta] \middle| X_l = i \right\} + E\left\{ g(X_{l+K}) | X_l = i \right\} \\
&= E\left\{ \sum_{k=0}^{K-1} [f(X_{l+k}) - \eta] + g(X_{l+K}) \middle| X_l = i \right\}.
\end{aligned}$$

Therefore, we can use $\sum_{k=0}^{K-1}[f(X_{l+k})-\eta]+g(X_{l+K})$ as an estimate of $g(X_l)$ and define the K-step temporal difference as

$$\delta_{l,K}=\left\{\sum_{k=0}^{K-1}[f(X_{l+k})-\eta]+g(X_{l+K})-g(X_l)\right\},\qquad l=0,1,\ldots.$$

Then, we obtain the following K-step TD(0) algorithm:

$$g(X_l):=g(X_l)+\kappa_l\delta_{l,K},\tag{6.26}$$

which updates $g(X_l)$ at $l+K$.

On the other hand, the Monte Carlo estimate of $g(i)$ is based on the approximation of (6.13)

$$g(i)\approx E\left\{\sum_{k=0}^{K-1}[f(X_{l+k})-\eta]\,\middle|\,X_l=i\right\},\tag{6.27}$$

with a large enough integer K. Using (6.12), we can take $\sum_{k=0}^{K-1}[f(X_{l+k})-\eta]$ as an estimate of $g(X_l)$ and update it as

$$g(X_l):=g(X_l)+\kappa_l\left\{\sum_{k=0}^{K-1}[f(X_{l+k})-\eta]-g(X_l)\right\}.$$

The difference between this algorithm and (6.26) is only one term $g(X_{l+K})$. This difference is very small for large K because $E\left[g(X_{l+K})|X_l=i\right]\to\pi g=0$ as $K\to\infty$. Thus, for a large K, the above K-step TD(0) method is almost the same as the stochastic approximation approach based on (6.27).

Next, we can get a mixture of the K-step TD algorithms (6.26). First, we choose a real number $0<\lambda<1$ and define

$$\delta_{l,\lambda}:=(1-\lambda)\sum_{K=1}^{\infty}\lambda^{K-1}\delta_{l,K}.$$

Then we have

$$\begin{aligned}\delta_{l,\lambda}&=(1-\lambda)\sum_{K=1}^{\infty}\left\{\lambda^{K-1}\left\{\sum_{k=0}^{K-1}[f(X_{l+k})-\eta]+g(X_{l+K})-g(X_l)\right\}\right\}\\&=\sum_{k=0}^{\infty}\left\{\lambda^k\left\{[f(X_{l+k})-\eta]+(1-\lambda)g(X_{l+k+1})\right\}\right\}-g(X_l)\\&=\sum_{k=0}^{\infty}\left\{\lambda^k\left\{[f(X_{l+k})-\eta]+g(X_{l+k+1})\right\}\right\}-\sum_{k=0}^{\infty}\lambda^{k+1}g(X_{l+k+1})-g(X_l)\\&=\sum_{k=0}^{\infty}\lambda^k\left\{[f(X_{l+k})-\eta]+g(X_{l+k+1})-g(X_{l+k})\right\}.\end{aligned}$$

Finally, we have the updating algorithm (known as the *TD*(λ) *algorithm*) as follows.

$$g(X_l) := g(X_l) + \kappa_l \delta_{l,\lambda}.$$

TD(0) is a special case of TD (λ) when $\lambda = 0$.

The TD (λ) method for problems with the discounted reward criterion is discussed in [236, 238]. TD (λ) algorithms with linearly parameterized function approximations are presented in [244].

In a further extension of the K-step TD (λ) approach, we set the number of steps K to be a random variable. We formulate the idea in a special case in Problem 6.7, also see the next subsection.

6.2.2 Q-Factors and Other Extensions

TD Methods for Q-factors

When the system structure, i.e., the transition probability matrix $P^d = [p^{d(i)}(j|i)]$, is completely unknown, we cannot apply policy iteration directly by using the potentials. (As shown in Example 5.1, we need to know at least the parts in the transition probabilities that are related to the actions, i.e., $b^d(n)$, to implement policy iteration with the estimated potentials.) In this case, we may need to estimate the Q-factor defined as [25, 238]

$$Q^d(i,\alpha) = \left\{ \sum_{j=1}^{S} p^\alpha(j|i) g^d(j) \right\} + f(i,\alpha) - \eta^d, \qquad \alpha \in \mathcal{A}(i), \tag{6.28}$$

for every state-action pair (i,α). In (6.28), $g^d(j)$, $j \in \mathcal{S}$, and η^d are the potentials and the average reward associated with policy d, and $p^\alpha(j|i)$ and $f(i,\alpha)$ depend on action α, which can be any one in $\mathcal{A}(i)$ (may not be $d(i)$). Thus, $Q^d(i,\alpha)$ is the expected potential if, at the current time, action α is taken and, at the other times, the system follows policy d. As shown in (6.28), Q-factors are also only determined up to an additive constant; i.e., if $Q^d(i,\alpha)$, $\alpha \in \mathcal{A}(i)$, $i \in \mathcal{S}$, are Q-factors of policy d, then for any constant c, $Q^d(i,\alpha)+c$, $\alpha \in \mathcal{A}(i)$, $i \in \mathcal{S}$, are also Q-factors of the same policy. From (4.5), if we can estimate $Q^d(i,\alpha)$ for all $\alpha \in \mathcal{A}(i)$, with d being the policy used in the current iteration, then we can choose

$$\alpha \in \arg\left\{ \max_{\alpha' \in \mathcal{A}(i)} \left[Q^d(i,\alpha') \right] \right\}, \tag{6.29}$$

as the action taken in state i in the next iteration. Since a sample path also contains information on the transition probabilities, it might be a good idea to estimate $Q^d(i,\alpha)$ for all state-action pairs rather than to estimate $g^d(i)$ and $p^\alpha(j|i)$'s separately, when the $P^\alpha(j|i)$'s are unknown for all $\alpha \in \mathcal{A}(i)$, $i,j \in \mathcal{S}$.

Conceptually, (6.28) is well defined. However, one problem arises immediately in sample-path-based implementation: It is impossible to estimate $Q^d(i, \alpha')$ on a sample path if the pair (i, α') does not appear on the path at all. Therefore, if a sample path is under a deterministic policy $d = d(i)$, $d(i) \in \mathcal{A}(i)$, $i \in \mathcal{S}$, which maps a state to one action, this approach does not work because the sample path only contains one state-action pair for each state. That is, this approach applies only to random policies that pick up all possible actions randomly.

However, we may implement the idea in policy iteration by slightly modifying the deterministic policy to a so-called ϵ-*greedy* policy [238]: With probability $1 - \epsilon$, $0 < \epsilon < 1$, we choose actions according to (6.29), and with probability ϵ we choose randomly (probably with an equal probability) other actions. When ϵ is small enough, this policy is close to the deterministic policy produced by (6.29), yet it visits all the possible state-action pairs. The deviation from the deterministic policy by a small probability ϵ is the price to pay for exploring the behavior of other policies. An ϵ-greedy policy generated from a greedy policy d (cf. (4.5)) will be denoted as d_ϵ:

$$d_\epsilon(i) = \begin{cases} d(i), & \text{with probability } 1 - \epsilon, \\ \alpha, \ \alpha \neq d(i) \ , \alpha \in \mathcal{A}(i), & \text{with probability } \frac{\epsilon}{|\mathcal{A}(i)|-1}, \end{cases} \tag{6.30}$$

where $|\mathcal{A}(i)|$ is the number of actions in $\mathcal{A}(i)$, and we assume that $|\mathcal{A}(i)| > 1$.

Now, let us consider a sample path that visits all the state-action pairs, denoted as $\{X_0, A_0, \ldots, X_l, A_l, \ldots,\}$, where A_l denotes the action taken at l. Such a sample path may be generated by an ϵ-greedy policy. For simplicity, we will drop the superscript "d_ϵ" in $Q^{d_\epsilon}(i, \alpha)$. From (6.28) and (6.13), we have

$$Q(i, \alpha) = E\left\{ \sum_{l=0}^{\infty} [f(X_l, A_l) - \eta] \middle| X_0 = i, A_0 = \alpha \right\}. \tag{6.31}$$

The TD(0) approach for $Q(i, \alpha)$ is similar to that for estimating potentials $g(i)$, $i \in \mathcal{S}$, except that the Q-factor has two variables $\alpha \in \mathcal{A}(i)$ and $i \in \mathcal{S}$. At the lth transition, if $X_l = i$ and $A_l = \alpha$, we have

$$Q(i, \alpha) = [f(i, \alpha) - \eta] + E\left[Q(X_{l+1}, A_{l+1}) | X_l = i, A_l = \alpha \right]. \tag{6.32}$$

Upon observing the transition from (X_l, A_l) to (X_{l+1}, A_{l+1}), we can use $[f(X_l, A_l) - \eta] + Q(X_{l+1}, A_{l+1})$ as an estimate of $Q(X_l, A_l)$. Thus, from (6.12), we obtain the following TD(0) algorithm:

$$\begin{aligned} Q(X_l, A_l) &:= Q(X_l, A_l) - \kappa_l \{Q(X_l, A_l) - [f(X_l, A_l) - \eta + Q(X_{l+1}, A_{l+1})]\} \\ &= Q(X_l, A_l) + \kappa_l \delta_l, \end{aligned} \tag{6.33}$$

$$\delta_l = [f(X_l, A_l) - \eta] + Q(X_{l+1}, A_{l+1}) - Q(X_l, A_l), \quad l = 0, 1, \ldots. \tag{6.34}$$

where δ_l is the *temporal difference* (*TD*) at time l. Note that the Q-factors thus obtained are for the ϵ-greedy policy d_ϵ, and they can be used as an approximation for the Q-factors of policy d if ϵ is small. In particular, the actions A_{l+1}, $l = 0, 1, \ldots$, in (6.33) also reflect the randomness of the ϵ-greedy policy, as shown in (6.30). The procedure of (6.33) and (6.34) is illustrated in Figure 6.6.

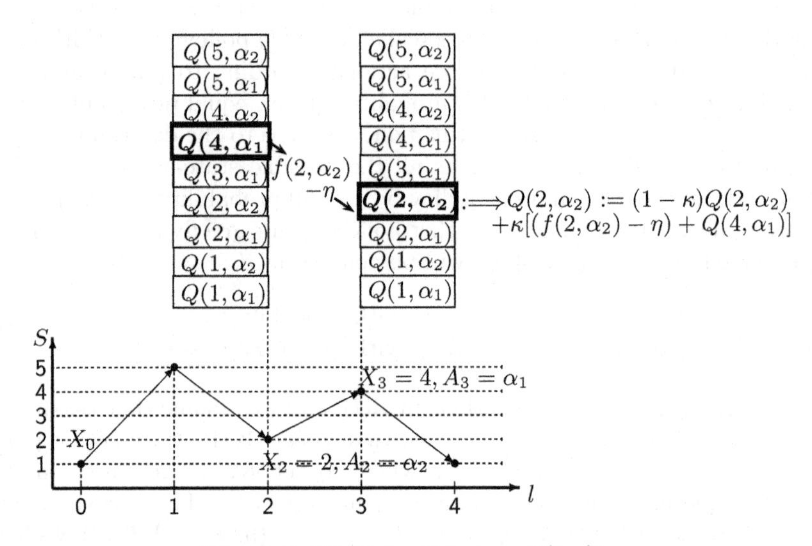

Fig. 6.6. TD(0) for Q-factors $Q(i, \alpha)$'s

A few observations are in order.

1. Because the action taken at each step depends only on the current state, for the ϵ-greedy policy, we have

$$E\left[Q(X_{l+1}, A_{l+1}) | X_{l+1} = i\right] = g(i).$$

Then,

$$E\left[Q(X_{l+1}, A_{l+1}) | X_l = i, A_l = \alpha\right] = E\left[g(X_{l+1}) | X_l = i, A_l = \alpha\right].$$

Thus, (6.32) becomes

$$Q(i, \alpha) = [f(i, \alpha) - \eta] + E\left[g(X_{l+1}) | X_l = i, A_l = \alpha\right]. \tag{6.35}$$

We can replace the $Q(X_{l+1}, A_{l+1})$ in (6.33) and (6.34) with $g(X_{l+1})$. Because $g(X_{l+1})$ is an average of $Q(X_{l+1}, A_{l+1})$, (6.35) may be more accurate than (6.33). However, we need to estimate, or to calculate, and to store $g(i)$, $i \in \mathcal{S}$, in addition to $Q(i, \alpha)$, $i \in \mathcal{S}$ and $\alpha \in \mathcal{A}(i)$.

2. If we only wish to estimate $Q(i,\alpha)$ for one particular state i, then we can develop TD(0) algorithms for only $Q(i,\alpha)$, with $\alpha \in \mathcal{A}(i)$, and $g(j)$, $j \neq i$ (without estimating the other Q-factors $Q(j,\beta)$, $\beta \in \mathcal{A}(j)$, $j \neq i$, or $\beta \neq \alpha$, $j = i$).
3. η does not need to be estimated. If we apply the G-adjustment step (6.22) in the algorithm, we can delete the item η in (6.33) and set

$$Q(X_l, A_l) := Q(X_l, A_l) - \kappa_l \left\{Q(X_l, A_l) - [f(X_l, A_l) + Q(X_{l+1}, A_{l+1})]\right\}.$$

Then, we fix a state $i^* \in \mathcal{S}$ and an action $\alpha^* \in \mathcal{A}(i)$ and choose a large number, $G > 0$, and add the following step to the algorithm:

$$\text{If} \quad \min_{i\in\mathcal{S},\alpha\in\mathcal{A}(i)} |Q(i,\alpha)| > G,$$
$$\text{set } Q(i,\alpha) := Q(i,\alpha) - Q(i^*,\alpha^*), \qquad \text{for all } i \in \mathcal{S},\ \alpha \in \mathcal{A}(i). \quad (6.36)$$

As shown in Example 6.5, the convergence rate may be slow, however.
4. Also, we may fix a reference state-action pair (i^*,α^*) and replace η by $Q(i^*,\alpha^*)$ in (6.33) and obtain

$$\delta_l = f(X_l, A_l) + Q(X_{l+1}, A_{l+1}) - Q(X_l, A_l) - Q(i^*,\alpha^*), \quad l = 0, 1, \ldots. \quad (6.37)$$

We expect that, with this algorithm, $Q(i^*,\alpha^*)$ indeed converges to η.

Extensions

There are a number of possible extensions to the TD method. Suppose that we only need to estimate the potentials for a subset of the state space, denoted as $\mathcal{S}_0 \subset \mathcal{S}$. (For example, in Example 5.1, only $g^\alpha(n,1)$'s need to be estimated.) We can develop an algorithm that is similar to the K-step TD method, except that the number of steps between two updates is random. On a sample path $\{X_0, X_1, \ldots\}$, we define $K_0, K_1, \ldots$ to be the sequence of time instants such that $X_{K_l} \in \mathcal{S}_0$, $l = 0, 1, \ldots$. For any $X_{K_l} = i \in \mathcal{S}_0$, we have

$$g(i) = E\left\{\sum_{k=0}^{\infty} [f(X_{K_l+k}) - \eta] \middle| X_{K_l} = i\right\}.$$

Therefore,

$$g(i) = E\left\{\sum_{k=K_l}^{K_{l+1}-1} [f(X_k) - \eta] + g(X_{K_{l+1}}) \middle| X_{K_l} = i\right\}.$$

The TD(0) algorithm is then

$$g(X_{K_l}) := g(X_{K_l}) + \kappa_l \delta_{l,K_l},$$

$$\delta_{l,K_l} = \left\{ \sum_{k=K_l}^{K_{l+1}-1} [f(X_k) - \eta] + g(X_{K_{l+1}}) - g(X_{K_l}) \right\}, \quad l = 0, 1, \ldots, \tag{6.38}$$

where δ_{l,K_l} is the temporal difference generated in the random period.

A close examination of δ_{l,K_l} indicates that this approach in fact estimates the difference of the potentials $g(X_{K_l}) - g(X_{K_{l+1}})$ by $\sum_{k=K_l}^{K_{l+1}-1}[f(X_k) - \eta]$. Therefore, this algorithm can be viewed as estimating the perturbation realization factors $\gamma(X_{K_{l+1}}, X_{K_l})$ by the stochastic approximation approach.

This TD method does not work if $\mathcal{S}_0$ contains only one state i. In this case, at all the embedded points, the system visits the same state $X_{K_l} = i$, $l = 0, 1, \ldots$. The periods between two embedded points K_l and K_{l+1} are regenerative points and therefore the mean of $\sum_{k=K_l}^{K_{l+1}-1}[f(X_k) - \eta]$ is zero and we would get $g(i) = 0$. Indeed, as we already know that TD methods for the average-reward problem only estimate the differences of the potentials, one cannot estimate the absolute value of $g(i)$ if $\mathcal{S}_0 = \{i\}$, $i \in \mathcal{S}$.

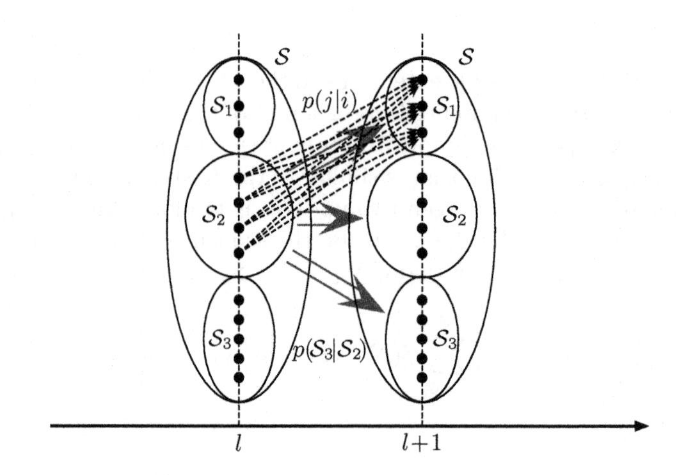

Fig. 6.7. The State Aggregation

In the Q-factor approach, the potential in a state i, $g(i)$, is “split” into a number of $Q(i, \alpha)$'s. From (6.31) and (6.13), we have

$$g(i) = \sum_{\alpha \in \mathcal{A}(i)} p_i(\alpha) Q(i, \alpha),$$

where $p_i(\alpha)$ is the probability that action α is taken in state i.

A dual approach is that we can aggregate a few potentials together. For example, we may partition the state space $\mathcal{S}$ into S_0 subsets: $\mathcal{S} = \cup_{k=1}^{S_0} \mathcal{S}_k$, $\mathcal{S}_k \cap \mathcal{S}_{k'} = \emptyset$, $k, k' = 1, 2, \ldots, S_0$. Each subset represents an aggregated state (see Figure 6.7). Assume that the aggregated states form a Markov chain. (This is a *very strong assumption*!) Suppose that the aggregated Markov chain is in the steady state. We may define the aggregated potentials as

$$g(\mathcal{S}_k) := E\left\{ \sum_{l=0}^{\infty} [f(X_l) - \eta] \middle| X_0 \in \mathcal{S}_k \right\}.$$

Apparently, we have

$$g(\mathcal{S}_k) = \sum_{i \in \mathcal{S}_k} \pi(i|\mathcal{S}_k) g(i), \tag{6.39}$$

where $\pi(i|\mathcal{S}_k) = \frac{\pi(i)}{\sum_{j \in \mathcal{S}_k} \pi(j)}$ is the conditional steady-state probability of i given that $i \in \mathcal{S}_k$ and $\pi(i)$ is the steady-state probability of state i. For the special case of $\mathcal{S}_k = \mathcal{S}$, we have $g(\mathcal{S}) = \sum_{j \in \mathcal{S}} \pi(j) g(j) = 0$, which is true because $g(\mathcal{S}) = E\left\{\sum_{l=0}^{\infty} [f(X_l) - \eta] | X_0 \in \mathcal{S}\right\} = 0$ in the steady state.

Consider a Markov chain $\{X_0, X_1, \ldots\}$. Set $S(i) = \mathcal{S}_k$ if $i \in \mathcal{S}_k$. Then, we have the *TD(0) algorithm for the aggregated potentials*:

$$g(S(X_l)) := g(S(X_l)) + \kappa_l \delta_l, \tag{6.40}$$

$$\delta_l = [f(X_l) - \eta + g(S(X_{l+1})) - g(S(X_l))], \qquad l = 0, 1, \ldots, \tag{6.41}$$

where δ_l is the *temporal difference*, expressed in the reward function of the original Markov chain.

Further extensions are possible. For example, we can combine the above two approaches, i.e., the state aggregation and sub-state-space approaches, together. That is, we may choose a subset of the aggregated states and combine (6.38) and (6.41) together. Another possible extension is to the event-based optimization problem discussed in Chapter 8. An event is defined as a set of state transitions. We can define the potential associated with an event a in the steady state, $g(a)$, as $g(a) := E\left\{\sum_{l=0}^{\infty} [f(X_l) - \eta] | a\right\}$. However, developing the corresponding TD algorithms like (6.40) and (6.41) for $g(a)$ is a challenging task. See Chapter 8 and Problem 6.17 for more discussion.

6.2.3 TD Methods for Performance Derivatives

As shown in Equations (3.36), (3.39), (3.44), and (3.45), the performance derivatives of η in the direction of $\Delta P = P' - P$ can be expressed as the mean of some random variables. Therefore, TD algorithms can be developed for estimating performance derivatives using these formulas. This is a new research area and few results exist. We only discuss a few cases to introduce the ideas.

Markov Chains

As an example, we study the performance derivative formula (3.45)

$$\frac{d\eta_\delta}{d\delta} = \frac{E\left\{\sum_{k=u_m}^{u_{m+1}-1}\left[f(X_k)-\eta\right]\hat{r}_k\right\}}{E\left[u_{m+1}-u_m\right]} \tag{6.42}$$

$$= E\left(\left[f(X_k)-\eta\right]\hat{r}_k\right), \tag{6.43}$$

where

$$\hat{r}_k = \sum_{l=u_{m(k)}}^{k} \frac{\Delta p(X_l|X_{l-1})}{p(X_l|X_{l-1})},$$

and u_m, $m = 0, 1, \ldots$, are a sequence of regenerative points with $X_0 = i^* \in \mathcal{S}$, $u_0 = 0$, and $u_{m+1} = \min\{n : n > u_m, X_n = i^*\}$, and for any integer $k \geq 0$, we define an integer $m(k)$ such that $u_{m(k)} \leq k < u_{m(k)+1}$.

Two algorithms can be developed based on (6.42) and (6.43). With (6.42), we can develop an algorithm that updates at every regenerative period. Note that the denominator $E\left[u_{m+1}-u_m\right]$ is easy to estimate with any standard algorithm. We develop an algorithm for the numerator $v_\delta := E\left\{\sum_{k=u_m}^{u_{m+1}-1}\left[f(X_k)-\eta\right]\hat{r}_k\right\}$ as follows:

$$v_\delta := v_\delta + \kappa_l\delta_l, \qquad l = 0, 1\ldots,$$

$$\delta_l = \sum_{k=u_l}^{u_{l+1}-1}\left[f(X_k)-\eta\right]\hat{r}_k - v_\delta. \tag{6.44}$$

This follows the standard algorithm (6.12) for estimating the mean of a random variable.

With (6.43), we can develop an algorithm that updates at every state transition.

$$\frac{d\eta_\delta}{d\delta} := \frac{d\eta_\delta}{d\delta} + \kappa_l\delta_l, \qquad l = 0, 1, \ldots,$$

$$\delta_l = \left[f(X_l)-\eta\right]\hat{r}_l - \frac{d\eta_\delta}{d\delta}. \tag{6.45}$$

This algorithm differs from (6.12) in that the observations $[f(X_l) - \eta]\,\hat{r}_l$ are not independent. The proof of the convergence of this algorithm is yet to be done.

Other algorithms can be developed based on other performance derivative formulas, e.g., (3.36), (3.39), and (3.44).

Queueing Systems

To develop TD-based algorithms for performance derivatives of queueing systems, we need to use performance derivative formulas expressed in the form of a mean of a random variable. In this subsection, we develop such a formula for an M/G/1 queue and use it as an example to illustrate the idea. (Problem 6.21 contains more discussion on the performance derivative formulas of queueing networks expressed in the form of a mean.)

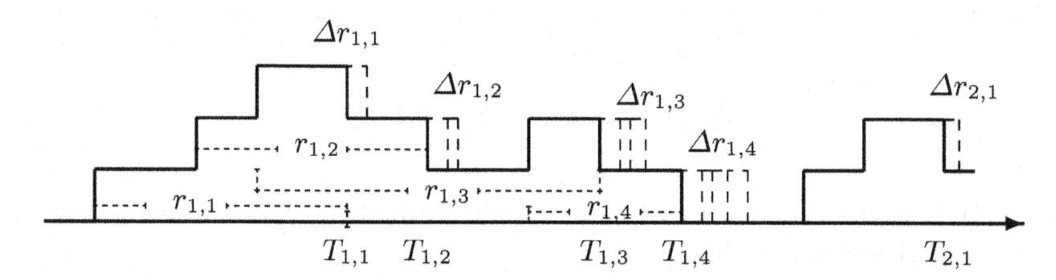

Fig. 6.8. The Response Times in an M/G/1 Queue

We take the mean response time (the length of the time period between the arrival and the departure times of a customer) as the system performance, and a sample path of an M/G/1 queue is shown in Figure 6.8. We apply PA to the sample path to obtain the desired performance derivative formula.

Let the service time be an i.i.d. random variable, and denote it as $s = F^{-1}(\xi, \theta)$, with ξ being a uniformly distributed random variable in $[0, 1)$ and θ being a parameter. Figure 6.8 illustrates the first busy period in which there are four customers served. Let $T_{k,i}$, $k, i = 1, 2, \ldots$, be the departure time of the ith customer in the kth busy period, and $s_{k,i}$ be its service time and $r_{k,i}$ be its response time. We have $s_{k,i} = F^{-1}(\xi_{k,i}, \theta)$, where $\xi_{k,i}$, $k, i = 1, 2, \ldots$, are i.i.d. random variables uniformly distributed in $[0, 1)$. Let n_k be the number of customers served in the kth busy period, and consider a sample path with K busy periods. The state process $\{X_t, t \geq 0\}$ is a regenerative process with the starting points of each busy period as the regenerative points. Thus, the mean response time is

$$\bar{r}(\theta) = \lim_{K \to \infty} \frac{\sum_{k=1}^{K} \sum_{i=1}^{n_k} r_{k,i}}{\sum_{k=1}^{K} n_k} = \frac{E\left[\sum_{i=1}^{n_k} r_{k,i}\right]}{E\left[n_k\right]}$$

$$= \lim_{K\to\infty} \frac{1}{L_K} \sum_{k=1}^{K} \sum_{i=1}^{n_k} r_{k,i}$$
$$= \lim_{K\to\infty} r_K(\theta),$$

where $L_K = \sum_{k=1}^{K} n_k$ is the number of customers served in the first K busy periods and

$$r_K(\theta) = \frac{1}{L_K} \sum_{k=1}^{K} \sum_{i=1}^{n_k} r_{k,i}. \tag{6.46}$$

The perturbation in $s_{k,i}$ due to the change in θ can be obtained by the inverse transform method (2.98):

$$\Delta s_{k,i} = \left. \frac{\partial F^{-1}(\xi_{k,i}, \theta)}{\partial \theta} \right|_{\xi_{k,i}=F(s_{k,i},\theta)} \Delta\theta.$$

The perturbed sample path can be constructed by PA (see Example 2.4), and, as shown in Figure 6.8, the perturbation of the response time $r_{k,i}$ is

$$\Delta r_{k,i} = \Delta T_{k,i} = \sum_{l=1}^{i} \Delta s_{k,l}.$$

The partial sample derivative of $r_{k,i}$ (which implicitly depends on $\xi_{k,i}$) with respect to θ (keeping $\xi_{k,i}$ fixed) is

$$\frac{\partial r_{k,i}}{\partial \theta} = \sum_{l=1}^{i} \frac{\partial s_{k,l}}{\partial \theta} = \sum_{l=1}^{i} \left\{ \left. \frac{\partial F^{-1}(\xi_{k,l}, \theta)}{\partial \theta} \right|_{\xi_{k,l}=F(s_{k,l},\theta)} \right\}.$$

Thus, from (6.46) we have

$$\frac{d r_K(\theta)}{d\theta} = \frac{1}{L_K} \sum_{k=1}^{K} \sum_{i=1}^{n_k} \sum_{l=1}^{i} \frac{\partial s_{k,l}}{\partial \theta}.$$

This is the PA-based sample derivative of $r_K(\theta)$ with respect to θ based on K busy periods. It is proved in [51] (see Problem 2.32) that this sample derivative is strongly consistent, i.e., we can exchange the order of "lim" and "$\frac{d}{d\theta}$" in the following equation:

$$\frac{d\bar{r}(\theta)}{d\theta} = \frac{d}{d\theta} \left\{ \lim_{K\to\infty} r_K(\theta) \right\} = \lim_{K\to\infty} \frac{d r_K(\theta)}{d\theta}$$
$$= \lim_{K\to\infty} \left\{ \frac{1}{L_K} \sum_{k=1}^{K} \sum_{i=1}^{n_k} \sum_{l=1}^{i} \frac{\partial s_{k,l}}{\partial \theta} \right\} \tag{6.47}$$
$$= \frac{E\left[\sum_{i=1}^{n_k} \sum_{l=1}^{i} \frac{\partial s_{k,l}}{\partial \theta} \right]}{E\left[n_k \right]}.$$

Next, we specifically assume that θ is the mean service time. By the standard queueing theory [169] (cf. Problem C.2), we have $E(n_k) = 1/(1 - \lambda\theta)$, with λ being the arrival rate. Thus, we have

$$\frac{d\bar{r}(\theta)}{d\theta} = E\left\{(1 - \lambda\theta)\left(\sum_{i=1}^{n_k}\sum_{l=1}^{i}\frac{\partial s_{k,l}}{\partial\theta}\right)\right\}. \tag{6.48}$$

From (6.48), we can use

$$h_k^{(r)} := (1 - \lambda\theta)\left(\sum_{i=1}^{n_k}\sum_{l=1}^{i}\frac{\partial s_{k,l}}{\partial\theta}\right) \tag{6.49}$$

obtained in any busy period (any $k = 1, 2, \ldots$) as an unbiased estimate of $\frac{d\bar{r}(\theta)}{d\theta}$. If λ is unknown, it can be easily estimated by the average of the inter-arrival times. With $h_k^{(r)}$, we may develop a TD algorithm estimating the performance derivative as follows:

$$\begin{aligned}\frac{d\bar{r}(\theta)}{d\theta} &:= \frac{d\bar{r}(\theta)}{d\theta} + \kappa_k\delta_k, \qquad k = 1, 2, \ldots,\\ \delta_k &= h_k^{(r)} - \frac{d\bar{r}(\theta)}{d\theta}.\end{aligned}$$

This algorithm updates in every busy period and the observations $h_k^{(r)}$s, $k = 1, 2, \ldots$, are independent of each other.

Equation (6.47) shows that the average of

$$h_{k,i}^{(r)} := \sum_{l=1}^{i}\frac{\partial s_{k,l}}{\partial\theta} \tag{6.50}$$

calculated at the ith service completion time of the kth busy period, converges to $\frac{d\bar{r}(\theta)}{d\theta}$ as the number of the customers served goes to infinity (strong consistency). Therefore, at each service completion time, we can simply use $h_{k,i}^{(r)}$ as an estimate of $\frac{d\bar{r}(\theta)}{d\theta}$. With $h_{k,i}^{(r)}$, we may develop a TD algorithm estimating the performance derivative as follows:

$$\begin{aligned}\frac{d\bar{r}(\theta)}{d\theta} &:= \frac{d\bar{r}(\theta)}{d\theta} + \kappa_{k,i}\delta_{k,i}, \qquad k = 1, 2, \ldots, \quad i = 1, 2, \ldots, n_k,\\ \delta_{k,i} &= h_{k,i}^{(r)} - \frac{d\bar{r}(\theta)}{d\theta}.\end{aligned}$$

In the algorithm, the transition times are indexed by both k and i, with k indexing the busy period and i indexing the departures in each busy period. The algorithm updates at every departure time; however, the observations $h_{k,i}^{(r)}$s in the same busy period k are not independent.

6.3 TD Methods and Performance Optimization

The TD algorithms introduced in the previous section enable us to estimate the performance potentials and Q-factors of a policy, as well as the performance derivatives. These quantities can be used in performance optimization, as shown in the previous chapters. We can also apply stochastic approximation methods to find the optimal policy directly. In PA-based algorithms, local optimal policies can be obtained by finding the zeros of the performance derivatives; and for Markov decision processes, optimal policies can be obtained by finding the solutions to the optimality equations.

6.3.1 PA-Based Optimization

PA-based gradient estimates can be used together with stochastic approximation methods to find the local optimal policies (parameters). Any of the PA algorithms in Section 3.2.2 and Chapter 2 can be used in optimization. As shown in Section 3.3.2, if we run the system long enough to get an accurate estimate of the gradient, the approach is similar to the standard gradient-based methods for deterministic systems [23]. In this section, we will focus on the approaches based on short-term estimates and stochastic approximation techniques.

Markov Chains

We consider the same parameterized space of transition probability matrices as discussed in Algorithm 3.4 in Section 3.2.2. We recall that the transition probability matrices are denoted as $P_\theta = [p_\theta(j|i)]$, $i, j \in \mathcal{S}$, and their corresponding long-run average rewards are $\eta_\theta = \pi_\theta f$. We choose a reference state i^*, set $g(i^*) = 0$ and $X_0 = i^*$, and define a sequence of regenerative points by $u_0 = 0$, and $u_{k+1} = \min\{n : n > u_k, X_n = i^*\}$. Furthermore, for any integer $n \geq 0$, we define an integer $m(n)$ such that $u_{m(n)} \leq n < u_{m(n)+1}$.

With this setting, (3.49) and (3.48) take the following form

$$\frac{d\eta_\theta}{d\theta} = E\left\{\left[\frac{\frac{d}{d\theta}p_\theta(X_{n+1}|X_n)}{p_\theta(X_{n+1}|X_n)}\right]\hat{w}_{n+1}\right\} \tag{6.51}$$

$$= E\left\{[f(X_n) - \eta_\theta]\,\hat{r}_n\right\}. \tag{6.52}$$

$$\hat{w}_{n+1} = \sum_{l=n+1}^{u_{m(n+1)+1}-1} [f(X_l) - \eta_\theta],$$

and

$$\hat{r}_n = \sum_{l=u_{m(n)}}^{n} \frac{\frac{d}{d\theta}p_\theta(X_l|X_{l-1})}{p_\theta(X_l|X_{l-1})}.$$

From Equations (6.51) and (6.52), at each transition time, we can use either $\left[\frac{\frac{d}{d\theta}p_\theta(X_{n+1}|X_n)}{p_\theta(X_{n+1}|X_n)}\right]\hat{w}_{n+1}$ or $[f(X_n)-\eta]\,\hat{r}_n$ as an unbiased estimate of $\frac{d\eta_\theta}{d\theta}$. Then, we may apply the Robbins-Monro algorithm (6.1) or (6.3) with step sizes (6.10) to find the zero points of the gradient (as a function of θ), i.e., the local optimal points of η_θ. For example, from (6.52), we can design a recursive algorithm that updates θ at every state transition time n as follows

$$\theta_{n+1} = \theta_n + \kappa_n\left[f(X_n) - \eta_\theta\right]\hat{r}_n, \tag{6.53}$$

where θ_n is the value of the system parameter θ used at the nth transition. In (6.53), we need to estimate η_θ too. Because $E\left[f(X_n)\right] = \eta_\theta$, η_θ can be estimated by the standard stochastic approximation algorithm (see (6.17)). This leads to the following algorithm [197]:

$$\begin{cases} \theta_{n+1} = \theta_n + \kappa_n\left[f(X_n) - \hat{\eta}_n\right]\hat{r}_n, \\ \hat{\eta}_{n+1} = \hat{\eta}_n + \varrho\kappa_n\left[f(X_n) - \hat{\eta}_n\right], \end{cases} \tag{6.54}$$

where $\hat{\eta}_n$ is the estimate of η_θ at the nth transition, and ϱ adjusts the step sizes for estimating η_θ. It is proved in [197] that, under some conditions, this algorithm converges to a point at which $\frac{d}{d\theta}\eta_\theta = 0$.

In algorithm (6.54), we update the parameter θ at each transition. With such an algorithm, the system never runs under the same policy (parameter) even for a few transitions. We may also develop an algorithm that updates the parameter in each regenerative period (between two visits to state i^*). This depends on the second equation of (3.48):

$$\frac{d\eta_\theta}{d\theta} = \frac{E\left\{\sum_{k=u_m}^{u_{m+1}-1}\left(\frac{\frac{d}{d\theta}p_\theta(X_{k+1}|X_k)}{p_\theta(X_{k+1}|X_k)}\hat{w}_{k+1}\right)\right\}}{E\left[u_{m+1}-u_m\right]}.$$

Thus, we can use $\sum_{k=u_m}^{u_{m+1}-1}\left(\frac{\frac{d}{d\theta}p_\theta(X_{k+1}|X_k)}{p_\theta(X_{k+1}|X_k)}\hat{w}_{k+1}\right)$ as an estimate of $\left\{E\left[u_{m+1}-u_m\right]\frac{d\eta_\theta}{d\theta}\right\}$. Note that $E(u_{m+1}-u_m)$ is a positive number. Therefore, the zeros of $\left\{E\left[u_{m+1}-u_m\right]\frac{d\eta_\theta}{d\theta}\right\}$ are the same as those of $\frac{d\eta_\theta}{d\theta}$. We have the following algorithm, which updates θ at the end of the mth regenerative period:

$$\begin{cases} \theta_{m+1} = \theta_m + \kappa_m\sum_{k=u_m}^{u_{m+1}-1}\left(\frac{\frac{d}{d\theta}p_\theta(X_{k+1}|X_k)}{p_\theta(X_{k+1}|X_k)}\hat{w}_{k+1}\right), \\ \hat{\eta}_{m+1} = \hat{\eta}_m + \varrho\kappa_m\sum_{k=u_m}^{u_{m+1}-1}\left[f(X_k) - \hat{\eta}_m\right], \end{cases} \tag{6.55}$$

where for $u_m \leq k < u_{m+1}$,

$$\hat{w}_{k+1} = \sum_{l=k+1}^{u_{m+1}-1} [f(X_l) - \hat{\eta}_{m+1}] ;$$

m is the index of the regenerative period; θ_m is the value of the parameter used in the mth period; and $\hat{\eta}_m$ is the estimated average reward used in the mth period, which is, in fact, obtained in the $(m-1)$th period. The second equation in (6.55) is a variant of the standard stochastic approximation, because for the average reward of the Markov chain in the mth period, η_{θ_m}, we have

$$\eta_{\theta_m}(\approx \hat{\eta}_m) = \frac{E\left\{\sum_{k=u_m}^{u_{m+1}-1} f(X_k)\right\}}{E\left[u_{m+1} - u_m\right]}.$$

Again, it is proved in [197] that, under some conditions, this algorithm also converges to the point at which $\frac{d}{d\theta}\eta_\theta = 0$.

Queueing Systems

As discussed in Section 6.2.3, in performance optimization, we need to express performance derivatives in the form of an expectation. As an example, we discuss the optimization of an M/G/1 queue with the performance derivative estimates $h_k^{(r)}$ in (6.49) and $h_{k,i}^{(r)}$ in (6.50). In general, we consider the performance function

$$\eta(\theta) = \bar{r}(\theta) + C(\theta),$$

with $\bar{r}(\theta)$ being the mean response time and $C(\theta)$ being a known function representing the associated cost. The estimate of $\frac{d\eta(\theta)}{d\theta}$ obtained in the kth busy period is

$$h_k = h_k^{(r)} + \frac{dC(\theta_k)}{d\theta},$$

and the estimate of $\frac{d\eta(\theta)}{d\theta}$ obtained at the ith service completion time in the kth busy period is

$$h_{k,i} = h_{k,i}^{(r)} + \frac{dC(\theta_{k,i})}{d\theta}.$$

It is not difficult to develop TD algorithms for finding the local optimal points of $\eta(\theta)$ with h_k or $h_{k,i}$. With h_k, θ is updated in every busy period, and with $h_{k,i}$, it is updated at every customer completion. This problem was studied in [82, 83, 84]. It is proved in those papers that, under some conditions, such algorithms indeed converge to local optimal points with probability 1. In [84], the same framework is extended to the optimization of other types of systems with a regenerative structure using PA-based gradient estimates.

6.3.2 Q-Learning

One of the most important developments in reinforcement learning is the *Q-learning* method for finding an optimal policy (see [238, 253] for problems with discounted-reward criteria and [3, 25] for problems with average-reward criteria.) We will briefly introduce the approach with the average reward problem.

The approach is based on the optimality equation in MDPs (4.7), which is re-stated as follows. Consider a policy $\widehat{d}$ with average reward $\eta^{\widehat{d}}$ and potential $g^{\widehat{d}}$. Let η^* be the optimal average reward. Then, $\widehat{d}$ is a gain-optimal policy (i.e., $\eta^{\widehat{d}} = \eta^*$) if and only if $\eta^{\widehat{d}}$ and $g^{\widehat{d}}$ satisfy the Bellman equation:

$$\eta^{\widehat{d}} + g^{\widehat{d}}(i) = \max_{\alpha \in \mathcal{A}(i)} \left\{ \sum_{j=1}^{S} p^{\alpha}(j|i) g^{\widehat{d}}(j) + f(i,\alpha) \right\}, \qquad i \in \mathcal{S}. \tag{6.56}$$

From the definition of the Q-factor (6.28), given any policy $d \in \mathcal{D}$ with potentials $g^d(j)$, $j \in \mathcal{S}$, and average reward η^d, for every state-action pair (i,α) we have

$$\eta^d + Q^d(i,\alpha) = \left\{ \sum_{j=1}^{S} p^{\alpha}(j|i) g^d(j) \right\} + f(i,\alpha), \qquad \alpha \in \mathcal{A}(i), \quad i \in \mathcal{S}. \tag{6.57}$$

Taking the maximum on both sides over the action space $\mathcal{A}(i)$, we get

$$\eta^d + \max_{\alpha \in \mathcal{A}(i)} Q^d(i,\alpha) = \max_{\alpha \in \mathcal{A}(i)} \left\{ \sum_{j=1}^{S} p^{\alpha}(j|i) g^d(j) + f(i,\alpha) \right\}.$$

Applying this equation to the optimal policy $\widehat{d}$, we have

$$\eta^{\widehat{d}} + \max_{\alpha \in \mathcal{A}(i)} Q^{\widehat{d}}(i,\alpha) = \max_{\alpha \in \mathcal{A}(i)} \left\{ \sum_{j=1}^{S} p^{\alpha}(j|i) g^{\widehat{d}}(j) + f(i,\alpha) \right\}.$$

Comparing this equation with (6.56), we conclude that, for the optimal policy $\widehat{d}$, we have

$$g^{\widehat{d}}(i) = \max_{\alpha \in \mathcal{A}(i)} Q^{\widehat{d}}(i,\alpha).$$

Substituting this into (6.57) indicates that a gain-optimal policy $\widehat{d}$ satisfies the *optimality equation for Q-factors*:

$$\eta^{\widehat{d}} + Q^{\widehat{d}}(i,\alpha) = \left\{ \sum_{j=1}^{S} p^{\alpha}(j|i) \left[\max_{\beta \in \mathcal{A}(j)} Q^{\widehat{d}}(j,\beta) \right] \right\} + f(i,\alpha), \qquad \alpha \in \mathcal{A}(i), \quad i \in \mathcal{S}. \tag{6.58}$$

Note that α can be any action in $\mathcal{A}(i)$. Equation (6.58) is the optimality equation for Q-factors. There are $\sum_{i=1}^{S} |\mathcal{A}(i)|$ equations, each for one state-action pair.

If $p^{\alpha}(j|i)$, $i, j \in \mathcal{S}$ and $\alpha \in \mathcal{A}(i)$, are known to us, then in principle (6.58) can be solved numerically by iteration. For example, applying the stochastic approximation algorithm (6.4) to equation (6.58), we may obtain an iterative algorithm as follows.

$$Q_{k+1}(i,\alpha) := Q_k(i,\alpha) + \kappa_k \delta_k, \qquad \alpha \in \mathcal{A}(i),$$
$$\delta_k = \sum_{j=1}^{S} p^{\alpha}(j|i) \left[\max_{\beta \in \mathcal{A}(j)} Q_k(j,\beta) \right] - Q_k(i,\alpha) + f(i,\alpha) - \bar{\eta}, \quad (6.59)$$

where $\bar{\eta}$ is an estimate of the optimal average reward. We can estimate it iteratively using the average reward of the "greedy" policy determined by the actions with $\max_{\alpha \in \mathcal{A}(i)} Q_k(i,\alpha)$. Furthermore, we may simply remove $\bar{\eta}$ and apply the G-adjustment method to keep $Q_k(i,\alpha)$ finite, because only the relative values of the Q-factors matter. We may also set a particular $Q_k(i^*,\alpha^*)$ to be zero or $\bar{\eta}$, as we did for the Q-factors in Section 6.2.2. We expect that the algorithm in (6.59) with any of these modifications converges to the solution to the optimality equations of Q-factors (6.58).

When we do not know the $p^{\alpha}(j|i)$'s exactly, we need to observe the state transitions on a sample path to get the information about these transition probabilities. The sample path can be obtained by running a real system, or by simulation. In simulation, we may use the system structure (e.g., for queueing systems) to simulate the system, without knowing the exact values of $p^{\alpha}(j|i)$'s. The TD approach combines the stochastic approximation approach (6.59) with sample-path information to find the solution to (6.58).

First, we note that in the optimality equation (6.58) and in the iterative algorithm (6.59) we need the Q-factors for all the state-action pairs (not only the action with the maximum Q-factor for a state); thus, we need to follow a sample path that visits all the state-action pairs to estimate the Q-factors for the optimal policy.

The basic idea is as follows. Suppose that α is taken at time l with $X_l = i$. Then,

$$E\left\{ \left[\max_{\beta \in \mathcal{A}(X_{l+1})} Q(X_{l+1},\beta) \right] \middle| X_l = i, A_l = \alpha \right\} = \sum_{j=1}^{S} p^{\alpha}(j|i) \left[\max_{\beta \in \mathcal{A}(j)} Q(j,\beta) \right].$$

Therefore, we may use $\max_{\beta \in \mathcal{A}(X_{l+1})} Q(X_{l+1},\beta)$ as a sample-path-based estimate of the right-hand side of the above equation. The sample-path-based version of (6.59) is

$$Q(X_l, A_l) := Q(X_l, A_l) + \kappa_l \delta_l,$$
$$\delta_l = \max_{\beta \in \mathcal{A}(X_{l+1})} Q(X_{l+1}, \beta) - Q(X_l, A_l) + f(X_l, A_l) - \bar{\eta}. \quad (6.60)$$

That is, at time l, we update one Q-factor $Q(X_l, A_l)$ according to (6.60); this is illustrated in Figure 6.9.

The above general discussion leads to a number of TD algorithms discussed below; they differ in the way of handling the constant term $\bar{\eta}$ in (6.60).

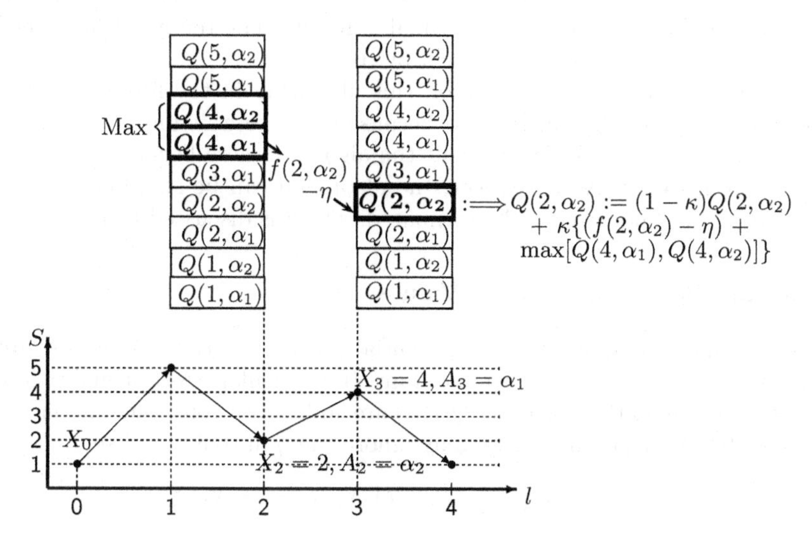

Fig. 6.9. Q-Learning for the Q-factors of the Optimal Policy

R-Learning

The R-learning algorithm proposed in [195] is one of the algorithms based on (6.60). (The word "R-learning" may be chosen to distinguish it from the Q-learning method, which is essentially based on the same idea but for discounted reward problems.) In R-learning, we observe a sample path with any policy that involves all the state-action pairs. (Such a policy is called a *behavior policy*; an example is the ϵ-greedy policy.) The Q-factors are updated according to (6.60). The optimal average reward $\bar{\eta}$ is also estimated from the sample path recursively. Because we expect that the algorithm converges to the optimal policy, we should count only these steps that eventually constitute the optimal policy. That is, we only update the average reward when the action taken is based on the "greedy" policy. Thus, it is proposed in [195] that we should update $\bar{\eta}$ as follows:

$$\begin{aligned}
\text{when } A_l &= \arg\left\{\max_{\alpha\in\mathcal{A}(X_l)} Q(X_l,\alpha)\right\} \\
\text{set } \bar{\eta}_{k+1} &:= \bar{\eta}_k + \kappa'_k\delta_k, \\
\delta_k &= f(X_l,A_l) - \bar{\eta}_k + \max_{\beta\in\mathcal{A}(X_{l+1})} Q(X_{l+1},\beta) - Q(X_l,A_l),
\end{aligned}$$

where κ'_k, $k = 1,2,\ldots$, are the step sizes, the subscript k indicates that the indexes of the iterations in $\bar{\eta}$ are different from those of the state transitions. When the Q-factors approach the solution to the optimality equation, the corresponding "greedy" policy approaches the optimal policy, and the estimated average reward $\bar{\eta}_k$ approaches its optimal value η^*. The temporal difference δ_k used in the update can be explained by (6.58).

There has been not much work on this algorithm in the literature. The convergence of $\bar{\eta}$ to η^* might be slow because it has to wait until a "greedy" action to make an update. We present the algorithm here because it follows naturally from (6.60). As we expect, the algorithm can be simplified in the sense that η^* does not need to be estimated, which is discussed below.

Q-Learning Based on Relative Values

As we have always emphasized, the potentials, or their variant Q-factors, are meaningful only up to an additive constant. In particular, the optimal policy does not change if all the Q-factors $Q(i,\alpha)$ are added by a constant. Therefore, similar to (6.21) for potentials, we may ignore the $\bar{\eta}$ in (6.60) and set

$$\delta_l = \max_{\beta\in\mathcal{A}(X_{l+1})} Q(X_{l+1},\beta) - Q(X_l,A_l) + f(X_l,A_l),$$

and

$$Q(X_l,A_l) := Q(X_l,A_l) + \kappa_l\delta_l. \tag{6.61}$$

Note that we need to add the G-adjustment step (6.36) to the numerical algorithm to avoid the estimates going to infinity.

Another way to keep the Q-factors finite is to use a reference state-action pair (i^*,α^*) and set $Q(i^*,\alpha^*) = \eta^*$. We have (cf. (6.24) for potentials)

$$\delta_l = \max_{\beta\in\mathcal{A}(X_{l+1})} Q(X_{l+1},\beta) - Q(X_l,A_l) + f(X_l,A_l) - Q(i^*,\alpha^*). \tag{6.62}$$

A similar algorithm is suggested in [25]: We only pick up one reference state i^* (not action), and at every step we subtract a constant $\max_{\alpha\in\mathcal{A}(i^*)} Q(i^*,\alpha)$. In other words, we set $\max_{\alpha\in\mathcal{A}(i^*)} Q(i^*,\alpha) = \eta^*$. Thus, the temporal difference is

$$\delta_l = \max_{\beta\in\mathcal{A}(X_{l+1})} Q(X_{l+1},\beta) - Q(X_l,A_l) + f(X_l,A_l) - \max_{\alpha\in\mathcal{A}(i^*)} Q(i^*,\alpha). \tag{6.63}$$

The convergence of the algorithm (6.61) and (6.63) to the optimal policy is given in [3]. (The minimization problem is considered there, but the principles are the same.)

Discussion

Compared with the numerical iterative algorithm (6.59), the Q-learning algorithms combine numerical methods with simulation (or the sample-path-based observations). Specifically, the $\max_{\beta \in \mathcal{A}(i)} Q(i, \beta)$ part in (6.60) is still based on numerical calculation, while the effects of the transition probabilities $p^\alpha(j|i)$'s are estimated on a sample path. Strictly speaking, the Q-learning algorithm is not a "sample-path-based learning" algorithm, or only a "partially sample-path-based" learning algorithm, because the sample path is only used in estimating the transition probabilities, and the update of Q-factors (the "learning" part) is based on the numerical method. In implementing the algorithm, the behavior policy used on the sample path is fixed, and it can be any one that visits all the state-action pairs. The sample path is only used to provide the information for transition probabilities associated with all the actions; it is not used to gather information for performance comparison.

When $p^\alpha(j|i)$'s are known, the optimal Q-factors can be obtained recursively by the numerical method (6.59). However, this numerical algorithm only serves illustrative purposes, because a more efficient numerical algorithm (the value iteration) based on performance potentials exists (see Section 6.3.4). In summary, Q-learning is useful when the system transition probabilities under different actions are unknown but a sample path of the system visiting all state-action pairs is available.

In both R-learning and Q-learning with relative values, the sample path can be under any behavior policy. The convergence property and speed depend heavily on the sample path. The policy may be even *time varying* as long as the sample path visits all the state-action pairs. As we want to concentrate on important states and actions, we may use the *ϵ-greedy policy*. That is, at the lth step, with probability $1-\epsilon$, $0 < \epsilon < 1$, we choose the action

$$\alpha_l \in \arg\left\{ \max_{\beta \in \mathcal{A}(X_l)} Q(X_l, \beta) \right\}, \tag{6.64}$$

where $Q(X_l, \beta)$ is the estimate of the Q-factor at the lth step, and with probability ϵ we choose all the other actions randomly with an equal probability. In this way, the policies on the sample path vary and converge to an ϵ-greedy optimal policy. In this sense, the algorithm is of an "on-line learning" nature (see the discussion in the next section). The policy used on the sample path improves as the algorithm is being implemented.

6.3.3 Optimistic On-Line Policy Iteration

In this subsection, we introduce two "on-line" optimization algorithms based on the TD methods. They are similar to the two Q-learning algorithms presented in the last subsection. However, the "on-line" nature implies that the sample path used in the learning process eventually converges to the optimal policy (in SARSA, this is true except for a small ϵ probability). In other words, the performance of the sample path gradually improves towards the optimal value.

With Q-factors: SARSA

The SARSA (stands for "State-action-reward-state-action") algorithm for discounted reward problems was proposed in [223, 237], and was further discussed in, e.g., [238, 25]. Here, we will discuss SARSA for average reward problems. Few results exist and the discussion is brief and of an exploratory nature.

This approach is based on the idea that if the sample path is generated according to the (*time varying*) ϵ-greedy policies following (6.64), the policies may converge to the optimal one if the Q-factors are accurately estimated, and there is no need to implement $\max_{\beta \in \mathcal{A}(j)} Q(j, \beta)$ in (6.60). (The Q-learning method is to find the solution to the optimality equation for Q-factors (6.58).) The Q-factors of the policies can be estimated by the TD algorithm (6.33) and (6.34) or (6.37). This approach is similar to the on-line potential-based policy iteration except that it is based on the estimated Q-factors $Q(i, \alpha)$, $\alpha \in \mathcal{A}(i)$, $i \in \mathcal{S}$, rather than on the estimated potentials $g(i)$ and the transition probabilities $p^{\alpha}(j|i)$, $i, j \in \mathcal{S}$. Of course, we still need to use a random policy that visits all the state-action pairs to generate the sample path in order to estimate the $Q(i, \alpha)$ for all $i \in \mathcal{S}$ and $\alpha \in \mathcal{A}(i)$.

SARSA is a very aggressive approach in implementing the above ideas. This approach may start with any initial values of $Q(i, \alpha)$'s, $i \in \mathcal{S}$ and $\alpha \in \mathcal{A}(i)$. At time l, $l = 0, 1 \ldots$, an action A_l is chosen from the ϵ-greedy policy based on (6.64). This corresponds to the "policy improvement" step in policy iteration. This action determines the transition to X_{l+1}; after the transition, the Q-factor $Q(X_l, A_l)$ is updated according to

$$\begin{aligned} Q(X_l, A_l) &:= Q(X_l, A_l) + \kappa_l \delta_l = Q(X_l, A_l) \\ &+ \kappa_l \left\{ Q(X_{l+1}, A_{l+1}) - Q(X_l, A_l) + f(X_l, A_l) - Q(i^*, \alpha^*) \right\}, \end{aligned} \tag{6.65}$$

with (i^*, α^*) being a pair of the reference state-action. This corresponds to the "policy evaluation" step in policy iteration procedures.

What this algorithm essentially does is as follows: At every step, it learns and updates the values of Q-factors; then, it uses the updated estimation of the Q-factors immediately in determining the greedy action. That is, it updates the policy every step on-line. Therefore, it is very "optimistic" in terms of learning. The convergence of such an aggressive algorithm is not guaranteed. If it does converge, the sample path reaches the optimal deterministic policy, perturbed by a small probability ϵ. One reason for SARSA to converge is that the update of the Q-factor at a transition usually changes the Q-factor only by a very small amount; therefore, this update usually will not change the greedy policy at the next transition. Thus, we may expect that, with SARSA, a policy is in fact used for a relatively long period before it changes to another one, and the Q-factors can be estimated with a reasonable accuracy.

With Potentials

If we know the transition probabilities (or their relative values, as shown in Example 5.1), then we may only need to estimate the potentials rather than the Q-factors to implement optimization. This results in an algorithm that is a combination of the TD(0) for potentials (6.15) and SARSA.

The approach starts with any initial values of $g(i)$'s, $i \in \mathcal{S}$. At time l, $l = 0, 1 \ldots$, an action A_l is chosen from the greedy policy (no need to be ϵ-greedy) according to $\max_{\alpha \in \mathcal{A}(i)} \sum_{j \in \mathcal{S}} p^{\alpha}(j|i)g(j) + f(i, \alpha)$. After a transition to X_{l+1}, the potential is updated according to

$$g(X_l) := g(X_l) + \kappa_l \left\{ g(X_{l+1}) - g(X_l) + f(X_l, A_l) - g(i^*) \right\}, \quad (6.66)$$

where i^* is a chosen reference state.

Essentially, this optimization algorithm applies the TD(0) algorithm (6.15) to update the potentials; however, it does not wait until accurate estimates are obtained before it implements policy improvement; instead, it updates the policy at every step, immediately after one update is made for one potential. Thus, this algorithm is also very "optimistic". Again, the convergence is not guaranteed. Similar to the situation of SARSA, hopefully, the algorithm is stable, because at each step only one potential is updated slightly, and in most cases the "greedy" policy obtained after one step may not change at all.

Obviously, there are a number of variants of this algorithm and SARSA, which may improve the convergence property. For example, we may use the K-step TD(0), or we may update the potentials a few steps before implementing policy iteration.

6.3.4 Value Iteration

As we mentioned, Q-learning algorithms rely on sample paths to obtain information about the transition probabilities $p^{\alpha}(j|i)$'s; these algorithms obtain the temporal differences and use them in stochastic approximation to approach the solution to the optimality equation of the Q-factors (6.58). If we know $p^{\alpha}(j|i)$'s, then we can develop pure numerical methods for the solution to the optimality equations. Now, we briefly discuss the ideas without exploring the convergence nature of the numerical algorithms.

If we set $\kappa_k = 1$ in (6.59), we get

$$Q(i,\alpha) := \left\{ \sum_{j=1}^{S} p^{\alpha}(j|i) \left[\max_{\beta \in \mathcal{A}(j)} Q(j,\beta) \right] \right\} + f(i,\alpha) - \bar{\eta}. \tag{6.67}$$

Because $\bar{\eta}$ is unknown, we need to use the Q-factors in relative values. Thus, we pick up a particular state-action pair (i^*, α^*) and set $Q(i^*, \alpha^*) = \eta^*$. We have

$$Q(i,\alpha) := \left\{ \sum_{j=1}^{S} p^{\alpha}(j|i) \left[\max_{\beta \in \mathcal{A}(j)} Q(j,\beta) \right] \right\} + f(i,\alpha) - Q(i^*, \alpha^*). \tag{6.68}$$

If the algorithm converges, we may expect that it converges to a solution to the optimality equation (6.58) with $Q(i^*, \alpha^*) = \eta^*$.

The algorithm similar to (6.63) is (or equivalently we set $\max_{\alpha \in \mathcal{A}(i^*)} Q(i^*, \alpha) = \bar{\eta}$ in (6.67)):

$$Q(i,\alpha) := \left\{ \sum_{j=1}^{S} p^{\alpha}(j|i) \left[\max_{\beta \in \mathcal{A}(j)} Q(j,\beta) \right] \right\} + f(i,\alpha) - \max_{\alpha \in \mathcal{A}(i^*)} Q(i^*, \alpha), \tag{6.69}$$

in which we only choose a reference state i^*.

However, there are $\sum_{i \in \mathcal{S}} |\mathcal{A}(i)|$ Q-factors. In fact, when $p^{\alpha}(j|i)$'s are known, we may work directly on the optimality equation for potentials (6.56). Applying the numerical recursive algorithm (6.3) to (6.56) yields an algorithm similar to (6.59):

$$g_{k+1}(i) := g_k(i) + \kappa_k \left\{ \max_{\alpha \in \mathcal{A}(i)} \left\{ \sum_{j=1}^{S} p^{\alpha}(j|i) g_k(j) + f(i,\alpha) \right\} - g_k(i) - \bar{\eta} \right\}, \tag{6.70}$$

where $\bar{\eta}$ is the average reward of the greedy policy determined with the current values of $g(i)$'s by using $\max_{\alpha \in \mathcal{A}(i)} \left\{ \sum_{j=1}^{S} p^{\alpha}(j|i) g(j) + f(i,\alpha) \right\}$, $i \in \mathcal{S}$. We may set $\kappa_k = 1$ for all k and get

$$g(i) := \max_{\alpha \in \mathcal{A}(i)} \left\{ \sum_{j=1}^{S} p^{\alpha}(j|i) g(j) + f(i,\alpha) \right\} - \bar{\eta},$$

Again, it is inconvenient to determine $\bar{\eta}$, and we need to use the relative values. Thus, we may pick any state as a reference state i^* and set $g(i^*) = \eta^*$. Then, we obtain the following algorithm

$$g(i) := \max_{\alpha \in \mathcal{A}(i)} \left\{ \sum_{j=1}^{S} p^{\alpha}(j|i) g(j) + f(i,\alpha) \right\} - g(i^*), \qquad i \in \mathcal{S}, \tag{6.71}$$

where $g(i^*)$ is updated in the same way:

$$g(i^*) := \max_{\alpha \in \mathcal{A}(i^*)} \left\{ \sum_{j=1}^{S} p^{\alpha}(j|i^*) g(j) + f(i^*, \alpha) \right\} - g(i^*).$$

This is called the *value iteration* algorithm in the literature. It is the same as the one discussed in [21], which takes the form

$$\begin{aligned} h(i) := & \max_{\alpha \in \mathcal{A}(i)} \left\{ \sum_{j=1}^{S} p^{\alpha}(j|i) h(j) + f(i, \alpha) \right\} \\ & - \max_{\alpha \in \mathcal{A}(i^*)} \left\{ \sum_{j=1}^{S} p^{\alpha}(j|i^*) h(j) + f(i^*, \alpha) \right\}, \\ & \qquad h(i) \equiv g(i) - g(i^*), \qquad i \in \mathcal{S}, \end{aligned}$$

with "max" changed to "min", because performance minimization is discussed there. The proof of convergence for the algorithm is not difficult and we refer readers to any existing text book on MDPs or dynamic programming. Different versions of value iteration exist (e.g., [216]).

It is non-traditional to introduce value iteration together with reinforcement learning. We, however, do so because both are methods for finding solutions to the optimality equations. In addition, we can see that the value iteration algorithm is a special case of the stochastic approximation algorithm (6.70) with step sizes $\kappa_k = 1$. In a sense, Q-learning is half numerical and half sample path based (for information on transition probabilities).

The algorithm indeed converges although the standard condition (6.11), or (6.10), does not hold (no random noise in this case, however). Of course, a more general value iteration algorithm for (6.71) is

$$g_{k+1}(i) := g_k(i) + \kappa_k \left\{ \max_{\alpha \in \mathcal{A}(i)} \left\{ \sum_{j=1}^{S} p^{\alpha}(j|i) g_k(j) + f(i, \alpha) \right\} - g_k(i^*) - g_k(i) \right\}$$

for all $i \in \mathcal{S}$, with the step sizes κ_k chosen properly, which may affect the convergence rate of the value iteration algorithm.

The computation complexity of both (6.68) and (6.71) is about the same: For each state, we need to do one "max" operation and for each state-action pair, we need to do one summation $\sum_{j=1}^{S}$. The difference is that, in (6.68), the "max" operation is implemented before the summation $\sum_{j=1}^{S}$, and in (6.71) the order is reversed. Both algorithms are equivalent. In fact, if we set $g(i) := \max_{\alpha \in \mathcal{A}(i)} Q(i, \alpha)$, $i \in \mathcal{S}$, in (6.69), we get

$$Q(i, \alpha) := \left\{ \sum_{j=1}^{S} p^{\alpha}(j|i) g(j) \right\} + f(i, \alpha) - g(i^*).$$

Taking $\max_{\alpha \in \mathcal{A}(i)}$ on both sides of the above equation, we obtain (6.71). Therefore, the value iteration algorithm based on potentials is simply a special version of the value iteration algorithm based on Q-factors with a clever way of saving memory space in which only $\max_{\alpha \in \mathcal{A}(i)} Q(i, \alpha)$ is recorded for each state $i \in \mathcal{S}$.

6.4 Summary of the Learning and Optimization Methods

By and large, there are two approaches to performance optimization, as illustrated in Table 6.3. The first one is to move step by step in the policy (parameter) space towards an optimal policy (or optimal parameters). This approach includes the policy iteration method for discrete policy spaces and the gradient-based methods for policy spaces with continuous parameters. The second approach is to find a solution to the optimality equations, or to find a zero point of the performance gradients. These two approaches are closely related because the first approach can be viewed as a special way of finding a solution to the optimality equations or finding a zero point of the gradients. The difference is that in the first approach, at every step we are working on a real policy in the policy space, while in the second approach, the intermediate steps may not correspond to any policy.

	Discrete Policy Spaces	Continuous Policy Spaces
Moving Iteratively	D1: Policy Iteration Based	C1: Gradient Based
Finding Zeros	D2: Opt. Eq. (6.56) and (6.58)	C2: $\frac{d\eta_\theta}{d\theta} = 0$

Table 6.3. Different Learning and Optimization Approaches

If we have complete knowledge about the transition probabilities and the reward functions for all the actions (or parameters), we may implement the above two approaches analytically and/or numerically. If the transition probabilities and/or the reward functions are not known, then different sample-path-based or simulation-based learning methods have to be used. Most methods combine the analytical and learning features together.

Let us first consider the problems with discrete policy spaces. We summarize the methods in the first approach (D1 in Table 6.3):

1. When the transition probabilities and the reward functions are known, we can use the standard policy iteration method discussed in Chapter 4; the potentials can be calculated by using the numerical methods discussed in Section 3.1.1; and the policy iteration algorithms produce a sequence of policies that moves towards an optimal policy.

2. Next, suppose that the matrix inversion in solving the Poisson equation is not feasible, or the transition probability matrix is not completely known (cf. Example 5.1 in Section 5.1), or the reward function is unknown but the rewards can be observed.
   a) We may first estimate the potentials of a policy accurately on a sample path and then use them to implement policy iteration.
      i. We may use the sample-path-based (Monte Carlo) algorithms discussed in Chapter 3 to estimate the potentials and then implement policy iteration.
      ii. We may use any of the TD methods discussed in Section 6.2.1 to estimate the potentials and then implement the policy iteration.
   b) We may update the policy while estimating the potentials before an accurate estimate is obtained. (This is called "generalized policy iteration" in [238].)
      i. The sample-path-based (Monte Carlo) optimization algorithms can be improved by using potential estimates with short sample paths; because of the random errors in the estimates, stochastic approximation techniques are used to make sure that the iterations converge to an optimal policy. This leads to the "faster" algorithms discussed in Section 5.3.2. This algorithm updates potentials and policies in every regenerative period.
      ii. A more aggressive algorithm is to apply the optimistic on-line policy iteration algorithm with potentials (6.66). This algorithm updates one potential estimate at every transition and uses the potential estimates to update the action at every transition.
3. Now, suppose that the transition probability matrix is completely unknown and the reward function is also unknown but the rewards can be observed. In this case, we need to use the Q-factors (6.28).
   a) We may first estimate the Q-factors of a policy accurately on a sample path and then use them to implement the policy iteration. To do so, we need to perturb the system a little bit to obtain a sample path visiting all the state-action pairs (e.g., the ϵ-greedy policy).
      i. We may use (6.31) to develop a Monte Carlo algorithm to estimate $Q(i, \alpha)$, $\alpha \in \mathcal{A}(i)$, $i \in \mathcal{S}$.
      ii. We may use any of the TD methods discussed in Section 6.2.2 to estimate the Q-factors and then implement the policy iteration.
   b) We may update the policy while estimating the Q-factors before an accurate estimate is obtained. (That is, we may use the generalized policy iteration.)
      i. We may develop "faster" algorithms similar to those in Section 5.3.2 for potential-based policy iterations.
      ii. A more aggressive algorithm is the SARSA algorithm, which updates the Q-factors at every transition according to (6.65) and updates the action according to the ϵ-greedy policy using these Q-factor estimates in every transition.

These policy-iteration-based learning and optimization methods are summarized in Figure 6.10.

Analytical (P,f known)	Learning $g(i)$ (No matrix inversion, etc)		Learning $Q(i,\alpha)$ (P completely unknown)	
Policy Iteration (Ch.4) Solving Poisson Eq. or by numerical methods for g (Sec. 3.1.1)	Monte Carlo		Monte Carlo	
	Long run accurate est. + PI (Sec. 3.1.2)	Short run noised est. +SA +GPI (Sec. 5.3)	Long run accurate est. + PI (Eq. (6.31))	Short run noised est. +SA +GPI (to be done)
	Temporal Difference		Temporal Difference	
	Long run accurate est. + PI (Sec. 6.2.1)	Short run noised est. +SA +GPI (Sec. 6.3.3)	Long run accurate est. + PI (Sec. 6.2.2)	Short run noised est. +SA +GPI (SARSA)

GPI: Generalized Policy Iteration
PI: Policy Iteration
SA: Stochastic Approximation

Fig. 6.10. Policy-Iteration Based Learning and Optimization Methods

The following methods belong to the second approach (i.e., D2 in Table 6.3):

1. When the transition probabilities and the reward functions are known, we can use the value iteration method discussed in Section 6.3.4, which is a numerical method that produces a solution to the optimality equation (6.56) iteratively.
2. When the transition probabilities are unknown, we can use the Q-learning algorithms discussed in Section 6.3.2, which provide a solution to the optimality equations for Q-factors (6.58). This requires a sample path that visits all the state-action pairs.

These methods are summarized in Table 6.11.

Now, we consider the policy spaces with continuous parameters, denoted as θ. PA-based optimization algorithms can be used.

P, f known (Numerically)	P unknown, f known (Simulation + numerically)
Value Iteration (Section 6.3.4)	Q-Learning (Section 6.3.2)

Fig. 6.11. Finding a Solution to the Optimality Equations

1. When the transition probabilities and the reward functions are known, we can first use formula (2.26) or (2.58) to calculate the performance derivatives, and then we use them in the deterministic gradient-based algorithm (3.51) to obtain a sequence of parameters that converge to the optimal values.
2. Suppose that the system structure is not completely known. Then, sample-path-based methods can be used.
   a) The potentials can be estimated by the Monte Carlo method or the TD methods; this is the same as what is discussed above in 2.a.i and 2.a.ii for the discrete state spaces.
   b) The performance derivatives can also be estimated directly on a sample path by using equations (3.46), (3.47), (3.48), and (3.49).
   c) The performance derivatives can be estimated by using the TD methods developed in Section 6.2.3.
   Because of the estimation errors, the stochastic approximation-based Robbins Monro algorithm has to be used in the gradient-based optimization algorithm.
3. In 2.a, 2.b, and 2.c, the parameters θ are updated after an accurate estimate of the performance gradient is obtained on a long sample path. The method may be improved if we update the parameters based on noisy estimates of the gradients obtained from short sample paths, with the help of stochastic approximation techniques. This leads to the two Robbins Monro algorithms in Section 6.3.1: (6.55), which updates the parameters in every regenerative period, and (6.54), which updates the parameters at every transition. These algorithms can also be viewed as the Robbins Monro algorithms for finding the zeros of the performance gradient function.

The PA-gradient-based learning and optimization methods are summarized in Figure 6.12.

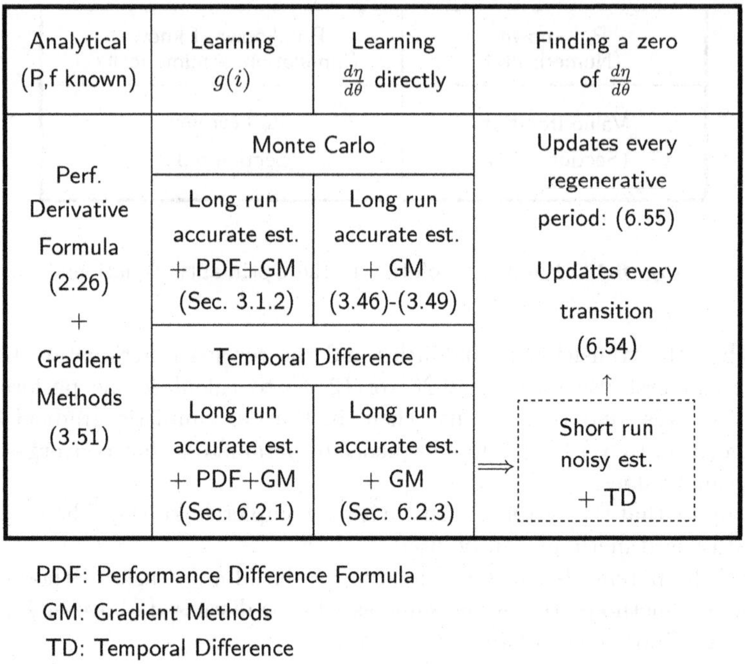

<table>
<tr><td>Analytical
(P,f known)</td><td>Learning
$g(i)$</td><td>Learning
$\frac{d\eta}{d\theta}$ directly</td><td>Finding a zero
of $\frac{d\eta}{d\theta}$</td></tr>
<tr><td rowspan="4">Perf.
Derivative
Formula
(2.26)
+
Gradient
Methods
(3.51)</td><td colspan="2">Monte Carlo</td><td rowspan="2">Updates every
regenerative
period: (6.55)

Updates every
transition
(6.54)</td></tr>
<tr><td>Long run
accurate est.
+ PDF+GM
(Sec. 3.1.2)</td><td>Long run
accurate est.
+ GM
(3.46)-(3.49)</td></tr>
<tr><td colspan="2">Temporal Difference</td><td rowspan="2">↑
⟹ Short run
noisy est.
+ TD</td></tr>
<tr><td>Long run
accurate est.
+ PDF+GM
(Sec. 6.2.1)</td><td>Long run
accurate est.
+ GM
(Sec. 6.2.3)</td></tr>
</table>

PDF: Performance Difference Formula

GM: Gradient Methods

TD: Temporal Difference

Fig. 6.12. PA-Gradient-Based Learning and Optimization Methods

PROBLEMS

6.1. Let us revisit the stochastic approximation algorithm (6.1) for the case with the function $f(x)$ known. In the proof of convergence, we have assumed that the function is convex and $\frac{df(x)}{dx} > 0$. Consider the convex function $f(x) = x^2$ with a zero at $x = 0$ at which $\frac{d(x^2)}{dx} = 0$. Modify the proof in the text to fit this case.

6.2. Study the convergence property of the sequence x_k, $k = 0, 1, \ldots$, in Example 6.1, for the following cases $1 > \kappa > 0$, $2 > \kappa > 1$, $\kappa = 2$, and $\kappa > 2$, respectively, by using the algorithm illustrated in Figure 6.1.

6.3. The algorithm in (6.12) can be used to estimate the mean of a random variable w. This has been verified for step sizes $\kappa_k = \frac{1}{k+1}$, $k = 0, 1, \ldots$, in Section 6.1.2.

a. Study the case for step sizes $\kappa_k = \frac{1}{2(k+1)}$, $k = 0, 1, \ldots$.

b. Choose a few sequences of κ_k, $k = 0, 1, \ldots$, that satisfy conditions (6.11) or (6.10) and run a simulation to see if the sequences of x_k, $k = 0, 1, \ldots$, converge and compare their convergence speeds, if possible.

6.4. Let us revisit Section 6.1.2 "Estimating Mean Values". Assume that the step sizes satisfy $\sum_{k=0}^{\infty} \kappa_k = \infty$ and $\sum_{k=0}^{\infty} \kappa_k^2 < \infty$. Working on (6.12) recursively, we may obtain

$$x_{k+1} = a_k x_0 + \xi_k.$$

a. Derive an expression of a_k and ξ_k in terms of $\kappa_0, \ldots, \kappa_k$ and $w_0, \ldots, w_k$.
b. Prove $\lim_{k\to\infty} a_k = 0$, $\lim_{k\to\infty} E(\xi_k) = E(w)$, and $\lim_{k\to\infty} \mathrm{var}(\xi_k) = 0$.

6.5. Consider the estimation of the average reward of a continuous time Markov process. Let $\{T_0, T_1, \ldots, T_l, \ldots\}$ be the sequence of transition times of the continuous Markov process with $T_0 = 0$. The state in the time period $[T_l, T_{l+1})$ is X_l, $l = 0, 1, \ldots$, and set $\tau_l = T_{l+1} - T_l$, $l = 0, 1, \ldots$. The reward function is $f(X_l)$ and the average reward is defined as

$$\eta = \lim_{l\to\infty} \eta_l, \qquad \text{w.p.1}, \qquad \eta_l := \frac{1}{T_l} \int_0^{T_l} f\left[X(t)\right] dt.$$

We wish to develop a recursive formula for η_l as follows:

$$\eta_{l+1} = \eta_l + \kappa_l \left[f(X_l) - \eta_l\right], \qquad l = 0, 1, \ldots, \quad \text{with } \eta_0 = 0.$$

Please find the value of κ_l, $l = 0, 1 \ldots$, in terms of T_l, etc.

6.6. Derive the TD(0) algorithm for the discounted performance criterion:

$$\eta_\beta(i) = E\left\{ \sum_{l=0}^{\infty} \beta^l f(X_l) \Big| X_0 = i \right\}, \qquad 1 > \beta > 0.$$

6.7. *TD (0) with random steps:* For any two states $i, j \in \mathcal{S}$, set $\mathcal{S}_0 = \{i, j\}$. Consider a sample path of a Markov chain $\{X_0, \ldots, X_l, \ldots\}$. Denote the time sequence at which the Markov chain is in $\mathcal{S}_0$ as $l_0, l_1, \ldots, l_k, \ldots$. We may set $g(i) = 0$.

a. Develop a TD(0) algorithm for estimating $g(j)$ by using the temporal differences obtained in the periods from $l_k + 1$ to l_{k+1}, $k = 0, 1, \ldots$.
b. Explain that the algorithm converges to the correct value, compare it with the realization factor $\gamma(i, j) = g(j) - g(i)$.

6.8. Consider a two-state Markov chain with transition probability matrix

$$P = \begin{bmatrix} 0.5 & 0.5 \\ 0.5 & 0.5 \end{bmatrix}$$

and reward function $f(1) = 1$ and $f(0) = 0$. We have $\eta = \frac{1}{2}$.

a. What are the potentials for the two states?
b. Write a computer program applying algorithms (6.15), (6.22), and (6.24), and observe the trends of the convergence of the sequences generated by these algorithms. (For Algorithm (6.22), observe the trend of convergence of $g(1) - g(0)$.)

6.9. The TD(0) algorithm (6.15) and (6.16) can only determine the potentials up to an additive constant. That is, starting from different initial values, the algorithm converges to different sets of potentials that have the same perturbation realization factors $\gamma(i,j)$, $i,j \in \mathcal{S}$.

a. Can we fix a reference state i^* and set $g(i^*) = 0$ in the TD(0) algorithm (6.15) and (6.16)?
b. If so, modify the algorithm.
c. Explain your algorithm using

$$g(i) = \gamma(i^*, i) = E\left\{ \sum_{l=0}^{L(i^*|i)-1} [f(X_l) - \eta] \,\middle|\, X_0 = i \right\}.$$

d. Apply this algorithm to the Markov chain in Example 6.5.

6.10. Consider the modified algorithm (6.21).

a. Can we fix a reference state i^* and set $g(i^*) = 0$ in (6.21), as we considered in Problem 6.9? (To find the answer, apply it to the Markov chain in Example 6.5.)
b. If not, why?

6.11. Derive an iterative numerical algorithm similar to the algorithm in (6.20) for potentials by using Equation (3.4).

6.12. Consider a finite state discrete-time birth-death problem: The state space is $\mathcal{S} = \{0, 1, 2, \ldots, S\}$. The state is the population $n \in \mathcal{S}$. The transition probability from state n to $n+1$ (the birth rate) is $p(n+1|n) = a$, $n = 1, \ldots, S-1$, and the death rate is $p(n-1|n) = b$, $n = 1, 2, \ldots, S-1$, $a + b = 1$; and $p(1|0) = p(S-1|S) = 1$. Let the reward function be $f(n) = n$. The performance is defined as the long-run average of the population $\eta = \sum_{n=0}^{S} \pi(n) f(n)$, where $\pi(n)$ denotes the steady-state probability of state n, $n = 0, 1, \ldots, S$.

a. Derive a formula expressing the performance η as a function of the birth rate a.
b. Set $a = \frac{1}{2}$. Use the derivative formula (6.43) to derive the performance derivative $\left.\frac{d\eta}{da}\right|_{a=\frac{1}{2}}$.
c. Develop a TD(0) algorithm for estimating $\left.\frac{d\eta}{da}\right|_{a=\frac{1}{2}}$.

6.13. Consider a randomized policy d_r. Denote $\mathcal{A}(i) := \{\alpha_{i,1}, \ldots, \alpha_{i,|\mathcal{A}(i)|}\}$, where $|\mathcal{A}(i)|$ is the number of actions in $\mathcal{A}(i)$, $i \in \mathcal{S}$. In state i, the system takes action $\alpha_{i,k} \in \mathcal{A}(i)$ with probability $p_{i,k}$, $k = 1, 2, \ldots, |\mathcal{A}(i)|$, and $\sum_{k=1}^{|\mathcal{A}(i)|} p_{i,k} = 1$, $i \in \mathcal{S}$. If action $\alpha \in \mathcal{A}(i)$ is taken in state i, then the transition probabilities are $p^{\alpha}(j|i)$, $j \in \mathcal{S}$, and the performance function is $f(i, \alpha)$, $i \in \mathcal{S}$. The Q-factors are defined in (6.28) as follows

$$Q^{d_r}(i, \alpha) = \left\{ \sum_{j=1}^{S} p^{\alpha}(j|i) g^{d_r}(j) \right\} + f(i, \alpha) - \eta^{d_r}, \qquad \alpha \in \mathcal{A}(i), \quad i \in \mathcal{S},$$

where $g^{d_r}(i)$, $i \in \mathcal{S}$, are the performance potentials of the system under this randomized policy d_r.

a. Determine the performance function and transition probabilities for the system under this randomized policy; derive the Poisson equation for it.
b. Prove that $g^{d_r}(i) = \sum_{k=1}^{|\mathcal{A}(i)|} p_{i,k} Q^{d_r}(i, \alpha_{i,k})$.
c. Given a deterministic policy $d(i) = \alpha_i^* \in \mathcal{A}(i)$, $i \in \mathcal{S}$, we define an ϵ-randomized policy: With probability $1 - \epsilon$ the system takes action α_i^*, and with probability $\epsilon/(|\mathcal{A}(i)| - 1)$ it takes any other actions in $\mathcal{A}(i)$, $i \in \mathcal{S}$. Let $g(i)$ be the potentials of the deterministic policy d, and $g_\epsilon(i)$, and $Q_\epsilon(i, \alpha)$, $\alpha \in \mathcal{A}(i)$, $i \in \mathcal{S}$, be the potentials and Q-factors of the ϵ-randomized policy. Prove

$$\lim_{\epsilon \to 0} g_\epsilon(i) = g(i), \qquad i \in \mathcal{S},$$

$$\lim_{\epsilon \to 0} Q_\epsilon(i, \alpha_i^*) = g(i), \qquad i \in \mathcal{S},$$

and

$$\lim_{\epsilon \to 0} Q_\epsilon(i, \alpha) = \left\{ \sum_{j=1}^{S} p^{\alpha}(j|i) g(j) \right\} + f(i, \alpha) - \eta, \qquad \alpha \neq \alpha_i^*, \quad i \in \mathcal{S}.$$

6.14. Suppose that we can only control the actions in the states in a subset of the state space $\mathcal{S}_0 \subset \mathcal{S}$ of a Markov chain, which is under a randomized policy that visits all the state-action pairs when the state is in $\mathcal{S}_0$. Denote the time sequence at which the Markov chain is in $\mathcal{S}_0$ as $l_0, l_1, \ldots, l_k, \ldots$; i.e., $X_{l_k} \in \mathcal{S}_0$, $k = 0, 1, \ldots$. Develop a TD(0) algorithm for Q-factors $Q(i, \alpha)$, $i \in \mathcal{S}_0$, with random steps K.

6.15. Develop a K-step algorithm for estimating the Q-factors (cf., (6.33) and (6.34)).

6.16. In (6.33) and (6.34), we may set the Q-factor of a pair of the reference state-action to be zero; i.e., $Q(i^*, \alpha^*) = 0$. Develop a TD(0)-learning algorithm.

6.17. We partition the state space $\mathcal{S}$ into S_0 subsets: $\mathcal{S} = \cup_{k=1}^{S_0} \mathcal{S}_k$, $\mathcal{S}_k \cap \mathcal{S}_{k'} = \emptyset$, $k, k' = 1, 2, \ldots, S_0$. Let $\pi(i)$, $i \in \mathcal{S}$, be the steady-state probability, and $\pi(i|\mathcal{S}_k) = \frac{\pi(i)}{\sum_{j \in \mathcal{S}_k} \pi(j)}$ be the conditional steady-state probability of i given that $i \in \mathcal{S}_k$. The potential associated with the aggregation $\mathcal{S}_k$ is defined as (6.39):

$$g(\mathcal{S}_k) = \sum_{i \in \mathcal{S}_k} \pi(i|\mathcal{S}_k) g(i).$$

We wish to establish a Poisson equation for the aggregations:

$$g(\mathcal{S}_k) = \sum_{k'=1}^{S_0} p(\mathcal{S}_{k'}|\mathcal{S}_k) g(\mathcal{S}_{k'}) + f(\mathcal{S}_k) - \eta, \qquad k = 1, 2, \ldots, S_0. \tag{6.72}$$

a. According to their physical meanings, determine the transition probabilities $p(\mathcal{S}_{k'}|\mathcal{S}_k)$ and the performance function $f(\mathcal{S}_k)$, $k, k' = 1, 2, \ldots, S_0$.
b. Prove that the Poisson equation (6.72) holds for the aggregations if, for any $\mathcal{S}_{k'}$, $k' = 1, \ldots, S_0$, and any $j \in \mathcal{S}_{k'}$, we have

$$\frac{\pi(j)}{\sum_{j' \in \mathcal{S}_{k'}} \pi(j')} = \frac{\sum_{i \in \mathcal{S}_k} \pi(i) p(j|i)}{\sum_{j' \in \mathcal{S}_{k'}} \sum_{i \in \mathcal{S}_k} \pi(i) p(j'|i)}, \qquad k = 1, \ldots, S_0. \tag{6.73}$$

c. Set

$$\pi(j|\mathcal{S}_{k'}, \mathcal{S}_k) = \frac{\sum_{i \in \mathcal{S}_k} \pi(i) p(j|i)}{\sum_{j' \in \mathcal{S}_{k'}} \sum_{i \in \mathcal{S}_k} \pi(i) p(j'|i)}.$$

Then (6.73) becomes $\pi(j|\mathcal{S}_{k'}, \mathcal{S}_k) = \pi(j|\mathcal{S}_{k'})$. Prove that (6.73) is equivalent to the following condition:

$$\pi(j|\mathcal{S}_{k'}, \mathcal{S}_k) \text{ is independent of } k. \tag{6.74}$$

d. Explain the meaning of $\pi(j|\mathcal{S}_{k'}, \mathcal{S}_k)$ and condition (6.74).
e. Derive a TD(0) algorithm for $g(\mathcal{S}_k)$, $k = 1, \ldots, S_0$.
f. Explain that the algorithm developed in e) may not work if the condition (6.74) does not hold.

6.18. In perturbation analysis of Markov chains, we have two Markov chains with transition probability matrices P and P', respectively. Let $\Delta P = P' - P$ and $P_\delta = P + \delta \Delta P$. Let η_δ be the long-run average reward of the Markov chain with transition probability matrix P_δ. Assume that the reward function f_δ is the same as f for all $0 \le \delta \le 1$. Let π and g be the steady-state probability and performance potential of the Markov chain with transition probability matrix P. Then the directional derivative of η_δ is (2.23):

$$\frac{d\eta_\delta}{d\delta} = \pi(\Delta P) g.$$

a. Write the performance derivative in the form of Q-factors.

b. Suppose that we do not know the values of P and P' and only know the corresponding actions. Develop a TD(0)-learning algorithm for the performance derivative.

6.19. Develop two TD(0) algorithms, similar to (6.44) and (6.45), based on the performance derivative formula (3.44).

6.20. Suppose that algorithms (6.62) and (6.63) converge to optimal Q-factors. Are the following statements true? If so, please explain

a. With (6.62), when the algorithm converges, we have $Q(i^*, \alpha^*) = \eta^*$, the optimal performance.
b. With (6.63), when the algorithm converges, we have $\max_{\alpha \in \mathcal{A}(i^*)} Q(i^*, \alpha) = \eta^*$.

6.21.* In this problem, we derive a performance derivative formula for closed Jackson networks in the form of sample path expectation. Consider a closed Jackson network consisting of M servers and N customers. The service times of server i are exponentially distributed with mean $\bar{s}_i = 1/\mu_i$, $i = 1, 2, \ldots, M$. The state of the system is $\boldsymbol{n} = (n_1, \ldots, n_M)$, n_i is the number of customers in server i, $\sum_{i=1}^M n_i = N$. Suppose that the system is in the steady state, and let $\pi(\boldsymbol{n})$ be the steady-state probability of state $\boldsymbol{n}$. Denote $\mu(\boldsymbol{n}) = \sum_{i=1}^M \epsilon(n_i)\mu_i$, with $\epsilon(n) = 1$ if $n > 0$ and 0 if $n = 0$. The system throughput is $\eta = \sum_{\text{all } \boldsymbol{n}} \pi(\boldsymbol{n})\mu(\boldsymbol{n})$, its derivative with respect to $\bar{s}_v$, $v = 1, 2, \ldots, N$, is (2.109):

$$\frac{\bar{s}_v}{\eta} \frac{\partial \eta}{\partial \bar{s}_v} = - \sum_{\text{all } \boldsymbol{n}} \pi(\boldsymbol{n}) c(\boldsymbol{n}, v),$$

where $c(\boldsymbol{n}, v)$ is the realization probability of a perturbation of server v when the system is in state $\boldsymbol{n}$.

a. Consider a sample path of the system. Denote the sequence of transition times as $T_0, T_1, \ldots, T_l, \ldots$. Suppose that the system is in state $\boldsymbol{n}$ in $[T_l, T_{l+1})$; i.e., $X_l = \boldsymbol{n}$. Assume that, in this period, server v obtains an (infinitesimal) perturbation. We define a *perturbation realization index* for this perturbation as follows:

$$RI(l, X_l, v) = \begin{cases} 1, & \text{if the perturbation is realized on the sample path,} \\ 0, & \text{otherwise;} \end{cases}$$

and set $\varsigma(t) = RI(l, X_l, v)$ for $t \in [T_l, T_{l+1})$. Then, by definition, we have

$$E\left[RI(l, X_l, v) | X_l = \boldsymbol{n}\right] = c(\boldsymbol{n}, v),$$

where "E" denotes the expectation with respect to the probability space generated by all the sample paths. Explain the following equation:

$$\frac{\bar{s}_v}{\eta}\frac{\partial \eta}{\partial \bar{s}_v} = -E\left[RI(l, X_l, v)\right] = -E\left[\varsigma(t)\right]$$
$$= \lim_{L\to\infty} \frac{1}{T_L}\int_0^{T_L} \varsigma(t)dt, \qquad \text{w.p.1.} \qquad (6.75)$$

b. Can you determine the function $\varsigma(t)$ based on the sample path in Figure 2.18 of Chapter 2? *(Note: $\varsigma(t)$ depends not only on the current state $\boldsymbol{n}$, but also on the future behavior of the system.)*
c. Derive a sample-path-based estimate of the performance derivative by using the above result.
d. Apply this equation (6.75) to a two-server closed Jackson network and verify the results. Can this be extended to networks with non-exponentially distributed service times?
e. Derive a recursive algorithm (cf. Problem 6.5).
f. Discuss and compare your results with other algorithms.

Never follow the beaten track, it leads only where others have been before.

Alexander Graham Bell,
American (Scottish-born)
scientist and inventor,
(1847 - 1922)

7 Adaptive Control Problems as MDPs

Adaptive control and identification theory for stochastic systems was developed in the last few decades and is now very mature. Many excellent textbooks exist, see e.g., [9, 165, 192, 193, 206]. There has been a continuing discussion of what adaptive control is. In general, the problems studied in this area involve systems whose structures and/or parameters are unknown and/or are time-varying, However, to precisely define adaptive control is not an easy job [9, 206]. In [9], adaptive control is viewed as a special type of feedback control in which the states of the process can be separated into two categories that change at different rates. The slow-changing states are viewed as a part of the system parameters. Thus, the goal of adaptive control is to design a control scheme that works well for systems with time-varying parameters.

Adaptive control can be categorized as direct and indirect. With an indirect adaptive control scheme, the system parameters are first estimated and the controller is then determined using the estimated values as the true ones for the parameters. On the other hand, with a direct adaptive control scheme, parameters of the controller are directly identified without knowing and/or estimating all the system parameters.

We take a slightly wider point of view. We view any control problem with unknown parameters as an adaptive control problem. Thus, a time-invariant system is simply a special case of adaptive control. We determine or identify, directly or indirectly, the optimal control law frequently even if the system is time-invariant. Indeed, even for time-invariant systems, we always need to continuously track the system to adjust our estimates for the parameters or the optimal controllers. Indirect adaptive control is usually equivalent to system identification plus the standard control problem. If we assume that the identification process is much faster than the parameter changes, then the identification-plus-control approach works well for time-varying systems.

The purpose of this chapter is not to provide an overall review of the adaptive control theory, but rather to apply one, perhaps non-standard, way to view the adaptive control problem and explore the possibility of solving it with this view. The fundamental idea is that a stochastic system can be viewed as a Markov system and therefore the approaches developed in the previous chapters, such as perturbation analysis (PA), Markov decision processes (MDPs), and reinforcement learning (RL), can be applied to develop learning and optimization algorithms for stochastic systems. This learning-and-optimization-based approach is similar to direct adaptive control because it focuses on the optimal policy and the system structures and parameters may not be estimated. The approach has been studied previously by many authors, see, e.g, [5, 30, 89, 124, 255]. We will introduce the possible advantages of this approach and the difficulties we may encounter with this approach.

We consider only the long-run average reward and, in most cases, we need to assume that the system under consideration is stable. For simplicity, we study only discrete-time systems.

In Section 7.1, we explain how a control problem is modelled as an MDP problem and discuss the basic differences of the two approaches: dynamic programming, which is widely used in adaptive control, and policy iteration in MDPs. In Section 7.2, we present the MDP theory for systems with continuous state spaces [24, 135, 136, 203]; this is necessary because, in most adaptive control problems, the system states are continuous. In Section 7.3, we first discuss the linear-quadratic (LQ) problem and show that we can derive the standard Riccati equation in adaptive control theory via policy iteration. We further discuss the Markov jump linear-quadratic (JLQ) problem and obtain the coupled Riccati equation. This section serves as a test showing that the proposed approach with policy iteration based on performance potentials works well and produces the same results as those in the literature on adaptive control for the LQ and JLQ problems. In Section 7.4, we study the possible applications of the sample-path-based policy iteration to the adaptive control of non-linear systems and discuss the difficulties and advantages of this approach.

7.1 Control Problems and MDPs

7.1.1 Control Systems Modelled as MDPs

Linear Systems

Consider a linear system with an additive noise:

$$X_{l+1} = AX_l + \xi_l, \qquad l = 0, 1, \ldots, \tag{7.1}$$

where $l = 0, 1, \ldots$ denotes the discrete time, $X_l \in \Re^n$, $\Re = (-\infty, +\infty)$, is an n-dimensional vector representing the system state at time l, A is an $n \times n$ square matrix, and ξ_l is an n-dimensional random vector representing the noise at l. We assume that the system is stable so that it returns to the vicinity of the origin infinitely often.

To illustrate the idea, we first study the one-dimensional case with $n = 1$. Thus, $A = a$ is a real constant and ξ_l is a random variable. The system is stable when $|a| < 1$. We further assume that the random noise ξ_l, at $l = 0, 1, \ldots,$ is independent and identically distributed with a distribution function denoted as $P_\xi(dx) = \nu(x)dx$, $x \in \Re$.

Now, suppose that at some time l, $X_l = x \in \Re$. Then, we have $X_{l+1} = ax + \xi_l$. As shown in Figure 7.1, X_{l+1} is distributed around ax, and $y := X_{l+1} - ax$ has a distribution function $P_\xi(dy)$, $y \in \Re$. This can be viewed as a state transition from $X_l = x$ to $X_{l+1} \in \Re$ with the *transition probability function* (or simply the transition function)

$$P(dy|x) = \nu(y - ax)dy. \tag{7.2}$$

By (7.2), the linear system behaves as a Markov process with a continuous state space.

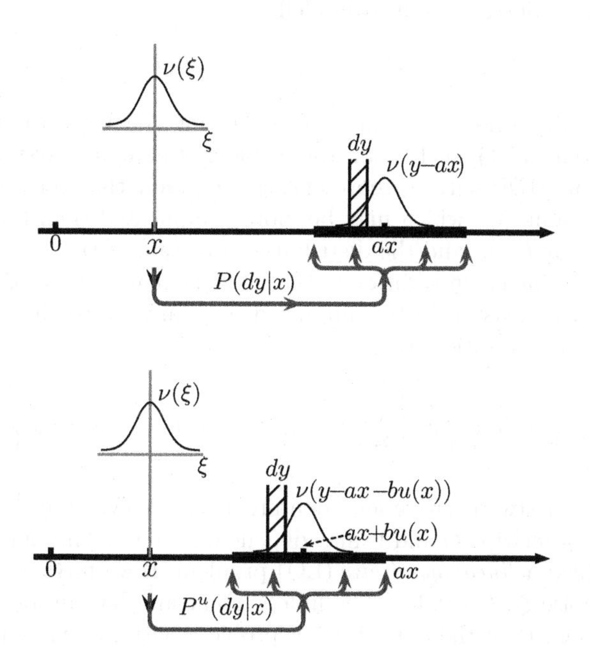

Fig. 7.1. The State Transition in a Linear Stochastic System

Linear Control Systems

Now, we introduce control to the linear system (7.1). Consider

$$X_{l+1} = AX_l + Bu_l + \xi_l, \qquad l = 0, 1, \ldots, \tag{7.3}$$

where u_l is an m-dimensional vector representing the control applied to the system at time l and B is an $n \times m$ matrix. The feedback control law is an m-dimensional function of the state denoted as $u_l(X_l) : \Re^n \to \Re^m$, $l = 0, 1, \ldots$. If it is independent of l, the control low is called *stationary* and is denoted as $u(x) : \Re^n \to \Re^m$. With a given $u(x)$, the system becomes

$$X_{l+1} = AX_l + Bu(X_l) + \xi_l, \qquad l = 0, 1, \ldots.$$

When $n = m = 1$, $A = a$ and $B = b$ are two constants. With a given control law $u(x) : \Re \to \Re$, the transition probability function (7.2) at x becomes

$$P^u(dy|x) = \nu(y - ax - bu(x))dy. \tag{7.4}$$

The superscript $u(x)$ indicates the dependency on the control variable. This is also shown in Figure 7.1. Clearly, $u(x)$ plays the same role as the actions do in MDPs, and the control function u is the same as a policy.

In general, a control system is modelled as

$$X_{l+1} = h(X_l, u_l, \xi_l), \qquad l = 0, 1, \ldots, \tag{7.5}$$

where h is usually a nonlinear function. Following the same argument as for the linear systems (7.1), such a control problem of nonlinear systems can also be viewed as an MDP with u_l representing the policy that controls the state transition probabilities, which may be more complicated than (7.4), but can be determined by h, u_l, and the distribution function of ξ_l.

In control problems, in addition to the system equation (7.3) or (7.5), there is a performance measure to be minimized. In many cases, the performance measure takes a quadratic form:

$$\eta^u(x) = \lim_{L \to \infty} \frac{1}{L} E \left\{ \sum_{n=0}^{L-1} \left[X_l^T Q X_l + u_l^T V u_l \right] \Big| X_0 = x \right\}, \tag{7.6}$$

where Q and V are two positive semi-definite matrices. (We only study stationary problems; otherwise, Q and V may depend on time l.) This optimal control problem is called a *linear-quadratic* (LQ) problem. If we further assume that the random noise ξ_l, $l = 0, 1, \ldots$, is independent and has an identical Gaussian distribution, then the optimization problem with system equation (7.3) and performance criterion (7.6) is called a *linear-quadratic-Gaussian (LQG)* problem.

In summary, a linear control system (7.3) with performance measure (7.6) can be modelled as an MDP problem with transition probability function (7.4) and performance measure (7.6). Non-linear control systems can be modelled in a similar way with the transition probability functions properly determined by (7.5). The state spaces are usually continuous, and the relevant results will be discussed in Section 7.2. However, for practical purposes, we can always discretize a continuous state space to obtain a finite space approximation.

7.1.2 A Comparison of the Two Approaches

The standard approach to control problems (7.3) and (7.6) is based on dynamic programming. In this section, we make a comparison between dynamic programming and the policy iteration in MDPs. To simplify the explanation, we consider systems with finite state spaces.

The finite-state MDP problem with stationary policies is described in Chapter 4. For dynamic programming, we need to consider non-stationary policies for finite-horizon problems; therefore, we need to modify the terminology slightly. Basically, we have a state space $\mathcal{S} = \{1, 2, \ldots, S\}$. In every state $i \in \mathcal{S}$, we can choose an action α in a subset $\mathcal{A}(i) \subseteq \mathcal{A}$, where $\mathcal{A}$ is the action space, and the action determines the state transition probabilities from state i, denoted as $p^{\alpha}(j|i)$, $j \in \mathcal{S}$, $\alpha \in \mathcal{A}(i)$. A *decision rule* is a mapping $d : \mathcal{S} \to \mathcal{A}$ with $d(i) \in \mathcal{A}(i)$ for all $i \in \mathcal{S}$. The transition probability matrix depends on d and is denoted as P^d. Associated with each decision rule d is a performance (reward or cost) function (vector) $f^d = (f(1, d(1)), \ldots, f(S, d(S))^T$. Therefore, a decision rule d corresponds to the pair (P^d, f^d).

Suppose that a decision rule d_l is used at time $l = 0, 1, 2, \ldots$. Then the sequence $\boldsymbol{d} := \{d_0, d_1, \ldots\}$ is called a *policy*. The sequence can be either finite or infinitely long. The quantities associated with a policy $\boldsymbol{d}$ are indicated with superscript $\boldsymbol{d}$, e.g., the steady-state probability vector, if it does exist, is denoted as $\pi^{\boldsymbol{d}}$. If $d_l \equiv d$ for all $l = 0, 1, \ldots$, i.e., $\boldsymbol{d} = \{d, d, \ldots\}$, we call $\boldsymbol{d}$ a stationary policy and simply denote it as d. All the policies studied in Chapter 4 are stationary policies and therefore are denoted in the same notation as the decision rules.

Assume that the system is ergodic under every stationary policy. The goal of an MDP problem is to find a stationary policy that maximizes (or minimizes) the long-run average reward

$$\eta^d = \lim_{L \to \infty} \left\{ \frac{1}{L} \sum_{l=0}^{L-1} f(X_l, d(X_l)) \right\}, \qquad \text{w.p.1.}$$

Because of the ergodicity, the limit exists and equals

$$\eta^d = \lim_{L \to \infty} \left\{ \frac{1}{L} \sum_{l=0}^{L-1} E\left[f(X_l, d(X_l)) \middle| X_0 = i \right] \right\}, \tag{7.7}$$

which is independent of the initial state i.

Dynamic Programming for Finite-Step Optimization Problems

As discussed in Chapter 4, this problem can be solved by policy iteration. Now we show that, in addition to the policy iteration approach, the problem can also be solved by taking the limit of the solutions to finite-step optimization problems described as follows. For any finite integer $L > 0$, define the L-step average reward in the period $[0, L-1]$ as

$$\eta_L^{\boldsymbol{d}}(i) = \frac{1}{L}\sum_{l=0}^{L-1} E\left[f(X_l, d_l(X_l))|X_0 = i\right], \qquad i \in \mathcal{S}. \tag{7.8}$$

The goal of the L-step optimization problem is to maximize this L-step performance by choosing a proper decision rule at every step. It is obvious that for a finite L the optimal policy may not be a stationary policy; so we denote the L-step policy in (7.8) as $\boldsymbol{d} := \{d_0, d_1, \ldots, d_{L-1}\}$.

If we only wish to optimize $\eta_L^{\boldsymbol{d}}(i)$ for a particular state i in $\mathcal{S}$, we only need to determine the optimal action $d_0(i)$, at $l = 0$. We will see that, because of the Markov property, the initial state i does not affect the optimal decision rules at times $l = 1, \ldots, L-1$; that is, the optimal decision rules, denoted as $\hat{d}_1, \ldots, \hat{d}_{L-1}$, are the same for different initial states $i \in \mathcal{S}$.

First, we observe that maximizing the L-step average (7.8) is the same as maximizing the L-step sum $\sum_{l=0}^{L-1} E\left[f(X_l, d_l(X_l))|X_0 = i\right]$. This simplifies the notation and therefore we change the performance criterion to

$$\eta_L^{\boldsymbol{d}}(i) = \sum_{l=0}^{L-1} E\left[f(X_l, d_l(X_l))|X_0 = i\right], \qquad i \in \mathcal{S}. \tag{7.9}$$

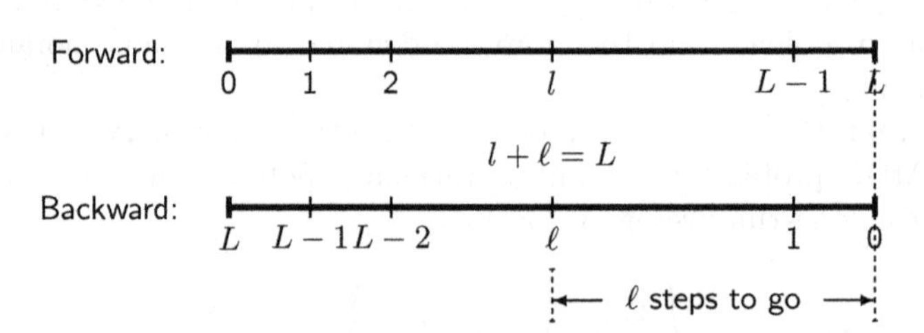

Fig. 7.2. Forward and Backward Indexes

This finite step optimization problem is typically solved by *dynamic programming*. We first solve the one-step problem at $l = L-1$, then the two-step problem at $l = L-2$, and so on. This iterative process goes backward in time. (A typical example of an L-step optimal problem (7.9) is the shortest-path problem illustrated in Problem 1.2.) Therefore, it is convenient to use an index to denote the time backwards. As shown in Figure 7.2, we denote the time epoch $l = L$ as $\ell = 0$, $l = L-1$ as $\ell = 1$, and so on, with $l + \ell = L$. The initial time $l = 0$ corresponds to $\ell = L$, and ℓ represents the number of steps to go in a finite-step optimization problem. With this notation, the state sequence is denoted as $\boldsymbol{X} = \{X_\ell, \ \ell = L, L-1, \ldots, 1, 0\} = \{X_{\ell=L}, X_{\ell=L-1}, \ldots, X_{\ell=1}, X_{\ell=0}\}$, with $X_{\ell=l} := X_{L-l}$, $l = 0, 1, \ldots, L$. The corresponding decision rule sequence is denoted as $\boldsymbol{d} = \{d_{\ell=L}, d_{\ell=L-1}, \ldots, d_{\ell=1}, d_{\ell=0}\}$, with $d_{\ell=l} := d_{L-l}$, $l = 0, 1, \ldots, L$. Furthermore, (7.9) becomes

$$\eta_{\ell=L}^{\boldsymbol{d}}(i) = \sum_{l=L}^{1} E\left[f(X_{\ell=l}, d_{\ell=l}(X_{\ell=l}))|X_{\ell=L} = i\right], \qquad i \in \mathcal{S}. \tag{7.10}$$

In the L-step optimization problem, we wish to choose $d_{\ell=1}, d_{\ell=2}, \ldots, d_{\ell=L}$ to maximize $\eta_{\ell=L}^{\boldsymbol{d}}(i)$. From (7.10), we have

$$\eta_{\ell=L+1}^{\boldsymbol{d}}(i) = f(i, d_{\ell=L+1}(i)) + \sum_{j=1}^{S} p^{d_{\ell=L+1}(i)}(j|i)\eta_{\ell=L}^{\boldsymbol{d}}(j).$$

Let $\eta_{\ell=L}^*(i)$ be the optimal L-step performance. Then, for $L = 1, 2, \ldots,$ we have

$$\eta_{\ell=L+1}^*(i) = \max_{\alpha \in \mathcal{A}(i)} \left\{ f(i,\alpha) + \sum_{j=1}^{S} p^\alpha(j|i)\eta_{\ell=L}^*(j) \right\}, \tag{7.11}$$

and the optimal decision rule is

$$\hat{d}_{\ell=L+1}(i) \in \arg\left[\max_{\alpha \in \mathcal{A}(i)} \left\{ f(i,\alpha) + \sum_{j=1}^{S} p^\alpha(j|i)\eta_{\ell=L}^*(j) \right\} \right]. \tag{7.12}$$

The L-step optimization problem can be solved iteratively by using (7.11) and (7.12) for $L = 1, 2, \ldots$ (backwards in time). The initial value for the optimal performance in this iterative procedure is

$$\eta_{\ell=1}^*(i) = \max_{\alpha \in \mathcal{A}(i)} f(i,\alpha), \tag{7.13}$$

and

$$\hat{d}_{\ell=1}(i) \in \arg\left\{ \max_{\alpha \in \mathcal{A}(i)} f(i,\alpha) \right\}.$$

From (7.12) and the backward nature of the optimization procedure, we can observe that, in the L-step optimization problem, the optimal decision rules at step l, $\hat{d}_{\ell=l}$, $l = 1, 2, \ldots, L-1$, in fact depend only on l and do not depend on L and the initial state $X_{\ell=L} = i$. In other words, if $\left\{\hat{d}_{\ell=L}, \ldots, \hat{d}_{\ell=1}, \hat{d}_{\ell=0}\right\}$ is the optimal policy for the L-step optimization problem, then for any $l < L$, $\left\{\hat{d}_{\ell=l}, \ldots, \hat{d}_{\ell=1}, \hat{d}_{\ell=0}\right\}$ is the optimal policy for the l-step optimization problem.

Finally, as $L \to \infty$, the L-step average (7.8) becomes the long-run average (7.7), and the L-step problem becomes the long-run average-reward optimization problem. Thus, if as $L \to \infty$, $\hat{d}_{\ell=L}(i)$ converges to a decision rule $\hat{d}$, then the stationary policy $\hat{d}$ must be the optimal policy for the long-run average-reward problem. That is, the long-run average-reward optimization problem can be solved by taking the limit of the solutions to the finite step average-reward problems.

Relation to Policy Iteration

The approach sketched above is different from the policy iteration approach. In fact, the two approaches can be viewed as two different ways to solve the optimality equation, and we can prove that they are equivalent. In (7.11), the optimal L-step performance is

$$\eta^*_{\ell=L}(i) = \sum_{l=L}^{1} E\left[f(X_{\ell=l}, \hat{d}_{\ell=l}(X_{\ell=l})) \middle| X_{\ell=L} = i \right], \tag{7.14}$$

where $\hat{d}_{\ell=l}$, $l = 1, 2, \ldots, L$, are the optimal decision rules. As $L \to \infty$, this quantity may not be bounded. However, as we observed many times in the previous chapters, because $\sum_{j=1}^{S} p^{\alpha}(j|i) = 1$, (7.12) remains the same if we add a constant to $\eta^*_{\ell=L}(j)$ for all $j \in \mathcal{S}$, $L = 1, 2, \ldots$. Therefore, we may choose any state i^* as a reference state and re-define

$$\begin{aligned} g^*_{\ell=L}(i^*) &:= 0, \qquad L = 1, 2, \ldots, \\ g^*_{\ell=L}(i) &:= \eta^*_{\ell=L}(i) - \eta^*_{\ell=L}(i^*), \qquad i \neq i^*, \quad L = 1, 2, \ldots. \end{aligned} \tag{7.15}$$

This is the same concept as the perturbation realization factor. Subtracting $\eta^*_{\ell=L}(i^*)$ from both sides of (7.11), we obtain

$$g^*_{\ell=L+1}(i) + \{\eta^*_{\ell=L+1}(i^*) - \eta^*_{\ell=L}(i^*)\} = \max_{\alpha \in \mathcal{A}(i)} \left\{ f(i, \alpha) + \sum_{j=1}^{S} p^{\alpha}(j|i) g^*_{\ell=L}(j) \right\}. \tag{7.16}$$

The decision rule (7.12) becomes

$$\hat{d}_{\ell=L+1}(i) \in \arg\left[\max_{\alpha \in \mathcal{A}(i)} \left\{ f(i, \alpha) + \sum_{j=1}^{S} p^{\alpha}(j|i) g^*_{\ell=L}(j) \right\} \right]. \tag{7.17}$$

As $L \to \infty$, $g^*_{\ell=L}(j)$ converges (see Problem 7.4). Let it converge to $g^*(j)$. Then from (7.17) the optimal decision rule converges to a stationary rule

$$\hat{d}(i) \in \arg\left[\max_{\alpha \in \mathcal{A}(i)} \left\{ f(i,\alpha) + \sum_{j=1}^{S} p^{\alpha}(j|i) g^*(j) \right\}\right].$$

In other words, $\lim_{L\to\infty} \hat{d}_{\ell=L} = \hat{d}$. Furthermore, from (7.10) and by the fact that as L goes to infinity, the L-step optimal policy converges to a long-run average optimal policy, we have

$$\lim_{L\to\infty} \frac{\eta^*_{\ell=L+1}(i^*)}{L+1} = \lim_{L\to\infty} \frac{\eta^*_{\ell=L}(i^*)}{L} = \eta^*,$$

where η^* is the steady-state average reward of the stationary policy $\hat{d}$. Thus,

$$\lim_{L\to\infty} \left[\eta^*_{\ell=L+1}(i^*) - \eta^*_{\ell=L}(i^*)\right] = \eta^*.$$

Finally, (7.16) becomes

$$g^*(i) + \eta^* = \max_{\alpha \in \mathcal{A}(i)} \left\{ f(i,\alpha) + \sum_{j=1}^{S} p^{\alpha}(j|i) g^*(j) \right\}.$$

This is the optimality equation for the long-run average-reward problem with η^* and g^* being the optimal average reward and the potential of the optimal policy, respectively. Note that the iteration procedure in (7.16) is equivalent to the value iteration method shown in (6.71). They differ only by a constant due to different normalization schemes: in (6.71), we set $g(i^*) = \eta^*$ and in (7.16) we have $g_{\ell=L}(i^*) = 0$.

Before further discussion, let us first study an example.

Example 7.1. Consider a system of three states $\mathcal{S} = \{1, 2, 3\}$. In each state i, two actions can be taken, denoted as $\alpha_{i,1}$ and $\alpha_{i,2}$, $i = 1, 2, 3$. The corresponding transition probabilities associated with every action and the reward function are listed in Table 7.1. As it is shown, the reward function does not depend on the actions.

We first solve the finite-step optimization problem by using (7.12). The values of $\eta^*_{\ell=L}(i)$, $\hat{d}_{\ell=L}(i)$, and $g^*_{\ell=L}(i)$ in (7.11), (7.12), and (7.15) for $L = 1, 2, 3, 4$, are listed in Table 7.2, with $i^* = 3$. As shown in the table, the optimal decision-rule sequence $\hat{d}_{\ell=L}$ converges to

$$\hat{d} = \left\{\hat{d}(1) = \alpha_{1,1}, \hat{d}(2) = \alpha_{2,2}, \hat{d}(3) = \alpha_{3,2}\right\}.$$

After only two iterations, i.e., when $L = 2$, the L-step optimization problem already yields the same optimal decision rule as the long-run average problem. $\eta^*_{\ell=L}$ increases as L increases; however, $g^*_{\ell=L}$ converges to finite numbers. The data in the table also show that the sequence of the optimal decision rules may reach its limit earlier than the sequence of $\eta^*_{\ell=L}$, $L = 1, 2, \ldots$.

state i	action	Transition prob. $p^{\alpha}(j\|i)$ $j=1$	2	3	Perf. func.
1	$\alpha_{1,1}$	0.5	0.4	0.1	10
	$\alpha_{1,2}$	0.5	0.1	0.4	
2	$\alpha_{2,1}$	0.6	0	0.4	8
	$\alpha_{2,2}$	0.4	0.3	0.3	
3	$\alpha_{3,1}$	0.2	0.2	0.6	0
	$\alpha_{3,2}$	0.3	0.4	0.3	

Table 7.1. The Actions and Performance Function in Example 7.1

L	1			2		
state i	$\eta^*_{\ell=1}(i)$	$g^*_{\ell=1}(i)$	$\hat{d}_{\ell=1}(i)$	$\eta^*_{\ell=2}(i)$	$g^*_{\ell=2}(i)$	$\hat{d}_{\ell=2}(i)$
1	10	10	$\alpha_{1,1}$, $\alpha_{1,2}$	18.2	12.0	$\alpha_{1,1}$
2	8	8	$\alpha_{2,1}$, $\alpha_{2,2}$	14.4	8.2	$\alpha_{2,2}$
3	0	0	$\alpha_{3,1}$, $\alpha_{3,2}$	6.2	0	$\alpha_{3,2}$
L	3			4		
state i	$\eta^*_{\ell=3}(i)$	$g^*_{\ell=3}(i)$	$\hat{d}_{\ell=3}(i)$	$\eta^*_{\ell=4}(i)$	$g^*_{\ell=4}(i)$	$\hat{d}_{\ell=4}(i)$
1	25.48	12.40	$\alpha_{1,1}$	32.632	12.48	$\alpha_{1,1}$
2	21.46	8.38	$\alpha_{2,2}$	28.554	8.402	$\alpha_{2,2}$
3	13.08	0	$\alpha_{3,2}$	20.152	0	$\alpha_{3,2}$

Table 7.2. The Results for the L-step Optimization Problems in Example 7.1

Next, let us solve the problem by policy iteration. Following the policy iteration algorithm in Section 4.1.1, we first choose an initial stationary policy d_0 and then determine the potentials of the Markov system under this policy. Suppose that we pick up $d_0 = \{\alpha_{1,1}, \alpha_{2,1}, \alpha_{3,1}\}$. The transition probability matrix is

$$P^{d_0} = \begin{bmatrix} 0.5 & 0.4 & 0.1 \\ 0.6 & 0 & 0.4 \\ 0.2 & 0.2 & 0.6 \end{bmatrix}.$$

From (3.5), we have the following approximation (we will omit the superscript d_0 in both P and g for simplicity, if there is no confusion)

$$\bar{g}_n \approx \sum_{k=0}^{n} P_-^k f_-,$$

where $P_- = P - ep_{S*}$, p_{S*} is the Sth row of P, and $f_-(i) = f(i) - f(S)$, $i = 1, 2, \ldots, S$. In this example, we have $S = 3$. Thus, we have

$$P_-^{d_0} = \begin{bmatrix} 0.3 & 0.2 & -0.5 \\ 0.4 & -0.2 & -0.2 \\ 0 & 0 & 0 \end{bmatrix},$$

and it happens that $f_- = f$. Thus, its potential g can be calculated iteratively:

$$\bar{g}_0 = f_-, \qquad \bar{g}_k = P_- \bar{g}_{k-1} + f_-, \qquad k = 1, 2, \ldots. \tag{7.18}$$

This calculation is similar to that in (7.11), except that in the latter case the matrix multiplication is carried out for every action, and the maximal value is picked up by comparison. The values for $\bar{g}_0$ to $\bar{g}_5$ are:

i	$\bar{g}_0(i)$	$\bar{g}_1(i)$	$\bar{g}_2(i)$	$\bar{g}_3(i)$	$\bar{g}_4(i)$	$\bar{g}_5(i)$
1	10	14.6	16.46	17.29	17.63	17.78
2	8	10.4	11.76	12.23	12.47	12.57
3	0	0	0	0	0	0.

We can see that $\bar{g}_k$ almost reaches the limit at $k = 5$, i.e., $g \approx \bar{g}_5$. Now, we apply (4.5) and obtain $d_1 = \{\alpha_{1,1}, \alpha_{2,2}, \alpha_{3,2}\}$. We have

$$P^{d_1} = \begin{bmatrix} 0.5 & 0.4 & 0.1 \\ 0.4 & 0.3 & 0.3 \\ 0.3 & 0.4 & 0.3 \end{bmatrix},$$

and therefore

$$P_-^{d_1} = \begin{bmatrix} 0.2 & 0 & -0.2 \\ 0.1 & -0.1 & 0 \\ 0 & 0 & 0 \end{bmatrix}.$$

Using (7.18), we have the following results for policy d_1:

i	$\bar{g}_0(i)$	$\bar{g}_1(i)$	$\bar{g}_2(i)$	$\bar{g}_3(i)$	$\bar{g}_4(i)$	$\bar{g}_5(i)$
1	10	12	12.4	12.48	12.496	12.499
2	8	8.2	8.38	8.402	8.408	8.409
3	0	0	0	0	0	0

Applying (4.5) again, we obtain the same policy $d_2 = d_1 = \{\alpha_{1,1}, \alpha_{2,2}, \alpha_{3,2}\}$. Thus, this policy is optimal.

Incidently, it is interesting to note that the values of $g^*_{\ell=L}$ in Table 7.2 are the same as those of $\bar{g}_k$ calculated in the numerical iterations for the potentials of the optimal policy (e.g., $\bar{g}_2(i) = g^*_{\ell=3}(i)$, $i = 1, 2, 3$). This is because $\hat{d}_{\ell=2} = \hat{d}_{\ell=3} = \hat{d}_{\ell=4}$ is the optimal policy. Therefore, the calculations for $\bar{g}_k(i)$, $k = 1, 2, 3$, $i \in \mathcal{S}$, are exactly the same as that for $g^*_{\ell=L+1}(i)$, $L = 1, 2, 3$, $i \in \mathcal{S}$. □

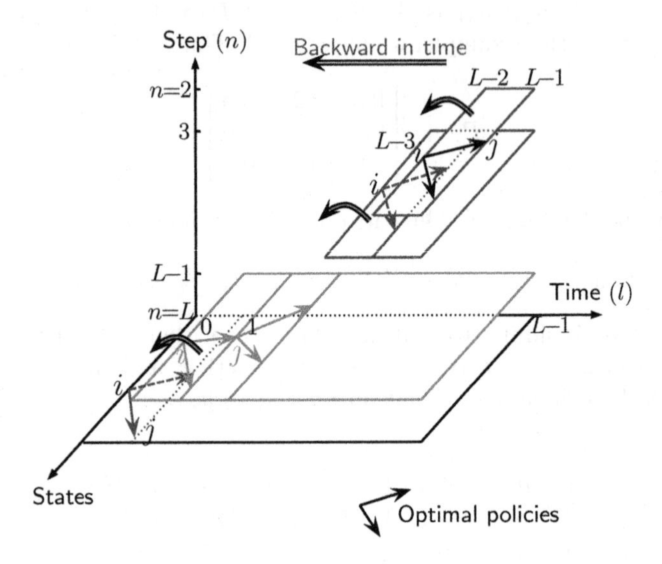

Fig. 7.3. The Dynamic Programming Approach

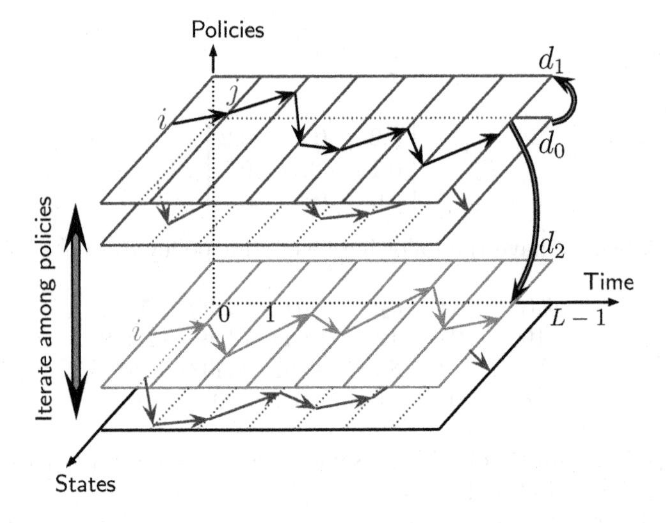

Fig. 7.4. The Policy Iteration Approach

Comparison: Backward and Forward in Time

Figures 7.3 and 7.4 illustrate the procedures of the dynamic programming and the policy iteration approaches, respectively. In Figure 7.3, each rectangle represents an L-step optimization problem, with the vertical axis denoting the steps L. Thus, the top rectangle represents a two-step problem from time $\ell = 2$ to $\ell = 1$; the one just below the top one represents a three-step problem from $\ell = 3$ to $\ell = 1$; and so on. With dynamic programming, we first solve for an optimal solution to the one-step problem, leading to the initial conditions (7.13). Then, we solve a two-step problem by (7.11) and (7.12) with $L = 1$ and the initial value (7.13). With the two-step optimal policy, we then solve the three-step problem and obtain the optimal decision rule at $\ell = 3$, by (7.11) and (7.12) with $L = 2$. This process goes on until it converges to an optimal decision rule, which is the optimal stationary policy for the long-run average-reward optimal problem. As is clearly shown in the figure, this procedure goes *backward* in time, and an optimal policy is obtained at each iteration for a finite-step problem. The long-run average-reward problem is treated as the limit of the finite-step problem when the number of steps, L, goes to infinity.

On the contrary, with the policy iteration approach shown in Figure 7.4, at each iteration, we deal with a (stationary) policy (not necessarily an optimal one) with an infinite horizon. First, we work *forward* in time to obtain the performance potentials $g(i)$, $i \in \mathcal{S}$, of the policy, either by calculation following (7.18) or by estimation following a sample path. Then, we find a better policy by using the potentials, and in this way we iterate in the policy space to reach an optimal policy. Figure 7.4 illustrates the first two iterations from d_0 to d_1 and to d_2.

There is another difference between the two approaches: The dynamic programming approach requires us to know the transition probabilities for all the actions. Because working backward is not realistic in practice, it cannot be implemented on a real system. On the other hand, as discussed in Chapter 5, policy iteration can be implemented on line, the potentials can be learned on a sample path, and in some cases, the exact transition probabilities may not need to be known when updating the policies.

7.2 MDPs with Continuous State Spaces

As noted in Section 7.1.1, the first problem in applying policy iteration to control problems is that the state spaces are generally continuous. Fortunately, the theory of MDPs and policy iteration has been extended to systems with continuous state spaces. The transition probability in a continuous state space can be described by an operator (integration) on a function space (cf. the matrix operator on a vector space for discrete state spaces). A rigorous treatment of the topic is beyond the scope of this book, and we will present some relevant results, in a self-contained way but with less rigor, to illustrate the main ideas; the readers are referred to [24, 135, 136, 203] for more details.

7.2.1 Operators on Continuous Spaces

Transition Probability Functions Operate on Functions

Consider a linear system with an additive noise described in (7.1):

$$X_{l+1} = AX_l + \xi_l, \qquad l = 0, 1, \ldots,$$

The state space $\mathcal{S}$ is the n-dimensional real space $\Re^n$, i.e., $\mathcal{S} = \Re^n$. As illustrated in Figure 7.1, given the current state $x \in \Re^n$, the transition probability function can be denoted as $P(dy|x)$, $y \in \Re^n$ (cf. (7.2)). Let $\mathcal{B}$ be the collection of sets in $\mathcal{S}$ such that $(\mathcal{S}, \mathcal{B}, P)$ forms a probability space as defined in Appendix A.1. $\mathcal{B}$ can be determined by the probability distribution function of the noise ξ_l via (7.4). A transition function is a function from $\Re^n \times \mathcal{B}$ to $[0, 1]$ such that, for any given $x \in \Re^n$, $P(R|x)$, $R \in \mathcal{B}$, is a probability distribution; and for any $R \in \mathcal{B}$, $P(R|x)$ is a function on $\Re^n$. Given the current state $x \in \Re^n$, the probability that the next state lies in $R \in \mathcal{B}$ is $P(R|x)$.

For any transition function $P(R|x)$ and any function $h(x)$ on $\Re^n$, we define their product, with P on the left, as a function, denoted as $(Ph)(x)$, defined as follows:

$$(Ph)(x) := \int_{\Re^n} h(y)P(dy|x). \tag{7.19}$$

We can also view (7.19) as the definition of an operator P: $h \to Ph$. Thus, P is also called a *transition operator*. We assume that, for any $x \in \Re^n$, $(P|h|)(x)$ exists; i.e., $|h(y)|$ is integrable with respect to $P(R|x)$. Although in many cases, $(Ph)(x)$ is finite for any $x \in \Re^n$; there is no conceptual difficulty if $(Ph)(x)$ is infinite for some x. From now on, we assume that all the functions $h(y)$ are integrable in the sense that (7.19) exists. In particular, for a set $R \in \mathcal{B}$, $P(R|y)$ as a function of y is integrable; i.e., $\int_{\Re^n} P(R|y)P(dy|x)$ exists.

The meaning of (7.19) is clear: for any initial state x, we have

$$(Ph)(x_0) = E\left[h(X_1)|X_0 = x_0\right]. \tag{7.20}$$

Define $e(x) = 1$ for all $x \in \Re^n$. This function corresponds to the vector $e = (1, 1, \ldots, 1)^T$ for finite-state MDPs. For any transition function P, we have $(Pe)(x) = 1$ for all $x \in \Re^n$. That is,

$$Pe = e.$$

Next, we define the identity transition function $I(R|x)$: For any $R \in \mathcal{B}$, we define

$$I(R|x) = \begin{cases} 1, & \text{if } x \in R, \\ 0, & \text{otherwise.} \end{cases}$$

For any function h and $x \in \Re^n$, we have $(Ih)(x) = h(x)$. I corresponds to the n-dimensional δ-function: for any function $h(x)$, we have $\int_{\Re^n} h(y)I(dy|x) = h(x)$.

For any two operators P_1 and P_2 corresponding to two transition functions $P_1(R|x)$ and $P_2(R|x)$, we define naturally

$$[(P_1 + P_2)h](x) := (P_1 h)(x) + (P_2 h)(x)$$

and

$$[(P_1 - P_2)h](x) := (P_1 h)(x) - (P_2 h)(x).$$

The product of two transition functions $P_1(R|x)$ and $P_2(R|x)$ is defined as a transition function $(P_1P_2)(R|x)$:

$$(P_1P_2)(R|x) := \int_{\Re^n} P_2(R|y)P_1(dy|x), \qquad x \in \mathcal{S}, \quad R \in \mathcal{B}. \tag{7.21}$$

In general, $P_1P_2 \neq P_2P_1$.

For any transition function P, we define the kth power, $k = 0, 1, \ldots$, as $P^0 = I$, $P^1 = P$, and $P^k = PP^{k-1}$, $k = 2, \ldots$. As an exercise, it can be verified that for any three operators P_1, P_2, and P_3, we have $(P_1P_2)P_3 = P_1(P_2P_3)$, and therefore $P^k = (P^{k-1})P = P(P^{k-1})$ (cf. Problem 7.2) .

Suppose that $\boldsymbol{X} := \{X_0, X_1, \ldots, X_l, \ldots\}$ is a time-homogeneous Markov chain with transition function $P(R|x)$, $x \in \Re^n$, $R \in \mathcal{B}$. The k-step transition probability functions, denoted as $P^{(k)}(R|x)$, $k = 1, 2, \ldots$, are given by the 1-step transition function defined as $P^{(1)}(R|x) = P(R|x)$ and

$$P^{(k)}(R|x) = \int_{\Re^n} P(dy|x)P^{(k-1)}(R|y), \qquad k \geq 2.$$

For any function $h(x)$, we have

$$\begin{aligned}(P^{(k)}h)(x) &= \int_{\Re^n} h(y)P^{(k)}(dy|x) = \int_{\Re^n} h(y)\int_{\Re^n} P(dz|x)P^{(k-1)}(dy|z) \\ &= \int_{\Re^n} P(dz|x)\left\{(P^{(k-1)}h)(z)\right\} = P(P^{(k-1)}h)(x).\end{aligned}$$

That is, as an operator, we have $P^{(k)} = P(P^{(k-1)})$, Recursively, we can prove that $P^{(k)} = P^k$. We have

$$(P^k h)(x_0) = E\left[h(X_k)|X_0 = x_0\right].$$

Transition Probability Functions Operate on Probability Distributions

For any transition function $P(R|x)$, $R \in \mathcal{B}$, $x \in \Re^n$, and probability distribution $\nu(R)$ on $\mathcal{B}$, we define their product, with $P(R|x)$ on the right, as a probability distribution, denoted as νP, defined as follows:

$$(\nu P)(R) := \int_{\Re^n} \nu(dx)P(R|x), \qquad R \in \mathcal{B}.$$

If ν is the initial distribution at time $l = 0$, then νP is the state probability distribution at time $l = 1$.

The product of a probability distribution ν and a function $h(x)$, with ν on the left, is defined as

$$\nu h := \int_{\Re^n} h(x)\nu(dx),$$

which is the mean of h under the probability distribution ν. We have $\nu e = 1$. Furthermore, for any probability distribution ν, transition distribution P, and real function h, we have

$$\nu(Ph) = \int_{\Re^n} \left\{ \int_{\Re^n} h(y)P(dy|x) \right\} \nu(dx).$$

We can verify that (see Problem 7.3)

$$\nu(Ph) = (\nu P)h. \tag{7.22}$$

A probability distribution ν defines a transition function denoted as $P(R|x) = \nu(R)$ for all $x \in \Re^n$. In fact, with the function $e(x) \equiv 1$ and any probability distribution ν, we may define their product, with ν on the right, as a transition function, denoted as $e\nu$ and defined as:

$$(e\nu)(R|x) = \nu(R), \qquad \text{for all } x \in \Re^n. \tag{7.23}$$

We can easily verify that

$$(e\nu)h = e(\nu h) := (\nu h)e$$

and that

$$(e\nu)^k = e\nu.$$

From (7.21), we further have

$$[P(e\nu)]\,(R|x) = \int_{\Re^n} \{P(dy|x)e(y)\nu(R)\} = e(x)\nu(R).$$

Thus,

$$P(e\nu) = (Pe)\nu = e\nu.$$

Similarly, we can verify that

$$(e\nu)P = e(\nu P) = e(x) \int_{\Re^n} \nu(dy)P(R|y).$$

Also, for any probability distribution π, we have

$$\begin{aligned} [\pi(e\nu)]\,(R) &:= \int_{\Re^n} \pi(dx)\nu(R) \\ &= \nu(R) \equiv (\pi e)\nu(R), \qquad R \in \mathcal{B}. \end{aligned} \tag{7.24}$$

Steady-State Probability Distributions and Steady-State Performance

Let $f(x)$ be a reward function. The long-run average reward is defined as

$$\eta(x) = \lim_{L\to\infty} \frac{1}{L} E\left\{ \sum_{l=0}^{L-1} f(X_l) \Big| X_0 = x \right\}, \qquad x \in \Re^n,$$

which is assumed to exist. By (7.19) and (7.20), we have

$$Pf(x) = \int_{\Re^n} f(y)P(dy|x) = E\left[f(X_1)|X_0 = x\right],$$

and because $P^l(R|x)$, $R \in \mathcal{B}$, $l = 1, 2, \ldots$, is the l-step transition function, we have

$$P^l f(x) = \int_{\Re^n} f(y)P^l(dy|x) = E\left[f(X_l)|X_0 = x\right], \qquad l = 1, 2, \ldots.$$

Thus, it is clear that

$$\eta(x) = \lim_{L\to\infty} \frac{1}{L} \left\{ \sum_{l=0}^{L-1} (P^l f)(x) \right\}. \tag{7.25}$$

The steady-state probability distribution of a transition function $P(R|x)$, $R \in \mathcal{B}$ and $x \in \Re^n$, is defined as a probability distribution π satisfying

$$\pi = \pi P.$$

Normally, we hope that as $l \to \infty$, P^l will converge, in some sense, to the transition function $e\pi$. However, with a continuous state space, there are many ways to define the convergence of P^l, $l = 1, 2, \ldots$. The convergence is related to the topic of ergodicity. For the analysis in this chapter, we only need to assume that for any reward function $f(x)$, we have

$$\lim_{k\to\infty} (P^k f)(x) = \left[(e\pi)f\right](x) = (\pi f)e(x), \qquad \forall x \in \Re^n. \tag{7.26}$$

Therefore, with the assumption (7.26), from (7.25), we have

$$\eta(x) = \lim_{k\to\infty} P^k f(x) = (\pi f)e(x),$$

with $\pi f = \int_{\Re^n} f(x)\pi(dx)$. With a slightly abused notation, we also use $\eta := \pi f$ as a constant. Thus, we have $\eta(x) = \eta e(x)$.

Ergodicity*

The convergence in (7.26) is closely related to the ergodicity. There are a number of ways to define the ergodicity of a Markov chain with a continuous state space. Some definitions of ergodicity impose more restrictive conditions on the transition probability functions and hence lead to strong convergence properties with a wider class of reward functions; and others may be less restrictive on the transition functions, but may require strong conditions on the reward functions to obtain nice convergence properties for the system performance and performance potentials.

It goes beyond the scope of this book to discuss the details of the theory of ergodicity. We, however, describe one definition of ergodicity to help obtain the flavor of this subject.

w-geometrically ergodic [203]

For a given real-valued function $\infty > w(x) \geq 1$ on $\Re^n$, any function $h(x)$ on $\Re^n$ is called *w-bounded* if its *w-norm* defined as

$$\|h\|_w := \sup_{x \in \Re^n} \frac{|h(x)|}{w(x)}$$

is finite. We will refer to such a w as a *weight function*, and denote by $\mathcal{B}_w$ the (Banach) space of all w-bounded functions on $\Re^n$. If $w(x) = e(x)$, then an e-bounded function is simply a bounded function.

The *w-norm* of a linear operator P on $\mathcal{B}_w$, $\|P\|_w$, is defined as

$$\|P\|_w := \sup_{u \in \mathcal{B}_w} \{\|Pu\|_w : \|u\|_w \leq 1 \text{ (i.e., } |u| \leq w)\}.$$

We assume that P has a finite $w-$norm.

We can easily verify that $\|Ph\|_w \leq \|P\|_w \|h\|_w$ for all $h \in B_w$. Thus, if $h \in \mathcal{B}_w$, then $Ph \in \mathcal{B}_w$. In addition, for any two operators P_1 and P_2, we have $\|P_1 P_2\|_w \leq \|P_1\|_w \|P_2\|_w$, and $\|P^n\|_w \leq \|P\|_w^n$, for $n = 1, 2, \ldots$.

Now we consider a Markov chain $\{X_l, l = 0, 1, \ldots\}$ with transition probability function P. It is said to be *w-geometrically ergodic* if there is a probability distribution π and nonnegative constants κ and ρ, with $0 < \rho < 1$, such that

$$\|P^n - e\pi\|_w \leq \kappa \rho^n, \qquad n = 0, 1, \ldots. \tag{7.27}$$

If the transition function P is w-geometrically ergodic, then for any w-bounded continuous reward function f, (7.26) holds.

For a finite-state ergodic Markov chain, P is an ergodic matrix and it is well known that it is e-geometrically ergodic (see, for example, Corollary 4.1.5 of [167]).

7.2.2 Potentials and Policy Iteration

Performance Potentials

Suppose that a Markov chain $\{X_l,\ l = 0, 1, \ldots\}$ with (P, f) defined in a continuous state space $\Re^n$ has a steady-state probability distribution π. As discussed above, the convergence of P^k, $k = 1, 2, \ldots$, depends on the definitions of ergodicity. As we will see, instead of verifying the ergodic condition, it would be easier to directly assume that

$$\lim_{k\to\infty} (P^k f)(x) = [(e\pi)f](x) = (\pi f)e(x), \qquad x \in \Re^n. \tag{7.28}$$

The performance potential g is a function $g(x)$, $x \in \Re^n$, that satisfies the Poisson equation

$$(I - P)g(x) + \eta(x) = f(x), \tag{7.29}$$

where I and P are two transition functions and $\eta(x) = (\pi f)e(x) = \eta e(x)$. The solution to the Poisson equation is unique up to an additive constant, i.e., if g is a solution to (7.29), then so is $g + ce$ with any constant c. Therefore, we will not distinguish these different versions of potentials with only a constant difference.

With a continuous state space, we have to be careful in exchanging the order of mathematical operations such as integration and limit. First, we define

$$g_K := \left\{ I + \sum_{k=1}^{K} (P^k - e\pi) \right\} f, \tag{7.30}$$

where $e\pi$ is a transition function defined according to (7.23), and set

$$g(x) := \lim_{K\to\infty} g_K(x), \qquad x \in \Re^n,$$

assuming that the limit exists. Next, we observe that $Pg_K(x) = \int_{\Re^n} g_K(y)P(dy|x)$. Therefore, if this integration uniformly converges for all K, we can exchange the order of the "$\lim_{K\to\infty}$" and the integration "$\int_{\Re^n}$" and obtain

$$\lim_{K\to\infty} Pg_K = Pg.$$

Lemma 7.1. *For any transition function P and performance function $f(x)$, if*

$$\lim_{k\to\infty} P^k f = (e\pi)f = \eta e, \ \lim_{K\to\infty} g_K = g, \ \textit{and} \ \lim_{K\to\infty} Pg_K = Pg \tag{7.31}$$

hold for every $x \in \Re^n$, then

$$g = \left\{ I + \sum_{k=1}^{\infty} (P^k - e\pi) \right\} f \tag{7.32}$$

is a solution to the Poisson equation (7.29).

This lemma can be established by directly verifying that the g in (7.32) satisfies (7.29). In addition, if we assume that $\lim_{K\to\infty} \pi g_K = \pi g$, then we can easily verify that

$$\pi g = \pi f = \eta. \tag{7.33}$$

This is a normalization condition of the potential g in (7.32). With (7.33), the Poisson equation (7.29) becomes

$$(I - P + e\pi) g(x) = f(x). \tag{7.34}$$

Equation (7.33) can also be verified by left-multiplying (7.34) with π and using (7.22), (7.24) and $\pi = \pi P$, $\pi e = e$ (also see Problem 7.3). Note that the g in (7.32) is one of the solutions, and for any constant c, $g + ce$ is also a solution to (7.29) (but not (7.34)).

As we have explained, the three equations in (7.31) hold under different ergodicity conditions for different types of reward functions. For example, if the Markov chain is w-geometrically ergodic and f is w-bounded, then we can verify that these equations hold and $g = \sum_{n=0}^{\infty} (P - e\pi)^n f$ is w-bounded and satisfies the Poisson equation (7.29). However, the ergodicity conditions are not necessary and therefore in Lemma 7.1 we prefer to use (7.31) as the general conditions. For a particular problem, sometimes it is more convenient to verify (7.31) directly than to find the right ergodicity conditions, as we will see for the LQ and JLQ problems discussed later.

Similar to the finite-state case, the Poisson equation plays a crucial role in performance optimization. Later in this chapter, when we mention the potential g, we always assume implicitly that the Poisson equation and (7.31) hold.

Performance Optimization

Now we can develop the MDP theory for problems with continuous state spaces by simply translating, with some care of handling the transition functions and probability distributions, the corresponding results from the finite state MDPs.

First, we modify the definition of the relations $=$, $\leq$, $<$, and $\preceq$ for two functions in $\Re^n$. Given a probability distribution ν on $\Re^n$, for two functions $h(x)$ and $h'(x)$, $x \in \Re^n$, we define $h' =_\nu h$, $h' \leq_\nu h$, and $h' <_\nu h$, respectively, if $h'(x) = h(x)$, $h'(x) \leq h(x)$, and $h'(x) < h(x)$, respectively, for all $x \in \Re^n$ except on a set R with $\nu(R) = 0$. We further define $h'(x) \preceq_\nu h(x)$ if $h'(x) \leq_\nu h(x)$ and $h'(x) < h(x)$ on a set R with $\nu(R) > 0$. Similar definitions are used for the relations $>_\nu$, $\succeq_\nu$, and $\geq_\nu$. With these definitions, we have the following theorem.

Let (P, f) and (P', f') be the transition functions and reward functions of two Markov chains with the same state space $\mathcal{S} = \Re^n$. Let η, g, π and η', g', and π' be their corresponding long-run average reward, performance potential functions, and steady-state probability distributions, respectively. Then, we have the following results.

The Average-Reward Difference Formula with Continuous Spaces:

$$\eta' - \eta = \pi' [(f' + P'g) - (f + Pg)] . \tag{7.35}$$

Comparison Lemma:

$$\text{If } f' + P'g \succeq_{\pi'} f + Pg, \text{ then } \eta' > \eta. \tag{7.36}$$

$$\text{If } f' + P'g \geq_{\pi'} f + Pg, \text{ then } \eta' \geq \eta.$$

The difficulty in verifying the condition $\succeq_{\pi'}$ lies in the fact that we may not know π', so we may not know for which set R we have $\pi'(R) > 0$. Fortunately, in many cases we can show that $\pi'(R) > 0$ if and only if R is a subset of $\mathcal{S} = \Re^n$ with a positive volume (formally called a Lebesgue measure). See the LQ and JLQ problems for example.

To develop the optimality equation, we need to further restrict the policy space. First, we use $u(x)$ to denote a policy. This is consistent with the notation used in control theory. In general, $u(x)$, $x \in \Re^n$, represents an action that determines the transition function. In control problems, $u(x)$ is a function $\Re^n \to \Re^m$. The policy $u(x)$ also determines the steady-state probability distribution through the transition function. We use the superscript u to denote the dependency on the policy. Thus, the transition function under policy $u(x)$ is denoted as $P^u(R|x)$, $R \in \mathcal{B}$, $x \in \Re^n$; and its corresponding steady-state probability distribution is denoted as $\pi^u(R)$, $R \in \mathcal{B}$, etc.

We say that two probability distributions π and π' have the same *support*, if $\pi(R) > 0$ then $\pi'(R) > 0$ for any $R \in \mathcal{B}$, and vice versa. Two policies u and u' have the same support, if their corresponding steady-state probability distributions have the same *support*. We assume that *all the policies in the policy space have the same support*. In control problems, if, for example, the noise has a Gaussian distribution, or any distribution that is over the whole space $\Re^n$, then all the policies have the same support. However, for problems where the steady-state probability distribution has discrete masses, the situation may be different. A special case is discussed in Problem 7.7.

If all the policies have the same support, we will drop the subscript π^u in the relationship notations such as $\geq$ and $\preceq$, etc. Now we assume that, for a Markov system, all the policies have the same support. Then, we have

Optimality Condition:
A policy $\hat{u}(x)$ is optimal, if and only if

$$f^{\hat{u}} + P^{\hat{u}} g^{\hat{u}} \geq f^u + P^u g^{\hat{u}}, \qquad \text{for all policies } u. \tag{7.37}$$

From the Optimality Condition (7.37), the optimality equation is

Optimality Equation:
A policy $\hat{u}(x)$ is optimal, if and only if

$$P^{\hat{u}} g^{\hat{u}} + f^{\hat{u}} = \max_u \left\{ P^u g^{\hat{u}} + f^u \right\}. \tag{7.38}$$

This equation holds with probability 1 with respect to the steady-state probability distribution of any policy. Under some conditions, the maximum in (7.38) can be reached. It is beyond the scope of this book to present further formulation in this direction.

With the Comparison Lemma (7.36), policy iteration algorithms can be designed. Roughly speaking, we may start with any policy $u_0(x)$. At the kth step with policy $u_k(x)$, $k = 0, 1, \ldots$, we set

$$u_{k+1}(x) \in \arg \left\{ \max_u \left[f^u + P^u g^{u_k} \right] \right\}, \qquad x \in \Re^n, \tag{7.39}$$

with g^{u_k} being any solution to the Poisson equation (7.29) for P^{u_k}. The iteration stops if $u_{k+1}(x)$ equals $u_k(x)$ almost everywhere under u_k; and in this case, $\eta_{k+1} = \eta_k$. By the performance difference formula (7.35), the average reward improves at each step. The Optimality Equation (7.37) indicates that the maximum is reached when no performance improvement can be further achieved. If the policy space is finite, the policy iteration will stop in a finite number of steps. However, in general, the iteration scheme may not stop at a finite number of steps, although the sequence of the performance η_k is increasing and hence converges.

All the above results hold if we change the problem to that of minimizing the performance criterion, by changing $\geq$ to $\leq$, max to min, etc.

7.3 Linear Control Systems and the Riccati Equation

7.3.1 The LQ Problem

In an LQ problem [39], we consider a system with a dynamic behavior described by

$$X_{l+1} = AX_l + Bu_l(X_l) + \xi_l, \qquad l = 0, 1, \ldots,$$

in which $u_l(x) \in \Re^m$ is a feedback control law at time l, A is an $n \times n$ matrix, and B is an $n \times m$ matrix. We assume that the system with A and B is controllable (in control terminology, this means that, starting from any initial position $X_0 \in \Re^n$, it is possible to control the system to the origin $X_l = 0$ when no noise is present). For each l, ξ_l denotes an n-dimensional noise with mean zero. ξ_l and ξ_k, $l \neq k$, are independent. (We do not require them to be Gaussian distributed.) Let $P_\xi(y)$ denote the probability distribution function of the random noise $\xi \in \Re^n$. $P_\xi(dy) = p_\xi(y)dy$ if the distribution density function $p_\xi(y)$ exists.

The goal of optimization is to minimize the performance criterion in a quadratic form:

$$\eta^u(x) = \lim_{L\to\infty} \frac{1}{L} E\left\{ \sum_{l=0}^{L-1} \left[X_l^T Q X_l + u_l^T V u_l\right] \Big| X_0 = x \right\}, \tag{7.40}$$

where Q and V are two $n \times n$ and $m \times m$ positive semi-definite matrices, respectively. (An $n \times n$ matrix Q is *positive semi-definite*, if for any $x \in \Re^n$ we have $x^T Q x \geq 0$.) Thus, the performance function is

$$f^u(x) = x^T Q x + u^T V u. \tag{7.41}$$

We assume that (7.40) exists; in particular, this requires that both integrations $\int_{\Re^n} \left[x^T Q x\right] P_\xi(dx)$ and $\int_{\Re^n} \left[u(x)^T V u(x)\right] P_\xi(dx)$ exist. A linear control system with this performance criterion (7.40) is called a *linear-quadratic (LQ) problem.*

For a stationary control law $u_l = u(x)$, we have

$$X_{l+1} = AX_l + Bu(X_l) + \xi_l, \qquad l = 0, 1, \ldots.$$

The transition function of this system under policy $u(x)$ is

$$P^u(dy|x) = p_\xi\left\{y - [Ax + Bu(x)]\right\} dy. \tag{7.42}$$

For any quadratic function $h(x) = x^T W_0 x$, we have

$$\begin{aligned}(P^u h)(x) &= \int_{\Re^n} h(y) P^u(dy|x) = \int_{\Re^n} (y^T W_0 y) p_\xi\left\{y - [Ax + Bu(x)]\right\} dy \\ &= \int_{\Re^n} \left\{z + [Ax + Bu(x)]\right\}^T W_0 \left\{z + [Ax + Bu(x)]\right\} p_\xi(z) dz.\end{aligned}$$

For zero-mean noise, we have $\int_{\Re^n} p_\xi(z)dz = 1$ and $\int_{\Re^n} zp_\xi(z)dz = 0$. Therefore,

$$(P^u h)(x) = \int_{\Re^n} z^T W_0 z p_\xi(z)dz + [Ax + Bu(x)]^T W_0 [Ax + Bu(x)]$$
$$= c_1 e(x) + [Ax + Bu(x)]^T W_0 [Ax + Bu(x)], \qquad (7.43)$$

where $c_1 := \int_{\Re^n} z^T W_0 z p_\xi(z)dz$ is a constant.

For stationary linear controllers, we set $u(x) = -Dx$, where D is an $m \times n$ matrix. With this controller, the system equation becomes

$$X_{l+1} = CX_l + \xi_l, \qquad l = 0, 1, \ldots, \qquad (7.44)$$

where $C = A - BD$. The performance measure becomes

$$\eta^u(x) = \lim_{L\to\infty} \frac{1}{L} E\left\{ \sum_{l=0}^{L-1} X_l^T W X_l \,\middle|\, X_0 = x_0 \right\}, \qquad (7.45)$$

where $W = Q + D^T V D$. Thus, the performance function is $f(x) = x^T W x$. It is easy to verify that W is also positive semi-definite. We assume that the system is stable, so the spectral radius of C, $\rho(C) < 1$.

Finally, if the Markov system (7.44) is stable, it is proved in [135, Example 7.4.4] or [203, Proposition 12.5.1 and Theorem 17.6.2] that this Markov chain is w-geometrically ergodic, with the weight function $w(x) = \sqrt{x_1^2 + \cdots + x_n^2} + 1$ or $w(x) = \sum_{i=1}^n x_i^2 + 1$.

Potentials for Linear Quadratic Systems

The transition function of the linear system (7.44) is $P(dy|x) = p_\xi(y - Cx)dy$. Denote $W_0 = W$. Then, the performance function is $f(x) = x^T W_0 x$. From (7.43), the transition operator is (we drop the superscript "u" for this particular controller $u = -Dx$)

$$(Pf)(x) = c_1 e(x) + x^T W_1 x,$$

where $c_1 := \int_{\Re^n} z^T W_0 z p_\xi(z)dz$ is a constant and $W_1 := C^T W_0 C$. From $Pe = e$ and $P^2 f = P(Pf)$, we have

$$(P^2 f)(x) = c_1 e(x) + \int_{\Re^n} (y^T W_1 y) p_\xi(y - Cx)dy$$
$$= c_2 e(x) + x^T W_2 x,$$

where $c_2 = \int_{\Re^n} z^T (W_0 + W_1) z p_\xi(z)dz$ and $W_2 = C^T W_1 C$. Continuing this process, we get

$$(P^k f)(x) = c_k e(x) + x^T W_k x, \qquad (7.46)$$

where

$$c_k = \int_{\Re^n} z^T(W_0 + W_1 + \cdots + W_{k-1}) z p_\xi(z) dz$$
$$= \int_{\Re^n} z^T S_{k-1} z p_\xi(z) dz$$

with

$$S_k := \sum_{i=0}^{k} W_i,$$

and

$$W_k = C^T W_{k-1} C = (C^k)^T W_0 C^k,$$

with $W_0 = W$.

Because $\rho(C) < 1$, we have

$$\lim_{k \to \infty} W_k = 0,$$

and S_k converges as $k \to \infty$. Set

$$S := \lim_{k \to \infty} S_k = \sum_{k=0}^{\infty} W_k.$$

Then,

$$S - W = C^T S C. \tag{7.47}$$

From (7.46), we have

$$\lim_{k \to \infty} (P^k f)(x) = \eta e(x),$$

and $\eta(x) = \eta e(x)$, where

$$\eta = \lim_{k \to \infty} c_k = \int_{\Re^n} (z^T S z) p_\xi(z) dz.$$

Thus, we have proved that, for the LQ problem, the steady-state performance $\eta(x)$ exists.

Finally, from (7.30), we have

$$g_K(x) = \left[\sum_{k=1}^{K} (c_k - \eta) \right] e(x) + x^T S_K x.$$

We can prove that $\lim_{K \to \infty} g_K(x) = g(x)$ (cf. Problem 7.10), where

$$g(x) = \left[\sum_{k=1}^{\infty} (c_k - \eta) \right] e(x) + x^T S x. \tag{7.48}$$

We can ignore the first constant term and simply use

$$g(x) = x^T S x$$

as the potential function. Therefore, the potential of a linear quadratic system is a quadratic function with the positive semi-definite matrix S.

Finally, we have

$$\begin{aligned} P g_K &= \int_{\Re^n} g_K(y) P(dy|x) \\ &= \left[\sum_{k=1}^{K} (c_k - \eta) \right] + \int_{\Re^n} \left[y^T S_K y \right] p_\xi(y - Cx) dy. \end{aligned}$$

Because all W_k, $k = 0, 1, \ldots$, are positive semi-definite, we have $y^T S_K y \leq y^T S y$ for all $y \in \Re^n$. Therefore, if we assume that

$$\int_{\Re^n} \left[y^T S y \right] p_\xi(y - Cx) dy$$

exists, then $\int_{\Re^n} \left[y^T S_K y \right] p_\xi(y - Cx) dy$ uniformly converges for all K. Thus, we have $\lim_{K \to \infty} P g_K = P g$. From Lemma 7.1, (7.48) is the potential that satisfies the Poisson equation (7.29).

The Optimal Policy

Before we apply the Optimality Equation (7.38), we need to verify that all the policies have the same support. To see this, we first assume that the noise distribution $P_\xi(dy)$ is supported on the entire space $\Re^n$; i.e., $P_\xi(R) > 0$ for all $R \in \mathcal{B}$ with a positive volume (Lebesgue measure). This is common, e.g., the Gaussian distribution has this property. Now, for any $R \in \mathcal{B}$ with a positive Lebesgue measure, if π is the steady-state probability distribution of the transition function P, we have

$$\pi(R) = (\pi P)(R) = \int_{\Re^n} \pi(dx) P(R|x).$$

In the LQ problem, for any $x \in \Re^n$ we have $P(R|x) = P_\xi(R - Cx)$, [1] where C depends on the control u. Because $P_\xi(R - Cx) > 0$, from the above equation, we can easily conclude that $\pi(R) > 0$. That is, all the policies have the same support of $\Re^n$, except for a set with a zero Lebesgue measure. Of course, the condition that $P_\xi(R) > 0$ for all set with a positive Lebesgue measure is not a necessary condition.

Now suppose that we are given a linear policy $u(x) = -Dx$, and we want to apply (7.39) to obtain a better policy. We consider any policy $\widetilde{u}(x)$ (not necessary a linear one). Setting $h = g = x^T S x$ in (7.43), we have

[1] For $R \subset \Re^n$ and $x \in \Re^n$, we define $R - Cx := \{y : y = x' - Cx, \text{ for all } x' \in R\}$.

$$(P^{\widetilde{u}}g)(x) = c_1 e(x) + [Ax + B\widetilde{u}(x)]^T S [Ax + B\widetilde{u}(x)],$$

where $c_1 := \int_{\Re^n} z^T S z p_\xi(z) dz$. In addition, from (7.41), we have

$$f^{\widetilde{u}}(x) = x^T Q x + \widetilde{u}(x)^T V \widetilde{u}(x).$$

Ignoring the two terms $c_1 e(x)$ and $x^T Qx$, in $(P^{\widetilde{u}}g)(x)$ and $f^{\widetilde{u}}(x)$, that are independent of $\widetilde{u}$, we have

$$\begin{aligned}
u'(x) &= \arg\left\{\min_{\widetilde{u}}\left[(P^{\widetilde{u}}g)(x) + f^{\widetilde{u}}(x)\right]\right\} \\
&= \arg\left\{\min_{\widetilde{u}}\left\{[Ax + B\widetilde{u}(x)]^T S [Ax + B\widetilde{u}(x)] + \widetilde{u}(x)^T V \widetilde{u}(x)\right\}\right\} \\
&= -(B^T S B + V)^{-1} B^T S A x \\
&= -D'x,
\end{aligned}$$

With

$$D' = (B^T S B + V)^{-1} B^T S A.$$

We conclude that if the original policy is a linear policy $u(x) = -Dx$, then we can find an improved policy $u'(x) = -D'x$, which is also linear.

Next, if $D' = D$ then the policy $u(x) = -Dx$ satisfies the Optimality Equation (7.38) and thus it is an optimal policy. Denote it as $\hat{u}(x) = -\hat{D}x$ and its corresponding quantity as $\hat{S}$, etc. We have

$$\hat{D} = (B^T \hat{S} B + V)^{-1} B^T \hat{S} A. \tag{7.49}$$

On the other hand, from (7.47) we have (C and W are defined in (7.44) and (7.45))

$$\begin{aligned}
\hat{S} &= \hat{C}^T \hat{S} \hat{C} + \hat{W} \\
&= (A - B\hat{D})^T \hat{S} (A - B\hat{D}) + (\hat{D})^T V \hat{D} + Q.
\end{aligned} \tag{7.50}$$

From (7.49) and (7.50) and after some calculation, we have

$$\hat{S} = A^T \hat{S} A - A^T \hat{S} B \left[B^T \hat{S} B + V\right]^{-1} B^T \hat{S} A + Q. \tag{7.51}$$

This is called the *Riccati equation.* In summary, we have shown that if $\hat{S}$ satisfies the Riccati equation, then the linear control $\hat{u}(x) = -\hat{D}x$, with $\hat{D}$ determined by (7.49) is an optimal policy.

Now, we can easily understand the policy iteration procedure. We start with a linear policy $u_0(x) = -D^{\{0\}}x$. Set $C^{\{0\}} = A - BD^{\{0\}}$, $W^{\{0\}} = Q + {D^{\{0\}}}^T V D^{\{0\}}$, and $S^{\{0\}} - W^{\{0\}} = {C^{\{0\}}}^T S^{\{0\}} C^{\{0\}}$. After one iteration, we have $u_1(x) = -D^{\{1\}}x$, with

$$D^{\{1\}} = (B^T S^{\{0\}} B + V)^{-1} B^T S^{\{0\}} A.$$

Continuing the iteration, we have $u_k(x) = -D^{\{k\}}x$, $C^{\{k\}} = A - BD^{\{k\}}$, $W^{\{k\}} = Q + {D^{\{k\}}}^T V D^{\{k\}}$, and $S^{\{k\}} - W^{\{k\}} = {C^{\{k\}}}^T S^{\{k\}} C^{\{k\}}$, and

$$D^{\{k\}} = (B^T S^{\{k-1\}} B + V)^{-1} B^T S^{\{k-1\}} A.$$

On the other hand, we have

$$\begin{aligned} S^{\{k-1\}} &= {C^{\{k-1\}}}^T S^{\{k-1\}} C^{\{k-1\}} + W^{\{k-1\}} \\ &= \left[A - BD^{\{k-1\}}\right]^T S^{\{k-1\}} \left[A - BD^{\{k-1\}}\right] + Q + {D^{\{k-1\}}}^T V D^{\{k-1\}}. \end{aligned}$$

We can see that when this iterative procedure continues, the above two equations converge to (7.49) and (7.50), which lead to the Riccati equation (7.51).

The above policy iteration procedure can be viewed as a numerical approach to solving the Riccati equation iteratively.

7.3.2 The JLQ Problem*

In a discrete-time *jump linear quadratic (JLQ) problem*, we consider a two-level stochastic control system. The system state at time l, $l = 0, 1, \ldots$, is denoted as (M_l, X_l), where $M_l \in \mathcal{M} := \{1, 2, \ldots, M\}$ represents the *mode* (high level) that the system is in, and $X_l \in \Re^n$ denote the continuous part of the state (low level). The system mode changes according to a finite-state ergodic Markov chain $\{M_l, l = 0, 1, \ldots\}$ with transition probabilities $p(j|i)$, $j, i = 1, 2, \ldots, M$. When the system is in mode $M_l = i$, $l = 0, 1, \ldots$, the continuous part X_l, evolves as

$$X_{l+1} = A_{M_l} X_l + B_{M_l} u_l + \xi_l, \qquad M_l = i \in \mathcal{M}, \tag{7.52}$$

where u_l is a control variable depending on M_l and X_l, and ξ_l, $l = 0, 1, \ldots$, is independent zero-mean noise. The dimensions of the A_m's and B_m's, $m \in \{1, 2, \ldots, M\}$, are $n \times n$ and $n \times m$, respectively. We assume that the transition among the modes at any time l is independent of the continuous state X_l.

The performance criterion is defined as

$$\eta^u(i, x) = \lim_{L \to \infty} \frac{1}{L} E \left\{ \sum_{l=0}^{L-1} \left[X_l^T Q_{M_l} X_l + u_l^T V_{M_l} u_l\right] \Big| X_0 = x, M_0 = i \right\}, \tag{7.53}$$

where Q_i and V_i, $i = 1, 2, \ldots, M$, are $n \times n$ and $m \times m$ positive semi-definite matrices, respectively. Our goal is to find a stationary control law $u = u(i, x)$ that minimizes $\eta^u(i, x)$. Note that the control does not affect the transition probabilities among the modes. For a feedback control law $u_l = u(M_l, X_l)$, from (7.53), the reward function for the JLQ problem is

$$f(i, x) = x^T Q_i x + u(i, x)^T V_i u(i, x). \tag{7.54}$$

The approach used to solve this problem is the same as that for the LQ problem, and we therefore only briefly present the results.

The Transition Operator

Denoting the transition function of the JLQ problem as $P(j, R|i, x)$, we have

$$P(j, R|i, x) = p(j|i)P_i(R|x), \qquad i, j \in \mathcal{M}, \quad x \in \Re^n, \quad R \in \mathcal{B}.$$

$P_i(R|x)$ is the transition function in mode i, which can be obtained from (7.52). For any function $h(i, x)$, $i \in \mathcal{M}$ and $x \in \Re^n$, we define the transition operator P as

$$(Ph)(i, x) := \sum_{j \in \mathcal{M}} \int_{\Re^n} h(j, y)P(j, dy|i, x) = \sum_{j \in \mathcal{M}} \left\{ p(j|i) \int_{\Re^n} h(j, y)P_i(dy|x) \right\}.$$

For any two operators P_1 and P_2, we define

$$(P_1 P_2)(j, R|i, x) := \sum_{k \in \mathcal{M}} \int_{\Re^n} P_1(k, dy|i, x)P_2(j, R|k, y),$$
$$i, j \in \mathcal{M}, \quad x \in \Re^n, \quad R \in \mathcal{B}.$$

Set $P^k = PP^{k-1} = P^{k-1}P$, $k \geq 2$, and $P^0 = I$, with

$$I(j, R|i, x) := \begin{cases} 1, & \text{if } i = j, \text{ and } x \in R, \\ 0, & \text{otherwise.} \end{cases}$$

And for any function $h(i, x)$, we have

$$(Ih)(i, x) = \sum_{j \in \mathcal{M}} \int_{\Re^n} h(j, y)I(j, dy|i, x) = h(i, x).$$

The unit function is defined as $e(i, x) = 1$ for all $i \in \mathcal{M}$ and $x \in \Re^n$. We have $Pe = e$ for any P. A probability distribution on $\mathcal{M} \times \Re^n$ is denoted as $\nu(i, R)$, with $\sum_{i \in \mathcal{M}} \int_{\Re^n} \nu(i, dy) = 1$. For any function $h(i, x)$, $i \in \mathcal{M}$, $x \in \Re^n$, define

$$\nu h := \sum_{k \in \mathcal{M}} \int_{\Re^n} h(k, x)\nu(k, dx),$$

provided it exists. We have $\nu e = 1$. The product of ν and P is defined as

$$(\nu P)(i, R) = \sum_{k \in \mathcal{M}} \int_{\Re^n} \nu(k, dy) P(i, R|k, y).$$

A steady-state probability distribution of P satisfies $\pi = \pi P$. For any ν, we define a transition function

$$(e\nu)(j, R|i, x) := \nu(j, R), \qquad i \in \mathcal{M}, \quad x \in \Re^n.$$

We have $(e\nu)^k = e\nu$.

Performance Potentials

The long-run average reward of the jump linear system is defined as

$$\eta(i, x) = \lim_{L \to \infty} \frac{1}{L} E\left\{ \sum_{l=0}^{L-1} f(M_l, X_l) \middle| M_0 = i, X_0 = x \right\}.$$

We have

$$\eta(i, x) = \lim_{L \to \infty} \frac{1}{L} \sum_{l=0}^{L-1} (P^l f)(i, x),$$

where $P(j, R|i, x)$ is the transition function. Similar to (7.28), with the steady-state probability distribution of P, π, we have

$$\lim_{k \to \infty} P^k f = (\pi f) e = \eta e. \tag{7.55}$$

Thus, we have

$$\eta(i, x) = \lim_{L \to \infty} (P^k f)(i, x) = \eta e(i, x).$$

The performance potential g satisfies the Poisson equation

$$(I - P) g + \eta = f.$$

Similar to the LQ problem, (7.32) holds if the convergence conditions (7.31) in Lemma 7.1 hold.

For any set of quadratic functions $h(i, x) = x^T W_i x^T$, where W_i, $i = 1, 2, \ldots, M$, are positive semi-definite matrices, and a control law $u(i, x)$, we have

$$(P^u h)(i, x) = \sum_{j \in \mathcal{M}} \left\{ p(j|i) \int_{\Re^n} h(j, y) P_i^u(dy|x) \right\}$$

$$= \sum_{j \in \mathcal{M}} \left\{ p(j|i) \int_{\Re^n} y^T W_j y p_\xi \left\{ y - [A_i x + B_i u(i, x)] \right\} dy \right\}$$

$$= \sum_{j \in \mathcal{M}} \left\{ p(j|i) \int_{\Re^n} \{ z + [A_i x + B_i u(i, x)] \}^T W_j \{ z + [A_i x + B_i u(i, x)] \} p_\xi(z) dz \right\}.$$

Again, for zero-mean distributions, we have $\int_{\Re^n} p_\xi(z)dz = 1$ and $\int_{\Re^n} z p_\xi(z)dz = 0$. Therefore,

$$(P^u h)(i,x) = c_1(i)e(x) + \sum_{j\in\mathcal{M}} p(j|i)\left[A_i x + B_i u(i,x)\right]^T W_j \left[A_i x + B_i u(i,x)\right], \tag{7.56}$$

where $c_1(i) := \sum_{j\in\mathcal{M}} c_0(j)p(j|i)$ with

$$c_0(j) := \int_{\Re^n} \left[z^T W_j z\right] p_\xi(z)dz.$$

For a linear control $u_l = -D_{M_l} X_l$, the system equation (7.52) becomes

$$X_{l+1} = C_{M_l} X_l + \xi_l,$$

where

$$C_i = A_i - B_i D_i, \qquad i = 1, 2, \ldots, M.$$

The performance function (7.54) becomes $f(i,x) = x^T W_i x$, with $W_i = Q_i + D_i^T V_i D_i$.

Now, we set $W_{i,0} := W_i$, $i = 1, 2, \ldots, M$, in the performance function $f(i,x)$ and set $h(i,x) = f(i,x)$ in (7.56). Then, for this linear jump system, from (7.56), we have (we drop the superscript "u" for simplicity)

$$\begin{aligned}(Pf)(i,x) &= c_{1,0}(i)e(x) + \sum_{j\in\mathcal{M}} p(j|i)x^T(C_i^T W_{j,0} C_i)x \\ &= c_{1,0}(i)e(x) + x^T W_{i,1} x,\end{aligned}$$

where

$$W_{i,1} = \sum_{j\in\mathcal{M}} p(j|i)(C_i^T W_{j,0} C_i),$$

$$c_{1,0}(i) := \sum_{j\in\mathcal{M}} c_{0,0}(j)p(j|i),$$

with

$$c_{0,0}(j) := \int_{\Re^n} \left[z^T W_{j,0} z\right] p_\xi(z)dz.$$

Next, we have

$$(P^2 f)(i,x) = \left[c_{2,0}(i) + c_{1,1}(i)\right] e(x) + x^T W_{i,2} x,$$

where

$$W_{i,2} = \sum_{j\in\mathcal{M}} p(j|i)(C_i^T W_{j,1} C_i),$$

$c_{2,0}(i) := \sum_{j\in\mathcal{M}} c_{1,0}(j)p(j|i)$, and $c_{1,1}(i) := \sum_{j\in\mathcal{M}} c_{0,1}(j)p(j|i)$, and

$$c_{0,1}(j) := \int_{\Re^n} z^T W_{j,1} z p_\xi(z) dz.$$

Continuing this process, we have

$$(P^k f)(i,x) = c_k(i)e(x) + x^T W_{i,k} x, \tag{7.57}$$

with

$$W_{i,k} = \sum_{j \in \mathcal{M}} p(j|i)(C_i^T W_{j,k-1} C_i)$$

and

$$c_k(i) := c_{k,0}(i) + c_{k-1,1}(i) + \cdots + c_{1,k-1}(i),$$

with the items $c_{k,0}$, $c_{k-1,1}$, ..., and $c_{1,k-1}$ properly defined. We will not give the explicit form of these items since they are not relevant to our main results.

For stable systems with $\rho(C_i) < 1$ for all $i \in \mathcal{M}$, we have $W_{i,k} \to 0$ as $k \to \infty$. Therefore, from (7.57) and (7.55) we have $c_k(i) \to \eta$, for all $i \in \mathcal{M}$. Furthermore, the following sum exists:

$$S_i := \sum_{k=0}^{\infty} W_{i,k}.$$

Following the same argument as for the LQ problem, we obtain the performance potentials

$$g(i,x) = \left[\sum_{k=1}^{\infty} (c_k(i) - \eta) \right] e(x) + x^T S_i x. \tag{7.58}$$

The first term depends on the mode i and, therefore, is not a constant. In fact, $x^T S_i x$ corresponds to the potentials of the states in the same mode i, while $\sum_{k=1}^{\infty}(c_k(i) - \eta)$ reflects the difference in the potentials of the states in different modes.

Let $H_i := \sum_{j \in \mathcal{M}} S_j p(j|i)$. We can easily verify that

$$S_i - W_i = C_i^T H_i C_i, \qquad i \in \mathcal{M}. \tag{7.59}$$

Thus, we have

$$\begin{aligned} S_i &= C_i^T H_i C_i + W_i \\ &= (A_i - B_i D_i)^T H_i (A_i - B_i D_i) + Q_i + D_i^T V_i D_i, \qquad i \in \mathcal{M}. \end{aligned} \tag{7.60}$$

The Optimal Policy

We may apply the optimality equation (7.38) to determine an optimal policy. Again, we assume that $P_\xi(x)$ has full support on $\Re^n$ and hence all policies have the same support $\Re^n$. We start with a linear policy $u(i,x) = -D_i x$.

The performance potential is (7.58). From (7.56), for any other policy $\widetilde{u}(i,x)$, $i \in \mathcal{M}$, $x \in \Re^n$, which may not be linear, we have

$$\begin{aligned}(P^{\widetilde{u}}g)(i,x) + f^{\widetilde{u}}(i,x) &= c(i)e(x) + x^T Q_i x + \widetilde{u}(i,x)^T V_i \widetilde{u}(i,x) \\ &\quad + \sum_{j \in \mathcal{M}} p(j|i) \left[A_i x + B_i \widetilde{u}(i,x)\right]^T S_j \left[A_i x + B_i \widetilde{u}(i,x)\right] \\ &= c(i)e(x) + x^T Q_i x + \widetilde{u}(i,x)^T V_i \widetilde{u}(i,x) \\ &\quad + \left[A_i x + B_i \widetilde{u}(i,x)\right]^T H_i \left[A_i x + B_i \widetilde{u}(i,x)\right],\end{aligned}$$

in which $c(i)$ is some constant and $c(i)e(x) + x^T Q_i x$ does not depend on $\widetilde{u}$. The improved policy is the one that minimizes the above quantity shown as follows

$$u'(i,x) = -D_i' x,$$

where

$$D_i' = (B_i H_i B_i + V_i)^{-1} B_i^T H_i A_i.$$

When $D_i = D_i'$, $i \in \mathcal{M}$, the policy $u(i,x) = -D_i x$ is optimal. Denote the optimal policy as $\hat{u}(i,x) = -\hat{D}_i x$ and the corresponding quantities as $\hat{S}_i$ and $\hat{H}_i$. We have

$$\hat{D}_i = (B_i \hat{H}_i B_i + V_i)^{-1} B_i^T \hat{H}_i A_i, \tag{7.61}$$

where

$$\hat{H}_i = \sum_{j \in \mathcal{M}} \hat{S}_j p(j|i), \tag{7.62}$$

and $\hat{S}_i$ satisfies (7.59) and (7.60).

Substituting (7.61) into (7.60), we have

$$\hat{S}_i = A_i^T \hat{H}_i A_i - A_i^T \hat{H}_i B_i (V_i + B_i^T \hat{H}_i B_i)^{-1} B_i^T \hat{H}_i A_i + Q_i. \tag{7.63}$$

This is the *coupled Riccati equation* for the optimal policy $\hat{u}(i,x) = -\hat{D}_i x$, where $\hat{D}_i$ can be determined by (7.61), (7.62), and (7.63).

The results in this section is reported in [265].

7.4 On-Line Optimization and Adaptive Control

In the previous sections, we showed that the MDP policy iteration approach leads to the same results as those in the literature for standard LQ and JLQ problems. The power of this approach lies in its on-line implementation feature. For complex systems, the system parameters and even the system structures may not be known, and yet we may be able to estimate the performance potentials on sample paths and to implement policy iteration to improve the system performance.

With the on-line MDP approach, a system is modelled by a Markov chain in general and the specific system structure is usually not identified. The parameters are the transition probabilities or transition functions, and in many cases these parameters need not be completely estimated. Instead, we can directly determine the control law by estimating the performance potentials or the Q-factors (on-line learning). Thus, the approach is in the spirit of direct adaptive control.

7.4.1 Discretization and Estimation

Discretization of State Spaces

The first step towards an on-line learning algorithm is the discretization of the continuous state space. To simplify the discussion, we consider a one-dimensional state space $\Re$. We divide $\Re$ into S intervals by $x_1 = -\infty, x_2, \ldots, x_S, x_{S+1} = \infty$. Set $\Delta x_i := x_{i+1} - x_i$, $i = 1, 2, \ldots, S$. The values of Δx_i for $i = 1$ and $i = S$ are both infinity. We call the interval $(x_i, x_i + \Delta x_i]$, $i = 2, \ldots, S-1$, the ith interval and denote it simply as Δ_i, and the first interval is denoted as $\Delta_1 = (-\infty, x_2]$ and the Sth interval is denoted as $\Delta_S = (x_S, \infty)$. We assume that the "effects" of the states in each interval Δ_i are almost the same, either because the potentials in these states are very close, or because the steady-state probability of the interval is close to zero and thus the "effects" of the states in the interval are negligible. The former is the case when Δx_i is very small, and the latter is for Δ_1 and Δ_S (cf. the performance difference formulas).

The ith interval represents a discretized state denoted as i, and we aggregate all the continuous states in Δ_i, $i = 1, 2, \ldots, S$, into a discrete state i. Thus, we have a finite state space $\mathcal{S} = \{1, 2, \ldots, S\}$ to approximate the continuous state space $\Re$.

Potentials of the Approximate Chains

Given a system $(P(dy|x), f(x))$ with potentials $g(x)$, $x \in \Re$, we wish to determine the corresponding transition probability matrix $P := [p(j|i)]$ and the potentials $g(i)$, $i, j \in \mathcal{S}$, for the approximately discretized finite-state Markov chain. For simplicity, we use the same notation of P and g for both continuous and discretized versions. We distinguish them by the arguments x, y and i, j in the expressions. First, we consider the steady-state expected value of g. For the continuous model, we have

$$\pi g = (\pi P)g = \int_{x\in\Re}\int_{y\in\Re} g(y)P(dy|x)\pi(dx). \tag{7.64}$$

For the discrete version, we need to determine the $\pi(i)$'s, $p(j|i)$'s, and $g(j)$'s such that

$$\pi g = \sum_{i\in\mathcal{S}}\sum_{j\in\mathcal{S}} \pi(i)p(j|i)g(j). \tag{7.65}$$

From (7.64), we have

$$\begin{aligned}\pi g &= \sum_{i\in\mathcal{S}}\sum_{j\in\mathcal{S}}\left\{\int_{x\in\Delta_i}\int_{y\in\Delta_j} g(y)P(dy|x)\pi(dx)\right\}\\ &= \sum_{i\in\mathcal{S}}\sum_{j\in\mathcal{S}}\left\{\left(\int_{y\in\Delta_j} g(y)\frac{\int_{x\in\Delta_i} P(dy|x)\pi(dx)}{\int_{x\in\Delta_i}\int_{y\in\Delta_j} P(dy|x)\pi(dx)}\right)\right.\\ &\qquad\left.\left(\frac{\int_{x\in\Delta_i}\int_{y\in\Delta_j} P(dy|x)\pi(dx)}{\int_{x\in\Delta_i}\pi(dx)}\right)\left(\int_{x\in\Delta_i}\pi(dx)\right)\right\}.\end{aligned}$$

Comparing this with (7.65), we should have

$$\pi(i) = \int_{x\in\Delta_i}\pi(dx), \tag{7.66}$$

$$p(j|i) = \frac{\int_{x\in\Delta_i}\int_{y\in\Delta_j} P(dy|x)\pi(dx)}{\int_{x\in\Delta_i}\pi(dx)}. \tag{7.67}$$

These two equations can be explained intuitively, since

$$\int_{x\in\Delta_i}\pi(dx) = \pi(\Delta_i) \text{ and } \int_{x\in\Delta_i}\int_{y\in\Delta_j} P(dy|x)\pi(dx) = \pi(\Delta_i,\Delta_j),$$

where $\pi(\Delta_i,\Delta_j)$ is the steady-state probability of $X_l \in \Delta_i$ and $X_{l+1} \in \Delta_j$. The item in the first bracket, however, has a form depending on both i and j, so we need to denote it as

$$g(i,j) = \int_{y\in\Delta_j} g(y)\frac{\int_{x\in\Delta_i} P(dy|x)\pi(dx)}{\int_{x\in\Delta_i}\int_{y\in\Delta_j} P(dy|x)\pi(dx)} = \int_{y\in\Delta_j} g(y)\mu_i(dy),$$

where

$$\mu_i(dy) = \frac{\int_{x\in\Delta_i} P(dy|x)\pi(dx)}{\int_{x\in\Delta_i}\int_{y\in\Delta_j} P(dy|x)\pi(dx)} = \frac{\pi(\Delta_i, dy)}{\pi(\Delta_i,\Delta_j)}$$

is a conditional probability of dy given that $x \in \Delta_i$ and $y \in \Delta_j$. From the Markov property, $g(y)$ does not depend on i. In addition, $\int_{y\in\Delta_j}\mu_i(dy) = 1$. Therefore, if $g(y)$ are very close to each other in $y \in \Delta_j$ (e.g., Δ_j is very small), then $g(i,j) \approx g(y)$, $y \in \Delta_j$, are also very close to each other for all i.

Furthermore, if the $g(i,j)$'s are almost independent of i, we have the following approximation

$$\begin{aligned} g(i,j) &= \frac{\int_{y\in\Delta_j} g(y) \int_{x\in\Delta_i} P(dy|x)\pi(dx)}{\int_{x\in\Delta_i}\int_{y\in\Delta_j} P(dy|x)\pi(dx)} \\ &\approx \frac{\sum_{i\in\mathcal{S}}\left\{\int_{y\in\Delta_j} g(y)\int_{x\in\Delta_i} P(dy|x)\pi(dx)\right\}}{\sum_{i\in\mathcal{S}}\left\{\int_{x\in\Delta_i}\int_{y\in\Delta_j} P(dy|x)\pi(dx)\right\}} \\ &= \frac{\int_{y\in\Delta_j} g(y)\int_{x\in\Re} P(dy|x)\pi(dx)}{\int_{y\in\Delta_j}\int_{x\in\Re} P(dy|x)\pi(dx)}. \end{aligned}$$

For the steady-state probability distribution $\pi(dx)$, we have $\int_{x\in\Re} P(dy|x)\pi(dx) = \pi(dy)$ and $\int_{y\in\Delta_j}\int_{x\in\Re} P(dy|x)\pi(dx) = \int_{y\in\Delta_j}\pi(dy) = \pi(\Delta_j)$. Denote

$$\pi(dy|\Delta_j) := \frac{\pi(dy)}{\pi(\Delta_j)},$$

which is a steady-state conditional probability distribution. Then, we have $g(i,j) \approx g(j)$ for all i, where

$$g(j) := \int_{y\in\Delta_j} g(y)\pi(dy|\Delta_j) \tag{7.68}$$

is the mean of the potentials in the interval Δ_j. The approximation in (7.68) may not hold if Δ_j is not small, e.g., for $\Delta_1 = (-\infty, x_2]$ and $\Delta_S = [x_S, \infty)$. However, the probabilities that the system is in these intervals are almost zero, so the effect is negligible.

Finally, with the $\pi(i)$, $p(j|i)$, and $g(j)$, $i,j \in \mathcal{S}$, defined in (7.66), (7.67), and (7.68), respectively, the average potentials in (7.64) and (7.65) are approximately the same for both the continuous and the discretized Markov chains. Therefore, they are the values for the approximate discretized Markov chains.

Estimation of Potentials and Parameters

For any interval $\Delta \subset \Re$ and two intervals $\Delta_1, \Delta_2 \subset \Re$, we define $I_\Delta(x) = 1$ if $x \in \Delta$, and $I_\Delta(x) = 0$ otherwise; and $I_{\Delta_1,\Delta_2}(x,y) = 1$ if $x \in \Delta_1$ and $y \in \Delta_2$, and $I_{\Delta_1,\Delta_2}(x,y) = 0$, otherwise. Given a sample path $\{X_0, X_1, \ldots, X_{L-1}\}$ of a Markov chain with state space $\Re$, it is clear from (7.66), (7.67), and (7.68) that

$$\pi(i) = \lim_{L\to\infty}\frac{1}{L}\sum_{l=0}^{L-1} I_{\Delta_i}(X_l), \qquad \text{w.p.1},$$

$$p(j|i) = \lim_{L \to \infty} \frac{\sum_{l=0}^{L-1} I_{\Delta_i,\Delta_j}(X_l, X_{l+1})}{\sum_{l=0}^{L-1} I_{\Delta_i}(X_l)}, \qquad \text{w.p.1}, \tag{7.69}$$

and

$$g(j) = \lim_{L \to \infty} \frac{\sum_{l=0}^{L-1} I_{\Delta_j}(X_l) g(X_l)}{\sum_{l=0}^{L-1} I_{\Delta_j}(X_l)}, \qquad \text{w.p.1}.$$

In practice, to estimate S (say 10,000) potentials $g(j)$, $j \in \mathcal{S}$, is usually feasible; however, to estimate S^2 transition probabilities $p(j|i)$, $i, j \in \mathcal{S}$, may not be desirable. Furthermore, based on observing a sample path of a system under one policy, it may not be possible to estimate the transition probabilities of other policies. We have the following alternatives:

1. Reinforcement learning methods may be used to estimate Q-factors with different actions. This requires us to estimate $\sum_{i=1}^{S} |\mathcal{A}(i)|$ items, and it also requires to "perturb" the system with an ϵ-greedy policy that visits all the state-action pairs (see Section 6.3.2), and this may not be practically desirable.
2. The number of the transition probabilities under different actions to be estimated can be reduced if the system possesses some special structures.

To illustrate the second idea, we assume that all the intervals (except those with probabilities almost zero) are of an equal size, i.e., $\Delta x_i = \Delta$ for all $i \in \mathcal{S}$. For one-dimensional linear systems, with a control policy $u(x)$, the transition function (7.42) takes the following form:

$$P^u(dy|x) = p_\xi\left[y - (ax + bu)\right] dy.$$

If we assume that Δ is very small, from (7.67) we have

$$p^u(j|i) \approx p_\xi\left[y - (ax + bu)\right] \Delta, \qquad y \in \Delta_j, \quad x \in \Delta_i. \tag{7.70}$$

Thus, if we know the distribution function $p_\xi(y)$, then we can calculate the transition probabilities $p^u(j|i)$, $i, j \in \mathcal{S}$, for any $u(x)$. This means that we can convert a problem of estimating a two-dimensional matrix P to a problem of estimating a one-dimensional vector $p_\xi(y)$, $y \in \mathcal{S}$.

We use some simple examples to illustrate the main ideas.

Example 7.2. We consider a one-dimensional control system. Suppose that the distribution density function $p_\xi(y)$ concentrates around the origin $y = 0$. We may take $\Delta = 0.1$ and divide $\Re$ with the points $-9\Delta, -8\Delta \ldots, -\Delta, 0, \Delta, \ldots, 9\Delta$. We have $S = 20$ discrete states, $1, 2, \ldots, 20$, corresponding to intervals $\Delta_1 = (-\infty, -9\Delta]$, $\Delta_2 = (-9\Delta, -8\Delta]$, $\ldots$, $\Delta_{19} = (8\Delta, 9\Delta]$, and $\Delta_{20} = (9\Delta, \infty)$. We assume that the probability that the random noise ξ lies in Δ_1 and Δ_{20} is very small.

Let $a = 1.5$ and $b = 2$. If we take action $u = 0$, then by (7.70) the transition probabilities are

$$p^{\{0\}}(j|i) \approx p_\xi(y - 1.5x)\Delta, \qquad y \in \Delta_j, \quad x \in \Delta_i.$$

As an example, let $i = 3$ correspond to $x = -7\Delta$. Therefore, $p^{\{0\}}(19|3) = p_\xi(9\Delta + 1.5 \times 7\Delta)\Delta = p_\xi(19.5\Delta)\Delta$, $p^{\{0\}}(18|3) = p_\xi(8\Delta + 1.5 \times 7\Delta)\Delta = p_\xi(18.5\Delta)\Delta$, $\ldots$, $p^{\{0\}}(2|3) = p_\xi(-8\Delta + 1.5 \times 7\Delta)\Delta = p_\xi(2.5\Delta)\Delta$. Thus, $p^{\{0\}}(j|3)$, $j = 2, 3, \ldots, 19$, which can be estimated by (7.69), correspond to the probabilities of the random noise ξ in the intervals $(2\Delta, 3\Delta]$, $\ldots$, $(18\Delta, 19\Delta]$, and $(19\Delta, 20\Delta]$. Similarly, $p^{\{0\}}(j|2)$, $j = 2, 3, \ldots, 19$, correspond to the probabilities of the random noise ξ in the intervals $(3\Delta, 4\Delta]$, $\ldots$, $(17\Delta, 18\Delta]$, and $(20\Delta, 21\Delta]$; and $p^{\{0\}}(j|19)$, $j = 2, 3, \ldots, 19$, correspond to the probabilities of the random noise ξ in the intervals $(-23\Delta, -22\Delta]$, $\ldots$, $(-3\Delta, -4\Delta]$, and $(-4\Delta, -5\Delta]$. Overall, the transition probabilities $p^{\{0\}}(j|i)$, $i, j = 2, \ldots, 19$, correspond to the probabilities of the random noise ξ in the intervals $(-23\Delta, -22\Delta]$, $(-22\Delta, -21\Delta]$, $\ldots$, and $(20\Delta, 21\Delta]$; there are altogether 44 intervals. That is, we need only to estimate 44, rather than 18×18, values. Transition probabilities $p^{\{0\}}(j|1)$ and $p^{\{0\}}(j|20)$ have to be dealt with separately by using (7.69). Hopefully, these transition probabilities are not so important because the steady-state probabilities $\pi^u(1)$ and $\pi^u(20)$ under any policy u are very small.

Now, suppose that we apply a feedback control law with $u = -x$. Then, the system is

$$y = ax + bu + \xi = 1.5x + 2(-x) + \xi = -0.5x + \xi,$$

and we have

$$p^{-x}(j|i) \approx p_\xi(y + 0.5x)\Delta, \qquad y \in \Delta_j, \quad x \in \Delta_i.$$

As an example, consider $i = 6$ corresponding to $x = -4\Delta$. We have $p^{-x}(19|6) = p_\xi(9\Delta - 0.5 \times 4\Delta)\Delta = p_\xi(7\Delta)\Delta$, $p^{-x}(18|6) = p_\xi(8\Delta - 0.5 \times 4\Delta)\Delta = p_\xi(6\Delta)\Delta$, $\ldots$, $p^{-x}(2|6) = p_\xi(-8\Delta - 0.5 \times 4\Delta)\Delta = p_\xi(-10\Delta)\Delta$. Overall, the transition probabilities $p^{-x}(j|i)$, $i, j = 2, \ldots, 19$, correspond to the probabilities of the random noise ξ in the intervals $(-13\Delta, -12\Delta]$, $(-12\Delta, -11\Delta]$, $\ldots$, and $(12\Delta, 13\Delta]$; there are altogether 26 intervals. All these intervals are among the previous intervals for determining $p^{\{0\}}(j|i)$, $i, j = 2, \ldots, 19$; thus, all $p^{-x}(j|i)$ are known once $p^{\{0\}}(j|i)$, $i, j = 2, \ldots, 19$, are known. □

Example 7.3. Consider a JLQ system. At the high level, there are two modes, $\mathcal{M} = \{1, 2\}$, which form a Markov chain with the following transition probability matrix

$$P = \begin{bmatrix} 0.5 & 0.5 \\ 0.5 & 0.5 \end{bmatrix}.$$

The system parameters are: $A_1 = 1,\ B_1 = 1,\ Q_1 = 1, V_1 = 1;\ A_2 = 2,\ B_2 = -1,\ Q_2 = 2, V_2 = 2$. Let the noise ξ_k be a normally distributed random variable $N(0, 1)$.

The coupled Riccati equation (7.63) is

$$\begin{cases} \hat{S}_1 = \hat{H}_1 - \hat{H}_1(1+\hat{H}_1)^{-1}\hat{H}_1 + 1, \\ \hat{S}_2 = 4\hat{H}_2 - 4\hat{H}_2(2+\hat{H}_2)^{-1}\hat{H}_2 + 2, \end{cases} \tag{7.71}$$

where $\hat{H}_1 = \hat{H}_2 = 0.5\hat{S}_1 + 0.5\hat{S}_2$.

Solving the coupled Riccati equation yields the theoretical values for the optimal control law as $\hat{D}_1 = 0.8253$ and $\hat{D}_2 = -1.405$.

We simulated the system and applied the learning algorithm to estimate $\hat{D}_1$ and $\hat{D}_2$. We took $\Delta = 0.1$ and divided $\Re$ by the points $-20\Delta, -19\Delta, \dots, -\Delta, 0, \Delta, \dots, 20\Delta$. We ran the system for $L = 10,000$ transitions for the first five iterations, and $L = 50,000$ for other iterations. The initial policy was chosen to be $D_1^{\{0\}} = 1$ and $D_2^{\{0\}} = -1$.

We applied the on-line learning algorithm to update control policies. The performance of each iteration is shown in Figure 7.5. The figure shows that the performance improves rapidly, and after seven iterations, the performance is almost the same as the optimal one. We stop the iteration when the differences between the control values $u(x)$ in two consecutive iterations for all x are smaller than $\varepsilon = 0.03$. The iterations terminate after 50 iteration steps. Figures 7.6 and 7.7 illustrate the optimal control laws $u = -\hat{D}_i x$, for modes $i = 1$ and $i = 2$, respectively. The straight lines are theoretical optimal control laws while the two wavy ones are obtained from the on-line algorithm. In the on-line learning approach, we estimate the potentials; but we assume that the parameters are known, and we apply (7.70) to estimate the transition probabilities $p^u(j|i)$, $i, j \in \mathcal{S}$, and use them to implement policy iteration without solving the coupled Riccati equation. □

7.4.2 Discussion

One advantage of the sample-path-based learning and optimization approach is that principally it applies to both linear and non-linear systems in the same way. The system structure affects only the transition probability matrix. However, determining the transition probability matrices for different control parameters might be a difficult task.

Another difficulty one may encounter is that, because the state and action spaces in such problems are usually continuous, the theoretical results on convergence, etc, may be limited.

More research in this direction is certainly needed.

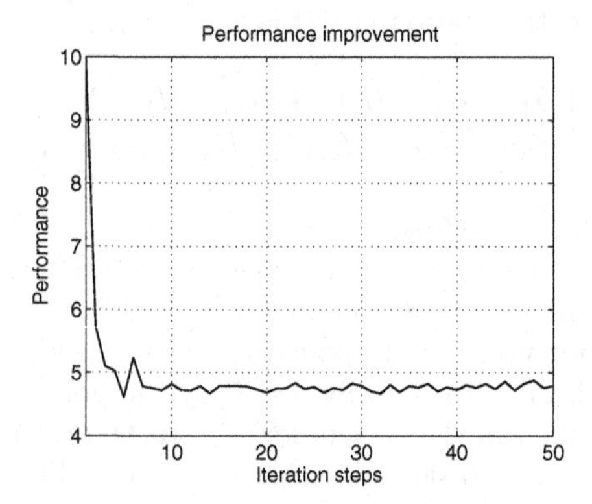

Fig. 7.5. Performance Improvement

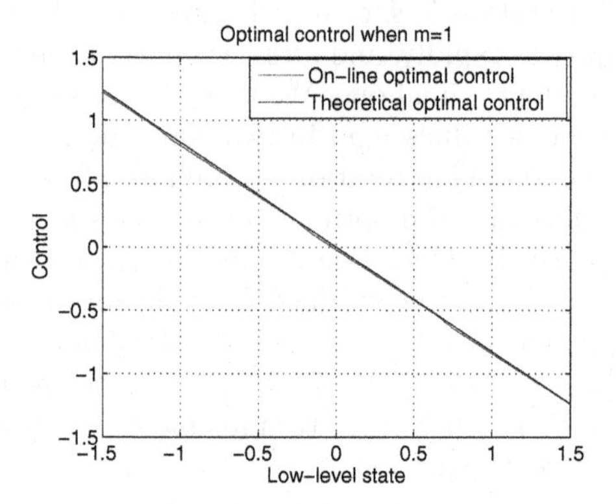

Fig. 7.6. Optimal Control Law for Mode 1

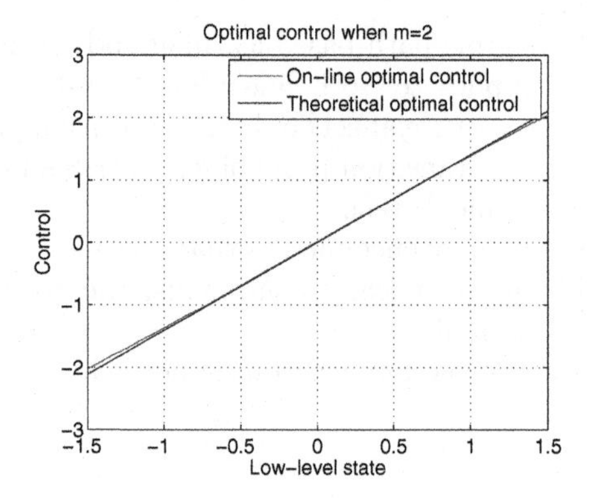

Fig. 7.7. Optimal Control Law for Mode 2

PROBLEMS

7.1. Repeat Example 7.1 with the data listed in Table 7.3.

state		Transition prob. $p^{\alpha}(j\|i)$			
i	action	$j=1$	2	3	Perf. func.
	$\alpha_{1,1}$	0.3	0.6	0.1	
1	$\alpha_{1,2}$	0.4	0.2	0.4	10
	$\alpha_{1,3}$	0.2	0.3	0.5	
	$\alpha_{2,1}$	0.6	0	0.4	
2	$\alpha_{2,2}$	0.4	0.3	0.3	0
	$\alpha_{3,1}$	0.4	0.2	0.4	
3	$\alpha_{3,2}$	0.3	0.5	0.2	-5
	$\alpha_{3,3}$	0.2	0.1	0.7	

Table 7.3. The Actions and Performance Function in Problem 7.1

7.2. For any three operators P_1, P_2, and P_3, prove

a. For any function $h(x)$ on $\Re^n$, we have $(P_1P_2)h(x) = P_1[P_2h(x)]$ (assuming the integrations exist); and
b. $(P_1P_2)P_3 = P_1(P_2P_3)$; and
c. $P^k = PP^{k-1} = P^{k-1}P$.

7.3. For any probability distribution ν, transition function P, and any function h, prove $\nu(Ph) = (\nu P)h$. Explain the meaning of both sides.

7.4. With the forward-time index used in (7.9), from (7.14) and (7.15), we can define the finite-step perturbation realization factor for any policy $\boldsymbol{d} = \{d_0, d_1, \ldots, d_{L-1}\}$ as follows:

$$g^{\boldsymbol{d}}_{\ell=L}(i) = \sum_{l=0}^{L-1} E\{[f(X_l, d_l(X_l)) - f(X'_l, d_l(X'_l))]| X_0 = i, X'_0 = i^*\},$$

where $\boldsymbol{X} = \{X_0, X_1, \ldots\}$ and $\boldsymbol{X}' = \{X'_0, X'_1, \ldots\}$ are two independent sample paths with initial states $X_0 = i$ and $X'_0 = i^*$, respectively. Note that the decision rules d_l may be different for different $l = 0, 1, \ldots$. Let L_{ii^*} be the time at which the two sample paths merge together, i.e., $X_{L_{ii^*}} = X'_{L_{ii^*}}$.

a. Prove that if $E(L_{ii^*}) < \infty$, then $\lim_{L\to\infty} g^{\boldsymbol{d}}_{\ell=L}(i)$ exists.
b. Find a condition under which $E(L_{ii^*}) < \infty$.

7.5. Prove Lemma 7.1.

7.6.* For any bounded function $f(x)$, $x \in \Re$, we define the e-norm $\|f(x)\| = \sup_x |f(x)|$. The e-norm of a linear operation $P(R|x)$ is defined as $\|P\| := \sup\{\|Pu\| : \|u\| \leq 1\}$. A transition function P is called e-ergodic, if

$$\lim_{k \to \infty} \|(P^k - e\pi)\| = 0.$$

Prove: if P is e-ergodic, then

$$\lim_{k \to \infty} g_k = g, \text{ and } \lim_{k \to \infty} Pg_k = Pg,$$

where $g_k := \left\{I + \sum_{l=1}^{k}(P^l - e\pi)\right\} f$, for any bounded function f.

7.7. Consider the two steady-state probability distributions π and π' defined as shown in Figure 7.8. The two distributions have discrete masses as follows: $\pi(-0.2) = \pi(-0.4) = \pi(-0.6) = \pi(-0.8) = \pi(-1) = 0.1$, and $\pi'(0.2) = \pi'(0.4) = \pi'(0.6) = \pi'(0.8) = \pi'(1) = 0.1$. The total probabilities on these discrete points are $\frac{1}{2}$ for both distributions. The other $\frac{1}{2}$ is evenly distributed on the interval $[-1, 1]$. Explain that these two distribution functions have the same state space, but they do not have the same support.

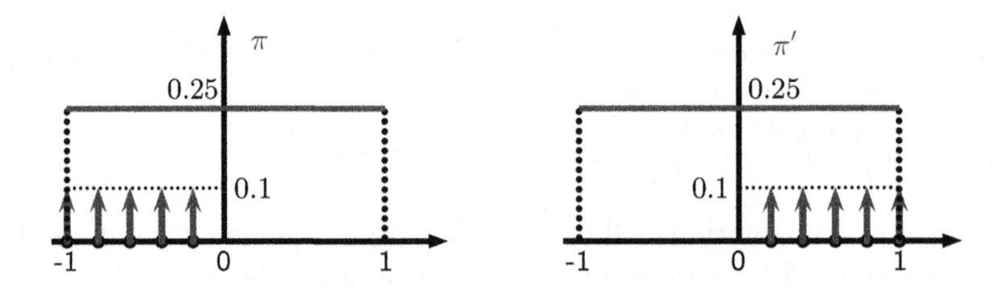

Fig. 7.8. The Two Steady-State Probability Distributions in Problem 7.7

7.8. Consider a non-linear control system

$$X_{l+1} = uX_l + \xi_l, \qquad l = 0, 1, \ldots,$$

where u is a control variable. Let $p_\xi(\bullet)$ be the distribution density function of the independent and identically distributed random noise ξ_l, $l = 0, 1, \ldots$.

a. Derive the transition probability function $P^u(dy|x)$.
b. How do we estimate the discrete approximation $p(j|i)$, $i, j = 1, 2, \ldots, S$? Can we reduce the number of transition probabilities to be estimated?

7.9. Consider a JLQ problem.

a. Suppose that the modes change slowly. That is, $p(i|i) \approx 1$ and $p(j|i) \approx 0$ for $j \neq i$. Show that the coupled Riccati equation is de-coupled into M Riccati equations corresponding to M LQ problems.
b. We consider another extreme case: the mode changes rapidly. As an example, we consider a 2-mode system ($M = 2$). Suppose $p(2|1) = p(1|2) \approx 1$ and $p(1|1) = p(2|2) \approx 0$. What is the coupled Riccati equation in this case? Explain your results.

7.10. Prove that in (7.48),

$$\sum_{k=1}^{\infty}(c_k - \eta) = -\int_{\Re^n} \left[z^T U z\right] p_\xi(z)dz,$$

with $U = \sum_{k=1}^{\infty} kW_k$. Prove that

$$U - C^T UC = C^T SC.$$

7.11. Consider a linear system

$$X_{l+1} = CX_l + \xi_l, \qquad l = 0, 1, \ldots,$$

with a discounted quadratic performance criterion

$$\eta(x) = \lim_{L\to\infty} E\left\{\sum_{l=0}^{L} \beta^l (X_l^T W X_l) \middle| X_0 = x\right\}, \qquad 0 < \beta < 1,$$

with W being a positive semi-definite matrix. Determine the performance potentials of this *LDQ (linear-discounted-quadratic)* problem.

7.12. Consider a linear control problem

$$X_{l+1} = AX_l + Bu(X_l) + \xi_l, \qquad l = 0, 1, \ldots,$$

with a discounted quadratic performance criterion

$$\eta(x) = \lim_{L\to\infty} E\left\{\sum_{l=0}^{L} \beta^l \left[X_l^T Q X_l + u_l^T V u_l\right] \middle| X_0 = x\right\}, \qquad l = 0, 1, \ldots.$$

Apply policy iteration to this LDQ control problem to derive the (discounted) Riccati equation for the optimal policy.

7.13. Consider the JLQ problem

$$X_{l+1} = A_{M_l} X_l + B_{M_l} u_l + \xi_{M_l,l},$$

in which the noise $\xi_{M_l,l}$, $M_l = 1, 2, \ldots, M$, has different probability distributions $P_{\xi_i}(y)$, $i = 1, 2, \ldots, M$, $y \in \Re^n$. Derive the solution to this problem.

Part II

The Event-Based Optimization - A New Approach

Imagination is more important than knowledge.

Albert Einstein
American (German born)
physicist (1879 - 1955)

8 Event-Based Optimization of Markov Systems

In the previous chapters, we developed a sensitivity-based approach that provides a unified framework for learning and optimization. We have shown that the two performance sensitivity formulas are the bases for learning and optimization of stochastic systems. The performance derivative formula leads to the gradient-based optimization approach, and the performance difference formula leads to the policy iteration approach to the standard MDP-type of problems.

However, the standard Markov-model-based formulation suffers from a number of drawbacks. First and foremost, the state space is usually too large for practical systems. Thus, the number of potentials to be calculated or estimated is too large for most problems. Second, the generally applicable Markov model does not reflect any special structure of a particular problem. Thus, from the Markov model alone, it is not clear whether or not, or how, potentials can be aggregated to save computation by exploiting the special structure of the system. The third issue is related to policy iteration: It requires the actions at different states to be chosen independently (the *independent-action assumption*, see Chapter 4). In many practical problems, however, these actions may have to be correlated; the standard policy iteration cannot handle such problems properly.

A natural question is, can we apply the sensitivity-based approach to overcome the aforementioned drawbacks? In other words, with the sensitivity-based approach, can we develop optimization methods to the problems with special features that may not fit the standard MDP formulation? And can we utilize the special features of a problem to save computation, or even to achieve a better performance than the standard MDP formulation?

To answer these questions, we start with the characterization of the special features of an optimization problem. In many problems, such special features can be captured by the concept of "event". For example, in an admission control problem (cf. Example 8.4), an event may be a customer arrival. In a two-level hierarchical control problem, an event may be a transition between two higher-level modes. In a traffic control problem of a large communication network, an event may be a packet transmission between two subnetworks. In these problems, control can only be applied when some events occur.

With the above observation, we propose an event-based optimization approach. Specifically, we will study what the main features of the event-based optimization are, how to formally characterize events, how to utilize the special features captured by events to derive performance sensitivity formulas, how to aggregate the performance potentials using the special features, how to further develop learning and optimization approaches based on these sensitivity formulas and aggregated potentials, and what the advantages and disadvantages of this event-based optimization approach are.

8.1 An Overview

8.1.1 Summary of Previous Chapters

In the previous chapters, we introduced the different areas in the learning and optimization of stochastic dynamic systems using a unified framework based on a sensitivity point of view. Among these areas are perturbation analysis (PA) [42, 51, 62, 72, 141, 142], Markov decision processes (MDPs) [6, 21, 25, 202, 216], reinforcement learning (RL) [17, 18, 161, 170, 171, 229, 230, 236, 238, 244], and direct adaptive control [9, 206]. This unified framework was illustrated by the "map of the learning and optimization world" in Figure 1.19, which is reproduced here as Figure 8.1.

A Map with a Sensitivity-Based View

At the center of the world are the two types of performance sensitivity formulas: When the system parameters are continuous variables, the sensitivity is the gradient (derivatives) of the performance with respect to the parameters; when the system is characterized by discrete quantities (e.g., policies), the sensitivity is the difference between the performance of the system with two different sets of parameters (e.g., under two different policies). The fundamental concept for both sensitivities is the performance potential of a Markov process. Both PA and MDP can be explained from a performance sensitivity point of view: PA gives the performance gradients (or policy gradients as referred in the RL literature), and policy iteration can be derived naturally from

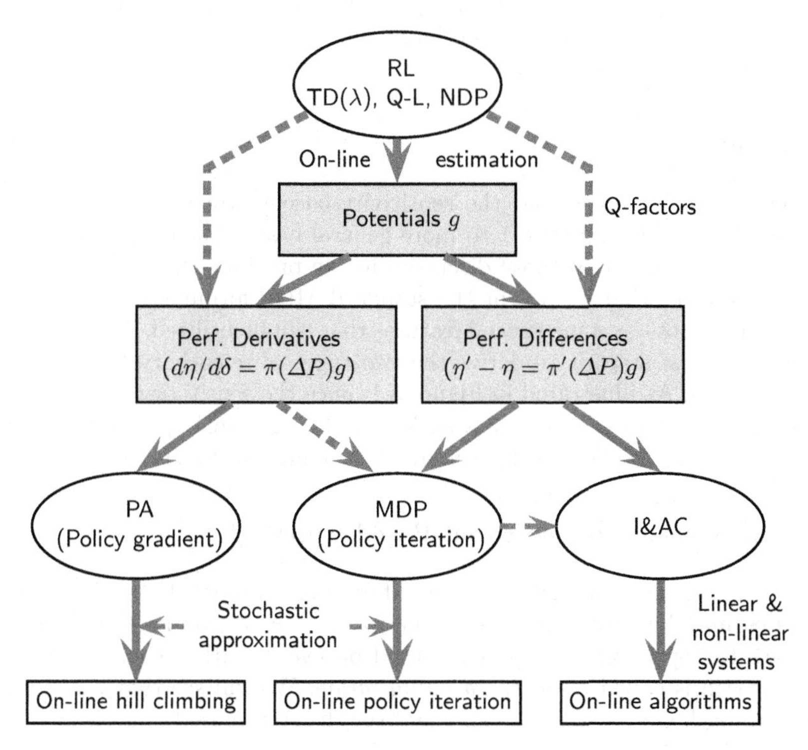

Fig. 8.1. A Map of the Learning and Optimization World

the performance difference formula [53, 67, 97] by utilizing its special form. Both of them can be implemented on a single sample path; Q-learning, TD(λ), neuro-dynamic programming, etc., provide sample-path-based efficient methods of estimating the performance potentials and/or other related quantities (e.g., Q-factors). As shown in the figure, performance optimization can be achieved either by using the performance gradients combined with stochastic approximation methods, see, e.g., [82, 83, 84, 103, 197, 198, 246] (in each step, performance improves by a small amount), or by applying policy iteration algorithms in MDPs (in each iteration, policy jumps to another one with a better performance). In recent years, there has been a considerable amount of effort to bring the researchers in different areas together to develop multi-disciplinary approaches to this common subject of performance optimization [58, 207, 208, 228].

Figure 8.1 describes the relations among the different optimization areas and illustrates how the two sensitivity formulas lead to different optimization approaches. However, as explained previously, because these two sensitivity formulas are derived with the standard MDP formulation, the results presented in the previous chapters do not cover non-standard problems; and it is well known that the standard MDP approach suffers from two major drawbacks: the state space is usually too large, and the actions have to be chosen independently at different states.

Extensions

Our general goal is to extend the sensitivity-based framework and the approaches depicted in Figure 8.1 to more general cases. The extension allows the approaches similar to those discussed in the previous chapters to be applied to problems that do not fit the standard MDP formulation well; these problems may possess some special features that can be utilized. The essential work (as the first step) is to derive the two types of sensitivity formulas for these problems. As illustrated in Figure 8.1, once the sensitivity formulas are obtained, optimization approaches such as policy gradient, policy iteration, and RL, etc., may be derived for these non-standard problems.

8.1.2 An Overview of the Event-Based Approach

In this chapter, we show that, in a class of optimization problems, the special structures may be captured by the concept of *"event"*; and we propose to formulate the optimization problem based on events rather than on states. The approach is called *event-based optimization.* We will see that the event-based approach enjoys some advantages over the state-based approach.

Events and Event-Based Policies

An *event* is defined as a set of state transitions. In a real world system, a physical event that happens at a particular time instant can be characterized by the state transition at that instant; e.g., if a customer arrives at a network and is accepted at a particular instant, then the population of the network increases by one at that instant. Therefore, the event corresponding to a customer arrival can be defined as a set of state transitions that increase the network population by one. Furthermore, in many systems, actions can be taken only when some events occur; e.g., in the admission control problem, actions (accept or reject) can be taken only when a customer arrives. Thus, policies can be defined on the event space instead of on the state space. Such policies are called *event-based policies.*

We first continue Example 1.5 in Chapter 1 to illustrate the main ideas in problem formulation.

Example 8.1. **(Moving Robot)** Let us first give the transition diagram in Figure 1.15 a "physical" meaning. A robot takes a random walk among six rooms, denoted as 1, 2, 3, 4, 5, and 6, as shown in Figure 8.2. There are two special passages, which connect rooms 1, 2, 3, and 4, and rooms 1, 2, 5, and

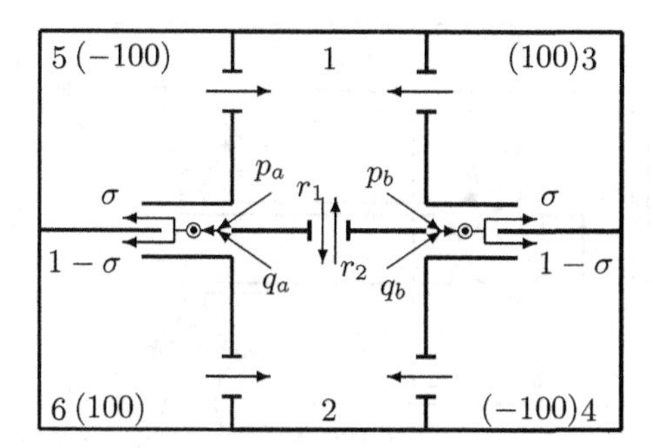

Fig. 8.2. A Simple Example for Event-Based Optimization

6, together, respectively, as shown in Figure 8.2. There is one traffic light, denoted as $\odot$, in each passage.

When the robot is in room 3 or 5, in the next step, it moves to room 1. When it is in room 4 or 6, in the next step, it moves to room 2. When the robot is in room 1, in the next step, it moves to room 2 with probability r_1, or enters the right passage with probability p_b, or enters the left passage with probability p_a, where $r_1 + p_a + p_b = 1$. When the robot is in room 2, in the next step, it moves to room 1 with probability r_2, or enters the right passage with probability q_b, or enters the left passage with probability q_a, where $r_2 + q_a + q_b = 1$. The robot meets the traffic light after it enters a passage; if it is green, it moves to the top room (3 on the right or 5 on the left); if it is red, it moves to the bottom room (4 on the right or 6 on the left).

Suppose that we can only turn on both lights together with the same color (either both red or both green). Denote the probability of the light being green and red as σ and $1-\sigma$, respectively. We may choose σ from $[0, 1]$. We, however, cannot distinguish room 1 or 2; i.e., we only know that the robot is in either room 1 or 2, but do not know which one it is exactly in.

When the robot is in room i, it receives a reward $f(i)$, $i = 1, 2, 3, 4, 5, 6$, with $f(1) = f(2) = 0$, $f(3) = f(6) = 100$, and $f(4) = f(5) = -100$. Our goal is to determine the probability σ so that the long-run average reward is maximal.

We denote the room number i as the system state, $i = 1, 2, 3, 4, 5, 6$. The system can be modelled as a Markov chain. The state transition diagram corresponding a fixed σ is shown in Figure 8.3, and the state transition probabilities are listed in Figure 8.4. The transition diagram 1.15 in Example 1.5 is a part of Figure 8.3.

The problem can be solved by the event-based approach. As discussed in Example 1.5 of Chapter 1, we can define two events

$$a = \{\langle 1, 5\rangle, \langle 1, 6\rangle, \langle 2, 5\rangle, \langle 2, 6\rangle\}$$

and

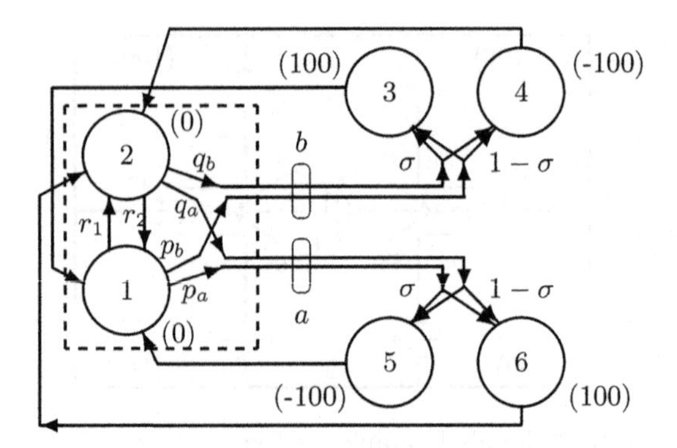

Fig. 8.3. The State Transition Diagram for a Given σ

Event b Event a

	1	2	3	4	5	6
1	0	r_1	$p_b\sigma$	$p_b(1-\sigma)$	$p_a\sigma$	$p_a(1-\sigma)$
2	r_2	0	$q_b\sigma$	$q_b(1-\sigma)$	$q_a\sigma$	$q_a(1-\sigma)$
3	1	0	0	0	0	0
4	0	1	0	0	0	0
5	1	0	0	0	0	0
6	0	1	0	0	0	0

Fig. 8.4. The Events Illustrated by State Transitions

$$b = \{\langle 1,3\rangle, \langle 1,4\rangle, \langle 2,3\rangle, \langle 2,4\rangle\}.$$

They are illustrated by the two small ovals in Figure 8.3. From Figure 8.2, it is clear that event a represents the "physical" event of the robot moving to the front of the left light, and event b represents the "physical" event of the robot moving to the front of the right light. Therefore, we may call a an "arrival to the left light", and b an "arrival to the right light". The event of "not an arrival" is denoted as $\overline{a \cup b}$. of course, other events can also be defined.

The action space is defined as $\mathcal{A} = \{r, g, \emptyset\}$, with r and g denoting "turning the red and the green lights on", respectively, and $\emptyset$ denoting "doing nothing".

Let $\langle X_l, X_{l+1}\rangle$ denote the state transition (which is also called a single event) at time l, $l = 0, 1, \ldots$. In the event-based approach, we assume that we cannot observe X_l and X_{l+1}, but we can observe the event happening at time l; and we denote it as E_l, with $\langle X_l, X_{l+1}\rangle \in E_l$, $l = 0, 1, \ldots$. We may have $E_l = a$, $E_l = b$, or $E_l = \overline{a \cup b}$. Let $\boldsymbol{E}_l := (E_0, E_1, \ldots, E_{l-1}, E_l)$ be the history of the events. When $E_l = \overline{a \cup b}$, we do nothing ($A_l = \emptyset$); when $E_l = a$, or b, we take an action r, or g ($A_l = r$ or g), with probability $1-\sigma_l$ or σ_l. We may choose σ_l according to a randomized policy (see Page 12) denoted as $\sigma_l = \sigma(\boldsymbol{H}_l)$, with $\boldsymbol{H}_l = (\boldsymbol{E}_l, \boldsymbol{A}_{l-1})$. As shown later, in this example, a stationary history-independent optimal policy $\sigma(E_l)$ is good enough. We denote $\sigma_a = \sigma(a)$ and $\sigma_b = \sigma(b)$, i.e., when event a (or b) occurs, we choose the probability of g to be σ_a (or σ_b). We wish to determine an event-based policy $\sigma(E_l)$ to achieve the best performance.

For convenience, sometimes we also say "take an action σ", meaning "take an action g with probability σ", as we did in Example 1.5 in Chapter 1. □

Other Approaches Do Not Perform Well

Before further introducing the event-based approach, let us first try to address this problem with other approaches in the literature. For illustrative purposes, we further assume that $p_a > p_b$ but $q_a < q_b$.

Example 8.2. (**Moving Robot**, continued)

A. MDP

This is not an MDP problem because we cannot exactly observe state 1 or state 2.

The problem can be modelled and solved by the MDP approach if we assume that the state is completely observable. With the MDP model, when the robot is in room 1 or 2, we may choose a σ from the set $[0, 1]$, denoted as σ_1 or σ_2. We may choose $\sigma_1 \neq \sigma_2$. When the robot is in room 1 and σ_1 is chosen, the transition probabilities are shown in the first row of Figure 8.4 with σ replaced by σ_1; and when the robot is in room 2 and σ_2 is chosen, the transition probabilities are shown in the second row of Figure 8.4 with σ replaced by σ_2.

First, from Figure 8.3, it is easy to see that if a occurs we should take $\sigma = 0$ (the smallest value) so that the robot can move to room 6 and get a reward of 100 with the largest probability; and similarly, if b occurs we should take $\sigma = 1$ (the largest value). Let $\sigma^*(1)$ and $\sigma^*(2)$ be the optimal policy with the MDP formulation. When the robot is in room 1, the probability of event a, p_a, is larger than that of event b, p_b, and thus we should set $\sigma^*(1) = 0$. Similarly, we have $\sigma^*(2) = 1$.

It is clear that the optimal performance with the MDP model may not be very good because for any σ, the two events lead to two opposite values. When $p_a = p_b = q_a = q_b$, the average reward is zero.

B. POMDP

If we cannot distinguish state 1 and state 2, the problem can be modelled as POMDP. Set $0 = \{1, 2\}$; i.e., when $X_l = 1$ or 2, we observe an aggregated state $Y_l = 0$. The observation history looks like $\boldsymbol{Y}_l = \{0, 5, 0, 4, 0, 0, 6, 0\}$. When the state is 1 or 2, σ depends on the conditional distribution $\nu(1)$ and

$\nu(2)$ under the given information history. The probabilities of events a and b are

$$\mathcal{P}(a) = \nu(1)p_a + \nu(2)q_a,$$

and

$$\mathcal{P}(b) = \nu(1)p_b + \nu(2)q_b.$$

If $\mathcal{P}(a) > \mathcal{P}(b)$, we take $\sigma = 0$; otherwise, $\sigma = 1$.

A special case of the POMDP is the history-independent POMDP in which we can only observe the current aggregated state. At steady state, we know

$$\nu(1) = \pi(1|0) = \frac{\pi(1)}{\pi(1) + \pi(2)},$$

and

$$\nu(2) = \pi(2|0) = \frac{\pi(2)}{\pi(1) + \pi(2)},$$

where π denotes the steady-state probability distribution. The solution is: If $\nu(1)p_a + \nu(2)q_a > \nu(1)p_b + \nu(2)q_b$, then we set $\sigma = 0$; otherwise, set $\sigma = 1$. (π, and hence ν, depends on σ, though.)

Note that in this special case, the same σ is applied to both states 1 and 2.

This approach suffers from the same drawback as MDPs: For the same $\sigma \neq \frac{1}{2}$, the two events lead to two opposite values of the reward. When $p_a = p_b = q_a = q_b$, the average reward is always zero.

C. POMDP with $\boldsymbol{E}_l$ as Observation Histories

In addition to $\boldsymbol{Y}_l$, we may also take $\boldsymbol{E}_l$, $l = 0, 1, \ldots$, as observations and use them to estimate the states. Indeed, this may help to improve the estimation of the conditional probabilities of states 1 and 2. However, as explained above, the performance of this approach will not be better than the MDP approach.

D. The Equivalent Aggregated Markov Chain (EAMC)

We can aggregate two states 1 and 2 into a big state 0. However, the resulting chain $\boldsymbol{Y}$ is not Markov. Suppose that the system currently is in state 0. If the previous state is 5, then the current state must be 1; and if the previous state is 4, then the current state must be 2. Note that states 1 and 2 have different probabilities for events a and b.

We can form a Markov chain that is equivalent to the aggregated chain. The problem with this EAMC approach is that, generally, such aggregation takes average and some structural information is lost. More specifically, let us reset $f(2) = 20$ and $f(1) = 0$. With the POMDP model (even the history-independent case) if the system moves from state 4 to state 2, we know it receives a reward of 20; but in the EAMC model, it receives a reward of weighted average of 20 and 0. This will affect the potential $g(4)$. Thus, with the EAMC model, the potential of the equivalent chain may be different from the original one.

Another difficulty with this approach is that the transition probabilities of the equivalent chain may depend on σ.

E. MDP with Correlated Actions

In this model, we assume that the states are completely observable, but we require that the probabilities of the red and green lights in states 1 and 2 must be the same. The problem is to choose a σ in Figure 8.4 so that the long-run performance is the best.

This problem is the same as the history-independent POMDP.

F. The Event-Based Approach

As explained in Example 1.5 in Chapter 1, and it is clear from the reward structure shown in Figure 8.3, we may design a myopic policy: If a occurs we choose $\sigma_a = 0$, which leads to state 6 and the reward at the next step is 100; and similarly, if b occurs, we choose $\sigma_b = 1$, and the reward at the next step is also 100. This myopic event-based policy is much better than the optimal MDP policy. □

Discussion on Problem Formulation

We have shown that the other approaches discussed above may not provide a satisfactory optimal solution, and we hope that with the event-based formulation we may utilize the structure better and obtain a better solution. We will formally define events in Section 8.2 and provide a solution to the event-based optimization problem in Section 8.3. Now, let us first study more aspects of the moving robot problem to obtain a better understanding,

Example 8.3. (**Moving Robot**, Continued) Logically, the process of the robot passing through a passage consists of three phases: First, the robot moves to the front of a light, either on the left or the right. (The robot moving to the front of the left light is called event a, and the robot moving to the front of the right light is called event b.) Second, an action is taken (turning on the red or the green light). We can control the probabilities of the actions (red or green), by using the information obtained in the first phase (i.e., the robot moves to the front of the left, or the right, light). Third, the robot moves on to its destination following the instruction of the light.

These three phases can be modelled as three types of events, as shown in Section 8.2. The three phases (events) have a logical order in timing but happen simultaneously in the Markov model, and these three events, together, determine a state transition. □

In the above example, when an action is taken, the robot already moves to the front of a light. In other words, when an action is taken, the state transition already passed its first phase. Therefore, knowing an event implies knowing something about the future ("half" of the transition).

In optimization, to obtain a better performance, we wish to control a system's future behavior by taking proper actions (we can do nothing about what happened in the past). In MDPs, we assume that the state is known; in order to choose the right action, we predict and compare the future behavior under different actions by using their corresponding transition probabilities. In POMDP, we do our best to get an estimate of the state by using its conditional distribution given all the information observed. With this distribution, we predict and compare the future behavior under different actions.

In the MDP model, nothing beyond X_l is known. At time l, we determine an action A_l based on X_l, then the transition takes place instantly. In the event-based approach, a time instant l is split into three phases: An event E_l happens in the first phase, and we wait until E_l happened then take an action. Because E_l contains some information about the state transition from X_l to X_{l+1}, it contains some information about X_{l+1}. In event-based approach, we use this information directly in optimization; while in MDP, future information is predicted indirectly from X_l using the transition probabilities $p(X_{l+1}|X_l)$. In POMDP, it is even worse: we need to estimate X_l before predicting X_{l+1}.

In summary, in the event-based model, we assume that we know something about the next state. Information about future is more directly related to optimization than that about the current state. All the information about the current state is for predicting the future behavior. Therefore, as we see in Example 8.2, knowing an event might be much better than knowing exactly the current state.

Sensitivity Formulas and Potential Aggregation

The performance sensitivity formulas can be constructed for event-based policies, and optimization can be implemented based on these sensitivity formulas in a way similar to the standard MDPs. The potentials of the states associated with an event can be aggregated, and computation is reduced. In aggregating the potentials, we may exploit the special system structure (e.g., the queueing structure), and may not need the explicit form of the transition probabilities of the underlying Markov system. With the formulation of event-based policies, the assumption that the actions can be chosen independently at different states is not required; for example, in the moving robot example, the same σ is used for both rooms 1 and 2. In the admission control problem (cf. Example 8.4), we often accept an arriving customer when the network population is less than a certain number, which may correspond to many different states.

Finally, the performance sensitivity formulas can be derived analytically from the general formulas (2.26) and (2.27). They can also be constructed by using the special features and with potentials as building blocks. The construction approach for performance derivatives is similar to that described in Section 2.1, and the construction approach for performance differences is introduced in Chapter 9. Compared with the analytical method, the construction approach is more flexible and intuitive. When we do not know the form of the sensitivity formulas, the construction approach is extremely useful because it helps us "guess" the formulas before we prove it analytically.

Advantages and Limitations

In summary, there are a number of advantages to event-based optimization:

1. In many real world problems, actions are taken only after some events occur. MDP and POMDP do not model the effect of the events well, and the important information about the next state, which are available when the decision is made, may be lost. The event-based approach directly utilizes this future information as well as the structure properties captured by the events; therefore, in some cases, this approach may lead to a better optimal solution than the approaches with policies depending only on the current state.
2. Using the sensitivity-based approach discussed in this book, we can develop general solution methods to these problems (either policy iteration when possible, or gradient-based approach).
3. We will see later, the approach aggregates the potentials staring from the next state after an event. The number of aggregated potentials is the number of events, which may scale to the system size. This may reduce the number of potentials to be estimated in the learning process and significantly save computation.

The limitation of the approach is that in many problems the aggregated potentials in the performance difference formula may depend on both policies under comparison; this may prevent the aggregated potentials from being used in policy iteration. However, gradient-based approach can always be developed. We will discuss this issue later.

Chapter Organization

We first formally introduce the concept of events, the event space, and the probability measure on the event space in Section 8.2. We use two examples, the moving robot and the admission control, to illustrate the basic concepts and ideas, including three types of events, in Section 8.2.3. We classify these three types of events in Section 8.2.4.

We then discuss event-based optimization in Section 8.3. We derive two fundamental performance sensitivity formulas for event-based policies in Sections 8.3.2 and 8.3.3. These two formulas have a similar structure as those with the standard MDP, except i) the steady-state probabilities of events (instead of states) are used, ii) actions depend on events (instead of states), and iii) the potentials are generally aggregated. With these two formulas, event-based optimization (gradient-based optimization in general and policy iteration in some special cases) is introduced in Section 8.3.4.

In Section 8.4, we show that an aggregated potential can be estimated on a sample path with the same computation and same accuracy as estimating a potential of a state.

The two problems, the moving robot and the admission control, are used as examples in all these sections.

In Sections 8.5.1 and 8.5.2, we give two more examples, one in manufacturing and the other in queueing systems, to illustrate the application of the event-based approach. In Section 8.5.3, we show that the event-based approach provides a unified framework for many other approaches in performance optimization and we provide a summary of the chapter.

8.2 Events Associated with Markov Chains

In many problems, the special feature related to a system's structural or parameter's changes can be characterized by "events". In a real world system, the system behavior is usually modelled as a Markov chain, and a physical event that happens at a particular time instant can be characterized by the state transition of the Markov chain at that instant; e.g., if a customer arrives at a network and is accepted at a particular instant, then the population of the network increases by one at that instant. Therefore, an *event* is defined as a set of state transitions that satisfy some common properties. Throughout this chapter, we will use the moving robot problem and the admission control problem as examples to illustrate the main ideas.

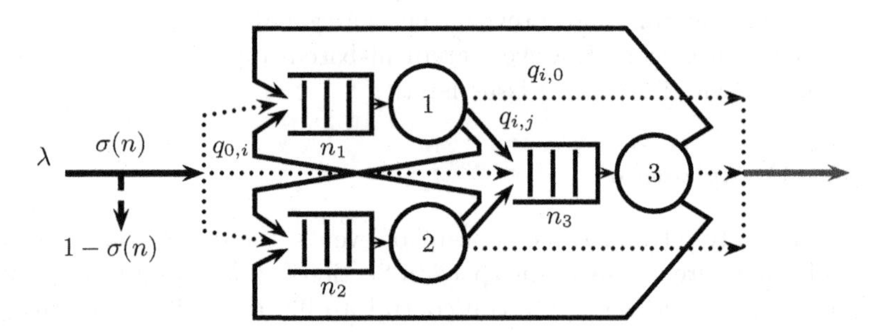

Fig. 8.5. The Admission Control Problem in Example 8.4

Example 8.4. (Admission Control) Consider the admission control problem in a communication system modelled as a variant of an open Jackson network [96] shown in Figure 8.5. The network consists of M servers; the service time of server i is exponentially distributed with mean $1/\mu_i$, and μ_i is called the service rate, $i = 1, 2, \dots, M$. After being served at server i, a customer will join the queue at server j with probability $q_{i,j}$, and will leave the network with probability $q_{i,0}$, $\sum_{j=0}^{M} q_{i,j} = 1$, $i = 1, 2, \dots, M$. Let n_i be the number of customers at server i, and $n = \sum_{i=1}^{M} n_i$ be the number of all customers in the system; n is also called the *population* of the network.

The customers arrive at the network in a Poisson process with rate λ. If an arriving customer finds n customers in the network, the customer will be admitted to the system with probability $\sigma(n)$ and will be rejected with probability $1 - \sigma(n)$. The system has a capacity of N; i.e., $\sigma(N) = 0$; thus, an arriving customer finding N customers in the system will be rejected. An admitted customer will join queue i with probability $q_{0,i}$, $\sum_{i=1}^{M} q_{0,i} = 1$. For simplicity, we assume $q_{i,i} = 0$ for all i. As explained later, this assumption is not restrictive.

We model the system with the discrete-time Markov chain embedded at the transition instants (If necessary, apply the uniformization technique, cf. Problem A.8)). The system state is $\boldsymbol{n} = (n_1, n_2, \ldots, n_M)$, and the state space is $\mathcal{S} := \left\{\text{all } \boldsymbol{n} : \sum_{i=1}^{M} n_i \leq N\right\}$. Let $\mathcal{S}_n := \left\{\text{all } \boldsymbol{n} : \sum_{i=1}^{M} n_i = n\right\}$, $n = 0, 1, \ldots, N$, be the set of states with population n. We have $\mathcal{S} = \cup_{n=0}^{N} \mathcal{S}_n$. In the embedded chain, there is only one customer transition at each time instant, because the probability of two transitions occurring at the same time instant is zero. We assume that at each transition instant, the state of the embedded chain takes the value BEFORE the transition. This convention is made only for convenience.

Note that in this problem we assume that the admission probability depends only on the population n and not on the state $\boldsymbol{n}$. This is similar to the partially observable Markov decision processes (POMDPs) in which only partial information about the state is observable.

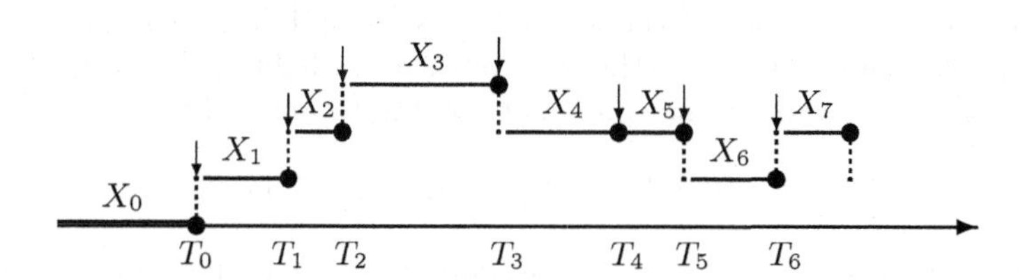

Fig. 8.6. The Embedded Markov Chain for a Single-Server Admission System

Figure 8.6 illustrates a continuous-time sample path $\{X(t), t \geq 0\}$ for a single server system, in which $T_0, T_1, \ldots,$ are transition instants, and the embedded chain is $X_0, X_1, \ldots, X_l, \ldots$. Thus, with our convention, the continuous-time sample is left-continuous. The arrows indicate the embedded points. As shown in the figure, the transition from state $X_0 = 0$ to state $X_1 = 1$ happens instantly at T_0+ in the continuous-time model, but this transition is reflected by two discrete points $X_0 = 0$ and $X_1 = 1$ in the discrete-time model. In general, we have $X(T_l) = X_l$ and $X(T_l+) = X_{l+1}$, $l = 0, 1, \ldots$. At T_4, an arriving customer is rejected. □

8.2.1 The Event and Event Space

We first formally define the events. Consider an ergodic finite Markov chain defined on state space $\mathcal{S} = \{1, 2, \ldots, S\}$ with transition probability matrix $P = [p(j|i)]_{i,j=1}^{S}$.

Definition 8.1. A single event, denoted as $\langle i, j \rangle$, is a state transition from i to j, $i, j \in \mathcal{S}$. The space of all the single events is denoted as $\mathcal{E} = \{\emptyset, \langle i, j \rangle : i, j \in \mathcal{S}\}$, with $\emptyset$ being a null event. A set of single events is called an *event.*

Consistent with the convention of X_l being the state before the transition at T_l, we say that a Markov chain $\boldsymbol{X} = \{X_l, l \geq 0\}$ makes a transition at time l from X_l to X_{l+1} (not from X_{l-1} to X_l). Thus, a single event $\langle i, j \rangle$ happens at time l if $X_l = i$ and $X_{l+1} = j$. The null event $\emptyset$ is defined purely for logical purposes and is different from any real event. Any event a is a subset of $\mathcal{E}$: $a \subseteq \mathcal{E}$. Thus, all the set operations apply to events. For any $a, b, c \subseteq \mathcal{E}$, we can write $c = a \cap b$, $c = a \cup b$, or $c = \bar{a} = \mathcal{E} - a$. Also, we may have $a \subseteq b$, which indicates that if event a happens, then so does b.

Although an event is mathematically defined as a set of state transitions, it usually has a physical interpretation for many real world problems. Therefore, we will refer to an event by its physical meaning. For example, we may call an event (a set of state transitions) "a customer arrival".

Example 8.5. (**Moving Robot**, Continued) Apparently, the event "the robot moving to the front of the left light" is $a = \{\langle 1, 5 \rangle, \langle 1, 6 \rangle, \langle 2, 5 \rangle, \langle 2, 6 \rangle\}$, the event "the robot moving to the front of the right light" is $b = \{\langle 1, 3 \rangle, \langle 1, 4 \rangle, \langle 2, 3 \rangle, \langle 2, 4 \rangle\}$, the event of the robot not moving to any light is

$$c := \overline{a \cup b} = \{\langle 2, 1 \rangle, \langle 1, 2 \rangle, \langle 3, 1 \rangle, \langle 4, 2 \rangle, \langle 5, 1 \rangle, \langle 6, 2 \rangle\}$$

(excluding the transitions with zero probability). Furthermore, the events of the robot moving to a green, or a red, light on the left, and moving to a green, or a red, light on the right, are

$$a_g = \{\langle 1, 5 \rangle, \langle 2, 5 \rangle\}, \quad a_r = \{\langle 1, 6 \rangle, \langle 2, 6 \rangle\},$$

and

$$b_g = \{\langle 1, 3 \rangle, \langle 2, 3 \rangle\}, \quad \text{and} \quad b_r = \{\langle 1, 4 \rangle, \langle 2, 4 \rangle\},$$

respectively.

We have $a = a_g \cup a_r$, $b = b_g \cup b_r$, $a_g, a_r \subset a$, $b_g, b_r \subset b$ and $\mathcal{E} = a \cup b \cup c$. □

Example 8.6. (**Admission Control**, Continued) In the admission control problem, a state transition is denoted as $\langle \boldsymbol{n}, \boldsymbol{n}' \rangle$, $\boldsymbol{n}, \boldsymbol{n}' \in \mathcal{S}$. With the convention $q_{i,i} = 0$ for all $i = 1, 2, \ldots, M$, a transition $\langle \boldsymbol{n}, \boldsymbol{n} \rangle$ clearly indicates that an arriving customer is rejected by the system. For any state $\boldsymbol{n}$, denote the state after a customer joins server i as $\boldsymbol{n}_{+i} = (n_1, \ldots, n_{i-1}, n_i + 1, n_{i+1}, \ldots, n_M)$.

Let $a_{n,+}$, $n < N$, be the event representing that *an arriving customer is accepted AND there are a total of n customers in the network before the arrival.* Then, we have

$$a_{n,+} := \{\langle \boldsymbol{n}, \boldsymbol{n}_{+i} \rangle : \boldsymbol{n} \in \mathcal{S}_n, i = 1, \ldots, M\} .$$

Figure 8.6 shows that at $l = 0$ we have $X_0 = 0$ and event $a_{0,+}$ happens, thus we have $X_1 = 1$ and $\langle X_0, X_1 \rangle \in a_{0,+}$.

Let $a_{n,-}$, $n \leq N$, be the event representing that *an arrival customer is rejected AND there are a total of n customers before the arrival*; we have

$$a_{n,-} := \{\langle \boldsymbol{n}, \boldsymbol{n} \rangle \ : \boldsymbol{n} \in \mathcal{S}_n\} .$$

Figure 8.6 shows that at $l = 4$ we have $X_4 = 2$ and event $a_{2,-}$ happens; thus, we have $X_5 = 2$ and $\langle X_4, X_5 \rangle \in a_{2,-}$.

The event representing *a customer arrival when there are a total of n customers before the arrival* is

$$a_n := a_{n,+} \cup a_{n,-}, \qquad n \leq N,$$

with $a_{N,+} = \emptyset$. The event representing a *customer arrival* is

$$a := \cup_{n=0}^{N} a_n .$$

The event representing that there is *no customer arrival* (including internal transitions and customer departures) is

$$b := \mathcal{E} - a. \tag{8.1}$$

The event representing that *an arriving customer is accepted* is

$$a_+ = \cup_{n=0}^{N-1} a_{n,+} .$$

The event representing that *an arriving customer is rejected* is

$$a_- = \cup_{n=0}^{N} a_{n,-} .$$

Furthermore, the event representing *an arriving customer joining server i when there are n customers before the arrival* is

$$a_{n,+i} := \{\langle \boldsymbol{n}, \boldsymbol{n}_{+i} \rangle, \ \boldsymbol{n} \in \mathcal{S}_n\} ;$$

and the event representing *an arrival customer joining server i* is

$$a_{+i} = \cup_{n=0}^{N-1} a_{n,+i} = \{\langle \boldsymbol{n}, \boldsymbol{n}_{+i} \rangle, \ \boldsymbol{n} \in \mathcal{S} \text{ and } n < N\} .$$

From the above definitions, if a state transition $\langle \boldsymbol{n}, \boldsymbol{n}' \rangle \in a_n \cap a_+ \cap a_{+i}$ (equivalently $\langle \boldsymbol{n}, \boldsymbol{n}' \rangle \in a_n \cap a_{n,+} \cap a_{n,+i}$), then $\langle \boldsymbol{n}, \boldsymbol{n}' \rangle \in a_{n,+i} =$

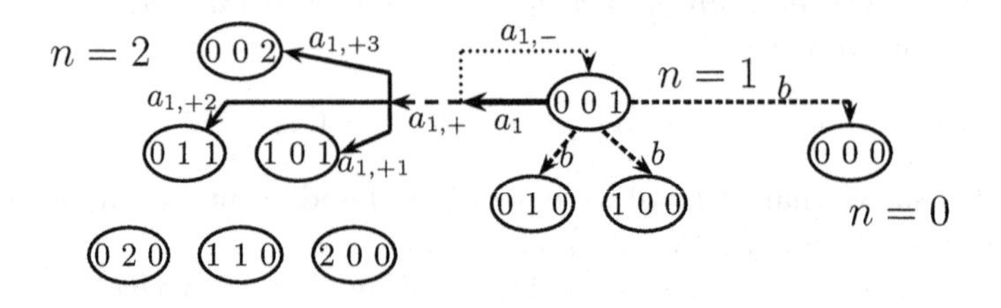

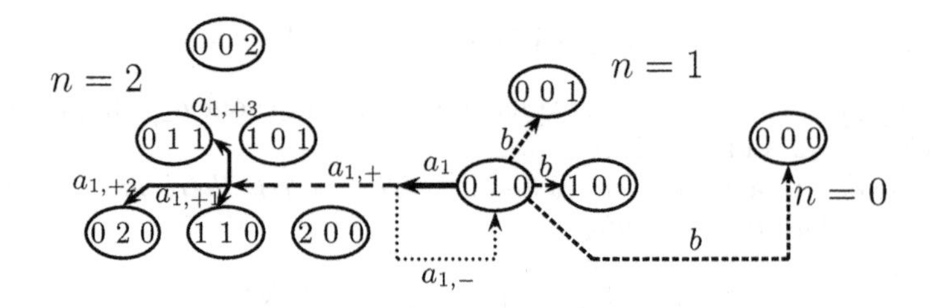

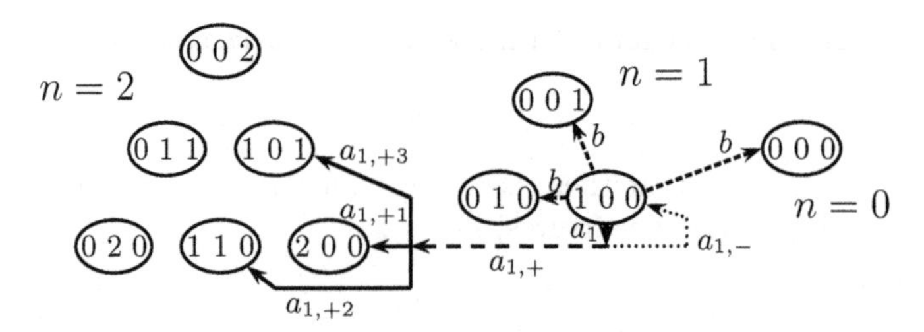

Fig. 8.7. The Events in Example 8.4 with $M = 3$, $n = 1$

$\left\{\langle \boldsymbol{n}, \boldsymbol{n}_{+i}\rangle \text{ with } \sum_{k=1}^{M} n_k = n\right\}$, i.e., a customer, arriving when the system population is n, is accepted and joins server i.

Figure 8.7 illustrates all the events that may happen when $n = 1$ in the admission control problem with $M = 3$. There are three states corresponding to $n = 1$: $(0,0,1)$, $(0,1,0)$, and $(1,0,0)$. The top graph in the figure illustrates the events that may happen when the system is in state $(0,0,1)$. The three dashed arrows represent event b which contains internal transitions to states $(0,1,0)$ and $(1,0,0)$, and the customer-departure transition to state $(0,0,0)$. The thick line represents the arrival event a_1, which contains both $a_{1,-}$ (rejection), denoted as the dotted line pointing to the state $(0,0,1)$ itself, and $a_{1,+}$ (accept), denoted as the long-dashed line, which splits into three sub-events $a_{1,+1}$, $a_{1,+2}$, and $a_{1,+3}$, denoted as the three thin arrows pointing to $(0,0,2)$,

$(0, 1, 1)$, and $(1, 0, 1)$, in the figure, respectively. Event $a_{1,+i}$ represents the accepted customer joining server i, $i = 1, 2, 3$.

The middle and bottom graphs illustrate, in a similar way, the events that may happen in states $(0, 1, 0)$ and $(1, 0, 0)$, respectively. It is shown that event a_1 may happen when the system is in either state $(0, 0, 1)$, or $(0, 1, 0)$, or $(1, 0, 0)$. Overall, a_1 contains all the transitions represented by the three thick lines in the three graphs in the figure. The situation for other values of n is similar. Overall, we have $a_+ = \cup_{n=0}^{N-1} a_{n,+}$ and $a_- = \cup_{n=0}^{N} a_{n,-}$, etc., and event b may happen for any $n > 0$.

We have assumed $q_{i,i} = 0$ for convenience. This is not restrictive. If $q_{i,i} \neq 0$, then a transition $\langle \boldsymbol{n}, \boldsymbol{n} \rangle$ corresponds to two situations: An arriving customer is rejected, or a customer finishes its service at a server and returns to the same server (with probability $q_{i,i} \neq 0$). In a real system, one can observe the difference between these two situations; however, the Markov chain $\boldsymbol{X}$ does not reflect this difference since the state transition is the same for both situations. From the learning point of view, we need to introduce an additional index to distinguish these two situations. This will make the notations more complicated but will not change the concepts and the results. □

As shown in Examples 8.5 and 8.6, if we know that an event (other than a single event, i.e., a state transition) happens, we may not know the exact state that the system is in, but we know that the system is in a state which belongs to a particular subset of the state space. In addition to this partial information about the current state, we also have some knowledge about the state transition at this moment. In Example 8.5, if we observe an arrival to the left light, we know that the state can be either 1 or 2, and we also know that the next state must be either 5 or 6. In the case of Figure 8.7, if we know that event a_1 happens, we do not know whether the state is $(0, 0, 1)$, or $(0, 1, 0)$, or $(0, 0, 1)$. However, in addition to the partial information about the state (i.e., the population is 1), we do know some partial information about the transition: After the event a_1, the population of the system either increases (accept), or remains the same (reject); it cannot decrease. That is, the next state cannot be $(0, 0, 0)$. If we know that $a_{1,+}$ happens, then the next state cannot be $(0, 0, 0)$, $(0, 0, 1)$, $(0, 1, 0)$, or $(0, 0, 1)$. Therefore, *an observation of an event may contain more information than a partial observation of the system state.*

8.2.2 The Probabilities of Events

Observe a sample path of a Markov chain with L consecutive transitions, denoted as $\langle X_l, X_{l+1} \rangle$, $l = 0, 1, \ldots, L - 1$. This sample path $\{X_0, X_1, \ldots, X_L\}$ consists of $L + 1$ states. The space spanned by all such sample paths is denoted as $\mathcal{S}^{L+1}$. Each sample path $\{X_0, X_1, \ldots, X_L\}$ is represented by a point in $\mathcal{S}^{L+1}$, which corresponds to a set of points in $\mathcal{S}^{l+1}$, $l > L$. In other words, $\mathcal{S}^{L+2}$ is "finer" than $\mathcal{S}^{L+1}$, $L \geq 0$, etc. The initial distribution

$\pi_0 = (\pi_0(1), \ldots, \pi_0(S))$ and the transition probability matrix $P = [p(j|i)]_{i,j=1}^S$ determine a probability measure $\mathcal{P}_L$ on $\mathcal{S}^{L+1}$, $L = 1, 2, \ldots$, and $\mathcal{P}_\infty$ (hereafter denoted as $\mathcal{P}$ for simplicity) on $\mathcal{S}^\infty$ in the standard way.

Recall that $\langle X_l, X_{l+1} \rangle$ is the transition at time instant l. Let $\pi_l = (\pi_l(1), \ldots, \pi_l(S))$ be the probability vector at time instant l, $l = 0, 1, \ldots$. The probability of event $\langle i, j \rangle$ happening at $l = 0$ is

$$\mathcal{P}(\langle X_l, X_{l+1} \rangle = \langle i, j \rangle) = \pi_0(i) p(j|i).$$

A feasible sequence of single events must take the form $\{\langle i_0, i_1 \rangle, \langle i_1, i_2 \rangle, \ldots, \langle i_{L-1}, i_L \rangle\}$, and its probability is

$$\mathcal{P}\{\langle i_0, i_1 \rangle, \langle i_1, i_2 \rangle, \ldots, \langle i_{L-1}, i_L \rangle\} = \pi_0(i_0) \prod_{l=0}^{L-1} p(i_{l+1}|i_l). \tag{8.2}$$

If the initial state probability is the steady-state probability, i.e., $\pi_0(i) = \pi(i)$ for all $i \in \mathcal{S}$, (8.2) becomes the steady-state probability of the sequence of single events. The steady-state probability of event $\langle i, j \rangle$ is

$$\pi(\langle i, j \rangle) = \pi(i) p(j|i), \tag{8.3}$$

which defines a probability measure on $\mathcal{E}$.

Since events are defined as sets in $\mathcal{E}$, the standard probability laws on sets apply to them. For example, if $c = a \cup b$ and $a \cap b = \emptyset$, then

$$\mathcal{P}(\langle X_l, X_{l+1} \rangle \in c) = \mathcal{P}(\langle X_l, X_{l+1} \rangle \in a) + \mathcal{P}(\langle X_l, X_{l+1} \rangle \in b).$$

This equation holds for any $l = 0, 1, 2, \ldots$. In such cases, we will simplify the notation by dropping the subscript l and setting $\mathcal{P}(a) := \mathcal{P}(\langle X_l, X_{l+1} \rangle \in a)$, and thus,

$$\mathcal{P}(c) = \mathcal{P}(a) + \mathcal{P}(b).$$

In addition, we can define the conditional probability

$$\mathcal{P}(b|a) := \mathcal{P}(\langle X_l, X_{l+1} \rangle \in b | \langle X_l, X_{l+1} \rangle \in a) = \frac{\mathcal{P}(a \cap b)}{\mathcal{P}(a)}.$$

Furthermore, events a and b are said to be *independent* if and only if $\mathcal{P}(a \cap b) = \mathcal{P}(a)\mathcal{P}(b)$.

The steady-state probability of an event defined in (8.3) depends on the steady-state probability $\pi(i)$, which depends on the transition probability matrix in a complicated way. The conditional probability, however, may depend only on the transition probabilities directly in a simple way. For example, define

$$\langle i, \bullet \rangle := \{\langle i, j \rangle, \text{ for all } j \in \mathcal{S}\}$$

be the event of state transitions going out of state i. Then

$$\mathcal{P}(\langle i,j\rangle|\langle i,\bullet\rangle) = p(j|i).$$

This property of independence on the steady-state probability is important in optimization problems. Since we can only directly control the transition probabilities, not the steady-state probabilities, we can only directly control the conditional probabilities, not the event probabilities.

Example 8.7. (**Moving Robot**, Continued) In the example, we have

$$p(a_g|a) = \sigma_a, \qquad p(b_g|b) = \sigma_b,$$

and

$$p(a_r|a) = 1 - \sigma_a, \qquad p(b_r|b) = 1 - \sigma_b.$$

These equations for conditional probabilities can be intuitively derived from the meaning of the events. □

Example 8.8. (**Admission Control**, Continued) In the example, we have

$$\mathcal{P}(a_{n,+}|a_n) = \sigma(n), \tag{8.4}$$

and

$$\mathcal{P}(a_{n,-}|a_n) = 1 - \sigma(n).$$

□

Conditional Probabilities of States Given an Event Sequence

In the event-based approach, we assume that we can only observe the events, but not the states. Given a sequence of events $\boldsymbol{E}_l = \{E_0, E_1, \ldots, E_l\}$, we have $\boldsymbol{E}_l = \{\boldsymbol{E}_{l-1}, E_l\}$. We wish to obtain $\mathcal{P}(X_l|\boldsymbol{E}_{l-1})$ and $\mathcal{P}(X_l|\boldsymbol{E}_l)$. Note that the event sequence is not Markov. This situation is similar to the partially observable MDPs. We first define

Definition 8.2. An input set of event a is

$$I(a) := \{\text{all } i \in \mathcal{S} : \langle i,j\rangle \in a \text{ for some } j\}.$$

An output set of event a is

$$O(a) := \{\text{all } j \in \mathcal{S} : \langle i,j\rangle \in a \text{ for some } i\}.$$

An input set of state j in event a is

$$I_j(a) = \{\text{all } i \in \mathcal{S} : \langle i,j\rangle \in a\}.$$

An output set of state i in event a is

$$O_i(a) = \{\text{all } j \in \mathcal{S} : \langle i,j \rangle \in a\}.$$

The states in $I(a)$ are called the input states of event a, and the states in $O(a)$ are called the output states of event a.

Apparently, if $i \notin I(a)$, then $O_i(a) = \emptyset$, and if $j \notin O(a)$, then $I_j(a) = \emptyset$. For any two events $a, b \subseteq \mathcal{E}$ and $i \in I(a)$, $i \in I(b)$, we have

$$O_i(a \cup b) = O_i(a) \cup O_i(b). \tag{8.5}$$

Likewise, if $j \in O(a)$ and $j \in O(b)$, then

$$I_j(a \cup b) = I_j(a) \cup I_j(b).$$

We consider the event E_l observed at time instant l. Obviously, we have

$$\mathcal{P}(E_l | X_l = i) = \sum_{j \in O_i(E_l)} p(j|i),$$

which is zero if event E_l cannot happen when $X_l = i$, i.e., if $O_i(E_l) = \emptyset$. Suppose that the *a priori* probability of X_l (before knowing whether E_l occurs) is $\pi^\dagger(X_l)$, $X_l = 1, 2, \ldots, S$. Then the conditional probability of X_l given that E_l occurs is

$$\begin{aligned} \mathcal{P}(X_l = i | E_l) &= \frac{\mathcal{P}(E_l | X_l = i)\pi^\dagger(i)}{\sum_{k \in \mathcal{S}} \mathcal{P}(E_l | X_l = k)\pi^\dagger(k)} \\ &= \frac{\left[\sum_{j \in O_i(E_l)} p(j|i)\right] \pi^\dagger(i)}{\sum_{k \in \mathcal{S}} \left\{\left[\sum_{j \in O_k(E_l)} p(j|k)\right] \pi^\dagger(k)\right\}}. \end{aligned} \tag{8.6}$$

Therefore, if we know $\mathcal{P}(X_l | \boldsymbol{E}_{l-1})$, we can obtain $\mathcal{P}(X_l | \boldsymbol{E}_l)$ by setting $\pi^\dagger(i)$ in (8.6) to be $\mathcal{P}(X_l = i | \boldsymbol{E}_{l-1})$. That is,

$$\mathcal{P}(X_l = i | \boldsymbol{E}_l) = \frac{\left[\sum_{j \in O_i(E_l)} p(j|i)\right] \mathcal{P}(X_l = i | \boldsymbol{E}_{l-1})}{\sum_{k \in \mathcal{S}} \left\{\left[\sum_{j \in O_k(E_l)} p(j|k)\right] \mathcal{P}(X_l = k | \boldsymbol{E}_{l-1})\right\}}.$$

We also have

$$\begin{aligned} &\mathcal{P}(X_l = i | \boldsymbol{E}_{l-1}) \\ = &\sum_{k \in I_i(E_{l-1})} [\mathcal{P}(X_{l-1} = k | \boldsymbol{E}_{l-1}) \mathcal{P}(X_l = i | X_{l-1} = k, \boldsymbol{E}_{l-1})] \\ = &\sum_{k \in I_i(E_{l-1})} [\mathcal{P}(X_{l-1} = k | \boldsymbol{E}_{l-1}) \mathcal{P}(X_l = i | X_{l-1} = k, \langle X_{l-1}, X_l \rangle \in E_{l-1})] \\ = &\sum_{k \in I_i(E_{l-1})} \left[\mathcal{P}(X_{l-1} = k | \boldsymbol{E}_{l-1}) \frac{p(i|k)}{\sum_{j \in O_k(E_{l-1})} p(j|k)}\right]. \end{aligned}$$

All the above equations can be obtained by the basic probability theorems. With these equations, $\mathcal{P}(X_l = i|\boldsymbol{E}_{l-1})$ and $\mathcal{P}(X_l = i|\boldsymbol{E}_l)$ can be obtained step by step recursively, starting from the initial probability $\pi_0(i)$ and

$$\mathcal{P}(X_0 = i|E_0) = \frac{\left[\sum_{j \in O_i(E_0)} p(j|i)\right] \pi_0(i)}{\sum_{k \in \mathcal{S}} \left\{\left[\sum_{j \in O_k(E_0)} p(j|k)\right] \pi_0(k)\right\}},$$

which follows from (8.6).

8.2.3 The Basic Ideas Illustrated by Examples

The event-based optimization approach depends on the logical relations among different types of events. In this subsection, we first illustrate the main ideas with the moving robot and the admission control examples.

Example 8.9. (**Moving Robot**, continued) In this problem, we first observe whether the robot arrives to a light (i.e., whether event a, or b, happens) at an instant. If not, we do nothing. Now, suppose that the robot arrives at the left light, i.e., event a occurs, we can either turn on the green or the red light, with probability σ_a or $1 - \sigma_a$. If the green light is given, the event at this instant belongs to a_g; and if the red light is given, the event at this instant belongs to a_r. We can control the conditional probabilities of a_g and a_r given the event a occurring. The same discussion can be applied when b occurs.

We call a and b the *observable events*, because we can observe whether they occurred, and a_r, a_g, b_r, and b_g the *controllable events*, because we can control the probabilities of their occurrence. After the red or the green light is turned on, the robot moves to its destination, either room 5 or 6, if event a occurs, or room 3 or 4, if event b occurs. These are called the *natural transition events.* (In this example, after an action is taken, there is only one destination; there might be more than one destination, as shown in Problem 8.9.)

In summary, the state transition at an arrival instant may belong to three types of events, the observable, the controllable, and the natural transition events (which will be discussed further later); the probabilities of the controllable events can be controlled by the action taken after the observation is made. The three types of events and the action taken at the arrival instant happen simultaneously in the Markov model, but they have a logical order in timing, as shown in Figure 8.8. □

Example 8.10. (**Admission Control,** continued) In controlling the system, we first observe whether a customer arrives at an instant. If not, we do nothing. Assume that when a customer arrives, we also know the network population n. Thus, suppose that a_1 is observed at an instant, which means that a customer arrives and the system population is 1 at this instant. We can either accept the arriving customer, or reject it. That is, we can control the probabilities $\sigma(1)$ and $1 - \sigma(1)$, which are the conditional probabilities of the state transition

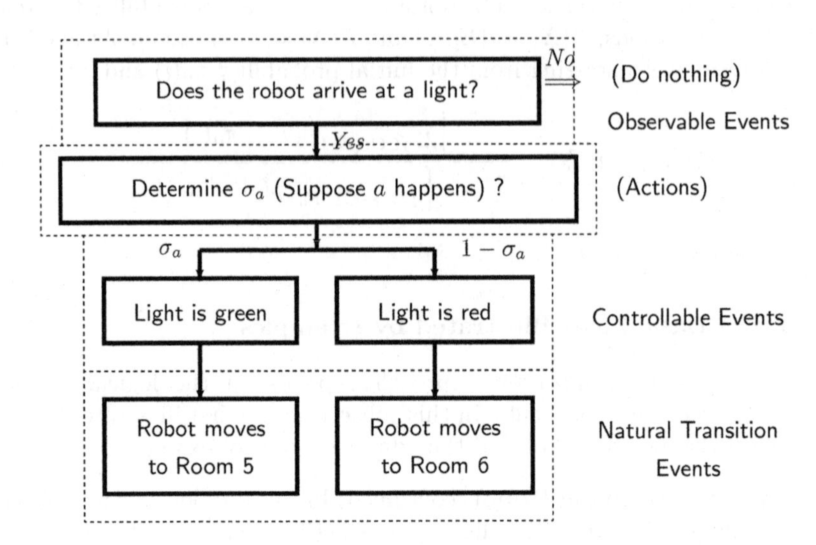

Fig. 8.8. The Timing Logic of the Three Types of Events When a Occurs in Example 8.9

belonging to a_+ or a_- (equivalently, $a_{1,+}$ or $a_{1,-}$). We call a_n, $n = 0, 1, \ldots, N$, the *observable events* and a_+ and a_- (or $a_{1,+}$ or $a_{1,-}$) the *controllable events.* Finally, after a customer is accepted, the nature determines which server it joins. That is, the nature randomly chooses which sub-event a_{+i} (equivalently, $a_{1,+i}$), $i = 1, 2, \ldots M$, the transition at this instant belongs to. These events a_{+i} (or $a_{1,+i}$), $i = 1, 2, \ldots, M$, are called the *natural transition events.*

In summary, the state transition at a customer arrival instant may belong to three types of events, the observable, the controllable, and the natural transition events; the probabilities of the controllable events can be controlled by the action taken after the customer arrives. The three types of events and the action taken at the arrival instant happen simultaneously in the Markov model, but have a logical order in timing, as shown in Figure 8.9.

A state transition representing an arriving customer being accepted and joining server i when the system state is $\boldsymbol{n}$ can be expressed as the event

$$\langle \boldsymbol{n}, \boldsymbol{n}_{+i} \rangle \in a_n \cap a_+ \cap a_{+i}.$$

We can also write

$$\langle \boldsymbol{n}, \boldsymbol{n}_{+i} \rangle \in a_n \cap a_{n,+} \cap a_{n,+i}, \tag{8.7}$$

in which $a_{n,+} = a_n \cap a_+ \subset a_n$ is a subset of a_n, and $a_{n,+i} = a_{n,+} \cap a_{+i} \subset a_{n,+}$ is a subset of $a_{n,+}$. Although (8.7) looks mathematically redundant (in the sense that $a_n \cap a_{n,+} \cap a_{n,+i} = a_{n,+i}$), it helps in understanding the logic among the observable, the controllable, and the natural transition events. For example, in Figure 8.7, we have

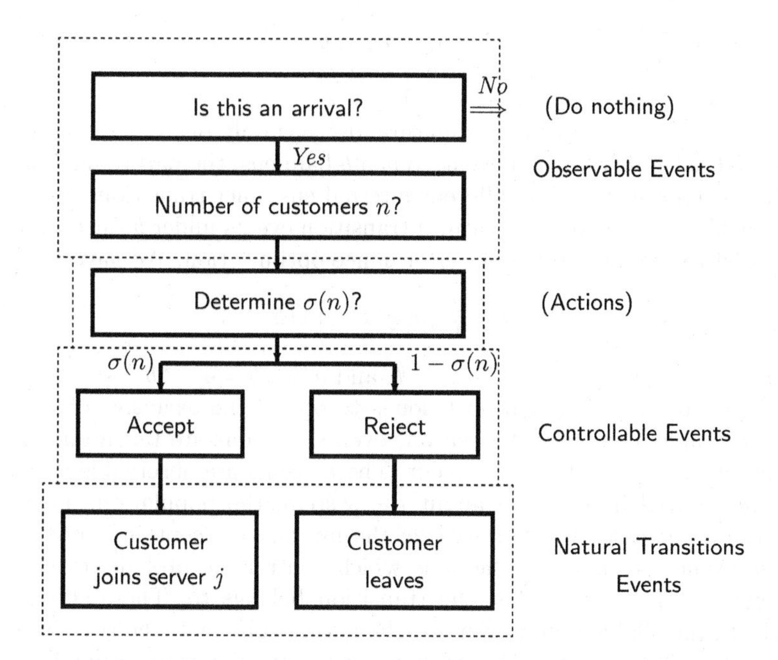

Fig. 8.9. The Timing Logic of the Three Types of Events in Example 8.10

$$\langle (0,0,1),(0,0,2)\rangle \in a_1 \cap a_{1,+} \cap a_{1,+3}.$$

An arriving customer being rejected when the system is in state $\boldsymbol{n}$ can be expressed as the event

$$\langle \boldsymbol{n},\boldsymbol{n}\rangle \in a_n \cap a_- = a_n \cap a_{n,-}, \qquad a_{n,-} = a_n \cap a_-,$$

with $\sum_{i=1}^{M} n_i = n$. No natural transition event explicitly appears in this expression; or the nature has only one choice in this case. In Figure 8.7, we have

$$\langle (0,0,1),(0,0,1)\rangle \in a_1 \cap a_{1,-}.$$

Now, let us decompose the event space $\mathcal{E}$. Note that the event b defined in (8.1), consisting of all internal customer transitions and departures from the network, can also be treated as an observable event. Thus, $\mathcal{E}$ can be decomposed into a set of mutually exclusive observable events:

$$\mathcal{E} = \left\{\cup_{n=0}^{N} a_n\right\} \cup b,$$

with $a_i \cap a_j = \emptyset$, $i \neq j$, $a_i \cap b = \emptyset$. Figure 8.7 shows that all the transitions from the states with $n = 1$ belong to either a_1 or b.

Next, when b is observed, there is only one choice for control: do nothing. Thus, b can also be viewed as a special controllable event representing only one action corresponding to "do nothing". Therefore, we have the mutually exclusive decomposition of $\mathcal{E}$ with the controllable events:

$$\mathcal{E} = a_+ \cup a_- \cup b,$$

with $a_+ \cap a_- = \emptyset$, $a_+ \cap b = a_- \cap b = \emptyset$.

Lastly, when a_- happens, the nature also has only one choice. Thus, a_- is a special natural transition event. When b happens, the nature may have a few choices representing the different internal customer transitions; however, we will not elaborate on these natural transition events under b since they are not related to the main topic of performance optimization. We have

$$\mathcal{E} = \{\cup_{i=1}^{M} a_{+i} \cup a_-\} \cup b,$$

with $a_{+i} \cap a_{+j} = \emptyset$, $i \neq j$, $a_{+i} \cap a_- = \emptyset$, and $a_{+i} \cap b = a_- \cap b = \emptyset$.

In summary, every transition belongs to one of the exclusive observable (or, controllable, or natural transition) events. Actions are taken only when some events a_n, $n = 0, 1, \ldots, N$, occur. These events are observable and contain some partial information about the system, the population. They are called the observable events. Based on the information contained in the observable events, we may take actions which control the probabilities of the sub-events, $a_{n,+}$ or $a_{n,-}$, that the transition belongs to. These sub-events are called controllable events. Finally, the nature completes the transition by choosing the destination of the arriving customer. These correspond to the natural transition events, $a_{n,+i}$.

The three types of events, the observable, the controllable, and the natural transition events happen at the same time in the Markov model, and together they determine a transition in the Markov model. These three types of events have a logical timing order, which is not captured by the standard MDP formulation. □

8.2.4 Classification of Three Types of Events

The logical structure demonstrated in the moving robot and the admission control examples is applicable to many problems. We now provide a general formulation of this structure in the event space $\mathcal{E}$. In our formulation, the system evolves according to the Markov rule; however, in the approach, we mainly deal with events; the system state is only a hidden concept that helps in the analysis.

Consider a Markov chain with transition probability matrix P and state space $\mathcal{S} = \{1, 2, \ldots, S\}$. The single-event space is $\mathcal{E} = \mathcal{S} \times \mathcal{S}$. There are $S \times S$ single events $\langle i, j \rangle$, $i, j \in \mathcal{S}$, and $2^{S \times S}$ possible events (subsets of $\mathcal{E}$).

Three Types of Events

The first type of event is the *observable event*.

An Observable Event Has Two Features:

1. We can tell whether the event occurs at any time instant from the system behavior.
2. The event contains some information about the system, which can be used to determine the control actions.

Because the information carried in different observable events is different, these events are mutually exclusive. In addition, we assume that every transition belongs to one observable event. (This is not restrictive since we may group all the "unobservable" transitions and/or the transitions containing no useful information into one event; this event can be viewed as an observable event, e.g., the event b in Example 8.6.) Thus we have

$$\mathcal{E} = \cup_{k=1}^{k_o} e_o(k), \quad e_o(k) \cap e_o(k') = \emptyset, \quad k \neq k', \quad k, k' \in \{1, 2, \ldots, k_o\}, \tag{8.8}$$

where $e_o(k)$, $k = 1, 2, \ldots, k_o$, are the observable events and k_o is the number of observable events.

The second type of event is the *controllable event.*

A controllable event is an event that we can control the probability of its occurrence by taking actions based on the information obtained from an observable event that has just happened (see (8.4) and (8.13)).

We have

$$\mathcal{E} = \cup_{k=1}^{k_c} e_c(k), \quad e_c(k) \cap e_c(k') = \emptyset, \quad k \neq k', \quad k, k' \in \{1, 2, \ldots, k_c\}, \tag{8.9}$$

where $e_c(k)$, $k = 1, 2, \ldots, k_c$, are the controllable events and k_c is the number of controllable events.

Suppose that $e_o(k)$ is the event we observed, then from (8.9) we have

$$e_o(k) = \cup_{k_2=1}^{k_c} \{e_o(k) \cap e_c(k_2)\}.$$

With this form, we can take actions to assign probabilities to those controllable events $e_c(k_2)$ for which $e_o(k) \cap e_c(k_2) \neq \emptyset$, $k_2 = 1, 2, \ldots, k_c$. In particular, if for an observable event $e_o(k_1)$ there is only one controllable event $e_c(k_2)$ such that $e_o(k_1) \cap e_c(k_2)$ is non-null, then at $e_o(k_1)$ there is only one choice of controllable event, namely $e_c(k_2)$. That is, at such an observable event $e_o(k_1)$, we can take only one action. In most cases, this unique action corresponds to "do nothing", and therefore at such an event the system is customarily said to be not controllable (e.g., the event b in Example 8.6).

The third type of event is the *natural transition event.*

A natural transition event is an event whose corresponding transitions are governed by nature; thus, the probability of the occurrence of a natural transition event cannot be controlled.

Generally, we have

$$\mathcal{E} = \cup_{k=1}^{k_t} e_t(k), \quad e_t(k) \cap e_t(k') = \emptyset, \quad k \neq k', \quad k, k' \in \{1, 2, \ldots, k_t\}, \quad (8.10)$$

where $e_t(k)$, $k = 1, 2, \ldots, k_t$, are the natural transition events and k_t is the number of natural transition events. Again, after the event $e_o(k_1)$ was observed and the controllable event $e_c(k_2)$ occurred, only those $e_t(k)$ with $\{e_o(k_1) \cap e_c(k_2)\} \cap e_t(k) \neq \emptyset$ will have non-zero natural transition probabilities.

As explained in Example 8.10, there is a logical timing order among the different types of events: At any instant, an observable event occurs first, and when it happens, the exact state transition is not determined. One needs to take an action to determine the probabilities of the controllable events, which is followed by a natural transition event. These three types of events happen in a logical sequence simultaneously; together they determine the exact transition from a state. Many real systems possess such a property. This logical and structural property represented by events, however, is lost in the standard MDP formulation.

The standard MDP can be viewed as a special case if we properly (and probably artificially) define the observable and controllable events as follows: For any state $i \in \mathcal{S}$, we define an observable event as $e_o(i) := \{\langle i, k \rangle : \text{ all } k \in \mathcal{S}\}$, and for any $j \in \mathcal{S}$, we define a controllable event $e_c(j) = \{\langle k, j \rangle : \text{ all } k \in \mathcal{S}\}$. We have $\mathcal{E} = \cup_{i \in \mathcal{S}} e_o(i) = \cup_{j \in \mathcal{S}} e_c(j)$, and $\langle i, j \rangle = e_o(i) \cap e_c(j)$, for all $i, j \in \mathcal{S}$.

Composition of a Single Event

Because the decompositions of the event space $\mathcal{E}$, (8.8), (8.9), and (8.10), are mutually exclusive, for any single event (a state transition) $\langle i, j \rangle \in \mathcal{E}$, there is a unique set of integers k_1, k_2, and k_3 such that

$$\langle i, j \rangle \in e_o(k_1) \cap e_c(k_2) \cap e_t(k_3) =: e(k_1, k_2, k_3), \quad (8.11)$$

with $k_1 \in \{1, \ldots, k_o\}, k_2 \in \{1, \ldots, k_c\}, k_3 \in \{1, \ldots, k_t\}$. We can always write (8.11) as

$$\begin{aligned} e(k_1, k_2, k_3) &= e_o(k_1) \cap e_c(k_2) \cap e_t(k_3) \\ &= e_o(k_1) \cap \{e_o(k_1) \cap e_c(k_2)\} \cap \{e_o(k_1) \cap e_c(k_2) \cap e_t(k_3)\} \\ &= e_o(k_1) \cap e'_c(k_2) \cap e'_t(k_3) \end{aligned} \quad (8.12)$$

where $e'_c(k_2) = e_o(k_1) \cap e_c(k_2) \subseteq e_o(k_1)$ and $e'_t(k_3) = e_o(k_1) \cap e_c(k_2) \cap e_t(k_3) = e'_c(k_2) \cap e_t(k_3) \subseteq e'_c(k_2)$. That is, the controllable and the natural transition events can be further partitioned into small events such that the controllable events are subsets of the observable events and the natural transition events are subsets of the controllable events. This view helps us to think: When an observable event is observed, we may control the occurrence of its sub-events by taking actions; and afterwards, nature finalizes the state transition by further determining the natural transition sub-event. However, the model in (8.11) is more general and more concise than (8.12).

The three events $e_o(k_1)$, $e_c(k_2)$, and $e_t(k_3)$ in (8.11) may not specify a single event; i.e., $e_o(k_1) \cap e_c(k_2) \cap e_t(k_3)$ may not be a singleton. However, we wish that starting from any state i, if a single event $\langle i, j \rangle$ belongs to $e_o(k_1) \cap e_c(k_2) \cap e_t(k_3)$, then j is uniquely determined. For instance, in Example 8.10, assume that we have $e_o(1) = a_n$, $e_c(2) = a_+$, and $e_t(3) = a_{+i}$. Then the event $e_o(1) \cap e_c(2) \cap e_t(3) = a_{n,+i}$ denotes that an arrival customer is accepted and joins server i; this event corresponds to many state transitions. However, for any particular $\boldsymbol{n}$, if we know $\langle \boldsymbol{n}, \boldsymbol{n}' \rangle \in a_{n,+i}$, then $\boldsymbol{n}' = \boldsymbol{n}_{+i}$ is uniquely determined. To be more precise, we make the following definition.

Definition 8.3. *An event a is said to be deterministic, if for any $i \in I(a)$, the output set $O_i(a)$ contains only one state.* □

Therefore, if a is deterministic and $i \in I(a)$, $\langle i, j \rangle \in a$, then j is uniquely determined. In other words, in a deterministic event a, a state cannot move to more than one state. In this case, we write $j = O_i(a)$ for convenience. However, in a deterministic event, two or more input states can move to the same output state.

Example 8.11. (**Moving Robot**, Continued) In the moving robot example, we have $I(a) = I(b) = \{1, 2\}$, $O(a) = O_1(a) = O_2(a) = \{5, 6\}$, $O(b) = O_1(b) = O_2(b) = \{3, 4\}$, $O_1(a_g) = O_2(a_g) = \{5\}$, $O_1(a_r) = O_2(a_r) = \{6\}$, $O_1(b_g) = O_2(b_g) = \{3\}$, and $O_1(b_r) = O_2(b_r) = \{4\}$. The events a_r, a_g, b_r, and b_g are deterministic.

The decomposition for the observable events is $\mathcal{E} = a \cup b \cup c$. When $c = \overline{a \cup b}$ occurs, there is only one action: "do nothing". Therefore, c can be viewed as a special controllable event. Since in this example, nature has only one choice after every controllable event, the natural transition events are the same as the controllable events. □

In general, we can always make the natural transition event decomposition "fine" enough to make sure that the transitions starting from any state i in event $e(k_1, k_2, k_3)$ in (8.11) is uniquely determined by $e_o(k_1) \cap e_c(k_2) \cap e_t(k_3)$. For example, if

$$O_i[e_o(k_1) \cap e_c(k_2) \cap e_t(k_3)] = \{j_1, j_2\}$$

contains two states j_1 and j_2, or equivalently

$$\langle i, j_1 \rangle, \langle i, j_2 \rangle \in e_o(k_1) \cap e_c(k_2) \cap e_t(k_3),$$

we can always split $e_t(k_3)$ into two sub-events $e_t(k_3) = e_t(k_3, 1) \cup e_t(k_3, 2)$, $e_t(k_3, 1) \cap e_t(k_3, 2) = \emptyset$, such that $\langle i, j_1 \rangle \in e_t(k_3, 1)$ and $\langle i, j_2 \rangle \in e_t(k_3, 2)$, and therefore,

$$\langle i, j_1 \rangle \in e_o(k_1) \cap e_c(k_2) \cap e_t(k_3, 1)$$

and

$$\langle i, j_2 \rangle \in e_o(k_1) \cap e_c(k_2) \cap e_t(k_3, 2).$$

Thus, by splitting a natural transition event into a few sub-natural transition events, we can always assume that the event $e(k_1, k_2, k_3)$ in (8.11) is deterministic and denote

$$j = O_i \{e_o(k_1) \cap e_c(k_2) \cap e_t(k_3)\}.$$

In summary, the event-based approach applies to systems in which the event space can be decomposed into mutually exclusive subsets of observable events $e_o(k)$, $k = 1, 2, \ldots, k_o$, mutually exclusive subsets of controllable events $e_c(k)$, $k = 1, 2, \ldots, k_c$, and mutually exclusive subsets of natural transition events $e_t(k)$, $k = 1, 2, \ldots, k_t$. Every single event (a state transition) belongs to one event in each type. In addition, there is a logical timing order among the three types of events: At any time instant, an observable event happens first, followed by a controllable event whose probability is determined by the action taken after the observable event is observed, then followed by a natural transition event. As a special case, which happens often, for some observable events only one action (usually "do nothing") is available. When such an observable state is observed, the system is not controllable at that time instant. We can also assume that, for any k_1, k_2, and k_3, the composition event $e(k_1, k_2, k_3) = e_o(k_1) \cap e_c(k_2) \cap e_t(k_3)$ in (8.11) is deterministic.

8.3 Event-Based Optimization

8.3.1 The Problem Formulation

The mechanism of event-based optimization is different from that of the standard MDP formulation. With the basic concepts introduced in the last section, we can now give a mathematical model for the event-based optimization and describe the system evolution with this model.

We assume that the classification of events described in Section 8.2.4 does not depend on any policy. In other words, the classification of events is determined only by the system. This assumption is satisfied by many real systems.

The mechanism of the event-based optimization is as follows. At time l the system is in state X_l, $l = 0, 1, \ldots$. However, X_l is not observed, and instead we observe an observable event $E_l = e_o(k_1) \subseteq \mathcal{E}$, with a probability distribution $\mu(e_o(k_1)|X_l)$, $k_1 = 1, 2, \ldots, k_o$. We assume that the system is time homogeneous and therefore $\mu(e_o(k_1)|X_l)$ is independent of time l.

In addition to some knowledge about the current state, the observable event also contains some information about the next state after the transition at time l; it, however, does not completely specify the transition.

Based on the information contained in the observable event $e_o(k_1)$, we take an action $\alpha \in \mathcal{A}(k_1)$, where $\mathcal{A}(k_1)$ is the set of actions that are available when $e_o(k_1)$ is observed. Once this action is taken, a controllable event $e_c(k_2)$ will follow with probability $p^\alpha[e_c(k_2)|e_o(k_1)]$, $k_2 = 1, 2, \ldots, k_c$. The superscript "α" indicates the dependence of the probabilities on the action taken. After the controllable event $e_c(k_2)$ occurs, nature chooses a natural transition event $e_t(k_3)$, $k_3 = 1, 2, \ldots, k_t$, which, together with $e_o(k_1)$ and $e_c(k_2)$, finally determines the state transition at time l, $\langle X_l, X_{l+1}\rangle \in e_o(k_1) \cap e_c(k_2) \cap e_t(k_3)$.

State Transition Probabilities

More precisely, let us assume that $X_l = i$ and $X_{l+1} = j$ and denote the transition at time l as $\langle i, j\rangle$. Because we observe the event $e_o(k_1)$, $k_1 = 1, 2, \ldots, k_o$, we have $\langle i, j\rangle \in e_o(k_1)$, but both i and j may not be known. If action $\alpha \in \mathcal{A}(k_1)$ is taken, then the conditional probability of $\langle i, j\rangle \in e_c(k_2)$, $k_2 = 1, 2, \ldots, k_c$, given that $\langle i, j\rangle \in e_o(k_1)$ is controlled by α and can be denoted as

$$p^\alpha[\langle i, j\rangle \in e_c(k_2)|\langle i, j\rangle \in e_o(k_1)], \qquad k_1 = 1, 2, \ldots, k_o, \quad k_2 = 1, 2, \ldots, k_c. \tag{8.13}$$

By convention, $p^\alpha[\langle i, j\rangle \in e_c(k_2)|\langle i, j\rangle \in e_o(k_1)] = 0$ for all $\alpha \in \mathcal{A}(k_1)$, if $e_c(k_2) \cap e_o(k_1) = \emptyset$. We make the following assumption.

Assumption 8.1. *The conditional probability in (8.13) depends only on $e_o(k_1)$ and $e_c(k_2)$; i.e., it is the same for all $i \in I[e_o(k_1)]$.*

Assumption 8.1 is a restriction on the effect of control actions. It is reasonable because we may not be able to observe i. It is not a restriction on the system structure. Under Assumption 8.1, we may denote (8.13) as

$$p^\alpha[e_c(k_2)|e_o(k_1)] := p^\alpha[\langle i, j\rangle \in e_c(k_2)|\langle i, j\rangle \in e_o(k_1)].$$

Then, we have

$$\sum_{k_2=1}^{k_c} p^\alpha[e_c(k_2)|e_o(k_1)] = 1.$$

The natural transition probability given the observable event $e_o(k_1)$ and controllable event $e_c(k_2)$ is denoted as

$$p[e_t(k_3)|e_c(k_2), e_o(k_1)] := p[\langle i, j\rangle \in e_t(k_3)|\langle i, j\rangle \in e_c(k_2) \cap e_o(k_1)],$$
$$k_1 = 1, 2, \ldots, k_o, \quad k_2 = 1, 2, \ldots, k_c, \quad k_3 = 1, 2, \ldots, k_t.$$

They are determined by nature. As shown later, for our analysis we do not require this probability to be independent of i. As discussed in the last section, we may assume that the event $e_o(k_1) \cap e_c(k_2) \cap e_t(k_3)$ is deterministic. Hence, the three events $e_o(k_1)$, $e_c(k_2)$, and $e_t(k_3)$ uniquely determine an output state $j = O_i[e_o(k_1) \cap e_c(k_2) \cap e_t(k_3)]$.

Let us consider the stationary deterministic policies that depend on the current observable events. Such a policy is a mapping from the set of observable events $\mathcal{E}_o$ to the action set $\mathcal{A} = \cup_{k_1=1}^{k_o} \mathcal{A}(k_1)$, denoted as $d : \mathcal{E}_o \to \mathcal{A}$, which specifies the action $d[e_o(k_1)] \in \mathcal{A}(k_1)$ taken when the observable event $e_o(k_1)$ is observed, where $\mathcal{E}_o = \{e_o(k_1) : k_1 = 1, \ldots, k_o\}$. Denote $\mathcal{D}_e$ as the set of all the stationary policies that depend only on the current observable events. (The subscript "e" indicates that the policies are event based.)

From Section 8.2.4, for any transition $\langle i, j \rangle$, $i, j \in \mathcal{S}$, there exists a unique set of integers, k_1, k_2, and k_3, $k_1 \in \{1, 2, \ldots, k_o\}$, $k_2 \in \{1, 2, \ldots, k_c\}$ and $k_3 \in \{1, 2, \ldots, k_t\}$, such that

$$\langle i, j \rangle \in e_o(k_1) \cap e_c(k_2) \cap e_t(k_3), \tag{8.14}$$

and $j = O_i[e_o(k_1) \cap e_c(k_2) \cap e_t(k_3)]$. From (8.14) and the mathematical model of the event-based optimization, the state transition probabilities from state i under an event-based policy d is

$$p^d(j|i) = \mu(e_o(k_1)|i) p^{d[e_o(k_1)]} [e_c(k_2)|e_o(k_1)] \, p[e_t(k_3)|e_c(k_2), e_o(k_1)], \tag{8.15}$$

where $j = O_i[e_o(k_1) \cap e_c(k_2) \cap e_t(k_3)]$. This holds for all output states of $e_o(k_1)$ (cf. (8.5)):

$$j \in O_i[e_o(k_1)] = \cup_{k_2=1}^{k_c} \cup_{k_3=1}^{k_t} O_i \{e_o(k_1) \cap e_c(k_2) \cap e_t(k_3)\}.$$

As assumed, the probability distribution $\mu(e_o(k_1)|i)$ is independent of the policy d. We denote the state transition probability matrix under the event-based policy d as $P^d = [p^d(j|i)]_{i,j \in \mathcal{S}}$. From (8.15), the process $\{X_l, l = 0, 1, \ldots\}$ under any event-based policy d is indeed a time-homogenous Markov chain.

From (8.15), in event-based optimization, we decompose the state transition probability into the controllable part, $p^{d[e_o(k_1)]}[e_c(k_2)|e_o(k_1)]$, and the two uncontrollable parts, $\mu(e_o(k_1)|i)$ and $p[e_t(k_3)|e_c(k_2), e_o(k_1)]$; each of them has a clear physical meaning. The decomposition utilizes the special features of the problem. With this formulation, actions depend on events and therefore the same action can be taken at different states. Furthermore, only the controllable part in the transition probabilities contains important parameters, and as we shall see later, the other parts may be "aggregated".

Average Reward and Optimization

Let $f(i, \alpha)$, $i \in \mathcal{S}$, $\alpha \in \mathcal{A}$, be a reward function. Generally, the long-run average reward of event-based policy d is defined as

$$\eta^d(i,\alpha) := \lim_{L\to\infty} \frac{1}{L} \sum_{l=0}^{L-1} E\left[f(X_l, A_l)|X_0 = i, A_0 = \alpha\right], \qquad (8.16)$$

where A_l is the action taken at time l according to policy d, $l = 0, 1, \ldots$. (8.16) exists for all stationary event-based policies. The goal is to find an event-based policy $d \in \mathcal{D}_e$ that maximizes the average reward or other performance criteria (e.g., the discounted reward).

8.3.2 Performance Difference Formulas

For performance optimization, we study only ergodic policies in this chapter. An event-based policy $d\,[e_o(k)]$, $k = 1, \ldots, k_o$, is said to be *ergodic*, if the Markov chain with transition probabilities $p^d(j|i)$, $i, j = 1, 2, \ldots, S$, as shown in (8.15), is ergodic. From now on, we assume that all the policies in $\mathcal{D}_e$ are ergodic.

For any policy $d \in \mathcal{D}_e$, there always exists a steady-state probability, denoted as $\pi^d = (\pi^d(1), \pi^d(2), \ldots, \pi^d(S))$. To simplify the analysis, we first assume that the reward function f does not depend on the actions. Thus, $f(i,\alpha) = f(i)$ for any $i \in \mathcal{S}$. The average reward is $\eta^d = \pi^d f$ and the performance potential g^d is determined by the Poisson equation (2.12) with transition probability matrix P^d. The extension to f depending on actions will be given later.

Let $\pi^d(e_o(k))$ be the steady-event probability of event $e_o(k)$, $k = 1, 2, \ldots, k_o$, under policy $d \in \mathcal{D}_e$. We have

$$\pi^d(e_o(k_1)) = \sum_{i \in I[e_o(k_1)]} \pi^d(i)\mu(e_o(k_1)|i), \qquad k_1 = 1, 2, \ldots, k_o. \qquad (8.17)$$

In addition, we can write

$$\pi^d(i) = \sum_{k_1=1}^{k_o} \pi^d(e_o(k_1))\pi^d(i|e_o(k_1)), \qquad i \in \mathcal{S},$$

where the steady-state conditional probability

$$\pi^d(i|e_o(k_1)) = \frac{\pi^d(i)\mu(e_o(k_1)|i)}{\pi^d(e_o(k_1))} = \frac{\pi^d(i)\mu(e_o(k_1)|i)}{\sum_{j \in I[e_o(k_1)]} \pi^d(j)\mu(e_o(k_1)|j)}, \qquad (8.18)$$

and $\sum_{i \in I[e_o(k_1)]} \pi^d(i|e_o(k_1)) = 1$.

By the standard MDP average-reward difference formula (4.1), for any two event-based policies $d \in \mathcal{D}_e$ and $h \in \mathcal{D}_e$, by (8.15), (8.17), and (8.18), we have

$$\begin{aligned} &\eta^h - \eta^d \\ &= \pi^h(P^h - P^d)g^d \end{aligned}$$

$$= \sum_{i=1}^{S} \pi^h(i) \sum_{j=1}^{S} \left[p^h(j|i) - p^d(j|i)\right] g^d(j)$$

$$= \sum_{i=1}^{S} \pi^h(i) \sum_{k_1=1}^{k_o} \sum_{k_2=1}^{k_c} \sum_{k_3=1}^{k_t} \mu(e_o(k_1)|i)$$

$$\left\{p^{h[e_o(k_1)]}\left[e_c(k_2)|e_o(k_1)\right] - p^{d[e_o(k_1)]}\left[e_c(k_2)|e_o(k_1)\right]\right\}$$

$$p\left[e_t(k_3)|e_c(k_2), e_o(k_1)\right] g^d(O_i\left[e_o(k_1) \cap e_c(k_2) \cap e_t(k_3)\right])$$

$$= \sum_{k_1=1}^{k_o} \sum_{k_2=1}^{k_c} \sum_{k_3=1}^{k_t} \sum_{i=1}^{S} \pi^h(i)\mu(e_o(k_1)|i)$$

$$\left\{p^{h[e_o(k_1)]}\left[e_c(k_2)|e_o(k_1)\right] - p^{d[e_o(k_1)]}\left[e_c(k_2)|e_o(k_1)\right]\right\}$$

$$p\left[e_t(k_3)|e_c(k_2), e_o(k_1)\right] g^d(O_i\left[e_o(k_1) \cap e_c(k_2) \cap e_t(k_3)\right])$$

$$= \sum_{k_1=1}^{k_o} \sum_{k_2=1}^{k_c} \sum_{k_3=1}^{k_t} \sum_{i=1}^{S} \pi^h(e_o(k_1))\pi^h(i|e_o(k_1))$$

$$\left\{p^{h[e_o(k_1)]}\left[e_c(k_2)|e_o(k_1)\right] - p^{d[e_o(k_1)]}\left[e_c(k_2)|e_o(k_1)\right]\right\}$$

$$p\left[e_t(k_3)|e_c(k_2), e_o(k_1)\right] g^d(O_i\left[e_o(k_1) \cap e_c(k_2) \cap e_t(k_3)\right])$$

$$= \sum_{k_1=1}^{k_o} \pi^h(e_o(k_1)) \sum_{k_2=1}^{k_c} \left\{p^{h[e_o(k_1)]}\left[e_c(k_2)|e_o(k_1)\right] - p^{d[e_o(k_1)]}\left[e_c(k_2)|e_o(k_1)\right]\right\}$$

$$\sum_{i \in I[e_o(k_1)]} \sum_{k_3=1}^{k_t} \pi^h(i|e_o(k_1))p\left[e_t(k_3)|e_c(k_2), e_o(k_1)\right] g^d(j),$$

$$\text{with } j = O_i\left[e_o(k_1) \cap e_c(k_2) \cap e_t(k_3)\right]).$$

Thus, we have

The Average-Reward Difference Formula with Event-Based Policies:

$$\eta^h - \eta^d = \sum_{k_1=1}^{k_o} \pi^h(e_o(k_1))$$

$$\sum_{k_2=1}^{k_c} \left\{p^{h[e_o(k_1)]}\left[e_c(k_2)|e_o(k_1)\right] - p^{d[e_o(k_1)]}\left[e_c(k_2)|e_o(k_1)\right]\right\} g^{d,h}(k_1, k_2), \quad (8.19)$$

where

$$g^{d,h}(k_1,k_2) = \sum_{i\in I[e_o(k_1)]} \sum_{k_3=1}^{k_t} \left\{\pi^h(i|e_o(k_1))p\left[e_t(k_3)|e_c(k_2),e_o(k_1)\right] g^d(j)\right\}, \tag{8.20}$$

with $j = O_i\left[e_o(k_1) \cap e_c(k_2) \cap e_t(k_3)\right]$ is the aggregated potential depending on both policy d and policy h.

Equation (8.19) is the average-reward difference formula for event-based policies. The aggregated potential (8.20) can be extended further by allowing the natural transition probability depending on state i, which is denoted as $p_i\left[e_t(k_3)|e_c(k_2),e_o(k_1)\right]$. In this case, (8.19) remains the same and (8.20) becomes

$$g^{d,h}(k_1,k_2) = \sum_{i\in I[e_o(k_1)]} \sum_{k_3=1}^{k_t} \left\{\pi^h(i|e_o(k_1))p_i\left[e_t(k_3)|e_c(k_2),e_o(k_1)\right] g^d(j)\right\}, \tag{8.21}$$

with $j = O_i\left[e_o(k_1) \cap e_c(k_2) \cap e_t(k_3)\right]$.

Finally, it should be noted that although the derivation of the difference formula in (8.19) is simple and direct, this formula was first "constructed" intuitively with performance potentials as building blocks, by following the method in Chapter 9.

When f Depends on Policy

Now, we assume that the reward functions for the two policies in comparison are different and are denoted as f^h and f^d, respectively, where $f^h = (f^h(1),\dots,f^h(S))^T$ and $f^d = (f^d(1),\dots,f^d(S))^T$, and in (8.16), we have $f^h(i) = f(i,h[e_o(k_1)]), i \in \mathcal{S}$. In this case, we need to add a term $\pi^h(f^h - f^d)$ on the right-hand side of (8.19). We have

$$\begin{aligned}
\pi^h f^h &= \sum_{i=1}^{S} \pi^h(i) f^h(i) \\
&= \sum_{i=1}^{S} \left\{ \sum_{k_1=1}^{k_o} \left[\pi^h(i|e_o(k_1))\pi^h(e_o(k_1))\right] f^h(i) \right\} \\
&= \sum_{k_1=1}^{k_o} \left\{ \pi^h(e_o(k_1)) \left[\sum_{i=1}^{S} \pi^h(i|e_o(k_1)) f^h(i) \right] \right\} \\
&= \sum_{k_1=1}^{k_o} \left[\pi^h(e_o(k_1)) f^h(k_1)\right],
\end{aligned}$$

where

$$f^h(k_1) := \sum_{i=1}^{S} \left[\pi^h(i|e_o(k_1)) f^h(i) \right]. \tag{8.22}$$

Similarly, we have

$$\pi^h f^d = \sum_{i=1}^{S} \pi^h(i) f^d(i)$$
$$= \sum_{k_1=1}^{k_o} \left[\pi^h(e_o(k_1)) f^{d,h}(k_1) \right],$$

where

$$f^{d,h}(k_1) := \sum_{i=1}^{S} \left[\pi^h(i|e_o(k_1)) f^d(i) \right]. \tag{8.23}$$

Therefore, with different reward functions f^d and f^h, we have

$$\begin{aligned}
&\eta^h - \eta^d \\
&= \sum_{k_1=1}^{k_o} \pi^h(e_o(k_1)) \sum_{k_2=1}^{k_c} \Big\{ p^{h[e_o(k_1)]} \left[e_c(k_2)|e_o(k_1) \right] \\
&\qquad - p^{d[e_o(k_1)]} \left[e_c(k_2)|e_o(k_1) \right] \Big\} g^{d,h}(k_1, k_2) \\
&\quad + \sum_{k_1=1}^{k_o} \left\{ \pi^h\left(e_o(k_1)\right) \left[f^h(k_1) - f^{d,h}(k_1) \right] \right\} \\
&= \sum_{k_1=1}^{k_o} \pi^h(e_o(k_1)) \left\{ \left[f^h(k_1) + \sum_{k_2=1}^{k_c} \left\{ p^{h[e_o(k_1)]} \left[e_c(k_2)|e_o(k_1) \right] g^{d,h}(k_1, k_2) \right\} \right] \right. \\
&\quad \left. - \left[f^{d,h}(k_1) + \sum_{k_2=1}^{k_c} \left\{ p^{d[e_o(k_1)]} \left[e_c(k_2)|e_o(k_1) \right] g^{d,h}(k_1, k_2) \right\} \right] \right\}.
\end{aligned} \tag{8.24}$$

This is similar to (4.1).

8.3.3 Performance Derivative Formulas

In the average-reward difference formula (8.19), the "aggregated" potentials (8.20) depend on the steady-state conditional probability of the perturbed system $\pi^h(i|e_o(k_1))$, and in general, they cannot be estimated on a sample path of the original system. The average-reward gradient along any direction, however, can be expressed in aggregated potentials that depend only on the original systems.

To study the average-reward gradient, we assume that the conditional transition probabilities of the controllable events and the reward function depend on a continuous parameter $\theta \in \Theta \subseteq \Re$ and are denoted as $p_\theta\left[e_c(k_2)|e_o(k_1)\right]$, $k_1 = 1, 2, \ldots, k_o$, $k_2 = 1, 2, \ldots, k_c$, and $f_\theta(i)$, $i \in \mathcal{S}$, respectively. Taking $p_{\theta_2}\left[e_c(k_2)|e_o(k_1)\right]$ as $p^h\left[e_c(k_2)|e_o(k_1)\right]$ and $p_{\theta_1}\left[e_c(k_2)|e_o(k_1)\right]$ as $p^d\left[e_c(k_2)|e_o(k_1)\right]$ in (8.24), where $\theta_1, \theta_2 \in \Theta$, we have

$$\begin{aligned}
&\eta(\theta_2) - \eta(\theta_1) \\
&= \sum_{k_1=1}^{k_o} \pi_{\theta_2}(e_o(k_1)) \left\{ \left[f_{\theta_2}(k_1) - f_{\theta_2,\theta_1}(k_1)\right] \right. \\
&\quad \left. + \sum_{k_2=1}^{k_c} \left\{ p_{\theta_2}\left[e_c(k_2)|e_o(k_1)\right] - p_{\theta_1}\left[e_c(k_2)|e_o(k_1)\right] \right\} g_{\theta_1,\theta_2}(k_1, k_2) \right\},
\end{aligned}$$

where

$$g_{\theta_1,\theta_2}(k_1, k_2) = \sum_{i \in I[e_o(k_1)]} \sum_{k_3=1}^{k_t} \left\{ \pi_{\theta_2}(i|e_o(k_1)) p_i\left[e_t(k_3)|e_c(k_2), e_o(k_1)\right] g_{\theta_1}(j) \right\},$$

with $j = O_i\left[e_o(k_1) \cap e_c(k_2) \cap e_t(k_3)\right]$, and

$$f_{\theta_2}(k_1) := \sum_{i=1}^{S} \pi_{\theta_2}(i|e(k_1)) f_{\theta_2}(i),$$

$$f_{\theta_2,\theta_1}(k_1) := \sum_{i=1}^{S} \pi_{\theta_2}(i|e(k_1)) f_{\theta_1}(i).$$

Dividing both sides with $\theta_2 - \theta_1$ and letting $\theta_2 \to \theta_1$ and assuming that the derivatives exist, we obtain

$$\begin{aligned}
\left.\frac{d\eta(\theta)}{d\theta}\right|_{\theta=\theta_1} &= \sum_{k_1=1}^{k_o} \pi_{\theta_1}(e_o(k_1)) \left\{ \sum_{i=1}^{S} \pi_{\theta_1}(i|e(k_1)) \left.\frac{df_\theta(i)}{d\theta}\right|_{\theta=\theta_1} \right. \\
&\quad \left. + \sum_{k_2=1}^{k_c} \left.\left\{ \frac{d}{d\theta} p_\theta\left[e_c(k_2)|e_o(k_1)\right] \right\}\right|_{\theta=\theta_1} g_{\theta_1}(k_1, k_2) \right\} \\
&= \sum_{i=1}^{S} \pi_{\theta_1}(i) \left.\frac{df_\theta(i)}{d\theta}\right|_{\theta=\theta_1} + \sum_{k_1=1}^{k_o} \pi_{\theta_1}(e_o(k_1)) \\
&\quad \left\{ \sum_{k_2=1}^{k_c} \left.\left\{ \frac{d}{d\theta} p_\theta\left[e_c(k_2)|e_o(k_1)\right] \right\}\right|_{\theta=\theta_1} g_{\theta_1}(k_1, k_2) \right\}, \quad (8.25)
\end{aligned}$$

where

$$g_{\theta_1}(k_1, k_2) = \sum_{i \in I[e_o(k_1)]} \sum_{k_3=1}^{k_t} \left\{ \pi_{\theta_1}(i|e_o(k_1)) p_i\left[e_t(k_3)|e_c(k_2), e_o(k_1)\right] g_{\theta_1}(j) \right\}. \quad (8.26)$$

The aggregated potential $g_{\theta_1}(k_1, k_2)$ depends only on the original system with parameter θ_1.

The Examples

In these examples, we discuss both the performance difference and the performance derivative formulas of the event-based policies.

Example 8.12. (**Moving Robot**, Continued) Now, we derive the performance sensitivity formulas for the moving robot problem. Consider two different policies (σ_a, σ_b) and (σ'_a, σ'_b). We assume that the reward functions are the same for both policies. Let $\pi(i)$ and $\pi'(i)$ denote the steady-state probabilities of state i, $i = 1, 2, 3, 4, 5, 6$, and $\pi(a)$, $\pi(b)$, and $\pi'(a)$, $\pi'(b)$ be the steady-state probabilities of events a and b under two policies (σ_a, σ_b) and (σ'_a, σ'_b), respectively.

We derive the performance difference formula by using (8.19). There are two observable events a and b, and corresponding to each observable event, there are two controllable events: a_r and a_g for a, and b_r and b_g for b. The conditional probabilities of the controllable events are $p(a_g|a) = \sigma_a$ for policy (σ_a, σ_b), and $p'(a_g|a) = \sigma'_a$ for policy (σ'_a, σ'_b). Let η and η' be the long-run average performance of these two policies, respectively. By construction, or by applying (8.19), we have

$$\begin{aligned}\eta' - \eta &= \pi'(a)\left\{[p'(a_g|a) - p(a_g|a)]\, g'(a_g) + [p'(a_r|a) - p(a_r|a)]\, g'(a_r)\right\} \\ &\quad + \pi'(b)\left\{[p'(b_g|b) - p(b_g|b)]\, g'(b_g) + [p'(b_r|a) - p(b_r|a)]\, g'(b_r)\right\} \\ &= \pi'(a)\left\{(\sigma'_a - \sigma_a)\,[g'(a_g) - g'(a_r)]\right\} \\ &\quad + \pi'(b)\left\{(\sigma'_b - \sigma_b)\,[g'(b_g) - g'(b_r)]\right\},\end{aligned}$$

where (cf. (8.38))

$$g'(a_g) = \sum_{i=1}^{2} [\pi'(i|a) g(5)] = g(5),$$

$$g'(a_r) = \sum_{i=1}^{2} [\pi'(i|a) g(6)] = g(6).$$

$$g'(b_g) = \sum_{i=1}^{2} [\pi'(i|b) g(3)] = g(3),$$

and

$$g'(b_r) = \sum_{i=1}^{2} [\pi'(i|b) g(4)] = g(4).$$

Therefore, we have

$$\begin{aligned}\eta' - \eta &= \pi'(a)\left\{(\sigma'_a - \sigma_a)\,[g(5) - g(6)]\right\} \\ &\quad + \pi'(b)\left\{(\sigma'_b - \sigma_b)\,[g(3) - g(4)]\right\}. \end{aligned} \tag{8.27}$$

Each term on the right-hand side of the performance difference formula (8.27) is factorized into three factors. For example, in the first term, the first factor is $\pi'(a)$, which only depends on policy (σ'_a, σ'_b); the second factor is $\sigma'_a - \sigma_a$, which is under our control; and the third factor is $g(5) - g(6)$, which only depends on policy (σ_a, σ_b), As we have seen in Chapter 4, this particular form is the basis for policy iteration.

Suppose that the probabilities depend on a parameter θ: $\sigma_a = \sigma_{a,\theta}$ and $\sigma_b = \sigma_{b,\theta}$. Then η and π, etc., also depend on θ. We use a subscript "θ" to denote the dependency of a quantity on θ. From (8.27), we have

$$\frac{d\eta_\theta}{d\theta} = \pi_\theta(a)\frac{d\sigma_{a,\theta}}{d\theta}\left[g(5) - g(6)\right] + \pi_\theta(b)\frac{d\sigma_{b,\theta}}{d\theta}\left[g(3) - g(4)\right]. \qquad (8.28)$$

The potentials $g(a_g)$, etc., have a clear physical meaning. For example, $g(a_g)$ is the aggregated potential after event a occurs and the robot passes the green light, which happens to be the same as $g(5)$. (See the extended example discussed later.) □

Example 8.13. (**Admission Control**, Continued) Now, we derive the performance sensitivity formulas for the admission control problem. Consider two admission policies $\sigma^d(n)$ and $\sigma^h(n)$ for $n = 0, 1, \ldots, N$. For simplicity, we assume that the reward functions are the same for both policies. This is usually the case, since we are most likely concerned about the same physical quantities such as mean waiting times, etc. Let $\pi^h(n) := \pi^h(a_n)$ denote the steady-state probability of event a_n under policy h, i.e., the probability that a customer arrives and finds n customers in the system. Let $\pi^h(\boldsymbol{n}|n)$ be the conditional steady-state probability that the system is in state $\boldsymbol{n}$ when event a_n happens. Then

$$\pi^h(\boldsymbol{n}, n) = \pi^h(n)\pi^h(\boldsymbol{n}|n)$$

is the probability that a_n occurs and at the same time the system state is $\boldsymbol{n}$, $n_1 + \cdots + n_M = n$.

We derive the performance difference formula by using (8.19). The observable events are $e_o(k_1) = a_{k_1}$, $k_1 = 0, 1, \ldots, N$. There are only two controllable events a_+ and a_-. We have (by changing the index from k_1 to n): $\sigma^d(N) = \sigma^h(N) = 0$, and

$$p^d(a_{n,+}|a_n) = \sigma^d(n), \quad \text{and} \quad p^h(a_{n,+}|a_n) = \sigma^h(n).$$

$$p^d(a_{n,-}|a_n) = 1 - \sigma^d(n), \quad \text{and} \quad p^h(a_{n,-}|a_n) = 1 - \sigma^h(n).$$

The natural transition probabilities are

$$p\left(a_{n,+i}|a_n, a_+\right) = q_{0,i}, \quad \text{and} \quad p\left(a_{n,-}|a_n, a_-\right) = 1.$$

Therefore, from (8.19), we have

$$\eta^h - \eta^d = \sum_{n=0}^{N-1} \Big\{ \pi^h(n) \left\{ \left[p^h(a_{n,+}|a_n) - p^d(a_{n,+}|a_n) \right] g^{d,h}(a_{n,+}) \right.$$
$$\left. + \left[p^h(a_{n,-}|a_n) - p^d(a_{n,-}|a_n) \right] g^{d,h}(a_{n,-}) \right\} \Big\}$$
$$= \sum_{n=0}^{N-1} \left\{ \pi^h(n) \left[\sigma^h(n) - \sigma^d(n) \right] \left[g^{d,h}(a_{n,+}) - g^{d,h}(a_{n,-}) \right] \right\}$$
$$= \sum_{n=0}^{N-1} \left\{ \pi^h(n) \left[\sigma^h(n) - \sigma^d(n) \right] \gamma^{d,h}(n) \right\}, \tag{8.29}$$

where (cf. (8.20))

$$g^{d,h}(a_{n,+}) = \sum_{\boldsymbol{n} \in \mathcal{S}_n} \left\{ \pi^h(\boldsymbol{n}|n) \sum_{i=1}^{M} \left[q_{0,i} g^d(\boldsymbol{n}_{+i}) \right] \right\},$$
$$g^{d,h}(a_{n,-}) = \sum_{\boldsymbol{n} \in \mathcal{S}_n} \left[\pi^h(\boldsymbol{n}|n) g^d(\boldsymbol{n}) \right],$$

and

$$\gamma^{d,h}(n) := g^{d,h}(a_{n,+}) - g^{d,h}(a_{n,-})$$
$$= \sum_{\boldsymbol{n} \in \mathcal{S}_n} \left\{ \pi^h(\boldsymbol{n}|n) \sum_{i=1}^{M} \left\{ q_{0,i} \left[g^d(\boldsymbol{n}_{+i}) - g^d(\boldsymbol{n}) \right] \right\} \right\}, \tag{8.30}$$
$$n = 0, 1, \ldots, N-1,$$

where $\mathcal{S}_n = \left\{ \boldsymbol{n} : \sum_{i=1}^{M} n_i = n \right\}$.

From the product-form solution of queueing networks [96], we can prove that for any two policies h and d, we have (Problem 8.6)

$$\pi^h(\boldsymbol{n}|n) = \pi^d(\boldsymbol{n}|n). \tag{8.31}$$

Thus, (8.29)-(8.30) become

$$\eta^h - \eta^d = \sum_{n=0}^{N-1} \left\{ \pi^h(n) \left[\sigma^h(n) - \sigma^d(n) \right] \gamma^d(n) \right\}. \tag{8.32}$$

$$g^d(a_{n,+}) = \sum_{\boldsymbol{n} \in \mathcal{S}_n} \left\{ \pi^d(\boldsymbol{n}|n) \sum_{i=1}^{M} \left[q_{0,i} g^d(\boldsymbol{n}_{+i}) \right] \right\}, \tag{8.33}$$

$$g^d(a_{n,-}) = \sum_{\boldsymbol{n} \in \mathcal{S}_n} \left[\pi^d(\boldsymbol{n}|n) g^d(\boldsymbol{n}) \right], \tag{8.34}$$

and

$$\gamma^d(n) := g^d(a_{n,+}) - g^d(a_{n,-})$$
$$= \sum_{\boldsymbol{n}\in\mathcal{S}_n} \pi^d(\boldsymbol{n}|n) \left\{ \sum_{i=1}^{M} q_{0,i} \left[g^d(\boldsymbol{n}_{+i}) - g^d(\boldsymbol{n}) \right] \right\}, \tag{8.35}$$
$$n = 0, 1, \ldots, N-1.$$

Both $g^d(a_{n,+})$ and $g^d(a_{n,-})$ have a clear physical meaning. $g^d(a_{n,+})$ is the aggregated potential of a system with population n after an arriving customer is accepted, and $g^d(a_{n,-})$ is the aggregated potential of a system with population n after an arriving customer is rejected. Because of the PASTA property, $g^d(a_{n,-})$ equals the aggregated potential of the set of states with the same population n. $\gamma^d(n)$ is the aggregated perturbation realization factors with $\gamma^d(\boldsymbol{n}, \boldsymbol{n}_{+i}) = g^d(\boldsymbol{n}_{+i}) - g^d(\boldsymbol{n})$ and

$$\gamma^d(n) = \sum_{\boldsymbol{n}\in\mathcal{S}_n} \left\{ \pi^d(\boldsymbol{n}|n) \sum_{i=1}^{M} \left[q_{0,i} \gamma^d(\boldsymbol{n}, \boldsymbol{n}_{+i}) \right] \right\}, \qquad n = 0, 1, \ldots, N-1.$$

The form of the difference formula (8.29) and the aggregate potentials (8.30) depends only on the system parameters $q_{0,i}$, $i = 1, \ldots, M$, and does not depend on the state transition probabilities of the underlying Markov process, nor on the customer routing probabilities $q_{i,j}$, $i, j = 1, 2, \ldots, M$. Thus, this approach maintains the structural property of the system and avoids the tedious effort in finding and storing the large matrix of transition probabilities.

For performance derivatives, we assume that the policy changes from $\sigma^d(n)$ to $\sigma^d(n) + \delta_n$ for a fixed n. From (8.32), we can easily derive

$$\left. \frac{\partial \eta(\delta_n)}{\partial \delta_n} \right|_{\delta_n = 0} = \pi^d(n) \gamma^d(n), \qquad n = 0, 1, \ldots, N-1,$$

where $\gamma^d(n)$ is the aggregated potential in (8.35). This is the derivative with respect to the admission probability at one population. In general, suppose that the policy $\sigma(n)$ depends on a parameter θ and is denoted as $\sigma_\theta(n)$, $n = 0, 1, \ldots, N-1$. Then, from (8.32), we have

$$\left. \frac{d\eta(\theta)}{d\theta} \right|_{\theta=\theta_1} = \sum_{n=0}^{N-1} \pi_{\theta_1}(n) \left. \frac{d\sigma_\theta(n)}{d\theta} \right|_{\theta=\theta_1} \gamma_{\theta_1}(n),$$

in which $\gamma_{\theta_1}(n)$ and $\pi_{\theta_1}(n)$ depend only on the original policy $\sigma_{\theta_1}(n)$, $n = 0, 1, \ldots, N-1$. □

8.3.4 Optimization

Both the performance difference and the performance derivative formulas (8.19), (8.24), and (8.25) have a similar form as those for the standard MDPs (2.27) and (2.26). However, there is a slight but crucial difference between the two cases: The aggregated potential $g^{d,h}(k_1, k_2)$ (8.20) and the aggregated performance function $f^{d,h}(k_1)$ in the performance difference formula (8.24) depend not only on the original policy d but also on the perturbed one h. As a result, policy iteration type of algorithms can be only developed under some conditions, by using the performance difference formula. On the other hand, the aggregated potential $g_{\theta_1}(k_1, k_2)$ (8.26) in the performance derivative formula (8.25) depends only on the original system with parameter θ_1. Therefore, the gradient-based type of optimization approaches can be developed, by using the performance derivative formula. In this section, we will provide a brief discussion on these issues.

Policy Iteration

We first consider the simple case where f does not depend on actions. The aggregated potential (8.20) (or (8.21)) contains items from both policies, $\pi^h(i|e_o(k_1))$ from the perturbed system and $g^d(j)$ from the original one. Such a quantity depending on both policies cannot be used in policy iteration. However, under some special situations, $g^{d,h}(k_1, k_2)$ in (8.20) depends only on the original system and therefore can be obtained by analyzing only the original system, or can be estimated from a sample path of the original system.

First, for some systems (e.g., the system in Example 8.4), the following equation holds for any $h, d \in \mathcal{D}_e$:

$$\pi^h(i|e_o(k_1)) = \pi^d(i|e_o(k_1)), \qquad \text{for all } i \in I\left[e_o(k_1)\right], \quad k_1 = 1, 2, \ldots, k_o. \tag{8.36}$$

In such cases, (8.21) becomes

$$g^d(k_1, k_2) := \sum_{i \in I[e_o(k_1)]} \sum_{k_3=1}^{k_t} \left\{ \pi^d(i|e_o(k_1)) p_i\left[e_t(k_3)|e_c(k_2), e_o(k_1)\right] g^d(j) \right\}, \tag{8.37}$$

which depends only on the original policy d.

Another situation under which the aggregated potential depends only on the original policy d is as follows. Suppose that given $i \in I[e_o(k_1)]$, both $p_i\left[e_t(k_3)|e_c(k_2), e_o(k_1)\right]$ and $j = O_i\left[e_o(k_1) \cap e_c(k_2) \cap e_t(k_3)\right]$ do not explicitly depend on i. In this case, the $g^{d,h}(k_1, k_2)$ in (8.21) becomes

$$\begin{aligned} g^d(k_1, k_2) &:= \sum_{i \in I[e_o(k_1)]} \sum_{k_3=1}^{k_t} \left\{ \pi^h(i|e_o(k_1)) p\left[e_t(k_3)|e_c(k_2), e_o(k_1)\right] g^d(j) \right\} \\ &= \sum_{k_3=1}^{k_t} \left\{ \left[\sum_{\text{all } i:\langle i,j\rangle \in e_o(k_1)} \pi^h(i|e_o(k_1)) \right] p\left[e_t(k_3)|e_c(k_2), e_o(k_1)\right] g^d(j) \right\} \\ &= \sum_{k_3=1}^{k_t} \left\{ p\left[e_t(k_3)|e_c(k_2), e_o(k_1)\right] g^d(j) \right\}, \end{aligned} \tag{8.38}$$

which also depends only on the original policy d. Note that the statement "$j = O_i[e_o(k_1) \cap e_c(k_2) \cap e_t(k_3)]$ does not depend on i" means that for any state i in $I[e_o(k_1)]$ if $e_c(k_2)$ occurs (no matter what action is taken) and if the same natural transition event $e_t(k_3)$ is followed, then the next state will be the same state j. This essentially resets the system, randomly by nature, after the action is taken. See the discussion in the manufacturing example in Section 8.5.1 for an example.

When the aggregated potentials $g^d(k_1, k_2)$ depend only on the original policy d, they can be estimated on a sample path of the original system. Event-based policy iteration algorithms can be developed from the performance difference equation (8.19) following the same idea as the standard MDPs. In essence, at each iteration, one chooses the action α' in the available action set corresponding to the observed event $e_o(k_1)$ that leads to the largest value of the average aggregated potential; i.e.,

$$\alpha' = \arg\max_{\alpha \in \mathcal{A}(k_1)} \left\{ \sum_{k_2=1}^{k_c} \left\{ p^\alpha [e_c(k_2)|e_o(k_1)] - p^{d[e_o(k_1)]} [e_c(k_2)|e_o(k_1)] \right\} g^d(k_1, k_2) \right\}$$

$$= \arg\max_{\alpha \in \mathcal{A}(k_1)} \left\{ \sum_{k_2=1}^{k_c} p^\alpha [e_c(k_2)|e_o(k_1)] \, g^d(k_1, k_2) \right\}.$$

From the same principle as policy iteration in the standard MDPs, we know that the policy iteration procedure leads to an optimal policy among the policies in the event-based policy space in a finite number of steps.

In general, there are $k_o \times k_c$ aggregated potentials $g^d(k_1, k_2)$ in (8.37), compared with S potentials in the standard MDPs. In Example 8.4, $k_o = (N+1)$ and $k_c = 2$. There are $2(N+1)$ aggregated potentials in (8.33) and (8.34), compared with the number of potentials, $\sum_{n=0}^{N} \frac{(n+M-1)!}{n!(M-1)!}$ $(0! = 1)$, in the standard MDPs.

If in addition to the conditions in (8.38), $p[e_t(k_3)|e_c(k_2), e_o(k_1)]$ does not depend on $e_o(k_1)$, we have from (8.38)

$$g^d(k_1, k_2) = \sum_{k_3=1}^{k_t} \{ p[e_t(k_3)|e_c(k_2)] \, g^d(j) \} =: g^d(k_2). \tag{8.39}$$

In such cases, we have only k_c potentials to estimate, and (8.19) becomes

$$\eta^h - \eta^d = \sum_{k_1=1}^{k_o} \pi^h(e_o(k_1)) \sum_{k_2=1}^{k_c} \left\{ p^h [e_c(k_2)|e_o(k_1)] - p^d [e_c(k_2)|e_o(k_1)] \right\} g^d(k_2).$$

The situation is more complicated when f depends on actions. In such cases, the aggregated functions (8.22) and (8.23) also depend on the perturbed policy d. When condition (8.36) holds, policy iteration with action-dependent f can be developed.

Gradient-Based Optimization

The aggregated potential $g_{\theta_1}(k_1, k_2)$ in (8.26) can be estimated on a sample path of the original system with θ_1. The fundamental ideas and some algorithms for estimating the aggregated potentials will be discussed in Section 8.4. Other sample path-based algorithms are yet to be developed. As discussed above, the number of aggregated potentials is usually smaller than the number of states. Once these aggregated potentials are estimated, the performance gradients with respect to any parameter can be obtained by (8.25).

Examples

Example 8.14. (**Moving Robot**, Continued) As shown in Example 8.12, the aggregated potentials $g'(a_g) = g(5)$, $g'(a_r) = g(6)$, $g'(b_g) = g(3)$, and $g'(b_r) = g(4)$ depend only on the original policy. Therefore, a policy iteration algorithm can be derived from the performance difference formula (8.27). The algorithm is very simple because the parameter space is only two dimensional. We first choose any (σ_a, σ_b) in the feasible region $[0, 1]^2$ and calculate or estimate its potentials $g(3)$, $g(4)$, $g(5)$, and $g(6)$. From (8.27), if, for example, $g(5) - g(6) > 0$ and $g(3) - g(4) < 0$, then any policy (σ'_a, σ'_b) with $\sigma'_a > \sigma_a$ and $\sigma'_b < \sigma_b$ performs better than (σ_a, σ_b). It is interesting to note that one of the boundary points (with $\sigma_a = 0$ or 1, and $\sigma_b = 0$ or 1) must be an optimal policy,

Gradient-based optimization algorithms can be developed from the performance derivative formula (8.28). □

Example 8.15. (**Admission Control**, Continued) Policy iteration algorithms can be derived from the performance difference formula (8.32). Following the same reasoning as the standard MDPs, because $\pi^h(n) > 0$ for any policy $\sigma^h(n)$, we have $\eta^h > \eta^d$ if $\left[\sigma^h(n) - \sigma^d(n)\right] \gamma^d(n) \geq 0$ for all n with $\left[\sigma^h(n) - \sigma^d(n)\right] \gamma^d(n) > 0$ for at least one n, $n = 1, 2 \ldots, N - 1$. Thus, at each iteration we can improve the policy as follows: If $\gamma^d(n) > 0$ (or < 0), $n = 0, 1, \ldots, N-1$, for the current policy, then we choose the action α with the largest (or the smallest) $\sigma^\alpha(n)$ for the next iteration; and if $\gamma^d(n) = 0$ for some n, we keep the same action α at this n for the next iteration. Also, as in the standard MDPs, using the aggregated potentials $\gamma^d(n)$, $n = 1, 2, \ldots, N - 1$, (either calculated analytically, or estimated on a sample path of the current system), we can always find a better policy by following the above described policy improvement rule, if such better policies exist. This means that if the iteration stops (i.e., we cannot find a better policy), then the current policy is the best.

In addition, from the form of (8.32), we can conclude that the optimal policy $\hat{\sigma}(n)$ can be chosen at the corner of the feasible policy space: Let $\hat{\gamma}(n)$ is the aggregated potential under policy $\hat{\sigma}(n)$, $n = 0, 1, \ldots, N - 1$. If $\hat{\gamma}(n) > 0$ (or < 0) then $\hat{\sigma}(n)$ should have the largest (or the smallest) among all possible

$\sigma(n)$'s, and if $\hat{\gamma}(n) = 0$, we can choose any $\sigma(n)$ including the largest one at the corner.

From the physical meaning of $g^d(a_{n,+})$ and $g^d(a_{n,-})$, $\gamma^d(n) = g^d(a_{n,+}) - g^d(a_{n,-}) > 0$ implies that the aggregated potential after accepting a customer is larger than that after rejecting the customer, or is larger than the original aggregated potential at population n; in this case, accepting the arriving customer makes the performance better. □

Policies Based on a Sequence of Events

Because the event sequence is not Markov (cf. Section 8.2.2), we may define policies that depend on a sequence of events instead of on only the current event; and the policies depending on a sequence of events may perform better than those depending only on the current event. The policy space expands but the fundamental principles remain the same. We may define the *observable event sequences*, which may contain more information than a single observable event. For instance, in Example 8.6, the event sequence $\{a_{n-1}, a_n\}$ is different from $\{a_{n+3}, a_n\}$. The former indicates that there is no departure between the two arrivals, and the latter indicates that there are three customers departing from the system between the two arrivals. Thus, the corresponding inter-arrival time for $\{a_{n-1}, a_n\}$ is most likely shorter than that for $\{a_{n+3}, a_n\}$. In addition, this sequence may also give us some information about the possible state of the system, depending on the network topology. For example, if the network is a series of servers in tandem, after a sequence of events $\{a_{n+3}, a_n\}$, there will be more likely fewer customers in the exit server.

Not much work has been done in this direction. The problem is similar to partially observable MDPs; however, the event sequence approach maintains the system structure by defining events properly, and an event at time l consists of information about the state transition at time l.

8.4 Learning: Estimating Aggregated Potentials

8.4.1 Aggregated Potentials

The aggregated potentials such as $g^d(a_{n,+})$ and $g^d(a_{n,-})$ in (8.33) and (8.34) can be estimated on a sample path of the original system. We first give an intuitive explanation about how an event-based aggregated potential can be estimated on a sample path with the same amount of computation and with the same accuracy as estimating the potential of a single state.

Estimating a Potential

Let us first review how the potential of a state i, $g(i)$, can be estimated. The simplest way is shown in Figure 8.10A. After each visit to state i, we sum

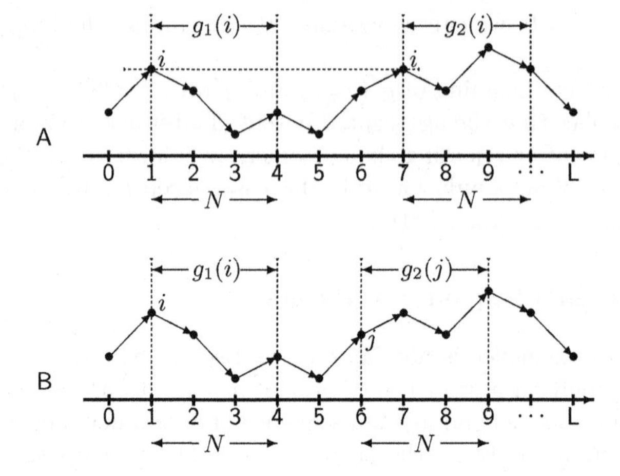

Fig. 8.10. Estimating the Aggregated Potentials Based on States

up the rewards for $N(>> 1)$ consecutive transitions to obtain estimates $g_1(i)$, $g_2(i), \ldots$. For example, in the figure $g_1(i) = \sum_{l=1}^{N} f(X_l)$, $g_2(i) = \sum_{l=7}^{6+N} f(X_l)$, etc. (in the figure, $N = 4$). In general, we define $\mathcal{T}_i = \{l_1, l_2, \ldots, l_k, \ldots, l_{K_i}\}$ as the set of consecutive time instants on the sample path $\{X_0, \ldots, X_L\}$ such that at l_k, $k = 1, 2, \ldots, K_i$, the state of the Markov chain is $X_{l_k} = i$, with K_i being the number of the visits to state i on the sample path. Let

$$g_k(i) = \sum_{l=0}^{N-1} f(X_{l_k+l}). \tag{8.40}$$

From (3.16), we have

$$g(i) \approx \frac{1}{K_i} \sum_{k=1}^{K_i} g_k(i) \quad \left(= \frac{1}{K_i} \sum_{k:\ l_k \in \mathcal{T}_i} g_k(i) \right). \tag{8.41}$$

More precisely, we need to define $g_k(i)$ as $g_k(i) = \sum_{l=0}^{N-1} [f(X_{l_k+l}) - \eta]$, with η being the long-run average reward. We have omitted the constant term $N\eta$ in (8.40) and (8.41).

If we make a slight change in the above expression, we can obtain an estimate of a potential weighted by the steady-state probability of the corresponding state. In fact, for a large integer L, we have

$$\frac{1}{L} \sum_{k=1}^{K_i} g_k(i) = \frac{K_i}{L} \frac{1}{K_i} \sum_{k=1}^{K_i} g_k(i) \approx \pi(i) g(i),$$

where $\frac{K_i}{L} \approx \pi(i)$ is the steady-state probability of i.

Estimating an Aggregated Potential Based on State Aggregations

With the same principle, we can estimate the potential aggregated at a subset of states. This is shown in Figure 8.10B. Let $\mathcal{I} \subseteq \mathcal{S}$ be a subset of the state space, which we call an *aggregated state.* Instead of collecting the sums of the reward functions g_k's starting from a particular state i, we collect the g_k's starting from any state in $\mathcal{I}$. Let $\mathcal{T}_\mathcal{I} := \cup_{i\in\mathcal{I}}\mathcal{T}_i = \{l_1, l_2, \ldots, l_K\}$ be the set of consecutive time instants such that $X_{l_k} \in \mathcal{I}$, and $K := \sum_{i\in\mathcal{I}} K_i$ be the number of visits to the set $\mathcal{I}$ on the sample path $\{X_0, \ldots, X_L\}$. Denote $g_k(X_{l_k}) = \sum_{l=0}^{N} f(X_{l_k+l})$. Then, we have

$$\frac{1}{K}\sum_{k:\ l_k\in\mathcal{T}_\mathcal{I}} g_k(X_{l_k}) = \sum_{i\in\mathcal{I}}\left[\frac{1}{K}\sum_{k:\ l_k\in\mathcal{T}_i} g_k(i)\right]$$
$$= \sum_{i\in\mathcal{I}}\left\{\frac{K_i}{K}\left[\frac{1}{K_i}\sum_{k:\ l_k\in\mathcal{T}_i} g_k(i)\right]\right\}.$$

Clearly, $\frac{K_i}{K} \approx \pi(i|\mathcal{I})$ is the steady-state conditional probability of i given it is in the aggregated state $\mathcal{I}$. Thus, from (8.41)

$$\frac{1}{K}\sum_{k:\ l_k\in\mathcal{T}_\mathcal{I}} g_k(X_{l_k}) \approx \sum_{i\in\mathcal{I}} \pi(i|\mathcal{I})g(i) =: \tilde{g}(\mathcal{I}), \tag{8.42}$$

where $\tilde{g}(\mathcal{I})$ denotes the potential aggregated at the aggregated state $\mathcal{I}$. In the same spirit, we have

$$\frac{1}{L}\sum_{k:\ l_k\in\mathcal{T}_\mathcal{I}} g_k(X_{l_k}) \approx \pi(\mathcal{I})\sum_{i\in\mathcal{I}} \pi(i|\mathcal{I})g(i) = \sum_{i\in\mathcal{I}} \pi(i,\mathcal{I})g(i),$$

where $\pi(\mathcal{I})$ is the steady-state probability of the aggregated state $\mathcal{I}$, and $\pi(i,\mathcal{I}) = \pi(i)$ is the steady-state joint probability of $\mathcal{I}$ and state i.

Estimating an Aggregated Potential Based on Events

Equation (8.42) illustrates how to estimate the potential aggregated on a set of states. There is, however, one difference for estimating a potential aggregated on an event: the potential is usually aggregated at the time instant after the event occurs. Let a be an event (it may not be deterministic) and $j \in O_i(a)$ be an output state of a with $i \in I(a)$ being an input state of a. Obviously, j is the state at the time instant following the instant at which event a occurs. Let $p(j|i,a)$ be the transition probability from i to j given that the event a happens. Define an aggregated potential of event a as

$$\tilde{g}(a) = \sum_{i\in I(a)}\sum_{j\in\mathcal{S}} \pi(i|a)p(j|i,a)g(j). \tag{8.43}$$

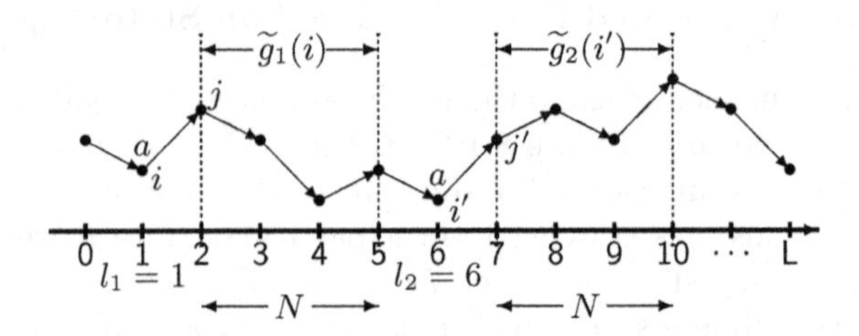

Fig. 8.11. Estimating the Aggregated Potentials Based on Events

Figure 8.11 shows how to estimate $\tilde{g}(a)$. In the figure, event a happens at $l = 1$ and $l = 6$, and $\tilde{g}_k(i)$, $i \in I(a)$, denotes the sum of the rewards of N transitions starting from the next state after the kth occurrence of event a:

$$\tilde{g}_k(X_{l_k}) = \sum_{l=1}^{N} f(X_{l_k+l}), \qquad X_{l_k} \in I(a).$$

Let $\mathcal{T}_a := \{l_1, \ldots, l_k, \ldots, l_K\}$ be the set of consecutive time instants on the sample path $\{X_0, X_1, \ldots, X_L\}$ such that $X_{l_k} \in I(a)$. We can easily prove that for a large N, we have

$$\tilde{g}(a) \approx \frac{1}{K} \sum_{k=1}^{K} \tilde{g}_k(X_{l_k}). \tag{8.44}$$

We have demonstrated that the event-based aggregated potentials (8.43) can be estimated on a sample path, as shown in (8.44). It is important to note that (8.44) has the same form as the estimate of the potential of a single state (8.41); thus estimating an aggregated potential requires the same amount of computation and has the same level of accuracy as estimating a potential of a state. Because the number of aggregated potentials is usually much less than that of the states, the event-based method saves considerable computation in the sample-path based approach for performance optimization.

8.4.2 Aggregated Potentials in the Event-Based Optimization

Many event-based aggregated potentials have the form (8.43). First, we observe that $g^d(k_1, k_2)$ in (8.37) is in the form of (8.43) with the event a being $e_o(k_1) \cap e_c(k_2)$. To see this, let us rewrite the transition probability $p_i\left[e_t(k_3)|e_o(k_1), e_c(k_2)\right]$ in (8.37) in the same form as in (8.43). Because the event $e_o(k_1) \cap e_c(k_2) \cap e_t(k_3)$ is deterministic, $j = O_i\left[e_o(k_1) \cap e_c(k_2) \cap e_t(k_3)\right]$ contains a single state. Next, the set $O_i\left[e_o(k_1) \cap e_c(k_2)\right]$ contains the states $j = O_i\left[e_o(k_1) \cap e_c(k_2) \cap e_t(k_3)\right]$ for all k_3, i.e.,

$$O_i\left[e_o(k_1) \cap e_c(k_2)\right] = \cup_{\text{all } k_3} O_i\left[e_o(k_1) \cap e_c(k_2) \cap e_t(k_3)\right].$$

Given that the event $e_o(k_1) \cap e_o(k_2)$ occurs and $i \in I\left[e_o(k_1) \cap e_c(k_2) \cap e_t(k_3)\right]$, the probability of event $e_t(k_3)$ is the same as the probability of the output state $j = O_i\left[e_o(k_1) \cap e_c(k_2) \cap e_t(k_3)\right]$, Therefore, in the notation of (8.43) we can write

$$p_i\left[e_t(k_3)|e_c(k_2), e_o(k_1)\right] = p\left[j|i; e_c(k_2), e_o(k_1)\right].$$

Thus, (8.37) becomes

$$g^d(k_1, k_2) = \sum_{i \in I[e_o(k_1)]} \sum_{\text{all } j} \left\{\pi^d(i|e_o(k_1)) p\left[j|i; e_c(k_2), e_o(k_1)\right] g^d(j)\right\},$$

where $j = O_i[e_o(k_1) \cap e_c(k_2) \cap e_t(k_3)]$. Because the choice of the control action does not depend on i, we have $\pi^d(i|e_o(k_1)) = \pi^d(i|e_o(k_1), e_c(k_2))$. Therefore, the above equation is exactly the same form as (8.43) with $a = e_o(k_1) \cap e_c(k_2)$.

In particular, $g^d(a_{n,+})$ in (8.33) and $g^d(a_{n,-})$ in (8.34) can be written in the form of (8.43) as $\tilde{g}(a_{n,+})$ and $\tilde{g}(a_{n,-})$, respectively. As an example, we now develop the details for the sample-path-based estimation of $g^d(a_{n,+})$ in the admission control problem.

Example 8.16. (**Admission Control**, Continued) Consider a sample path $\{X_0, X_1, \ldots, X_L\}$, with $L >> 1$. Denote the sequence of the time instants at which event $a_{n,+}$ happens (i.e., an arriving customer finding n customers in the system is accepted) on the sample path as $\mathcal{T}_{a_{n,+}} := \left\{l_1, \ldots, l_{L_{n,+}}\right\}$. Then at $l_k + 1$, $k = 1, 2, \ldots, L_{n,+}$, there are $n+1$ customers in the system. Choose a large integer N. Set

$$\tilde{g}_{l_k} = \sum_{l=1}^{N} f(X_{l_k + l}).$$

Next, we decompose the set $\mathcal{T}_{a_{n,+}}$ into sub-groups $\mathcal{T}_{a_{n,+}} = \cup_{\boldsymbol{n} \in \mathcal{S}_n} \mathcal{T}_{a_{\boldsymbol{n},+}}$, such that before the customer arriving at $l \in \mathcal{T}_{a_{\boldsymbol{n},+}}$ is accepted the system state is $\boldsymbol{n}$ with population n. Let $L_{\boldsymbol{n},+}$ be the number of instants in $\mathcal{T}_{a_{\boldsymbol{n},+}}$. We have $L_{n,+} = \sum_{\boldsymbol{n} \in \mathcal{S}_n} L_{\boldsymbol{n},+}$. We further decompose $\mathcal{T}_{a_{\boldsymbol{n},+}}$ into $\mathcal{T}_{a_{\boldsymbol{n},+}} = \cup_{i=1}^{M} \mathcal{T}_{a_{\boldsymbol{n}+i}}$; in $\mathcal{T}_{a_{\boldsymbol{n}+i}}$, the accepted customer joins server i, $i = 1, \ldots, M$. Let $L_{\boldsymbol{n}+i}$ be the number of instants in $\mathcal{T}_{a_{\boldsymbol{n}+i}}$. We have

$$L_{\boldsymbol{n},+} = \sum_{i=1}^{M} L_{\boldsymbol{n}+i}, \qquad L_{n,+} = \sum_{\boldsymbol{n} \in \mathcal{S}_n} \sum_{i=1}^{M} L_{\boldsymbol{n}+i}.$$

From the above definitions, we have

$$\begin{aligned}
\frac{1}{L_{n,+}} \sum_{k=1}^{L_{n,+}} \tilde{g}_{l_k} &= \frac{1}{L_{n,+}} \sum_{l_k \in \mathcal{T}_{a_{n,+}}} \tilde{g}_{l_k} \\
&= \frac{1}{L_{n,+}} \sum_{\boldsymbol{n} \in \mathcal{S}_n} \sum_{l_k \in \mathcal{T}_{a_{\boldsymbol{n},+}}} \tilde{g}_{l_k}
\end{aligned}$$

$$= \frac{1}{L_{n,+}} \sum_{n \in \mathcal{S}_n} \sum_{i=1}^{M} \sum_{l_k \in \mathcal{T}_{a_{n_{+i}}}} \tilde{g}_{l_k}$$

$$= \sum_{n \in \mathcal{S}_n} \left[\frac{L_{\boldsymbol{n},+}}{L_{n,+}} \sum_{i=1}^{M} \frac{L_{\boldsymbol{n}_{+i}}}{L_{\boldsymbol{n},+}} \left(\frac{1}{L_{\boldsymbol{n}_{+i}}} \sum_{k \in \mathcal{T}_{a_{\boldsymbol{n}_{+i}}}} \tilde{g}_{l_k} \right) \right]. \quad (8.45)$$

By definitions, when L is large enough, we have

$$\frac{1}{L_{\boldsymbol{n}_{+i}}} \sum_{k \in \mathcal{T}_{a_{\boldsymbol{n}_{+i}}}} \tilde{g}_{l_k} \approx g(\boldsymbol{n}_{+i}),$$

and

$$\frac{L_{\boldsymbol{n}_{+i}}}{L_{\boldsymbol{n},+}} \approx q_{0,i}, \qquad \frac{L_{\boldsymbol{n},+}}{L_{n,+}} \approx \pi(\boldsymbol{n}|n).$$

With these equations and (8.45), we obtain

$$\frac{1}{L_{n,+}} \sum_{k=1}^{L_{n,+}} g_{l_k} \approx g^d(a_{n,+}). \quad (8.46)$$

In (8.46), we ignored a constant $N\eta$. More precisely, we need to define

$$\tilde{g}_{l_k} = \sum_{l=1}^{N} \left[f(X_{l_k+l}) - \eta \right].$$

With this definition, we have

$$\lim_{N \to \infty} \lim_{L_{n,+} \to \infty} \frac{1}{L_{n,+}} \sum_{k=1}^{L_{n,+}} g_{l_k} = g^d(a_{n,+}), \qquad \text{w.p.1.} \quad (8.47)$$

Sample-path-based algorithms can be developed with (8.46) and (8.47). □

8.5 Applications and Examples

In this section, we first illustrate the applications of the event-based approach via two examples, and then discuss other possible applications in general.

8.5.1 Manufacturing

Consider a manufacturing system which produces M types of products. The system processes one product for a period of time and then switches to another product. When the system processes a product of type i, it operates as a Markov chain with different "phases" denoted as $\{1, 2, \ldots, N_i\}$,

$i = 1, 2, \ldots, M$. The system state is then denoted as (i, j), $i = 1, \ldots, M$, and $j = 1, \ldots, N_i$. In state (i, j), the system completes the product i with probability $\sigma_i(j)$, and continues with the same product i with probability $1 - \sigma_i(j)$. If the system continues with the same product, it chooses phase j' with probability $p_i(j, j')$ for $j' = 1, \ldots, N_i$. If the system completes processing on the product i, it picks up a product i' with probability $p(i, i')$, $i, i' = 1, \ldots, M$. Once the system picks up a product i', the initial phase is chosen according to the distribution $\rho_{i,i'}(j')$, $j' = 1, \ldots, N_{i'}$. For convenience, we will assume $p(i, i) = 0$. This is not conceptually restrictive. It is the same as the assumption of $q_{i,i} = 0$ in the admission control example.

Events

A single event in this system is a state transition $\langle (i, j), (i', j') \rangle$. The event representing that the system completes a product i is

$$a_{-i} := \{\langle (i, j), (i', j') \rangle,\ i' \neq i, i' = 1, \ldots, M, j = 1, \ldots, N_i, j' = 1, \ldots, N_{i'}\}.$$

The event representing that the system picks up a product i' is

$$a_{+i'} := \{\langle (i, j), (i', j') \rangle,\ i \neq i', i = 1, \ldots, M, j = 1, \ldots, N_i, j' = 1, \ldots, N_{i'}\}.$$

The event representing a shift from product i to i' is

$$a_{-i,+i'} := \{\langle (i, j), (i', j') \rangle,\ j = 1, \ldots, N_i, j' = 1, \ldots, N_{i'}\}, \qquad i' \neq i.$$

We have $a_{-i,+i'} = a_{-i} \cap a_{+i'}$. Finally, the event representing that the system enters phase j' of product i' from product i, $i \neq i'$, is

$$a_{-i;+i',j'} := \{\langle (i, j), (i', j') \rangle,\ j = 1, \ldots, N_i\}, \qquad i' \neq i.$$

When $i = i'$, we have

$$a_{i,j;j'} := \{\langle (i, j), (i, j') \rangle\},$$

which is a single event representing that the system continues with the same product i and enters phase j' from phase j.

The probabilities of events $\mathcal{P}(E_l = a_{-i,+i'})$ and $\mathcal{P}(E_l = a_{-i})$ can be obtained from the Markov model. But from the meaning of the events we can easily get the conditional probability

$$\mathcal{P}(a_{-i,+i'} | a_{-i}) = p(i, i'). \tag{8.48}$$

Now, suppose that we can control the probabilities $p(i, i')$, $i, i' = 1, 2, \ldots, M$, by actions. The control process can be described in terms of events: First, we observe whether an event a_{-i}, $i = 1, 2, \ldots, M$, occurs; if so, we apply an action which controls the conditional probabilities of events $a_{-i,+i'}$ that follow, see (8.48); afterwards, nature determines the event $a_{-i;+i',j'}$ with probability $\rho_{i,i'}(j')$ (i.e., the initial phase of product i' is determined by the nature of the product).

In this formulation, a_{-i}, $i = 1, 2, \ldots, M$, are the observable events; $a_{-i,+i'}$, $i, i' = 1, 2, \ldots, M$, are the controllable events; and $a_{-i;+i',j'}$ are the natural transition events. The logical relationship between these events are shown in Figure 8.12. The figure shows that a state transition from (i, j) to (i', j') is composed of three arrows: the solid arrow corresponding to the event a_{-i}, the dashed arrow corresponding to the event $a_{-i,+i'}$, and the dotted arrow corresponding to the event $a_{-i;+i',j'}$.

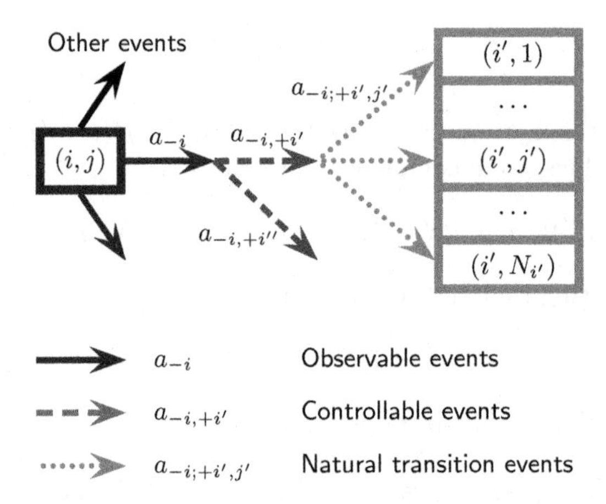

Fig. 8.12. The Logical Relation Among Three Types of Events in the Manufacturing System

The Event Space Decomposition

The space of single events is

$$\mathcal{E} := \{\langle (i,j), (i',j') \rangle, j = 1, \ldots, N_i, j' = 1, \ldots, N_{i'}, i, i' = 1, 2, \ldots, M\},$$

which can be decomposed. For observable events, the decomposition is

$$\mathcal{E} = \left\{\cup_{i=1}^{M} a_{-i}\right\} \cup b, \tag{8.49}$$

where we define

$$b = \cup_{i=1}^{M} \cup_{j=1}^{N_i} \cup_{j'=1}^{N_i} a_{i,j;j'}, \tag{8.50}$$

to represent all the transitions among the different phases in the same product. Event b is observable and when b happens, the system is not controllable (or we can do nothing). The information contained in a_{-i} is that a product i is completed. For controllable events, we have

$$\mathcal{E} = \left\{\cup_{j=1}^{M} a_{+j}\right\} \cup b = \left\{\cup_{i=1}^{M} \cup_{j=1, j \neq i}^{M} a_{-i,+j}\right\} \cup b. \tag{8.51}$$

For natural transition events, we have

$$\mathcal{E} = \left\{ \cup_{i=1}^{M} \cup_{i'=1}^{M} \cup_{j'=1}^{N_{i'}} a_{-i;+i',j'} \cup b \right\}. \tag{8.52}$$

The decomposition of the event space for the manufacturing system with $M = 3$ and $N_i = 3$, $i = 1, 2, 3$, according to (8.49), (8.51), and (8.52) is shown in Figure 8.13. The three upper horizontal rectangles represent events a_{-i}, $i = 1, 2, 3$, respectively; and the three vertical rectangles represent events a_{+i}, $i = 1, 2, 3$. The natural transition events $a_{-*;+*,j}$, $j = 1, 2, 3$, are illustrated by different grays; each such event consists of three narrow vertical rectangles. Event b consists of all the transitions among the different phases within the same product. When b is observed, there is only one control action: "do nothing", which is followed by a natural transition event expressed in (8.50). For clarity, this detailed composition of b is not shown in the figure because it only has a conceptual meaning and is not related to the control problem. For observable events, the decomposition is $\mathcal{E} = \cup_{i=1}^{3} a_{-i} \cup b$; for controllable events, we have $\mathcal{E} = \cup_{i=1}^{3} a_{+i} \cup b$; and for natural transition events, we have $\mathcal{E} = \cup_{j'=1}^{3} a_{-*;+*,j'} \cup b$, where $a_{-*;+*,j'} = \cup_{i=1}^{3} \cup_{i'=1,i'\neq i}^{3} a_{-i;+i',j'}$.

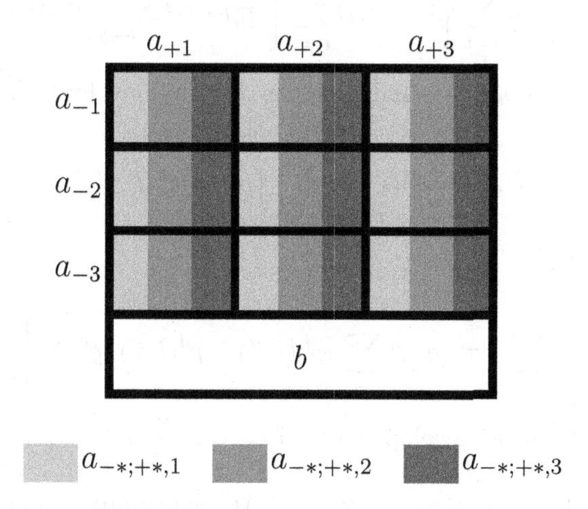

Fig. 8.13. The Decomposition of the Event Space $\mathcal{E}$ into Three Types of Events in the Manufacturing Example, with $M = 3$ and $N_i = 3$, $i = 1, 2, 3$

Performance Sensitivity Formulas

Denote the steady-state probability of a_{-i} as $\pi(a_{-i})$. The controllable events are $a_{+i'}$, $i' = 1, 2, \ldots, M$. When a_{-i} occurs, the conditional probability of $a_{+i'}$ is $\mathcal{P}(a_{+i'}|a_{-i}) = p(i, i')$, $i, i' = 1, 2, \ldots, M$. Given two policies $p^h(i, i')$ and $p^d(i, i')$, $i, i' = 1, 2, \ldots, M$, (8.19) becomes

$$\eta^h - \eta^d = \sum_{i=1}^{M} \left\{ \pi^h(a_{-i}) \sum_{i'=1}^{M} \left[p^h(i, i') - p^d(i, i') \right] g^{d,h}(i, i') \right\},$$

where

$$g^{d,h}(i,i') = \sum_{j=1}^{N_i} \sum_{j'=1}^{N_{i'}} \left\{ \pi^h \left[(i,j) | a_{-i} \right] \rho_{i,i'}(j') g^d(i',j') \right\}. \tag{8.53}$$

A close examination reveals that the condition for (8.38) does hold in this problem, and we can rewrite (8.53) as

$$g^{d,h}(i,i') := g^d(i,i') = \sum_{j'=1}^{N_{i'}} \left[\rho_{i,i'}(j') g^d(i',j') \right], \tag{8.54}$$

which depends only on the original system. Thus

$$\eta^h - \eta^d = \sum_{i=1}^{M} \left\{ \pi^h(a_{-i}) \sum_{i'=1}^{M} \left[p^h(i,i') - p^d(i,i') \right] g^d(i,i') \right\}. \tag{8.55}$$

The performance derivative (8.25), with $f_\theta = f$, becomes

$$\left. \frac{d\eta(\theta)}{d\theta} \right|_{\theta=\theta_1} = \sum_{i=1}^{M} \left\{ \pi_{\theta_1}(a_{-i}) \sum_{i'=1}^{M} \left[\left. \frac{d}{d\theta} p_\theta(i,i') \right|_{\theta=\theta_1} g_{\theta_1}(i,i') \right] \right\}.$$

When $\rho_{i,i'}(j') = \rho_{i'}(j')$ is independent of i, we have for all i,

$$g^d(i,i') = g^d(i') := \sum_{j'=1}^{N_{i'}} \left\{ \rho_{i'}(j') g^d(i',j') \right\}, \tag{8.56}$$

and

$$\eta^h - \eta^d = \sum_{i=1}^{M} \left\{ \pi^h(a_{-i}) \sum_{i'=1}^{M} \left[p^h(i,i') - p^d(i,i') \right] g^d(i') \right\}. \tag{8.57}$$

Policy iteration can be developed based on (8.55) and (8.54), or (8.57) and (8.56).

In this problem, there are $k_o \times k_c = M^2$ aggregated potentials $g^d(i,i')$, $i, i' = 1, 2, \ldots, M$. This is smaller than the number of states $\sum_{i=1}^{M} N_i$, if $M < N_i$, $i = 1, 2, \ldots, M$. If the natural transition probabilities $\rho_{i,i'}(j') = \rho_{i'}(j')$ do not depend on i, then (8.39) holds and there are only M aggregated potentials $g^d(i')$, $i' = 1, 2, \ldots, M$.

8.5.2 Service Rate Control

Consider a closed Jackson network consisting of N customers circulating among M servers (single-server stations). Let n_i be the number of customers in server i, $i = 1, 2, \ldots, M$, with $\sum_{i=1}^{M} n_i = N$. We assume that the service rate at server i may depend on the number of customers in the server n_i

(called a "load-dependent" service rate in the literature). Denote the service rate of server i with n_i customers as μ_{i,n_i}, and let τ_i be a customer's service requirement at server i. If n_i remains the same in the service period of a customer, then the service time of the customer is $s_i = \frac{\tau_i}{\mu_{i,n_i}}$. By convention, $\mu_{i,0} = 0$, $i = 1, \ldots, M$. We assume that all τ_i, $i = 1, \ldots, M$, are exponentially distributed. Without loss of generality, we may further assume that the means of all τ_i, $i = 1, 2, \ldots, M$, are equal to one. (We may adjust the service time by changing the service rate.)

After the completion of its service at server i, a customer moves to server j with probability $q_{i,j}$, $\sum_{j=1}^{M} q_{i,j} = 1$. Again, for simplicity, we assume $q_{i,i} = 0$ for all $i = 1, 2, \ldots, M$. For performance study, the network can be modelled as a continuous-time Markov process with state $\boldsymbol{n} = (n_1, n_2, \ldots, n_M)$. Let $\boldsymbol{n}_{-i,+j} = (n_1, \ldots, n_i - 1, \ldots, n_j + 1, \ldots, n_M)$ be a neighboring state of $\boldsymbol{n}$, $n_i \geq 1$. For example, in a three-server network if $\boldsymbol{n} = (2, 1, 3)$, then $\boldsymbol{n}_{-1,+2} = (1, 2, 3)$, or we write $(1, 2, 3) = (2, 1, 3)_{-1,+2}$. The transition rate from state $\boldsymbol{n}$ to state $\boldsymbol{n}_{-i,+j}$ in the continuous-time Markov process is $\mu_{i,n_i} q_{i,j}$.

The Embedded and Augmented Chain

Suppose that we may control the service rates μ_{i,n_i}, $n_i = 1, 2, \ldots, N$, $i = 1, 2, \ldots, M$, to optimize a long-run average performance with some reward function f. We first transfer the problem into a discrete-time model by the uniformization technique. We assume that the service rates are finite; thus, there is a $\mu > 0$ such that $0 < \mu_{i,n_i} < \mu$, $n_i = 1, 2, \ldots, N$, $i = 1, 2, \ldots, M$, for all possible service rates. It is well known (see Appendix C.3) that in a network with a single class of customers, a server with rate μ_{i,n_i} and routing probabilities $q_{i,j}$, $q_{i,i} = 0$, is equivalent to a server with service rate μ in which after the completion of its service, a customer will move back to the end of the queue of the same server (feedback) with probability $1 - p_{i,n_i}$ and will leave the server with probability p_{i,n_i}, $p_{i,n_i} = \frac{\mu_{i,n_i}}{\mu} < 1$. If the customer leaves the server, it will join server j with probability $q_{i,j}$. If we replace all the servers in the network with its equivalent server described above, we obtain a closed network in which all the service rates with any loads are equal to μ, and the effect of μ_{i,n_i} is reflected by the "feedback" probability $1 - p_{i,n_i}$. This network is uniformized and we may study its embedded chain, denoted as $\boldsymbol{X}$, which describes a discrete closed queueing network. In this discrete-time Markov chain, a state transition is $\langle \boldsymbol{n}, \boldsymbol{n}_{-i,+j} \rangle$, $i, j = 1, 2 \ldots, M$. With this discrete version, the problem becomes how to choose p_{i,n_i}, $i = 1, 2, \ldots, M$, in order to optimize a given performance measure. We assume that the performance function f is the same for all policies.

We first note that in this system, with the standard MDP formulation, a transition $\langle \boldsymbol{n}, \boldsymbol{n} \rangle$ may correspond to a few different events. For example, if at a time instant the system state is $\boldsymbol{n}$ and there is a customer at server i, $i = 1, 2, \ldots, M$, which completes its service and feeds back to the queue of the same server, then the transition is $\langle \boldsymbol{n}, \boldsymbol{n} \rangle$, regardless of which server the customer is in. Therefore, we need to modify the notation to make distinctions among these situations. To this end, we may artificially denote a load vector as $\boldsymbol{n} = \boldsymbol{n}_{-i,+i}$, $i = 1, 2, \ldots, M$. For example, we write $(2,1,3) = (2,1,3)_{-1,+1} = (2,1,3)_{-2,+2} = (2,1,3)_{-3,+3}$. With this notation, a transition $\langle \boldsymbol{n}, \boldsymbol{n}_{-i,+i} \rangle$ clearly represents the event that a customer feeds back at server i. Therefore, by using $\boldsymbol{n}_{-i,+i}$ to denote the system status if necessary, we can distinguish all events in the system. This modified notation only helps us to distinguish events and will not be needed for system states. For example, the potentials $g(\boldsymbol{n}_{-i,+i}) \equiv g(\boldsymbol{n})$ are the same for all i, $i = 1, 2, \ldots, M$.

Events

With the modified notation, the event representing a customer finishing a service at server i, including the feedback, can be represented by the set

$$a_{-i} := \{\langle \boldsymbol{n}, \boldsymbol{n}_{-i,+j} \rangle : \text{ all } \boldsymbol{n} \text{ with } n_i > 0, \text{ all } j = 1, 2, \ldots, M\}.$$

The event representing a customer joining server j, including the feedback, is

$$a_{+j} := \{\langle \boldsymbol{n}, \boldsymbol{n}_{-i,+j} \rangle : \text{ all } \boldsymbol{n} \text{ with } n_i > 0, \text{ all } i = 1, 2, \ldots, M\}.$$

The event representing a customer moving from server i to server j ($i = j$ implies a feedback at server i) is

$$a_{-i,+j} := \{\langle \boldsymbol{n}, \boldsymbol{n}_{-i,+j} \rangle : \text{ all } \boldsymbol{n} \text{ with } n_i > 0\}.$$

The event representing a customer feeding back to server i is

$$a_{-i,+i} := \{\langle \boldsymbol{n}, \boldsymbol{n}_{-i,+i} \rangle : \text{ all } \boldsymbol{n} \text{ with } n_i > 0\}.$$

The event representing a customer departing from server i is

$$a_{-i,*} := \cup_{j \neq i} a_{-i,+j} = \{\langle \boldsymbol{n}, \boldsymbol{n}_{-i,+j} \rangle : \text{ all } \boldsymbol{n} \text{ with } n_i > 0, \ j \neq i\}.$$

The event representing a customer finishing its service at server i when there are $n_i > 0$ customers in it is

$$a_{-i}(n_i) := \{\langle \boldsymbol{n}, \boldsymbol{n}_{-i,+j} \rangle : \text{ all } \boldsymbol{n} \text{ with } n_i, \text{ all } j = 1, 2, \ldots, M\}.$$

The event representing a customer moving from server i to server j when there are $n_i > 0$ customers in it before the departure is

$$a_{-i,+j}(n_i) := \{\langle \boldsymbol{n}, \boldsymbol{n}_{-i,+j} \rangle : \text{ all } \boldsymbol{n} \text{ with } n_i\}.$$

The event representing a customer feeding back to server i when there are n_i customers in it is

$$a_{-i,+i}(n_i) := \{\langle \boldsymbol{n}, \boldsymbol{n}_{-i,+i} \rangle : \text{ all } \boldsymbol{n} \text{ with } n_i\}.$$

The event representing a customer departing from server i when there are n_i customers in it is

$$a_{-i,*}(n_i) := \cup_{j \neq i} a_{-i,+j}(n_i) = \{\langle \boldsymbol{n}, \boldsymbol{n}_{-i,+j} \rangle : \text{ all } \boldsymbol{n} \text{ with } n_i, \; j \neq i\}.$$

In the system, the events $a_{-i}(n_i)$, $i = 1, 2, \ldots, M$ and $n_i = 1, 2, \ldots, N$, are the observable states. The controllable events are $a_{-i,+i}(n_i)$ and $a_{-i,*}(n_i)$. The conditional probabilities that we can control are

$$\mathcal{P}\left[a_{-i,*}(n_i) | a_{-i}(n_i)\right] = p_{i,n_i},$$

and

$$\mathcal{P}\left[a_{-i,+i}(n_i) | a_{-i}(n_i)\right] = 1 - p_{i,n_i}.$$

After the controllable event $a_{-i,*}(n_i)$ occurs, the nature chooses the destination server, denoted as j, with probability $q_{i,j}$ (i.e., $a_{-i,+j}(n_i)$ occurs). After the feedback event $a_{-i,+i}(n_i)$, nature has only one choice: to put the customer at the end of queue i.

Performance Sensitivity Formulas

Denote the steady-state probabilities of the observable events $a_{-i}(n_i)$ as $\pi\left[a_{-i}(n_i)\right]$, $n_i = 1, 2 \ldots, N$, $i = 1, 2, \ldots, M$. At any observable event $a_{-i}(n_i)$, there are two controllable events: departure $a_{-i,*}(n_i)$ and feedback $a_{-i,+i}(n_i)$. Consider two policies denoted as p^h_{i,n_i} and p^d_{i,n_i}, $n_i = 1, 2, \ldots, N$, $i = 1, 2, \ldots, M$. For convenience, instead of using the indexes k_1 and k_2, we directly use $a_{-i}(n_i)$ to indicate the observable events and use two letters, "fb" and "dp", to indicate the two controllable events, "feedback" and "departure", respectively. We have

$$\begin{aligned}
\eta^h - \eta^d = & \sum_{i=1}^{M} \sum_{n_i=1}^{N} \left\{ \pi^h \left[a_{-i}(n_i)\right] \left\{ \left(p^h_{i,n_i} - p^d_{i,n_i}\right) g^{d,h}\left[a_{-i}(n_i), fb\right] \right. \right. \\
& \left. \left. + \left[(1 - p^h_{i,n_i}) - (1 - p^d_{i,n_i})\right] g^{d,h}\left[a_{-i}(n_i), dp\right] \right\} \right\} \\
= & \sum_{i=1}^{M} \sum_{n_i=1}^{N} \left\{ \pi^h \left[a_{-i}(n_i)\right] \left(p^h_{i,n_i} - p^d_{i,n_i}\right) \right. \\
& \left. \left\{ g^{d,h}\left[a_{-i}(n_i), fb\right] - g^{d,h}\left[a_{-i}(n_i), dp\right] \right\} \right\}, \qquad (8.58)
\end{aligned}$$

where

$$g^{d,h}\left[a_{-i}(n_i), fb\right] = \sum_{\text{all } \boldsymbol{n}} \pi^h \left[\boldsymbol{n} | a_{-i}(n_i)\right] g^d(\boldsymbol{n}),$$

and

$$g^{d,h}\left[a_{-i}(n_i), dp\right] = \sum_{\text{all } \boldsymbol{n}} \sum_{j=1}^{M} \pi^h \left[\boldsymbol{n} | a_{-i}(n_i)\right] q_{i,j} g^d(\boldsymbol{n}_{-i,+j}).$$

In general, these two aggregated potentials depend on both the original system and the perturbed system. However, the performance derivatives depend only on the original system:

$$\left.\frac{d\eta}{d\theta}\right|_{\theta=\theta_1}=\sum_{i=1}^{M}\sum_{n_i=1}^{N}$$
$$\left\{\pi_{\theta_1}\left[a_{-i}(n_i)\right]\left.\frac{dp_{i,n_i}(\theta)}{d\theta}\right|_{\theta=\theta_1}\left\{g_{\theta_1}\left[a_{-i}(n_i),fb\right]-g_{\theta_1}\left[a_{-i}(n_i),dp\right]\right\}\right\},$$

with

$$g_{\theta_1}\left[a_{-i}(n_i),fb\right]=\sum_{\text{all }\boldsymbol{n}}\pi_{\theta_1}\left[\boldsymbol{n}|a_{-i}(n_i)\right]g_{\theta_1}(\boldsymbol{n}),$$

and

$$g_{\theta_1}\left[a_{-i}(n_i),dp\right]=\sum_{\text{all }\boldsymbol{n}}\sum_{j=1}^{M}\pi_{\theta_1}\left[\boldsymbol{n}|a_{-i}(n_i)\right]q_{i,j}g_{\theta_1}(\boldsymbol{n}_{-i,+j}).$$

Gradient-based optimization algorithms can be developed by using this performance derivative formula.

Because the aggregated potentials in (8.58) depend on both the original and the perturbed policies d and h, policy iteration cannot be developed based on this performance difference formula. However, all is not lost. Suppose that we change the service rates μ_{i,n_i} (this is equivalent to changing the feedback probabilities p_{i,n_i}), only for one server, denoted as k. That is, we set $p^h_{i,n_i}=p^d_{i,n_i}$ for all n_i and $i\neq k$, and only change the service rates of server k to $p^h_{k,n_k}\neq p^d_{k,n_k}$ for a particular k and all n_k. Then, by the product-form solution to this queueing network, we can prove

$$\pi^h\left[\boldsymbol{n}|a_{-k}(n_k)\right]=\pi^d\left[\boldsymbol{n}|a_{-k}(n_k)\right]. \tag{8.59}$$

Thus, we have

$$\eta^h-\eta^d=\sum_{n_k=1}^{N}\left\{\pi^h\left[a_{-k}(n_k)\right]\left(p^h_{k,n_k}-p^d_{k,n_k}\right)\left\{g^d\left[a_{-k}(n_k),fb\right]-g^d\left[a_{-k}(n_k),dp\right]\right\}\right\}, \tag{8.60}$$

with

$$g^d\left[a_{-k}(n_k),fb\right]=\sum_{\text{all }\boldsymbol{n}}\pi^d\left[\boldsymbol{n}|a_{-k}(n_k)\right]g^d(\boldsymbol{n}), \tag{8.61}$$

and

$$g^d\left[a_{-k}(n_k),dp\right]=\sum_{\text{all }\boldsymbol{n}}\sum_{j=1}^{M}\pi^d\left[\boldsymbol{n}|a_{-k}(n_k)\right]q_{k,j}g^d(\boldsymbol{n}_{-k,+j}). \tag{8.62}$$

Both depend only on the original system.

We may use (8.60)-(8.62) to develop a policy iteration algorithm that provides a local optimal policy.

Policy Iteration

Suppose that when event $a_{-i}(n_i)$, $i = 1, \ldots, M$ and $n_i = 1, \ldots, N$, happens, we can choose K_{i,n_i} different feedback probabilities in the set $\{p_{i,n_i}(1), p_{i,n_i}(2), \ldots, p_{i,n_i}(K_{i,n_i})\}$. We wish to find a policy that has the best performance.

The following policy iteration algorithm can be developed from (8.60), in which at each iteration the service rates of only one server are updated. At the kth iteration, the policy is denoted as

$$d_k := \left\{p_{i,n_i}^{d_k},\ i = 1, \ldots, M,\ n_i = 1, \ldots, N\right\}.$$

Algorithm 8.1. Policy Iteration Algorithm in Which Service Rates Are Updated One by One:

1. Guess an initial policy d_0, set $k := 0$.
2. (Policy evaluation) Estimate the aggregated potentials $g^{d_k}[a_{-i}(n_i), fb]$ and $g^{d_k}[a_{-i}(n_i), dp]$ for $i = 1, \ldots, M$ and $n_i = 1, \ldots, N$ defined in (8.61) and (8.62) on a sample path of the system under policy d_k .
3. (Policy improvement) Set $i := 1$, do
   (*a*) for $n_i = 1, \ldots, N$, do
      i. if $g^{d_k}[a_{-i}(n_i), fb] \geq g^{d_k}[a_{-i}(n_i), dp]$ then set
         $p_{i,n_i}^{d_{k+1}} = \max_{1 \leq l \leq K_{i,n_i}} p_{i,n_i}(l)$;
      ii. if $g^{d_k}[a_{-i}(n_i), fb] < g^{d_k}[a_{-i}(n_i), dp]$ then set
         $p_{i,n_i}^{d_{k+1}} = \min_{1 \leq l \leq K_{i,n_i}} p_{i,n_i}(l)$; and
   (*b*) if $p_{i,n_i}^{d_{k+1}} = p_{i,n_i}^{d_k}$ for all $n_i = 1, \ldots, N$,
         then if $i < M$ set $i := i + 1$ and go to step 3(a);
            if $i = M$, stop;
      otherwise go to step 4.
4. Set $k := k + 1$ and go to step 2.

Suppose that this algorithm stops at a policy $\hat{d}$, then at this point, no improvement can be made by changing the service rates of only one server, and the directional derivative along any direction in the policy space is non-positive. That is, this point is a local maximal point (see Problem 8.13).

In this problem, $k_o = M \times N$ and $k_c = 2$. The number of aggregated potentials is $k_o \times k_c = 2MN$, which is usually much smaller than the number of states, $\frac{(N+M-1)!}{N!(M-1)!}$.

From (8.60), if $g^d[a_{-k}(n_k), fb] - g^d[a_{-k}(n_k), dp] > 0$ (or < 0) for a particular n_k, then $\eta^h > \eta^d$ if $p_{k,n_k}^h > p_{k,n_k}^d$ (or $p_{k,n_k}^h < p_{k,n_k}^d$) and $p_{k,n}^h = p_{k,n}^d$ for all $n \neq n_k$. Also, if $g^d[a_{-k}(n_k), fb] - g^d[a_{-k}(n_k), dp] = 0$, then $\eta^h = \eta^d$ for any p^h_{k,n_k} if we have $p^h_{k,n} = p^d_{k,n}$ for all $n \neq n_k$. Therefore, we may conclude that there must be a local optimal policy that is at a corner of the feasible region in the policy space. In other words, at this local optimal policy the feedback rates at any server, denoted as $p^{\hat{d}}_{i,n_i}$, $i = 1, 2 \ldots, M$, are either the largest (when $g^{\hat{d}}[a_{-i}(n_i), fb] - g^{\hat{d}}[a_{-i}(n_i), dp] > 0$, with $\hat{d}$ denoting the optimal policy $p^{\hat{d}}_{i,n_i}$) or the smallest (when $g^{\hat{d}}[a_{-i}(n_i), fb] - g^{\hat{d}}[a_{-i}(n_i), dp] < 0$). Therefore, if at a local optimal policy $g^{\hat{d}}[a_{-k}(n_k), fb] \neq g^{\hat{d}}[a_{-k}(n_k), dp]$ for any i and n_i, this policy $\hat{d}$ must be at a corner.

8.5.3 General Applications

In the literature, there are a number of topics dealing with special features and computational issues in performance optimization of stochastic systems. We found that many of these topics may fit the event-based framework. A few examples are listed below. Further research is needed to formulate and to solve these problems with the event-based approach; here we will only briefly describe the problems.

A. *Multilevel control problem.* The two-level hierarchical control problem is similar to the manufacturing problem discussed in 8.5.1. We denote the high-level state (which changes with a slow time scale) as x and the lower-level state (which changes with a fast time scale) as y. The overall system state is (x, y). Any transition out from a high-level state x can be viewed as an observable event. The rest can be defined according to the specifics of the problem; a few of such specifics are described in [250].
B. *Time aggregation and options.* When control actions can only be applied to a subset of the state space $\mathcal{I} \in \mathcal{S}$, we can make the observable event as a transition out from a state in $\mathcal{I}$. As shown in Problem 8.8, the event-based approach may save computation in some special cases. Furthermore, the visits to the states in the subset $\mathcal{I}$ form an embedded Markov chain, and a "time aggregation" approach can be developed [67]. The time aggregation approach was first applied to estimate the performance gradients in [264]. Options [15] can be modelled in a similar way; see Problem 8.15.
C. *State aggregation.* Partition the state space into subsets $\mathcal{S} = \cup_{\text{all } l} \mathcal{S}_l$ and group the states in each subset as an aggregated state. The observation event can be defined as the transitions out from a subset.
D. *Singular perturbation.* In singular perturbed systems, the state space is partitioned into a few subsets, and the system stays in the same subset for a long period before it moves to other subsets. The observable events represent the transitions among different subsets.
E. *Queueing applications.* A large queueing network can be decomposed into a number of small sub-networks connected with each other. The customer transitions among different sub-networks can be viewed as observable events. Large communication networks are usually networks of sub-networks and can therefore be modelled by such queueing networks.

F. *Partially observable MDPs (POMDPs).* In POMDPs, the state x is not observable, but a random variable y, which is stochastically related to x, can be observed. We may use y, or the "belief state", or the "internal state" [159, 161, 188], to define the observable events and then apply the event-based approach. Problems 8.16 to 8.20 provide some introductory materials to POMDPs.

Once the above problems are precisely defined in the event-based framework, the performance difference and derivative formulas can be developed for them. With these formulas, we may aggregate potentials and develop gradient-based optimization algorithms. Under some conditions, we may also develop policy iteration algorithms for performance optimization. These are the future research topics.

Summary

In summary, with a sensitivity point of view of performance optimization, we proposed an event-based optimization approach. The approach is based on two formulas: The performance derivative formula for gradient-based optimization and the performance difference formula for policy iteration. The approach utilizes the special feature of a system, which is captured by the structure of events; performance potentials are aggregated according to events using these special features.

In performance derivative formulas, the aggregated potentials depend only on the original policy, and sample-path-based algorithms can be developed for gradient-based approaches. As shown in Section 8.3.4, under some special conditions, the aggregated potentials in performance difference formulas also depend only on the original policy, and therefore, policy iteration algorithms can be developed for these problems. However, the limitation of this approach is that the aggregated potentials in the performance difference formula may depend on the two policies under comparison. This prevents the aggregated potentials from being used in policy iteration. In this regards, the approach clearly indicates whether policy iteration can be implemented in optimization for a particular problem, and if not, why. In general, performance gradient-based optimization (with events) is more applicable than event-based policy iteration.

As shown in Section 8.4, estimating an aggregated potential on a sample path requires the same amount of effort as estimating a potential of a state. The number of aggregated potential depends on that of the events, which may scale to the system size, while the number of states usually grows exponentially with the system size. Thus, this approach may save significant computation.

The approach applies to many practical problems that do not fit well with the standard MDP formulation. It provides a unified view to a number of subjects including the POMDP problem. Applying the event-based approach to these different areas is a future research topic.

The performance difference formula (8.19) is derived in this chapter via the standard formula $\eta^h - \eta^d = \pi^h(P^h - P^d)g^d$. The structure of the formula (8.19) clearly reflects the special event-based feature of the system. It is important to note that the formula can be constructed by intuition using the potentials as building blocks, in a way similar to what we did in Section 2.1.1 for the performance derivative formula. In fact, (8.19) was first derived by construction, and was then verified by the standard formula, as we did on page 417. The construction approach utilizes the special features of the events; it is intuitive and leads directly to the final form. This construction approach can help us to quickly "guess" a final form of the difference formula for a particular problem, by using its special structure. The approach is also very flexible, and it can be applied, for example, to problems where the sizes of the state spaces of the two policies are different. We will introduce the construction method and discuss its advantages in the next chapter.

PROBLEMS

8.1. In a discrete-time birth-death process, the system moves from state n to $n+1$ with a birth probability p_n, $0 < p_n < 1$, $n = 0, 1, \ldots$, and from state n to $n-1$ with a death probability q_n, $0 < q_n < 1$, or stays in the same state n with probability $1 - p_n - q_n$, $p_n + q_n \leq 1$. When $n = 0$, the death probability is $q_0 = 0$. Define the events representing: A birth (denoted as event b), a death (denoted as event a), and no population change (denoted as event c), respectively.

8.2. In the discrete-time birth-death process considered in Problem 8.1, we set $p_n = p$ for all $n \geq 0$ and $q_n = q$ for all $n > 0$.

a. Find the steady-state probability $\pi(n)$, $n = 0, 1, \ldots$.
b. Suppose that we know a prior that at time l the system is at steady state, and we observed a birth event b at time $l-1$, what is the conditional distribution $\mathcal{P}(X_l|E_{l-1} = b)$?
c. What is the conditional probability of X_l if we have observed two consecutive birth events?
d. What if we observed a death event at steady state; i.e., what is $\mathcal{P}(X_l|E_{l-1} = a)$?

8.3. Please define the following events in Problem 4.2 (the state of the system is the stock in every evening before the order):

a. The retailer ordered more than the next day's demand.
b. The retailer ordered less than or equal to the next day's demand.
c. The retailer does not or may not have enough merchandise to sell.

8.4. We modify and restate the retailer's problem (Problem 4.2 and Problem 8.3) as follows: The system state x is the stock left every evening. We only consider threshold types of policies. That is, the state space $\{0, 1, 2, \ldots\}$ is divided into N intervals $I_1 := [0, n_1]$, $I_2 := [n_1 + 1, n_2]$, ..., $I_{N-1} := [n_{N-2}, n_{N-1}]$, $I_N := [n_{N-1}, \infty)$. The retailer is allowed to order M pieces of merchandise, or $2M$ pieces of merchandise, or not to order at all. Assume that we can only observe that the state is in a particular interval and cannot observe the state itself. Based on the observation $x \in I_i$, $i = 1, 2, \ldots, N$, the retailer may choose different probabilities of ordering 0, M, or $2M$ pieces of merchandise. Every day's demand on merchandise can be described by an integer random variable with distribution p_n, $n = 0, 1, \ldots$. Describe the three types of events: the observable, the controllable, and the natural transition events.

8.5. Suppose that the derivative $\left.\frac{df_\theta(i)}{d\theta}\right|_{\theta=0}$ is known. Derive a sample-path-based formula for the event-based average

$$\left.\frac{df_\theta(k_1)}{d\theta}\right|_{\theta=0} = \sum_{i=1}^{S} \left[\pi(i|e(k_1)) \left.\frac{df_\theta(i)}{d\theta}\right|_{\theta=0} \right].$$

8.6.* Derive equation (8.31), by using the arrival theorem and the steady state probabilities of the open Jackson networks.

8.7.* In Chapter 3, we derived a few sample-path-based direct-learning algorithms for the performance derivatives $\frac{d\eta_\delta}{d\delta}$, e.g., (3.30), (3.33), and (3.35). Derive similar direct-learning algorithms for the aggregated potentials (8.26) and the event-based performance derivatives by using formula (8.25).

8.8. Suppose that in an MDP problem, we can only apply control actions when the system is in a subset of state space, denoted as $\mathcal{I} \subset \mathcal{S}$. The observable events can be defined as when the system leaves any state $i \in \mathcal{I}$ or leaves the non-controllable set $\mathcal{S} - \mathcal{I}$.

a. Precisely define the observable events.
b. What are the controllable events?
c. Apply the event-based approach to this problem to derive the performance difference and derivative formulas for any two policies.

8.9. This problem is designed to further illustrate the ideas of natural transition events and potential aggregation. Compared with Example 8.1, there are two additional rooms 7 and 8, as shown in Figure 8.14. As in Example 8.1, after passing the green light on the right, the robot moves to the top; however, it will enter room 3 with probability u_1 and enter room 7 with probability u_2. Likewise, after passing the red light on the right, the robot will enter room 4 with probability v_1 and will enter room 8 with probability v_2.

a. Formulate this problem with the event-based approach, and define the observable, controllable, and natural transition events.

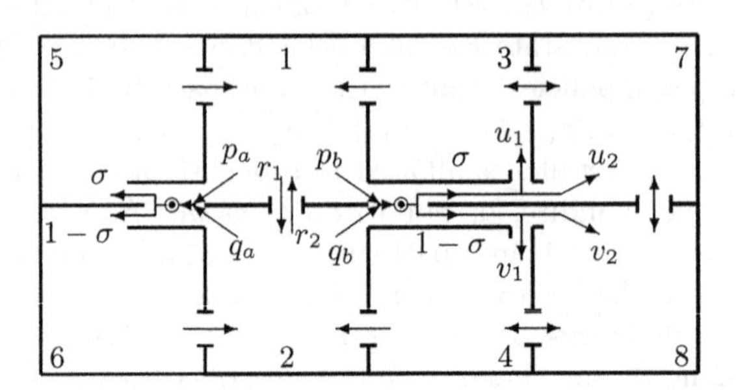

Fig. 8.14. Extended Moving Robot Problem

b. Derive the performance difference and derivative formulas.

8.10. A robot takes a random walk among four rooms, denoted as 1, 2, 3, and 4, as shown in Figure 8.15. When the robot is in room 3, in the next step, it moves to room 1. When it is in room 4, in the next step, it moves to rooms 2. There is a special passage that connects the four rooms as shown in the middle of Figure 8.15. When the robot is in room 1, in the next step, it moves to room 2 with probability $1-p_1$, or it tries to go through the passage with probability p_1. There is a traffic light, denoted as $\odot$ in the figure, in the passage. If it is red, the try fails and the robot moves back to room 1 in the next step; if the light is green, the robot passes the light and moves to room 3. The robot behaves in a similar way when it is in room 2: In the next step, it moves to room 1 with probability $1-p_2$, or it tries to go through the passage with probability p_2; and the robot moves back to room 2 in the next step if the light is red, and it passes the light and moves to room 4 in the next step, if the light is green. Denote the reward function as f.

Denote the probabilities of the light being green and red as σ and $1-\sigma$, respectively. We may control σ when we observe that the robot is in front of the light; we, however, do not know which room does the robot come from. Our goal is to determine the probability σ so that the long-run average reward is the maximum.

a. Formulate this problem with the event-based approach.
b. Derive the performance difference and derivative formulas.
c. Derive a policy iteration algorithm.
d. Show that one of the boundary points, $\sigma_{\max}$ or $\sigma_{\min}$, must be an optimal policy.

8.11.* Derive equation (8.59) by using the arrival theorem and the product-form solution to the steady-state probabilities of the closed Jackson networks.

8.12. Develop a sample-path-based estimation algorithm for $g\left[a_{-k}(n_k), fb\right]$ in (8.61) and $g\left[a_{-k}(n_k), dp\right]$ in (8.62).

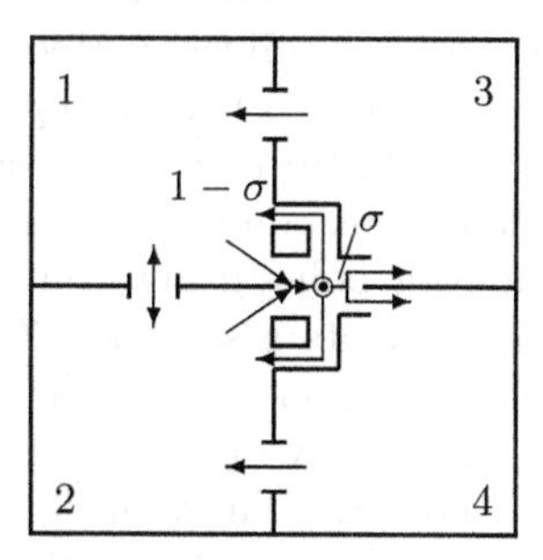

Fig. 8.15. The Moving Robot System in Problem 8.10

8.13. Consider the policy iteration Algorithm 8.1 in the service rate control problem in Section 8.5.2.

a. Prove that the algorithm reaches a local optimal policy in a finite number of iterations. Why is this policy not a "global" optimal policy?
b. If we change the policy improvement step to
   3. (Policy improvement) For $i = 1, \ldots, M$, do for $n_i = 1, \ldots, N$, do
      i. if $g^{d_k}[a_{-i}(n_i), fb] \geq g^{d_k}[a_{-i}(n_i), dp]$ then set

$$p_{i,n_i}^{d_{k+1}} = \max_{1 \leq l \leq K_{i,n_i}} p_{i,n_i}(l);$$

      ii. if $g^{d_k}[a_{-i}(n_i), fb] < g^{d_k}[a_{-i}(n_i), dp]$ then set

$$p_{i,n_i}^{d_{k+1}} = \min_{1 \leq l \leq K_{i,n_i}} p_{i,n_i}(l).$$

   If $p_{i,n_i}^{d_{k+1}} \neq p_{i,n_i}^{d_k}$ for any $i = 1, \ldots, M$, and $n_i = 1, \ldots, N$, then set $k := k + 1$ and go to step 2.
   If $p_{i,n_i}^{d_{k+1}} = p_{i,n_i}^{d_k}$ for all $i = 1, \ldots, M$, and $n_i = 1, \ldots, N$, then stop.

What is the difference that such a change makes to the algorithm? Will this algorithm stop? Will it reach a local optimal policy?

8.14. In the policy iteration algorithm in the service rate control problem in Section 8.5.2, at every iteration we always start from server 1, in the order of server 1, server 2, and so on, to update the service rates of the servers. We may try to update the service rates of the servers in a round-robin way: e.g., if server 1's service rates are updated at an iteration, then in the next iteration, we start from server 2 to update the service rates, etc. Develop such an algorithm and discuss its advantages, if any.

8.15.*(*Options* [15]) This problem is closely related to the time aggregation formulation. Consider a Markov process $\boldsymbol{X}$ with state space $\mathcal{S}$, and let $\mathcal{I} \subset \mathcal{S}$

be a subset of $\mathcal{S}$. As in Problem 8.8, we may define an observable event as when the system leaves a state in $\mathcal{I}$. Let us call the period between two consecutive events (i.e., two consecutive visits to $\mathcal{I}$) as an *option* period. The control problem is described as follows. There is a space, denoted as Π, of a finite number of options. An option corresponds to a state transition probability matrix in $\mathcal{S}$ (i.e., equivalent to a policy); however, it is only applied to an option period. After the system visits a state $i \in \mathcal{I}$, the system may evolve with any option in the available option set $\Pi_i \subseteq \Pi$ until it reaches the next state $j \in \mathcal{I}$. We assume that under any option in Π, the set $\mathcal{I}$ is reachable.

We consider randomized policies. Thus, in this problem for any given $i \in \mathcal{I}$ a policy specifies a probability distribution on Π_i. Precisely, let $o_{i,1}, o_{i,2}, \ldots, o_{i,n_i}$ be the options in Π_i. A policy d specifies a probability distribution $d(i) := (p_{i,1}, \ldots, p_{i,n_i})$. With policy d, after the system visits a state $i \in \mathcal{I}$, the system operates under option $o_{i,k}$ with probability $p_{i,k}$, $\sum_{k=1}^{n_i} p_{i,k} = 1$, for one option period until it visits another state $j \in \mathcal{I}$. Our goal is to determine the policy that achieves the maximum long-run average reward. For simplicity, we assume that the reward function f is the same for all policies.

The standard event-based optimization approach discussed in this chapter does not directly apply to this problem. However, the basic principles and concepts can be easily modified and extended to this problem. In the standard formulation, a control action taken at a time instant only affects the transition to the next state and therefore the controllable event can be defined. In the option problem, however, a control action affects the transitions in the entire option period.

Please formulate this problem in the framework of event-based optimization.

a. What are the observable events?
b. What are the aggregated potentials? (*Hint: it can be denoted as* $g(i, o_i)$.)
c. Derive the performance difference and derivative formulas for the two policies in the problem.
d. Comment on this event-option-based optimization approach.

8.16.* Consider a partially observable Markov chain with the structure shown in Figure 8.16. The fifteen states are grouped into three functionally similar groups. Group 1 consists of five states denoted as 1, 111, 112, 121, and 122; Group 2 consists of five states denoted as 2, 211, 212, 221, and 222; and Group 3 consists of five states denoted as 3, 311, 312, 321, and 322. States 1, 2, and 3 are completely observable. The other twelve states are grouped in to six super-states, denoted as 11, 12, 21, 22, 31, and 32 (which are denoted as dashed ovals in Figure 8.16); each consisting of two states as shown in the figure; e.g, the super-state 11 consists of two states 111 and 112. Only the super-states are observable; for example, after the system moves out from state 1, we only know that the system is in super-state 11 or 12 and cannot exactly know which state the system is in.

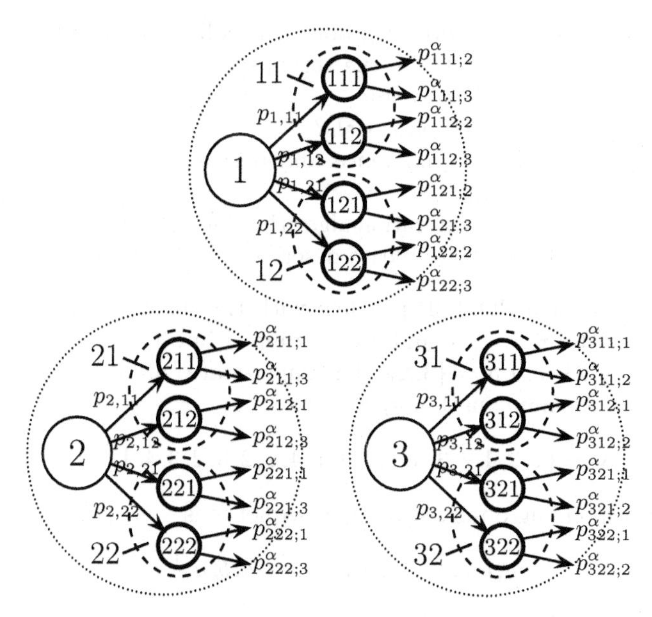

Fig. 8.16. The POMDP in Problem 8.16

The state transition probabilities are indicated in the figure. The transition probabilities from the observable states 1, 2, and 3, e.g., $p_{1,11}$, $p_{1,12}$, $p_{1,21}$, and $p_{1,22}$, are fixed and known. The transition probabilities from the non-observable states are controllable by actions and are denoted as $p^{\alpha}_{111;2}$, $p^{\alpha}_{111;3}$, $p^{\alpha}_{121;2}$, and $p^{\alpha}_{121;3}$, etc. The superscript α denotes any feasible action for the corresponding state. Because we cannot exactly determine the states in a super-state, we need to assume that the sets of the feasible actions for the two states in a super-state are the same. For example, if we know that the system is in super-state 12 and we decide to take action α, then this action must be feasible to both 121 and 122.

A sample path of the Markov chain may look like:

$$\boldsymbol{X} = \{2, 221, 1, 112, 2, 211, 3, 322, 1, 111, \ldots\},$$

with an observable state followed by a non-observable state and followed by another observable state, etc. The corresponding observed random sequence is

$$\boldsymbol{Y} = \{2, 22, 1, 11, 2, 21, 3, 32, 1, 11, \ldots\}.$$

Suppose that when the system is in state x, a random reward is received with $f(x)$ being its average. In addition, we assume that the function f is unknown but the reward at any time instant is observable. We consider the optimization of the long-run average reward. Please formulate this problem in the event-based formulation.

a. Explain that in this POMDP problem, a policy is a mapping from the space $\{11, 12, 21, 22, 31, 32\}$ to the action space.
b. What are the observable events?
c. What are the aggregated potentials?
d. Derive the performance difference and derivative formulas for the two policies in the problem.
e. Can we develop a policy iteration algorithm for the performance optimization of this problem? If so, please describe the algorithm in detail.

8.17.* We consider a POMDP problem with the structure shown in Figure 8.17. The four states 1, 2, 3, and 4 are grouped into two super-states a and b, with $a = \{1,2\}$ and $b = \{3,4\}$. The super-states are observable, but the states are not. A sample path may look like

$$\boldsymbol{X} = \{1,2,4,2,3,4,1,2,4,1,4,1,2,3,2,3,4,1,\ldots\},$$

and the corresponding observed random sequence is

$$\boldsymbol{Y} = \{a,a,b,a,b,b,a,a,b,a,b,a,a,b,a,b,b,a,\ldots\}. \tag{8.63}$$

Unlike in Problem 8.16 where a super-state completely determines the probability distribution of the system state, here the state distribution may depend on the history of the observed super-states. For example, if we observe two a's in a row, from Figure 8.17 we know that the system must be in state 2. Similarly, two consecutive observations of b lead to state 4. Therefore, after two consecutive a's or b's, denoted as (a,a) or (b,b), the system "regenerates" from state 2 or 4.

The regenerative property simplifies the analysis as well as the notation. Let x, or x', denote any sequence of super-states. Then an observation history (x', a, a, x) can be denoted as (a, a, x), and (x', b, b, x) can be denoted as (b, b, x), because the past history x' does not contain any extra information. Furthermore, if x is non-null, we may further omit the prefix (a,a) or (b,b) and simply denote them as x (if x starts with a, the prefix cannot be (a,a), and vise versa). Therefore, the observation histories correspond to the following cases: (a,a), (b,b), (a), (b), (a,b), (b,a), (a,b,a), (b,a,b), and (a,b,a,b), and so on. In general, the sequence alternates between a and b.

If at a time instant the observation history is $\boldsymbol{Y} = \{x', a, a, x\}$ or $\boldsymbol{Y} = \{x', b, b, x\}$, then x (or (a,a) and (b,b) if x is null) completely determines the probability distribution of the states at that time instant. For example, $x = (a)$ implies that the system just moves from state 4 to state 1 or 2. Thus, the state probability distribution is $p(3) = p(4) = 0$ and

$$p(1) = \frac{p^{\alpha}_{4,1}}{p^{\alpha}_{4,1} + p^{\alpha}_{4,2}} \quad \text{and} \quad p(2) = \frac{p^{\alpha}_{4,2}}{p^{\alpha}_{4,1} + p^{\alpha}_{4,2}}.$$

Therefore, in terms of the state probability distribution, the history $\boldsymbol{Y}$ in (8.63) is equivalent to

$$\{\bullet, 2, (b), (b,a), (b,a,b), 4, (a), 2, (b), (b,a),$$
$$(b,a,b), (b,a,b,a), 2, (b), (b,a), (b,a,b), 4, (a), \ldots, \}$$

where "$\bullet$" represents the initial probability.

Suppose that when the system is in state i, a random reward is received with $f(i)$ being its average. In addition, we assume that the function f is unknown but the reward at any time instant is observable. We consider the long-run average reward, its existence is guaranteed by the regenerative property.

a. Derive the state probability distributions corresponding to (b), (b,a) (a,b) and so on.
b. What are the observable events?
c. What are the aggregated potentials?
d. Derive the performance difference and derivative formulas for the two policies in the problem.
e. Can we develop a policy iteration algorithm for the performance optimization of this problem? If so, please describe the algorithm in details.

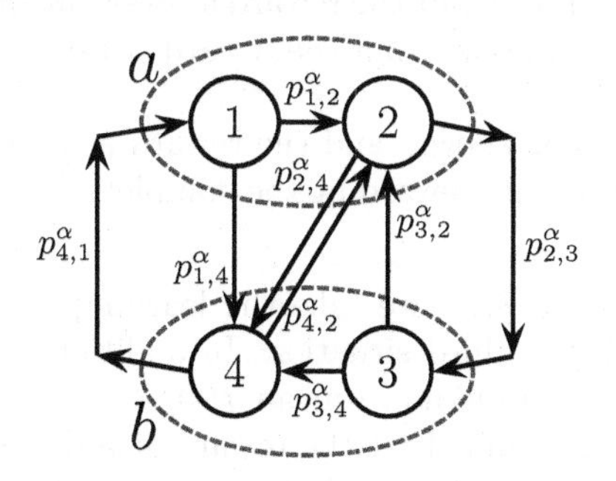

Fig. 8.17. The POMDP in Problem 8.17

8.18.* Suppose that in Problem 8.17, for simplicity we only take (a,a), (b,b), (a), (b), (a,b), and (b,a) as the possible events; i.e., we aggregate the histories according to the latest two super-states. For example, history (a,b,a,b,a) is aggregated into (b,a) and so on. In this formulation, the action taken at a time instant depends only on the last two super-states in the observation history.

a. Derive the performance difference formula.
b. Explain that in general, policy iteration cannot be developed from such a performance difference formula.
c. Do this problem and Problem 8.17 help you understand the POMDP problems?

8.19.* In Problem 8.17, if we can trace back from the observation history, we can estimate the earlier system state better. For example, as shown in (8.63), the observations from $l=0$ to $l=5$ are $\{a,a,b,a,b,b\}$. We know that at $l=1$, the system is at $X_1 = 2$, and the state probability distributions at times $l=3$,

$l = 4$, and $l = 5$ can be calculated, see Problem 8.17. However, at $l = 5$ we have observed (b, b) and therefore we know that the system state is $X_5 = 4$. Knowing so, from the structure shown in Figure 8.17, we may trace back to $l = 4$ and assert that $X_4 = 3$. Similarly, we can know for sure that $X_3 = 2$.

a. Update the state probability distribution at $l = 2$ after observing $\{a, a, b, a, b, b\}$ at $l = 5$.
b. Does this posterior information help in determining the optimal policy?

8.20.* In the analytical approach for MDPs, the reward function $f(i)$ is assumed to be known; and in the reinforcement learning approach, the reward at every time instant is assumed to be observable. In MDPs, the state i is assumed to be completely observable, therefore, both assumptions are equivalent. In POMDPs, however, the state is not observable; therefore, knowing the form of the function $f(i)$ does not allow us to know the actual reward at every instant. As such, we may have four different situations regarding the rewards:

i. The function f is known, and the reward at every instant is observable;
ii. The function f is known, but the reward at every instant is not observable;
iii. The function f is not known, but the reward at every instant is observable; and
vi. The function f is not known, and the reward at every instant is also not observable, but the final reward at the completion of each sample path is known.

In Problems 8.16 and 8.17, we take the learning approach and therefore we were dealing with the third situation. In addition, we assumed that the reward is random with an unknown mean $f(i)$.

Now, let us further assume that the reward at any state i is a fixed deterministic number $f(i)$, which is an unknown function but the reward received at every time instant is observable. In this case, we may determine the state i by the reward received. For instance, in Problem 8.16, when super-state 11 is observed, the system may be in either state 111 or 112 with probabilities $\sigma_{111} := \frac{p_{1,11}}{p_{1,11}+p_{1,12}}$ or $\sigma_{112} := \frac{p_{1,12}}{p_{1,11}+p_{1,12}}$, respectively. Thus, the reward received is either $f(111)$ or $f(112)$ with probabilities σ_{111} or σ_{112}, respectively. To be more precise, suppose $\sigma_{111} = 0.4$ and $\sigma_{112} = 0.6$. Let us observe the sample path for a while. We may find that when 11 is observed, we have 0.4 chance of obtaining a reward of 0 and 0.6 chance of obtaining a reward of 1. Then we can easily know that $f(111) = 0$ and $f(112) = 1$, and later on when 11 is observed, if we receive 0 we know that the state is 111 and if we receive 1, we know it is in 112. The following questions are for your further investigation:

a. Can we develop an algorithm from this reasoning?
b. Can we apply the same reasoning to Problem 8.17?
c. Can we apply the same reasoning to the general POMDPs?

Genius is one per cent inspiration, ninety-nine per cent perspiration.

Thomas A. Edison, American Inventor (1847 - 1931)

9 Constructing Sensitivity Formulas

9.1 Motivation

Although the two sensitivity formulas for Markov chains can be derived easily from the Poisson equation, this mathematical derivation lacks structural insights needed for deriving similar sensitivity formulas for other non-standard problems.

In Section 8.3.2, the performance difference formula for event-based policies is derived from the performance different formula for Markov chains. However, strictly speaking, the derivation in Section 8.3.2 is only a proof, or a verification of the formula, because it is "derived" with the particular formula in mind, which, in fact, was obtained first by the construction method introduced in this chapter.

As shown in Section 2.2, the performance derivative formula for Markov chains can be constructed using performance potentials by first principles. In this chapter, we show that the performance difference formulas can also be constructed, by first principles, using performance potentials as building blocks. We first develop such a construction approach for the standard MDP formulation, and then apply the same principles to construct performance difference formulas for other non-standard special systems.

To achieve our goal, we first study the structure of the performance difference. We show that we can build a sample path of a Markov chain under one policy (called a perturbed policy) by a sample path of the Markov chain under another policy (called an original policy) together with the sample paths that represent the performance potentials of the other policy (the original policy). The construction of the performance differences is then based on this decomposition of the sample path.

The difference in deriving the performance derivatives in Section 2.2 and the performance difference formulas in this chapter is as follows: In the former case, we can assume that the perturbations happen rarely because the parameter changes are infinitesimal; thus the effects of two perturbations on a sample path can be considered as decoupled. However, in the latter case, perturbations happen often and their effects are coupled. That is, before the effect of a perturbation on a sample path ends, another perturbation may occur. The extension of the construction approach from performance derivatives to performance differences represents a major effort in dealing with the coupling effect of two jumps that happen closely; in PA, this represents a change from infinitesimal perturbations to finite perturbations of the system parameters.

We then show that the construction approach can be applied to more general systems, including cases where the two systems under comparison have different state spaces but share a common subspace, systems with event-based policies, and parameterized systems. The construction approach is flexible and intuitive, and it utilizes the special features of a system. The performance potentials can be aggregated using the special features, and with this approach only the states that are affected by the parameter or structural changes need to be considered.

The approach clearly illustrates the physical meaning of potentials, or equivalently, realization factors, and their crucial role in performance optimization of discrete event dynamic systems. Compared with the traditional MDP solutions where the potentials of all states are treated as a vector (a solution to the Poisson equation) and considered as a group altogether, the construction approach offers a novel view to potentials by treating them separately and flexibly. Using this approach, we can flexibly derive formulas for performance sensitivities which may not be easy to conceive otherwise.

As shown in Chapter 8, the performance sensitivity formulas constructed play the central role in learning and optimization; gradient-based or, under some conditions, policy-iteration-based, optimization methods may be developed with these formulas.

To simplify notation, we use P and P' to denote the two policies and $\boldsymbol{X}$ and $\boldsymbol{X}'$ as their corresponding sample paths.

9.2 Markov Chains on the Same State Space

In this section, we show how we can "construct" performance difference formulas, by applying first principles to the case where the Markov chains under comparison are ergodic and defined on the same finite state space. Such a performance difference formula for ergodic Markov systems has been developed in Section 2.1.3 by using the Poisson equation.

We first give a brief review of terminologies and main concepts in sample-path-based sensitivity analysis. Following the terminology of PA, we refer to the two systems under comparison as the original system and the perturbed one, respectively, and their sample paths as the original path and the perturbed path, respectively. With this terminology, a Markov system under two different policies are referred to as two Markov chains. For performance derivatives with respect to a parameter θ, the original system is the one with θ, and the perturbed one, with $\theta + \Delta\theta$.

The main idea of perturbation analysis is as follows: Any change in a system parameter is reflected by "jumps" on the system's sample path; a jump here refers to the case where from the same state, the original path moves to state i, while the perturbed one moves to state j. The effect of such a single jump from i to j on the system performance can be measured by the *realization factor* $\gamma(i,j)$, which equals the difference of the performance potentials, $g(j)-g(i)$. Both $\gamma(i,j)$ and $g(i)$ can be estimated on sample paths. Finally, the performance derivative with respect to a system parameter can be decomposed into the sums of the effects of many single jumps, induced by the parameter change, on the system's sample path and can therefore be constructed by using realization factors or potentials as building blocks.

Now, we show how we can use realization factors, or potentials, as building blocks to construct the difference in the performance of the two Markov chains with two different transition probability matrixes P and P' (2.25).

Consider the simulation of two sample paths, one for the Markov chain with P and the other for the Markov chain with P'; both are defined on the same state space $\mathcal{S} = \{1, 2, \ldots, S\}$. We first assume that the two Markov chains have the same performance function, i.e., $f' = f$. As we see in Section 2.1.3, for $P_\delta = P + \delta(P' - P) = P + \delta\Delta P$, $\Delta P = P' - P$, with a small δ, if we use the same sequence of uniformly distributed $[0, 1)$ random variables to determine the state transitions for both chains, then the two sample paths $\boldsymbol{X}_\delta$ and $\boldsymbol{X}$ are very close, and the jumps happen rarely on $\boldsymbol{X}_\delta$ and their effects can be treated separately. However, when we consider $P' = P + \Delta P$ (corresponding to $\delta = 1$), two sample paths $\boldsymbol{X}'$ and $\boldsymbol{X}$ are completely different and the effects of the jumps may be coupled. That is, after a jump of $\boldsymbol{X}'$, another jump may occur before $\boldsymbol{X}'$ and $\boldsymbol{X}$ merge together, as we say it in PA.

The Effect of Two Coupled Perturbations

We first show how to determine the effect of two "coupled" jumps. In Figure 9.1, $A - B - W - C$ denotes the original sample path $\boldsymbol{X}$ (with P), and $A - B - G - E - H - D$ denotes the perturbed path $\boldsymbol{X}'$ (with P'). Suppose that the sample path $\boldsymbol{X}'$ (**not** $\boldsymbol{X}$!) is generated with a sequence of independent and uniformly distributed $[0, 1)$ random variables $\xi_1, \xi_2, \ldots, \xi_l, \ldots$, by using (2.2). We use a similar terminology as for the performance derivative problem: If with ξ_{l-1} and by (2.2), from X'_{l-1} (which is most likely different from X_{l-1})

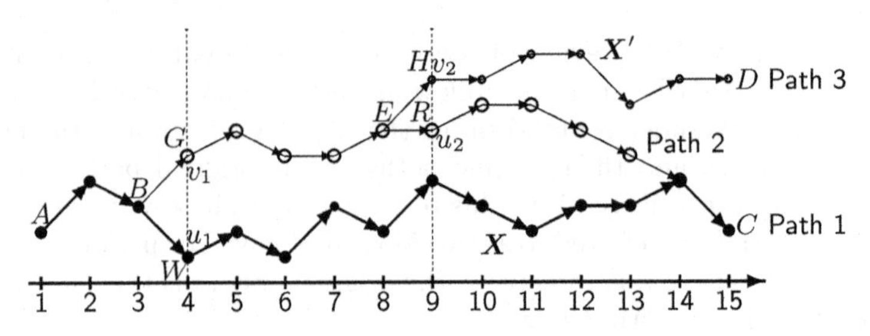

Fig. 9.1. The Effect of Two Perturbations

the Markov chain moves to the same state according to both P' and P, we say that the sample path $\boldsymbol{X}'$ does not have a jump at l. However, if with ξ_{l-1} and by (2.2), X'_{l-1} moves to state u according to P while it moves to state $X'_l = v$ according to P', we say that the perturbed chain $\boldsymbol{X}'$ has a jump (or a perturbation) from u to v at time l. Figure 9.1 illustrates two jumps on $\boldsymbol{X}'$, one at $l = 4$ from u_1 to v_1, the other at $l = 9$ from u_2 to v_2.

We cannot see u_2 on either $\boldsymbol{X}$ or $\boldsymbol{X}'$. In Figure 9.1, we have added a point R to illustrate the transition at $l = 8$ from point E to state u_2 if P (instead of P') were followed. Because there are no jumps on $G - E$, as we assumed, all the transitions on $G - E$ of $\boldsymbol{X}'$ are the same as if they followed the transition matrix P. Thus, all the transitions on $G - E - R$ are the same as if they followed the transition matrix P.

Now, after R, we add an auxiliary path that follows P until the auxiliary path merges with $\boldsymbol{X}$ at $l = 14$. Let us denote the path $A - B - W - C$ as path 1, $A - B - E - R - C$ as path 2, and $A - B - E - D$ as path 3. Path 1 follows P (hence it is $\boldsymbol{X}$), and path 3 follows P' (hence it is $\boldsymbol{X}'$) on which the segments $A - B$, $G - E$, and $H - D$ are the same as if they were generated according to P. With the auxiliary path, segment $G - E - R - C$ also follows P. Now it is clear that the effect of the jump from u_1 to v_1 can be measured by the performance on $G - E - R - C$ and $W - C$, and the effect of the jump from u_2 to v_2 can be measured by the performance on $H - D$ and $R - C$, all these segments follow the transition matrix P.

Let us make the above observation precise. We use superscripts to indicate the paths associated with a quantity. For example, the sequences of states on these three paths in the period from $l = 1$ to $l = 15$ are denoted as $X_1^{(1)}, X_2^{(1)}, \ldots$; $X_1^{(2)}, X_2^{(2)}, \ldots$; and $X_1^{(3)}, X_2^{(3)}, \ldots$; respectively. Of course, at some times the states may be the same on the three paths, or on any two of them, e.g., $X_1^{(1)} = X_1^{(2)} = X_1^{(3)}$, and $X_7^{(2)} = X_7^{(3)}$. It is clear from Figure 9.1 that for any $L > 9$, we have

$$F_L^{(2)} - F_L^{(1)} = \sum_{l=1}^{L} f(X_l^{(2)}) - \sum_{l=1}^{L} f(X_l^{(1)})$$

$$= \sum_{l=4}^{L} f(X_l^{(2)}) - \sum_{l=4}^{L} f(X_l^{(1)}),$$

$$F_L^{(3)} - F_L^{(2)} = \sum_{l=9}^{L} f(X_l^{(3)}) - \sum_{l=9}^{L} f(X_l^{(2)}).$$

Therefore, for any $L > 9$, we have

$$F_L^{(3)} - F_L^{(1)} = (F_L^{(3)} - F_L^{(2)}) + (F_L^{(2)} - F_L^{(1)})$$
$$= \left[\sum_{l=4}^{L} f(X_l^{(2)}) - \sum_{l=4}^{L} f(X_l^{(1)})\right] + \left[\sum_{l=9}^{L} f(X_l^{(3)}) - \sum_{l=9}^{L} f(X_l^{(2)})\right]. \quad (9.1)$$

Because both $G - E - C$ and $W - C$ follow transition probability matrix P, the expectation of $\sum_{l=4}^{L} f(X_l^{(2)}) - \sum_{l=4}^{L} f(X_l^{(1)})$ as $L \to \infty$ is $\gamma(u_1, v_1)$. Similarly, $H - D$ also follows transition probability matrix P. Because we assume that there are only two jumps on $\boldsymbol{X}'$, there will be no jumps after point H. Thus, both $H - D$ onwards and $R - C$ onwards follow P. Path 3 will eventually merge with path 2 (before or after $l = 14$), and the expectation of $\sum_{l=9}^{L} f(X_l^{(3)}) - \sum_{l=9}^{L} f(X_l^{(2)})$ as $L \to \infty$ is $\gamma(u_2, v_2)$. Finally, the effect of the two "coupled" jumps at $l = 4$ and $l = 9$ together is on average $\gamma(u_1, v_1) + \gamma(u_2, v_2)$.

The Effect of More Coupled Perturbations

The above observation for the two-jump case sheds light on the general case (Figure 9.2). Let P change to $P' = P + \Delta P$, and suppose that there are K jumps on $\boldsymbol{X}'$: $A - B - E - D - J$ (After the Kth jump, $\boldsymbol{X}'$ looks the same as if it followed P). Let $w_1, w_2, \dots, w_K$ be the instants at which jumps occur, and denote the jump at w_k as from state u_k to state v_k. Let $w_0 = 1$. In Figure 9.2, $K = 3$ and $w_1 = 4$, $w_2 = 9$, and $w_3 = 13$. By assumption, the segments from X'_{w_k} to $X'_{w_{k+1}-1}$, $k = 0, 1, 2, \dots, K$, look the same as if they were generated according to P. (It is possible that $w_{k+1} - 1 = w_k$, and in such cases the segment is null.) Similar to the two-jump case, we add an auxiliary path starting from each X'_{w_k-1} that exactly follows the transition matrix P (e.g., the paths $E - R - C$ and $D - M - C$ in Figure 9.2). Denote the auxiliary path starting from X'_{w_k-1} as AP_k. In Figure 9.2, AP_1 is $B - W - C$, which is the same as the original sample path $\boldsymbol{X}$, AP_2 is $E - R - C$, and AP_3 is $D - M - C$. Denote the path from $X'_{w_0} (= X'_1)$ to X'_{w_1-1} to AP_1 as path 1 and the path from X'_{w_0} via X'_{w_1-1}, X'_{w_2-1}, ... and X'_{w_k-1} and then to AP_k, as path k, etc., with path 1 being $\boldsymbol{X}$ (a sample path for P). We denote $\boldsymbol{X}'$ as path $K+1$. In Figure 9.2, path 1 is $A-B-W-C$, path 2 is $A-B-E-R-C$, path 3 is $A - B - E - D - M - C$, and path 4 is $A - B - E - D - J$.

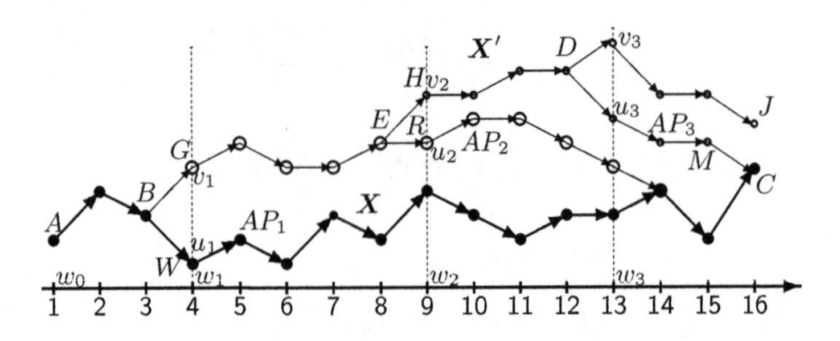

Fig. 9.2. The Potential Structure of a Sample Path

Applying the same reasoning as for the two-jump case illustrated in Figure 9.1, we can obtain an equation similar to (9.1) for the K-jump case (assuming $L > w_K$):

$$\begin{aligned} F'_L - F_L = & \left[\sum_{l=w_K}^{L} f(X_l^{(K+1)}) - \sum_{l=w_K}^{L} f(X_l^{(K)}) \right] \\ & + \left[\sum_{l=w_{K-1}}^{L} f(X_l^{(K)}) - \sum_{l=w_{K-1}}^{L} f(X_l^{(K-1)}) \right] \\ & + \cdots \\ & + \left[\sum_{l=w_1}^{L} f(X_l^{(2)}) - \sum_{l=w_1}^{L} f(X_l^{(1)}) \right] \end{aligned} \tag{9.2}$$

in which the expectation of $\sum_{l=w_k}^{L} f(X_l^{(k+1)}) - \sum_{l=w_k}^{L} f(X_l^{(k)})$ as $L \to \infty$ is $\gamma(u_k, v_k)$, $k = 1, 2, \ldots, K$.

Construction of the Perturbed Sample Path

Figure 9.2 illustrates that a sample path of a Markov chain with transition matrix P', $\boldsymbol{X}'$, can be decomposed into the sum of a sample path with transition matrix P, $\boldsymbol{X}$, and the differences of many segments, such as $G-E-R-C$ and $W-C$, $H-D-M-C$ and $R-C$, etc.; the effect of each such a difference can be measured on average by the perturbation realization factors of the Markov chain with P.

Pictorially, the perturbed sample path $\boldsymbol{X}'$ in Figure 9.2 starts from point A, then moves as if it followed the transition matrix P on the original path $\boldsymbol{X}$ until point B, at which $\boldsymbol{X}'$ jumps to point G according to P', and then moves as if it followed P again on another "original path" (with large circles) to point E, at which it jumps to point H according to P', and then moves as if it followed P again on another "original path" (with small circles) to point D, and so on.

The Performance Difference Formula

Now, let us observe a sample path of the perturbed system with transition probability matrix $P' = P + \Delta P$ for L transitions, denoted as $\{X'_0, \dots, X'_L\}$, $L >> 1$. Among the L states on this sample path, there are $L\pi'(i)$ states being state i on average. Suppose that after visiting state i, $\boldsymbol{X}'$ has a jump from u to v (we allow $u = v$). Denote the probability of a jump from u to v after visiting i as $p(u, v|i)$, with $\sum_{u=1}^{S} \sum_{v=1}^{S} p(u, v|i) = 1$. Then, on average, on the sample path there are $L\pi'(i)p(u, v|i)$ jumps from u to v that happen after visiting i. According to (9.2), each such jump has, on average, an effect of $\gamma(u, v)$ on F_L. Thus, on average, the total effect on F_L due to the change from P to P' is

$$\begin{aligned} E(F'_L - F_L) &\approx \sum_{i=1}^{S} \left[\sum_{u,v=1}^{S} L\pi'(i)p(u, v|i)\gamma(u, v) \right] \\ &= \sum_{i=1}^{S} \left\{ \sum_{u,v=1}^{S} L\pi'(i)p(u, v|i) \left[g(v) - g(u) \right] \right\}. \end{aligned} \tag{9.3}$$

Similar to (2.20) and (2.19), and by a probabilistic argument, we have $\sum_{u=1}^{S} p(u, v|i) = p'(v|i)$, and $\sum_{v=1}^{S} p(u, v|i) = p(u|i)$. Thus, (9.3) becomes

$$\begin{aligned} E(F'_L - F_L) &\approx \sum_{i=1}^{S} \left\{ L\pi'(i) \left[\sum_{j=1}^{S} \left[p'(j|i) - p(j|i)\right] g(j) \right] \right\} \\ &= L\pi' \left(P' - P\right) g = L\pi' \Delta P g, \end{aligned}$$

with $\Delta P = P' - P$. Finally, we have

$$\eta' - \eta = \lim_{L \to \infty} \frac{1}{L} E(F'_L - F_L) = \pi' \Delta P g. \tag{9.4}$$

This is the same as (2.25) derived directly from the Poisson equation.

Now, suppose that $f' \neq f$. Setting $h = f' - f$, we can easily obtain

$$\begin{aligned} \eta' - \eta &= \pi' f' - \pi f \\ &= (\pi' f - \pi f) + \pi'(f' - f) \\ &= \pi'(\Delta P g + h) \\ &= \pi' \left[(f' + P' g) - (f + P g) \right]. \end{aligned} \tag{9.5}$$

Example 9.1. Consider a one-dimensional random walk in a lattice consisting of $N + 1$ positions denoted as $0, 1, \dots, N$. At position $i = 1, 2, \dots, N - 1$, the walker moves to $i + 1$ or $i - 1$ with probabilities σ_i and $1 - \sigma_i$, respectively. Positions 0 and N are called walls. When the walker reaches wall 0, it stays there with probability α and leaves for position 1 with probability $1 - \alpha$.

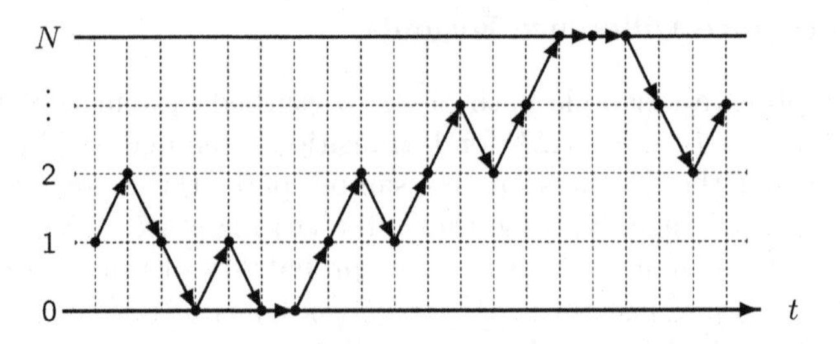

Fig. 9.3. A Sample Path of the Random Walk

When the walker reaches wall N, it stays there with probability β and leaves for position $N-1$ with probability $1-\beta$. The walker receives a reward $f(i)$ each time it is at position i. The long-run average reward is denoted by η. A sample path of this random walker is illustrated in Figure 9.3.

Now suppose that the transition probabilities α and β change to α' and β', respectively, and other transition probabilities remain the same. Let η' be the corresponding average reward and $\pi'(i)$ be the corresponding steady-state probability of the random walker being at position i, $i = 0, 1, \ldots, N$. To construct the performance difference $\eta' - \eta$, we investigate a perturbed sample path $\boldsymbol{X}'$ with α' and β' (without loss of generality, we assume $\alpha' > \alpha$, $\beta' > \beta$). It is clear that jumps occur only when the state is 0 or N. Let $g(i)$, $i = 0, 1, \ldots, N$, be the potentials of the original system. To construct a performance difference formula, we think as follows: Jumps occur on $\boldsymbol{X}'$ only when it visits the two walls 0 or N. In state 0, $\boldsymbol{X}'$ may jump from state 1 to 0 with probability $\alpha' - \alpha$, and in state N, it may jump from state $N-1$ to N with probability $\beta' - \beta$. Therefore, by construction, we have

$$\begin{aligned}\eta' - \eta = \pi'(0)\{(\alpha' - \alpha)[g(0) - g(1)]\} \\ +\pi'(N)\{(\beta' - \beta)[g(N) - g(N-1)]\}.\end{aligned} \tag{9.6}$$

Of course, this can also be derived from the standard formula (9.4) as follows:

$$\begin{aligned}\eta' - \eta = \pi'(0)\left\{(\alpha' - \alpha)\, g(0) + [(1-\alpha') - (1-\alpha)]\, g(1)\right\} \\ +\pi'(N)\left\{(\beta' - \beta)\, g(N) + [(1-\beta') - (1-\beta)]\, g(N-1)\right\}.\end{aligned}$$

However, the construction method (9.6) is more direct and reflects the meaning of the potentials. The potentials $g(0), g(1), g(N-1)$, and $g(N)$ can be estimated from a sample path of the original system. □

Policy Iteration

Next, we show, by an example, that the performance difference formula can be used to develop policy iteration procedure for performance optimization in problems that do not fit the standard MDP formulation.

Suppose that in Example 9.1, a number of actions can be taken in states 0 and N, resulting in different probabilities $\alpha_1, \alpha_2, \ldots, \alpha_M$ and $\beta_1, \beta_2, \ldots, \beta_M$, respectively. The standard policy iteration procedure for performance optimization in the policy space $\{(\alpha_i, \beta_j), i, j = 1, 2, \ldots, M\}$ can be derived from the performance difference formula (9.6). Moreover, (9.6) also holds if the actions taken in state 0 and N are correlated. For example, we may assume that when the retaining probability in state 0 is α_i, then that in state N must be β_i, $i = 1, 2, \ldots, M$. In the following example, we show that a policy iteration algorithm can be derived for this special non-standard problem as well.

Example 9.2. In this example, we assume that α_i and β_j cannot be chosen independently. Rather, they have to be chosen in pairs as (α_i, β_i), $i = 1, 2, \ldots, M$. We need to slightly re-structure the performance difference formula (9.6). First, we have

$$\pi'(0) = \pi'(0, N)\pi'\left[0|(0, N)\right],$$

and

$$\pi'(N) = \pi'(0, N)\pi'\left[N|(0, N)\right],$$

where $\pi'(0, N) = \pi'(0) + \pi'(N)$ is the steady-state probability that the random walker is at a wall, and $\pi'\left[0|(0, N)\right]$ and $\pi'\left[N|(0, N)\right]$ are the conditional steady-state probabilities of the random walker being at walls 0 and N, respectively, given that s/he is at a wall. We have

$$\pi'\left[0|(0, N)\right] = \frac{\pi'(0)}{\pi'(0) + \pi'(N)},$$
$$\pi'\left[N|(0, N)\right] = \frac{\pi'(N)}{\pi'(0) + \pi'(N)}.$$

These conditional distributions can be determined. We first observe that when state 0, or state N, is visited, the random walker stays at the wall for, on average, $\frac{1}{1-\alpha_i}$, or $\frac{1}{1-\beta_i}$, steps, respectively, before leaving it. Next, we construct another random walk which is the same as the random walk in this example except that when the walker hits one wall s/he is bounced back (to position 1 or $N-1$) immediately. This random walker behaves the same as if we shrink every period that the walker stays at the walls, 0 or N, in Figure 9.3 into one point. Denote the probabilities of states 0 and N in this constructed random walk as p_0 and p_N, respectively. Obviously, we have

$$\pi'\left[0|(0, N)\right] : \pi'\left[N|(0, N)\right] = \pi'(0) : \pi'(N) = \frac{p_0}{1-\alpha_i} : \frac{p_N}{1-\beta_i}.$$

Therefore, the performance difference formula (9.6) becomes

$$\eta' - \eta = \pi'(0, N)\kappa\left\{p_0 \frac{\alpha' - \alpha}{1-\alpha'}\left[g(0) - g(1)\right] + p_N \frac{\beta' - \beta}{1-\beta'}\left[g(N) - g(N-1)\right]\right\}, \tag{9.7}$$

where $\kappa > 0$ being a constant. The quantities p_0, p_N, $g(0)$, $g(1)$, $g(N-1)$, and $g(N)$ can be obtained analytically or by estimation from the original system. Furthermore, we can develop a policy iteration algorithm by using the difference formula (9.7):

1. Guess an initial policy $(\alpha_{i_0}, \beta_{i_0})$, $1 \le i_0 \le M$, set $k = 0$.
2. Run the system with $(\alpha_{i_k}, \beta_{i_k})$ and estimate its $p_0, p_N, g(0), g(1), g(N-1)$, and $g(N)$ on the sample path.
3. Choose

$$(\alpha_{i_{k+1}}, \beta_{i_{k+1}}) \in \arg \max_{(\alpha', \beta') \in \{(\alpha_i, \beta_i),\ i=1,2,\ldots,M\}}$$
$$\left\{ p_0 \frac{\alpha' - \alpha_{i_k}}{1-\alpha'} [g(0) - g(1)] + p_N \frac{\beta' - \beta_{i_k}}{1-\beta'} [g(N) - g(N-1)] \right\}.$$

4. If $(\alpha_{i_{k+1}}, \beta_{i_{k+1}}) = (\alpha_{i_k}, \beta_{i_k})$, then $(\alpha_{i_k}, \beta_{i_k})$ is the optimal policy; otherwise set $k := k+1$ and go to step 2. □

9.3 Event-Based Systems

9.3.1 Sample-Path Construction*

Now, we apply the construction approach to derive the difference formula (8.19). Recall that the conditional probabilities of the controllable events under the two policies in consideration are denoted as $p\left[e_c(k_2)|e_o(k_1)\right]$ and $p'\left[e_c(k_2)|e_o(k_1)\right]$, $k_1 = 1, 2, \ldots, k_o$, $k_2 = 1, 2, \ldots, k_c$, respectively; the natural transition probabilities are the same for both policies and are denoted as $p\left[e_t(k_3)|e_c(k_2), e_o(k_1)\right]$. For simplicity, we assume that these natural transition probabilities do not depend on the input state i, and that the performance functions for both policies are the same, i.e., $f'(i) = f(i)$, $i = 1, 2, \ldots, S$.

Now, we consider a sample path of the perturbed system (with transition probabilities $p'\left[e_c(k_2)|e_o(k_1)\right]$) denoted as $\boldsymbol{X}' = \{X'_0, X'_1, \ldots, X'_L\}$, with $L >> 1$, shown as path $A-C$, the top path in Figure 9.4. Assume that the perturbed sample path has reached its steady state. Let $\pi'(e_o(k_1))$ be the steady-state probability of the observable event $e_o(k_1)$ in the perturbed system. Among the L transitions $\langle X'_l, X'_{l+1}\rangle$, $l = 0, 1, \ldots, L-1$, on the sample path, there are approximately $L\pi'(e_o(k_1))$ transitions belonging to event $e_o(k_1)$. In Figure 9.4, the lower-case letters indicate the states the system is in; e.g., at time instants $l = 1$ and $l = 4$, the system is in state i and u, respectively. Let $\pi'(i|e_o(k_1))$ be the steady-state conditional probability of state i given that $e_o(k_1)$ is observed at an instant. Because

$$\sum_{\text{all } i \in I[e_o(k_1)]} \pi'[i|e_o(k_1)] = 1,$$

we can write

$$\pi'[e_o(k_1)] = \pi'[e_o(k_1)] \sum_{\text{all } i \in I[e_o(k_1)]} \pi'[i|e_o(k_1)].$$

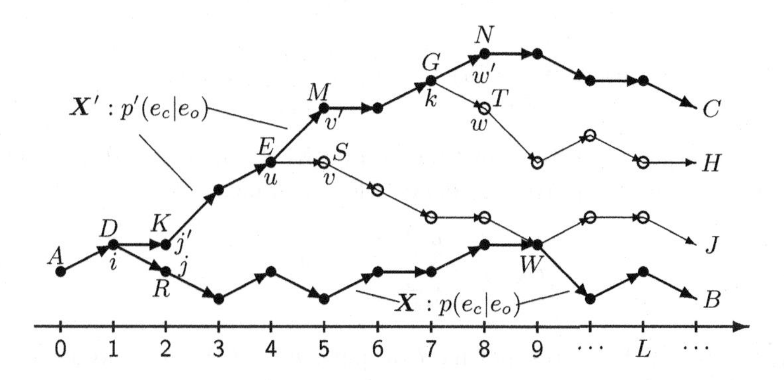

Fig. 9.4. The Performance Difference of Two Policies

In the perturbed system, from state i, given $i \in I\left[e_o(k_1)\right]$, the system will move to state $j' = O_i\left[e_o(k_1) \cap e_c(k_2) \cap e_t(k_3)\right]$ with probability

$$p'_{k_1}(j'|i) := p'\left[e_c(k_2)|e_o(k_1)\right] p\left[e_t(k_3)|e_c(k_2), e_o(k_1)\right], \quad k_2 \in \{1, \ldots, k_c\}, \ k_3 \in \{1, \ldots, k_t\}. \tag{9.8}$$

However, in the original system (i.e., with $p\left[e_c(k_2)|e_o(k_1)\right]$), from state i, given $i \in I\left[e_o(k_1)\right]$, the system will move to state $j = O_i\left[e_o(k_1) \cap e_c(k_2) \cap e_t(k_3)\right]$ with a different probability

$$p_{k_1}(j|i) := p\left[e_c(k_2)|e_o(k_1)\right] p\left[e_t(k_3)|e_c(k_2), e_o(k_1)\right], \quad k_2 \in \{1, \ldots, k_c\}, \ k_3 \in \{1, \ldots, k_t\}. \tag{9.9}$$

According to the construction method discussed in Section 9.2, we follow a perturbed sample path with transition probabilities $p'_{k_1}(j|i)$, $i, j \in \mathcal{S}$, shown as path $A - C$ in Figure 9.4. At every time instant $l = 0, 1, \ldots$, on the perturbed path, we determine, by using (2.2) with a uniformly distributed $[0, 1)$ random variable ξ_l, whether the state transition would be different if it followed transition probabilities $p_{k_1}(j|i)$, $i, j \in \mathcal{S}$, instead of $p'_{k_1}(j|i)$, $i, j \in \mathcal{S}$. The figure illustrates that the transitions in segments $A - D$, $K - E$, $M - G$, and $N - C$ happen to be the same for both p and p'. In other words, these segments can be viewed as a part of either an original path (with p) or a perturbed one (with p').

However, at Points D, E, and G, the transitions following p and p' are different. For example, at Point D, the perturbed system (with p') moves to state j', while the original system (with p) moves to state j. We say that a "jump" from j to j' occurs at $l = 2$. There is a jump from v to v' at $l = 5$ and a jump from w to w' at $l = 8$ in Figure 9.4.

Following the construction method, after each jump, we add an auxiliary path ($D-B$, $E-J$, or $G-H$) that follows the original transition probabilities p. Based on the construction, paths $A-B$, $K-E-J$, $M-G-H$, and $N-C$ can be viewed as original sample paths (following p). Let

$$F_L = \sum_{l=0}^{L-1} f(X_l), \qquad F'_L = \sum_{l=0}^{L-1} f(X'_l),$$

where X_l and X'_l are the states at time l on paths $A-B$ (original) and $A-C$ (perturbed), respectively. We also denote, for example,

$$F_{K-J} = \sum_{l=0}^{L-1} f(X_l^{K-J}),$$

with X_l^{K-J} being the states at time l on path $K-J$. Similar notations are used for other paths. We have

$$\begin{aligned}\Delta F_L &:= F'_L - F_L = F_{A-C} - F_{A-B} \\ &= (F_{A-C} - F_{A-H}) + (F_{A-H} - F_{A-J}) + (F_{A-J} - F_{A-B}) \\ &= (F_{N-C} - F_{T-H}) + (F_{M-H} - F_{S-J}) + (F_{K-J} - F_{R-B}).\end{aligned}$$

As shown in (2.6), when $L \to \infty$, the average of $F_{K-J} - F_{R-B}$, $E(F_{K-J} - F_{R-B})$, goes to $\gamma(j, j')$, and the average of $F_{M-H} - F_{S-J}$, $E(F_{M-H} - F_{S-J})$, goes to $\gamma(v, v')$, etc. In other words, each jump from j to j' contributes, on average, $\gamma(j, j')$ to the performance difference.

Let $p_{k_1}(j, j'|i)$ be the probability that after the system visits i, $i \in I[e_o(k_1)]$, a jump from j to j' occurs. We have

$$\sum_{j'=1}^{S} p_{k_1}(j, j'|i) = p_{k_1}(j|i), \qquad \sum_{j=1}^{S} p_{k_1}(j, j'|i) = p'_{k_1}(j'|i). \tag{9.10}$$

The number of the jumps from j to j' after visiting $i \in I[e_o(k_1)]$ when $e_o(k_1)$ is observed on the perturbed path is roughly

$$L\pi'[i, e_o(k_1)] p_{k_1}(j, j'|i),$$

where $\pi'[i, e_o(k_1)] = \pi'[e_o(k_1)]\pi'[i|e_o(k_1)]$ and L is the length of the perturbed path. Thus, adding up the contributions of all these jumps together over all possible observable events $e_o(k_1)$, $k_1 = 1, \ldots, k_o$, all possible input states i in each observable event, and all possible "jumps" from j to j' after visiting i, we have, on average and for a very large L, the following equation

$$\begin{aligned}E(\Delta F_L) &= E(F'_L) - E(F_L) \\ &\approx \sum_{k_1} \pi'[e_o(k_1)] \sum_{i \in I[e_o(k_1)]} \Bigg\{ \sum_{j \in O_i[e_o(k_1)]}\end{aligned}$$

$$\left.\sum_{j':\in O_i[e_o(k_1)]} \{L\pi'[i|e_o(k_1)]p_{k_1}(j,j'|i)\gamma(j,j')\}\right\}.$$

With $\eta = \lim_{L\to\infty} E(F_L)/L$, we have

$$\eta' - \eta = \sum_{k_1} \pi'[e_o(k_1)] \sum_{i\in I[e_o(k_1)]} \left\{ \sum_{j\in O_i[e_o(k_1)]} \right.$$

$$\left. \sum_{j':\in O_i[e_o(k_1)]} \{\pi'[i|e_o(k_1)]p_{k_1}(j,j'|i)\gamma(j,j')\} \right\}. \tag{9.11}$$

Finally, the difference formula (8.19) can be easily derived from (9.8), (9.9), (9.10), and (9.11), and $\gamma(j,j') = g(j') - g(j)$.

Remarks

Historically, the performance difference formula for event-based policies (8.19) was first discovered intuitively by the construction approach [59], and it was later verified analytically by using the standard performance difference formula with the MDP formulation. Furthermore, the event-based difference formula was first derived for the admission control problem before the general formulas were developed [60]. That is to say, the event-based approach was motivated first by practical problems and intuitive thinking. The construction approach helps us to derive the formulas from structural insights.

9.3.2 Parameterized Systems: An Example

In this section, we show that the construction approach can be applied in a more flexible way to obtain performance sensitivity formulas for systems that are parameterized (a special case of the event-based systems). Such a system has some special features: First, the value of a parameter usually determines the state transition probabilities of many states; thus the performance optimization problem cannot be viewed as a standard MDP problem. Second, in such a system, a parameter change usually affects only a part of the transition probability matrix, and the sample-path construction needs to be applied only to the parts of the sample path that are affected by the changes of the values of the parameters. For example, if a parameter change only affects the transition probabilities $p(i|k)$ and $p(j|k)$, $i, j, k \in \mathcal{S}$, then the jumps on the perturbed path are only from state i to j (or j to i) after visiting state k, and auxiliary paths need only to be added after such a jump following a visit to state k, and only the potentials $g(i)$ and $g(j)$ will appear in the performance sensitivity formulas. With the construction approach, we may obtain performance sensitivities by on-line estimation without estimating the potentials for all states and without knowing the entire transition probability matrix. Sample-path-based algorithms can be developed for these non-standard MDP problems. The ideas are illustrated by an example.

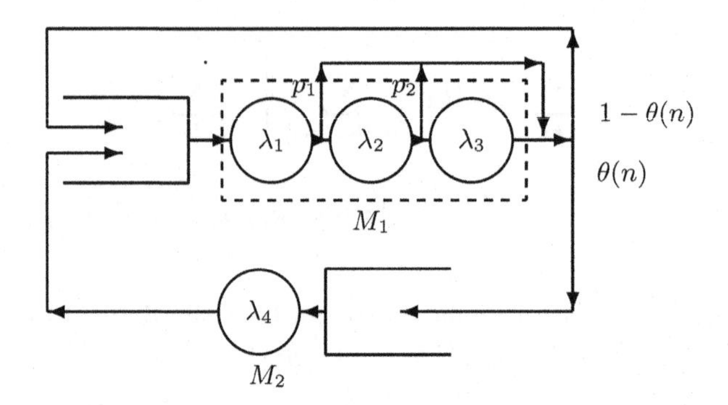

Fig. 9.5. A Manufacturing System

We consider a manufacturing system consisting of two machines and N work parts circulating between the two machines, as shown in Figure 9.5. Machine 1 (M_1) performs three operations (1, 2, and 3); their service times are exponentially distributed with rates λ_1, λ_2, and λ_3, respectively. Some parts only require operation 1, some others require two operations: 1 followed by 2, and the rest parts need to go through all three operations in the sequence of 1, 2, and 3. The probabilities that a part belongs to these three types are p_1, $(1-p_1)p_2$, $(1-p_1)(1-p_2)$, respectively, as shown in Figure 9.5. Machine 2 (M_2) has only one operation; its service time is exponential with rate λ_4. Each machine can only perform one operation on one part at a time. Machine 1 can also be viewed as having a Coxian distributed service time [90], see Figure A.1.

The system can be modelled as a Markov process with states denoted as (n, i), where $n = 0, 1, \ldots, N$ is the number of parts at M_1 and $i = 1, 2$, or 3 denotes the operation that M_1 is performing; when $n = 0$, we may simply write $(0, i) := (0)$. The state space is $\mathcal{S} = \{(n, i), n = 0, 1, \ldots, N, i = 1, 2, 3\}$. After the completion of its service at M_1, a part goes to M_2 with probability $\theta(n) \in [0, 1]$ (assumed to be independent of i, the operation that the part just finished), or immediately returns to M_1 with probability $1 - \theta(n)$. Let $f(n, i)$, $n = 1, 2, \ldots, N$, $i = 1, 2, 3$, and $f(0)$ be the reward function, and η_θ be the long-run average reward, which depends on the parameter vector $\theta := (\theta(1), \theta(2), \ldots, \theta(N))$.

We can use uniformization to convert this model to a discrete-time Markov chain so that we can apply the results for discrete-time Markov chains (A parallel theory can be developed for continuous time Markov processes, see [62] for performance derivatives of continuous-time Markov processes). The transition probability matrix of this Markov chain can be easily derived by using λ_i, $i = 1, 2, 3, 4$, and θ. We, however, will not do so because its explicit form is not needed in our approach.

Let us first construct the performance derivative formula. Following the same procedure as in Section 9.2, we consider a sample path with $L >> 1$ transitions. Let $\pi(n)$ be the steady-state probability of an event representing that a transition is due to a service completion of M_1 and meanwhile there are n customers in it. Now, suppose that $\theta(n)$ changes to $\theta(n) + \delta_n$, with $\delta_n > 0$ being a very small number, for a particular n, with $0 < n \leq N$. This change in the system parameter may cause "jumps" of the system state on the sample path from $(n, 1)$ to $(n-1, 1)$ (the original sample path moves to state $(n, 1)$ but the perturbed path moves to state $(n-1, 1)$). The conditional probability of such jumps at the service completion of M_1 with n customers in it is δ_n and the average effect of such a jump can be measured by the realization factor $\gamma\left[(n, 1), (n-1, 1)\right]$. The average number of transitions on the sample path corresponding to M_1's service completion with n customers in it is approximately $L\pi(n)$, and the average number of jumps after such service completions is approximately $L\pi(n)\delta_n$. Thus, we have

$$E(F_{\delta_n, L} - F_L) \approx L\pi(n)\delta_n\gamma\left[(n, 1), (n-1, 1)\right].$$

Therefore, we can obtain

$$\begin{aligned}\frac{d\eta_\theta}{d\theta(n)} &= \pi(n)\gamma\left[(n, 1), (n-1, 1)\right] \\ &= \pi(n)\left[g(n-1, 1) - g(n, 1)\right].\end{aligned} \tag{9.12}$$

Both $\pi(n)$ and $\gamma[(n, 1), (n-1, 1)]$ or $g(n-1)$ and $g(n)$ can be directly estimated on a sample path without knowing P and π. From (2.17), $\gamma\left[(n, 1), (n-1, 1)\right]$ can be estimated simply by averaging the sum of $f(X_l) - \eta_\theta$ over the periods starting from state $(n-1, 1)$ and ending in state $(n, 1)$.

From (9.12), to obtain $\frac{d\eta_\theta}{d\theta(n)}$ for a particular n, we need to estimate only $\gamma\left[(n, 1), (n-1, 1)\right]$. We can obtain $\frac{d\eta_\theta}{d\theta(n)}$ for all n if we estimate $g(n, 1)$ for all $n = 1, 2, \ldots, N$. These derivatives can be used in various optimization schemes (even with constraints). For example, if $\theta(n)$ changes to $\theta(n) + \alpha_n\delta$ for all n with a set of fixed α_n, then the performance derivative with respect to δ is

$$\begin{aligned}\frac{d\eta_\theta}{d\delta} &= \sum_{n=1}^{N} \alpha_n\pi(n)\gamma\left[(n, 1), (n-1, 1)\right] \\ &= \sum_{n=1}^{N} \alpha_n\pi(n)\left[g(n-1, 1) - g(n, 1)\right].\end{aligned}$$

Now, suppose that we have another system working under parameters $\theta'(n)$, $n = 1, 2, \ldots, N$. Following the same procedure as in Section 9.2, we may obtain the performance difference formula

$$\eta' - \eta = \sum_{n=1}^{N} \pi'(n) \left[\theta'(n) - \theta(n)\right] \gamma\left[(n,1),(n-1,1)\right]$$

$$= \sum_{n=1}^{N} \pi'(n) \left[\theta'(n) - \theta(n)\right] \left[g(n-1,1) - g(n,1)\right]. \qquad (9.13)$$

Note that $\pi'(n)$ denotes the steady-state probability of an event, not a particular state.

This example shows that by the performance sensitivity construction method, we can obtain the performance sensitivity formulas by analyzing a sample path. The formulas do not explicitly depend on the transition probability P and only involve the potentials $g(i)$ for some states (only $g(n,1)$, $n = 1, 2, \ldots, N$, are needed). Gradient-based optimization can be developed from (9.12) and policy iteration can be derived from (9.13). On-line algorithms can be implemented without knowing P and there is no need to estimate the potentials for all states. In this problem, the same value of $\theta(n)$ (viewed as an action, in the terminology of MDPs) determines the transition probabilities for different states $(n,1)$, $(n,2)$ and $(n,3)$; this violates the independent-action assumption, and the standard MDP formulation does not apply.

9.4 Markov Chains with Different State Spaces*

The advantage of the construction approach is that it can be applied flexibly to other general problems that may not fit the standard MDP framework. For these problems the Poisson equation may not exist. In this section, we apply this approach to two special problems to illustrate its flexibility.

9.4.1 One Is a Subspace of the Other*

We now construct the performance difference between two irreducible Markov chains defined on two different state spaces, with one being a subspace of the other. Let $\mathcal{S} = \{1, 2, \ldots, S\}$ and $\mathcal{S}' = \{1, 2, \ldots, S'\}$, $S < S'$, be two such state spaces, with $\mathcal{S} \subset \mathcal{S}'$. An example is the M/M/1/N and the M/M/1/N+1 queues. The state spaces of these two systems are $\mathcal{S} = \{0, 1, \ldots, N\}$ and $\mathcal{S}' = \{0, 1, \ldots, N+1\}$. Let P (an $S \times S$ matrix) and P' (an $S' \times S'$ matrix) be the (irreducible) transition probability matrices of the two Markov chains, respectively. We decompose P' into

$$P' = \begin{bmatrix} P'_1 & P'_{12} \\ P'_{21} & P'_2 \end{bmatrix},$$

where P'_1 is an $S \times S$ matrix corresponding to the states in $\mathcal{S}$.

Let $f = (f(1), \ldots, f(S))^T$ and $f' = (f'(1), \ldots, f'(S), f'(S+1), \ldots, f'(S'))^T$ be the two reward vectors, and let $\widetilde{f} = (f(1), \ldots, f(S), f'(S+1), \ldots, f'(S'))^T$.

For simplicity, we first assume $f'(i) = f(i)$, $i = 1, 2, \ldots, S$. Then we have $\widetilde{f} = f'$ and $\eta' = \pi' f' = \pi' \widetilde{f}$. As shown in (9.5), extension to $f(i) \neq f'(i)$, $i = 1, \ldots, S$, is straightforward.

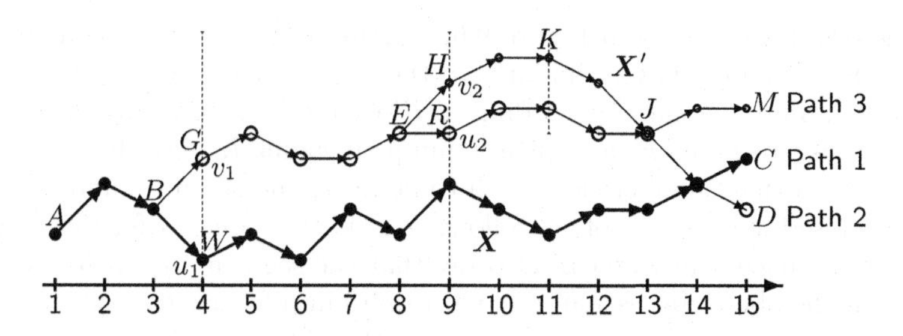

Fig. 9.6. Perturbation Between Two Different State Spaces

Construction of the Perturbed Sample Path

Figure 9.6 illustrates the sample paths of the two Markov chains, in which path 1 $(A - B - W - C)$ is viewed as the original path $\boldsymbol{X}$ with P, and path 3 $(A - B - G - E - H - J - M)$ is the perturbed path $\boldsymbol{X}'$ with P'. We assume that the initial states of both sample paths are the same (so they must start with a state in $\mathcal{S}$).

For any segment in which $\boldsymbol{X}'$ lies in $\mathcal{S}$, the situation is the same as the case of two Markov chains with the same state space, discussed in Section 9.2. For example, at $l = 4$, $\boldsymbol{X}'$ has a jump from state u_1 to state v_1. If both u_1 and v_1 are in $\mathcal{S}$, then after the jump, $\boldsymbol{X}'$ may follow the transition matrix P until at $l = 9$ it has another jump from u_2 to v_2. By adding an auxiliary path $E - R - J - D$ that follows P, we have a segment $G - E - R - J - D$, which is the same as if it follows transition matrix P. Thus, the jump at $l = 4$ from u_1 to v_1, $u_1, v_1 \in \mathcal{S}$, can be treated in the same way as in Section 9.2.

However, if a jump is from a state in $\mathcal{S}$ to a state outside of $\mathcal{S}$ (in $\mathcal{S}' - \mathcal{S}$), then after the jump, $\boldsymbol{X}'$ will follow the sub-matrix $[P'_{21}, P'_2]$ until it reaches $\mathcal{S}$ again. For example, in Figure 9.6 there is a jump from $u_2 \in \mathcal{S}$ to $v_2 \in \mathcal{S}' - \mathcal{S}$ at $l = 9$, and after the jump $\boldsymbol{X}'$ follows $[P'_{21}, P'_2]$ until at point K it reaches $\mathcal{S}$ again. (More precisely, $\boldsymbol{X}'$ follows $[P'_2]$ until at $l = 10$ it moves back into $\mathcal{S}$ at K following $[P'_{21}]$.) For simplicity, Figure 9.6 illustrates the situation where there is no jump on $\boldsymbol{X}'$ until it merges with path 2 at J. This does not lose any generality because once $\boldsymbol{X}'$ returns back to $\mathcal{S}$ at point K, if there is another jump on $\boldsymbol{X}'$ after K before it merges with path 2, we can always add an auxiliary path and denote it as $K - J$. In both cases, the effect of the jump from u_2 to v_2 can be measured by the difference between the two segments $H - J$ and $R - J$; $R - J$ follows P, while the first part of $H - J$, $H - K$, follows $[P'_{21}, P'_2]$, and the last part, $K - J$, follows P. (It in fact follows P'_1, but looks like following P if there is no jump, or if it is an auxiliary path.) That is, $H - J$ starts from a state in $\mathcal{S}' - \mathcal{S}$ and follows the following $S' \times S'$ transition matrix:

$$\widetilde{P} = \begin{bmatrix} P & 0 \\ P'_{21} & P'_2 \end{bmatrix}. \tag{9.14}$$

Pictorially, as shown in Figure 9.6, starting from a state in $\mathcal{S}$ at point A, $\boldsymbol{X}'$ follows P and behaves similarly to the original path $\boldsymbol{X}$ until point B, at which it jumps to point G according to P' then follows P again on the "large circles" path to point E, at which it jumps to point H according to P' then follows $\widetilde{P}$ on the "small circles" path to point J, and so on. In addition, since P is a closed sub-matrix of $\widetilde{P}$, following P is the same as following $\widetilde{P}$ in a large state space; thus, we can also say that the segment $R - J$ follows $\widetilde{P}$.

From the above discussion, the effect of a jump from u to v, $u \in \mathcal{S}, v \in \mathcal{S}'$, on the average reward can be measured by the difference of the two segments, both of them following the transition matrix $\widetilde{P}$. When $u, v \in \mathcal{S}$, the two paths (using an auxiliary path if necessary, e.g., $W - C$ and $G - E - R - J - D$) follow P, the top sub-matrix of $\widetilde{P}$; and when $u \in \mathcal{S}$ and $v \in \mathcal{S}' - \mathcal{S}$, one path (e.g., $R - J$) follows P and the other follows $\widetilde{P}$ (e.g., $H - K$ follows $[P'_{21}, P'_2]$ and $K - J$ follows P). Note that no jump may occur in $H - K$, which follows $[P'_{21}, P'_2]$ (no jump may occur after visiting a state in $\mathcal{S}' - \mathcal{S}$).

Potentials of $\widetilde{P}$

Now, we study the potentials of $\widetilde{P}$ with reward vector $\widetilde{f}$. Let $\widetilde{\Gamma} = [\widetilde{\gamma}(i,j)]_{i,j=1}^{S'}$ ($S' \times S'$) be its realization matrix, then

$$\widetilde{\Gamma} - \widetilde{P}\widetilde{\Gamma}\widetilde{P}^T = \widetilde{F}, \tag{9.15}$$

where $\widetilde{F} = e_{S'}\widetilde{f}^T - \widetilde{f}e_{S'}^T$. We have $\widetilde{\Gamma}^T = -\widetilde{\Gamma}$, $\widetilde{\Gamma} = e_{S'}\widetilde{g}^T - \widetilde{g}e_{S'}^T$, and $\widetilde{g}$ is the potential satisfying the Poisson equation

$$(I - \widetilde{P})\widetilde{g} + \widetilde{\eta}e_{S'} = \widetilde{f}. \tag{9.16}$$

Later, we will see that the solution to (9.15) or (9.16) exists. In the Markov chain with transition matrix $\widetilde{P}$, all recurrent states are in $\mathcal{S}$. Thus, its steady-state probability takes the form

$$\widetilde{\pi} = [\pi(1), \ldots, \pi(S), 0, \ldots, 0], \tag{9.17}$$

or in a vector form $\widetilde{\pi} = (\pi, 0)$, with $\pi = (\pi(1), \ldots, \pi(S))$ being the steady-state probability corresponding to P. We have $\widetilde{\pi}e_{S'} = \pi e_S = 1$, $\widetilde{\eta} = \widetilde{\pi}\widetilde{f} = \pi f = \eta$, and $\widetilde{\pi}(I - \widetilde{P}) = 0$. Left-multiplying both sides of (9.16) with $\widetilde{\pi}$, we indeed get

$$\widetilde{\eta} = \widetilde{\pi}\widetilde{f} = \pi f = \eta.$$

Recall that in (9.16) $\tilde{g}$ is determined only up to an additive constant. We may set the normalization condition for $\tilde{g}$ as

$$\tilde{\pi}\tilde{g} = \tilde{\eta}. \tag{9.18}$$

Denote

$$\tilde{g} = [g^T, g_2^T]^T, \tag{9.19}$$

where g is an S-dimensional vector. (9.18) becomes $\pi g = \eta$. Denote $\widetilde{\Gamma}$ as

$$\tilde{\Gamma} = \begin{bmatrix} \Gamma & \Gamma_{12} \\ \Gamma_{21} & \Gamma_2 \end{bmatrix}, \tag{9.20}$$

with Γ being an $S \times S$ matrix. Putting (9.14) and (9.20) into (9.15), we get the following three equations (9.21), (9.22), and (9.23)

$$\Gamma - P\Gamma P^T = F, \tag{9.21}$$

where $F = e_S f^T - f e_S^T$, which shows that the up-left sub-matrix of $\tilde{\Gamma}$ is the same as the realization factor matrix for the Markov chain with transition matrix P; and

$$\Gamma_{12} - (P\Gamma P_{21}^{'T} + P\Gamma_{12}P_2^{'T}) = F_{12}, \tag{9.22}$$

where $F_{12} = e_S f_2^T - f e_{S'-S}^T$ is an $S \times (S' - S)$ matrix, with $f_2 = (f'(S+1), \ldots, f'(S'))^T$ being an $(S' - S)$ dimensional vector, $\widetilde{f} = (f^T, f_2^T)^T$; and

$$\Gamma_2 - P_2'\Gamma_2 P_2^{'T} = F_2 + (P_{21}'\Gamma + P_2'\Gamma_{21})P_{21}^{'T} + P_{21}'\Gamma_{12}P_2^{'T}, \tag{9.23}$$

where $F_2 = e_{S'-S} f_2^T - f_2 e_{S'-S}^T$.

From (9.21), we have $\Gamma = e_S g^T - g e_S^T$ and g is the potential of P satisfying the Poisson equation $(I - P)g + \eta e_S = f$. Furthermore, from (9.19) and (9.20) we have

$$\Gamma_{12} = e_S g_2^T - g e_{S'-S}^T.$$

Substituting the above equation into (9.22) and using $Pe_S = e_S$, $P_2 e_{S'-S} + P_{21} e_S = e_{S'-S}$, we get

$$e_S g_2^T - e_S g_2^T P_2^{'T} = e_S f_2^T - f e_{S'-S}^T + g e_{S'-S}^T + e_S g^T P_{21}^{'T} - P g e_{S'-S}^T.$$

Left-multiplying the above equation with π and using $\pi e_S = 1$, $\pi P = \pi$, we obtain

$$g_2 - P_2' g_2 = f_2 - \eta e_{S'-S} + P_{21}' g, \tag{9.24}$$

or (the inverse $(1 - P_2')^{-1}$ exists for uni-chains in the form of (9.14), see Lemma B.2 in Appendix B or [216])

$$g_2 = (I - P_2')^{-1} (f_2 - \eta e_{S'-S} + P_{21}' g). \tag{9.25}$$

Thus, the $\tilde{g}$ defined in (9.19) satisfies (9.22). Let $\Gamma_2 = e_{S'-S} g_2^T - g_2 e_{S'-S}^T$. Substituting it into (9.23), we can verify that (9.24) also satisfies (9.23). We thus conclude that the solution to (9.15), or equivalently to (9.16), indeed exists as $\tilde{g} = (g^T, g_2^T)^T$, with g being the potential of P and g_2 satisfying (9.25).

Performance Difference and Derivative Formulas

After determining the realization factors $\tilde{\gamma}(i,j)$, $i,j = 1,\ldots,S$, and the potentials $\tilde{g}(j)$, $j = 1,\ldots,S$, which reflect the effect of a single jump on the average reward, the next step is to determine the total effect of all the jumps caused by the changes in the transition probability matrix as well as the change in the state space.

To this end, we observe a perturbed sample path on the state space $\mathcal{S}'$ following the transition probability matrix P' for L transitions $\{X'_1,\ldots,X'_L\}$, $L >> 1$. Recall that $p(u,v|i)$ is the probability that after visiting state i the chain jumps from u to v, $u = 1,\ldots,S$ and $v = 1,\ldots,S'$. We may follow the same procedure as described in Section 9.4.1 with only one exception: there would be jumps only when the system is in $\mathcal{S}$ (There is no jump between H and K in Figure 9.6). Therefore, corresponding to (9.3), we have

$$E(F'_L - F_L) \approx \sum_{i=1}^{S}\left\{\sum_{u=1}^{S}\sum_{v=1}^{S'}[L\pi'(i)p(u,v|i)\tilde{\gamma}(u,v)]\right\}, \tag{9.26}$$

where $\tilde{\gamma}(u,v) = \tilde{g}(v) - \tilde{g}(u)$ is the (u,v)th component of $\tilde{\Gamma}$. We have $\sum_{u=1}^{S} p(u,v|i) = p'(v|i)$ and $\sum_{v=1}^{S'} p(u,v|i) = p(u|i)$, for $i = 1,\ldots,S$. Thus,

$$E(F'_L - F_L) \approx \sum_{i=1}^{S}\left\{L\pi'(i)\left\{\sum_{v=1}^{S'}[p'(v|i)\tilde{g}(v)] - \sum_{u=1}^{S}[p(u|i)\tilde{g}(u)]\right\}\right\}.$$

Setting $p(u|i) = 0$ for $i = 1,\ldots,S$, $u = S+1,\ldots,S'$, we have

$$E(F'_L - F_L) \approx \sum_{i=1}^{S}\left\{L\pi'(i)\left\{\sum_{v=1}^{S'}[p'(v|i)\tilde{g}(v)] - \sum_{u=1}^{S'}[p(u|i)\tilde{g}(u)]\right\}\right\}.$$

Note that $\eta = \lim_{L\to\infty}\frac{1}{L}E(F_L) = \pi f$, and $\eta' = \lim_{L\to\infty}\frac{1}{L}E(F'_L) = \pi'\tilde{f}$. Finally, we have

The Average-Reward Difference Formula with $\mathcal{S} \subset \mathcal{S}'$:

$$\eta' - \eta = \pi'_{-}(\Delta P)_{-}\tilde{g}, \tag{9.27}$$

where

$$\pi'_{-} = [\pi'(1),\ldots,\pi'(S)],$$

and

$$(\Delta P)_{-} := [P'_1, P'_{12}] - [P, 0],$$

where "0" denotes an $S \times (S' - S)$ matrix in which all components are zeros.

In (9.27), we have $\widetilde{g} = (g^T, g_2^T)^T$, and g is determined by P. From (9.25), g_2 can be determined by g, f_2, P'_2, and P'_{21}. Thus, $\widetilde{g}$ is determined by P, P'_2, and P'_{21} and is independent of P'_1 and P'_{12}.

Now, suppose that $f'(i) \neq f(i)$, $i = 1, 2, \ldots, S$. (No value is assigned to $f(i)$, $i > S$.) Recall that $f' = (f'(1), \ldots, f'(S), f'(S+1), \ldots, f'(S'))^T$ and $\widetilde{f} = (f(1), \ldots, f(S), f'(S+1), \ldots, f'(S'))^T$. Let $f'_- = (f'(1), \ldots, f'(S))^T$ and $h_- = f'_- - f$. We have

$$\begin{aligned}\eta' - \eta &= \pi' f' - \pi f = (\pi' \widetilde{f} - \pi f) + \pi'(f' - \widetilde{f}) \\ &= \pi'_-(\Delta P)_- \tilde{g} + \pi'_- h_- = \pi'_-[(\Delta P)_- \tilde{g} + h_-].\end{aligned}$$

Therefore,

$$\eta' - \eta = \pi'_- \left[(f'_- + P'_- \tilde{g}) - (f + Pg)\right], \tag{9.28}$$

where $P'_- = [P'_1, P'_{12}]$.

The intuitively obtained equation (9.27) can be easily verified. Left-multiplying (9.16) by π' and using $\pi' e_{S'} = 1$, $\widetilde{\eta} = \widetilde{\pi}\widetilde{f} = \eta$, $\eta' = \pi'\widetilde{f}$, $\pi' P' = \pi'$, we have

$$\eta' - \eta = \pi' \widetilde{f} - \widetilde{\eta} = \pi'(I - \widetilde{P})\widetilde{g} = \pi'(P' - \widetilde{P})\widetilde{g} = \pi'_-(\Delta P)_- \tilde{g}.$$

For performance derivatives, we define

$$P_\delta = \widetilde{P} + \delta\left(P' - \widetilde{P}\right) = \begin{bmatrix} P + \delta(P'_1 - P) & \delta P'_{12} \\ P'_{21} & P'_2 \end{bmatrix}, \qquad 0 \le \delta \le 1.$$

Thus, $P_\delta|_{\delta=1} = P'$ and $P_\delta|_{\delta=0} = \tilde{P}$, which has the same steady-state performance as P. Subscript δ is added to the quantities associated with Markov chain P_δ. We discuss the case with $f(i) = f'(i)$, $i = 1, \ldots, S$. Applying (9.27) to P_δ and P, we obtain $\eta_\delta - \eta = \pi_{\delta -}(\Delta P)_- \delta \tilde{g}$, where $\pi_{\delta -} = (\pi_\delta(1), \ldots, \pi_\delta(S))$. Letting $\delta \to 0$, we get

$$\frac{d\eta_\delta}{d\delta} = \pi(\Delta P)_- \tilde{g}, \tag{9.29}$$

where $(\Delta P)_- := [P'_1, P'_{12}] - [P, 0]$.

Example 9.3. Suppose that in Example 9.1 the number of positions of the random walker increases from $N+1$ to $N+2$. The transition probability matrix of the random walk with $N+1$ positions takes the form

$$P = \begin{bmatrix} \alpha & 1-\alpha & 0 & 0 & \cdots & 0 & 0 & 0 \\ \sigma_1 & 0 & 1-\sigma_1 & 0 & \cdots & 0 & 0 & 0 \\ 0 & \sigma_2 & 0 & 1-\sigma_2 & \cdots & 0 & 0 & 0 \\ \cdots & \cdots & \cdots & \cdots & \cdots & \cdots & \cdots & \cdots \\ 0 & 0 & 0 & 0 & \cdots & \sigma_{N-1} & 0 & 1-\sigma_{N-1} \\ 0 & 0 & 0 & 0 & \cdots & 0 & 1-\beta & \beta \end{bmatrix}. \qquad (9.30)$$

The transition probability matrix P' for $N+2$ has the same form as (9.30) except that its size is larger by one. Comparing P and P', we can construct $\tilde{P}$ in (9.14). Indeed, we have

$$P'_{21} = [0, 0, \ldots, 1-\beta],$$

and $P'_2 = \beta$. Therefore, from (9.25) we have

$$g_2 = \frac{1}{1-\beta} \left[f(N+1) - \eta + (1-\beta) g(N) \right],$$

where $f(N+1)$ is the reward at the added position $N+1$ and $g(N)$ is the potential at position N in the original system with P.

Now, let us determine the performance difference. According to (9.27), we need to determine $(\Delta P)_{-}$. Comparing P and P', we find that $[P'_1, P'_{12}]$ and $[P, 0]$ differ only on their last rows. Thus, $(\Delta P)_{-}$ is zero everywhere expect its last row, which is

$$[0, \ldots, 0, \sigma_N - (1-\beta), -\beta, 1-\sigma_N],$$

in which the last three components are non-zeros. Finally, from (9.27) we have

$$\eta' - \eta = \pi'(N) \left\{ [\sigma_N - (1-\beta)] \, g(N-1) - \beta g(N) + (1-\sigma_N) g_2 \right\}.$$

Because $\pi'(N) > 0$, the performance improves (i.e., $\eta' > \eta$) if and only if

$$[\sigma_N - (1-\beta)] \, g(N-1) - \beta g(N) + (1-\sigma_N) g_2 > 0.$$

All the items g_2, $g(N-1)$, and $g(N)$ can be estimated from the original "smaller" system. Therefore, if the system parameters $f(N+1)$, β, and σ_N are known, we can determine, by analyzing a sample path of the random walk with $N+1$ positions, whether the performance improves when the number of positions increases from $N+1$ to $N+2$. □

An Alternative Approach: From Larger to Smaller

In the above analysis, we view the sample path associated with transition matrix P in the smaller state space $\mathcal{S}$ as the original one and that of the Markov chain with P' in the larger state space $\mathcal{S}'$ as the perturbed one. The role of the two sample paths can be reversed, i.e., we may view the sample path with P' as the original one and that with P as the perturbed one. In this way, we will follow the perturbed path and observe the jumps from states in $\mathcal{S}'$ to states in $\mathcal{S}$.

Recall $f' = (f'(1), \dots, f'(S), f'(S+1), \dots, f'(S'))^T$ and we denote $f'_- = (f'(1), \dots, f'(S))^T$. Again, for simplicity, we first assume $f(i) = f'(i)$, $i = 1, 2, \dots, S$. Thus, $f = f'_-$. Set $\eta' = \pi' f'$ and $\eta = \pi f = \pi f'_-$. The realization factor matrix $\Gamma' = [\gamma'(i,j)] = e_{S'} g'^T - g' e_{S'}^T$ is an $(S' \times S')$ matrix satisfying

$$\Gamma' - P' \Gamma' P'^T = F',$$

with $F' = e_{S'} f'^T - f' e_{S'}^T$, and g' satisfies

$$(I - P')g' + \eta' e_{S'} = f'. \tag{9.31}$$

To construct the performance difference, we consider a perturbed sample path with transition probability matrix P for $L >> 1$ transitions. Among them, on average $L\pi(i)$ transitions are from state i, $i \in \mathcal{S}$. After each visit to state i, a jump from $u \in \mathcal{S}'$ to $v \in \mathcal{S}$ happens with probability $p(u, v|i)$; and so on. Following the same reasoning as we did above, we eventually obtain

$$\eta - \eta' = \pi (\Delta P)'_- g', \tag{9.32}$$

where $(\Delta P)'_- = [P, 0] - [P'_1, P'_{12}] = -(\Delta P)_-$.

Equation (9.32) can be verified simply by left-multiplying both sides of (9.31) by $\widetilde{\pi} = (\pi(1), \pi(2), \dots, \pi(S), 0, \dots, 0) = (\pi, 0)$.

Next, suppose that $f'(i) \neq f(i)$, $i = 1, 2, \dots, S$. Set $f = (f(1), \dots, f(S))^T$ and $\eta = \pi f$. Recall $f'_- = (f'(1), \dots, f'(S))^T$ and define $h'_- = f - f'_- = -h_-$, we get

$$\begin{aligned} \eta - \eta' &= \pi f - \pi' f' = (\pi f'_- - \pi' f') + \pi (f - f'_-) \\ &= \pi [(\Delta P)'_- g' + h'_-]. \end{aligned}$$

Therefore,

The Average-Reward Difference Formula with $\mathcal{S} \subset \mathcal{S}'$:

$$\eta - \eta' = \pi \left[(f + P_+ g') - (f'_- + P'_- g') \right],$$

where $P_+ = [P, 0]$ and $P'_- = [P'_1, P'_{12}]$.

To study performance derivatives, we use $\tilde{P}$ defined in (9.14) again. Set

$$P_\delta = P' + \delta(\tilde{P} - P') = \begin{bmatrix} P'_1 + \delta(P - P'_1) & (1-\delta) P'_{12} \\ P'_{21} & P'_2 \end{bmatrix}, \qquad 0 \leq \delta \leq 1,$$

with $P_\delta|_{\delta=1} = \tilde{P}$, which has the same long-run average reward η as P, and $P_\delta|_{\delta=0} = P'$. In fact, in this case, the perturbed chain P_δ has the same state space as $\mathcal{S}'$. We can simply apply (9.32) and obtain $\eta_\delta - \eta' = \pi_{\delta-}(\Delta P)'_{-}\delta g'$. Therefore, letting $\delta \to 0$ we get

$$\left.\frac{d\eta_\delta}{d\delta}\right|_{\text{at } \eta'} = \pi'_{-}(\Delta P)'_{-}g', \tag{9.33}$$

where $\pi'_{-} = (\pi'(1), \ldots, \pi'(S))$.

Both g' and π'_{-} in (9.33) can be estimated with a single sample path of the Markov chain P'. Thus, when the original state space is larger, the performance derivative from a large state space to a small state space can be determined based on a sample path of the original Markov chain. However, in (9.29), $\widetilde{g}$ is determined by $\widetilde{P}$ in (9.14), which depends on $[P'_{21}, P'_2]$. Therefore, for performance derivatives from a small state space to a large state space, additional information is needed besides a sample path of the original Markov chain, which is in the small state space.

9.4.2 A More General Case

In this section, we study the case where two state spaces $\mathcal{S}$ and $\mathcal{S}'$ have a common subspace. For notational convenience, we order the states from the top downwards. Thus, we denote

$$\mathcal{S} = \{S, S-1, \ldots, 2, 1\}$$

and

$$\mathcal{S}' = \{S_0, \ldots, 1, 0, -1, \ldots, -S_{-1}+1, -S_{-1}\}.$$

The common subspace is $\mathcal{S}_0 = \{S_0, \ldots, 2, 1\}$, $S_0 < S$. $\mathcal{S}'$ has $S' = S_{-1} + S_0 + 1$ states. In addition, we denote $\mathcal{S}_1 = \{S, S-1, \ldots, S_0+1\}$ and $\mathcal{S}_{-1} = \{0, -1, \ldots, -S_{-1}\}$. We have $\mathcal{S} = \mathcal{S}_1 \cup \mathcal{S}_0$ and $\mathcal{S}' = \mathcal{S}_0 \cup \mathcal{S}_{-1}$. Let

$$\begin{aligned}\tilde{\mathcal{S}} &= \mathcal{S} \cup \mathcal{S}_{-1} = \mathcal{S}_1 \cup \mathcal{S}_0 \cup \mathcal{S}_{-1} \\ &= \{S, \ldots, 1, 0, -1, \ldots, -S_{-1}+1, -S_{-1}\},\end{aligned}$$

which has $\widetilde{S} := S + S_{-1} + 1$ states. The relations among the state spaces are illustrated in Figure 9.7.

The Augmented Matrix $\widetilde{P}$

Consider two Markov chains $\boldsymbol{X}$ and $\boldsymbol{X}'$ defined on $\mathcal{S}$ and $\mathcal{S}'$, respectively. Let P and P' be their transition probability matrices, respectively. Let

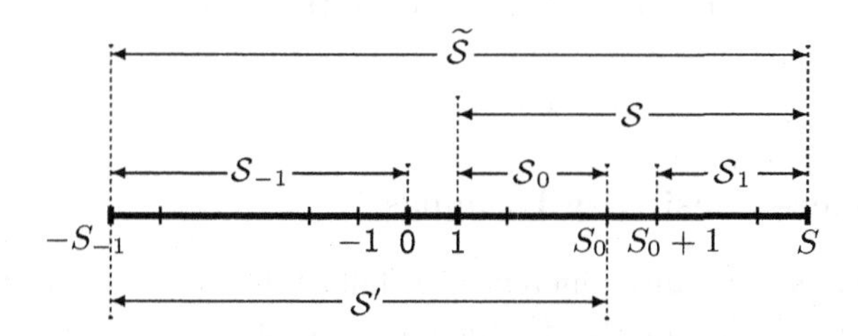

Fig. 9.7. Two Overlapped State Spaces $\mathcal{S}$ and $\mathcal{S}'$

$f = (f(S),\ldots,f(1))^T$ and $f' = (f'(S_0),\ldots,f'(1),f'(0),\ f'(-1),\ldots,f'(-S_{-1}))^T$ be the two reward vectors. Again, for simplicity we first assume that $f'(i) = f(i)$, for all $i \in \mathcal{S}_0$. Let π, η and π', η' be the steady-state probability vectors and the long-run average rewards of the two chains, respectively. Assume that both Markov chains are irreducible.

We decompose P and P' into

$$P = \begin{bmatrix} P_1 & P_{10} \\ P_{01} & P_0 \end{bmatrix},$$

and

$$P' = \begin{bmatrix} P'_0 & P'_{0-1} \\ P'_{-10} & P'_{-1} \end{bmatrix},$$

where $[P_1, P_{10}]$ in P corresponds to $\mathcal{S}_1$, $[P_{01}, P_0]$ in P and $\left[P'_0, P'_{0-1}\right]$ in P' correspond to subspace $\mathcal{S}_0$, and $\left[P'_{-10}, P'_{-1}\right]$ corresponds to $\mathcal{S}_{-1}$.

Without loss of generality, we assume that both $\boldsymbol{X}$ and $\boldsymbol{X}'$ start from the same state in $\mathcal{S}_0$. Following the same argument as in Section 9.4.1, we use Figure 9.6 to construct the performance difference, with path 1 $(A{-}B{-}W{-}C)$ as $\boldsymbol{X}$ and path 3 $(A-B-G-E-H-J-M)$ as $\boldsymbol{X}'$. We can see that by using auxiliary paths if necessary, the paths that determine the realization factors follow the transition matrix on the large state space $\tilde{\mathcal{S}}$ defined as (cf. (9.14)):

$$\tilde{P} = \begin{bmatrix} P_1 & P_{10} & 0 \\ P_{01} & P_0 & 0 \\ 0 & P'_{-10} & P'_{-1} \end{bmatrix}. \tag{9.34}$$

Similar to (9.15), we define

$$\tilde{\Gamma} - \tilde{P}\tilde{\Gamma}\tilde{P}^T = \tilde{F}, \tag{9.35}$$

where $\tilde{\Gamma}$ is an $\widetilde{S} \times \widetilde{S}$ realization matrix, $\tilde{F} = e_{\widetilde{S}}\tilde{f}^T - \tilde{f}e^T_{\widetilde{S}}$, and

$$\tilde{f} = [f(S),\ldots,f(1),f'(0),\ldots,f'(-S_{-1})]^T = [f^T, f^T_{-1}]^T,$$

with $f_{-1} = (f'(0), f'(-1),\ldots,f'(-S_{-1}))^T$.

We have $\tilde{\Gamma}^T = -\tilde{\Gamma}$, $\tilde{\Gamma} = e_{\tilde{S}}\tilde{g}^T - \tilde{g}e_{\tilde{S}}^T$, and $\tilde{g}$ is the potential vector satisfying

$$(I - \tilde{P})\tilde{g} + \tilde{\eta}e_{\tilde{S}} = \tilde{f}. \tag{9.36}$$

The Performance Sensitivity Formulas

To construct the performance difference, we follow the sample path $\boldsymbol{X}'$, which has transition probability matrix P', for L transitions $\{X_1', \ldots, X_L'\}$, $L >> 1$. Similar to (9.26), we have

$$E(F_L' - F_L) \approx \sum_{i \in \mathcal{S}_0} \left\{ \sum_{u \in \mathcal{S}} \sum_{v \in \mathcal{S}'} [L\pi'(i)p(u,v|i)\widetilde{\gamma}(u,v)] \right\},$$

where $\widetilde{\gamma}(u,v) = \tilde{g}(v) - \tilde{g}(u)$ is the (u,v)th component of $\widetilde{\Gamma}$. We have $\sum_{u \in \mathcal{S}} p(u,v|i) = p'(v|i)$ and $\sum_{v \in \mathcal{S}'} p(u,v|i) = p(u|i)$. Thus,

$$E(F_L' - F_L) \approx \sum_{i \in \mathcal{S}_0} \left\{ L\pi'(i) \left\{ \sum_{v \in \mathcal{S}'} [p'(v|i)\tilde{g}(v)] - \sum_{u \in \mathcal{S}} [p(u|i)\tilde{g}(u)] \right\} \right\}.$$

Setting $p(u|i) = 0$ for $i \in \mathcal{S}_0$, $u \in \mathcal{S}_{-1}$, and $p'(v|i) = 0$ for $i \in \mathcal{S}_0$, $v \in \mathcal{S}_1$, we have

$$E(F_L' - F_L) \approx \sum_{i \in \mathcal{S}_0} \left\{ L\pi'(i) \left\{ \sum_{v \in \widetilde{\mathcal{S}}} [p'(v|i)\tilde{g}(v)] - \sum_{u \in \widetilde{\mathcal{S}}} [p(u|i)\tilde{g}(u)] \right\} \right\}.$$

Finally, because $\eta = \lim_{L\to\infty} \frac{1}{L}E(F_L)$ and $\eta' = \lim_{L\to\infty} \frac{1}{L}E(F_L')$, we have

The Average-Reward Difference Formula with $\mathcal{S} \cap \mathcal{S}' = \mathcal{S}_0$:

$$\eta' - \eta = \lim_{L\to\infty} \frac{1}{L}E(F_L' - F_L) = \pi_-'(\Delta P)_- \tilde{g}, \tag{9.37}$$

where $\pi_-' = (\pi'(S_0), \ldots, \pi'(1))$ and

$$(\Delta P)_- := \left[0, P_0', P_{0-1}'\right] - \left[P_{01}, P_0, 0\right].$$

Now we consider the performance derivatives. To this end, we define

$$\tilde{P}_\delta = \begin{bmatrix} P_1 & P_{10} & 0 \\ (1-\delta)P_{01} & P_0 + \delta(P_0' - P_0) & \delta P_{0-1}' \\ 0 & P_{-10}' & P_{-1}' \end{bmatrix}.$$

Thus, $\tilde{P}_\delta|_{\delta=0} = \tilde{P}$, which has the same long-run average reward as P, and

$$\tilde{P}_\delta|_{\delta=1} = \begin{bmatrix} P_1 & P_{10} & 0 \\ 0 & P'_0 & P'_{0-1} \\ 0 & P'_{-10} & P'_{-1} \end{bmatrix} =: \tilde{P}',$$

which has the same long-run average reward as P'. We have $\tilde{P}_\delta = \tilde{P} + \delta(\tilde{P}' - \tilde{P})$. From (9.37), we have $\eta_\delta - \eta = \pi_{\delta-}(\Delta P)_- \delta \tilde{g}$, where $\pi_{\delta-} = (\pi_\delta(S), \ldots, \pi_\delta(1))$. Letting $\delta \to 0$, we get

$$\frac{d\eta_\delta}{d\delta} = \pi_-(\Delta P)_- \tilde{g},$$

where $\pi_- = (\pi(S_0), \ldots, \pi(1))$.

We have constructed performance sensitivity formulas with intuitions. The results need to be rigorously proved. First, we define an $S + S_{-1} + 1 = \widetilde{S}$ dimensional row vector

$$\tilde{\pi} = [\pi(S), \ldots, \pi(1), 0, \ldots, 0] = [\pi, 0].$$

The non-zero part is π with $\pi P = \pi$. Thus, $\tilde{\pi}\tilde{P} = \tilde{\pi}$. Left-multiplying both sides of (9.36) with $\tilde{\pi}$ and noting $\tilde{\pi} e_{\widetilde{S}} = 1$, we get

$$\tilde{\eta} = \tilde{\pi}\tilde{f} = \pi f = \eta.$$

Let $\tilde{\pi}' = (0, \pi')$ be an $S + S_{-1} + 1 = \widetilde{S}$ dimensional row vector with "0" denoting an $S - S_0 = S_1$ dimensional row vector with all components being zeros. We have $\tilde{\pi}'\tilde{P}' = \tilde{\pi}'$ and $\tilde{\pi}'\tilde{f} = \pi' f' = \eta'$. Left-multiplying both sides of (9.36) with $\tilde{\pi}'$, we get

$$\eta' - \eta = \tilde{\pi}'(I - \tilde{P})\tilde{g}.$$

Because $\pi' P' = P'$, we have

$$\tilde{\pi}'(I - \tilde{P}) = (0, \pi') \left\{ \begin{bmatrix} 0 & 0 & 0 \\ 0 & P'_0 & P'_{0-1} \\ 0 & P'_{-10} & P'_{-1} \end{bmatrix} - \begin{bmatrix} P_1 & P_{10} & 0 \\ P_{01} & P_0 & 0 \\ 0 & P'_{-10} & P'_{-1} \end{bmatrix} \right\}.$$

It is then clear that $\tilde{\pi}'(I - \tilde{P}) = \pi'_-(\Delta P)_-$, and (9.37) is proved.

Potentials of $\widetilde{P}$

Next, we calculate the potential vector $\tilde{g}$. Equations (9.17) to (9.25) hold with some minor modifications. For clarity, we repeat these equations with only notation changes to fit the general case. We rewrite (9.34) as

$$\tilde{P} = \begin{bmatrix} P & 0 \\ P'_{-1*} & P'_{-1} \end{bmatrix}, \tag{9.38}$$

where $P'_{-1*} = \left[0, P'_{-10}\right]$. Next, we denote

$$\tilde{g}^T = (g^T, g_{-1}^T)^T = (g_1^T, g_0^T, g_{-1}^T)^T, \tag{9.39}$$

where g is defined on $\mathcal{S}$, g_1, g_0, and g_{-1} are on $\mathcal{S}_1$, $\mathcal{S}_0$ and $\mathcal{S}_{-1}$, respectively. Denote $\tilde{\Gamma}$ as

$$\tilde{\Gamma} = \begin{bmatrix} \Gamma & \Gamma_{*-1} \\ \Gamma_{-1*} & \Gamma_{-1} \end{bmatrix}, \tag{9.40}$$

with Γ being an $S \times S$ matrix corresponding to $\mathcal{S}$. Putting (9.38) and (9.40) into (9.35), we get

$$\Gamma - P\Gamma P^T = F,$$

where $F = e_S f^T - f e_S^T$. This is the same for the Markov chain with transition matrix P; thus g in (9.39) satisfies the Poisson equation

$$(I - P)g + \eta e_S = f.$$

Finally, we can obtain (cf. (9.25))

$$\begin{aligned} g_{-1} &= (I - P'_{-1})^{-1} \left(f_{-1} - \eta e_{S_{-1}+1} + P'_{-1*} g\right) \\ &= (I - P'_{-1})^{-1} \left(f_{-1} - \eta e_{S_{-1}+1} + P'_{-10} g_0\right). \end{aligned}$$

9.5 Summary

We introduced an intuitive approach for constructing performance sensitivity formulas for Markov systems, including those that do not fit the standard MDP formulation. This approach utilizes the special structures of a system. The sensitivity formulas thus obtained have a clear meaning and may not be easily conceived otherwise. Only the potentials that are directly related to the changes in the system structure/parameters are involved in the formulas.

Specifically, we showed that a sample path of a Markov chain with transition probability matrix P' can be built upon a sample path of a Markov chain with transition probability matrix P on the same state space together with the segments that can be measured on average by the performance potentials of P (see Figure 9.2). We refer to this property as the *potential structure* of a sample path. We showed that this structure allows us to construct performance sensitivities, both performance derivatives and performances differences, by first principles with sample-path-based arguments. Based on this potential structure of a sample path, performance potentials, or performance realization factors, can be used as building blocks in the construction of performance sensitivities, and they can be estimated on sample paths of the Markov chain with transition probability matrix P.

The construction approach can be extended and applied flexibly to many problems with special non-standard features. For example, we can construct the performance sensitivity formulas for two Markov chains with different (but with some overlap) state spaces. When the original system has a larger state space, which contains the state space of the perturbed system as a subspace, the potentials used in the sensitivity formulas can be estimated on a single sample path of the original system. This is in the same spirit as perturbation analysis: one can obtain the performance sensitivity by analyzing only the original system. When the two systems have two different state spaces, or the perturbed system has a larger state space, the potentials can be estimated on sample paths with an augmented transition probability matrix (see (9.14) and (9.34)). For these systems, efficient methods in estimating potentials based on reinforcement learning should be developed. We also gave an example showing that in general we only need to estimate the potentials of the states that are related to the system parameter changes.

As shown in Figure 8.1, the performance sensitivity formulas serve as the bases for learning and optimization. Thus, the sensitivity formulas obtained by construction can be used to develop learning and optimization approaches for systems with special features. This is the topic of Chapter 8.

PROBLEMS

9.1. As explained in Section 9.2, in the performance difference construction approach shown in Figure 9.1, the construction is done in the following way:

i. On the perturbed sample path $A-B-E-D$, we use the same random variable ξ_l to determine whether or not there is a jump at each transition l; and
ii. when a jump is identified, we use another independent sequence of random variables to generate an auxiliary path, e.g., $W-C$.

While the above construction is convenient, it is not necessary. Convince yourself that we can derive the same results as those in Section 9.2 if we construct the sample paths in the following way:

i. On the perturbed sample path $A-B-E-D$, we use two independent random variable ξ_l and ξ_l' to determine whether or not there is a jump at each transition l; i.e., a jump from j to j' occurs if after visiting state i, the system moves to state j according to ξ_l and P, but it moves to state j' according to ξ_l' and P'; and
ii. we generate the auxiliary paths by using the same sequence of random variables as the perturbed path, e.g., we generate $W-C$ by using the same sequence as that used for generating the perturbed path $G-D$.

9.2. For two ergodic transition probability matrices P and P', set $P(\delta) := P+\delta(P'-P)$. Assume that δ is very small. Apply the construction approach described in Section 9.2 by following a sample paths of the Markov chains with $P(\delta)$. Show that this is equivalent to the performance derivative construction described in Section 2.1.3. (In Section 9.2, we follow the perturbed sample path, while in Section 2.1.3, we follow the original path.)

9.3. Suppose that the transition probability matrices of all the policies in an MDP problem are uni-chains on the same finite state space $\mathcal{S}$. (A uni-chain is a special case of a multi-chain defined in (B.1) with $m = 1$.)

a. Apply the construction approach shown in Section 9.2 to any two uni-chain policies and derive the performance difference formula. Show that it is a special case of the performance difference formula (4.36) in Chapter 4 for the multi-chain case.
b. Derive the Poisson equation for a uni-chain policy, prove that its solution exists, and express the potentials of the transient states in terms of those of the recurrent states.
c. Develop the policy iteration algorithm for uni-chain MDPs, and show that it is the same as that for ergodic chains.
d. Explain point c) using the policy iteration algorithm for the general case of multi-chain MDPs.

9.4. Prove that the policy iteration algorithm developed in Example 9.2 converges to an optimal policy.

9.5. In this exercise, we modify the random walk problem studied in Examples 9.1 and 9.2 as follows. First we simplify the problem by assuming that the random walker can take only $N + 1 = 5$ positions denoted as $0, 1, 2, 3$, and 4. When the walker hits the wall 0 or 4, s/he stays there with probability α_0, or α_4, respectively, and jumps to position 1, or 3, with probability $1 - \alpha_0$, or $1-\alpha_4$, respectively. Second, we assume that when the walker is at position 1, 2, or 3, s/he will also stay there with probability α_1, α_2, and α_3, respectively, and will leave the position with probability $1-\alpha_1$, $1-\alpha_2$, and $1-\alpha_3$, respectively. If s/he leaves position i, $i = 1, 2, 3$, s/he will have an equal probability of 0.5 to jump to one of its neighboring position $i - 1$ or $i + 1$, $i = 1, 2, 3$.

Now suppose that at each position i we may choose α_i from a finite set denoted as $\alpha_{i,1}, \alpha_{i,2}, \ldots, \alpha_{i,M}$, $i = 0, 1, \ldots, 4$.

a. Derive the performance difference formula (similar to (9.6)) and the policy iteration algorithm for this problem.
b. Furthermore, we assume that $\alpha_{0,i}$ and $\alpha_{4,i}$ (with the same i), $i = 1, 2, \ldots, M$, have to be chosen together, and $\alpha_{1,i}, \alpha_{2,i}$, and $\alpha_{3,i}$ (with the same i), $i = 1, 2, \ldots, M$, have to be chosen together. Derive a performance difference formula (similar to (9.7)) for this problem.
c. Based on the performance difference formula derived in b), develop a policy iteration algorithm for the optimization problem in which actions at different states cannot be chosen independently.

9.6. Study the random walk problem in Example 9.3 by using the system with $N + 2$ positions as the original system and the system with $N + 1$ positions as the perturbed one. Derive the performance difference formula similar to (9.32).

9.7. Extend the performance derivative formulas (9.29) and (9.33) to the case with $f(i) \neq f'(i)$, $i = 1, \ldots, S$.

9.8. In Section 9.4.2, suppose that $f'(i) \neq f(i)$, for $i = 1, \ldots, M$. Modify the performance difference formula (9.37) (i.e., derive the formula similar to (9.5) and (9.28)).

9.9. Draw a sample-path diagram to illustrate the effect of one jump in the example of the parameterized system in Section 9.3.2.

9.10. Consider a discrete-time Markov chain consisting of three super states denoted as 1, 2, and 3, respectively; each of them is further composed of three phases a, b, and c, as shown in Figure 9.8. Each phase represents a state of the Markov chain and thus it has altogether nine states denoted as $1a$, $1b$, $1c$; $2a$, $2b$, $2c$; and $3a$, $3b$, and $3c$. The transition probabilities between any two phases in the same super state are denoted by $p(1b|1a)$, $p(3a|3c)$, etc. When the system leaves a phase, it does not feed back immediately, i.e., $p(1a|1a) = 0$, etc. At each super state, phase a is an input phase, i.e., the system enters phase a to start its journey in the corresponding super state. Phase c is an exit phase, i.e, the system leaves a super state from phase c. At super state 1, for example, we have $p(1b|1a) + p(1c|1a) = 1$ and $p(1a|1b) + p(1c|1b) = 1$. At phase $1c$, there is a positive probability $p(0|1c)$ to leave the super state 1. Thus, $p(1a|1c) + p(1b|1c) + p(0|1c) = 1$. When a system leaves a super state i, $i = 1, 2, 3$, it moves to super state j and enters phase ja, $j = 1, 2, 3$, with probability $p(j|i)$, $\sum_{j=1}^{3} p(j|i) = 1$. The reward function is denoted as $f(1a)$, $f(1b)$, etc.

Suppose that the transition probabilities $p(j|i)$ depend on a parameter θ and are denoted as $p_\theta(j|i)$, $i, j = 1, 2, 3$. Construct the performance derivative and difference formulas for this system, similar to (9.12) and (9.13).

9.11. Consider a discrete-time M/M/1/N queue with capacity N. The system state is the number of customers in the system (in the queue plus in the server), denoted as n. The transition probabilities are $p(1|0) = p$, $p(0|0) = q$, $p(N-1|N) = q$, $p(N|N) = p$, and $p(n+1|n) = p$, $p(n-1|n) = q$, $p > 0$, $q > 0$, $p + q = 1$. Suppose that

a. the capacity changes to $N - 1$, or
b. the capacity changes to $N + 1$.

Construct the difference formula for the mean response time.

9.12. Suppose that we have two independent M/M/1/N queues with parameters p_1, q_1, N_1 and p_2, q_2, N_2, respectively, as explained in Problem 9.11. If we have one more buffer space, to which queue should we allocate this extra buffer space to maximally reduce the customers' mean response time? Please develop an on-line approach.

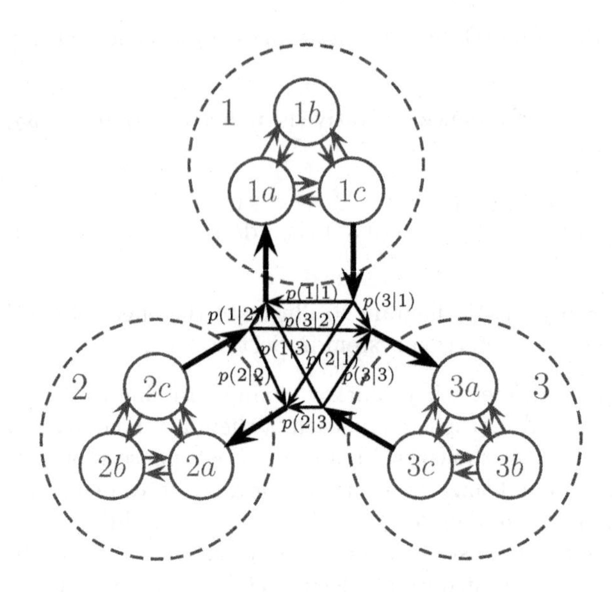

Fig. 9.8. The Transition Probabilities in Problem 9.10

9.13.* Extend the construction approach in Section 9.2 to (continuous-time) Markov processes. *(Hint: This extension is not as straightforward as what it may appear. To develop a construction approach to the changes in transition probabilities of the embedded Markov chain* $p(j|i)$, $i, j = 1, 2, \ldots, S$, *in (A.12) may be easy; the extension to the changes in transition rate* $\lambda(i)$ *may be more involved.)*

9.14.* Propose a construction approach for the performance differences and derivatives for a (continuous-time) closed Jackson (Gordon-Newell) network (Section C.2) with respect to the changes in routing probabilities. *(Hint: Use the results in Problem 9.13 for the transition probability matrix of the embedded chain.)*

Part III

Appendices: Mathematical Background

A carpenter who wishes to make good work must sharpen his axes first.

Confucius, Chinese thinker and social philosopher (551 BC - 479 BC)

The mathematical background required for this book includes some fundamental knowledge about probability theory, stochastic processes, and linear algebra. In addition, many application examples used in the book are drawn from communications and manufacturing, and these systems are generally modelled as queueing networks; and the contents in Section 2.4 on perturbation analysis are directly related to queueing systems. Therefore, some knowledge about queueing theory is also required to understand these parts of the book.

In this part, we briefly introduce the fundamentals in probability theory, Markov processes, stochastic matrices, and queueing theory that are related to the contents of this book, with an emphasis on the concepts and methodologies that are important to the subjects studied in the book. Some contents and problems (marked with asterisks) are designed to advance the readers' understanding of the main concepts. They may be quite difficult, and first time readers can ignore them.

The future depends on what we do in the present.

Mahatma Gandhi
(1869 - 1948)

Probability and Markov Processes

A.1 Probability

In this subsection, we review some concepts and results in probability theory [26, 28, 32] that are important to understanding the contents of this book.

Probability Spaces

A probability space consists of three elements:

1. A set, also called *a space*, Ω.
2. A collection $\mathcal{F}$ of subsets of Ω, called a σ-field, which satisfies the following properties:
   a) $\Omega \in \mathcal{F}$.
   b) If $A \in \mathcal{F}$, then $\bar{A} = \{\omega \in \Omega | \omega \notin A\} \in \mathcal{F}$.
   c) If $A_k \in \mathcal{F}$, $k = 1, 2, \ldots$, then $\cup_{k=1}^{\infty} A_k \in \mathcal{F}$.
   If a set $A \in \mathcal{F}$, A is called an *event*.
3. A set function $\mathcal{P}$ that assigns to each set $A \in \mathcal{F}$ a real number $\mathcal{P}(A)$, called the *probability* of the event A; $\mathcal{P}$ satisfies:
   a) $\mathcal{P}(A) \geq 0$, for any $A \in \mathcal{F}$.
   b) $\mathcal{P}(\Omega) = 1$.
   c) If $A_i \cap A_j = \emptyset$ for all $i \neq j$, then $\mathcal{P}(\cup_{k=1}^{\infty} A_k) = \sum_{k=1}^{\infty} \mathcal{P}(A_k)$.

The set function $\mathcal{P}$ is called a *probability measure*. Such a probability space is denoted as $(\Omega, \mathcal{F}, \mathcal{P})$.

If $A, B \in \mathcal{F}$ are two events, the *conditional probability* of event B given that event A occurs is defined as

$$\mathcal{P}(B|A) = \frac{\mathcal{P}(A \cap B)}{\mathcal{P}(A)}.$$

Random Variables

Let $(\Omega, \mathcal{F}, \mathcal{P})$ be a probability space and $\mathcal{R} = (-\infty, \infty)$ be the space of real numbers. A random variable X is a function, $\Omega \to \mathcal{R}$, such that for any real number $x \in \mathcal{R}$, the set $\{\omega \in \Omega, X(\omega) \leq x\} \subseteq \Omega$ belongs to $\mathcal{F}$ and thus has a probability with $\mathcal{P}$. The *distribution function* of X is defined as

$$\Phi(x) := \mathcal{P}\{\omega \in \Omega : X(\omega) \leq x\} := \mathcal{P}(X \leq x).$$

$\Phi(x)$ is nondecreasing; $\lim_{x \to -\infty} \Phi(x) = 0$; and $\lim_{x \to \infty} \Phi(x) = 1$. A *distribution density function* of a random variable X is a nonnegative function $\phi(x)$ on $\mathcal{R}$ such that

$$\Phi(x) = \int_{-\infty}^{x} \phi(y) dy, \qquad x \in \mathcal{R}.$$

If $\Phi(x)$ is differentiable at x, then $\phi(x) = \frac{d}{dx}\Phi(x)$. We have the *normalization condition*:

$$\int_{-\infty}^{\infty} \phi(x) dx = \Phi(\infty) = 1.$$

If X represents the lifetime of an event, then $\phi(x)$ is the rate at which the event will end in $[x, x + \Delta x)$ (i.e., $\phi(x)\Delta x$ is the probability that the event will end in $[x, x + \Delta x)$).

The *mean* (or the *expected value*) of a random variable X is defined as

$$E(X) := \int_{-\infty}^{\infty} x d\Phi(x) = \int_{-\infty}^{\infty} x\phi(x) dx.$$

The *variance* of X is

$$Var(X) := E[X - E(X)]^2 = E(X^2) - [E(X)]^2.$$

The *hazard rate* function is defined as

$$r(x) = \frac{\phi(x)}{1 - \Phi(x)}.$$

If $\Phi(x)$ is differentiable, then

$$\phi(x) = r(x) \exp\left(- \int_{-\infty}^{x} r(y) dy \right).$$

If X represents the lifetime of an event, then $r(x)$ is the "conditional" rate that the event will end in $[x, x + \Delta x)$ given that the event survives up to x.

If a random variable X is defined on a discrete space as $\Omega \to \mathcal{S}$, $\mathcal{S} = \{0, 1, 2, \ldots\}$, X is called a *discrete random variable*, which is characterized by the probabilities $p_n := \mathcal{P}(X = n)$, $n = 0, 1, \ldots$. $\{p_n, n = 0, 1, \ldots\}$ is called the *probability distribution* of X.

Let X and Y be two random variables defined on $(\Omega, \mathcal{F}, \mathcal{P})$. Their joint distribution function is defined as

$$\Phi(x, y) := \mathcal{P}\{\omega \in \Omega : X(\omega) \leq x, Y(\omega) \leq y\} := \mathcal{P}(X \leq x, Y \leq y).$$

Two random variables X and Y are said to be *independent*, if and only if

$$\Phi(x, y) = \Phi(x)\Phi(y).$$

The Memoryless Property of Exponential Distributions

A random variable X is said to have an *exponential distribution* if

$$\Phi(x) := \mathcal{P}(X \leq x) = \begin{cases} 1 - \exp\left(-x/\bar{x}\right), & x \geq 0, \\ 0, & x < 0. \end{cases}$$

We have $E(X) = \bar{x}$, $Var(X) = \bar{x}^2$. The hazard rate is a constant $r(x) = \lambda := \frac{1}{\bar{x}}$, which is called the *rate* of the exponential distribution. The conditional distribution function of X given $X \geq x_0 \geq 0$ is

$$\begin{aligned} \Phi(x|X \geq x_0) &:= \mathcal{P}(X \leq x|X \geq x_0) \\ &= 1 - \exp\left(-\frac{x - x_0}{\bar{x}}\right) = \Phi(x - x_0), \qquad x \geq x_0. \end{aligned} \tag{A.1}$$

Imagine X as the lifetime of an event. Equation (A.1) shows that if one knows that the event survives at $x_0 > 0$, then the residual lifetime of the event at x_0, $X - x_0$, has the same distribution as the lifetime itself, independent of x_0. This is called the *memoryless property* of the exponential distribution. This property is fundamental in the Markov property of continuous time processes, which will be discussed below in Appendix A.2. Basically, because of the memoryless property, if the event time is exponentially distributed, then at any time the future event time is independent of the elapsed time since the start of the event.

It can be shown that the exponential distribution is the only distribution that has the memoryless property [169]. For discrete random variables, the geometric distribution $p_n = q^n(1 - q)$, $0 < q < 1$, $n = 0, 1, \ldots$, is the only distribution that has the memoryless property.

The Coxian Distribution

Cox [90] proved that any distribution function whose Laplace transform is a rational function can be constructed by a series of stages, each of them having an exponential distribution. Figure A.1 shows the structure of such a construction with k stages. The idea can be explained with the example of a service station. Suppose that a customer arrives at a station for service. The customer first enters stage 1, which has a service time with an exponential distribution with mean $\bar{s}_1$. After the completion of the service at stage 1, the customer goes to stage 2 with probability p_1 and leaves the station with probability $q_1 = 1 - p_1$. The customer stays at stage 2 for an exponentially distributed service time with mean $\bar{s}_2$; then, the customer either enters stage 3 with probability p_2 or leaves the station with probability $q_2 = 1 - p_2$, and so on. If the customer enters the last stage k, he or she will leave the station with probability $q_k = 1$. The customer's service time at the station has a Coxian distribution.

If a customer's total service time has a general distribution, we need to use the elapsed service time (the amount of service time that the customer has received) to denote the status of the service: the probability distribution of the remaining service time depends on the elapsed time. Because of the memoryless property of exponential distributions, if a customer's service time has a Coxian distribution, then the probability distribution of the remaining service time depends only on the current stage. Thus, with the Coxian distribution, we may change a continuous variable, the elapsed service time, to a discrete one, the stage the customer is in, to denote the current status of the service.

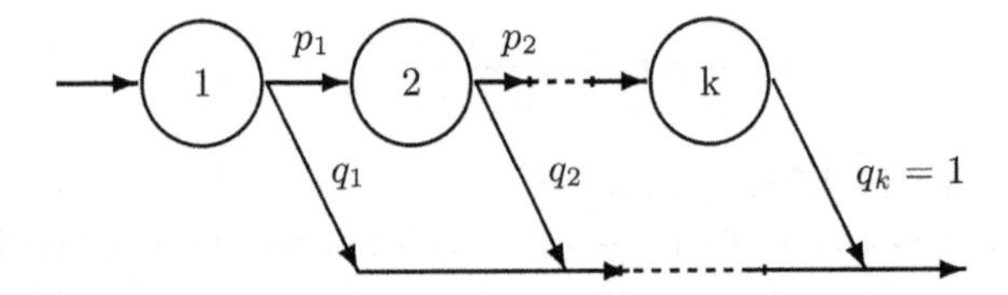

Fig. A.1. The Coxian Distribution

The Inverse Transform Method

This method is used to generate a random variable with any distribution function $\Phi(x)$ from a random variable that is uniformly distributed on $[0, 1)$ [12]. It also helps to understand the underlying probability space of a Markov process or a queueing network.

The method is based on the observation that for a random variable X with any distribution function $\Phi(x)$ the random variable $\xi = \Phi(X)$ is uniformly distributed on $[0, 1)$. Thus, to obtain a random variable X with a given distribution $\Phi(x)$, we first generate a random variable ξ, uniformly distributed on $[0, 1)$, and then set

$$X = \Phi^{-1}(\xi) = \sup\{x : \Phi(x) \leq \xi\}.$$

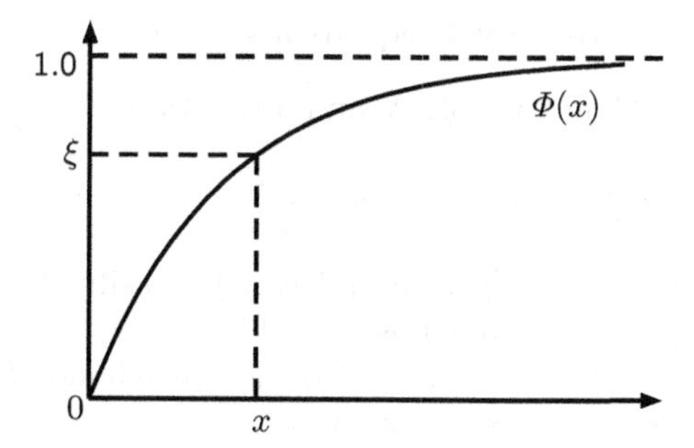

Fig. A.2. The Inverse Transform Method

This is illustrated in Figure A.2. The random variable is thus a function of ξ and can be denoted as $X(\xi)$, $\xi \in [0,1)$. Therefore, the underlying probability space of any random variable X, which represents the randomness of X, is the same space as ξ, $\Omega = [0,1)$.

Indeed, the random variable thus generated has the distribution $\Phi(x)$ as shown below:

$$\mathcal{P}(X \le x) = \mathcal{P}[\Phi^{-1}(\xi) \le x] = \mathcal{P}[\xi \le \Phi(x)] = \Phi(x).$$

The last equality holds because ξ is uniformly distributed on $[0,1)$ and $0 \le \Phi(x) < 1$.

The inverse transform method will be used in this book to determine the state transitions in a Markov process, to generate the service times, and to determine the customer routings in a queueing system.

For exponential distributions $\Phi(x) = 1 - \exp(-x/\bar{x})$, we have

$$X = -\bar{x}\ln(1-\xi).$$

To generate a discrete random variable K with probability distribution $\mathcal{P}(K = k) = p_k \ge 0$, $k = 1, 2, \ldots$, $\sum_{k=1}^{\infty} p_k = 1$, we partition the interval $[0,1)$ into small pieces: $[0,1) = [0,p_1) \cup [p_1, p_1+p_2) \cup \cdots \cup [\sum_{j=1}^{k-1} p_j, \sum_{j=1}^{k} p_j)$ $\cdots$. If $\xi \in [\sum_{j=1}^{k-1} p_j, \sum_{j=1}^{k} p_j)$, with $\sum_{j=1}^{0} p_j = 0$ and $\sum_{j=1}^{\infty} p_j = 1$, then we set $K = k$.

One important feature of the inverse transform method is that it separates the randomness involved in a random variable and the deterministic part. The randomness is represented by the uniformly distributed random variable ξ, and the deterministic part is described by function $\Phi(x)$. In the case of an exponential distribution, the method clearly separates the randomness ξ from the parameter $\bar{x}$. With this formulation, the correlation between different random variables can be expressed by that of different uniformly distributed random variables. Problem A.4 provides an example.

The Markov and Chebyshev Inequalities

For a non-negative random variable X and a constant $c > 0$, we have

$$\mathcal{P}(X \geq c) \leq \frac{E(X)}{c}.$$

This is called the *Markov inequality* and can be easily derived as follows. Consider a random variable defined as $Y = 0$ if $X < c$, and $Y = c$ if $X \geq c$. We have $Y \leq X$. Thus, $E(Y) \leq E(X)$. On the other hand, $E(Y) = c\mathcal{P}(Y = c) = c\mathcal{P}(X \geq c)$. Thus, $c\mathcal{P}(X \geq c) \leq E(X)$.

Next, for any random variable X with mean $E(X)$ and variance $Var(X)$, we have the *Chebyshev inequality*:

$$\mathcal{P}(|X - E(X)| \geq c) \leq \frac{Var(X)}{c^2}, \qquad \text{for any } c > 0.$$

This follows directly from the Markov inequality by using $[X - E(X)]^2$ as the nonnegative random variable and c^2 as the positive constant.

Convergence of Random Sequences

Let $X_1, X_2, \ldots, X_n, \ldots$ be a sequence of random variables defined on the same probability space $(\Omega, \mathcal{F}, \mathcal{P})$. There are four major concepts regarding the convergence of a random sequence. (The same definitions apply when the discrete index "n" is replaced by a continuous one, say "$t \in [0, \infty)$".)

i. *Convergence in probability.* The sequence of random variables $\{X_n, n = 1, 2, \ldots\}$ converges in probability to a random variable X, if for any $\epsilon > 0$,

$$\lim_{n\to\infty} \mathcal{P}(|X_n - X| \geq \epsilon) = 0.$$

ii. *Convergence with probability 1 (w.p.1).* The sequence of random variables $\{X_n, n = 1, 2, \ldots\}$ converges with probability 1 to a random variable X, if

$$\mathcal{P}\left(\omega : \lim_{n\to\infty} X_n = X\right) = 1,$$

or equivalently, for any $\epsilon > 0$,

$$\lim_{n\to\infty} \mathcal{P}(|X_k - X| > \epsilon \text{ for some } k \geq n) = 0.$$

iii. *Convergence in the mean or in the mean square.* The sequence of random variables $\{X_n, n = 1, 2, \ldots\}$ converges in the mean, or in the mean square, respectively, to a random variable X, if

$$\lim_{n\to\infty} E(|X_n - X|) = 0,$$

or

$$\lim_{n\to\infty} E(|X_n - X|^2) = 0,$$

respectively.

iv. *Convergence in the distribution and weak convergence.* The sequence of random variables $\{X_n, n = 1, 2, \ldots\}$ converges in the distribution to a random variable X, if

$$\lim_{n \to \infty} \Phi_n(x) = \Phi(x)$$

at every continuous point of Φ, where $\Phi_n(x)$ and $\Phi(x)$ are the distribution functions of X_n and X, respectively. The sequence of distribution functions $\{\Phi_n(x), n = 1, 2, \ldots\}$ is said to converge weakly to $\Phi(x)$.

Both convergence with probability 1 and convergence in the mean (or in the mean square) imply convergence in probability, which, in turn, implies convergence in the distribution (see, e.g., [28]). Convergence with probability 1 and convergence in the mean do not imply each other. However, if X_n are dominated by a random variable Y having a finite mean (i.e., $|X_n| \le Y$, w.p.1, $n \ge 1$, and $E(|Y|) < \infty$), then X_n converges to X with probability 1 implies that X_n converges to X in the mean.

The Law of Large Numbers

Suppose that $\{X_n, n = 1, 2, \ldots\}$ is a sequence of independent random variables with $E(X_n) = 0$, $n = 1, 2, \ldots$. If $\sum_{n=1}^{\infty} \frac{Var(X_n)}{n^2} < \infty$, then

$$\lim_{n \to \infty} \frac{1}{n} \sum_{k=1}^{n} X_k = 0, \qquad \text{w.p.1.}$$

The following result is more often used in practice: Suppose that $\{X_n, n = 1, 2, \ldots\}$ is a sequence of independent and identically distributed (i.i.d.) random variables and $E(X_n) = E(X)$ exists, then

$$\lim_{n \to \infty} \frac{1}{n} \sum_{k=1}^{n} X_k = E(X), \qquad \text{w.p.1.}$$

That is, the sample mean $\frac{1}{n} \sum_{k=1}^{n} X_k$ converges to the true mean $E(X)$ with probability 1. This is the *strong law of large numbers.* The *weak law of large numbers* is as follows: Suppose that $\{X_n, n = 1, 2, \ldots\}$ is a sequence of independent random variables with $E(X_n) = E(X) < \infty$ and $\sum_{n=1}^{\infty} \frac{Var(X_n)}{n^2} < \infty$. Then, for any $\epsilon > 0$, we have

$$\lim_{n \to \infty} \mathcal{P} \left[\left| \frac{1}{n} \sum_{k=1}^{n} X_k - E(X) \right| \ge \epsilon \right] = 0.$$

That is, the sample mean converges to the true mean in probability.

A.2 Markov Processes

A *stochastic process* is defined as a sequence of random variables, defined on the same probability space $(\Omega, \mathcal{F}, \mathcal{P})$ and indexed by discrete times $l = 0, 1, \ldots$, or continuous times $t \in [0, \infty)$, denoted as $\boldsymbol{X}(\omega) = \{X_0(\omega), X_1(\omega), \ldots, X_l(\omega), \ldots\}$ or $\boldsymbol{X}(\omega) = \{X_t(\omega), t \in [0, \infty)\}$, $\omega \in \Omega$, respectively. For every fixed point $\omega \in \Omega$, the sequence $\boldsymbol{X}(\omega)$ is called a *sample path* of $\boldsymbol{X}$. Thus, a point $\omega \in \Omega$ represents a sample path of $\boldsymbol{X}$, and Ω is the probability space generated by all the sample paths of $\boldsymbol{X}$. Generally, we will omit the argument ω if there is no confusion.

The Markov Property and System States

Naturally, the first step in modelling a system is to describe precisely the system's "behavior" as a history of the system state. The word "state" is used with different meanings in the literature. In [87], X_l or X_t in any process $\boldsymbol{X}$ is called the *state* of the process at time l or t. With this definition, any quantity in a system that depends on time can be defined as the state of a stochastic process; however, such a "state" may not represent the whole status of the system at any particular time.

To define a state that can completely capture the system status at any particular time, we need the Markov property. A stochastic process is said to possess the *Markov property*, if, given the current state, the process's future behavior is independent of its past history. In other words, by knowing the current state of a process, we can predict the future behavior of the process as well as if we know the entire history (the current state plus the past history); the past history of the process does not provide any additional information that may help with the prediction. In this book, we use the word "state" in the strict sense, i.e., a state is a random variable (or random vector) indexed by time that satisfies the Markov property. This is also called a *mathematical state*, while the state in the general sense, i.e., any random variable indexed by time, is called a *physical state*, which usually has a physical interpretation (e.g., the queueing length in queueing theory). A stochastic process that possesses the Markov property is called a *Markov process.*

In this book, we follow the terminology used in [87], call discrete-time Markov processes *Markov chains* and reserve the phrase "Markov processes" for continuous-time Markov processes.

Markov Chains

We first discuss Markov chains. Let $\mathcal{S}$ be a state space of a Markov chain $\boldsymbol{X}$. Although many concepts and results presented and discussed in this book apply to Markov chains and processes with infinitely many states as well, in this book we assume that the state space is finite and denote it as $\mathcal{S} = \{1, 2, \ldots, S\}$, if not otherwise mentioned. A Markov chain with a finite state space is called a finite Markov chain.

Definition A.1. [87] *A Markov chain* $\boldsymbol{X} = \{X_0, X_1, \ldots\}$ *is a sequence of random variables defined on* $(\Omega, \mathcal{F}, \mathcal{P})$ *with values on* $\mathcal{S}$ *such that*

$$\mathcal{P}(X_{l+1} = j | X_0 = i_0, \ldots, X_l = i_l) = \mathcal{P}(X_{l+1} = j | X_l = i_l) \tag{A.2}$$

for all $j, i_0, \ldots, i_l \in \mathcal{S}$ *and any* $l \in \{0, 1, \ldots\}$.

The property (A.2) is called the *Markov property.* Define

$$p_l(j|i) := \mathcal{P}(X_{l+1} = j | X_l = i)$$

as the *transition probabilities* of the Markov chain at time l, $l = 0, 1, \ldots$. If $p_l(j|i) =: p(j|i)$ for all l and $i, j \in \mathcal{S}$, the Markov chain is said to be *time homogeneous.* We shall restrict ourselves to time-homogeneous Markov chains in most part of this book.

The matrix $P = [p(j|i)]_{i,j \in \mathcal{S}}$ is called the *transition probability matrix.* We have $p(j|i) \geq 0$ and $\sum_{j \in \mathcal{S}} p(j|i) = 1$ for all i, or

$$Pe = e,$$

with $e = (1, 1, \ldots, 1)^T$ being a column (unit) vector whose all components are 1, where "T" denotes transpose. Sometimes, we use e_S to indicate that the vector e is S-dimensional. Such a matrix P is called a *Markov matrix* (or a *stochastic matrix*).

Let $p^{(k)}(j|i)$ be the (i, j)th element in P^k, i.e., $P^k := [p^{(k)}(j|i)]_{i,j \in \mathcal{S}}$. Then, it is easy to verify that

$$\mathcal{P}(X_{l+k} = j | X_l = i) = p^{(k)}(j|i), \qquad \text{for all } l; \tag{A.3}$$

thus, P^k is called the *k-step transition matrix.*

A set of states is said to be *closed* if no state outside the set can be reached from any state in the set; i.e., $\mathcal{S}_0 \subseteq \mathcal{S}$ is a closed set if $p(j|i) = 0$ for all $i \in \mathcal{S}_0$ and $j \in \mathcal{S} - \mathcal{S}_0$. A closed set containing one state i, (i.e., $p(i|i) = 1$) is called an *absorbing state.* A closed set is *irreducible* if no proper subset of it is closed. A Markov chain is said to be *irreducible* if its only non-empty closed set is $\mathcal{S}$. Naturally, a Markov chain is irreducible if and only if all its states can be reached from each other, either directly, or by going through other states. A Markov chain is reducible, if and only if, by re-labelling the states, its transition probability matrix can be written as

$$P = \begin{bmatrix} P_1 & 0 \\ R & P_2 \end{bmatrix},$$

where "0" represents a matrix whose elements are all zeros; the sub-matrix P_2 may be further reduced, see the canonical form (B.1) in Appendix B.1. A Markov chain is either irreducible or reducible.

Consider a Markov chain $\boldsymbol{X}$ starting from an initial state $X_0 = j$. Let L_j be the time of the first visit of $\boldsymbol{X}$ to state j, i.e., $L_j = \min\{l : X_l = j; X_n \neq$

$j, 0 < n < l\}$. A state j is said to be *recurrent* if $\mathcal{P}(L_j < \infty | X_0 = j) = 1$; otherwise, if $\mathcal{P}(L_j = \infty | X_0 = j) > 0$, then j is said to be *transient*. If j is a transient state, then for any $i \in \mathcal{S}$,

$$\lim_{k \to \infty} p^{(k)}(j|i) = 0.$$

A recurrent state j is said to be *periodic* with *period* κ, if there is an integer $\kappa \geq 2$, such that κ is the largest integer for which

$$\mathcal{P}(L_j = n\kappa \text{ for some integer } n \geq 1) = 1; \tag{A.4}$$

otherwise, if there is no such $\kappa \geq 2$, j is said to be *aperiodic*. It is known that if $\boldsymbol{X}$ is an irreducible and finite-state Markov chain and if one state is periodic, then all the states are periodic with the same period [87]. Such a Markov chain is called a *periodic Markov chain*; otherwise, it is called an *aperiodic Markov chain*. The transition probability matrix of a periodic Markov chain is called a *periodic matrix*.

A Markov chain is said to be *stationary* [87], if for any $k \geq 0$ and $i_0, i_1, \ldots \in \mathcal{S}$, it holds

$$\mathcal{P}(X_0 = i_0, X_1 = i_1, \ldots) = \mathcal{P}(X_k = i_0, X_{k+1} = i_1, \ldots). \tag{A.5}$$

By taking the marginal distribution for X_k, we get $\mathcal{P}(X_k = i) = \mathcal{P}(X_0 = i)$ for all $k > 0$ and $i \in \mathcal{S}$. That is, the distribution functions of the states of a stationary Markov chain at any time are the same. This distribution is called the *stationary distribution* and is denoted as $\pi(j)$, $j \in \mathcal{S}$. Let $\pi = (\pi(1), \pi(2), \ldots, \pi(S))$ be the (row) vector of the stationary probabilities of a Markov chain with transition probability matrix P. Then, π satisfies the following probability flow-balance equation:

$$\pi P = \pi, \qquad \text{with } \pi e = 1. \tag{A.6}$$

If the initial state distribution of a Markov chain is the stationary distribution, then the Markov chain is stationary. The stationary distribution π is a distribution defined on $\mathcal{S}$; it is also called the *steady-state probability* vector.

Let $f : \mathcal{S} \to \mathcal{R}$, $\mathcal{R} = (-\infty, +\infty)$ be a reward function defined on $\mathcal{S}$. Then, $f(X_l)$, $l = 0, 1, \ldots$, is a random variable defined on Ω. For parsimony, we also use f to denote the (column) vector $f = (f(1), f(2), \ldots, f(S))^T$. For stationary Markov chains, we have $E[f(X_l)] = E_\pi[f(X)] = \pi f$, where "$E$" denotes the expectation corresponding to the probability measure $\mathcal{P}$ on Ω, and "E_π" the expectation corresponding to π on $\mathcal{S}$, X_l is the state at time l, $l = 0, 1, \ldots$, and X denotes a generic random variable with probability distribution $\pi(i)$, $i \in \mathcal{S}$.

It is well known [87] that if a finite Markov chain (with $S < \infty$) is irreducible, then (A.6) has a unique solution, and all states are recurrent with $\pi(i) > 0$ for all $i \in \mathcal{S}$. If, in addition, it is aperiodic, then the Markov chain is *asymptotically stationary*; i.e.,

$$\lim_{k\to\infty} p^{(k)}(j|i) = \pi(j), \tag{A.7}$$

for all $i, j \in \mathcal{S}$, which does not depend on i. In a matrix form, this is

$$\lim_{k\to\infty} P^k = e\pi. \tag{A.8}$$

Also, the *ergodicity* holds; i.e., the long-run average reward equals its steady-state mean,

$$\eta := \lim_{L\to\infty} \left[\frac{1}{L} \sum_{l=0}^{L-1} f(X_l) \Big| X_0 = i \right] = E_\pi[f(X)] = \pi f, \qquad \text{w.p.1}, \tag{A.9}$$

which holds regardless of the initial state. Because of this, an irreducible and aperiodic finite Markov chain is called an *ergodic* Markov chain.

However, if P is periodic, the limit (A.7) or (A.8) does not exist. For example, if

$$P = \begin{bmatrix} 0 & 1 \\ 1 & 0 \end{bmatrix},$$

then $P^k = I$ if k is even and $P^k = P$ if k is odd. Thus, for periodic Markov chains, the asymptotical stationarity (A.7) does not hold. However, for finite irreducible periodic Markov chains, (A.6) has a unique solution π and the time average equation (A.9) holds.

Because the concepts and methodologies in learning and optimization can be clearly and concisely explained with the discrete-time model with a finite state space, most results in this book are stated in the discrete-time, finite-state version. In addition, we always assume that the Markov chain is ergodic to assure that the stationary distribution exists. This assumption can be easily relaxed by using the time averages expressed in the form of the Cesaro limit to replace the probabilities (see the discussion for the multi-chain case in Appendix B.3).

Markov Processes

A Markov process defined on a finite state space $\mathcal{S}$ and the continuous time domain $t \in [0, \infty)$ is denoted as $\boldsymbol{X} = \{X_t, t \in [0, \infty)\}$, $X_t \in \mathcal{S}$. Denote the underlying probability space as $(\Omega, \mathcal{F}, \mathcal{P})$. The Markov property becomes [87]

$$\mathcal{P}(X_{s+t} = j | X_u; u \le s) = \mathcal{P}(X_{s+t} = j | X_s), \qquad \text{for any } 0 < t, s < \infty. \tag{A.10}$$

For a time-homogeneous process, the transition probability function does not depend on s and is denoted as $\mathcal{P}(X_{s+t} = j | X_s = i) = p_t(j|i)$, for all $i, j \in \mathcal{S}$ and $t \ge 0$. We have $p_t(j|i) \ge 0$ and $\sum_{k\in\mathcal{S}} p_t(k|i) = 1$, for all $i, j \in \mathcal{S}$, $t \ge 0$. We discuss only time-homogeneous Markov processes.

Let $T_0 = 0$, T_1, T_2, ... be the state transition instants of the Markov process $\boldsymbol{X} = \{X_t\}$ and X_0, X_1, X_2, ... be the successive states visited by $\boldsymbol{X}$.

We assume that the sample paths are right continuous; i.e., $X_l = X_{T_l+0}$. Because of the Markov property of $\boldsymbol{X}$, $\boldsymbol{X}^\dagger := \{X_0, X_1, \ldots\}$ is a Markov chain, called the *embedded Markov chain* of $\boldsymbol{X}$. $T_{l+1} - T_l$ is called the *sojourn time* in state X_l.

Let $\tau \in [T_{l-1}, T_l)$ for some l. Because of the Markov property, the remaining sojourn time at τ, $R_\tau := T_l - \tau$, depends only on X_{l-1} and is independent of the elapsed time in state X_{l-1}, $\tau - T_{l-1}$. Hence, the random variable R_τ satisfies the "memoryless" property. Thus, R_τ has an exponential distribution with a mean depending on $X_\tau = i$, denoted as $1/\lambda(i)$:

$$\mathcal{P}(R_\tau \le t | X_\tau = i) = 1 - \exp\left[-\lambda(i)t\right], \qquad t \ge 0.$$

$\lambda(i)$ is called the *transition rate* of the Markov process in state i. Taking $\tau = T_{l-1}$, we obtain $\mathcal{P}(T_l - T_{l-1} \le t | X_{l-1} = i) = 1 - \exp[-\lambda(i)t]$, $t \ge 0$. Therefore, for any l, $j \in \mathcal{S}$, and $t \ge 0$, if $X_l = i$, then we have

$$\begin{aligned}&\mathcal{P}(X_{l+1} = j, T_{l+1} - T_l \le t | X_0, \ldots X_l = i; T_0, \ldots, T_l) \\ &= p(j|i)\left\{1 - \exp\left[-\lambda(i)t\right]\right\},\end{aligned} \tag{A.11}$$

where $p(j|i)$, $i, j \in \mathcal{S}$, are the transition probabilities of the embedded Markov chain $\boldsymbol{X}^\dagger$. We have $p(j|i) \ge 0$ and $\sum_{j \in \mathcal{S}} p(j|i) = 1$. By the definition of the embedded chain, if we cannot observe a transition from one state to itself, we have $p(i|i) = 0$. However, in a general setting, we allow the Markov process $\boldsymbol{X}$ to jump from one state to itself. In this case, we may have $p(i|i) > 0$.

Let

$$b(i,j) := \begin{cases} -\lambda(i)[1 - p(i|i)], & \text{if } i = j, \\ \lambda(i)p(j|i), & \text{if } i \ne j. \end{cases} \tag{A.12}$$

We have

$$\sum_{j \in \mathcal{S}} b(i,j) = 0, \qquad i \in \mathcal{S}.$$

That is, $Be = 0$, where $B := [b(i,j)]_{i,j \in \mathcal{S}}$. Let $P_t := [p_t(j|i)]_{i,j \in \mathcal{S}}$. Then, we have the following *Kolmogorov's* equation [87]:

$$\frac{d}{dt} P_t = BP_t = P_t B, \tag{A.13}$$

with the initial condition $P_0 = I$ being the identity matrix. The matrix B is called the *infinitesimal generator* of the Markov process $\boldsymbol{X}$. The solution to (A.13) is

$$P_t = \exp\left(tB\right) := \sum_{n=0}^{\infty} \frac{t^n}{n!} B^n. \tag{A.14}$$

If the embedded Markov chain is irreducible, then $\boldsymbol{X}$ is asymptotically stationary; i.e.,

$$\lim_{t \to \infty} \mathcal{P}(X_t = i | X_0 = j) = \pi(i),$$

independent of j, where $\pi(i)$, $i \in \mathcal{S}$, is the stationary (steady-state) distribution of the Markov process $\boldsymbol{X}$. Let $\pi = (\pi(1), \ldots, \pi(S))$. We have

$$\lim_{t \to \infty} P_t = e\pi. \tag{A.15}$$

From (A.13), and $\lim_{t \to \infty} \frac{d}{dt} P_t = 0$, we get

$$\pi B = 0.$$

If the embedded Markov chain is irreducible, we also have

$$\lim_{T \to \infty} \left[\frac{1}{T} \int_0^T f(X_t) dt \right] = E_\pi[f(X)] = \pi f, \qquad \text{w.p.1}, \tag{A.16}$$

for any initial condition, where $f : \mathcal{S} \to \mathcal{R}$ is a reward function and "E_π" is the expectation corresponding to the steady-state probability distribution π on $\mathcal{S}$.

Semi-Markov Processes

A stochastic process $\boldsymbol{X} = \{X_t, \ t \in [0, \infty)\}$ defined on state space $\mathcal{S}$ is called a *semi-Markov process* (SMP) if

$$\begin{aligned} &\mathcal{P}(X_{l+1}{=}j, T_{l+1} - T_l \le t | X_0, \ldots, X_l; T_0, \ldots, T_l) \\ &= \mathcal{P}(X_{l+1}{=}j, T_{l+1} - T_l \le t | X_l), \end{aligned}$$

for all $l = 0, 1, \ldots$, $j \in \mathcal{S}$, and $t \ge 0$, where $\boldsymbol{X}^\dagger := \{X_0, X_1, \ldots\}$ is an embedded chain. We assume that the process is time homogeneous and set $p(j;t|i) := \mathcal{P}(X_{l+1} = j, T_{l+1} - T_l \le t | X_l = i)$, which is called a *semi-Markov kernel* on $\mathcal{S}$. A semi-Markov process "partially" enjoys the Markov property: at any transition instant T_l, with the current state known, the future behavior is independent of the past history.

Let $p(j|i) = \lim_{t \to \infty} p(j;t|i)$. It is an easy exercise to prove that $\boldsymbol{X}^\dagger = \{X_l, \ l = 0, 1, \ldots\}$ is a Markov chain with transition matrix $P = [p(j|i)]$. This is called the *embedded Markov chain* of $\boldsymbol{X} = \{X_t, t \in [0, \infty)\}$. If

$$p(j;t|i) = p(j|i) \left\{1 - \exp\left[-\lambda(i)t\right]\right\},$$

then a semi-Markov process is a Markov process (cf. (A.11)).

Similar to the case of Markov processes, if s is a transition instant, we define $p_t(j|i) = \mathcal{P}(X_{s+t} = j | X_s = i)$, $0 < t, s < \infty$, which does not depend on s for time-homogeneous semi-Markov processes. Let $P_t = [p_t(j|i)]_{i,j \in \mathcal{S}}$. Finally, if the embedded Markov chain is irreducible, then the steady-state probabilities π can also be defined by (A.15), in the same way as for Markov processes, and (A.16) also holds for a semi-Markov process.

PROBLEMS

A.1. Consider the Coxian distribution shown in Figure A.1.

a. Derive the probability distribution density function for the Coxian distribution.
b. Derive the Laplace transform of the density function.
c. Construct a Coxian distribution such that the Laplace transform of its density function is the rational function given below:

$$\Phi^*(s) = \frac{2 + 1.08s + 0.2s^2}{2 + 5s + 4s^2 + s^3}.$$

A.2. Consider a sequence of independent random variables X_n with $\mathcal{P}(X_n = 1) = \frac{1}{n}$ and $\mathcal{P}(X_n = 0) = 1 - \frac{1}{n}$. Does the sequence converge in probability, with probability 1, in mean, or in mean square?

A.3. Consider a sequence of independent random variables X_n with $\mathcal{P}(X_n = 1) = \frac{1}{n^2}$ and $\mathcal{P}(X_n = 0) = 1 - \frac{1}{n^2}$. Does the sequence converge in probability, with probability 1, in mean, or in mean square?

A.4. * Let X and Y be two random variables with probability distributions $\Phi(x)$ and $\Psi(y)$, respectively. Their means are denoted as $\bar{x} = E(X)$ and $\bar{y} = E(Y)$. We wish to estimate $\bar{x} - \bar{y} = E(X - Y)$ by simulation. We generate random variables X and Y using the inverse transform method. Thus, we have $X = \Phi^{-1}(\xi_1)$ and $Y = \Psi^{-1}(\xi_2)$, where ξ_1 and ξ_2 are two uniformly distributed random variables in $[0, 1)$. Prove that if we choose $\xi_1 = \xi_2$, then the variance of $X - Y$, $Var[X - Y]$, is the smallest among all possible pairs of ξ_1 and ξ_2.

A.5. Consider a sequence of i.i.d. random variables $\{X_n, n = 1, 2, \ldots\}$ with mean $E(X_n) = E(X)$. Define another sequence of $0 - 1$ valued i.i.d. random variables $\{\chi_n, n = 1, 2, \ldots\}$ where $\chi_n = 1$ with probability p, $1 > p > 0$ and $\chi_n = 0$ with probability $1 - p$. Let

$$N_n = \sum_{k=1}^{n} \chi_k$$

be the number of 1's in the first n samples. Define

$$M_n := \frac{1}{N_n} \sum_{k=1}^{n} (\chi_k X_k).$$

Prove that M_n converges to $E(X)$ with probability 1 as $n \to \infty$, i.e.,

$$\lim_{n \to \infty} M_n = E(X), \qquad \text{w.p.1},$$

and M_n converges to $E(X)$ in probability as $n \to \infty$, i.e., for any $\epsilon > 0$,

$$\lim_{n \to \infty} \mathcal{P}[|M_n - E(X)| \geq \epsilon] = 0.$$

A.6. Consider a sequence of independent random variables $\{X_n, n = 1, 2, \ldots\}$. The mean value of X_n, $E(X_n)$, converges to a constant $\bar{X}$, $\lim_{n\to\infty} E(X_n) = \bar{X}$, and $Var(X_n) < \infty$. Prove that the mean sample $M_n = \frac{1}{n}\sum_{k=1}^{n} X_k$ converges to $\bar{X}$ both with probability 1 and in probability.

A.7. Let $\boldsymbol{X}$ be an irreducible and periodic Markov chain with transition probability matrix P. The asymptotic stationarity (A.8) does not hold. However, we may define $\pi(i)$ as the time average

$$\pi(i) = \lim_{L\to\infty} \frac{1}{L} E\left[\sum_{l=0}^{L-1} \chi_i(X_l)\middle| X_0 = j\right], \qquad i, j \in \mathcal{S}, \tag{A.17}$$

with $\chi_i(x) = 1$, if $x = i$, and $\chi_i(x) = 0$, otherwise. Prove the following results:

a. Prove that the $\pi(i)$ in (A.17) indeed does not depend on j.
b. Let $\pi = (\pi(1), \ldots, \pi(S))$, then

$$P^* := \lim_{L\to\infty} \frac{1}{L} \sum_{l=0}^{L-1} P^l = e\pi.$$

c. $\pi P = \pi$ and $\pi e = 1$. That is, the time average π plays the same role as the steady-state probability.
d. $\lim_{L\to\infty} \frac{1}{L}\left[\sum_{l=0}^{L-1} \chi_i(X_l)\right]$, $i \in \mathcal{S}$, converges with probability 1 to $\pi(i)$. Therefore, $\pi(i)$ can also be defined as the limit of the sample-path average of $\chi_i(X_l)$, $l = 0, 1, \ldots$.

A.8. *(Uniformization)* Consider a Markov process $\boldsymbol{X}$ with transition rates $\lambda(i)$, $i \in \mathcal{S} = \{1, 2, \ldots, S\}$. Let $P = [p(j|i)]$ be the transition probability matrix of the embedded Markov chain, with $p(i|i) = 0$. Define another Markov process $\boldsymbol{X}'$ as follows: the transition rate in state i changes to $\lambda'(i) = \frac{\lambda(i)}{1-c_i}$, where $c_i \in (0, 1)$ is a fixed number, $i \in \mathcal{S}$; the transition probabilities change to $p'(i|i) = c_i$ and $p'(j|i) = p(j|i)(1 - c_i)$, $i \neq j$.

a. Prove that the steady-state probabilities of both processes are equal; i.e., $\pi'(i) = \pi(i)$, $i \in \mathcal{S}$.
b. Explain the relation between the sample paths of both processes.
c. Find the values for c_i, $i \in \mathcal{S}$, such that the embedded Markov chain of $\boldsymbol{X}'$, $\boldsymbol{X}'^{\dagger}$, has the same steady-state probabilities as those of $\boldsymbol{X}'$ and $\boldsymbol{X}$; i.e., $\pi'^{\dagger}(i) = \pi'(i) = \pi(i)$, $i \in \mathcal{S}$.

A.9. Let $\boldsymbol{X}^{\dagger}$ be the embedded Markov chain of Markov process $\boldsymbol{X}$. Assume that $\boldsymbol{X}^{\dagger}$ is ergodic. Let $\lambda(i)$, $i \in \mathcal{S} = \{1, 2, \ldots, S\}$ be the transition rates of $\boldsymbol{X}$; and $\pi^{\dagger}(i)$, $\pi(i)$, $i \in \mathcal{S}$, be the steady-state probabilities of $\boldsymbol{X}^{\dagger}$ and $\boldsymbol{X}$, respectively. Prove

$$\pi(i) = c\frac{\pi^{\dagger}(i)}{\lambda(i)}$$

where

$$c = \sum_{i \in \mathcal{S}} \pi(i)\lambda(i) = \frac{1}{\sum_{i \in \mathcal{S}} \frac{\pi^{\dagger}(i)}{\lambda(i)}}.$$

A.10. Consider an ergodic Markov chain $\mathbf{X} = \{X_0, X_1, \ldots\}$ with transition probability matrix $P = [p(j|i)]$ on state space $\mathcal{S} = \{1, \ldots, S\}$. Let π be the steady-state probability vector. Define a performance function that depends on two consecutive states: $f(i,j)$, $i, j \in \mathcal{S}$. Prove that the following ergodicity equation holds:

$$\lim_{n \to \infty} \left[\frac{1}{L} \sum_{l=0}^{L-1} f(X_l, X_{l+1}) \right] = E_{\pi,P}[f(X_l, X_{l+1})]$$

$$:= \sum_{i=1}^{S} \sum_{j=1}^{S} [f(i,j)\pi(i)p(j|i)] = \sum_{i=1}^{S} [\bar{f}(i)\pi(i)], \qquad \text{w.p.1},$$

where $\bar{f}(i) = \sum_{j=1}^{S} [f(i,j)p(j|i)]$. Extend these results to function $f(X_l, X_{l+1}, \ldots, X_{l+N})$ for a finite integer N.

A.11. Prove that the sojourn time that a time-homogenous Markov process stays in a state i is exponentially distributed, by using the Markov property (A.10).

A.12. Is the following statement true? (NO!)

If the inter-transition times of a semi-Markov process are exponentially distributed, i.e., if $\mathcal{P}(T_{l+1} - T_l \le t | X_l = i) = 1 - \exp[-\lambda(i)t]$, $i \in \mathcal{S}$, *then the semi-Markov process is a Markov process.*

If your answer is "yes", prove it; if the answer is "no", explain why and give a counter example.

Life is good for only two things, discovering mathematics and teaching mathematics

Simon Poisson, French Mathematician (1781-1840)

B Stochastic Matrices

To be consistent with the notation used for the transition probability matrix, we use $p(j|i)$ to denote the element in the ith row and the jth column of a matrix P. A matrix P is called a *non-negative matrix* if $p(j|i) \geq 0$ for all i, j (i.e., $P \geq 0$), and it is called a *positive matrix* if $p(j|i) > 0$ for all i, j (i.e., $P > 0$). A square non-negative matrix P is called a *stochastic matrix* if $Pe = e$. Therefore, a transition probability matrix is a stochastic matrix. Many properties of Markov processes are related to the theory of stochastic matrices. In this section, we review some relevant results of stochastic matrices.

B.1 Canonical Form

A stochastic matrix P is said to be *reducible*, if, by permutation of the rows and columns in the same order (corresponding to relabelling the states of the Markov chain with the transition probability matrix P), it can be written as

$$P = \begin{bmatrix} P_1 & 0 \\ R & P_2 \end{bmatrix},$$

where 0 denotes a zero matrix with all the elements being zeros, and R may or may not be a zero matrix. Otherwise, it is called *irreducible*. This definition of irreducibility is consistent with that given in Appendix A.2 for Markov chains, i.e., a Markov chain is irreducible if and only if its transition probability matrix is irreducible. In an irreducible Markov chain, every state will be eventually visited as time goes on. The matrices P_1 and P_2 may be further reducible. Finally, a transition probability matrix takes the *canonical form*:

$$P = \begin{bmatrix} P_1 & 0 & 0 & \cdots & \cdot & 0 \\ 0 & P_2 & 0 & \cdots & \cdot & 0 \\ \cdot & \cdot & \cdot & \cdots & \cdot & \cdot \\ 0 & 0 & 0 & \cdots & P_m & 0 \\ R_1 & R_2 & R_3 & \cdots & R_m & R_{m+1} \end{bmatrix}, \tag{B.1}$$

where $P_1, P_2, \ldots, P_{m-1}$ and P_m are all irreducible square matrices. This form indicates that the state space of a Markov chain with the transition probability matrix P consists of m closed subsets of recurrent states; each subset corresponds to one of the sub-matrices P_k, $k = 1, 2, \ldots, m$. The states corresponding to the last row, $R_1, R_2, \ldots, R_{m+1}$, are transient; the Markov chain starting from any transient state will eventually reach one of the closed subsets of recurrent states.

If $m = 1$, the Markov chain is called a *uni-chain*, and if $m > 1$, it is called a *multi-chain.* A uni-chain with no transient states is an ergodic chain.

An irreducible transition probability matrix is called *aperiodic* if $P^k > 0$, componentwisely, for some integer k (therefore, for all k that are larger). If such k does not exist, it is called *periodic.* (This definition is equivalent to (A.4).) A periodic transition probability matrix has a period κ such that, by relabelling the states, P can be written as

$$P = \begin{bmatrix} 0 & R_1 & 0 & \cdots & 0 \\ 0 & 0 & R_2 & \cdots & 0 \\ \cdot & \cdot & \cdot & \cdots & 0 \\ 0 & 0 & 0 & \cdots & R_{\kappa-1} \\ R_\kappa & 0 & 0 & \cdots & 0 \end{bmatrix},$$

where the diagonal blocks are square zero matrices. A stochastic matrix is called *aperiodic* if all the irreducible matrices $P_1, \ldots, P_m$ in the canonical form are aperiodic.

B.2 Eigenvalues

Let I denote the identity matrix whose diagonal elements are ones and other elements are zeros. For a given S-dimensional real or complex square matrix A, the determinant $det(\lambda I - A)$ defines a polynomial of λ with coefficients determined by A. The equation $det(\lambda I - A) = 0$ has exactly S real or complex roots denoted as $\lambda_1, \lambda_2, \ldots, \lambda_S$; they are called *eigenvalues* of A [20, 109]. If an eigenvalue λ appears only once in $\lambda_1, \ldots, \lambda_S$, it is said to be *simple.* If an eigenvalue λ appears in the sequence $m \geq 2$ times, it is said to have a *multiplicity* of m.

For any eigenvalue λ of A, there exists at least one (nonzero) column vector v satisfying $Av = \lambda v$. Such a vector v is called an *eigenvector* of A corresponding to the eigenvalue λ. If v is an eigenvector, then so is cv for any constant $c \neq 0$. Up to this multiplication, there is only one eigenvector corresponding to a simple eigenvalue. There are at most m linearly independent eigenvectors corresponding to an eigenvalue with m multiplicity.

The largest absolute value of the eigenvalues of a matrix A is called the *spectrum radius* of A and is denoted as $\rho(A)$.

Some results about eigenvalues are important in proving and understanding the main results in our book, and we state them as lemmas.

Eigenvalues of Irreducible Matrices

Lemma B.1. [20, 87]

(*a*) If $P > 0$ is a positive square matrix, then $\rho(P) =: \lambda_1$ is a simple and positive eigenvalue, and for any other eigenvalues λ of P we have $|\lambda| < \lambda_1$.

(*b*) If $P \geq 0$ is an irreducible non-negative matrix, then $\rho(P) =: \lambda_1$ is a simple and positive eigenvalue. (By definition, for any other eigenvalues λ of P we have $|\lambda| \leq \lambda_1$.)

  i) If, in addition, P is aperiodic, then for any other eigenvalues λ of P we have $|\lambda| < \lambda_1$.

  ii) If P is periodic with period κ, then there are exactly κ simple eigenvalues $\lambda_1, \ldots, \lambda_\kappa$ with absolute values exactly equal $\rho := \rho(P)$. These eigenvalues are

$$\lambda_1 = \rho, \ \lambda_2 = \rho \exp\left(\frac{2\pi}{\kappa}\sqrt{-1}\right), \ldots, \ \lambda_\kappa = \rho \exp\left(\frac{2(\kappa-1)\pi}{\kappa}\sqrt{-1}\right).$$

(*c*) If P is stochastic, then $\lambda_1 = \rho(P) = 1$.

In the literature, this lemma is known as the *Perron-Frobeniu* theorem. The results in Lemma B.1(*b*.i) for aperiodic matrices follow directly from that in Lemma B.1(*a*) for positive matrices. Indeed, if P is irreducible and aperiodic, there is an integer k such that $P^k > 0$ is positive. The eigenvalues of P^k are λ_i^k with λ_i being the eigenvalues of P, $i \in \mathcal{S}$. By Lemma B.1(*a*), λ_1 is a simple eigenvalue and all the other eigenvalues satisfy $|\lambda_i|^k < \lambda_1^k$ (i.e., $|\lambda| < \lambda_1$), $i \neq 1$.

Eigenvalues of Reducible Matrices in the Canonical Form

Next, we study reducible matrices that take the canonical form (B.1), in which $P_1, \ldots, P_m$ are irreducible. R_{m+1} cannot be periodic, because the corresponding states are transient. We have [20, 87]

Lemma B.2. (*a*) In the canonical form (B.1), we have $\rho(R_{m+1}) < 1$.
(*b*) For any square matrix A, if $\rho(A) < 1$, then $\lim_{k\to\infty} A^k = 0$, $(I - A)$ is invertible and

$$(I-A)^{-1} = \sum_{k=0}^{\infty} A^k.$$

Consider an aperiodic stochastic matrix in the canonical form (B.1). Denote the dimensions of $P_1, \dots, P_m$ as $S_1, \dots, S_m$, $S_k > 1$, $k = 1, 2, \dots, m$ (no absorbing state), and the dimension of R_{m+1} as S_{m+1}. We have $\sum_{k=1}^{m+1} S_k = S$. Following Lemma B.1(*b*.i), we may denote the eigenvalues of P_i, $i = 1, 2, \dots, m$, as $\lambda_{i,1} = 1, \lambda_{i,2}, \dots, \lambda_{i,S_i}$, with $\lambda_{i,k} \neq 1$ and $|\lambda_{i,k}| < 1$ for $k = 2, \dots, S_i$. We observe that the S_i-dimensional vector $e_{S_i} = (1, 1, \dots, 1)^T$ is an eigenvector of P_i corresponding to eigenvalue $\lambda_{i,1} = 1$, because $P_i e_{S_i} = e_{S_i}$. From Lemma B.2(*a*) and (*b*), $(I - R_{m+1})^{-1}$ exists. Set

$$w_i = (I - R_{m+1})^{-1} R_i e_{S_i}, \qquad i = 1, 2, \dots, m,$$

(an S_{m+1}-dimensional column vector), and define

$$v_{i,1} = [0, \dots, 0, e_{S_i}^T, 0, \dots, 0, w_i^T]^T, \tag{B.2}$$

in which e_{S_i} is the ith block, w_i is the $(m+1)$th block, and the other $(m-1)$ 0's represent the zero vectors with dimensions $S_1, \dots, S_{i-1}$, $S_{i+1}, \dots, S_m$, respectively. We can easily verify that $Pv_{i,1} = v_{i,1}$. Therefore, 1 is an eigenvalue of P with (at least) m multiplicity. Corresponding to this eigenvalue, there are (at least) m linearly independent eigenvectors $v_{i,1}$, $i = 1, 2, \dots, m$.

Denote the eigenvalues of R_{m+1} as $\lambda_{m+1,k}$, $k = 1, 2, \dots, S_{m+1}$ (there may be multiple eigenvalues). Let $u_{m+1,k}$ be the corresponding eigenvectors, $k = 1, 2, \dots, S_{m+1}$. It is easy to verify that P has the same eigenvalues with the corresponding eigenvectors

$$v_{m+1,k} = [0, \dots, 0, u_{m+1,k}^T]^T, \qquad k = 1, 2, \dots, S_{m+1}. \tag{B.3}$$

Now let $u_{i,k}$ be the eigenvector of P_i, $i = 1, 2, \dots, m$, corresponding to the eigenvalue $\lambda_{i,k}$, $k = 2, \dots, S_i$. We have $P_i u_{i,k} = \lambda_{i,k} u_{i,k}$, $\lambda_{i,k} \neq 1$, $|\lambda_{i,k}| \leq 1$. In this case, if we choose

$$v_{i,k} = [0, \dots, 0, u_{i,k}^T, 0, \dots, 0, w_{i,k}^T]^T,$$

where the block structure is the same as $v_{i,1}$ and $w_{i,k}$ satisfies

$$(\lambda_{i,k} I - R_{m+1}) w_{i,k} = \lambda_{i,k} \Big(I - \frac{1}{\lambda_{i,k}} R_{m+1}\Big) w_{i,k} = R_i u_{i,k}, \tag{B.4}$$

then we have $Pv_{i,k} = \lambda_{i,k}v_{i,k}$. That is, P has the same eigenvalue $\lambda_{i,k}$, whose corresponding eigenvector is $v_{i,k}$. Note that the eigenvalues of $(I - \frac{1}{\lambda_{i,k}}R_{m+1})$ are $1 - \frac{\lambda_{m+1,l}}{\lambda_{i,k}}$, with $\lambda_{m+1,l}$, $l = 1, 2, \ldots, S_{m+1}$ being the eigenvalues of R_{m+1}. Thus, if $\lambda_{m+1,l} \neq \lambda_{i,k}$, $l = 1, 2, \ldots, S_{m+1}$, then $\left(I - \frac{1}{\lambda_{i,k}}R_{m+1}\right)^{-1}$ exists and (B.4) has a unique solution

$$w_{i,k} = \frac{1}{\lambda_{i,k}}\left(I - \frac{1}{\lambda_{i,k}}R_{m+1}\right)^{-1} R_i u_{i,k}.$$

On the other hand, if $\lambda_{m+1,l} = \lambda_{i,k}$ for some $l = 1, 2, \ldots, S_{m+1}$, then $(I - \frac{1}{\lambda_{i,k}}R_{m+1})$ is not of full rank, and there may be more than one $w_{i,k}$ satisfying (B.4). Indeed, in this case, if $v_{i,k}$ is an eigenvector of P corresponding to $\lambda_{i,k}$, and $v_{m+1,l}$ in (B.3) is an eigenvector corresponding to $\lambda_{m+1,l} = \lambda_{i,k}$, then for any constant c,

$$v_{i,k} + cv_{m+1,l} = [0, \ldots, 0, u_{i,k}^T, 0, \ldots, 0, w_{i,k}^T + cu_{m+1,l}^T]^T$$

is also an eigenvector of P corresponding to $\lambda_{i,k}$. A similar discussion applies when there is more than one eigenvalue of R_{m+1} equal to $\lambda_{i,k}$.

In summary, we have the following lemma.

Lemma B.3. Let P be a stochastic matrix in the canonical form (B.1) in which $P_1, \ldots, P_m$ are irreducible square matrices with dimensions $S_1, \ldots, S_m$, respectively, and let S_{m+1} be the dimension of R_{m+1}. Denote the set of eigenvalues of $P_1, \ldots, P_m$ as $\{\lambda_{i,1} = 1, \lambda_{i,2}, \ldots, \lambda_{i,S_i}\}$, $i = 1, 2, \ldots, m$, respectively, with $\lambda_{i,k} \neq 1$, $|\lambda_{i,k}| \leq 1$, $k = 2, \ldots, S_i$, $i = 1, 2, \ldots, m$, and the set of eigenvalues of R_{m+1} as $\{\lambda_{m+1,1}, \ldots, \lambda_{m+1,S_{m+1}}\}$, with $|\lambda_{m+1,k}| < 1$, $k = 1, 2, \ldots, S_{m+1}$. Then the set of eigenvalues of P is

$$\cup_{i=1}^{m}\{\lambda_{i,1} = 1, \lambda_{i,2}, \ldots, \lambda_{i,S_i}\} \cup \{\lambda_{m+1,1}, \ldots, \lambda_{m+1,S_{m+1}}\},$$

with $\lambda_{i,1} = 1$, $i = 1, \ldots, m$, being an eigenvalue of m multiplicity. If, furthermore, P is aperiodic, then the magnitude of all the other eigenvalues is less than one.

B.3 The Limiting Matrix

Recall that, from (A.8), if P is aperiodic and irreducible, then $\lim_{k\to\infty} P^k = e\pi$, where π is the steady-state probability vector satisfying $\pi = \pi P$. This asymptotical stationarity does not hold for periodic Markov chains.

The Limiting Matrix P^*

In general, for both aperiodic and periodic matrices (with multi-chains) we may use the *Cesaro limit*:

$$P^* = \lim_{N\to\infty} \frac{1}{N} \sum_{n=0}^{N-1} P^n. \tag{B.5}$$

This limit exists for any transition matrix (see [86, 87]). In component notation, this is

$$p^*(j|i) = \lim_{N\to\infty} \frac{1}{N} \sum_{n=0}^{N-1} p^{(n)}(j|i),$$

where $p^{(n)}(j|i)$ denotes the component of P^n, and $p^{(0)}(j|i) = 1$ if $i = j$, 0 otherwise. Clearly $p^*(j|i)$ is the long-run average of the number of visits to state j on a sample path starting from state i. When P is ergodic, $p^*(j|i) = \pi(j)$, independent of i. Therefore, P^* measures the sample average, which is equivalent to $e\pi$ for ergodic chains. If j is a transient state in a multi-chain matrix P, $p^*(j|i) = 0$ for all $i \in \mathcal{S}$. From this probabilistic meaning, we can be convinced that the Cesaro limit (B.5) exists.

From (B.5), it is easy to verify the following properties:

$$P^*e = e, \tag{B.6}$$

which corresponds to $\pi e = 1$, and

$$P^*P = PP^* = P^*P^* = P^*, \tag{B.7}$$

which corresponds to $\pi = \pi P$. Note that the solutions to both (B.6) and (B.7) are not unique for a multi-chain stochastic matrix P. For example, if

$$P = \begin{bmatrix} 1 & 0 & 0 \\ 0 & 1 & 0 \\ \frac{1}{3} & \frac{1}{3} & \frac{1}{3} \end{bmatrix},$$

then

$$P^* = \begin{bmatrix} c & 1-c & 0 \\ c & 1-c & 0 \\ c & 1-c & 0 \end{bmatrix},$$

for any $0 \le c \le 1$, satisfies both (B.6) and (B.7). This is different from the ergodic case where $\pi e = 1$ and $\pi = \pi P$ uniquely determine π. Indeed, (B.7) is a property that specifies the steady-state (stationary) distributions (cf. (A.6)). In the multi-chain case, we may still use (A.5) to define the steady-state distributions, but they are not unique, and the asymptotical stationarity

(A.7) do not hold. Furthermore, if $\pi P = \pi$, then $e\pi$ is a solution to (B.6) and (B.7). However, there may be more than one steady-state distribution π satisfying $\pi P = \pi$ for multi-chain stochastic matrices. In the example, $\pi = (c, 1-c, 0)$ for any $0 \le c \le 1$. On the other hand, P^* in (B.5) reflects the long-run average starting from a particular initial state. Every row of P^* can be viewed as a steady state distribution of P, satisfying $\pi P = \pi$. In addition, for any stochastic matrix P, there exists only one matrix P^* with the form of (B.1) that satisfies both (B.6) and (B.7) (see Problem B.2).

When P is aperiodic, $\lim_{N\to\infty} P^N$ exists. We have

$$P^* = \lim_{N\to\infty} \frac{1}{N} \sum_{n=0}^{N-1} P^n = \lim_{N\to\infty} P^N. \tag{B.8}$$

This is to say, the sample average equals the steady-state probabilities.

If P is aperiodic, then in the canonical form (B.1), we have [216]

$$P^* = \begin{bmatrix} P_1^* & 0 & 0 & \cdots & \cdot & 0 \\ 0 & P_2^* & 0 & \cdots & \cdot & 0 \\ \cdot & \cdot & \cdot & \cdots & \cdot & \cdot \\ 0 & 0 & 0 & \cdots & P_m^* & 0 \\ R_1^* & R_2^* & R_3^* & \cdots & R_m^* & 0 \end{bmatrix}, \tag{B.9}$$

where $P_i^* = \lim_{N\to\infty} P_i^k$, $i = 1, 2, \ldots, m$. Let π_i, $i = 1, 2, \ldots, m$, be the steady-state probability vector of the ith closed irreducible subset of the state space of the Markov chain. We have $\pi_i = \pi_i P_i$, $\pi_i e_{S_i} = 1$, $i = 1, 2, \ldots, m$, and $P_i^* = e_{S_i}\pi_i$, $i = 1, 2, \ldots, m$. From Lemma B.2(a), $\rho(R_{m+1}) < 1$. Thus, from Lemma B.2(b), $\lim_{k\to\infty} R_{m+1}^k = 0$ and $I - R_{m+1}$ is invertible, and from $PP^* = P^*$, we can verify that [216]

$$\begin{aligned} R_i^* &= (I - R_{m+1})^{-1} R_i P_i^* \\ &= (I - R_{m+1})^{-1} R_i e_{S_i} \pi_i \\ &= w_i \pi_i, \end{aligned} \tag{B.10}$$

where $w_i = (I - R_{m+1})^{-1} R_i e_{S_i}$. Denote $w_i = (w_i(1), \ldots, w_i(S_{m+1}))^T$, $i = 1, 2, \ldots, m$, where S_{m+1} is the number of the transient states of P. From $Pe = e$, we have $\sum_{i=1}^m R_i e_{S_i} + R_{m+1} e_{S_{m+1}} = e_{S_{m+1}}$. Therefore, $\sum_{i=1}^m w_i = e_{S_{m+1}}$. Finally, (B.9) takes the form

$$P^* = \begin{bmatrix} e\pi_1 & 0 & 0 & \cdots & \cdot & 0 \\ 0 & e\pi_2 & 0 & \cdots & \cdot & 0 \\ \cdot & \cdot & \cdot & \cdots & \cdot & \cdot \\ 0 & 0 & 0 & \cdots & e\pi_m & 0 \\ w_1\pi_1 & w_2\pi_2 & w_3\pi_3 & \cdots & w_m\pi_m & 0 \end{bmatrix}. \tag{B.11}$$

This form has a clear probabilistic meaning. First, we have

$$(I - R_{m+1})^{-1} R_i = R_i + R_{m+1} R_i + \cdots + R_{m+1}^k R_i + \cdots.$$

We note that the first component of the column vector $R_i e_{S_i}$ is the probability that, starting from the first transient state, the Markov chain enters the ith irreducible closed set of the recurrent states after one state transition; the first component of the column vector $R_{m+1} R_i e_{S_i}$ is the probability that, starting from the first transient state, the Markov chain enters the ith irreducible closed set of the recurrent states after two state transitions; and so on. Therefore, the first component of $w_i = (I - R_{m+1})^{-1} R_i e_{S_i}$, $w_i(1)$, is the total probability that starting from the first transient state the Markov chain eventually enters the ith irreducible closed set of the recurrent states. Once it enters the ith irreducible closed set, the steady-state probability, or the long-run average, will be π_i. Therefore, starting from the first transient state, the probability that the system visits the states in the ith irreducible closed set is $w_i(1)\pi_i$, $i = 1, 2, \ldots m$. The same explanation applies to other components. This also explains $\sum_{i=1}^m w_i = e_{S_{m+1}}$.

The Matrix $(I - P + P^*)^{-1}$

In performance analysis of Markov systems, we always encounter equations such as $\pi = \pi P$, or $\pi(I - P) = 0$. However, $(I - P)$ is not invertible. In many problems in performance sensitivity analysis, we need to study the matrix $(I - P + P^*)$. We first prove that it is invertible.

Let us examine its eigenvalues. From Lemma B.3, we define $\{1, 1, \ldots, 1, \lambda_{m+1}, \ldots, \lambda_S\}$ to be the set of the eigenvalues of P with "1" appearing m times (some of the other eigenvalues λ_k, $k = m+1, \ldots, S$, may also be repeated). We have $|\lambda_i| \leq 1$ for all $i = m+1, \ldots, S$ ($|\lambda_i| < 1$ if P is aperiodic). Now, we show that λ_i, $i = m+1, \ldots, S$, are also eigenvalues of $P - P^*$. To see this, let $v_i \neq 0$ be the eigenvector of P corresponding to λ_i; i.e., $P v_i = \lambda_i v_i$, $i = m+1, \ldots, S$. If $\lambda_i \neq 0$, then

$$P^* v_i = \frac{1}{\lambda_i}(P^*) P v_i = \frac{1}{\lambda_i} P^* v_i.$$

Thus $P^* v_i = 0$ since $\lambda_i \neq 1$. We have $(P - P^*) v_i = P v_i = \lambda_i v_i$. That is, λ_i is an eigenvalue of $P - P^*$ with the same eigenvector v_i. On the other hand, if $\lambda_i = 0$, then $P v_i = 0$ and $(P - P^*) v_i = (I - P^*) P v_i = 0$. That is, $\lambda_i = 0$ is also an eigenvalue with the same eigenvector v_i. Thus, we have proved that $\lambda_{m+1}, \ldots, \lambda_S$ are eigenvalues of $P - P^*$ with the same eigenvectors.

Next, we turn our attention to the m-multiple eigenvalues 1. Let $v_i \neq 0$, $i = 1, 2, \ldots, m$, be the corresponding eigenvectors. We have $P v_i = v_i$. From this, we get $P^k v_i = v_i$ for any integer k. From (B.5), we have $P^* v_i = v_i$. Thus, $(P - P^*) v_i = 0$. That is, v_i, $i = 1, 2, \ldots, m$, are eigenvalues of $(P - P^*)$ corresponding to the eigenvalue 0. As shown in the proof of Lemma B.3, the m vectors v_i, $i = 1, 2, \ldots, m$, (the $v_{i,1}$'s in (B.2)) are linearly independent. Therefore, $P - P^*$ has an eigenvalue 0 with m multiplicity.

Therefore, the set of eigenvalues of $(P - P^*)$ is $\{0, 0, \dots, 0, \lambda_{m+1}, \dots, \lambda_S\}$ with the same set of corresponding eigenvectors of P. The first m items are zeros and $|\lambda_i| \leq 1$, $\lambda_i \neq 1$ for all $i = m+1, \dots, S$. It is also possible that there are some more zeros among $\{\lambda_{m+1}, \dots, \lambda_S\}$.

It is easy to see that the set of eigenvalues of $I - P + P^*$ is $\{1, 1, \dots, 1, 1 - \lambda_{m+1}, \dots, 1 - \lambda_S\}$; none of them is zero. Therefore, $(I - P + P^*)$ is invertible and the eigenvalues of $(I - P + P^*)^{-1}$ are $1, \dots, 1, \frac{1}{1-\lambda_{m+1}}, \dots, \frac{1}{1-\lambda_S}$, with $|\lambda_i| \leq 1$, $i = m+1, \dots, S$.

If P is aperiodic, then $|\lambda_i| < 1$, $i = m+1, \dots, S$, and $\rho(P - P^*) < 1$. From Lemma B.2(b), we have

$$(I - P + P^*)^{-1} = I + \sum_{k=1}^{\infty} (P - P^*)^k.$$

It is easy to verify $(P - P^*)^k = P^k - P^*$. Therefore,

$$(I - P + P^*)^{-1} = I + \sum_{k=1}^{\infty} (P^k - P^*). \tag{B.12}$$

The matrix $(I - P + P^*)^{-1}$ is called the *fundamental matrix*.

However, if P is periodic, we have $\rho(P - P^*) = 1$, and the expansion (B.12) does not hold for the fundamental matrix.

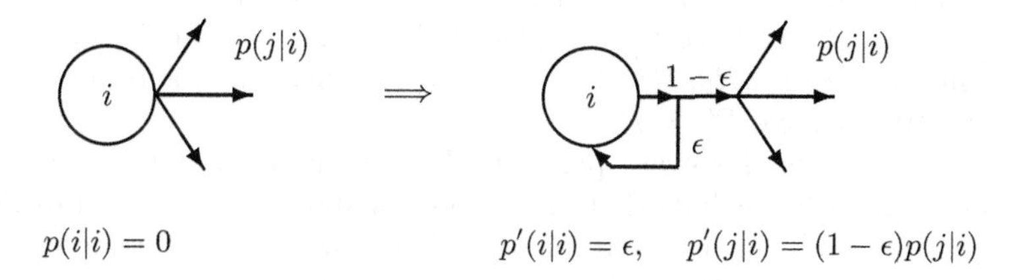

Fig. B.1. The Equivalent Aperiodic Matrix

Equivalent Aperiodic Markov Chains

We can see that the periodicity of a Markov chain causes some technical difficulties in analyzing its performance: The matrix expansion (B.12) and the asymptotic stationarity (A.7) do not hold, and the limit, $\lim_{k\to\infty} P^k$, has to be replaced by the Cesaro limit (B.5). However, for every periodic Markov chain, we can easily construct an aperiodic Markov chain equivalent to it in the sense that the steady-state probabilities of the aperiodic Markov chain equal the long-run average of the periodic one.

First, we note that $p(i|i) = 0$ for all $i \in \mathcal{S}$ in any periodic matrix. Now, suppose that P is irreducible and periodic and let $\pi = \pi P$ be its steady-state distribution, which represents the long-run average of the periodic Markov chain. For any $0 < \epsilon < 1$, we define a stochastic matrix $P' = \epsilon I + (1-\epsilon)P$, i.e., $p'(i|i) = \epsilon$, $p'(j|i) = (1-\epsilon)p(j|i)$, $j \neq i$, for all $i, j \in \mathcal{S}$ (see Figure B.1). P' is irreducible because P is irreducible, P' is aperiodic because $p'(i|i) \neq 0$. Let $\pi' = \pi' P'$ be the steady-state probability of P'. We have

$$\pi' = \pi' P' = \pi'[\epsilon I + (1-\epsilon)P]$$
$$= \epsilon\pi' + (1-\epsilon)\pi' P.$$

From this, we have $\pi' = \pi' P$, and therefore $\pi' = \pi$.

The results can be explained intuitively. Denote the two Markov chains with P and P' as $\boldsymbol{X}$ and $\boldsymbol{X}'$, respectively. $\boldsymbol{X}'$ evolves as follows: Whenever $\boldsymbol{X}'$ visits a state i, it remains visiting the state a few times, then it "jumps" to another state, following the transition probability matrix P, in the same way as $\boldsymbol{X}$. The number of times that $\boldsymbol{X}'$ stays in a same state before it jumps is identically distributed with a geometric distribution with the same mean $\frac{1}{1-\epsilon}$. Therefore, the long-run sample average of the number of visits to any particular state of $\boldsymbol{X}$ and $\boldsymbol{X}'$ must be the same. $\boldsymbol{X}'$ is aperiodic because the number of consecutive visits to any state is random.

The above results extend to reducible Markov chains. Suppose that, in a periodic stochastic matrix P in the canonical form (B.1), some blocks among $P_1, \ldots, P_m$, denoted as $P_{l_1}, \ldots, P_{l_k}$, $1 \leq l_1, \ldots, l_k \leq m$, are periodic. We define an equivalent stochastic matrix P' by replacing these blocks in P with $P'_{l_i} = \epsilon I + (1-\epsilon)P_{l_i}$, $i = 1, \ldots, k$. P' is aperiodic because every P'_k, $k = 1, \ldots, m$, is aperiodic. Applying the above result for irreducible matrices to P'_{l_i}, $i = 1, \ldots, k$, we have $\pi'_k = \pi_k$ for all $k = 1, 2 \ldots, m$ in the form (B.11). In addition, because R_k, $k = 1, \ldots, m+1$, are the same for both P' and P, from (B.10) we have $P'^* = P^*$.

Finally, because for every periodic Markov chain, we have an equivalent aperiodic one, we conclude that all the results about the steady-state performance of aperiodic chains hold for periodic chains as well if we interpret the steady-state probabilities as the long-run time averages.

PROBLEMS

B.1. In the canonical form (B.1), R_{m+1} may be further reducible.

a. Write R_{m+1} in a canonical form, and
b. Explain the meaning of this canonical form in terms of the transitions of the transient states.

B.2. Derive a general form for the solution to (B.6) and (B.7).

B.3. Many results for a series of real numbers have their counterparts in matrix form. For example, for a real number series, we have $\frac{1}{1-x} = 1 + x +$

$x^2+\dots$ if $|x|<1$; and for a matrix series we have $(I-P)^{-1}=I+P+P^2+\dots$ if $\rho(P)<1$. In real analysis we have the following Stolz theorem: For two series of real numbers x_n and y_n, $n=1,2,\dots$, if $y_{n+1}>y_n$, $n=1,2,\dots$, $\lim_{n\to\infty} y_n=\infty$, and $\lim_{n\to\infty}\frac{x_{n+1}-x_n}{y_{n+1}-y_n}$ exists, then

$$\lim_{n\to\infty}\frac{x_n}{y_n}=\lim_{n\to\infty}\frac{x_{n+1}-x_n}{y_{n+1}-y_n}.$$

a. Prove the Stolz theorem.
b. Prove that if $\lim_{n\to\infty} x_n$ exists, then $\lim_{n\to\infty}\frac{1}{n}\sum_{k=1}^{n} x_n=\lim_{n\to\infty} x_n$.
c. Prove the matrix formula (B.8).

B.4. Let P be an irreducible periodic stochastic matrix. We have $p(i|i)=0$ for all $i\in\mathcal{S}$. To break the periodicity, it is enough to introduce a "feedback probability" $p(i|i)=\epsilon$ for only one state i (but not for all states). Therefore, we define an aperiodic matrix by setting $p'(i|i)=\epsilon$, $p'(j|i)=(1-\epsilon)p(j|i)$, $j\neq i$ for one particular state i, and $p'(k|j)=p(k|j)$ for $k\in\mathcal{S}$, $j\neq i$.

a. Express the steady-state probabilities $\pi'(i)$ of P' in terms of ϵ and the steady-state probabilities $\pi(i)$ of P.
b. Let f denote the reward function and $\eta=\pi f$ be the long-run average reward for the Markov chain with transition probability matrix P. Define a reward function f' so that the long-run average reward of the Markov chain with transition probability matrix P', $\eta'=\pi' f'$, equals η.

Do not worry about your difficulties in mathematics. I can assure you mine are still greater.

Albert Einstein
American (German born)
physicist (1879 - 1955)

Queueing Theory

In this section, we review some material from the standard queueing theory [169] that is related to the topics in this book. This section also contains results about performance sensitivity formulas and their computational algorithms, which are particularly relevant to performance optimization.

C.1 Single-Server Queues

The Model

A single-station queue is the basic component of queueing systems. Figure C.1 illustrates a model of a single-station queue containing only one server. In the figure, the circle represents the server and the open box represents a buffer. Customers arrive at a sequence of random times denoted as $0 \leq t_1 \leq t_2 \leq \ldots$. The server serves one customer at a time, and the other customers have to wait in the queue held in the buffer. The nth customer, arriving at the server at t_n, requires a certain amount of service that will take the server s_n time units to process. A customer leaves the server once it finishes its service. The nth customer's departure time is denoted as t'_n, $n = 1, 2 \ldots$. Define $\tau_n := t_{n+1} - t_n$ as the *inter-arrival time* and $t'_{n+1} - t'_n$, as the *inter-departure time*. $\{t_1, t_2, \ldots\}$ is called an *arrival process* and $\{t'_1, t'_2, \ldots\}$ is called a *departure process*. A multi-server station contains multiple servers sharing a common queue.

A buffer may have an infinite capacity or a finite capacity accommodating K customers. A customer arriving at a full buffer will be lost (e.g., in the case of a single-server queue) or will have to wait elsewhere (e.g., in some other server in a network of queues).

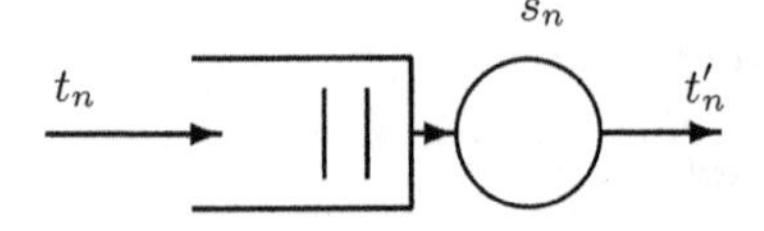

Fig. C.1. A Single-Server Queue

It is conventional to adopt the four-part description $A/B/m/K$ to specify a single-station queue. In this description, m denotes the number of servers in the station, K defines the buffer size, and A and B specify the types of the distributions of the inter-arrival times τ_n and the service times s_n.

The most common arrival process is the *Poisson process*, in which the inter-arrival times $\tau_n = t_{n+1} - t_n$, $n = 1, 2, \ldots$, are independent and exponentially distributed. The rate of the exponential distribution λ is called the rate of the Poisson process.

The most common single-server queue is the $M/M/1$ queue, where M/M indicates that both the inter-arrival times and the service times are independent and identically distributed (i.i.d.) with exponential distributions. (Hence the arrival process is a Poisson process.) The last letter $K = \infty$ is omitted for simplicity. Examples of other queues are the $M/G/1$ queue, where G indicates that the service times are independent and identically distributed with a general (non-exponential) distribution, and the $GI/M/1$ queue, where GI indicates that the arrival times form a renewal process (i.e., the inter-arrival times are i.i.d. with a general distribution). Other commonly used symbols for A and B are: D for deterministic distributions, E_r for r-stage Erlangian distributions, H_R for R-stage hyper-exponential distributions [169], and PH for phase-type distributions [209].

Service Disciplines

A service discipline determines which customer is served at any given time. We first assume that the *work-conservative* law holds; i.e., a server will provide service at its full capacity as long as its queue is not empty. Thus, if any server in the service station is idle, then an arriving customer gets served immediately.

Some commonly used service disciplines for a server are as follows.

1. *First come first served (FCFS):* The customer who arrives at the queue first is served first by the server.
2. *The priority scheme:* Customers are assigned different priorities and the customers with the highest priority in the queue get served first. A priority scheme may be either *preemptive* or *non-preemptive*. In a preemptive scheme, a customer being served is liable to be ejected from service when a customer with a higher priority enters the queue. In a non-preemptive scheme, the arriving customer has to wait in the queue until the server

completes its service to its current customer, even if the arriving customer has a higher priority. A preemptive scheme may allow service to resume or not to resume. In a preemptive resume scheme, the service that a preempted customer has received is not lost when the service is resumed. In a preemptive non-resume scheme, the preempted customer has to start the whole service again when it restarts service. The customers in the same priority class may follow the FCFS scheme, or the LCFS scheme introduced below, or any other possible schemes.

3. *Last come first served (LCFS):* The customer who arrives at the queue last receives the service first. This scheme has different versions: preemptive resume, preemptive non-resume, and non-preemptive.
4. *Processor sharing (PS):* The service power of the server is shared equally by all the customers in the queue. Thus, if the service rate of a server is μ and there are n customers in the queue, then the service rate for each customer is μ/n. In modelling computer systems, a *round-robin* scheme with a small quantum size can be approximated by a PS scheme.

The $M/M/1$ Queue

In an $M/M/1$ queue, the arrival process is Poisson and the service time distribution is exponential. Because of the memoryless property of the exponential distribution, the state of an $M/M/1$ queue can be simply chosen as the number of customers in the queue, denoted as n. The state process is a Markov process.

Let λ and $\mu(> \lambda)$ be the arrival and the service rate, respectively. The stationary distribution of n, $\pi(n)$, can be easily obtained by solving the probability flow-balance equations; we have

$$\pi(n) = \rho^n(1-\rho), \qquad \rho = \frac{\lambda}{\mu}, \quad n = 0, 1, \ldots.$$

Let $\bar{n} := \sum_{n=0}^{\infty} n\pi(n)$ be the average number of customers in the queue (including the customer receiving service), $\bar{n}_b$ be the average number of customers in the buffer (excluding the customer receiving service), $\bar{\tau}$ be the average time that the customers spend in the system (including the service time), and W be the average waiting time (excluding the service time) of the customers. Then,

$$\bar{n} = \frac{\rho}{1-\rho},$$

$$\bar{n}_b = \frac{\rho^2}{1-\rho},$$

$$\bar{\tau} = \frac{1}{\mu-\lambda},$$

and

$$W = \bar{\tau} - \frac{1}{\mu} = \frac{\rho}{\mu - \lambda} = \frac{\rho/\mu}{1-\rho}.$$

In this book, we mainly deal with discrete-time model. A *discrete $M/M/1$ queue* can be modelled as a Markov chain with $p(1|0) = p_a$, $p(0|0) = 1 - p_a$, $p(n-1|n) = p_d$, $p(n+1|n) = p_a$, and $p(n|n) = 1 - p_a - p_d$, with $0 < p_a, p_d < 1$, $p_a + p_d \leq 1$, $n = 1, 2, \ldots$. In the model, "p_a" denotes the arrival probability, and "p_d" denotes the departure probability.

The $M/D/1$ Queue

The arrival process is a Poisson process with rate λ and the service times are a constant $\bar{s}$. Set $\rho = \lambda\bar{s}$. We have

$$\bar{n} = \frac{\rho}{1-\rho} - \frac{\rho^2}{2(1-\rho)}$$

and

$$W = \frac{\rho\bar{s}}{2(1-\rho)}.$$

Thus, the average waiting time of a customer in an $M/D/1$ queue is only half of that in an $M/M/1$ queue. (Randomness increases the waiting time!)

The $M/M/1/K$ Queue

An $M/M/1/K$ queue is the same as an $M/M/1$ queue except its buffer size is finite, denoted as K. If a customer arrives when the system contains K customers (including the one being served), the arriving customer is simply lost. The steady-state probabilities are $\pi(n) = 0$ if $n > K$ and

$$\pi(n) = \frac{(1-\rho)\rho^n}{1-\rho^{K+1}}, \qquad 0 \leq n \leq K, \quad \rho = \frac{\lambda}{\mu}.$$

The $M/G/1$ Queue

The arrival process is a Poisson process with rate λ and let $b(s)$ be the probability density function of the service time. Define the mean and the second moment of the service time as

$$\bar{s} = \int_0^\infty s b(s) ds, \qquad \bar{s^2} = \int_0^\infty s^2 b(s) ds,$$

respectively. Set $\rho = \lambda\bar{s}$. We have

$$\bar{n} = \rho + \frac{\lambda^2 \bar{s^2}}{2(1-\rho)},$$

and

$$W = \frac{\rho\bar{s}}{2(1-\rho)}(1+C_b^2) = \frac{\rho\bar{s}}{2(1-\rho)}\frac{\bar{s^2}}{\bar{s}^2},$$

where $C_b^2 = \frac{\bar{s^2}-\bar{s}^2}{\bar{s}^2}$. We can see that both $\bar{n}$ and W are linear in $\bar{s^2}$.

Readers are referred to standard textbooks [110, 169] for formulas for other single-station queues.

Little's Law

We observe that for an $M/M/1$ queue, we have

$$\bar{n} = \lambda\bar{\tau} \tag{C.1}$$

and

$$\bar{n}_b = \lambda W. \tag{C.2}$$

These two equations are *Little's Law* for an $M/M/1$ queue.

In general, Little's law applies to any system, including queueing networks or subnetworks. It says that *the average number of customers in any system (or any subsystem) equals the product of the average time that a customer stays in the system (or the subsystem) and the average arrival rate of the customers to the system (or the subsystem).* The customer arrival process may be any point process, not necessarily a renewal process (in which the inter-arrival times are i.i.d.).

Applying Little's law to the subsystem consisting of only the buffer in the $M/M/1$ queue, we obtain (C.2). Applying it to the system consisting of both the buffer and the server, we have (C.1). This is shown in Figure C.2. In general, we may view the dashed box in the figure as a black box that may be any complex system and (C.1) still holds.

If there are different classes of customers, Little's law applies to each class as well as to all classes of customers together. Let λ_k be the arrival rate of the class k customers, $\bar{n}_k$ be the average number of the class k customers in a system, and $\bar{\tau}_k$ be the average time that the class k customers stay in the system, $k = 1, \ldots, K$. Furthermore, let $\lambda = \sum_{k=1}^{K} \lambda_k$ be the arrival rate of all classes of customers, $\bar{n} = \sum_{k=1}^{K} \bar{n}_k$ be the average number of all classes of customers in the system, and $\bar{\tau}$ be the average time that all classes of customers stay in the system. Then, by Little's law, we have

$$\bar{n}_k = \lambda_k \bar{\tau}_k, \qquad k = 1, 2, \ldots, K,$$

and

$$\bar{n} = \lambda\bar{\tau},$$

where

$$\bar{\tau} = \sum_{k=1}^{K} \frac{\lambda_k}{\left(\sum_{l=1}^{K} \lambda_l\right)} \bar{\tau}_k;$$

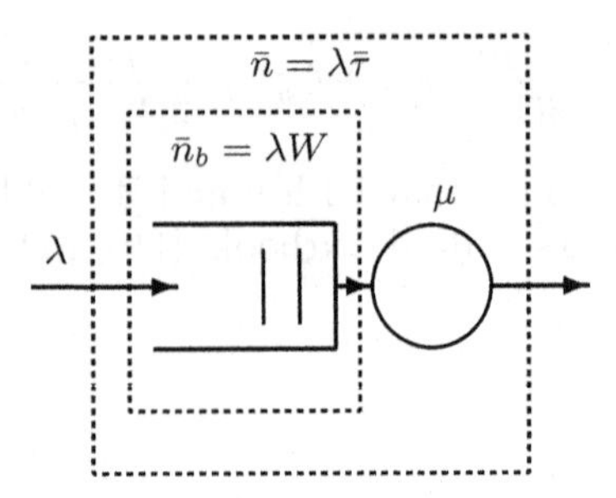

Fig. C.2. Little's Law for an $M/M/1$ Queue

this, of course, makes sense in terms of probability.

C.2 Queueing Networks

A queueing network is a system consisting of a number of service stations. Customers in a queueing network move among servers or leave the network according to certain routing mechanisms. Customers may belong to different classes, meaning that they may have different routing mechanisms, different service time distributions, or different service priorities. A queueing network may belong to one of the three types: open, closed, or mixed. In an open network, customers arrive at the network from outside and eventually leave the network; in a closed network, customers circulate among stations and no customer arrives or leaves the network; a mixed network is open to some classes of customers and is closed to other classes.

Jackson Networks

We consider an open network consisting of M single-server stations and N single-class customers. Each server has a buffer with an infinite capacity and the service discipline is FCFS. Customers arrive at server i in a Poisson process with a rate $\lambda_{0,i}$, $i = 1, 2, \ldots, M$. After the completion of the service at server i, a customer enters server j with probability $q_{i,j}$ and leaves the network with probability $q_{i,0}$. We have $\sum_{j=0}^{M} q_{i,j} = 1$, $i = 1, 2, \ldots, M$. The service time of server i is exponentially distributed with mean $\bar{s}_i = 1/\mu_i$, $i = 1, 2, \ldots, M$. Such a network is called an (open) *Jackson network* [160].

Because of the memoryless property of the exponential distribution, the system state can be denoted as $\boldsymbol{n} = (n_1, n_2, \ldots, n_M)$, where n_i is the number of customers in server i. Let λ_i be the overall arrival rate (including both external and internal arrivals) of the customers to server i. Then,

$$\lambda_i = \lambda_{0,i} + \sum_{j=1}^{M} \lambda_j q_{j,i}, \qquad i = 1, 2, \ldots, M.$$

It is known that in an acyclic open Jackson network (in which a customer does not visit the same server more than once), the overall arrival process to each server is a Poisson process. However, if there are feedback loops in the network, the overall arrival process to a server is generally not a Poisson process [166]. Thus, each server behaves differently from an $M/M/1$ queue. Nevertheless, the steady-state distribution, $\pi(\boldsymbol{n})$, of the network looks the same as if every server is an $M/M/1$ queue. Indeed, we have

$$\pi(\boldsymbol{n}) = \pi(n_1, n_2, \ldots, n_M) = \prod_{k=1}^{M} \pi(n_k), \tag{C.3}$$

with

$$\pi(n_k) = (1 - \rho_k)\rho_k^{n_k}, \qquad \rho_k = \frac{\lambda_k}{\mu_k}, \quad k = 1, 2, \ldots, M.$$

This shows that in an open Jackson network, each server behaves as if it is an independent $M/M/1$ queue with arrival rate λ_k and service rate μ_k, $k = 1, 2, \ldots, M$.

Closed Jackson (Gordon-Newell) Networks

In a closed Jackson (or Gordon-Newell) network [122], there are N customers circulating among M servers according to the routing probabilities $q_{i,j}$, $\sum_{j=1}^{M} q_{i,j} = 1$, $i = 1, 2, \ldots, M$; and no customers are allowed to enter the network from the outside or to leave the network. We have $\sum_{k=1}^{M} n_k = N$. We assume that the network is irreducible, in the sense that the routing probability matrix $Q := [q_{i,j}]$ is irreducible. In such a network, every customer will visit every server in the network.

We consider a more general model for the servers, called *load-dependent servers.* In this model, a customer requires a certain amount of service, r, from a server. The service time of a customer equals its service requirement divided by the service rate of the server. We assume that the service requirement r is exponentially distributed with a mean $\bar{r}$ equal to one. The service rate of a server may depend on the number of customers in the server (called the *"load"* of the server). Let μ_{i,n_i} be the service rate of server i when there are n_i customers in the server, $0 \leq \mu_{i,n_i} < \infty$, $n_i = 1, 2, \ldots, N$, $i = 1, 2, \ldots, M$. Note that the assumption of the mean service requirement being one does not lose generality, since a service requirement with mean $\bar{r} > 0$ and service rate μ_{i,n_i} is equivalent to a service requirement with mean one and a service rate $\mu_{i,n_i}/\bar{r}$, in the sense that they have the same service time distribution.

In a network with load-independent servers, $\mu_{i,n_i} \equiv \mu_i$ for n_i, $i = 1, 2, \ldots, M$. The service time at server i is exponentially distributed with mean $\bar{s}_i = 1/\mu_i$, $i = 1, 2, \ldots, M$.

Figure C.3 illustrates a three-server closed Jackson network with mean service times $\bar{s}_i$, and routing probabilities $q_{i,j}$, $i, j = 1, 2, 3$.

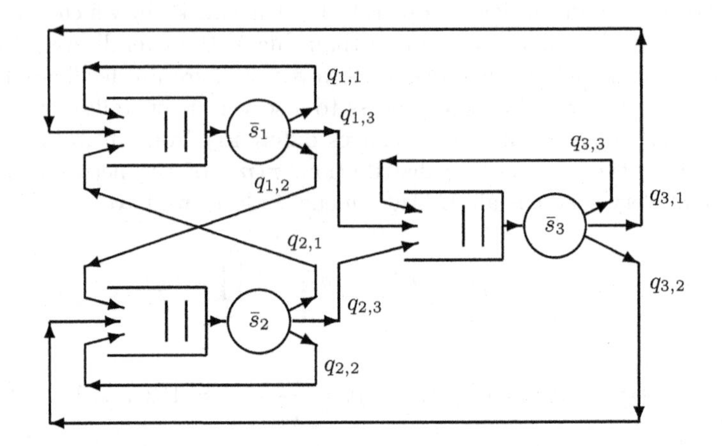

Fig. C.3. A Three-Server Closed Jackson Network

Because the capacity of any real-world system is always finite, open networks, even the $M/M/1$ queue, are always approximate in modelling and performance analysis of real systems. On the other hand, closed networks may provide exact models. For example, consider a manufacturing system consisting of M machines modelled as servers. Each workpiece has to be loaded on a pallet in order to be processed by the machines. Each workpiece moves among machines on a pallet, and when the workpiece finishes its service in the system it moves to a load-and-unload machine that unloads the workpiece from the pallet and loads another raw workpiece on it. There are N pallets moving around in the system. Assume that there are always more than N workpieces available, including those in the system and those waiting to be loaded at the load-and-unload station. Such a system clearly can be modelled as a closed network with M servers and N customers. Closed networks can also be used to model packet switches in communication [52, 68] (see Problem C.6).

The Product-Form Solution

The state of a closed Jackson network is $\boldsymbol{n} = (n_1, n_2, \dots, n_M)$. Let $\mathcal{S}$ denote the state space; it contains $\frac{(N+M-1)!}{N!(M-1)!}$ states. Let $\pi(\boldsymbol{n})$ be the steady-state probability of state $\boldsymbol{n}$. We use $\boldsymbol{n}_{-i,+j} = (n_1, \dots, n_i - 1, \dots, n_j + 1, \dots, n_M)$, $n_i > 0$, to denote a "neighboring" state of $\boldsymbol{n}$. Let

$$\epsilon(n_k) = \begin{cases} 1, & \text{if } n_k > 0, \\ 0, & \text{if } n_k = 0, \end{cases}$$

and define

$$\mu(\boldsymbol{n}) = \sum_{k=1}^{M} \epsilon(n_k)\mu_{k,n_k}.$$

Then the flow balance equation for the steady-state probabilities $\pi(\boldsymbol{n})$ is

$$\mu(\boldsymbol{n})\pi(\boldsymbol{n}) = \sum_{i=1}^{M}\sum_{j=1}^{M} \epsilon(n_j)\mu_{i,n_i+1}q_{i,j}\pi(\boldsymbol{n}_{-j,+i}). \tag{C.4}$$

Gordon and Newell [122] derived a solution to the above equations. Let $v_i > 0$, $i = 1, 2, \ldots, M$, be the *visit ratio* to server i, i.e., a solution (within a multiplicative constant) to the equations

$$v_i = \sum_{j=1}^{M} q_{j,i}v_j, \qquad i = 1, 2, \ldots, M. \tag{C.5}$$

The solution to these equations is not unique. In fact, if $v := (v_1, v_2, \ldots, v_M)$ is a vector of visit ratios, then for any $\kappa > 0$, $(\kappa v_1, \kappa v_2, \ldots, \kappa v_M)$ is also a set of visit ratios. Equation (C.5) is similar to the probability flow balance equation of a Markov chain $\pi = \pi P$. Indeed, written in a vector form, (C.5) is $v = vQ$.

Let $A_i(0) = 1$, $i = 1, 2, \ldots, M$, and

$$A_i(k) = \prod_{j=1}^{k} \mu_{i,j}, \qquad i = 1, 2, \ldots, M.$$

For every $n = 1, 2, \ldots, N$ and $m = 1, 2, \ldots, M$, let

$$G_m(n) = \sum_{n_1+\cdots+n_m=n} \prod_{i=1}^{m} \frac{v_i^{n_i}}{A_i(n_i)}. \tag{C.6}$$

Then, the solution to (C.4) satisfying the normalization condition $\sum_{\boldsymbol{n}\in\mathcal{S}} \pi(\boldsymbol{n}) = 1$ is [122]

$$\pi(\boldsymbol{n}) = \frac{1}{G_M(N)} \prod_{i=1}^{M} \frac{v_i^{n_i}}{A_i(n_i)}. \tag{C.7}$$

Equation (C.7) is often called a *product-form* solution.

For load-independent networks, $\mu_{i,n_i} \equiv \mu_i$, $i = 1, 2, \ldots, M$. The product-form solution becomes

$$G_m(n) = \sum_{n_1+\cdots+n_m=n} \prod_{i=1}^{m} x_i^{n_i},$$

and

$$\pi(\boldsymbol{n}) = \frac{1}{G_M(N)} \prod_{i=1}^{M} x_i^{n_i},$$

where $x_i = v_i/\mu_i = v_i\bar{s}_i$, $i = 1, 2, \ldots, M$.

Marginal Distributions

From (C.7) we can obtain many other steady-state probabilities. For example, the marginal distribution of the number of customers at server M is

$$\begin{aligned}\pi(n_M = k) &= \sum_{\text{all } \boldsymbol{n}:\, n_M=k} \pi(\boldsymbol{n}) \\ &= \frac{v_M^k}{A_M(k)} \frac{G_{M-1}(N-k)}{G_M(N)},\end{aligned} \tag{C.8}$$

where $G_{M-1}(N-k)$ is calculated for the first $M-1$ servers with the visit ratios $v_1, v_2, \ldots, v_{M-1}$.

The $G_m(n)$ in (C.6) is the standard short-hand notation used in the literature. To make it precise and general, we set $\Gamma = \{1, 2, \ldots, M\}$ and let $\Gamma_0 \subseteq \Gamma$ be a subset of Γ. Define

$$G_{\Gamma_0}(n) = \sum_{\sum_{i\in\Gamma_0} n_i = n} \prod_{i\in\Gamma_0} \frac{v_i^{n_i}}{A_i(n_i)}.$$

Thus, $G_M(N) = G_\Gamma(N)$, and in (C.8) $G_{M-1}(N-k) = G_{\Gamma_{-M}}(N-k)$ with $\Gamma_{-M} := \Gamma - \{M\} = \{1, 2, \ldots, M-1\}$.

With this notation, we can get the expression for the steady-state marginal distribution for any subnetwork. Let $\Gamma_0 := \{i_1, i_2, \ldots, i_K\} \subset \Gamma$ and $\boldsymbol{n}_{\Gamma_0} = \boldsymbol{n}_{\{i_1,\ldots,i_K\}} = (n_{i_1}, \ldots, n_{i_K})$ be the state of the subnetwork. Then, we have

$$\pi\left(\boldsymbol{n}_{\{i_1,\ldots,i_K\}}\right) = \frac{G_{\Gamma-\{i_1,\ldots,i_K\}}(N - \sum_{k=1}^{K} n_{i_k})}{G_\Gamma(N)} \prod_{i\in\{i_1,\ldots,i_K\}} \frac{v_i^{n_i}}{A_i(n_i)}. \tag{C.9}$$

The marginal distribution is

$$\pi\Big(\sum_{k=1}^{K} n_{i_k} = n\Big)$$

$$= \frac{G_{\Gamma-\{i_1,\ldots,i_K\}}(N - \sum_{k=1}^K n_{i_k})}{G_\Gamma(N)} \sum_{\sum_{k=1}^K n_{i_k}=n} \prod_{i\in\{i_1,\ldots,i_K\}} \frac{v_i^{n_i}}{A_i(n_i)}$$

$$= \frac{G_{\Gamma-\{i_1,\ldots,i_K\}}(N-n)}{G_\Gamma(N)} G_{\{i_1,\ldots,i_K\}}(n). \tag{C.10}$$

Thus, the conditional distribution of $\boldsymbol{n}_{\{i_1,\ldots,i_K\}}$ given $\sum_{k=1}^K n_{i_k} = n$ is

$$\pi\left(\boldsymbol{n}_{\{i_1,\ldots,i_K\}} \middle| \sum_{k=1}^K n_{i_k} = n\right) = \frac{1}{G_{\{i_1,\ldots,i_K\}}(n)} \prod_{i\in\{i_1,\ldots,i_K\}} \frac{v_i^{n_i}}{A_i(n_i)}. \tag{C.11}$$

For networks with load-independent servers, we have

$$G_{\{i_1,\ldots,i_K\}}(n) = \sum_{\sum_{i\in\{i_1,\ldots,i_K\}} n_i=n} \prod_{i\in\{i_1,\ldots,i_K\}} x_i^{n_i},$$

and (C.9) becomes

$$\pi(\boldsymbol{n}_{\{i_1,\ldots,i_K\}}) = \frac{G_{\Gamma-\{i_1,\ldots,i_K\}}(N - \sum_{k=1}^K n_{i_k})}{G_\Gamma(N)} \prod_{i\in\{i_1,\ldots,i_K\}} x_i^{n_i}.$$

The marginal distribution keeps the same form as (C.10), and the conditional distribution (C.11) becomes

$$\pi\left(\boldsymbol{n}_{\{i_1,\ldots,i_K\}} \middle| \sum_{k=1}^K n_{i_k} = n\right) = \frac{1}{G_{\{i_1,\ldots,i_K\}}(n)} \prod_{i\in\{i_1,\ldots,i_K\}} x_i^{n_i}.$$

It is clear that this conditional distribution remains the same as long as the relative ratio $x_{i_1} : x_{i_2} : \cdots : x_{i_K}$ is the same. In particular, the conditional distribution does not depend on the service times of the servers outside the subnetwork consisting of servers $i_1, \ldots, i_K$.

Buzen's Algorithm

Buzen [40] developed an efficient algorithm for calculating $G_M(N)$. The algorithm is stated below in (C.12) and will be referred to as Buzen's algorithm. First, from (C.6) we observe

$$G_m(n) = \sum_{k=0}^{n} \frac{v_m^k}{A_m(k)} G_{m-1}(n-k), \tag{C.12}$$

with

$$G_m(0) = 1, \qquad m = 1, 2, \ldots, M,$$

and

$$G_1(n) = v_1^n / A_1(n), \qquad n = 0, 1, \ldots, N.$$

Starting from $G_2(0) = 1$ and $G_1(1) = v_1/\mu_{1,1}$, from (C.12), we can calculate $G_2(1) = G_1(1) + \frac{v_2}{\mu_{2,1}} G_1(0) = \frac{v_1}{\mu_{1,1}} + \frac{v_2}{\mu_{2,1}}$, then we can get $G_2(2) = G_1(2) + \frac{v_2}{\mu_{2,1}} G_1(1) + \frac{v_2^2}{\mu_{2,1}\mu_{2,2}} G_1(0)$, and so on, up to $G_2(N)$. From $G_2(n)$, $n = 0, 1, \ldots, N$, and $G_3(0) = 1$, we can calculate $G_3(n)$, $n = 1, \ldots, N$, and so on, up to $G_M(N)$.

For networks with load-independent servers, Buzen's algorithm is

$$G_m(n) = G_{m-1}(n) + x_m G_m(n-1), \tag{C.13}$$

with

$$G_m(0) = 1, \qquad m = 1, 2, \ldots, M,$$

and

$$G_1(n) = (x_1)^n, \qquad n = 0, 1, \ldots, N.$$

Starting from $G_2(0) = 1$ and $G_1(1) = x_1$, from (C.13), we can calculate $G_2(1) = G_1(1) + x_2 G_2(0) = x_1 + x_2$, then we can get $G_2(2) = G_1(2) + x_2 G_2(1) = x_1^2 + x_2(x_1 + x_2)$, and so on, up to $G_2(N)$. From $G_2(n)$, $n = 0, 1, \ldots, N$, and $G_3(0) = 1$, we can obtain $G_3(n)$, $n = 1, \ldots, N$ and so on, up to $G_M(N)$.

For networks with load-independent servers, from (C.13), we have in (C.8) that $G_{M-1}(N-k) = G_M(N-k) - x_M G_M(N-k-1)$ for $0 \le k < N$. Note that, in this expression, $G_{M-1}(N-k)$ implicitly depends on the choice of server M via x_M, and both $G_M(N-k)$ and $G_M(N-k-1)$ do not depend on the order of the servers. Thus, we can choose any server i as the server M in (C.8) and obtain

$$\pi(n_i = k) = \frac{x_i^k}{G_M(N)} [G_M(N-k) - x_i G_M(N-k-1)], \qquad 0 \le k < N,$$

and

$$\pi(n_i = N) = \frac{1}{G_M(N)} x_i^N.$$

These equations hold for any $i = 1, 2, \ldots, M$. From this, we obtain

$$\pi(n_i \ge k) = \sum_{l=k}^{N} \pi(n_i = l) = x_i^k \frac{G_M(N-k)}{G_M(N)}. \tag{C.14}$$

Therefore, the mean queueing length of server i in an N-customer network, $\bar{n}_i(N)$, is

$$\bar{n}_i(N) := \sum_{k=1}^{N} k\pi(n_i = k) = \sum_{k=1}^{N} \pi(n_i \ge k)$$

$$= \sum_{k=1}^{N} x_i^k \frac{G_M(N-k)}{G_M(N)}. \tag{C.15}$$

From (C.14), the steady-state throughput of server i is

$$\eta_i = \mu_i \pi(n_i \geq 1) = v_i \frac{G_M(N-1)}{G_M(N)}. \tag{C.16}$$

The Closed Form of the Normalizing Constant

The most important step in calculating the steady-state probability $\pi(\boldsymbol{n})$ in (C.7) and other probabilities such as (C.14) and (C.15) is to calculate the normalizing constant of the form $G_M(N)$. Besides the computational algorithms (C.12) and (C.13), some closed-form expressions for $G_M(N)$ have been found [111, 121, 125]. Here we simply quote one result for single-server closed Jackson queueing networks. Recall that $x_i = \frac{v_i}{\mu_i}$, with v_i being the visit ratio to server i, $i = 1, 2, \ldots, M$. It is shown in [121, 125] that if all x_i, $i = 1, 2, \ldots, M$, are distinct, then

$$G_M(N) = \sum_{i=1}^{M} \frac{x_i^{N+M-1}}{\prod_{j \neq i}(x_i - x_j)}.$$

The form is more complicated if some x_i, $i = 1, 2, \ldots, M$, are equal [121]. For the normalizing constants of other types of networks such as multi-server and/or multi-class networks, see [111, 121].

Sensitivity Formulas

Now, we present sensitivity formulas for closed Jackson networks. Let f be a performance function, $f : \mathcal{S} \to \mathcal{R}$. The steady-state mean performance is

$$E(f) = \sum_{\boldsymbol{n} \in \mathcal{S}} f(\boldsymbol{n}) \pi(\boldsymbol{n}).$$

The elasticities of $E(f)$ with respect to $\mu_{i,k}$, $i = 1, 2, \ldots, M$, $k = 1, 2, \ldots, N$, can be obtained by taking the derivatives of the product-form formula. We have (see, e.g., [66, 190])

$$\frac{\mu_{i,k}}{E(f)} \frac{\partial E(f)}{\partial \mu_{i,k}} = \pi_N(n_i \geq k) - \frac{E[f\chi(n_i \geq k)]}{E(f)}, \tag{C.17}$$

where $\chi(n_i \geq k) = 1$ if $n_i \geq k$, and 0 otherwise; $\pi_N(n_i \geq k)$ is the steady-state probability of $n_i \geq k$ in a network with N customers; and

$$E[f\chi(n_i \geq k)] = \sum_{\boldsymbol{n} \in \mathcal{S}:\, n_i \geq k} f(\boldsymbol{n}) \pi(\boldsymbol{n}).$$

The system throughput is defined as

$$\begin{aligned}
\eta := \sum_{\boldsymbol{n} \in \mathcal{S}} \mu(\boldsymbol{n})\pi(\boldsymbol{n}) &= \sum_{\boldsymbol{n} \in \mathcal{S}} \left\{ \left[\sum_{i=1}^{M} \mu_{i,n_i} \epsilon(n_i) \right] \pi(\boldsymbol{n}) \right\} \\
&= \sum_{i=1}^{M} \left\{ \sum_{\boldsymbol{n} \in \mathcal{S}} \left[\mu_{i,n_i} \epsilon(n_i) \pi(\boldsymbol{n}) \right] \right\} \\
&= \sum_{i=1}^{M} \left[\sum_{k=1}^{N} \mu_{i,k} \sum_{\boldsymbol{n}:\ n_i = k} \pi(\boldsymbol{n}) \right] \\
&= \sum_{i=1}^{M} \left[\sum_{k=1}^{N} \mu_{i,k} \pi(n_i = k) \right] = \sum_{i=1}^{M} \eta_i,
\end{aligned}$$

where $\eta_i = \sum_{k=1}^{N} \mu_{i,k} \pi(n_i = k)$. We have [66]

$$\frac{\mu_{i,k}}{\eta} \frac{\partial \eta}{\partial \mu_{i,k}} = \pi_N(n_i \geq k) - \pi_{N-1}(n_i \geq k). \tag{C.18}$$

For networks with load-independent servers, from (C.16) we have

$$\eta = \frac{G_M(N-1)}{G_M(N)} \left(\sum_{i=1}^{M} v_i \right).$$

Because $\mu_{i,k} \equiv \mu_i$ for all $0 \leq k \leq N$, it is easy to derive

$$\frac{\mu_i}{\eta} \frac{\partial \eta}{\partial \mu_i} = \sum_{k=1}^{N} \frac{\mu_{i,k}}{\eta} \frac{\partial \eta}{\partial \mu_{i,k}}.$$

From (C.18), we have [66]

$$\frac{\mu_i}{\eta} \frac{\partial \eta}{\partial \mu_i} = \bar{n}_i(N) - \bar{n}_i(N-1), \tag{C.19}$$

where $\bar{n}_i(N)$ is the steady-state mean of the number of customers in server i in a network with N customers.

Other sensitivity formulas obtained from the product-form solution can be found in, for example, [65, 66, 190, 234, 242, 258].

Computational Algorithms for $E(f)$

In general, to calculate $E(f)$, we need to calculate $\pi(\boldsymbol{n})$ for every state $\boldsymbol{n}$. This is not efficient for large networks. However, for a large class of performance functions, algorithms based on Buzen's algorithm can be developed, and the computational effort can be significantly reduced [66].

Suppose that the performance function takes the following form:

$$f(\mathbf{n}) = f(n_1, \ldots, n_M) = \sum_{k=1}^{K} f_k(n_1, \ldots, n_M), \qquad K \geq 1,$$

where

$$f_k(n_1, \ldots, n_M) = \prod_{i=1}^{M} h_{k,i}(n_i), \qquad k = 1, 2, \ldots, K.$$

For each $i = 1, 2, \ldots, M$ and $k = 1, 2, \ldots, K$, $h_{k,i}$ is a function on $\{0, 1, \ldots, N\}$. We have

$$\begin{aligned} E(f) &= \frac{1}{G_M(N)} \sum_{k=1}^{K} \sum_{n_1+\cdots+n_M=N} \left[f_k(n_1, \ldots, n_M) \prod_{i=1}^{M} \frac{v_i^{n_i}}{A_i(n_i)} \right] \\ &= \frac{1}{G_M(N)} \sum_{k=1}^{K} \sum_{n_1+\cdots+n_M=N} \prod_{i=1}^{M} \frac{h_{k,i}(n_i) v_i^{n_i}}{A_i(n_i)}. \end{aligned} \tag{C.20}$$

For each $k = 1, 2, \ldots, K$, $m = 1, 2, \ldots, M$, and $n = 1, 2, \ldots, N$, define

$$G_{k,m}(n) := \sum_{n_1+\cdots+n_m=n} \prod_{i=1}^{m} \frac{h_{k,i}(n_i) v_i^{n_i}}{A_i(n_i)}.$$

Similar to (C.12), for each k, the $G_{k,m}(n)$'s can be recursively computed via the recursion

$$G_{k,m}(n) = \sum_{j=0}^{n} \frac{h_{k,m}(j) v_m^j}{A_m(j)} G_{k,m-1}(n-j),$$

with

$$G_{k,1}(n) = \frac{h_{k,1}(n) v_1^n}{A_1(n)}, \qquad n = 0, 1, \ldots, N,$$

and

$$G_{k,m}(0) = \prod_{i=1}^{m} h_{k,i}(0), \qquad m = 1, 2, \ldots, M.$$

This is exactly the same as Buzen's algorithm for $G_m(n)$, except with an additional weighting factor $h_{k,m}(j)$. Finally, from (C.20) we have

$$E(f) = \frac{1}{G_M(N)} \sum_{k=1}^{K} G_{k,M}(N). \tag{C.21}$$

With this algorithm, the computational complexity of calculating $E(f)$ can be reduced to the same as that of applying Buzen's algorithm $K+1$ times (for $G_M(N)$ and $G_{k,M}(N)$, $k = 1, 2, \ldots, K$). The algorithm (C.21) can also be used in (C.17) to calculate the performance sensitivities.

BCMP Networks

Another important class of networks whose steady-state distribution possesses a product-form is the *BCMP* network [16]. In a BCMP network, there are M service stations and K classes of customers. While moving from station to station, customers may change their classes. The customers arrive to the network in Poisson processes. The service discipline may be FCFS, PS, LCFS preemptive-resume, or IS (a service station that has an infinite number of servers). The service time distribution for any PS, IS, and LCFS station may be any Coxian distribution, and that for an FCFS station must be exponential with the same mean for all classes of customers. It is shown in [16] that the stationary distribution of a BCMP network has a product form.

The Arrival Theorem

As we know, the steady state probabilities of an embedded chain may not be equal to those of the original Markov process (see Problem A.9). By the same token, the probabilities of the states at selected time instants in a queueing system at the steady state, such as the arrival or departure times at a server in a queueing network, may not be necessarily equal to the steady-state probabilities of the system states. However, it has been shown, that for a large class of queueing networks having product-form steady-state probabilities, these two steady-state probabilities are in fact the same, see, e.g., [81, 201, 227]. This is called the *arrival theorem* and we state it below for open and closed Jackson networks.

Arrival Theorem

(*a*) In a closed Jackson network with a single class of customers and single-server stations, the steady-state distribution seen by the arriving customers to a server (after leaving another server), or the steady-state distribution left behind by the departing customers from a server (before joining another server), is the same as the steady-state probability distribution of a network with one less customer in it.

(*b*) In an open Jackson network with a single class of customers and single-server stations, the steady-state distribution seen by the arriving customers, or left behind by the departing customers, of a server is the same as the steady-state probability distribution of the network.

The words "seen by" and "left behind by" a customer mean that the customer itself is not counted as in the network.

Let us describe the theorem in a precise mathematical term. Consider a closed Jackson network with N customers. Let t_k, $k = 1, 2, \ldots$, be the arrival (departure) instants of the customers at server i, $i = 1, 2, \ldots, M$. Let $X(t)$ be the state of the network at time t. For arrival instants, we set $X_{t_k} := X(t_k+)$ to be the state of the network after the transitions, and for departure instants, we set $X_{t_k} := X(t_k-)$ to be the state of the network before the transitions. Note that, by definition, the ith components of X_{t_k}, $k = 1, 2, \ldots$, is always positive. Define $X_{t_k,-i} := X_{t_k} - e_{\cdot i}$, with $e_{\cdot i}$ being a vector whose components are all zeros except the ith component being one. $X_{t_k,-i}$ denotes the state "seen by" the arriving customer, or "left behind by" the departing customer (not including the customer itself). Then $\boldsymbol{X}_{-i} := \{X_{t_1,-i}, X_{t_2,-i}, \ldots, X_{t_k,-i}, \ldots\}$ is a Markov chain. The arrival theorem claims that the steady-state distribution of $\boldsymbol{X}_{-i}$ is the same as that of the same network having $N - 1$ customers.

The same notation can be applied to open Jackson networks. For instance, we may define t_k, $k = 1, 2, \ldots$, to be the overall arrival (or departure) instants to a server and $X_{t_k} := X(t_k+)$ to be the state of the network after the arrival instants. In addition, we can also define t_k, $k = 1, 2, \ldots$, only to be the external arrival instants to a server, or to the network, and $X_{t_k,-i} := X_{t_k} - e_{\cdot i}$, with i being the server where the arriving customer enters at t_k+, to be the state "seen by" the arriving customer. The arrival theorem claims that the steady-state distribution of $\boldsymbol{X}_{-i} := \{X_{t_1,-i}, X_{t_2,-i}, \ldots\}$ is the same as that of the network.

PASTA

The arrival theorem for external arrivals in an open network is a special case of a more general result, *Poisson Arrivals See Time Averages (PASTA)*. In fact, because the steady-state probability of a state equals the fraction of time that the network is in that state, the arrival theorem for the external arrivals in an open network claims that the steady-state probability of a system state seen by the arriving customers (in a Poisson process) equals the fraction of time that the network is in that state.

The general form of PASTA is as follows [259]. Consider a Poisson arrival process to any stochastic system such as a queueing network. The arriving customers may interact with the system; however, we assume that the future arrival times are independent of the past history of the system. Then, the steady-state probability of a system state seen by the arrival customers equals the fraction of time that the system is in that state.

Note, however, that an internal arrival process (customers arriving from other servers), or an internal departure process (customers leaving a server to go to other servers), is usually not a Poisson process, and is not even a renewal process. That is, the inter-arrival or departure times $t_k - t_{k-1}$, $k = 2, 3, \ldots$, may be dependent [166].

C.3 Some Useful Techniques

In this section, we describe some useful techniques for analyzing the performance of queueing systems.

Routing Feedback

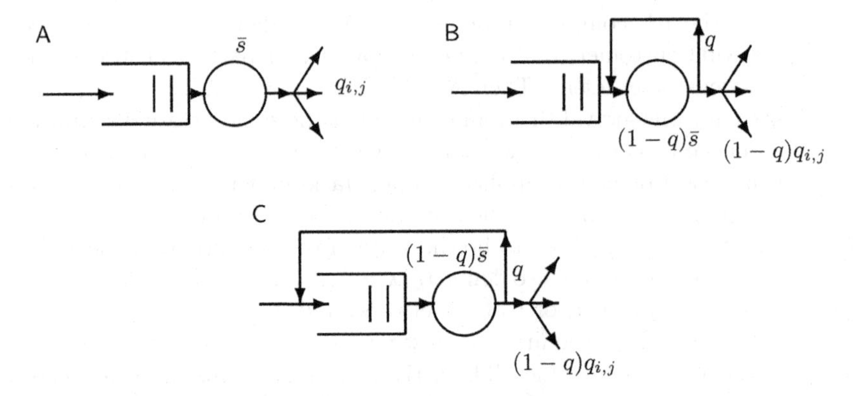

Fig. C.4. Equivalent Servers

Consider a single server in a queueing network, denoted as server i, shown in Figure C.4.A; the service times of this server are i.i.d. with an exponential distribution with mean $\bar{s}$, and the routing probabilities are $q_{i,j}$, with $q_{i,i} = 0$, $i, j = 1, 2, \ldots, M$.

Suppose that we add a feedback loop to the server; i.e., every time a customer finishes its service at the server, we let the customer, with probability $q_{i,i} = q > 0$, go back to the server immediately to receive another service, and with probability $1 - q$, leave the server. In this way, before leaving the server, a customer will receive service once from the server with probability $1 - q$, twice with probability $q(1 - q)$, and k times with probability $q^{k-1}(1 - q)$, $k = 1, 2, \ldots$. Next, we change the mean service time of the server to $(1 - q)\bar{s}$. This is shown in Figure C.4.B. It is easy to show that, with this feedback setting, the total service time that a customer receives from the server is still exponentially distributed with mean $\bar{s}$. Therefore, Figure C.4.B is equivalent to Figure C.4.A.

Next, if there is only one class of customers in the system, all the customers are identical. Thus, there is no distinction if we put the feedback customer at the end of the queue and pick up another customer in the queue, if any, to serve according to any service discipline. That is, the queue in Figure C.4.B behaves in the same way as if we put the feedback customer at the end of the queue, shown in Figure C.4.C.

In summary, the two queues in Figure C.4.A and Figure C.4.C are equivalent. That is, in a single-class queueing network, a server with exponentially distributed service time with mean $\bar{s}$ and routing probabilities $q_{i,j}$ ($q_{i,i} = 0$) is equivalent to a server with exponentially distributed service time with mean $(1-q)\bar{s}$ and routing probabilities $q'_{i,i} = q$ and $q'_{i,j} = (1-q)q_{i,j}$, $j \neq i$. The same argument holds if $q_{i,i} > 0$; the routing probabilities become $q'_{i,i} = q+(1-q)q_{i,i}$ and $q'_{i,j} = (1-q)q_{i,j}$, $j \neq i$.

Aggregation: Norton's Theorem

For networks having product-form steady-state distributions, there is a result similar to Norton's theorem for electrical circuits [11, 76]. That is, under some conditions, a subset of a network can be replaced by a single server without affecting the steady-state marginal distribution of the remainder of the network; the service rate of the single server can be determined by the throughput of a network obtained by "shorting" (i.e., setting the service time equal to zero) the complement of the subset in the original network.

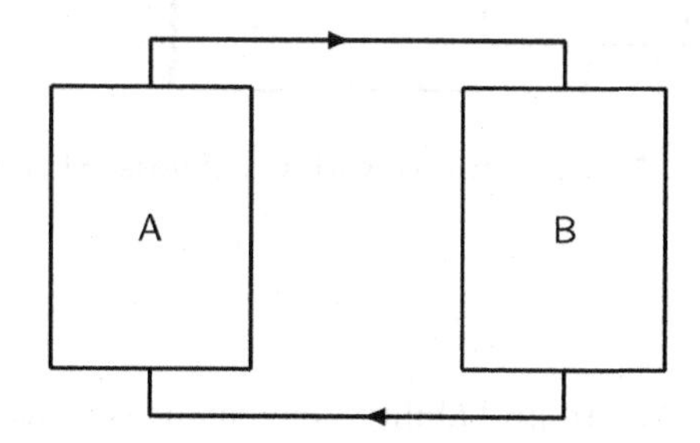

Fig. C.5. A Closed Jackson Network with Two Subsets

We explain the theorem by using Figures C.5-C.7. The network shown in Figure C.5 consists of two subnetworks, A and B. The customer leaving subnetwork A goes to subnetwork B, and vice versa. Using Norton's theorem, we can replace subnetwork A by a load-dependent server, resulting in the network shown in Figure C.6, in which the steady-state marginal distribution of subnetwork B is the same as that in the original network. The load-dependent service rate μ_k of the equivalent server with k customers in it equals the throughput on the shorted path in the closed network consisting of k customers, shown in Figure C.7. The shorted path is obtained by setting the service times of all the servers in subnetwork B to zero.

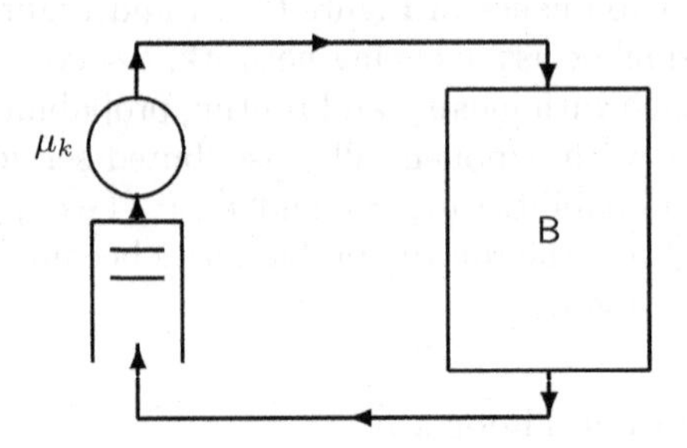

Fig. C.6. The Equivalent Network

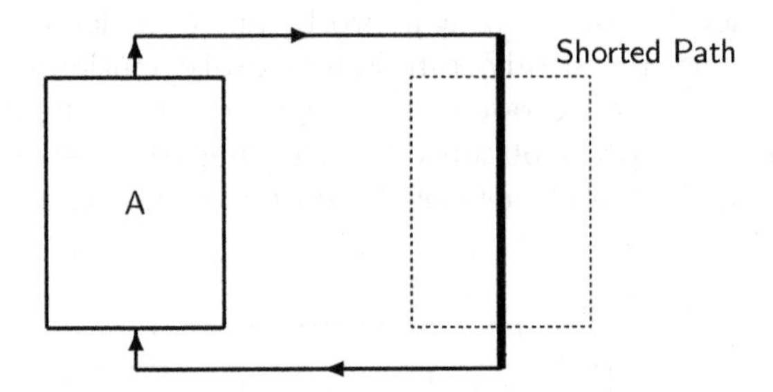

Fig. C.7. The Network with One Subset Shorted

PROBLEMS

C.1. Write the steady-state probability flow-balance equation for an $M/M/1$ queue.

C.2. Consider an $M/G/1$ queue with arrival rate λ and mean service time $\bar{s}$. Prove that the average of the number of customers served in a busy period is $\frac{1}{1-\lambda\bar{s}}$.

C.3. An $M/M/1$ queue with arrival rate λ and departure rate μ can be constructed as follows. Choose an initial state n_0 at time 0 and a rate $\sigma > \lambda + \mu$. Generate a Poisson process with rate σ, denoted as $t_0, t_1, \ldots, t_l, \ldots$. An instant t_l, $l = 0, 1, \ldots$, is chosen as an arrival point with probability $\frac{\lambda}{\sigma}$ and as a departure point with probability $\frac{\mu}{\sigma}$. At an arrival point, we increase the population by one: $n := n + 1$, and at a departure point if $n > 0$ then we decrease the population by one: $n := n - 1$, and at other points we keep the population unchanged. Prove that the discrete-time Markov chain embedded at t_l, $l = 0, 1, \ldots$, is the discrete $M/M/1$ queue described on Page 522. Determine its parameters p_a and p_d [148].

C.4. Many results in this book are stated only for discrete-time Markov models, but queueing systems are usually modelled by continuous-time Markov models. Therefore, we need to use the embedded Markov chains.

a. Find the transition probabilities of the Markov chain embedded at the arrival and departure instants of an $M/M/1$ queue with arrival rate λ and service rate μ.
b. If we use the reward function $f(n) = n$, does the long-run average of the embedded chain equal the mean length of the original $M/M/1$ queue?
c. If the answer to b) is "No", what can we do? (cf. Problem C.9)

C.5.* Consider the queueing system with an $M/M/1$ queue and a feedback loop shown in Figure C.8. This is the simplest non-acyclic open queueing network. The external arrival process to the system is a Poisson process. After the completion of its service at the server, a customer leaves the system with probability $1-q$ and returns back to the queue with probability q, $0 < q < 1$. The total arrival process to the queue at point A is a composition of both the external arrival process and the feedback process. Explain that this total arrival process at point A is not a renewal process. *(Hint: When the server is idle, the inter-arrival time is larger on average. Explain that the consecutive inter-arrival times at point A are not independent.)*

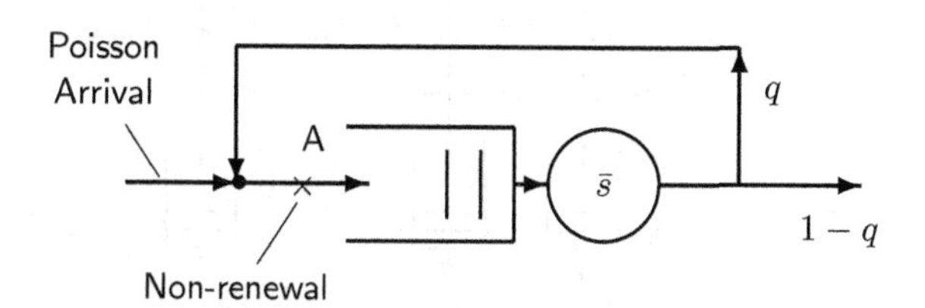

Fig. C.8. An $M/M/1$ Queue with Feedback

C.6. A nonblocking cross-bar switch can be modelled as a closed queueing network. Figure C.9 illustrates the structure of a nonblocking packet switch consisting of N input links and M output links. Packets arriving at each input queue are put in a buffer to wait to be transmitted. Suppose that all packets belong to the same class in terms of the statistics of their destinations: A packet arriving at any input has probability $q_{i,j}$ of being destined for output j given that the previous packet at that input was destined for output i, $i, j = 1, \ldots, M$. Every packet destined to output j requires an exponentially distributed transmission time with mean $\bar{s}_j$. At a particular time, only the head of line (HOL) packet (the first packet) in an input queue can be transmitted and the switch can only transmit one packet to every output queue at a time. The HOL packet of an input queue contends with the HOL packets of other input queues that have the same destination in a FCFS manner.

We wish to determine the maximum throughput of this $N \times M$ switch, i.e., how many packets can this switch transmit to their destinations per second if there are always packets at every input waiting for transmission. Develop a queueing model for this problem [52, 68]. *(Hint: The HOL packet of an input queue makes a request to the switch asking to be transmitted to its destination at the time when it moves to the head position. All the requests to the same destination output queue form a logical queue called a request queue. The M request queues constitute a closed queueing network.)*

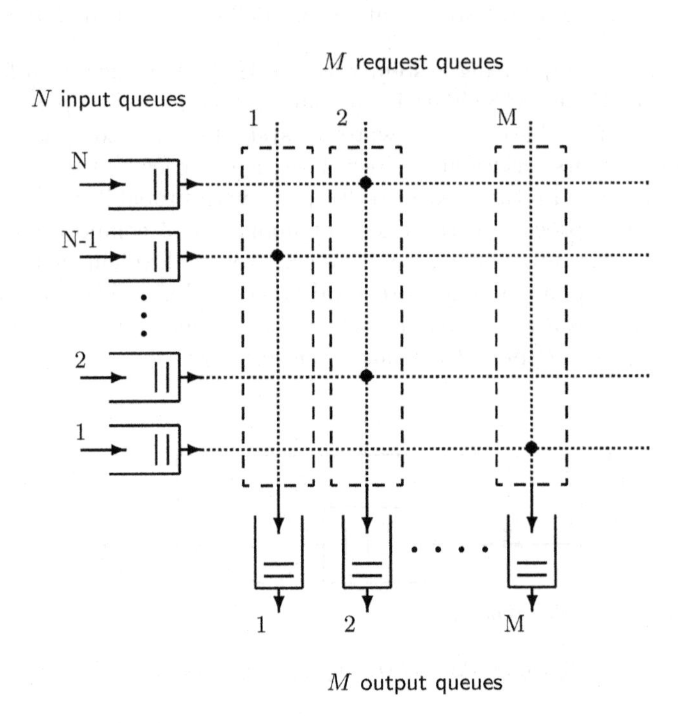

Fig. C.9. The Model of a Nonblocking Switch

C.7. A cyclic queueing network of M servers is a closed network that contains M servers connecting as a circle. A two-server cyclic network is a network of two servers with routing probabilities $q_{1,2} = q_{2,1} = 1$ and $q_{1,1} = q_{2,2} = 0$. Consider a two-server cyclic network with service rates λ and μ, and a population K. Show that this closed network is equivalent to an $M/M/1/K$ queue with arrival rate λ and service rate μ.

C.8. Consider an open Jackson network with M servers. The service times at server i are exponentially distributed with mean $\bar{s}_i$, $i = 1, 2, \ldots, M$; the routing probabilities are $q_{i,j}$, $i = 1, 2, \ldots, M$, $j = 0, 1, \ldots, M$, $\sum_{j=0}^{M} q_{i,j} = 1$, with $q_{i,0}$ being the probability that a customer leaves the network from server i after receiving service at the server, $i = 1, 2, \ldots, M$; and the external arrival rate to server i is $\lambda_{0,i}$, $i = 1, 2, \ldots, M$. The state of the network is $\boldsymbol{n} = (n_1, \ldots, n_M)$. Let $N := \sum_{k=1}^{M} n_k$.

a. Find the conditional steady-state probability $\pi(\boldsymbol{n}|N)$.
b. Show that this conditional probability is the same as an equivalent closed Jackson network with a population N.
c. Find the routing probabilities of this equivalent closed Jackson network and give an explanation for these routing probabilities.

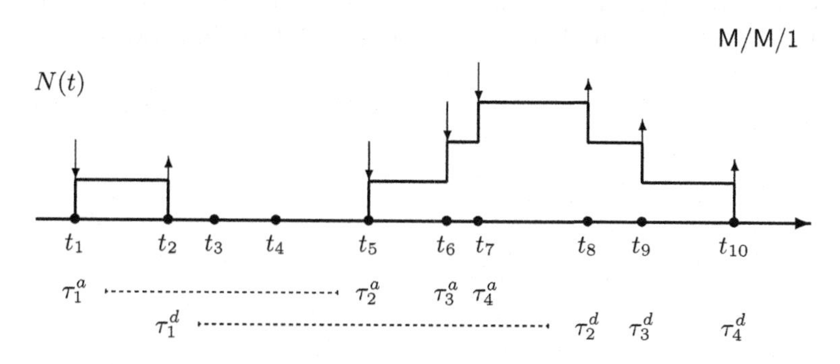

Fig. C.10. The Arrival Theorem and PASTA

C.9.* Figure C.10 illustrates a sample path $N(t)$ of an $M/M/1$ queue, in which the upward arrows indicate the departure instants and the downward arrows indicate the arrival instants. Let the arrival rate and service rate be λ and μ, respectively. We simulate the $M/M/1$ queue with the uniformization approach (cf. Problem A.8):

i. Generate a Poisson process with rate $\lambda + \mu$, shown in Figure C.10 as $\{t_1, t_2, t_3, \ldots\}$. Set $N(0) = n_0$ to be the initial state ($n_0 = 0$ in Figure C.10).
ii. At t_k, $k = 1, 2, \ldots$, generate an independent and uniformly distributed random variable $\xi_k \in [0, 1)$,
   (1) If $\xi_k < \frac{\lambda}{\lambda+\mu}$, then t_k is an arrival instant; set $N(t_k+) := N(t_k) + 1$.
   (2) If $\xi_k > \frac{\lambda}{\lambda+\mu}$ and $N(t_k) > 0$, then t_k is a departure instant; set $N(t_k+) := N(t_k) - 1$.
   (3) If $\xi_k > \frac{\lambda}{\lambda+\mu}$ and $N(t_k) = 0$, do nothing.

The process $N(t)$ thus generated is left-continuous. In Figure C.10, τ_k^a, $k = 1, 2, \ldots$, indicate the arrival instants and τ_k^d, $k = 1, 2, \ldots$, indicate the departure instants; at t_3 and t_4 the server is idle and nothing changes, these instants are called "dummy instants".

a. Explain why the process $N(t)$ generated by the above algorithm is indeed an $M/M/1$ queue with arrival rate λ and service rate μ (cf. Problem C.3).
b. Define $X_k := N(t_k)$. Prove that the embedded chain $\boldsymbol{X} := \{X_1, X_2, \ldots\}$ is a Markov chain and its steady-state distribution is the same as that of the $M/M/1$ queue (PASTA).
c. Prove that the average of the number of visits to any particular state n at the non-dummy instants $t_1, t_2, t_5, t_6, \ldots$, equals the steady-state probability of the state n, $n = 0, 1, \ldots$. Further, prove that the average of the number of visits to any particular state n at the arrival instants τ_k^a, $k = 1, 2, \ldots$, (or the departure instants τ_k^d, $k = 1, 2, \ldots$) equals the steady-state probability of the state n, $n = 0, 1, \ldots$ (the arrival theorem).
d. Extend this explanation to (open or closed) Jackson networks.

Notation

a	Event
α	Action
$\mathcal{A}$	Action space
$\mathcal{A}(i)$	Set of available actions in state i
A_l	Action taken at time l
$\boldsymbol{A}_l = (A_0, \ldots, A_l)$	Action history up to time l
β	Discount factor
B	Infinitesimal generator
$c(\boldsymbol{n}, i)$	Realization probability of a perturbation at server i when the queueing system is in state $\boldsymbol{n}$
$c^{(f)}(\boldsymbol{n}, i)$	Realization factor of a perturbation at server i when the queueing system is in state $\boldsymbol{n}$, for a performance function f
d, h	Policies
$\mathcal{D}$	Policy space
$\mathcal{D}_0$	Gain-optimal policy space
$\mathcal{D}_n$, $n = 1, 2, \ldots$	nth-bias optimal policy space
δ, θ	Parameters
$e = (1, \ldots, 1)^T$	A column vector with all components being one
T	D^T: transpose of matrix D
$e_{\cdot i}$	A column vector with all components being zero except its ith component being one
e_S	S-dimensional unit vector
$e_c(k_2)$	Controllable events, $k_2 = 1, \ldots, k_c$
$e_o(k_1)$	Observable events, $k_1 = 1, \ldots, k_o$
$e_t(k_3)$	Natural transition events, $k_3 = 1, \ldots, k_t$
$\mathcal{E}$	Set of all single events (state transitions)
E	Expectation
E_l	Event at time l
$\boldsymbol{E}_l = (E_0, \ldots, E_l)$	Event history up to time l
η	Performance measure (long-run average reward)
η_β	Discounted reward
η^*	Optimal gain
f	Performance (reward) function
f^d	Performance (reward) function of policy d
F_L	Equal to $\sum_{l=0}^{L-1} f(X_l)$ in discrete-time case, or $\int_0^{T_L} f(X_t)dt$ in continuous-time case
g	Performance potential function, vector; and bias
g_β	Discounted potential
g^d	Performance potential function of policy d
g_n, $n = 0, 1, \ldots$	nth potentials and nth bias
g_n^*, $n = 0, 1, \ldots$	Optimal nth bias
g_n^d, $n = 0, 1, \ldots$	nth potentials and nth bias of policy d

$\Gamma = [\gamma(i,j)]$	Performance realization matrix
$\boldsymbol{H}_l$	Information history up to time l, $\boldsymbol{H}_l = \{\boldsymbol{Y}_l, \boldsymbol{A}_{l-1}\}$
κ	Step size
$L(i\|j)$	First time a Markov chain reaches state i from initial state j
L^*_{ij}	Merging point of two Markov chains starting from initial states i and j, respectively
n_i	Number of customers in server i
$\boldsymbol{n} = (n_1, \ldots, n_M)$	State of a queueing network
$\boldsymbol{n}_{-i,+j}$	Neighboring state of $\boldsymbol{n}$
$P = [p(j\|i)]$	Transition probability matrix
P^d	Transition probability matrix under policy d
$\Delta P = (\Delta P)^{d,h}$	Equal to $P^h - P^d$: the direction from P^d to P^h
P_δ	Equal to $P + \delta\Delta P$ or $P^d + \delta(\Delta P)^{d,h}$
P^*	Cesaro limit of P^n
$\mathcal{P}$	Probability measure
π	Steady-state probability, row, vector
π^d	Steady-state probability of policy d
$Q = [q_{i,j}]$	Routing probability matrix in queueing networks
$Q(i,\alpha)$	Q-factor of state i and action α
$\Re$	Space of real numbers $(-\infty, \infty)$
ρ	Traffic intensity, $\rho = \frac{\lambda}{\mu}$ in an M/M/1 queue
$\rho(R)$	Spectrum radius of matrix R
$\bar{s}_i$	Mean service time of server i
S	Number of states in the state space
$\mathcal{S}$	State space
T_l	lth transition instant of a continuous-time Markov process
$T(i\|j)$	First time a Markov process reaches state i from initial state j
T^*_{ij}	Merging point of two Markov processes starting from initial states i and j, respectively
X_l	State of Markov chain $\boldsymbol{X}$ at time l
X_t	State of Markov process $\boldsymbol{X}$ at time t
$\boldsymbol{X}$	A Markov chain or Markov process, a sample path
$\boldsymbol{X}_l$	State history up to time l
$\boldsymbol{X}_\delta = \boldsymbol{X}^{d,h}_\delta$	Markov chain, a sample path for $P^d + \delta(\Delta P)^{d,h}$
Y_l	Observation at time l
$\boldsymbol{Y}_l$	Observation history up to time l
$\times$	The Cartesian product
$\otimes$	The Kronecker product
$\langle i, j \rangle$	Transition from state i to state j
$\geq (\leq)$	Two vectors $u \geq (\leq)\ v$ means $u(i) \geq (\leq)\ v(i)$ for all i
$\succeq (\preceq)$	Two vectors $u \succeq (\preceq)\ v$ means $u \geq (\leq)\ v$ and $u \neq v$
#	$B^\#$: group inverse of matrix B

Abbreviations

CR	Common realization
DEDS	Discrete event dynamic system
EAMC	Equivalent aggregated Markov chain
FCFS	First come first serve
GM	Gradient method
GPI	Generalized policy iteration
I&AC	Identification and adaptive control
i.i.d.	Identically and independently distributed
JLQ	Jump linear quadratic
LCFS	Last come first serve
LDQ	Linear discounted quadratic
LQ	Linear quadratic
LQG	Linear quadratic Gaussian
LR	Likelihood ratio
MDPs	Markov decision processes
NDP	Neuro-dynamic programming
PA	Perturbation analysis
PASTA	Poisson arrivals see time average
PDF	Performance difference formula
PE	Poisson equation
PG	Policy gradient
PI	Policy iteration
POMDPs	Partially observable Markov decision processes
PRF	Perturbation realization factor
PS	Processor sharing
Q-L	Q-learning
RL	Reinforcement learning
RM	Robins-Monro algorithm
SA	Stochastic approximation
SARSA	State-action-reward-state-action
SD	Standard deviation
SFM	Stochastic fluid model
SMP	Semi-Markov process
TD	Temporal difference
WD	Weak derivative
w.p.1	With probability 1

References

1. M. Abbad, J. A. Filar, and T. R. Bielecki, "Algorithms for Singularly Perturbed Limiting Average Control Problem," *IEEE Transactions on Automatic Control*, Vol. 37, 1421-1425, 1992.
2. D. Aberdeen and J. Baxter, "Scaling Internal-State Policy-Gradient Methods for Partially Observable Markov Decision Processes," *Proceedings of Nineteenth International Conference on Machine Learning*, Sydney, Australia, 3-10, July 2002.
3. J. Abounadi, D. Bertsekas, and V. S. Borkar, "Learning Algorithms for Markov Decision Processes with Average Cost," *SIAM Journal on Control and Optimization*, Vol. 40, 681-698, 2001.
4. R. W. Aldhaheri and H. K. Khalil, "Aggregation of the Policy Iteration Method for Nearly Completely Decomposable Markov Chains," *IEEE Transactions on Automatic Control*, Vol. 36, 178-187, 1991.
5. A. Al-Tamimi, F. L. Lewis and M. Abu-Khalaf, "Model-Free Q-Learning Designs for Linear Discrete-Time Zero-Sum Games with Application to H-Infinity Control," *Automatica*, Vol. 43, 473-481, 2007.
6. E. Altman, *Constrained Markov Decision Processes*, Chapman & Hall/CRC, Boca Raton, 1999.
7. E. Altman, K. E. Avrachenkov, and R. Núñez-Queija, "Perturbation Analysis for Denumerable Markov Chains with Applications to Queueing Models," *Advances in Applied Probability*, Vol. 36, 839-853, 2004.
8. A. Arapostathis, V. S. Borkar, E. Fernandez-Gaucherand, M. K. Ghosh, and S. I. Marcus, "Discrete-Time Controlled Markov Processes with Average Cost Criterion: A Survey," *SIAM Journal on Control and Optimization*, Vol. 31, 282-344, 1993.
9. K. J. Åström and B. Wittenmark, *Adaptive Control*, Addison-Wesley, Reading, Massachusetts, 1989.
10. M. Baglietto, F. Davoli, M. Marchese, and M. Mongelli, "Neural Approximation of Open-Loop Feedback Rate Control in Satellite Networks," *IEEE Transactions on Neural Networks*, Vol. 16, 1195-1211, 2005.
11. S. Balsamo and G. Iazeolla, "An Extension of Norton's Theorem for Queueing Networks," *IEEE Transactions on Software Engineering*, Vol. 8, 298-305, 1982.
12. J. Banks, J. S. Carson, B. L. Nelson, and D. M. Nicol, *Discrete-Event System Simulation*, Third Edition, Prentice Hall, New Jersey, 2000.
13. A. Y. Barraud, "A Numerical Algorithm to Solve $A^TXA - X = Q$," *IEEE Transactions on Automatic Control*, Vol. 22, 883-885, 1977.
14. R. H. Bartels and G. W. Stewart, "Solution of the Matrix Equation $AX+XB = C$," *Communications of the ACM*, Vol. 15, 820-826, 1972.

15. A. G. Barto and S. Mahadevan, "Recent Advances in Hierarchical Reinforcement Learning," Special Issue on Reinforcement Learning, *Discrete Event Dynamic Systems: Theory and Application*, Vol. 13, 41-77, 2003.
16. F. Baskett, K. M. Chandy, R. R. Muntz, and F. G. Palacios, "Open, Closed, and Mixed Networks of Queues with Different Classes of Customers," *Journal of the ACM*, Vol. 22, 284-260, 1975.
17. J. Baxter and P. L. Bartlett, "Infinite-Horizon Policy-Gradient Estimation," *Journal of Artificial Intelligence Research*, Vol. 15, 319-350, 2001.
18. J. Baxter, P. L. Bartlett, and L. Weaver, "Experiments with Infinite-Horizon, Policy-Gradient Estimation," *Journal of Artificial Intelligence Research*, Vol. 15, 351-381, 2001.
19. V. E. Benes, *Mathematical Theory of Connecting Networks and Telephone Traffic*, Academic Press, New York, 1965.
20. A. Berman and R. J. Plemmons, *Nonnegative Matrices in the Mathematical Sciences*, SIAM, Philadelphia, 1994.
21. D. P. Bertsekas, *Dynamic Programming and Optimal Control*, Volumes I and II. Athena Scientific, Belmont, Massachusetts, 1995, 2001, 2007.
22. D. P. Bertsekas, with A. Nedic and A. E. Ozdaglar, *Convex Analysis and Optimization*, Athena Scientific, Belmont, Massachusetts, 2003.
23. D. P. Bertsekas, *Nonlinear Programming*, Athena Scientific, Belmont, Massachusetts, 1995.
24. D. P. Bertsekas and S. E. Shreve, *Stochastic Optimal Control: The Discrete Time Case*, Academic Press, New York, 1978.
25. D. P. Bertsekas and T. N. Tsitsiklis, *Neuro-Dynamic Programming*, Athena Scientific, Belmont, Massachusetts, 1996.
26. D. P. Bertsekas and T. N. Tsitsiklis, *Introduction to Probability*, Athena Scientific, Belmont, Massachusetts, 2002.
27. T. R. Bielecki and J. A. Filar, "Singularly Perturbed Markov Control Problems: Limiting Average Cost," *Annals of Operations Research*, Vol. 28, 153-168, 1991.
28. P. Billingsley, *Probability and Measure*, John Wiley & Sons, New York, 1979.
29. J. P. C. Blanc, "A Numerical Approach to Cyclic-Server Queueing Models," *Queueing Systems: Theory and Applications*, Vol. 6, 173-188, 1990.
30. S. J. Bradtke, B. E. Ydestie and A. G. Barto, "Adaptive Linear Quadratic Control Using Policy Iteration," *Proceedings of the American Control Conference*, Baltimore, Maryland, U.S.A, 3475-3479, 1994.
31. P. Bratley, B. L. Fox, and L. E. Schrage, *A Guide to Simulation*, Second Edition, Springer-Verlag, New York, 1987.
32. L. Breiman, *Probability*, Addison-Wesley, Massachusetts, 1968.
33. P. Bremaud, "Maximal Coupling and Rare Perturbation Sensitivity Analysis," *Queueing Systems: Theory and Applications*, Vol. 11, 307-333, 1992.
34. P. Bremaud and W. B. Gong, "Derivatives of Likelihood Ratios and Smoothed Perturbation Analysis for Routing Problem," *ACM Transactions on Modeling and Computer Simulation*, Vol. 3, 134-161, 1993.
35. P. Bremaud, R. P. Malhame, and L. Massoulie, "A Manufacturing System with General Stationary Failure Process: Stability and IPA of Hedging Control Policies," *IEEE Transactions on Automatic Control*, Vol. 42, 155-170, 1997.
36. P. Bremaud and L. Massoulie, "Maximal Coupling and Rare Perturbation Analysis with a Random Horizon," *Discrete Event Dynamic Systems: Theory and Applications*, Vol. 5, 319-342, 1995.

37. P. Bremaud and F. J. Vazquez-Abad, "On the Pathwise Computation of Derivatives with Respect to the Rate of A Point Process: The Phantom RPA Method," *Queueing Systems: Theory and Applications*, Vol. 10, 249-269, 1992.
38. C. A. Brooks and P. Varaiya, "Using Augmented Infinitesimal Perturbation Analysis for Capacity Planning in Intree ATM Networks," *Discrete Event Dynamic Systems: Theory and Applications*, Vol. 7, 377-390, 1997.
39. A. E. Bryson and Y. C. Ho, *Applied Optimal Control: Optimization, Estimation, and Control*, Blaisdell, Waltham, Massachusetts, 1969.
40. J. P. Buzen, "Computational Algorithm for Closed Queueing Networks with Exponential Servers," *Communications of the ACM*, Vol. 16, 527-531, 1973.
41. E. Campos-Náñez and S. D. Patek, "Dynamically Identifying Regenerative Cycles in Simulation-Based Optimization Algorithms for Markov Chains," *IEEE Transactions on Automatic Control*, Vol. 49, 1022-1025, 2004.
42. X. R. Cao, "Convergence of Parameter Sensitivity Estimates in a Stochastic Experiment," *IEEE Transactions on Automatic Control*, Vol. 30, 845-853, 1985.
43. X. R. Cao, "First-Order Perturbation Analysis of a Single Multi-Class Finite Source Queue," *Performance Evaluation*, Vol. 7, 31-41, 1987.
44. X. R. Cao, "Sensitivity Estimates Based on One Realization of a Stochastic System," *Journal of Statistical Computation and Simulation*, Vol. 27, 211-232, 1987.
45. X. R. Cao, "Realization Probability in Closed Jackson Queueing Networks and Its Application," *Advances in Applied Probability*, Vol. 19, 708-738, 1987.
46. X. R. Cao, "A Sample Performance Function of Jackson Queueing Networks," *Operations Research*, Vol. 36, 128-136, 1988.
47. X. R. Cao, "Estimates of Performance Sensitivity of a Stochastic System," *IEEE Transactions on Information Theory*, Vol. 35, 1058-1068, 1989.
48. X. R. Cao, "A Comparison of the Dynamics of Continuous and Discrete Event Systems," *Proceedings of the IEEE*, Vol. 77, 7-13, 1989.
49. X. R. Cao, "Realization Probability and Throughput Sensitivity in a Closed Jackson Network," *Journal of Applied Probability*, Vol. 26, 615-624, 1989.
50. X. R. Cao, "Realization Factors and Sensitivity Analysis of Queueing Networks with State-Dependent Service Rates," *Advances in Applied Probability*, Vol. 22, 178-210, 1990.
51. X. R. Cao, *Realization Probabilities: The Dynamics of Queueing Systems*, Springer-Verlag, New York, 1994.
52. X. R. Cao, "The Maximum Throughput of a Nonblocking Space-Division Packet Switch with Correlated Destinations," *IEEE Transactions on Communications*, Vol. 43, 1898-1901, 1995.
53. X. R. Cao, "The Relations Among Potentials, Perturbation Analysis, and Markov Decision Processes," *Discrete Event Dynamic Systems: Theory and Applications*, Vol. 8, 71-87, 1998.
54. X. R. Cao, "Single Sample Path Based Optimization of Markov Chains," *Journal of Optimization Theory and Application*, Vol. 100, 527-548, 1999.
55. X. R. Cao, "A Unified Approach to Markov Decision Problems and Performance Sensitivity Analysis," *Automatica*, Vol. 36, 771-774, 2000.
56. X. R. Cao, "From Perturbation Analysis to Markov Decision Processes and Reinforcement Learning," *Discrete Event Dynamic Systems: Theory and Applications*, Vol. 13, 9-39, 2003.

57. X. R. Cao, "Semi-Markov Decision Problems and Performance Sensitivity Analysis," *IEEE Transactions on Automatic Control*, Vol. 48, 758-769, 2003.
58. X. R. Cao (eds.), "Introduction to the Special Issue on Learning, Optimization, and Decision Making in DEDS," *Discrete Event Dynamic Systems: Theory and Applications*, Vol. 13, 7-8, 2003.
59. X. R. Cao, "The Potential Structure of Sample Paths and Performance Sensitivities of Markov Systems," *IEEE Transactions on Automatic Control*, Vol. 49, 2129-2142, 2004.
60. X. R. Cao, "Basic Ideas for Event-Based Optimization of Markov Systems," *Discrete Event Dynamic Systems: Theory and Applications*, Vol. 15, 169-197, 2005.
61. X. R. Cao, "A Basic Formula for On-Line Policy-Gradient Algorithms," *IEEE Transactions on Automatic Control*, Vol. 50, 696-699, 2005.
62. X. R. Cao and H. F. Chen, "Perturbation Realization, Potentials and Sensitivity Analysis of Markov Processes," *IEEE Transactions on Automatic Control*, Vol. 42, 1382-1393, 1997.
63. X. R. Cao and X. P. Guo, "A Unified Approach to Markov Decision Problems and Sensitivity Analysis with Discounted and Average Criteria: Multichain Case," *Automatica*, Vol. 40, 1749-1759, 2004.
64. X. R. Cao and Y. C. Ho, "Estimating Sojourn Time Sensitivity in Queueing Networks Using Perturbation Analysis," *Journal of Optimization Theory and Applications*, Vol. 53, 353-375, 1987.
65. X. R. Cao and D. J. Ma, "New Performance Sensitivity Formulae for a Class of Product-Form Queueing Networks," *Discrete Event Dynamic Systems: Theory and Applications*, Vol. 1, 289-313, 1992.
66. X. R. Cao and D. J. Ma, "Performance Sensitivity Formulas, Algorithms, and Estimates for Closed Queueing Networks with Exponential Servers," *Performance Evaluation*, Vol. 26, 181-199, 1996.
67. X. R. Cao, Z. Y. Ren, S. Bhatnagar, M. C. Fu, and S. I. Marcus, "A Time Aggregation Approach to Markov Decision Processes," *Automatica*, Vol. 38, 929-943, 2002.
68. X. R. Cao and D. Towsley, "A Performance Model for ATM Switches with General Packet Length Distributions," *IEEE/ACM Transactions on Networking*, Vol. 3, 299-309, 1995.
69. X. R. Cao and Y. W. Wan, "Algorithms for Sensitivity Analysis of Markov Systems Through Potentials and Perturbation Realization," *IEEE Transactions on Control System Technology*, Vol. 6, 482-494, 1998.
70. X. R. Cao, X. M. Yuan, and L. Qiu, "A Single Sample Path-Based Performance Sensitivity Formula for Markov Chains," *IEEE Transactions on Automatic Control*, Vol. 41, 1814-1817, 1996.
71. X. R. Cao and J. Y. Zhang, "The nth-Order Bias Optimality for Multi-chain Markov Decision Processes," *IEEE Transactions on Automatic Control*, submitted.
72. C. G. Cassandras and S. Lafortune, *Introduction to Discrete Event Systems*, Kluwer Academic Publishers, Boston, 1999.
73. C. G. Cassandras and S. G. Strickland, "On-Line Sensitivity Analysis of Markov Chains," *IEEE Transactions on Automatic Control*, Vol. 34, 76-86, 1989.

74. C. G. Cassandras, G. Sun, C. G. Panayiotou, and Y. Wardi, "Perturbation Analysis and Control of Two-Class Stochastic Fluid Models for Communication Networks," *IEEE Transactions on Automatic Control*, Vol. 48, 770-782, 2003.
75. C. G. Cassandras, Y. Wardi, B. Melamed, G. Sun, and C. G. Panayiotou, "Perturbation Analysis for Online Control and Optimization of Stochastic Fluid Models," *IEEE Transactions on Automatic Control*, Vol. 47, 1234-1248, 2002.
76. K. M. Chandy, U. Herzog, and L. Woo, "Parametric Analysis of Queueing Networks," *IBM Journal on Research and Development*, Vol. 19, 36-42, 1975.
77. H. S. Chang, M. C. Fu, J. Hu and S. I. Marcus, *Simulation-Based Algorithms for Markov Decision Processes*, Springer, New York, 2007.
78. H. S. Chang, H. G. Lee, M. C. Fu, and S. I. Marcus, "Evolutionary Policy Iteration for Solving Markov Decision Processes," *IEEE Transactions on Automatic Control*, Vol. 50, 1804–1808, 2005.
79. H. F. Chen, *Stochastic Approximation and Its Applications*, Kluwer Academic Publishers, Dordrecht, 2002.
80. C. H. Chen, S. D. Wu, and L. Dai, "Ordinal Comparison of Heuristic Algorithms Using Stochastic Optimization," *IEEE Transactions on Robotics and Automation*, Vol. 15, 44-56, 1999.
81. H. Chen and D. D. Yao, *Fundamentals of Queuing Networks: Performance, Asymptotics and Optimization*, Springer-Verlag, New York, 2001.
82. E. K. P. Chong and P. J. Ramadge, "Convergence of Recursive Optimization Algorithms Using Infinitesimal Perturbation Analysis Estimates," *Discrete Event Dynamic Systems: Theory and Applications*, Vol. 1, 339-372, 1992.
83. E. K. P. Chong and P. J. Ramadge, "Optimization of Queues Using an Infinitesimal Perturbation Analysis-Based Stochastic Algorithm with General Update Times," *SIAM Journal on Control and Optimization*, Vol. 31, 698-732, 1993.
84. E. K. P. Chong and P. J. Ramadge, "Stochastic Optimization of Regenerative Systems Using Infinitesimal Perturbation Analysis," *IEEE Transactions on Automatic Control*, Vol. 39, 1400-1410, 1994.
85. E. K. P. Chong and S. H. Zak, *An Introduction to Optimization*, Second Edition, John Wiley & Sons, New York, 2001
86. K. L. Chung, *Markov Chains with Stationary Transition Probabilities*, Springer-Verlag, Berlin, 1960.
87. E. Çinlar, *Introduction to Stochastic Processes*, Prentice Hall, Englewood Cliffs, New Jersey, 1975.
88. W. L. Cooper, S. G. Henderson, and M. E. Lewis, "Convergence of Simulation-Based Policy Iteration," *Probability in the Engineering and Information Sciences*, Vol. 17, 213-234, 2003.
89. O. L. V. Costa and J. C. C. Aya, "Monte Carlo TD(λ)-Methods for the Optimal Control of Discrete-Time Markovian Jump Linear Systems," *Automatica*, Vol. 38, 217-225, 2002.
90. D. R. Cox, "A Use of Complex Probabilities in the Theory of Stochastic Processes," *Proceedings of Cambridge Philosophical Society*, Vol. 51, 313-319, 1955.
91. L. Dai, "Rate of Convergence for Derivative Estimation of Discrete-Time Markov Chains Via Finite-Difference Approximations with Common Random Numbers," *SIAM Journal on Applied Mathematics*, Vol. 57, 731-751, 1997.
92. L. Dai, "Perturbation Analysis via Coupling," *IEEE Transactions on Automatic Control*, Vol. 45, 614-628, 2000.

93. L. Dai and C. H. Chen, "Rates of Convergence of Ordinal Comparison for Dependent Discrete Event Dynamic Systems," *Journal of Optimization Theory and Applications*, Vol. 94, 29-54, 1997.
94. L. Dai, and Y. C. Ho, "Structural Infinitesimal Perturbation Analysis (SIPA) for Derivative Estimation of Discrete Event Dynamic Systems," *IEEE Transactions on Automatic Control*, Vol. 40, 1154-1166, 1995.
95. F. Davoli, M. Marchese, and M. Mongelli, "Resource Allocation in Satellite Networks: Certainty Equivalent Approaches Versus Sensitivity Estimation Algorithms," *International Journal of Communication Systems*, Vol. 18, 3-36, 2005.
96. N. V. Dijk, *Queueing Networks and Product Forms: A Systems Approach*, John Willey & Sons, Chichester, 1993.
97. H. T. Fang and X. R. Cao, "Potential-Based Online Policy Iteration Algorithms for Markov Decision Processes," *IEEE Transactions on Automatic Control*, Vol. 49, 493-505, 2004.
98. E. A. Feinberg and A. Shwartz (eds.), *Handbook of Markov Decision Processes: Methods and Application*, Kluwer Academic Publishers, Boston, 2002.
99. J. A. Filar, V. Gaitsgory, and A. B. Haurie, "Control of Singularly Perturbed Hybrid Stochastic Systems," *IEEE Transactions on Automatic Control*, Vol. 46, 179-190, 2001.
100. J. A. Filar and A. Haurie, "A Two-Factor Stochastic Production Model with Two Time Scales," *Automatica*, Vol 37, 1505-1513, 2001.
101. J. P. Forestier and P. Varaiya, "Multilayer Control of Large Markov Chains," *IEEE Transactions on Automatic Control*, Vol. 23, 298-305, 1978.
102. M. Freimer and L. Schruben, "Graphical Representation of IPA Estimation," *Proceedings of the 2001 Winter Simulation Conference*, Arlington, Virginia, U.S.A, Vol. 1, 422-427, December 2001.
103. M. C. Fu, "Convergence of a Stochastic Approximation Algorithm for the GI/G/1 Queue Using Infinitesimal Perturbation Analysis," *Journal of Optimization Theory and Applications*, Vol. 65, 149-160, 1990.
104. M. C. Fu and J. Q. Hu, "Consistency of Infinitesimal Perturbation Analysis for the GI/G/m Queue," *European Journal of Operational Research*, Vol. 54, 121-139, 1991.
105. M. C. Fu and J. Q. Hu, "Smoothed Perturbation Analysis Derivative Estimation for Markov Chains," *Operations Research Letters*, Vol. 15, 241-251, 1994.
106. M. C. Fu and J. Q. Hu, "Efficient Design and Sensitivity Analysis of Control Charts Using Monte Carlo Simulation," *Management Science*, Vol. 45, 395-413, 1999.
107. M. C. Fu and J. Q. Hu, *Conditional Monte Carlo: Gradient Estimation and Optimization Applications*, Kluwer Academic Publishers, Boston, 1997.
108. A. A. Gaivoronski, L. Y. Shi, and R. S. Sreenivas, "Augmented Infinitesimal Perturbation Analysis: An Alternate Explanation," *Discrete Event Dynamic Systems: Theory and Applications*, Vol. 2, 121-138, 1992.
109. F. R. Gantmacher, *The Theory of Matrices*, Volumes I and II, Chelsea, New York, 1959.
110. E. Gelenbe and G. Pujolle, *Introduction to Queueing Networks*, John Wiley & Sons, New York, 1987.
111. A. I. Gerasimov, "On Normalizing Constants in Multiclass Queueing Networks," *Operations Research*, Vol. 43, 704-711, 1995.

112. P. Glasserman, *Gradient Estimation Via Perturbation Analysis*, Kluwer Academic Publishers, Boston, 1991.
113. P. Glasserman, "The Limiting Value of Derivative Estimates Based on Perturbation Analysis," *Communications in Statistics: Stochastic Models*, Vol. 6, 229-257, 1990.
114. P. Glasserman and W. B. Gong, "Smoothed Perturbation Analysis for A Class of Discrete Event System," *IEEE Transactions on Automatic Control*, Vol. 35, 1218-1230, 1990.
115. P. W. Glynn, "Regenerative Structure of Markov Chains Simulated Via Common Random Numbers," *Operations Research Letters*, Vol. 4, 49-53, 1985.
116. P. W. Glynn, "Likelihood Ratio Gradient Estimation: An Overview," *Proceedings of the 1987 Winter Simulation Conference*, Atlanta, Georgia, U.S.A, 366-375, December 1987.
117. P. W. Glynn, "Optimization of Stochastic Systems Via Simulation," *Proceedings of the 1989 Winter Simulation Conference*, Washington, U.S.A, 90-105, December 1989.
118. P. W. Glynn and P. L'Ecuyer, "Likelihood Ratio Gradient Estimation for Stochastic Recursions," *Advances in Applied Probability*, Vol. 27, 1019-1053, 1995.
119. W. B. Gong and Y. C. Ho, "Smoothed Perturbation Analysis for Discrete Event Dynamic Systems," *IEEE Transactions on Automatic Control*, Vol. 32, 858-866, 1987.
120. W. B. Gong and J. Q. Hu, "The Maclaurin Series for the GI/G/1 Queue," *Journal of Applied Probability*, Vol. 29, 176-184, 1992.
121. J. J. Gordon, "The Evaluation of Normalizing Constants in Closed Queueing Networks," *Operations Research*, Vol. 38, 863-869, 1990.
122. W. J. Gordon and G. F. Newell, "Closed Queueing Systems with Exponential Servers," *Operations Research*, Vol. 15, 254-265, 1967.
123. X. H. Guan, C. Song, Y.C. Ho and Q. C. Zhao, "Constrained Ordinal Optimization - A Feasibility Model Based Approach," *Discrete Event Dynamic Systems: Theory and Applications*, Vol. 16, 279-299, 2006.
124. S. Hagen and B. Krose, "Linear Quadratic Regulation Using Reinforcement Learning," *Proceedings of 8th Belgian-Dutch Conference on Machine Learning*, Wageningen, The Netherlands, 39-46, 1998.
125. P. G. Harrison, "On Normalizing Constants in Queueing Networks," *Operations Research*, Vol. 33, 464-468, 1985.
126. P. Heidelberger, X. R. Cao, M. A. Zazanis, and R. Suri, "Convergence Properties of Infinitesimal Perturbation Analysis Estimates," *Management Science*, Vol. 34, 1281-1302, 1988.
127. P. Heidelberger and D. L. Iglehart, "Comparing Stochastic Systems Using Regenerative Simulation with Common Random Numbers," *Advances in Applied Probability*, Vol. 11, 804-819, 1979.
128. B. Heidergott, "Infinitesimal Perturbation Analysis for Queueing Networks with General Service Time Distributions," *Queueing Systems: Theory and Applications*, Vol. 31, 43-58, 1999.
129. B. Heidergott, "Customer-Oriented Finite Perturbation Analysis for Queueing Networks," *Discrete Event Dynamic Systems: Theory and Applications*, Vol. 10, 201-232, 2000.

130. B. Heidergott and X. R. Cao, "A Note on the Relation Between Weak Derivatives and Perturbation Realization," *IEEE Transactions on Automatic Control*, Vol. 47, 1112-1115, 2002.
131. B. Heidergott and A. Hordijk, "Taylor Series Expansions for Stationary Markov Chains," *Advances in Applied Probability*, Vol. 35, 1046-1070, 2003.
132. B. Heidergott and A. Hordijk, "Single-Run Gradient Estimation Via Measure-Valued Differentiation," *IEEE Transactions on Automatic Control*, Vol. 49, 1843-1846, 2004.
133. B. Heidergott and A. Hordijk, "Taylor Series Expansions for Stationary Markov Chains (correction)," *Advances in Applied Probability*, Vol. 36, 1300-1300, 2004.
134. B. Heidergott, A. Hordijk, and H. Weisshaupt, "Measure-Valued Differentiation for Stationary Markov Chains," *Mathematics of Operations Research*, Vol. 31, 154-172, 2006.
135. O. Hernández-Lerma and J. B. Lasserre, *Discrete-Time Markov Control Processes: Basic Optimality Criteria*, Springer-Verlag, New York, 1996.
136. O. Hernández-Lerma and J. B. Lasserre, "Policy Iteration for Average Cost Markov Control Processes on Borel Spaces," *Acta Appliandae Mathematicae*, Vol. 47, 125-154, 1997.
137. O. Hernández-Lerma and J. B. Lasserre, *Markov Chains and Invariant Probabilities*, Birkhäuser, Basel, 2003.
138. Y. C. Ho (eds.), *Discrete-Event Dynamic Systems: Analyzing Complexity and Performance in the Modern World*, IEEE Press, New York, 1992.
139. Y. C. Ho, "Heuristics, Rules of Thumb, and the 80/20 Proposition," *IEEE Transactions on Automatic Control*, Vol. 39, 1025-1027, 1994.
140. Y. C. Ho, "On the Numerical Solution of Stochastic Optimization Problems," *IEEE Transactions on Automatic Control*, Vol. 42, 727-729, 1997.
141. Y. C. Ho and X. R. Cao, "Perturbation Analysis and Optimization of Queueing Networks," *Journal of Optimization Theory and Applications*, Vol. 40, 559-582, 1983.
142. Y. C. Ho and X. R. Cao, *Perturbation Analysis of Discrete-Event Dynamic Systems*, Kluwer Academic Publisher, Boston, 1991.
143. Y. C. Ho, X. R. Cao, and C. Cassandras, "Infinitesimal and Finite Perturbation Analysis for Queueing Networks," *Automatica*, Vol. 19, 439-445, 1983.
144. Y. C. Ho, M. A. Eyler, and T. T. Chien, "A Gradient Technique for General Buffer Storage Design in A Production Line," *International Journal of Production Research*, Vol. 17, 557-580, 1979.
145. Y. C. Ho, M. A. Eyler, and T. T. Chien, "A New Approach to Determine Parameter Sensitivities of Transfer Lines," *Management Science*, Vol. 29, 700-714, 1983.
146. Y. C. Ho and J. Q. Hu, "Infinitesimal Perturbation Analysis Algorithm for A Multiclass G/G/1 Queue," *Operations Research Letters*, Vol. 9, 35-44, 1990.
147. Y. C. Ho and S. Li, "Extensions of Infinitesimal Perturbation Analysis," *IEEE Transactions on Automatic Control*, Vol. 33, 427-438, 1988.
148. Y. C. Ho, S. Li, and P. Vakili, "On the Efficient Generation of Discrete Event Sample Paths Under Different Parameter Values," *Mathematics and Computers In Simulation*, Vol. 30, 347-370, 1988.
149. Y. C. Ho and D. L. Pepyne, "Simple Explanation of the No Free Lunch Theorem and its Implications," *Journal of Optimization Theory and Applications*, Vol. 115, 549-570, 2002.

150. Y. C. Ho, R. Sreenivas, and P. Vakili, "Ordinal Optimization of Discrete Event Dynamic Systems," *Discrete Event Dynamic Systems: Theory and Applications*, Vol. 2, 61-88, 1992.
151. Y. C. Ho, Q. C. Zhao, and Q. S. Jia, *Ordinal Optimization: Soft Optimization for Hard Problems*, Springer, 2007, to appear.
152. Y. C. Ho, Q. C. Zhao, and D. Pepyne, "The No Free Lunch Theorem, Complexity and Computer Security," *IEEE Transactions on Automatic Control*, Vol. 48, 783-793, 2003.
153. G. Hooghiemstra, M. Keane, and S. Van De Ree, "Power Series for Stationary Distribution of Coupled Processor Models," *SIAM Journal of Applied Mathematics*, Vol. 48, 1159-1166, 1988.
154. A. Hordijk, *Dynamic Programming and Markov Potential Theory*, Mathematisch Centrum, Amsterdam, 1974.
155. R. A. Horn and C. R. Johnson, *Matrix Analysis*, Cambridge University Press, Cambridge, 1985.
156. J. Q. Hu, "Convexity of Sample Path Performance and Strong Consistency of Infinitesimal Perturbation Analysis Estimates," *IEEE Transactions on Automatic Control*, Vol. 37, 258-262, 1992.
157. J. Q. Hu, M. C. Fu, V. R. Ramezani, and S. I. Marcus, "An Evolutionary Random Search Algorithm for Solving Markov Decision Processes," *INFORMS Journal on Computing*, to appear, 2006.
158. J. Q. Hu, S. Nananukul, and W. B. Gong, "A New Approach to (s, S) Inventory Systems," *Journal of Applied Probability*, Vol. 30, 898-912, 1993.
159. T. Jaakkola, S. P. Singh, and M. I. Jordan, "Reinforcement Learning Algorithm for Partially Observable Markov Decision Problems," In G. Tesauro, D. S. Touretzky, and T. K. Leen (eds.), *Advances in Neural Information Processing Systems 7: Proceedings of the 1994 Conference*, MIT Press, Cambridge, Massachusetts, 345-352, 1995.
160. J. R. Jackson, "Networks of Waiting Lines," *Operations Research*, Vol. 5, 518-521, 1957.
161. L. P. Kaelbling, M. L. Littman, and A. R. Cassandra, "Planning and Acting in Partially Observable Stochastic Domains," *Artificial Intelligence*, Vol. 101, 99-134, 1998.
162. T. Kailath, *Linear Systems*, Prentice Hall, Englewood Cliffs, New Jersey, 1980.
163. L. C. M. Kallenberg, *Linear Programming and Finite Markovian Control Problems*, Mathematisch Centrum, Amsterdam, 1983.
164. R. Kapuscinski, and S. Tayur, "A Capacitated Production-inventory Model with Periodic Demand," *Operations Research*, Vol. 46, 899-911, 1998.
165. H. Kaufman, I. Bar-Kana, and K. Sobel, *Direct Adaptive Control Algorithms - Theory and Applications*, Springer-Verlag, Noew York, 1994.
166. F. P. Kelly, *Reversibility and Stochastic Networks*, John Wiley & Sons, Chichester, 1979.
167. J. G. Kemeny and J. L. Snell, *Finite Markov Chains*, Van Nostrand, Princeton, New Jersey, 1960.
168. J. G. Kemeny and J. L. Snell, "Potentials for Denumerable Markov Chains," *Journal of Mathematical Analysis and Applications*, Vol. 3, 196-260, 1960.
169. L. Kleinrock, *Queueing Systems, Volume I: Theory*, John Wiley & Sons, New York, 1975.

170. V. R. Konda and V. S. Borkar, "Actor-Critic Type Learning Algorithms for Markov Decision Processes," *SIAM Journal on Control and Optimization*, Vol. 38, 94-123, 1999.
171. V. R. Konda and J. N. Tsitsiklis, "On Actor-Critic Algorithms," *SIAM Journal on Control and Optimization*, Vol. 42, 1143-1166, 2003.
172. P. R. Kumar, "Re-Entrant Lines," *Queueing Systems: Theory and Applications*, Special Issue on Queueing Networks, Vol. 13, 87-110, 1993.
173. H. J. Kushner and G. Yin, *Stochastic Approximation Algorithms and Applications*, Springer-Verlag, New York, 1997.
174. P. Lancaster and M. Tismenetsky, *The Theory of Matrices with Applications*, Second Edition, Academic Press, Orlando, 1985.
175. T. W. E. Lau and Y. C. Ho, "Universal Alignment Probabilities and Subset Selection for Ordinal Optimization", *Journal of Optimization Theory and Applications*, Vol. 93, 455-489, 1997.
176. P. L'Ecuyer, "A Unified View of the IPA, SF, and LR Gradient Estimation Techniques," *Management Science*, Vol. 36, 1364-1383, 1990.
177. P. L'Ecuyer, "Convergence Rate for Steady-State Derivative Estimators," *Annals of Operations Research*, Vol. 39, 121-136, 1992.
178. P. L'Ecuyer, "On the Interchange of Derivative and Expectation for Likelihood Ratio Derivative Estimators," *Management Science*, Vol. 41, 738-748, 1995.
179. P. L'Ecuyer and G. Perron, "On the Convergence Rates of IPA and FDC Derivative Estimators," *Operations Research*, Vol. 42, 643-656, 1994.
180. D. C. Lee, "Applying Perturbation Analysis to Traffic Shaping," *Computer Communications*, Vol. 24, 798-810, 2001.
181. L. H. Lee, T. W. E. Lau, and Y. C. Ho, "Explanation of Goal Softening in Ordinal Optimization," *IEEE Transactions on Automatic Control*, Vol. 44, 94-99, 1999.
182. M. E. Lewis and M. L. Puterman, "A Probabilistic Analysis of Bias Optimality in Unichain Markov Decision Processes," *IEEE Transactions on Automatic Control*, Vol. 46, 96-100, 2001.
183. M. E. Lewis and M. L. Puterman, "Bias Optimality," in E. A. Feinberg and A. Shwartz (eds.), *The Handbook of Markov Decision Processes: Methods and Applications*, Kluwer Academic Publishers, Boston, 89-111, 2002.
184. D. Li, L. H. Lee, and Y. C. Ho, "Constraint Ordinal Optimization," *Information Sciences*, Vol. 148, 201-220, 2002.
185. Q. L. Li, and L. M. Liu, "An Algorithmic Approach on Sensitivity Analysis of Perturbed QBD Processes," *Queueing Systems*, Vol. 48, 365-397, 2004.
186. G. Liberopoulos and M. Caramanis, "Infinitesimal Perturbation Analysis for Second Derivative Estimation and Design of Manufacturing Flow Controllers," *Journal of Optimization Theory and Applications*, Vol. 81, 297-327, 1994.
187. G. Liberopoulos and M. Caramanis, "Dynamics and Design of A Class of Parameterized Manufacturing Flow Controllers," *IEEE Transactions on Automatic Control*, Vol. 40, 1018-1028, 1995.
188. M. L. Littman, A. R. Cassandra, and L. P. Kaelbling, "Learning Policies for Partially Observable Environments: Scaling up," *Proceedings of the Twelfth International Conference on Machine Learning*, Tahoe City, California, U.S.A, 362-370, July 1995.
189. Y. Liu and W. B. Gong, "Perturbation Analysis for Stochastic Fluid Queueing Systems," *Discrete Event Dynamic Systems: Theory and Applications*, Vol. 12, 391-416, 2002.

190. Z. Liu and P. Nain, "Sensitivity Results in Open, Closed and Mixed Product Form Queueing Networks," *Performance Evaluation*, Vol. 13, 237-251, 1991.
191. L. Ljung, G. Pflug, and H. Walk, *Stochastic Approximation and Optimization of Random Systems,* Birkhäuser, Basel, 1992.
192. L. Ljung and T. Söderström, *Theory and Practice of Recursive Identification*, MIT Press, Cambridge, Massachusetts, 1983.
193. L. Ljung, *System Identification - Theory for the User*, PTR Prentice Hall, 1999.
194. S. Mahadevan, "Sensitive Discount Optimality: Unifying Discounted and Average Reward Reinforcement Learning," *Proceedings of Thirteenth International Conference on Machine Learning*, Bari, Italy, 328-336, 1996.
195. S. Mahadevan, "Average Reward Reinforcement Learning: Foundations, Algorithms, and Empirical Results," *Machine Learning*, Vol. 22, 159-196, 1996.
196. N. B. Mandayam and B. Aazhang, "Gradient Estimation for Sensitivity Analysis and Adaptive Multiuser Interference Rejection in Code-Division Multiple-Access Systems," *IEEE Transactions on Communications*, Vol. 45, 848-858, 1997.
197. P. Marbach and T. N. Tsitsiklis, "Simulation-Based Optimization of Markov Reward Processes," *IEEE Transactions on Automatic Control*, Vol. 46, 191-209, 2001.
198. P. Marbach and T. N. Tsitsiklis, "Approximate Gradient Methods in Policy-Space Optimization of Markov Reward Processes," *Discrete Event Dynamic Systems: Theory and Applications*, Vol. 13, 111-148, 2003.
199. M. Marchese, A. Garibbo, F. Davoli, and M. Mongelli, "Equivalent Bandwidth Control for the Mapping of Quality of Service in Heterogeneous Networks," *IEEE International Conference on Communications*, Vol. 4, 1948-1952, 2004.
200. M. Marchese and M. Mongelli, "On-Line Bandwidth Control for Quality of Service Mapping over Satellite Independent Service Access Points," *Computer Networks*, Vol. 50, 2088-2111, 2006.
201. B. Melamed, "On Markov Jump Process Imbedded at Jump Epochs and Their Queueing-Theoretic Applications," *Mathematics of Operations Research*, Vol. 7, 111-128, 1982.
202. C. D. Meyer, "The Role of the Group Generalized Inverse in the Theory of Finite Markov Chains," *SIAM Review*, Vol. 17, 443-464, 1975.
203. S. P. Meyn and R. L. Tweedie, *Markov Chains and Stochastic Stability*, Springer-Verlag, London, 1993.
204. N. Miyoshi, "Application of IPA to the Sensitivity Analysis of the Leaky-Bucket Filter with Stationary Gradual Input," *Probability in the Engineering and Informational Sciences*, Vol. 14, 219-241, 2000.
205. M. K. Nakayama and P. Shahabuddin, "Likelihood Ratio Derivative Estimation for Finite-Time Performance Measures in Generalized Semi-Markov Processes," *Management Science*, Vol. 44, 1426-1441, 1998.
206. K. S. Narendra and A. M. Annaswamy, *Stable Adaptive Systems*, Prentice Hall, Englewood Cliffs, New Jersey, 1989.
207. NSF (USA) Workshop on Learning and Approximate Dynamic Programming, Playacar, Mexico, April 8-10, 2002.
208. NSF (USA) Workshop and Outreach Tutorials on Approximate Dynamic Programming, Cocoyoc, Mexico, April 3-6, 2006
209. M. F. Neuts, *Matrix-Geometric Solutions in Stochastic Models: An Algorithmic Approach*, The John Hopkins University Press, Baltimore, 1981.

210. C. Panayiotou and C. G. Cassandras, "Infinitesimal Perturbation Analysis and Optimization for Make-to-Stock Manufacturing Systems Based on Stochastic Fluid Models," *Discrete Event Dynamic Systems: Theory and Applications*, Vol. 16, 109-142, 2006.
211. C. Panayiotou, C. G. Cassandras, G. Sun, and Y. Wardi, "Control of Communication Networks Using Infinitesimal Perturbation Analysis of Stochastic Fluid Models," *Advances in Communication Control Networks, Lecture Notes in Control and Information Sciences*, Vol. 308, 1-26, 2004.
212. G. Pflug, *Optimization of Stochastic Models: The Interface between Simulation and Optimization*, Kluwer Academic Publishers, Boston, Massachusetts, 1996.
213. G. Pflug and X. R. Cao, unpublished manuscript.
214. E. L. Plambeck, B. R. Fu, S. M. Robinson, and R. Suri, "Sample-Path Optimization of Convex Stochastic Performance Functions," *Mathematical Programming*, Vol. 75, 137-176, 1996.
215. J. M. Proth, N. Sauer, Y. Wardi, and X. L. Xie, "Marking Optimization of Stochastic Timed Event Graphs Using IPA," *Discrete Event Dynamic Systems: Theory and Applications*, Vol. 6, 221-239, 1996.
216. M. L. Puterman, *Markov Decision Processes: Discrete Stochastic Dynamic Programming*, John Wiley & Sons, New York, 1994.
217. M. I. Reiman and A. Weiss, "Sensitivity Analysis for Simulations Via Likelihood Ratios," *Operations Research*, Vol. 37, 830-844, 1989.
218. D. Revuz, *Markov Chains*, North-Holland, Amsterdam, 1984.
219. Derek J. S. Robison, *A Course in the Theory of Groups*, Springer-Verlag, New York, 1993.
220. S. Roman, *Field Theory*, Springer-Verlag, New York, 1995.
221. R. V. Rubinstein, *Monte Carlo Optimization, Simulation, and Sensitivity Analysis of Queueing Networks*, John Wiley & Sons, New York, 1986.
222. R. V. Rubinstein and A. Shapiro, *Sensitivity Analysis and Stochastic Optimization by the Score Function Method*, John Wiley & Sons, New York, 1993.
223. G. A. Rummery and M. Niranjan, "On-Line Q-Learning Using Connectionist Systems," Technical Report CUED/F-INFENG/TR 166, Engineering Department, Cambridge University, 1994.
224. H. Salehfar and S. Trihadi, "Application of Perturbation Analysis to Sensitivity Computations of Generating Units and System Reliability," *IEEE Transactions on Power Systems*, Vol. 13, 152-158, 1998.
225. U. Savagaonkar, E. K. P. Chong, and R. L. Givan, "Online Pricing for Bandwidth Provisioning in Multi-class Networks," *Computer Networks*, Vol. 44, 835-853, 2004.
226. A. Schwartz, "A Reinforcement Learning Method for Maximizing Undiscounted Rewards," *Proceedings of the Tenth International Conference on Machine Learning*, Amherst, Massachusetts, U.S.A, 298-305, June 1993.
227. K. C. Sevcik and I. Mitrani, "The Distribution of Queuing Network States at Input and Output Instants," *Journal of the ACM*, Vol. 28, 358-371, 1981.
228. J. Si, A. G. Barto, W. B. Powell, and D. Wunsch, (eds) *Handbook of Learning and Approximate Dynamic Programming*, Wiley-IEEE Press, 2004.
229. S. P. Singh, "Reinforcement Learning Algorithms for Average-Payoff Markovain Decision Processes," *Proceedings of the Twelfth National Conference on Artificial Intelligence*, Seattle, Washington, U.S.A, 700-705, July-August 1994.

230. W. D. Smart and L. P. Kaelbling, "Practical Reinforcement Learning in Continuous Spaces," *Proceedings of the Seventeenth International Conference on Machine Learning*, Stanford, California, U.S.A, 903-910, June-July 2000.
231. G. Sun, C. G. Cassandras, and C. G. Panayiotou, "Perturbation Analysis of Multiclass Stochastic Fluid Models," *Discrete Event Dynamic Systems: Theory and Applications*, Vol. 14, 267-307, 2004.
232. R. Suri, "Infinitesimal Perturbation Analysis for General Discrete Event Systems," *Journal of the ACM*, Vol. 34, 686-717, 1987.
233. R. Suri and J. W. Dille, "A Technique for On-Line Sensitivity Analysis of Flexible Manufacturing Systems," *Annals of Operations Research*, Vol. 3, 381-391, 1985.
234. R. Suri and Y. T. Leung, "Single Run Optimization of Discrete Event Simulations - An Empirical Study Using the M/M/1 Queue," *IIE Transactions*, Vol. 21, 35-49, 1989.
235. R. Suri and M. A. Zazanis, "Perturbation Analysis Gives Strong Consistent Sensitivity Estimates for the M/G/1 Queue," *Management Science*, Vol. 34, 39-64, 1988.
236. R. S. Sutton, "Learning to Predict by the Methods of Temporal Differences," *Machine Learning*, Vol. 3, 9-44, 1988.
237. R. S. Sutton, "Generalization in Reinforcement Learning: Successful Examples Using Sparse Coarse Coding," in D. S. Touretzky, M. C. Mozer and M. E. Hasselmo (eds.), *Advances in Neural Information Processing Systems 8: Proceedings of the 1995 Conference*, MIT Press, Cambridge, Massachusetts, 1038-1044, 1996.
238. R. S. Sutton and A. G. Barto, *Reinforcement Learning: An Introduction*, MIT Press, Cambridge, Massachusetts, 1998.
239. R. S. Sutton, D. Precup, and S. Singh, "Between MDPs and Semi-MDPs: A Framework for Temporal Abstraction in Reinforcement Learning," *Artificial Intelligence*, Vol. 112, 181-211, 1999.
240. Q. Y. Tang and E. K. Boukas, "Adaptive Control for Manufacturing Systems Using Infinitesimal Perturbation Analysis," *IEEE Transactions on Automatic Control*, Vol. 44, 1719-1725, 1999.
241. Q. Y. Tang, P. L'Ecuyer, and H. F. Chen, "Central Limit Theorems for Stochastic Optimization Algorithms Using Infinitesimal Perturbation Analysis," *Discrete Event Dynamic Systems: Theory and Applications*, Vol. 10, 5-32, 2000.
242. Y. C. Tay and R. Suri, "Error Bounds for Performance Prediction in Queueing Networks," *ACM Transactions on Computer Systems*, Vol. 3, 227-254, 1985.
243. H. C. Tijms, *Stochastic Models: An Algorithmic Approach*, John Wiley & Sons, New York, 1994.
244. J. N. Tsitsiklis and B. Van Roy, "Average Cost Temporal-Difference Learning," *Automatica*, Vol. 35, 1799-1808, 1999.
245. S. Uryasev, "Analytic Perturbation Analysis for DEDS with Discontinuous Sample Path Functions," *Communications in Statistics: Stochastic Models*, Vol. 13, 457-490, 1997.
246. F. J. Vazquez-Abad, C. G. Cassandras, and V. Julka, "Centralized and Decentralized Asynchronous Optimization of Stochastic Discrete Event Systems," *IEEE Transactions on Automatic Control*, Vol. 43, 631-655, 1998.

247. F. J. Vazquez-Abad and J. H. Kushner, "Estimation of the Derivative of a Stationary Measure with Respect to a Control Parameter," *Journal of Applied Probability*, Vol. 29, 343-352, 1992.
248. A. F. Veinott, "On Finding Optimal Policies in Discrete Dynamic Programming with No Discounting," *The Annals of Mathematical Statistics*, Vol. 37, 1284-1294, 1966.
249. A. F. Veinott, "Discrete Dynamic Programming with Sensitive Discount Optimality Criteria," *The Annals of Mathematical Statistics*, Vol. 40, 1635-1660, 1969.
250. Y. W. Wan and X. R. Cao, "The Control of a Two-Level Markov Decision Process by Time Aggregation," *Automatica*, Vol. 43, 393-403, 2006.
251. Y. Wardi, M. W. McKinnon, and R. Schuckle, "On Perturbation Analysis of Queueing Networks with Finitely Supported Service Time Distributions," *IEEE Transactions on Automatic Control*, Vol. 36, 863-867, 1991.
252. Y. Wardi, B. Melamed, C. G. Cassandras, and C. G. Panayiotou, "Online IPA Gradient Estimators in Stochastic Continuous Fluid Models," *Journal of Optimization Theory and Applications*, Vol. 115, 369-405, 2002.
253. C. J. C. H. Watkins, *Learning from Delayed Rewards*, Ph.D thesis, Cambridge Uiniversity, 1989.
254. C. J. C. H. Watkins and P. Dayan, Q-Learning, *Machine Learning*, Vol. 8, 279-292, 1992.
255. P. J. Werbos, "Consistency of HDP applied to a simple reinforcement learning problem," *Neural Networks*, Vol. 3, 179-189, 1990.
256. P. J. Werbos, "Approximate Dynamic Programming for Real-Time Control and Neural Modeling," In D. A. White and D. A. Sofge (eds.), *Handbook of Intelligent Control: Neural, Fuzzy, and Adaptive Approaches*, Van Nostrand Reinhold, New York, 493-525, 1992.
257. W. Whitt, "Bivariate Distributions with Given Marginals," *Annals of Statistics*, Vol. 4, 1280-1289, 1976.
258. A. C. Williams and R. A. Bhandiwad, "A Generating Function Approach to Queueing Network Analysis of Multiprogrammed Computers," *Networks*, Vol. 6, 1-22, 1976.
259. R. W. Wolff, "Poisson Arrivals See Time Averages," *Operations Research*, Vol. 30, 223-231, 1982.
260. L. Xia and X. R. Cao, "Relationship Between Perturbation Realization Factors with Queueing Models and Markov Models," *IEEE Transactions on Automatic Control*, Vol. 51, 1699-1704, 2006.
261. H. Yu, and C. G. Cassandras, "Perturbation Analysis for Production Control and Optimization of Manufacturing Systems," *Automatica*, Vol. 40, 945-956, 2004.
262. H. Yu and C. G. Cassandras, "Perturbation Analysis of Feedback-Controlled Stochastic Flow Systems," *IEEE Transactions on Automatic Control*, Vol. 49, 1317-1332, 2004.
263. H. Yu and C. G. Cassandras, "Perturbation Analysis of Communication Networks with Feedback Control Using Stochastic Hybrid Models," *Nonlinear Analysis - Theory Methods and Applications*, Vol. 65, 1251-1280, 2006.
264. B. Zhang and Y. C. Ho, "Performance Gradient Estimation for Very Large Finite Markov Chains," *IEEE Transactions on Automatic Control*, Vol. 36, 1218-1227, 1991.

265. K. J. Zhang, Y. K. Xu, X. Chen and X. R. Cao, "Policy iteration based feedback control", submmited to *Automatica.*
266. Q. C. Zhao, Y. C. Ho, and Q. S. Jia, "Vector Ordinal Optimization," *Journal of Optimization Theory and Applications*, Vol. 125, 259-274, 2005.
267. Y. Zhu and H. Li, "The MacLaurin Expansion for a G/G/1 Queue with Markov-Modulated Arrivals and Services," *Queueing Systems: Theory and Applications*, Vol. 14, 125-134, 1993.

Index